AAPG Treatise of Petroleum Geology Reprint Series

The American Association of Petroleum Geologists
gratefully acknowledges and appreciates the leadership and support
of the AAPG Foundation in the development of the
Treatise of Petroleum Geology.

RESERVOIRS III

CARBONATES

COMPILED BY
EDWARD A. BEAUMONT
AND
NORMAN H. FOSTER

TREATISE OF PETROLEUM GEOLOGY
REPRINT SERIES, NO. 5

PUBLISHED BY
THE AMERICAN ASSOCIATION OF PETROLEUM GEOLOGISTS
TULSA, OKLAHOMA 74101, U.S.A.

Library of Congress Cataloging-in-Publication Data

Reservoirs.

(Treatise of petroleum geology reprint series; no. 5)
Bibliography: v. 2,p.
Contents: 1. Properties. 2. Sandstones
1. Petroleum—Geology. 2. Gas, Natural—Geology.
I. Foster, Norman H. II. Beaumont, E.A. (Edward A.) III.
American Association of Petroleum Geologists.
TN870.5.R43 1988 553.2'82 87-19599
ISBN 0-89181-404-3 (v.3)

Table of Contents

Reservoirs III
Carbonates

Carbonates

Carbonate Depositional Environments

Carbonate Diagenesis

Table of Contents

Reservoirs I

Properties

General

Pore Geometry, Permeability and Fluid Interaction

Pressures

Fractures

Production Geology

Table of Contents

Reservoirs II
Sandstones

Sandstones

Sandstone Depositional Environments

Sandstone Diagenesis

INTRODUCTION

The reprint volumes (I) Reservoirs: Properties, (II) Reservoirs: Sandstones, and (III) Reservoirs: Carbonates, belong to a series that is part of the *Treatise of Petroleum Geology*. The *Treatise of Petroleum Geology* was born during a discussion we had at the Annual AAPG Meeting in San Antonio in 1984. When our discussions ended, we had decided to write a state-of-the-art textbook in petroleum geology, directed not at the student, but at the practicing petroleum geologist. The project to put together one textbook gradually evolved into a series of three different publications: the Reprint Series, the Atlas of Oil and Gas Fields, and the Handbook of Petroleum Geology; collectively these publications are known as the *Treatise of Petroleum Geology*. Using input from the Advisory Board of the Treatise, we designed this entire effort so that the set of publications will represent the state-of-the-art in petroleum exploration knowledge and application. The Reprint Series collects the most up-to-date, previously published literature; the Atlas is a collection of detailed field studies to illustrate all the various ways oil and gas are trapped; and the Handbook is a professional explorationist's guide to the latest knowledge in the various areas of petroleum geology and related fields.

Papers in the various volumes of the Reprint Series are meant to complement the chapters of the Handbook. Papers were selected mainly on the basis of their usefulness today in petroleum exploration and development. Many "classic papers" that led to our present state of knowledge are not included because of space limitations.

We divided the general topic of reservoirs into three separate volumes: (I) properties, (II) sandstones, and (III) carbonates. We grouped papers that discuss properties common to all reservoirs in the volume entitled *Reservoirs: Properties*. The volumes, *Reservoirs: Carbonates* and *Reservoirs: Sandstones* contain papers dealing specifically with those two general types of reservoirs. Originally, the subject of reservoirs was intended to be one volume, but the number of papers suggested for it became so large that it had to be divided into three separate volumes. Therefore, if you examine the content of the papers in all three volumes you will notice that there is much overlap in content.

Volume I, *Reservoirs: Properties*, comprises five parts: (1) General (papers dealing with general reservoir properties), (2) Pore geometry, permeability, and fluid interactions, (3) Pressures, (4) Fractures, and (5) Production geology. Papers in this volume generally describe properties or production techniques that are common to all reservoirs.

Volumes II and III, *Reservoirs: Sandstones* and *Reservoirs: Carbonates*, are each divided into three sections: (1) a section of general papers, (2) depositional environments, and (3) diagenesis.

Edward A. Beaumont
Tulsa, Oklahoma

Norman H. Foster
Denver, Colorado

Treatise of Petroleum Geology
Advisory Board

Carbonates

The American Association of Petroleum Geologists Bulletin
V. 54, No. 2 (February, 1970), P. 207-250, 13 Figs., 3 Tables

Geologic Nomenclature and Classification of Porosity in Sedimentary Carbonates[1]

PHILIP W. CHOQUETTE[2] and LLOYD C. PRAY[3]
Littleton, Colorado 80121, and Madison, Wisconsin 53706

TABLE OF CONTENTS

Abstract Pore systems in sedimentary carbonates are generally complex in their geometry and genesis, and commonly differ markedly from those of sandstones. Current nomenclature and classifications appear inadequate for concise description or for interpretation of porosity in sedimentary carbonates. In this article we review current nomenclature, propose several new terms, and present a classification of porosity which stresses interrelations between porosity and other geologic features.

The time and place in which porosity is created or modified are important elements of a genetically oriented classification. Three major geologic events in the history of a sedimentary carbonate form a practical basis for dating origin and modification of porosity, independent of the stage of lithification. These events are (1) creation of the sedimentary framework by clastic accumulation or accretionary precipitation (final deposition), (2) passage of a deposit below the zone of major influence by processes related to and operating from the deposition surface, and (3) passage of the sedimentary rock into the zone of influence by processes operating from an erosion surface (unconformity). The first event, final deposition, permits recognition of predepositional, depositional, and postdepositional stages of porosity evolution. Cessation of final deposition is the most practical basis for distinguishing primary and secondary (postdepositional) porosity. Many of the key postdepositional changes in sedimentary carbonates and their pore systems occur near the surface, either very early in burial history or at a penultimate stage associated with uplift and erosion. Porosity created or modified at these times commonly can be differentiated. On the basis of the three major events heretofore distinguished, we propose to term the early burial stage "eogenetic," the late stage "telogenetic," and the normally very long intermediate stage "mesogenetic." These new terms are also applicable to process, zones of burial, or porosity formed in these times or zones (e.g., eogenetic cementation, mesogenetic zone, telogenetic porosity).

The proposed classification is designed to aid in geologic description and interpretation of pore systems

[1] Manuscript received, December 31, 1968; revised, July 16, 1969; accepted, July 31, 1969. Published with permission of Marathon Oil Company.

[2] Denver Research Center, Marathon Oil Company.

[3] Department of Geology and Geophysics, University of Wisconsin.

This article was largely formulated and written while both writers were part of a continuing research program on carbonate facies at the Denver Research Center of Marathon Oil Company. We are pleased to acknowledge the very appreciable help received from our colleagues, both within and outside Marathon, in evolving concepts expressed in this article. We extend special thanks to D. H. Craig, D. B. MacKenzie, P. N. McDaniel, and R. D. Russell of Marathon, R. G. C. Bathurst of the University of Liverpool, and P. O. Roehl of Union Oil Company for critical reviews of drafts of the manuscript and to A. S. Campbell of the Oasis Oil Company of Libya, Inc., for stimulating discussions.

and their carbonate host rocks. It is a descriptive and genetic system in which 15 basic porosity types are recognized: seven abundant types (interparticle, intraparticle, intercrystal, moldic, fenestral, fracture, and vug), and eight more specialized types. Modifying terms are used to characterize genesis, size and shape, and abundance of porosity. The genetic modifiers involve (1) process of modification (solution, cementation, and internal sedimentation), (2) direction or stage of modification (enlarged, reduced, or filled), and (3) time of porosity formation (primary, secondary, predepositional, depositional, eogenetic, mesogenetic, and telogenetic). Used with the basic porosity type, these genetic modifiers permit explicit designation of porosity origin and evolution. Pore shapes are classed as irregular or regular, and the latter are subdivided into equant, tubular, and platy shapes. A grade scale for size of regular-shaped pores, utilizing the average diameter of equant or tubular pores and the width of platy pores, has three main classes: micropores (< 1/16 mm), mesopores (1/16–4 mm), and megapores (4–256 mm). Megapores and mesopores are divided further into small and large subclasses. Abundance is noted by percent volume and/or by ratios of porosity types.

Most porosity in sedimentary carbonates can be related specifically to sedimentary or diagenetic components that constitute the texture or fabric (fabric-selective porosity). Some porosity cannot be related to these features. Fabric selectivity commonly distinguishes pore systems of primary and early postdepositional (eogenetic) origin from those of later (telogenetic) origin that normally form after extensive diagenesis has transformed the very porous assemblage of stable and unstable carbonate minerals into a much less porous aggregate of ordered dolomite and/or calcite. Porosity in most carbonate facies, including most carbonate petroleum reservoir rocks, is largely fabric selective.

Introduction

Most sedimentary carbonate rocks have very little porosity[4], but the minority that does contain more than a few percent pore space is collectively of immense economic importance. Porous limestone and dolomite facies contain about one half the world's known reserves of oil and gas and are sites of many significant ore deposits. Programs of exploration for oil and gas and mineral deposits based on an understanding of the geometry and genesis of the pore systems in such exploration targets are becoming increasingly practical. Sedimentary carbonates form aquifers in many regions, and a knowledge of the nature of their pore systems is useful for hydrologic interpretations. Even the utility of carbonate rocks for building stone, concrete aggregate, or other such purposes can be influenced by the nature and amount of their porosity. Hence, economics alone amply justifies a concern with the pore systems of this group of sedimentary rocks.

Carbonate-rock pore systems also can help to elucidate geologic history, for carbonate sediments and rocks and their pore systems are relatively sensitive indicators of physicochemical events. Much of the record of diagenesis is intertwined inextricably with the creation, modification, and obliteration of porosity. Understanding the evolution of porosity from that of an original sediment to that of an ancient carbonate rock can contribute to the unraveling of both depositional and postdepositional events, and can aid in interpreting the history of interaction between sedimentary strata and their contained fluids.

Because of the economic and interpretive value of porosity studies, it is unfortunate that porosity has been neglected in much of the research on sedimentary carbonates published in the past two decades. In many works it is ignored or is described in terms of volume percent, with little other description, illustration, or interpretation. Where aspects of porosity other than pore volume are described, the terminology commonly lacks descriptive clarity or genetic meaning, or both.

Evaluation of the literature relevant to carbonate porosity reveals the need for a more rigorous nomenclature and a more systematic, genetically oriented classification of carbonate pores and pore systems. An integrated system of nomenclature and classification fostering more precise descriptions and a more direct focus on pore genesis would aid interpretation not only of porous carbonates, but also of the many limestones and dolomite facies from which porosity has been essentially eliminated (Pray and Choquette, 1966). Though such systems may have been devised, they either have not been published or have not come to our attention. In any case, they are not generally available.

The purpose of this article is to present a comprehensive system of geologic nomenclature and classification that can be applied to the porosity of sedimentary carbonates. The goal of the system is to facilitate geologic description of porosity, and to assist in improving genetic interpretations of porosity across the broad spectrum of carbonate facies.

[4] In this article, diagenesis is used to encompass "those natural changes which occur in sediments or sedimentary rocks between the time of initial deposition and the time—if ever—when the changes created by elevated temperature, or pressure, or other conditions can be considered to have crossed the threshold into the realm of metamorphism" (Murray and Pray, 1965). It is independent of the process of lithification and may include weathering (*cf.*, Chilingar *et al.*, 1967; Fairbridge, 1967).

This article is in three main parts, each somewhat dependent on the others. Part 1 provides perspectives on the nature of porosity in sedimentary carbonates. It stresses the genetic and geometric complexity of carbonate pore systems, the distinctiveness of most carbonate porosity in comparison to that of porous sandstones, and the importance of ascertaining relations of porosity to fabric elements of carbonate rocks—the concept of fabric selectivity. Part 2 presents the more general aspects of porosity nomenclature. The general terms *porosity, pore, pore system,* and *pore interconnections* are reviewed. Terminology relating to the time of porosity origin and modification is discussed, and new terms and related concepts useful in designating time (and place) of porosity origin are presented. Part 3 presents the classification we propose. The major elements of the system are summarized in Figure 2 and illustrations of its use are given in Figures 3–12. Following the text is a glossary and discussion of most geologic terms that have been applied to the porosity of sedimentary carbonates in the past several decades (Appendix A). Most terms are defined briefly as befits a glossary, but more extended discussions are provided for important and much-used terms such as "vug", for which we believe that clarified definitions and more consistent usage are needed. The glossary is intended to serve both as a general reference and as the main source for definitions and usages of the terms employed in our proposed classification.

Considerations of nomenclature and classification in any scientific field present a reviewer with two end-member alternatives: either adapt preexisting terminology to the present state of knowledge of the field under review, or create a new system with a new nomenclature. Despite some distinct advantages in the second alternative for the description, classification, and interpretation of pore systems in sedimentary carbonates, we do not believe that wholesale changes in the current body of terms are justified by the present "state of the art." For the present, it seems more practical to use current terms as much as possible, sharpening or restricting usage where current concepts suggest that this will improve the precision or clarity of the term.

Part 1. Perspectives on Porosity in Sedimentary Carbonates

Sedimentary carbonates are being recognized increasingly as a complex and distinctive rock family. Their porosity is likewise complex and distinctive. Awareness of the many possible stages in porosity evolution is essential for geologists concerned with studies of carbonate facies, whether porous or not.

Although the origins of porosity are reasonably well understood, many modifications of porosity in carbonates are still inadequately known. For example, it long has been recognized that much pore space in sedimentary carbonates is created after deposition, and attention has been given to the processes of solution and dolomitization believed to have created most of this porosity. But much less attention seems to have been given to the dominant process in porosity evolution, which is the wholesale obliteration of both primary and secondary porosity that has occurred in most ancient carbonates. Newly deposited carbonate sediments commonly have porosity of 40–70 percent; ancient carbonates with more than a few percent porosity are unusual. The volume of pore-filling cement in ancient carbonates commonly may approach or exceed the volume of the initial sediment (Pray and Choquette, 1966). Most porous ancient carbonates are regarded more correctly as representing arrested stages in the normal trend toward obliteration of porosity than as examples of enhanced porosity in formerly less porous facies. Even the creation of molds by solution of aragonite particles, widely regarded as increasing rock porosity, may not involve much net change in pore volume, and the change may be a slight diminution rather than an increase in the pore volume (Harris and Matthews, 1967; Land *et al.*, 1967). The long-claimed increase of porosity that occurs during dolomitization is quantitatively minor compared to the overall porosity decrease which must have occurred in nearly all ancient dolomites. Processes causing this large decrease, however, have been largely neglected. Clearly, the evolution of porosity (both its genesis and modification) in sedimentary carbonates not only is commonly complex, but also records a very important part of the formation of ancient carbonate facies.

The discussion that follows stresses features of porosity in sedimentary carbonates that provide useful perspective for the consideration of nomenclature and the classification presented in Parts 2 and 3.

Complexity of Carbonate Pore Systems

The pores and pore systems of sedimentary carbonates are normally complex both physically and genetically. The pore space of some sedimentary carbonates consists almost entirely of interparticle (intergranular) openings between nonporous sediment grains of relatively uniform size and shape. Porosity of this kind may be relatively simple in geometry. If it formed at the time of deposition and was little modified by later diagenesis, the resulting pore system may closely resemble that of many well-sorted sandstones. It represents a physical and genetic simplicity that is unusual in sedimentary carbonates; much greater complexity is the rule.

It is not surprising that geologists generally have not attempted to describe quantitatively the geometry of pore openings; their sizes, shapes, and the nature of their boundaries commonly show extreme variability. The three-dimensional physical complexity can be visualized readily in some carbonates, but in many it is appreciated best by injecting plastic into the pore system, dissolving the rock with acid after the plastic has hardened, and directly observing the pore system. Illustrations of the results of this technique (Nuss and Whiting, 1947) are provided in articles by Imbt and Ellison (1946) and Etienne (1963).

The size and shape complexity of pores in carbonate rocks is caused by many factors. It relates partly to the wide range in size and shape of sedimentary carbonate particles, which create pores either by their packing or by their solution. It also relates partly to the size and shape variation of pores created within sedimentary particles by skeletal secretion. Extensive size and shape variation relates in part to the filling of former openings by carbonate cement or internal sediment.

The physical complexity of porosity in carbonate rocks is increased greatly by solution processes, which may create pore space that precisely mimics the size and shape of depositional particles or form pores that are independent of both depositional particles and diagenetic crystal textures. Fracture openings also are common in carbonate rocks and can strongly influence solution. Pores range in size from openings 1μ or less in diameter (if a single linear measure is applicable) to openings hundreds of meters across like the "Big Room" at Carlsbad Caverns, New Mexico, termed a "macropore" by Adams and Frenzel (1950, p. 305). Size complexity, in addition to a wide range in possible pore sizes, may involve juxtaposition of large and minute openings in the same rock unit or single sample. Size and shape complexity applies equally well to all openings, whether pores or pore interconnections.

Pores in sedimentary carbonates are fully as complex genetically as they are geometrically. Carbonate porosity is polygenetic in the sense of both time and modes of origin. Although interparticle porosity created at the time of final sediment deposition or accretion is important in many carbonate rocks, porosity created in sedimentary particles either before their final deposition or after deposition commonly ranks in importance with interparticle porosity of depositional origin.

Processes by which porosity is created and modified are greatly varied. Porosity can form by secretion of skeletal carbonate that creates cells, chambers, or other openings; it can form by sediment packing, sediment shrinkage, sediment distention as by gas evolution, or rock fracturing; and it can form by selective solution of sedimentary particles or indiscriminate solution of the rock, organic boring or burrowing, organic decomposition, or in other ways. Though not all of these genetic types of porosity may be present in the same sediment or rock, several generations of porosity and several types of processes may be involved in the genesis of pores in the same facies or even the same sample. Pore-modification processes also are diverse. The many possible processes of porosity creation and modification, operating through a long depositional and diagenetic history, make genetic as well as physical complexity the norm in sedimentary carbonates.

Comparison of Porosity in Sandstones and Sedimentary Carbonates

Many geologists interpret carbonate rocks and their pore systems by analogy with detrital sandstone, siltstone, and shale. Valid analogies can be drawn, but generally it is more useful to focus on the distinctiveness of carbonate pore systems and on their many differences from pores in detrital analogs. The most common analogy, particularly in analysis of petroleum reservoirs, is made between the pore systems of sandstones and those of carbonate rocks.

To illustrate the danger in reasoning by analogy, consider a comparison between the pore system of a sucrose dolomite composed of silt-size rhombic dolomite crystals and that of a slightly cemented quartz siltstone. The size, shape, and sorting of component crystals or particles may be very similar in the two rocks, and superficial analysis would suggest that their pore geometries and hence their fluid-flow properties might be comparable. But there must be basic differences in these pore systems. Few quartz siltstones are oil productive, and those which do constitute oil reservoirs have grain sizes near the upper end of the silt size range (0.06 mm). However, many petroleum reservoirs produce from dolomites with intercrystal porosity that is no more abundant, and superficially is no coarser, than that of nonproductive siltstone. In fact, several dolomite petroleum reservoirs produce from intercrystal porosity in which the crystals are in the smaller silt sizes, some as small as 0.01 mm or less. The good reservoir qualities of such microcrystalline dolomites might not be anticipated if one relied solely on knowledge of porosity characteristics of their apparent textural analogs, the siltstones.

A general comparison of carbonate rocks and sandstones shows a major difference between them in porosity evolution. Both families of rocks have evolved from sediments that had high initial porosity. Even though newly deposited carbonate sand and silt commonly are more porous than newly deposited detrital sand, ancient carbonate rocks commonly have much less pore space than sandstone, and may have none. Porosity reduction is much more extensive in carbonate facies. The normal post-depositional modification of porosity in sandstone is by cementation and minor compaction. This generally causes only modest reduction in original interparticle pore volume without substantial changes in the kind or position of pores. In most carbonates, much or all original pore space is obliterated during diagenesis, new voids are created, and these in turn may be partly or completely filled.

The more significant differences in porosity between sandstones and sedimentary carbonate rocks are summarized in Table 1. The many differences suggest not only that reasoning by analogy has pitfalls, but also that a nomenclature and classification of porosity adequate for sandstones is not adequate for sedimentary carbonates.

Table 1. Comparison of Porosity in Sandstone and Carbonate Rocks

Aspect	*Sandstone*	*Carbonate*
Amount of primary porosity in sediments	Commonly 25–40%	Commonly 40–70%
Amount of ultimate porosity in rocks	Commonly half or more of initial porosity; 15–30% common	Commonly none or only small fraction of initial porosity; 5–15% common in reservoir facies
Type(s) of primary porosity	Almost exclusively interparticle	Interparticle commonly predominates, but intraparticle and other types are important
Type(s) of ultimate porosity	Almost exclusively primary interparticle	Widely varied because of postdepositional modifications
Sizes of pores	Diameter and throat sizes closely related to sedimentary particle size and sorting	Diameter and throat sizes commonly show little relation to sedimentary particle size or sorting
Shape of pores	Strong dependence on particle shape—a "negative" of particles	Greatly varied, ranges from strongly dependent "positive" or "negative" of particles to form completely independent of shapes of depositional or diagenetic components
Uniformity of size, shape, and distribution	Commonly fairly uniform within homogeneous body	Variable, ranging from fairly uniform to extremely heterogeneous, even within body made up of single rock type
Influence of diagenesis	Minor; usually minor reduction of primary porosity by compaction and cementation	Major; can create, obliterate, or completely modify porosity; cementation and solution important
Influence of fracturing	Generally not of major importance in reservoir properties	Of major importance in reservoir properties if present
Visual evaluation of porosity and permeability	Semiquantitative visual estimates commonly relatively easy	Variable; semiquantitative visual estimates range from easy to virtually impossible; instrument measurements of porosity, permeability and capillary pressure commonly needed
Adequacy of core analysis for reservoir evaluation	Core plugs of 1-in. diameter commonly adequate for "matrix" porosity	Core plugs commonly inadequate; even whole cores (~3-in. diameter) may be inadequate for large pores
Permeability-porosity interrelations	Relatively consistent; commonly dependent on particle size and sorting	Greatly varied; commonly independent of particle size and sorting

Concept of Fabric Selectivity

The relation between pore space and solid depositional and diagenetic constituents of a sediment or rock, though commonly ignored, is important for geologic interpretation of sedimentary carbonates. The solid constituents include the various types of primary sediment particles and later formed diagenetic elements such as carbonate cement or recrystallization products, dolomite crystals, and gypsum and anhydrite crystals. These primary and secondary solid constituents, including their textural and smaller structural features, are here collectively referred to as the "fabric elements." The relation between the pore space and fabric elements of carbonate facies ranges from complete dependence, as for primary interparticle porosity, to complete independence, as for tectonic fractures in nonporous limestone. If a dependent relation can be discerned between porosity and fabric elements, we refer to the porosity as *fabric selective*. If such a relation cannot be established, we classify the porosity as *not fabric selective*. Most primary porosity and a large proportion of secondary porosity in carbonate facies are fabric selective. Fabric selectivity is especially useful in determining the time of origin of pores in relation to other events in the diagenetic evolution.

For better observation, description, and interpretation of carbonate porosity it is important to assess fabric selectivity, which we incorporate into our classification (Fig. 2). Interparticle, intercrystal, moldic, fenestral, shelter, and growth-framework porosity are almost invariably fabric selective. Most intraparticle porosity in carbonates is a primary constructional fabric and hence is fabric selective. But like "interparticle," the term refers to position, and some internal pores may form by nonselective solution, boring, or other processes. Fracture, vug, and channel porosity generally are not fabric selective in terms of their origins or by definition. Other types of porosity—boring, burrow, breccia, and shrinkage—are classed as "fabric selective or not" as there is no consistent relation for these types.

Two factors are involved in establishing fabric selectivity: the configuration of the pore boundary and the position of the pore relative to fabric elements. For most primary porosity, both boundary configuration and position of the pore are determined completely by the fabric elements. Thus, the boundaries of interparticle pores in an unconsolidated sediment are shaped by the depositional particles and the pore position is determined by these primary fabric elements. The same dependence of pore boundary and position occurs with internal pores formed by the growth of a skeletal organism. With secondary porosity, the degree of fabric selectivity ranges from complete dependence of both pore boundaries and position upon the fabric elements to apparent independence. For example, the selective removal of aragonite oöliths or gastropod shells from a calcitic rock, or secondary anhydrite crystals from a dolomite, results in pores the size and shape of the dissolved constituents. These molds are fabric selective in both boundaries and positions. However, the tortuous mazes of many phreatic caverns commonly show no positional relation to the depositional or diagenetic fabric elements of the rock (they commonly reflect joint systems). The commonly smooth walls of such caverns further demonstrate the independence of the pore boundary from small-scale fabric elements. The same indiscriminate relation of position and boundary can characterize smaller scale pores—pores we classify as vugs or channels. Such pore systems are not fabric selective. Fabric selectivity is independent of scale of pore.

Two types of fabric selectivity can be usefully differentiated: depositional and diagenetic. Depositional fabric selectivity shows a dependence on either the primary fabric elements or later features that reflect the primary fabric elements. This type of fabric selectivity would include primary interparticle porosity, fenestral porosity that formed shortly after sedimentation, and molds formed by selective removal of depositional particles. Depositional fabric selectivity also can apply to pores of more complex origin. For example, consider a pore occupying the site of a shell that was dissolved selectively to form a mold, which then was filled completely with sparry calcite cement that resisted subsequent dolomitization of the matrix rock but later was dissolved selectively to form a second-generation mold. The final pore still reflects a positional and configuration control by the original shell, a primary fabric element.

Diagenetic fabric selectivity shows a dependence of pore location and configuration upon postdepositional features. Thus, molds may be examples of diagenetic fabric selectivity if they formed by selective removal of diagenetic gypsum or anhydrite crystals. The intercrystal porosity of dolomites is determined by shape and arrangement of the constituent crystals, which normally were created during diagenesis. Their individual position does not coincide with recognizable primary elements in the sediment; hence, most dolomite porosity is diagenetic fabric selective. Many dolomites show evidence of another aspect of selectivity that should be distinguished from the diagenetic fabric selectivity. It relates to the broader scale influence of heterogeneities in depositional fabric, such as "lime mud" versus grains. The importance of fabric heterogeneities like these has been stressed repeatedly by R. C. Murray (*cf.* 1960, 1964; Murray and Lucia, 1967; Lucia and Murray, 1967) and others.

As expected in natural phenomena, there is a spectrum of fabric selectivity; pores and pore systems may have intermediate fabric-selectivity characteristics, but most carbonate porosity can be classified without ambiguity as fabric selective or not. Intermediate types are far less prevalent than the end members of the spectrum, and most of them are solution-enlarged pore systems. Some are pores which occupy positions related to former fabric elements, but have boundaries that cut indiscriminately across the diverse fabric elements. Others are pores that have fabric-selective boundaries, but are so large that their positional relation to smaller scale fabric elements or to former pores cannot be determined. Some caverns have these characteristics. Some pores show gradations in degree of fabric selectivity along their boundaries. Many pores of gradational or intermediate fabric selectivity result from two or more periods of pore development.

Characterization of secondary porosity as fabric selective or not assists in dating the origin of porosity in relation to the diagenetic evolution of a sediment or rock. Two major diagenetic changes of carbonates from newly deposited sediment to typical ancient rock are the change from a very porous sediment to a rock with little porosity, and the change from a mineral assemblage including unstable forms such as aragonite, high-magnesium calcite, and disordered forms of dolomite into a stable assemblage of low-magnesium calcite and/or ordered dolomite (*e.g.*, Land *et al.*, 1967). In essence, very porous, heterogeneous mixtures of stable and unstable carbonate minerals become nonporous rocks composed of calcite or dolomite.

As a broad generalization, most fabric-selective porosity of secondary origin formed before pervasive cementation (the predominant porosity-reducing process), and/or it formed at a time when unstable mineral components were present to influence porosity development. The selective solution of "lime mud" from between the coarser grains to form secondary interparticle porosity almost surely occurs before cementation. Most, if not almost all, of the many porosity fabrics that can be related to selective removal or other diagenetic modification of aragonitic components in sediments must have occurred before or during the elimination of aragonite. At the other extreme, most secondary porosity that is not fabric selective probably developed after pervasive cementation and mineralogic stabilization. The porosity normally associated with tectonic fracturing and with the late-stage solution that may accompany development of major unconformities is rarely fabric selective. Localization of solution porosity that is not fabric selective seems related chiefly to access of permeating water, and to the indiscriminate solution of calcium carbonate along the walls of major flow passageways. The lack of response to fabric elements suggests that the rock material acted largely as a homogeneous source of calcium and carbonate ions. Differences in size, shape, internal porosity, or minor compositional or structural aspects of the component particles or crystals being dissolved apparently had minor influence at this stage.

Other concepts somewhat related to fabric selectivity are also useful in interpreting porosity of carbonate facies. Adams (1953) stressed the importance of preexisting porosity in the development of secondary porosity by solution processes in the rock underlying an erosion surface. His valuable perspectives lead to a concept that can be termed "facies-selective porosity." A spectacular example is present in the Cretaceous El Abra formation of northeastern Mexico (Bonet, 1952). There, unconformity-related cavern systems have been developed by extensive solution within formerly porous rudistid-rich carbonate buildups, but are absent from the enclosing carbonate facies which presumably had little porosity and permeability at the time the unconformity was created. The El Abra example is only one type of facies-selective porosity. Many petroleum reservoirs are found in only one or a very few specific facies of an entire carbonate complex, the other facies being essentially nonporous or impermeable. Such reservoirs exemplify the "specific-reservoir concept" (Pray and Choquette, 1966). Pore systems in most specific reservoirs are fabric-selective types.

In summary, porosity is rarely fortuitous; usually "there is a reason" (Rittenhouse, 1959). Critical study of pore space in relation to the solid constituents of the rock too commonly is neglected. It can aid in interpreting the geologic history, as well as in exploiting the fluid contents of carbonate facies. The concept of fabric selectivity is useful in interpreting the origin of porosity and in relating its occurrence to specific facies. We believe most petroleum reservoirs consist predominantly of fabric-selective pore systems. As with carbonate rocks in general, "there is no substitute for looking at the rocks themselves" (Ham and Pray, 1962, p. 19), and with pore systems in carbonate rocks the need is not only to see the porosity, but to discern its relation to the enclosing solid rock.

One needs, in the American idiom, "to see both the doughnut and the hole."

PART 2. GENERAL CONSIDERATIONS OF POROSITY NOMENCLATURE

The nomenclature applied to porosity in sedimentary carbonates is not extensive, but a review of current terms and their usages suggests that it is inadequate to permit geologists to describe porosity succinctly or to interpret its evolution. Many terms have become part of the nomenclature haphazardly, as by adaptation from common to technical usage. Moreover, some terms have had more than one definition or usage, and others have been used with neither explicit definition nor clear connotations. An example can be seen in the usage of the venerable term "vug," which is discussed in some detail in the glossary. Some of the terms for porosity fail to convey effectively either descriptive geologic attributes or specific genetic information. The imprecise, varied, and changing usage of nomenclature for sedimentary carbonates has made much of the descriptive and interpretive literature on carbonate porosity less useful than it might have been. Because of these considerations and because recent advances in knowledge of sedimentary carbonates have shed new light on the origins of porosity, a reappraisal of the nomenclature applied to porosity seems both timely and needed.

Our views on porosity nomenclature are given in this section and in the glossary. The glossary provides definitions and discussion of most of the terms that have been applied to porosity in carbonates, including the several proposed herein. In this section two aspects of nomenclature are presented. First, the definitions and a brief discussion of the most general terms—*pore, pore interconnection, porosity,* and *pore system*—are given to avoid possible confusion in succeeding parts of this article. Second, an extensive discussion is provided relating to the nomenclature and concepts pertaining to the time at which porosity is formed or modified. Existing porosity-time nomenclature is reviewed, and several new terms are proposed to recognize important stages of porosity evolution of sedimentary carbonates. The new terms also can be used to denote the zones in which porosity evolution takes place. These terms and the ideas they embody require an integrated discussion beyond that feasible for a glossary.

Definitions of General Porosity Terms

The most general terms—*pore, pore interconnection, porosity,* and *pore system*—are usually employed in a way that leaves little or no doubt about their meanings.

The term "pore" was derived from the Greek word, *poros,* meaning passageway. Although it is used in this general sense, more often it is used restrictively to designate local swellings or enlargements along or within a pore system, as well as apparently discrete individual openings within a sediment or rock. In this usage, such discrete pores are distinguished from the more constricted openings that serve as "pore interconnections" or "pore throats." The distinction is useful and we use pore in this manner. Pores so defined constitute most of the porosity and fluid storage volume of the rock. The interconnections or throats generally contribute a small fraction of the total pore volume, but exert decisive control on the permeability and hence the fluid "deliverability" of the rock. The size, shape, and distribution of pores can be described with relative accuracy by macroscopic and microscopic visual observations. Pore interconnections are more difficult or impossible to assess visually, and capillary pressure measurements commonly are used to assess them quantitatively.

The term "porosity" long has been applied by geologists to all openings in the mineral framework of sediments or rocks. It is used in this broadest and most general sense in this article. "Porosity" also is used commonly to denote the amount of a sediment or rock composed of openings, generally expressed as a percent of bulk volume. Collectively, all the openings of sediments or rocks are usefully referred to as the "pore system." The term "pore system" can also denote each of several types of associated porosity, as in "the reservoir porosity consists of an interparticle pore system supplemented by a very permeable pore system of open fractures."

The classification of porosity presented in Part 3 is concerned primarily with the size, shape, and distribution of pores rather than with pore interconnections, and with pore systems in fractured and other carbonates where a distinction between pores and interconnections is either impractical or impossible. Although pore interconnections are not classified or evaluated directly in this article, for some types of porosity descriptions of the types and sizes of pores allow inferences about the nature and effectiveness of pore interconnections.

Porosity Terms of Time Significance

Introduction and summary.—Pore space can be created, modified, or destroyed at many stages in the history of a sedimentary carbonate. A nomenclature for the important phases of porosity evolution is an essential part of a genetically oriented system of classification. This section is concerned with the terminology pertaining to the time of porosity origin and time of significant porosity modification. Processes involved in creating and modifying pore space and the zones in which they operate are considered in relation to the time terms.

The traditional time terms for porosity in carbonates are "primary" and "secondary." These terms are firmly entrenched in the literature, and, despite some variation in meaning, are useful. Their general time significance is clear and generally is sufficiently explicit. As used by most geologists, "primary porosity" includes all pore space present immediately after final deposition, and "secondary porosity" designates any created after final deposition. This usage is still appropriate.

For detailed consideration of porosity evolution, however, such a simple two-fold time classification is inadequate. One carbonate sedimentary cycle begins with the secretion or precipitation of carbonate particles or growth frameworks and ends with metamorphism or erosion of long-buried carbonate rock. Important changes in porosity can occur between the time when sedimentary particles first form (commonly with intraparticle pore space) and the time when these particles, or aggregates of fragments derived from them, come to rest at their final site of deposition and later burial. Afterward, creation, modification, and/or elimination of porosity can occur at any time or continuously during the postdepositional period. This period is generally long, and the porosity history may be complex.

For a somewhat more detailed time characterization than is possible with "primary" and "secondary," three major time periods can be identified in their relation to the most easily defined and recognizable event in porosity evolution, final deposition. "Final deposition" here means the deposition just preceding final burial of the sediment or rock, in contrast to possible earlier phases of deposition, erosion and reworking, and redeposition. The three major stages of porosity evolution are: (1) *predepositional,* (2) *depositional,* and (3) *postdepositional.* To subdivide further the long and significant postdepositional period, we propose three formal subdivisions related to burial stages. Two of these are stages of shallow burial, when the major porosity-influencing processes operate from or are closely related to the surface—a subaerial or subaqueous interface. The third stage is the commonly long intermediate period of deeper burial. We refer to the time of early burial as *eogenetic,* the time of deeper burial as *mesogenetic,* and the late stage associated with erosion of long-buried carbonates as *telogenetic.* These terms also can be applied to porosity created during the three stages, to processes acting during the stages, or to the respective burial zones. Our usage of the major time-porosity terms is summarized in Figure 1. The major time terms and some aspects pertaining to the nature of porosity related to time, the processes important in porosity evolution, and the sites at which these processes operate are discussed in more detail hereafter.

Primary and secondary porosity.—The general meaning of "primary" and "secondary," simple and long-used terms for designating time of porosity genesis, is clear from the normal connotation of these words. The terms are most commonly and better used to refer to the porosity, as in "secondary porosity," than to a time period, such as "secondary time" (Fig. 1, top). Nomenclature problems with these terms arise from minor differences in their definitions and usages. We consider *primary porosity* to be any which was present in the sediment or rock immediately after final deposition. It includes any pore space present within individual sediment or rock components at the time of final deposition, or any which formed during the process of final deposition. This definition appears to be consistent with most current usage (*e.g.,* Illing *et al.,* 1967, p. 488; Murray, 1960, p. 61). It conflicts, however, with usage by Hohlt (1948, p. 6) and others who regard primary porosity as any which developed either before or during lithification. We believe a definition based on the time of final deposition to be much more useful, precise, and practical than one based on lithification. *Secondary porosity* is considered here to be any porosity created in a sediment or rock after final deposition.

Primary porosity and secondary porosity, as used here, depend only on the time when the pore opening was created; their definitions are independent of both the mode of origin of the pore space and the state of lithification. Any porosity formed before or during deposition is

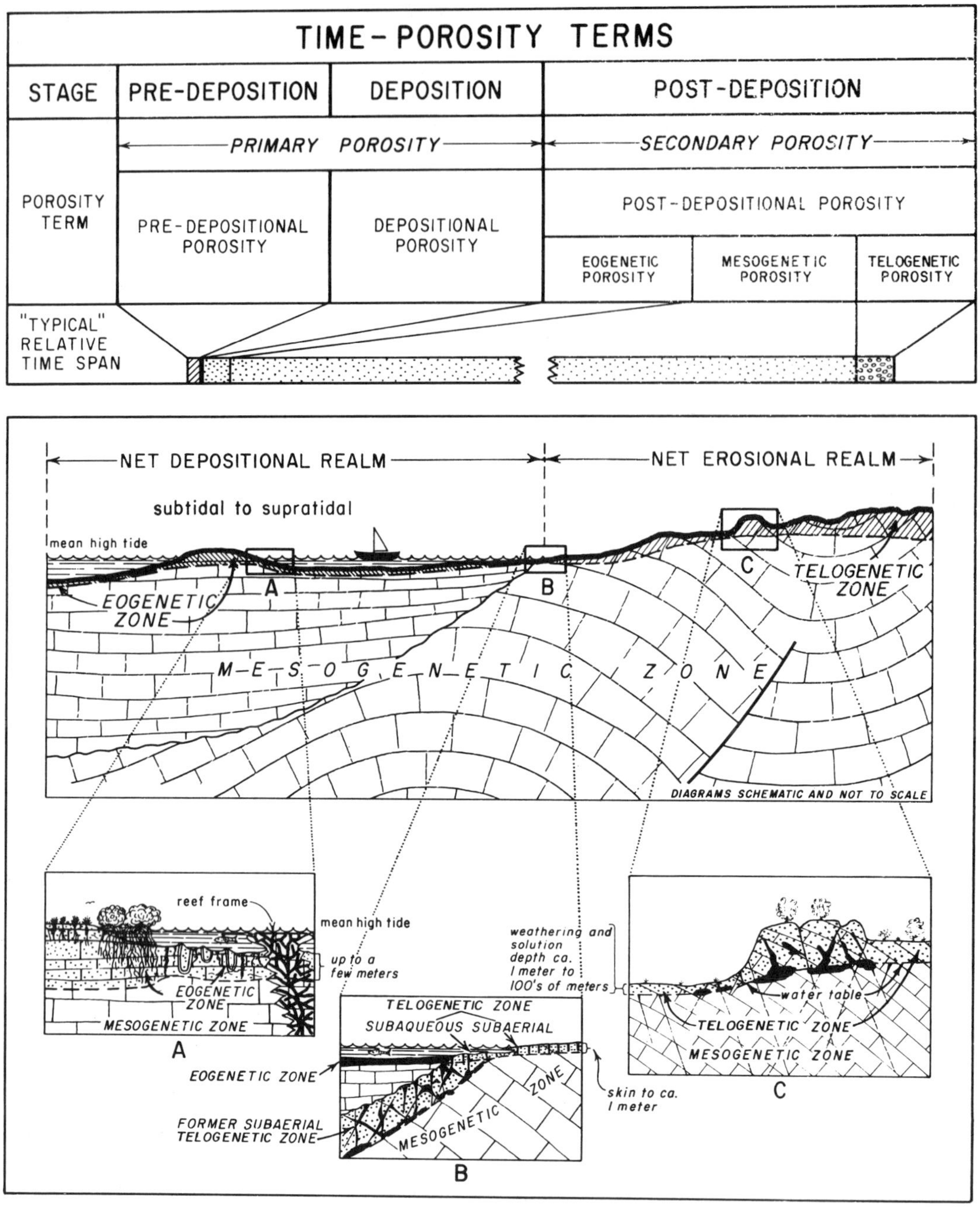

Fig. 1.—Time-porosity terms and zones of creation and modification of porosity in sedimentary carbonates.

Upper diagram: Interrelation of major time-porosity terms. *Primary* porosity either originates at time of deposition (*depositional* porosity) or was present in particles before their final deposition (*predepositional* porosity). *Secondary* or *postdepositional* porosity originates after final deposition and is subdivided into *eogenetic, mesogenetic,* or *telogenetic* porosity depending on stage or burial zone in which it develops (see lower diagram). Bar diagram depicts our concept of "typical" relative durations of stages.

Lower diagram: Schematic representation of major surface and burial zones in which porosity is created or modified. Two major surface realms are those of net deposition and net erosion. Upper cross section and enlarged diagrams **A, B,** and **C** depict three major postdepositional zones. Eogenetic zone extends from surface of newly deposited carbonate to depths where processes genetically related to surface become ineffective. Telogenetic zone extends from erosion surface to depths at which major surface-related erosional processes become ineffective. Below a subaerial erosion surface, practical lower limit of telogenesis is at or near water table. Mesogenetic zone lies below major influences of processes operating at surface. The three terms also apply to time, processes, or features developed in respective zones.

primary regardless of reduction later by cementation or partial filling by internal sediment. However, if primary pores were filled and later processes developed new openings at the sites of these pores, the new porosity would be classed as secondary or postdepositional. The time of origin of the void space is all important. Similarly, if primary pores were enlarged by solution, the resulting space would be partly primary and partly secondary.

A minor classification problem in applying the terms relates to the space occupied in newly deposited carbonates by organic matter that decomposes after burial. We arbitrarily consider pore space created after deposition by decay of buried particles of wood, fibrous roots, and the like to be secondary porosity of a moldic type. However, appreciable pore space in sedimentary carbonates could have been filled at the time of burial with living matter, mucilaginous slimes, or related partly decomposed organic matter, and this space commonly cannot be distinguished from similar pore space which was empty at the time of final deposition. Therefore it is generally practical to classify porosity of these latter kinds, as well as the pore space of body cavities, as primary porosity.

The complexity of carbonate porosity evolution and of differentiation of primary and secondary space becomes apparent if the porosity of individual particles or intraclasts is considered independent of the porosity of the final sedimentary aggregate. By definition all pore space present in a sediment at final deposition is primary porosity. But primary porosity of the sedimentary aggregate may be either primary or secondary in regard to the individual clasts. For example, intraclasts normally contain primary depositional porosity of an interparticle nature that was formed when the source material of the intraclast was deposited. Later, secondary porosity could have been developed in such an intraclast by boring, solution, or some other means, before the final (recycled) deposition of the intraclast. Likewise, porosity "inherited" in clasts derived by erosion of older rocks contributes to the primary porosity of the host sediment, but in relation to the clast itself such inherited porosity may be of primary or secondary origin. Thus, the terms *primary* and *secondary* can by applied usefully to each of the depositional events in which carbonate constituents take part enroute to their final sedimentation, but for these special applications the context in which the terms are used must make clear the cycle of deposition to which they refer.

Predepositional stage, predepositional porosity.—The *predepositional stage* begins at the time when sedimentary material first is formed and ends with the final deposition and burial of that material or of sedimentary particles derived from it. Porosity created in this stage is *predepositional porosity*. The duration of predepositional time can range from perhaps thousands of years in areas with low sedimentation rates accompanied by processes that permit intermittent reworking of bottom deposits to virtually no time at all where there is direct secretion or precipitation of calcium carbonate into an accreting growth framework.

Although a wide variety of porosity types contribute to predepositional porosity, most are formed by the creation of internal cells, chambers, or other openings within individual skeletal organisms, and by porosity within multiparticle individual grains such as pellets. The volumetric significance of conspicuous cellular openings in skeletal grains can hardly be missed. But nonskeletal grains also contain much primary predepositional porosity. For example, pellets can contain more than 50 percent pore space. Individual oöliths have appreciable porosity, and in many of the solid-appearing parts of skeletal organisms there is some porosity between or within the crystals.

The extent to which predepositional porosity of both skeletal and nonskeletal particles is modified prior to final deposition is poorly known and needs much research. Illing (1954) called attention to cementation of individual "grains-of-matrix" and composite grains on the Bahama sea floor. More recent work, such as that of Purdy (1963) and of Bathurst (1964, 1966) on Bahama sediments, suggests both the importance and the complexity of porosity genesis and modification in the predepositional stage.

Depositional stage, depositional porosity.—The *depositional stage* is the time period involved in the final deposition, at its site of ultimate burial, of a sediment or an accreting growth framework. Porosity created at this stage is *depositional porosity*. The importance of this stage in porosity formation is out of proportion to its short duration. Depositional porosity may form two thirds or more of the volume of many carbonate muds. In coarser, well-sorted carbonate sediments, depositional porosity commonly forms more than a third of the bulk volume. Most depositional porosity in carbonates is of an interparticle nature; a small amount is created by the growth frameworks of organic or, rarely, inorganic (*e.g.*, travertine) boundstones, but some porosity in boundstones is formed by boring, solution, or other processes at the depositional interface.

Postdepositional stage, postdepositional (secondary) porosity.—The postdepositional stage encompasses all the time after final deposition (Fig. 1). The term "postdepositional" can be used interchangeably with "secondary" in referring to porosity, to processes of porosity origin or evolution, or to the stage in which they occur. The postdepositional time interval is generally long, and there are significant differences in the time-related processes involved in creating and modifying pore space, as well as in the nature of this pore space. Moreover, the geologic zones in which these postdepositional processes operate also differ significantly. There are many ways to subdivide postdepositional time. A practical time subdivision of the postdepositional stage should distinguish among times of major "bursts" in porosity evolution, and should be based on criteria that can be recognized in many ancient carbonate facies. Our approach involves burial stages that commonly can be distinguished by the study of the pore systems of ancient carbonates. The three subdivisions we propose (*eogenetic, mesogenetic*, and *telogenetic*) are based on the importance in porosity evolution of very early and very late postdepositional events. These can be related to surficial processes in contrast to those occurring at greater depths. This approach also is based on the recognition that porosity created or modified early ("eo-") in postdepositional time commonly is very different from porosity created or modified in the late or final ("telo-") stage associated with an erosional surface carved into long-buried rock. The early porosity generally has fabric-selective characteristics and the late does not.

Alternative approaches to porosity evolution relate to the evolution of mineral assemblages or crystal fabrics, or both; to changing conditions of temperature or pressure during a burial cycle; to changes in formation fluids or the origin of formation water involved in diagenesis; and to the evolution of porosity in relation to lithification.

The responses of mineral assemblages and crystal fabrics to changing physicochemical conditions are clearly important and gradually are becoming better understood, as are the changes in fluid composition that occur during diagenesis of carbonates. However, at least at present, these considerations do not serve to establish general subdivisions of the burial cycle that are useful for porosity evolution.

A diagenetic classification (Fairbridge, 1967) proposed for all sediments emphasizes the origin, nature, and direction of movement of formation water. It recognizes that the impact of early and late surface-related processes, and the zones of syndiagenesis, anadiagenesis, and epidiagenesis, are broadly comparable with the zones of eogenesis, mesogenesis, and telogenesis proposed herein. A major difference is that the epidiagenetic zone extends to the recognizable limit of meteoric water, and appears to exclude surficial weathering; the telogenetic zone includes surficial weathering and extends downward through the vadose zone to about the water table. Other contrasts exist. We prefer the system proposed herein for porosity studies of carbonates, but the two approaches are not mutually exclusive.

Lithification is used as a definitive element in several approaches to diagenesis in carbonates (*e.g.*, Chilingar *et al.*, 1967; Hohlt, 1948). But lithification is an impractical criterion. One reason is, "What constitutes lithification?" Is lithification achieved when a granular sediment is cemented just enough to be coherent? Or when a rock is cemented sufficiently to be truly rigid even though rather porous? Or is lithification completed only when all porosity has been obliterated? More fundamental difficulties in using lithification, whatever definition is used, arise from the wide range in time of lithification, even for contemporaneous carbonate deposits of a very small area, and from the problems of determining that time. Some carbonates are essentially "born lithified," such as coral-algal reef frameworks or travertine. Closely associated clastic limestones may be lithified much later. Some carbonates, composed initially of loose aggregates, become lithified while still in the depositional environment, such as carbonate beachrock or supratidal dolomite crusts. Sea-floor cementation of carbonate deposits in subtidal environments is becoming recognized as a significant process as more examples are reported from ancient and Holocene carbonate studies (*e.g.*, Fischer and Garrison, 1967; Friedman, 1965; Ginsburg *et al.*, 1967; Lees, 1964; Lindstrom, 1963; Milliman, 1966; Pray, 1964, 1965; Taft and Harbaugh, 1964). At the other end of the time spectrum some ancient limestones, notably chalks and some oölitic carbonates, are essentially unlithified. Hence, the range in time required for lithification, the complexity of lithification in carbonates, and the difficulty of deciding when lithification has been achieved severely limit the use of lithification as a time reference to which porosity can be related.

Eogenetic stage, eogenetic porosity.—The *eogenetic stage* applies to the time interval between final deposition and burial of the newly deposited sediment or rock below the depth of significant influence by processes that either operate from the surface or depend for their effectiveness on proximity to the surface. *Eogenetic porosity* is that formed in the eogenetic stage. The term can also be used to designate processes occurring during this stage and the zone in which they occur.

The upper limit of the eogenetic zone is the surface (Fig. 1). This is normally a depositional interface, but it may be a surface of temporary nondeposition, or a surface of erosion during a temporary interruption in sedimentation. The upper surface of the eogenetic zone can be subaqueous or subaerial. The lower boundary is gradational with the underlying mesogenetic zone and is less clearly definable, because the effectiveness of some surface-related processes may diminish gradually with depth, and different processes can operate down to different depths. In many carbonate depositional environments, the boundary of the eogenetic zone may be only a few meters below the surface. Most organic influences, such as burrowing, boring, and root penetration, generally are limited to a few meters or less, and most organic decomposition and gas evolution are also likely to be accomplished at shallow depths. In the low-lying terrane of temporarily emergent carbonate deposits and in areas of flat supratidal surfaces that are only occasionally inundated, the effective range of these and other eogenetic processes probably is mainly in the uppermost few meters of the sediment, or perhaps is in a thickness comparable to the "tidal" range. Unfortunately, information is scanty on processes operating more than a meter or two below modern low-lying surfaces of Holocene carbonates, such as in western Andros, Bahamas (Roehl, 1967; Shinn *et al.*, 1965), and on carbonate sabkhas such as those described by Illing *et al.* (1965).

A problem in determining the lower limit of the eogenetic zone is related to the circulation of fluids at depth, which could cause solution, cementation, replacement (including dolomitization), or other diagenetic effects. If reflux of heavier brines occurred only in a shallow zone of sediments below the depositional interface, it would be classified as eogenetic. But if reflux extends to depths of hundreds of meters, as inferred for the Capitan Reef complex in the pioneering paper of Adams and Rhodes (1960), should all this be considered eogenetic? Or if a freshwater lens is present in newly deposited strata below an oceanic islet, does the eogenetic zone include the entire lens, which may extend many tens of meters below other surface-related processes? These unanswered questions demonstrate the difficulty of applying a single simple classification across the broad range of places and processes involved in carbonate deposition and diagenesis, but hardly prove that the concept of an eogenetic zone or process is without merit.

Although the time of eogenesis may be geologically brief, and the zone thin, eogenetic processes can be of extreme importance in diagenesis and in porosity evolution. The processes that effect these changes or the environments in which they occur are varied, and it is commonly useful to denote them more specifically; hence such terms as "freshwater," "subaerial," or "submarine eogenesis," or more explicit terms such as "vadose eogenetic solution," "submarine eogenetic cementation," or "eogenetic sabkha dolomitization." Very significant porosity evolution may occur during eogenesis of carbonate facies and retain its identity in ancient carbonates. In unconsolidated carbonate sediments, eogenetic porosity can be created by burrowing, sediment shrinkage, sediment distension caused by gas evolution from decomposing organic matter, or root penetration and subsequent decomposition. In more rigid carbonates, boring and root penetration, aided by solution, can create and modify porosity. Eogenetic solution of carbonate or evaporite minerals from a less soluble host, particularly selective solution of aragonite by fresh water, is probably the most important process creating porosity in this zone. However, the major change in the eogenetic zone probably is not creation of new or additional porosity, but reduction of porosity, chiefly through cementation by carbonate or, less commonly, by evaporite minerals. Even the creation of molds from aragonite constituents, though it adds new pores, may reduce porosity either by introduction of cement from another source or by reprecipitation of $CaCO_3$ derived from aragonite in the less dense form, calcite (*cf.* Land, 1967). Internal sedimentation, a porosity-reducing process, most generally occurs in the eogenetic zone. The relation between cement and internal sediment can help in diagnosis of eogenetic cementation, but in many carbonates eogenetic and mesogenetic cementations are difficult or impossible to differentiate.

Mesogenetic stage, mesogenetic porosity.—The term *mesogenetic* is here proposed for the time interval or stage in which the sediments or rocks are buried at depth below the major influence of processes directly operating from or closely related to the surface. It constitutes the entire time between the geologically brief early stage of burial and a final phase of imminent erosion. *Mesogenetic porosity* is that created during the mesogenetic stage. Mesogenetic also applies to the processes occurring within the mesogenetic stage and to the zone in which they occur (Fig. 1).

Porosity changes in sedimentary carbonates within the zone of mesogenesis can be presumed to be extensive. Cementation is probably the major process affecting porosity in the mesogenetic zone, whereas solution is probably minor. Both processes commonly begin in the eogenetic stage. Porosity obliteration may persist through much of mesogenesis. Filling of the innermost parts of formerly larger pore spaces by mosaics of coarsely crystalline calcite may reflect long-continued mesogenetic cementation in rock becoming progressively less permeable. Although many carbonate facies show little evidence of compaction (Beales, 1965; Pray, 1961), marked physical compaction and pressure-solution effects probably can be related to the higher pressures of the mesogenetic zone.

Subdivision of the broad realm of mesogenesis would be desirable, and may become more feasible as future research provides more interpretive leverage. The evolution of the mineral assemblages or of the fabrics undoubtedly will be useful.

Telogenetic stage, telogenetic porosity.—The term *telogenetic* is here proposed for the time interval or stage during which long-buried carbonate rocks are influenced significantly by processes associated with the formation of an unconformity. *Telogenetic porosity* is that formed during the telogenetic stage, and the term can be applied to processes operating at that time, or to the zone influenced by these late-stage processes. The term applies specifically to the erosion of old rocks, in contrast to erosion of newly deposited sediments or rocks during temporary interruptions of a long period of deposition. The prefix "telo-," from the Greek *telos* meaning end or final, is used to denote a final or very late effect in contrast to the early effects connoted by "eo-" (Greek *eos,* dawn or early). Telogenetic porosity is unconformity-developed porosity, but not all porosity associated with an unconformity is of this type. The upper limit of the telogenetic zone is the subaerial or subaqueous erosional interface. The lower boundary is gradational and is placed at the depth at which the erosional processes become insignificant or unrecognizable. The depth of the telogenetic zone and the sharpness of the lower boundary can range widely, depending on the nature of the erosional processes, the environment of erosion, and the nature of the substrate. Although the zone can range in thickness up to hundreds of meters in high-standing areas of karst topography, it is generally a thin zone in the geologic record. In some areas of low relief or of marine planation across massive, nonporous carbonates, it may be less than a meter, and beneath many unconformities no telogenetic zone is detectable.

The lower limit of the telogenetic zone extends to or somewhat below the water table. The telogenetic zone thus encompasses the vadose zone below subaerial surfaces and may include rock affected by phreatic solution associated with the water table. The water table commonly appears to serve as an effective lower limit of many weathering processes including solution. Although Davis (1930) and Bretz (1942) have shown the importance of phreatic solution in cavern development, fortunately for the purpose of locating a practical geologic boundary, most phreatic solution may occur not far below the water table (Thrailkill, 1968).

Solution by meteoric water is undoubtedly the major porosity-forming process of the telogenetic zone. Processes other than solution that create porosity in the telogenetic zone are many, but minor. Borings into a subaqueous erosion surface and fracture and breccia porosity consequent upon solution-collapse may be locally important. Porosity-reduction processes are largely those of filling by (1) internal sedimentation and (2) precipitation from solution in either the vadose zone below a subaerial surface or the subaqueous zone beneath a submarine erosion surface.

In ancient carbonates a variety of features can be used to identify the telogenetic zone. The criteria of fabric selectivity, discussed in Part 1, are very useful in distinguishing telogenetic secondary pore systems from eogenetic pore systems. If the pore boundaries are not fabric selective, the porosity is likely to be telogenetic. Features suggestive of the vadose zone below subaerial erosion surfaces include typical cave deposits, the subtler but distinctive forms of internal sediment (Dunham, 1963), dedolomitization structures shown to be products of weathering (Evamy, 1967), and cementing minerals such as limonite and goethite whose presence similarly reflects oxidizing conditions.

It was long believed that much porosity in carbonates formed by solution below major unconformities. Without doubt, some important petroleum reservoirs and aquifers were created in this manner. The oil fields of the "Golden Lane" in the El Abra Limestone of Mexico (Bonet, 1952) are excellent examples of spectacular porosity (cavern) development associated with an unconformity. But the importance commonly attributed to this late-stage porosity development (*e.g.*, Levorsen, 1967; Murray, 1930) seems overrated. Telogenetic porosity influence on petroleum reservoir quality seems much better demonstrated in the significant enhancement of permeability by fracturing and subsequent solution enlargement than in the development of large volumes of new pore space.

As the length of the mesogenetic stage diminishes, both the distinction between telogenesis and eogenesis and the usefulness of trying to make this distinction diminish. Telogenesis grades into eogenesis somewhat as unconformities grade into diastems. How much older must the eroded strata be than the erosion surface to qualify for telogenesis? Where the age difference spans geologic periods or epochs, the usual diagnosis would be telogenesis; for shorter intervals of time the decision is commonly more arbitrary. Elements besides actual age difference that might be pertinent in distinguishing temporary eogenetic erosion from telogenetic erosion could be the amount of uplift, the amount of eroded strata, or the degree of mineral stabilization or cementation. We would classify as eogenetic the zones of minor erosion and temporary uplift that are common in each of a series of cycles. Thus the intertidal to supratidal cycles described by Laporte (1967) and Roehl (1967) would represent eogenetic erosion, and not telogenesis. The erosion surface that separates Pleistocene from Holocene carbonates in south Florida and the Bahamas could be telogenetic or eogenetic, depending on one's purpose in classification.

Part 3. Classification of Carbonate Porosity

Classifications of porosity in sedimentary carbonates, like most systems of ordering natural phenomena, vary in objectives, emphasis, and detail. No single published classification we know of has the major goal of aiding the geologic interpretation of porosity, or is structured to permit the succinct recording of both physical and genetic characteristics. The literature bearing on the classification of porosity in carbonates can be grouped into two general types, one primarily concerned with the physical properties of principal utility for evaluating or exploiting the fluid content of the rocks, and the other with more geologic or genetic emphasis.

Several classifications in the petroleum geology literature focus on purely descriptive, physical properties of pore systems—the volume percent porosity or other physical properties such as size, shape, and distribution of pores. The main objective of these classifications is correlation of these physical properties with such petrophysical properties as permeability, relative permeability, capillary pressure, fluid saturation, and electrical resistivity. Classifications of this type include those of Teodorovich (*in* Chilingar, 1957; Aschenbrenner and Chilingar, 1960), Archie (1952), Stout (1964), and Jodry (1966).

Classifications of carbonate porosity with more geologic or genetic emphasis are of various types. Some of the earlier published articles relating rock textures or structures to porosity are the very generalized classifications of Howard (1928), Murray (1930), and Howard and David (1936), and the more detailed classification of Imbt and Ellison (1946). A detailed system with particular emphasis on fracture openings is that of Waldschmidt *et al.* (1956). Although the literature on carbonate-rock classification has increased significantly in the past decade, in most of it porosity classification has been ignored or treated only incidentally. Thus, of the articles in a symposium volume on carbonate rock classification edited by Ham (1962), only those by Leighton and Pendexter, Powers, and Thomas consider porosity in any detail. The treatment by Thomas is the most detailed and genetic. It provides illustrations and some discussion of a variety of types of porosity, and also calls attention to the effect of carbonate matrix sorting on porosity, a consideration generally overlooked. More recent classifications of carbonate rocks, however, treat porosity briefly and in general terms (*e.g.*, Sander, 1967; Todd, 1966).

Some of the most valuable recent literature on porosity of sedimentary carbonates is concerned primarily with geologic interpretation and not with problems of classification as such. Thus von Engelhardt's (1960) treatise on porosity in sedimentary rocks affords useful insight on carbonate porosity but does not give a detailed classification. Other articles that add to an understanding of porosity, but emphasize the geologic characteristics, occurrence, and origin of pore space rather than classification, are by Illing *et al.* (1967), Lucia (1962), Lucia and Murray (1967), R. C. Murray (1960), R. C. Murray and Lucia (1967), Roehl (1967), and Schmidt (1965). The general approach in these articles is well expressed by Murray's statement (1960, p. 61) that the ". . . general categories [of porosity] by no means form a suggested classification, but only provide a framework for examining processes and mechanisms of porosity formation and destruction." A recent article on carbonate reservoir rocks by Harbaugh (1967) uses with only slight modification the "general categories" of Murray (1960).

There long has been a need for a comprehensive geologic classification of porosity in sedimentary carbonates—a classification specific enough to permit detailed description, yet genetically oriented enough to foster and guide the kinds of observations needed to understand the origins and modifications of porosity. The system we propose has these goals.

Our approach is to identify porosity with other sediment or rock constituents, textures, and structures, focusing on those features to which the origin and later evolution of pores and pore systems seem most commonly related. In taking this approach we have drawn heavily on current terminology and concepts. The result is a broadly applicable system which is sufficiently "open" to allow the incorporation of other descriptive or interpretive terms. The classification was designed primarily for use in porosity study of hand specimens and cores, but also is adaptable to both microscopic and field studies.

The four main elements of the proposed classification are shown in Figures 2 and 3. These elements include what we call "basic porosity types" (the principal group of elements), and three sets of porosity modifiers that concern (1) the time of porosity origin and the process and direction or stage of porosity modification, aspects we designate with genetic modifiers, (2) pore size (and shape), and (3) pore abundance. The modifiers may be used singly or in combinations, either with a basic porosity type, or with the more general designations of pore or porosity. These features provide much of the descriptive and interpretive value of the classification and adapt it for different levels of detail.

Basic Porosity Types

A principal element of the proposed classification is the characterization of a given pore or pore system as one or more of several basic porosity types. Each type is simply a physically or genetically distinctive kind of pore or pore system that can be defined by such attributes as pore size, pore shape, genesis, and position or association relative to either particular constituents or overall fabric.

We recognize 15 basic types of porosity which are listed in the summary chart of Figure 2 and are defined in the glossary. Most of them are illustrated in Figures 4 through 12. Other basic types could be recognized, but these 15, in combination with the other elements of the classification, permit detailed geologic characterization of almost all porosity in sedimentary carbonates. Recognition of these 15 types appears to be the most practical way to cope with the extreme physical and genetic diversity of pore systems in carbonates and still use current terms.

The 15 basic types vary considerably in volumetric abundance and commonness of occurrence. Seven of them are extremely common and volumetrically important, probably forming the bulk of pore space in most sedimentary carbonates. These types are interparticle, intraparticle, intercrystal, moldic, fenestral, fracture, and vug porosity. The rest appear to be less abundant, although some are common (*e.g.*, shelter porosity), and their separate designation and recognition can be useful for porosity characterization and facies interpretation.

Most of the basic types recognized here have been identified and named by previous workers, and their names conform as closely as possible to current usage and generally identify the main characteristics. Definitions of some types, particularly vug, channel, and growth framework, have been restricted or modified from current ill-defined, broad, and/or variable usage, both to make them more meaningful as porosity terms and to integrate them better into the system.

The attributes that characterize the 15 basic porosity types, such as size, shape, genesis, or position with respect to fabric elements of the rock, merit further attention to clarify the nature of these types. Table 2 summarizes the main attributes involved in the definitions. The single element in defining three porosity types —interparticle, intraparticle, and intercrystal —is the position of the pore with respect to the fabric elements; pore size, shape, and origin are involved either secondarily or not at all in the definitions. One basic type, cavern porosity, is defined solely on the basis of size. Others such as moldic, boring, and shrinkage are defined solely on the basis of origin. Still others such as vug, channel, and various minor types are more complexly defined on the basis of several attributes.

Determining which of the basic porosity types are present in a sediment or rock is not only a matter of identification and interpretation; it also involves judgments as to which of the types best serve the classifier's needs. Several of the porosity types are not mutually exclusive. Thus, the porosity within a pellet composed of aragonite crystals may be interparticle porosity with reference to the component crystals, but on a larger scale it is intraparticle porosity. In Figure 8F the porosity of the sediment within the gastropod shell is primary interparticle porosity, but is also part of the intraparticle porosity in the gastropod. Likewise, dolomitized sediment in a burrow may contain intercrystal porosity that could be designated as burrow porosity, as intercrystal porosity, or as both. Some of the basic types of porosity are little more than physical or genetic varieties of other basic types; for example, fenestral, shelter, and breccia are all varieties of interparticle porosity. Clearly, classification cannot be separated from one's objectives. Deciding which basic porosity types are to be used relates to one's purpose.

The interrelations of the basic porosity types with the time of their origin relative to final deposition and with their mode of origin are summarized in Table 3. It is important to recognize that many of the types can be created at different times and by different processes. For example, although interparticle porosity commonly forms during the process of final deposition of the carbonate sediment, some can form prior to final deposition. Of more significance, interparticle porosity can also form after deposition by selective solution of matrix particles from between larger particles. Thus, the practice of equating the interparticle (intergranular) porosity of carbonates with primary or depositional porosity (*e.g.*, Levorsen, 1967, p. 113) is an unfortunate simplification, though it may be satisfactory for most sandstones. The complexity introduced into the interpretation of carbonate porosity by multiple modes and times of origins is a prime reason for using genetic modifiers with basic porosity types.

Compound and gradational basic porosity types.—Many carbonate facies contain two or more basic types of porosity that are easily differentiated. *Compound* pore systems are those composed of two or more basic types of pores, each type physically somewhat discrete and easily distinguishable. Common examples are those composed of both interparticle and intraparticle porosity, of moldic and intercrystalline porosity, or of any fabric-selective type of porosity combined with fracture porosity (Fig. 12).

Gradational pores or pore systems cannot be clearly differentiated physically and/or genetically. They may be intermediate in characteristics between two basic types; or they may intergrade in very short distances within a thin section, hand specimen, or small part of an exposure; or they may interconnect in a manner that makes separate recognition difficult. As an example, fracture porosity commonly grades both spatially and genetically into breccia porosity, a situation approximated in Figure 11A and B. In some sucrose dolomites where depositional texture is poorly preserved, the larger intercrystal porosity may grade into, and be indistinguishable from, that of small molds or vugs. As another example, carbonates may have some porosity that is both interparticle and moldic, but much of the porosity may not be resolvable into one or the other type; their porosity is gradational.

Gradational porosity also may be designated in the many instances in which fabric-selective porosity becomes nonselective within very short distances. For example, a pore of channel shape may have margins that are locally fabric selective; the channel may have begun to form by solution enlargement of interparticle voids, some of whose edges are preserved along its margins.

Another type of gradation of basic types occurs between interparticle and shelter porosity or interparticle and fenestral porosity. Shelter and fenestral are varieties of interparticle porosity, and distinguished from it partly by the larger size of the pore in relation to the associated particles. As pore size diminishes in relation to these fabric elements, the distinctions also diminish and pores can be classified as of either type or as gradational between them.

Fabric selectivity of basic porosity types.—

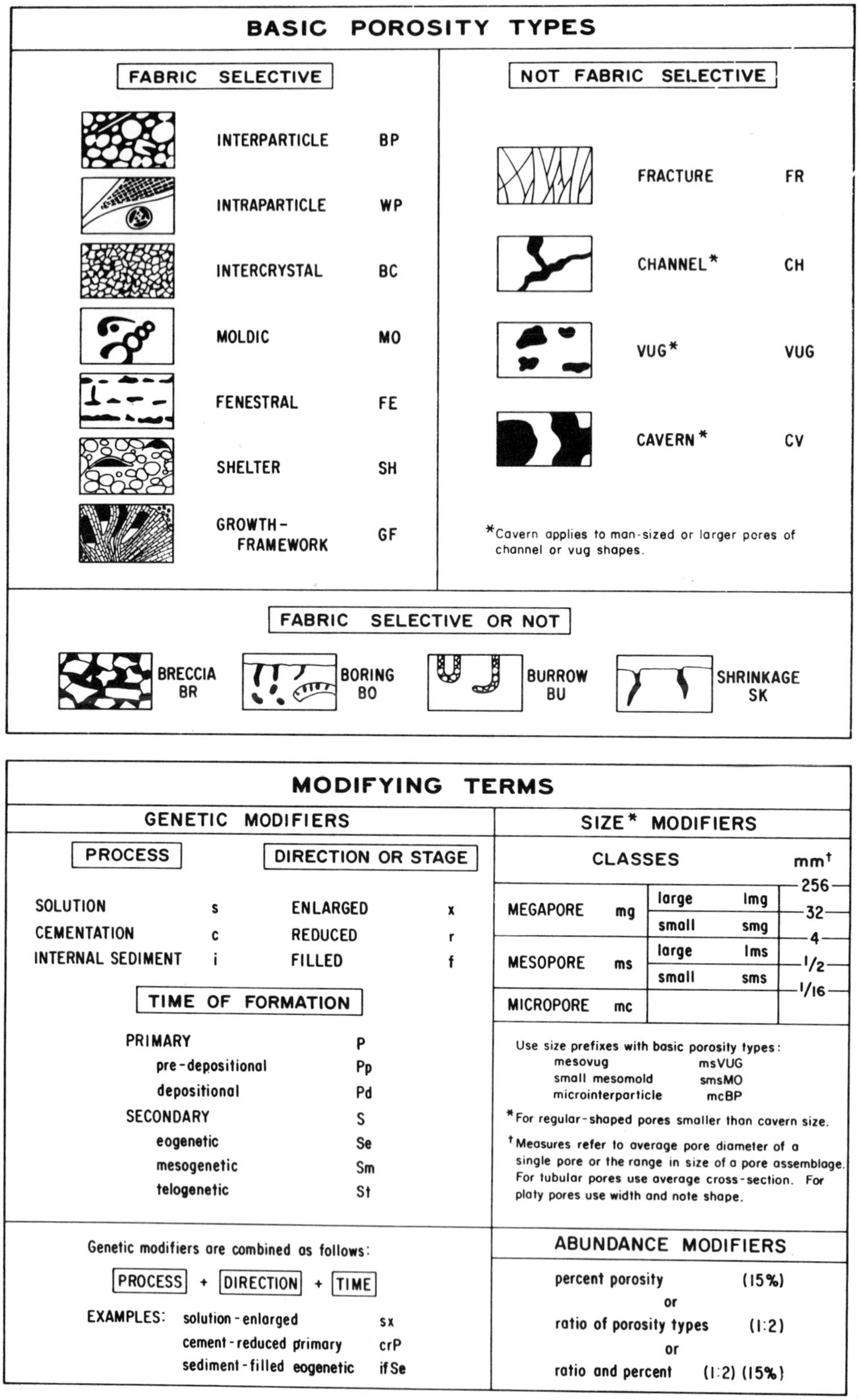

FIG. 2.—Geologic classification of pores and pore systems in carbonate rocks.

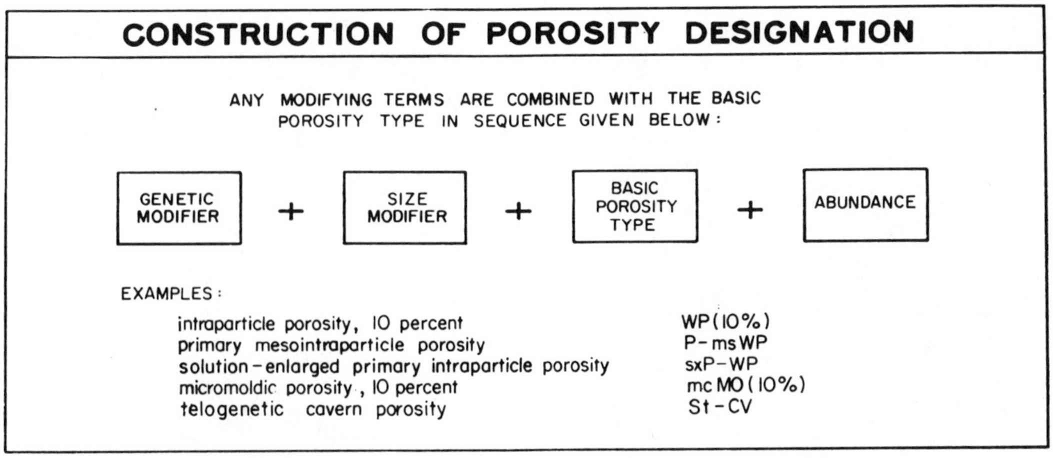

FIG. 3.—Format for construction of porosity name and code designations. Additional examples are shown in captions of Figures 5–12.

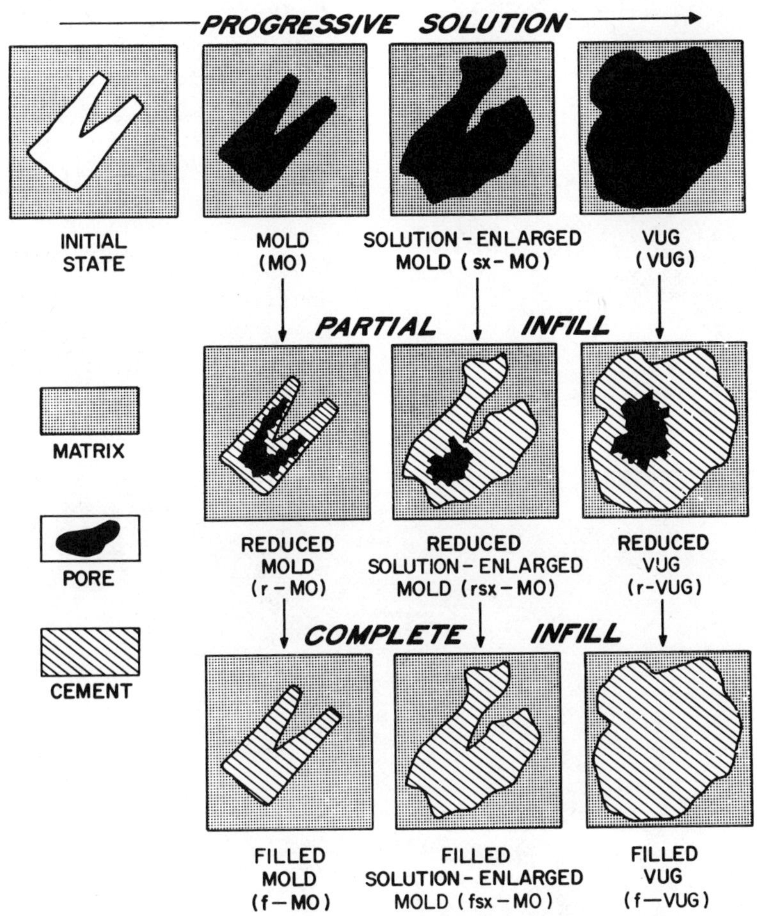

FIG. 4.—Common stages in evolution of one basic type of pore, a mold, showing applications of genetic modifiers and classification code. Starting material is crinoid columnal (top left). It, and matrix adjoining it, then may be dissolved in varying degrees. Depending on extent of solution (top row), resulting pore is classed as mold, solution-enlarged mold, or vug if precursor's identity is lost. Filling by cement could occur after each solution stage.

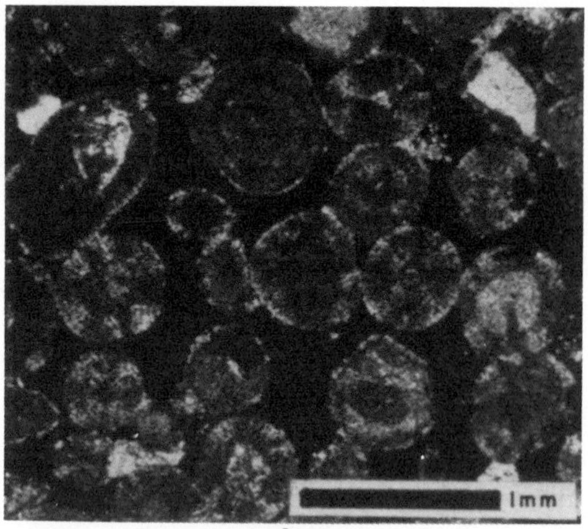

A

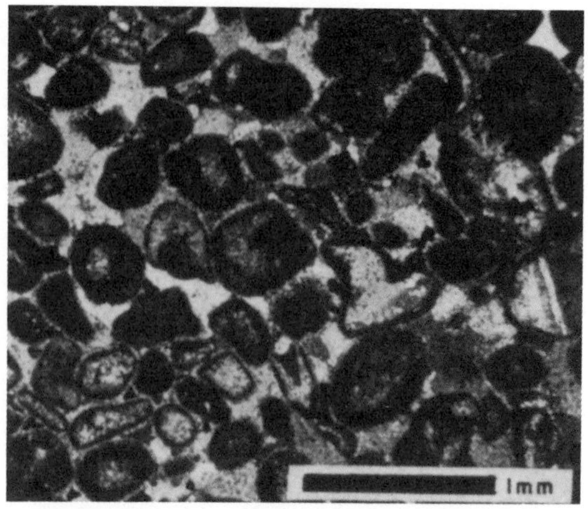

B

C

D

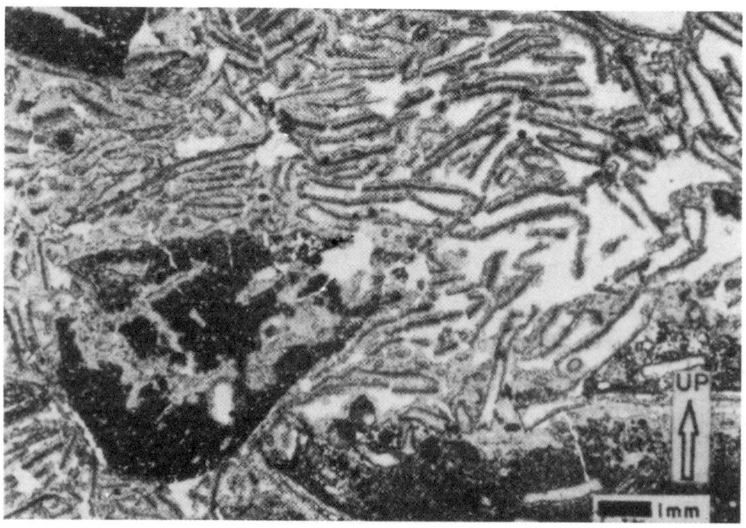

E

The basic porosity types can be characterized usefully on the basis of fabric selectivity (see Fig. 2 and discussion in Part 1), a property which stresses relations between pore space and other constituents. The two fabric-selectivity criteria of pore position and pore-boundary configuration help both in identifying basic porosity types and in interpreting their times of origin. Interparticle, intercrystal, moldic, and fenestral pores have both their positions and their boundaries determined by the fabric elements, hence are fabric selective. Most intraparticle porosity also is fabric selective and is so classed in Figure 2. Some nonselective intraparticle porosity may be present, however, such as a vug within a clast. By definition, vugs and channels are not fabric selective. Fracture porosity is generally insensitive to the smaller scale features of the rock, and hence is considered not fabric selective. But where pore systems that would be classified as vugs, channels, or fractures on the basis of their indiscriminate position relative to fabric elements show fabric selectivity along part or all of their boundaries, they may have formed before complete cementation of the rock. Thus, the abnormal fabric selectivity of some basic pore types has genetic significance.

Genetic Modifiers

Genetic information is implicit in some of the designations of basic porosity types, but other basic-type terms supply little or no genetic information (Table 2). Many types of porosity can originate at different times in relation to sediment deposition or burial, and by several processes. And, once created, porosity can be modified by various processes. These processes, and the direction and extent of porosity evolution, are significant descriptive and interpretive elements. The genetic modifers in this classification provide a way to designate such elements. They can be used either with the basic porosity types or independently.

We recognize 13 genetic modifiers of three types which denote (1) the *time* of origin of the porosity, (2) the *process* involved in its subsequent modification, and (3) the *direction* and extent of such modification(s), herein referred to simply as the "direction." The three types are listed in the summary diagram of Figure 2 with examples showing how individual genetic modifiers are used, singly or in combinations. The complete genetic-modifier term coupled with the basic porosity type provides a definitive porosity characterization (Fig. 3).

Time modifiers.—The seven modifers relating to time of origin consist of the two most general time terms, *primary* and *secondary,* and five more detailed time terms that are subdivisions of the two general terms. Specifically these are *predepositional* and *depositional,* both subdivisions of primary; and *eogenetic, mesogenetic,* and *telogenetic,* all subdivisions of secondary. These terms are defined in the glossary and discussed in Part 2. The five more explicit time terms can be combined directly with the basic porosity type (*e.g.,* depositional interparticle porosity, eogenetic moldic porosity, or telogenetic vug), but the full designation using pri-

←

FIG. 5.—Examples of interparticle porosity

A. Interparticle porosity in oölitic grainstone. Grainstone is well sorted and free of interparticle matrix. Little of its depositional porosity (black) has been filled. Classification: primary depositional interparticle porosity (Pd-BP). Ste. Genevieve Limestone (Mississippian), Bridgeport field, Illinois. Thin section, cross-polarized light.

B. Reduced primary interparticle porosity (black) in oölitic grainstone. Calcite cement, some as syntaxial rims on crinoid columnals, has filled most pore space. Classification: cement-reduced primary interparticle porosity (crP-BP). Remaining voids are classified as small mesopores (sms) in contrast to large mesopores (lms) of **A.** Ste. Genevieve Limestone (Mississippian), Bridgeport field, Illinois. Thin section, cross-polarized light.

C. Solution interparticle porosity in foraminiferal packstone. Pores are white. Note irregular, erratically distributed pores and finely particulate matrix (dark gray) within and between forams. Porosity appears to have resulted from solution of matrix. Tertiary, Libya. Thin section, plane light.

D. Solution interparticle porosity in crinoid-fusulinid packstone. Pores (black) were created largely by solution of matrix, in places with partial corrosion of large particles (arrow). Classification code: s-BP. Pennsylvanian, Hulldale field, Texas. Polished core surface.

E. Primary and reduced primary interparticle and intraparticle porosity in phylloid algal grainstone. In places, as on right side of photograph, some pores may have been solution enlarged. Some algal plates have trapped fine sediment. Classification code: rP-WP/P-BP. Paradox Formation (Pennsylvanian), Honaker Trail, San Juan River Canyon, Utah. Thin section, plane light.

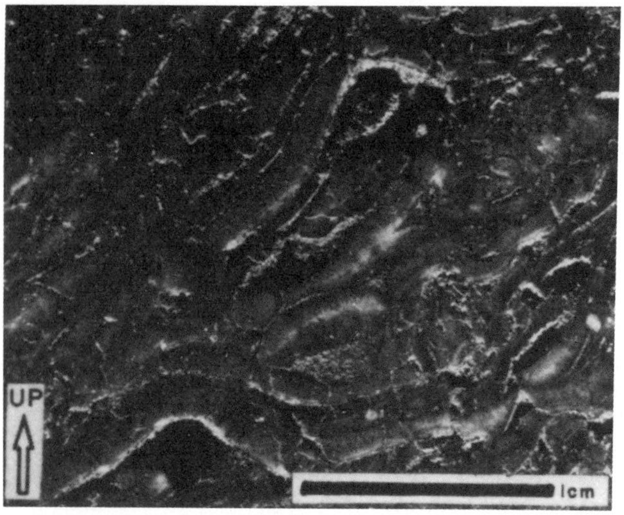

A

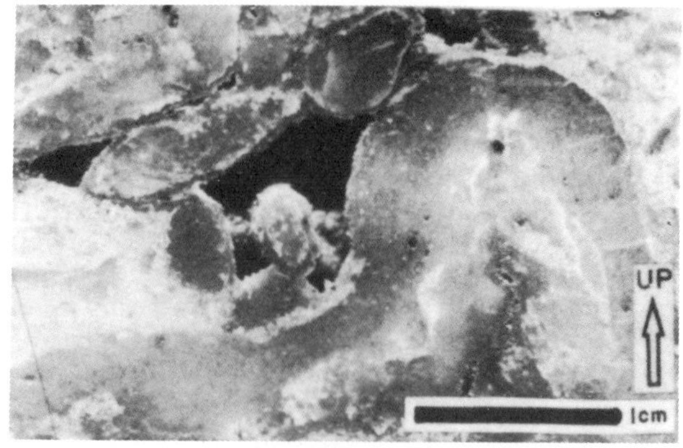

C

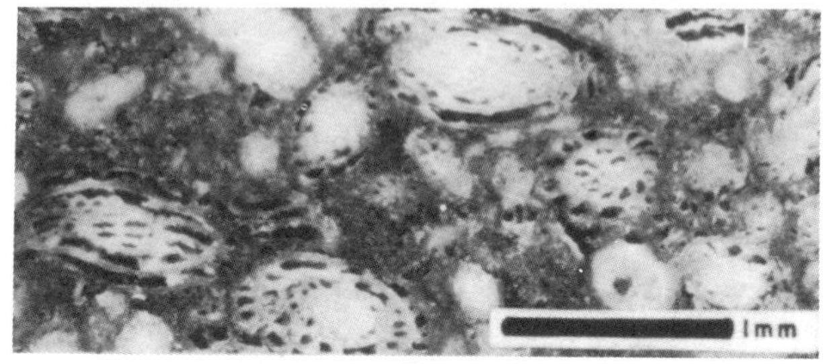

D

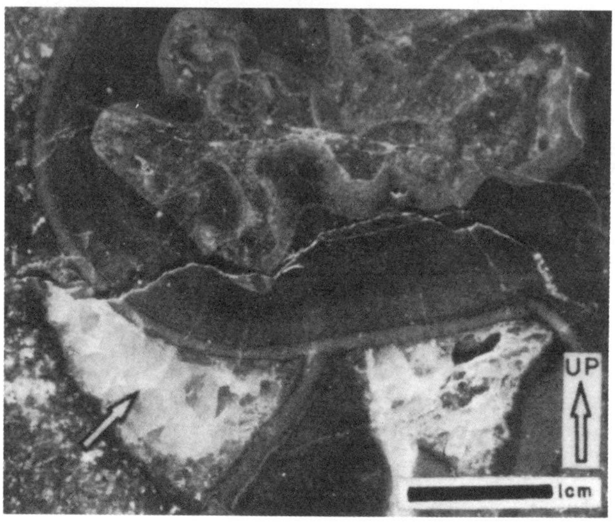

B

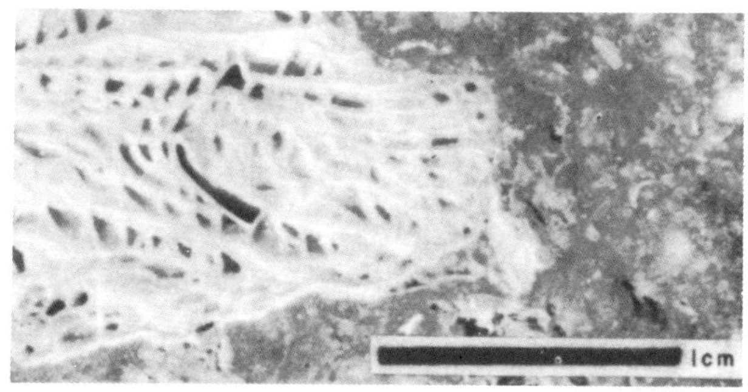

E

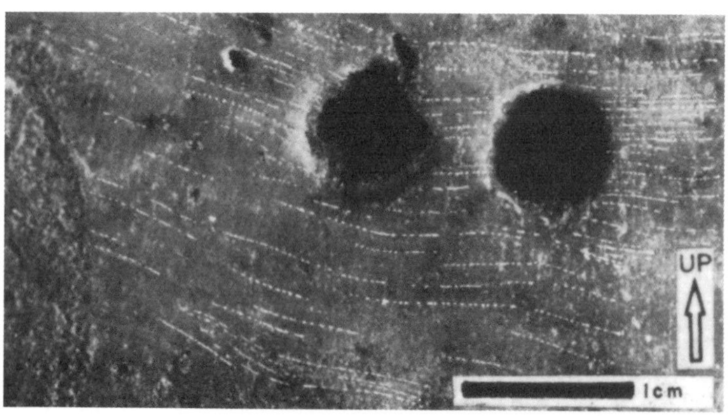

F

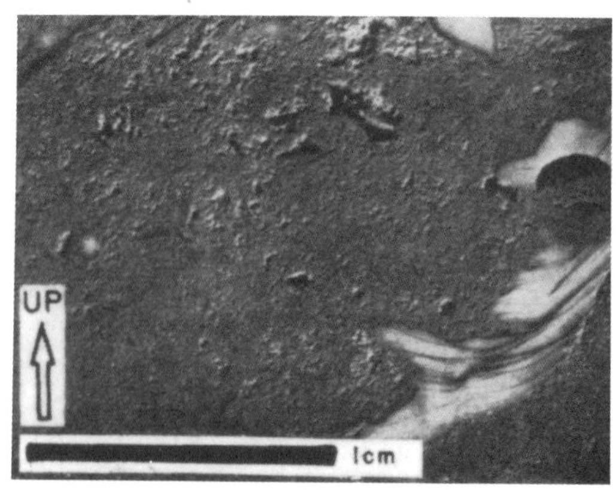

G

mary or secondary together with the more explicit terms may be desirable (*e.g.*, primary depositional or secondary eogenetic).

Modification process and direction.—Six of the 13 genetic modifiers relate to the processes by which pores are modified, and to the direction(s) taken by modifications. An example of some of the possible modification effects is shown diagrammatically in Figure 4 for a crinoid mold.

Of a variety of modification processes, *solution, cementation,* and *internal sedimentation* are recognized in this classification. *Solution* processes both create and modify pores. The main use of "solution" as a genetic modifier is to note solution enlargement of basic types of porosity and to designate a solution origin for those basic types that designate position in a fabric, namely interparticle and intercrystal. The designation of solution is not required for moldic porosity, a basic type defined as solution created. As solution normally is assumed to have been the genetic mechanism in vug, channel, and cavern porosity, with these terms it need not be specified. *Cementation,* used here in the broad sense for the filling of voids by precipitation of mineral matter from solutions, probably accounts for most of the wholesale reduction of porosity from newly deposited sediments to ancient rocks. The reduction process can be noted specifically by the modifier term "cementation." The quantitative importance of internal sedimentation as a porosity-reducing process in carbonates is still being debated, but the process is being recognized increasingly as a useful indicator of special postdepositional events, such as vadose circulation (Dunham, 1963). Normally it occurs as particle-by-particle deposition within the interstices of a porous sediment or rock. Processes of mass sediment injection associated with limestone clastic dikes (Pray, 1964) also may eliminate some porosity.

The porosity of some carbonates may be reduced by changes in packing consequent upon physical compaction. This is not provided for in our system, because it is believed to be uncommon and is difficult to recognize or record on the basis of pore characteristics. Other processes of porosity modification, such as gas distension or mineral-volume change, likewise are not considered feasible to note for the purpose of this classification.

The direction or extent of the porosity modification is noted by three modifiers, *enlarged* and *reduced* as the main direction terms, and *filled* for the commonly encountered end stage of porosity reduction. These are used best with the notation of process, but can be used independently. The modifier "enlarged" normally is used to denote enlargement by solution. It is applied only to modifications that do not obliterate the identity of the original pore. "Reduced" is used for stages of porosity reduction between the initial state and the end stage of "filled." In view of the almost universal reduction of pores by some cementation or other form of filling, the modifier "reduced" is used, as in "reduced primary interparticle porosity," normally only if the volumetric reduction is appreciable, perhaps 30–50 percent or more. Examples of reduced porosity are shown in

---

←««

FIG. 6.—Examples of intraparticle, boring, and shelter porosity.

A. Shelter porosity (SH), a type of interparticle porosity, in algal packstone. Large pores (black) were sheltered from sediment beneath "umbrellalike" phylloid algae. Paradox Formation (Pennsylvanian), Ratherford field, Utah. Polished core surface.

B. Primary (depositional) shelter porosity below reef framework (megabreccia clast) virtually filled by white sparry calcite cement. Classification: cement-filled depositional shelter porosity (cfPd-SH). Upper Bone Spring Limestone (Permian), Guadalupe Mountains, Texas. Polished surface.

C. Shelter porosity between coarse stromatoporoid fragments in coarse-textured part of fine-grained, nearly nonporous packstone. Loosely packed, relatively coarse debris prevented infilling by finer contemporaneous sediment (white). Leduc Formation (Devonian), Redwater field, Alberta, Canada. Polished core surface.

D. Intraparticle porosity within fusulinids. Classification: primary mesointraparticle porosity (P-msWP). Lansing Group (Pennsylvanian), Kansas. Polished core surface.

E. Intraparticle porosity in horn coral. Pennsylvanian, Hulldale field, Texas. Polished core surface.

F. Boring porosity (BO) of large-mesopore size which truncates growth laminations (accented by retouching on photo) in massive stromatoporoid. Matrix at left is fine-grained packstone. Leduc Formation (Devonian), Redwater field, Alberta, Canada. Polished core surface.

G. Boring in thick-shelled pelecypod. Note partial filling by internal sediment, which suggests that shell was bored before its final deposition. Classification: sediment-reduced, predepositional boring porosity (irPp-BO). Matrix surrounding shell and internal sediment are porous, dolomitic, bioclastic packstone. Tertiary, Libya. Polished core surface.

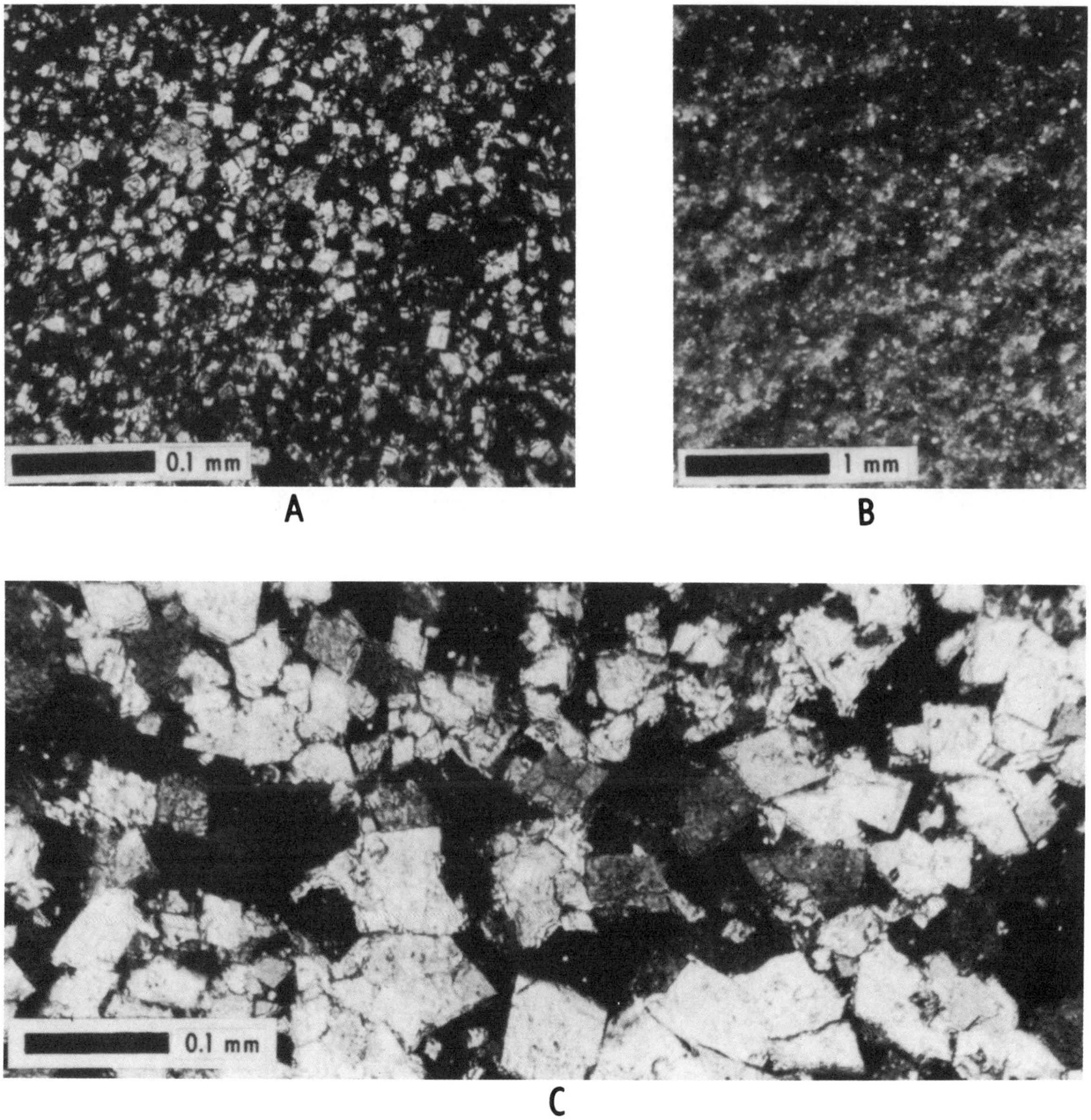

FIG. 7.—Intercrystal porosity in dolomites.

A. Microcrystal porosity (black) in sucrose dolomite of very fine texture and high porosity (37 percent). Micropores are well connected and form very permeable pore system. Classification code: mcBC (37%). Ste. Genevieve Formation (Mississippian), Bridgeport field, Illinois. Thin section, cross-polarized light.

B. Broken surface of rock shown in **A,** illustrating finely sucrose texture as highlighted by small reflecting crystal surfaces.

C. Mesointercrystal porosity (black) in sucrose dolomite. Apparent "diameters" of pores that look discrete in two dimensions range from about 0.02 to about 0.1 mm, but are dominantly of small-mesopore sizes. Note that photos **A** and **C** are at same scale. Porosity is 14 percent. Classification code: smsBC (14%). Madison Group (Mississippian), Oregon Basin field, Wyoming.

Figures 5B, 8E, and 10C. If greater detail is needed, the amount of reduction can be stated, as in "reduced 40 percent" or "reduced one half," and a corresponding code notation of "$r_{40\%}$" or "$r_{1/2}$" can be used. The designation "filled" is commonly very useful in description and interpretation of the porosity evolution of nonporous carbonate rocks, which are much more common than porous carbonates. It also permits treatment of obliterated voids in rocks retaining some porosity. Filled former voids may involve more than half the volume of the rock!

The direction term *incomplete* can be useful for designating incompletely formed molds, but for simplicity and because the term is not as widely applicable as the others, it is not a formal part of the classification.

Pore Size and Pore-Size Modifiers

The size of pores in sedimentary carbonates is an important descriptive parameter, but one difficult to treat. The distinction has been made between pores and pore throats or interconnections. Some quantitative visual characterization of pore size generally can be made without undue difficulty, but determination of pore-throat size by direct observation is generally difficult or impossible. Pore-throat size can be determined indirectly by observation in those unusual carbonates that consist of grains of uniform size, shape, and packing; but uncemented, well-sorted oölites and their textural analogs are rare! The vast preponderance of ancient carbonates and most modern carbonate sediments require capillary-pressure measurements combined with other mass-response petrophysical data to characterize pore-throat size quantitatively. Such characterizations, though important for an understanding of reservoir behavior, are outside the scope of this classification; here we are concerned primarily with pore-system attributes that can be directly and readily observed. How can the size of the pores, as differentiated from pore throats, best be characterized? What level of precision is feasible for most geologic description and interpretation? Our system is summarized in Figure 2.

Several factors make it difficult or impossible to be precise in a visual characterization of pore size. One is that the physical boundaries of an individual pore are arbitrary if, as is normal, the pore is part of a continuous pore system. A second limiting factor is the difficulty (and even impossibility) of observing the three-dimensional shape of pores. Shapes generally must be visualized from a two-dimensional surface in an opaque rock. A third factor is the irregular shape of pores. But the main control on useful precision relates, not to these difficulties in determining the size of an individual pore, but to the normal range in sizes and shapes of all the pores in a reservoir, outcrop, hand specimen, or even thin section. The common need for geologic description is an expression of average size or average size range as determined by quick visual inspection. Practicality thus dictates a need for broad size (and shape) classes. If size is expressed by a diameter measure, the question of which diameter is somewhat academic if many pores are considered. A pore-size grade scale as detailed as the Wentworth scale for grain sizes, utilizing a class interval ratio of two, or even of four (Todd, 1966), is usually much too refined for carbonate porosity.

Some need for expressing pore size in carbonates can be eliminated by careful lithologic description, coupled with a notation of the basic porosity type. Thus, the interparticle porosity in a slightly cemented, well-sorted, medium-grained oölite rarely needs a direct pore-size description. Likewise, the description, "fusulinid moldic porosity," may convey adequately both size and shape of pore. Another way of simplifying pore-size expressions is to describe the pore size of each porosity type individually, rather than the whole pore-size spectrum collectively.

To designate pore size quantitatively, pore shape first must be considered, and in carbonates the shapes can be extremely diverse. We divide pore shapes into two broad categories: *regular,* with shapes that can be characterized by one-, two-, or three-diameter measures, and *irregular,* with shapes so complex they cannot be described adequately by a few measurements. It is impractical to subdivide the many possible shapes of irregular pores. Regular pores, however, can be classified on the basis of their diameters and pore shapes: *equant, tubular,* and *platy*. Tubular and platy pores are notably elongate in one or two directions or diameters, in comparison to the short diameter. The equant class includes pores whose three diameters are about the same and pores that are not so distinctly elongate in one or two dimensions as to be called tubular or platy. The range in shape of closely associated pores makes unnecessary much concern with precise boundaries between these three regular-shape categories.

For size classification, equant pores can be characterized adequately by a single measure, an average diameter. Sizes of tubular and platy pores can be characterized adequately by an average cross-section diameter or width. In this pore-size classification, if shape is not specified, it can be assumed that essentially equant-shaped pores are referred to. Shape should be specified explicitly where pores are tubular or platy, unless shape is implicit in the porosity-

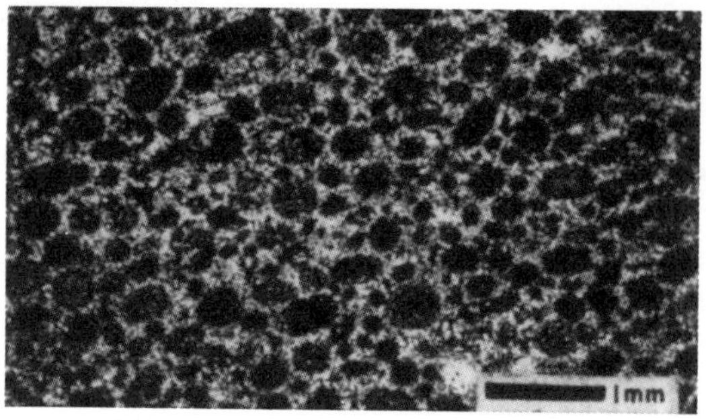

A

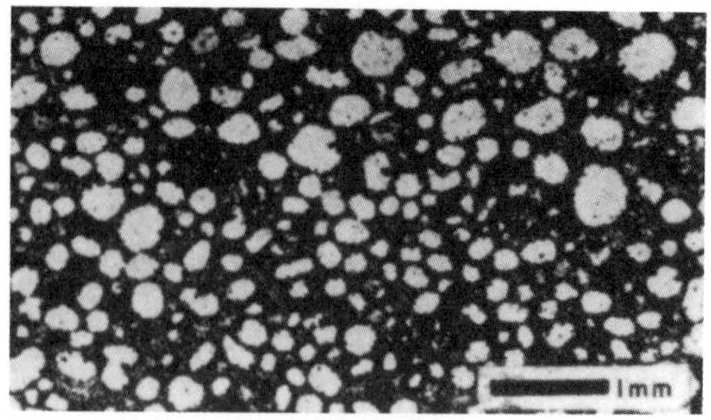

B

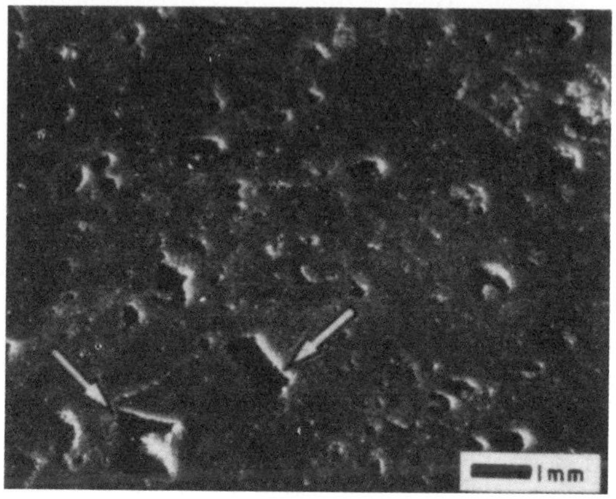

C

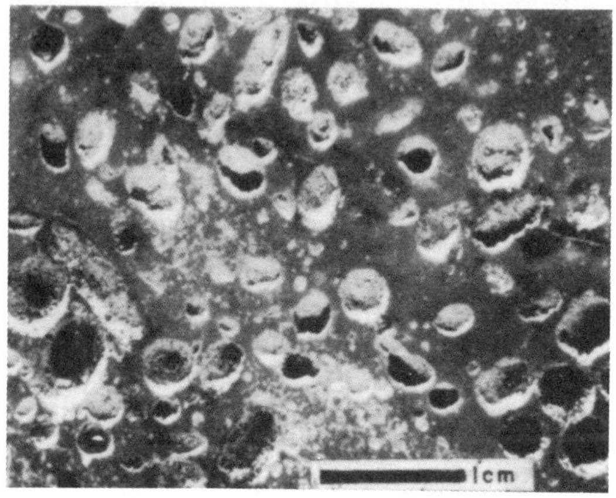

D

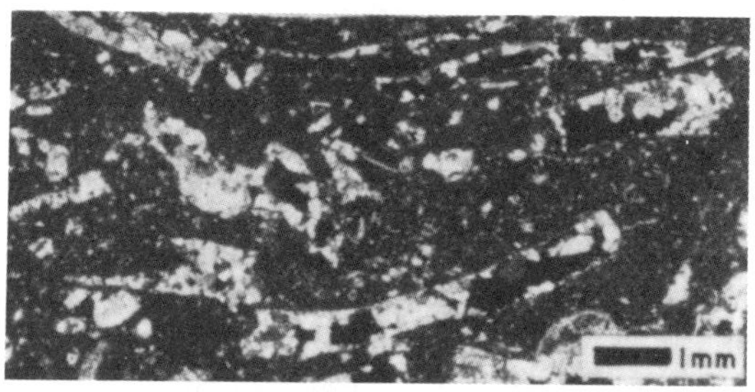

E

F

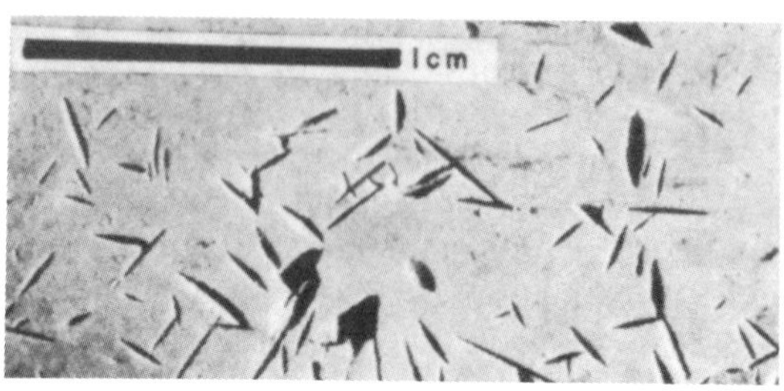

G

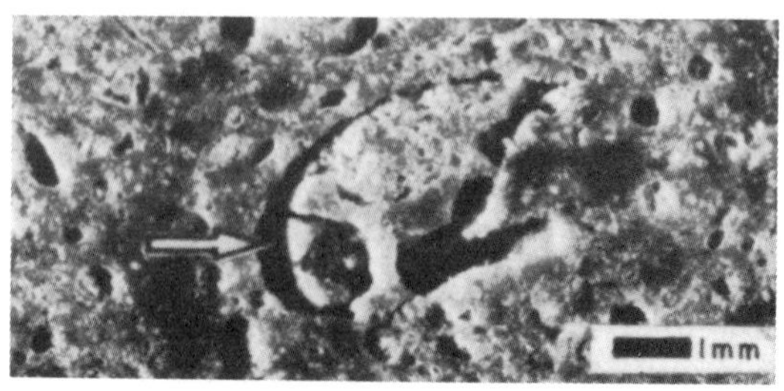

H

type term, such as fracture porosity where measurement could be assumed to mean width. In practice, size classification is made easier by the fact that average diameters of equant pores and cross-section diameters of tubular and platy openings are the most easily observed measures.

The pore-size classification is shown in the bottom part of Figure 2. We propose three main size classes for regular-shaped pores smaller than cavern size (diameter exceeds 256 mm): *micro-* for pores smaller than 1/16 mm, *meso-* for pores 1/16–4 mm, and *mega-* for pores larger than 4 mm. These class terms are used as prefixes to the word "pore" itself (*e.g.*, micropore or megapore) if only size is to be designated, or as prefixes to the basic porosity type if this is designated (*e.g.*, micromoldic, mesointercrystal, or megavug).

Two subdivisions of the mega- and meso-size ranges are useful, based on subclass interval ratios of eight. Subdivision of micropores is more difficult and has limited value in visual work. Subdivisions of the meso- size are *small* (1/16–½ mm) and *large* (½–4 mm). The two subclasses of megapores are likewise *small* (4–32 mm) and *large* (32–256 mm). Openings larger than about 256 mm (roughly "man size") in average diameter, cross section, or width are classified simply as "caverns" (see glossary).

There are significant advantages of such a grade scale permitting size prefixes to be combined with the word "pore" to form a complete size term, as in "micropore," "mesopore," or "megapore." Size characterization can be independent of judgments as to type of porosity. The term "vug" can be restricted more readily to the geologically more meaningful sense proposed in this classification, and terms like "micropore" and "small mesopore" can be used to replace the ill-defined and loosely used term "pinpoint" (see glossary).

Choices of boundaries for the size classes were guided in part by the natural selection or geologic breaks they achieve, and the manner in which the porosity can be observed. Micropores can be assessed only with a hand lens or microscope. Mesopores are normally observable in hand specimens and subsurface cores, and megapores are so large that generally they can be observed and described only in outcrop, or inferred from tool drops during the drilling of wells. These broadest size categories have significance as geologic fences across the wide range of pore sizes in carbonates. Porosity in chalks and earthy dolomites is dominantly microporosity. As shown by Archie (1952), carbonates of chalky or earthy texture generally differ significantly from coarser pore systems in such petrophysical attributes as porosity-permeability characteristics, fluid saturations, capillary-pressure behavior, or electrical resistivity. Most of the intraparticle, moldic, and fenestral porosity and the coarser interparticle and intercrystal porosity in limestones and dolomites are

←

FIG. 8.—Examples of moldic porosity.

A. Moldic porosity in pellet packstone. Many molds (black) are incomplete. Paradox Formation (Pennsylvanian), Ismay field, Utah. Thin section, cross-polarized light.

B. Moldic porosity in dolomitized pellet packstone. Molds (white) are dominantly of small mesopore sizes. Classification: small mesomoldic porosity (smsMO). Smackover Limestone (Jurassic), Texas. Thin section, plane light.

C. Moldic porosity in dolomite due largely to solution of calcitic crinoid columnals. Madison Group (Mississippian), Oregon Basin field, Wyoming. Polished core surface.

D. Fusulinid moldic or fusumoldic porosity in dolomitized wackestone. Some fusulinids were incompletely dissolved. San Andres Formation (Permian), Andrews Co., Texas. Polished core surface.

E. Reduced moldic porosity in phylloid algal wackestone. Molds of fragmented platy algae have been filled partly to completely by sparry calcite. Classification: cement-reduced moldic porosity (cr-MO). Paradox Formation (Pennsylvanian), Ismay field, Utah. Thin section, cross-polarized light. From Choquette and Traut (1963).

F. Gastropod mold in dolomitic packstone. Shell has been dissolved, leaving intraparticle filling of internal sediment and coarse calcite. Internal sediment and surrounding packstone matrix contain compound interparticle and intercrystal (in dolomitized micrite) porosity. Tertiary, Libya. Polished core surface.

G. Gypsum crystal molds in unfossiliferous, microcrystalline dolomite. Greenfield Formation (Silurian), Ohio. Polished surface. Specimen courtesy of C. H. Summerson (see Summerson, 1966).

H. Pelecypod (arrow) in dolomite containing abundant moldic porosity. Precursor responsible for molds can be noted by symbols such as MO_{pe} for this example; MO_{fu} for example in D; MO_{ga} for F; MO_{gy} for G; *etc.* Tertiary, Libya. Polished core surface.

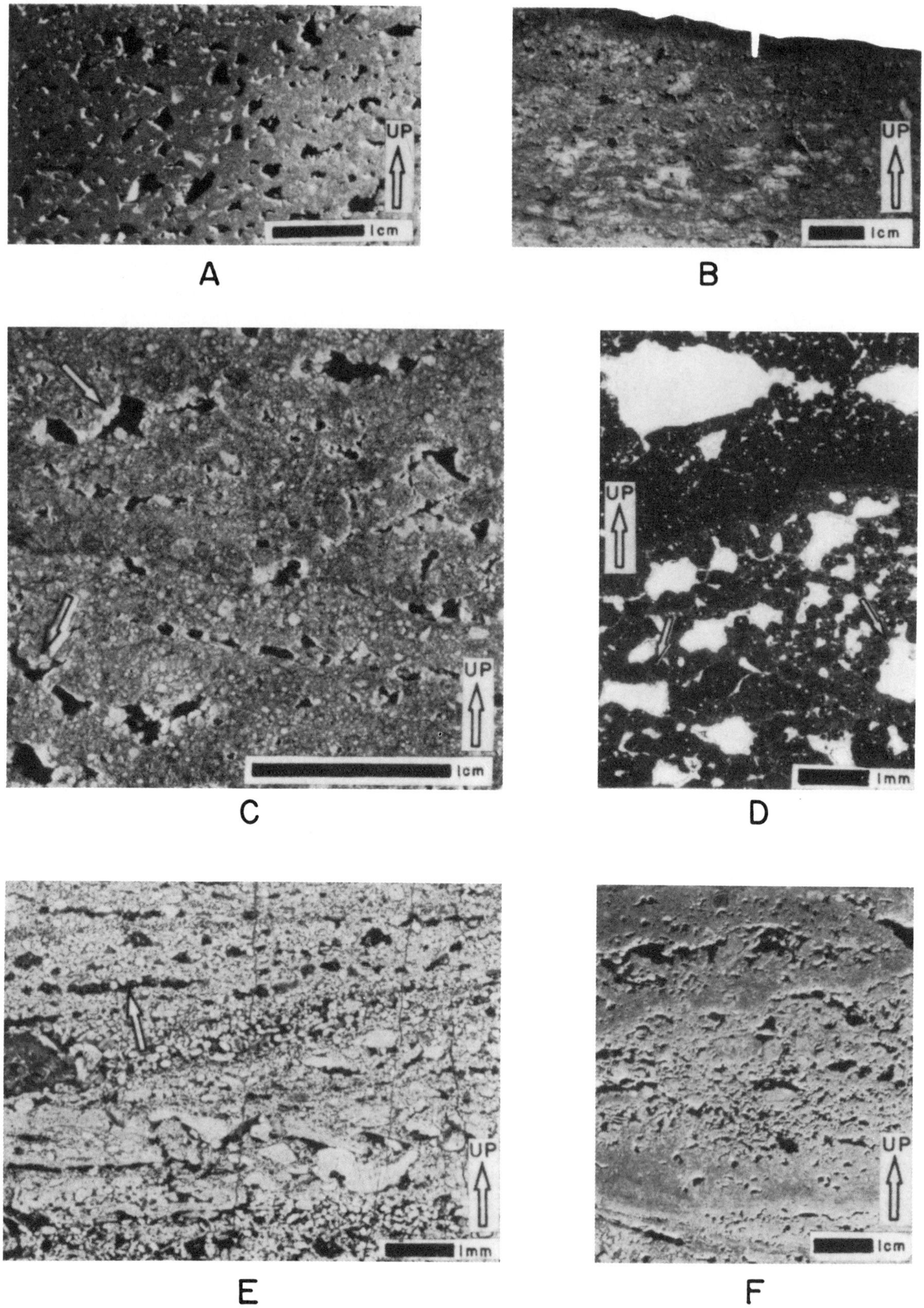
UP
1cm
A
UP
1cm
B
UP
1cm
C
UP
1mm
D
UP
1mm
E
UP
1cm
F

in the mesopore range. Most mesopores are fabric-selective types of pores, whereas many of the megapores represent types that are not fabric selective (channels, vugs, and caverns) and were formed by solution beneath erosion surfaces in the telogenetic zone.

Subclass boundaries within the mesopore range also correspond in part to natural groupings. The interparticle porosity of many pisolitic limestones, much moldic porosity due to solution of bioclastic debris, much interparticle porosity of coarse bioclastic limestones, and most fenestral porosity generally are predominantly in the large-mesopore size range (½–4 mm). Pore sizes in most oölitic limestones and significant amounts of the intercrystal pore space in sucrose dolomites that is coarser than microporosity (<1/16 mm) are in the small-mesopore size range (1/16–½ mm).

Porosity Abundance

Porosity abundance is a useful descriptive parameter that is hard to determine visually with accuracy. Thus, there is much use of qualitative terms such as "fair," "good," and "excellent," with varied amounts of porosity qualifying for each, depending on the perspective or purpose of the observer. This approach no doubt will continue. Our concern, however, is with more quantitative ways to express porosity abundance. Accurate visual estimation of porosity abundance is difficult with most porous carbonates, for several reasons: very large pores commonly cause sampling problems; a multiplicity of porosity types and complexities of size and shape make estimates inaccurate; and micropore systems, such as those found in many marls, chalks, and very finely crystalline dolomite, are more difficult to estimate than porosity of rocks with mesopores. Study of thin sections impregnated with colored plastic helps in determining porosity abundance in uniform types of mesopore and micropore systems. Experience with rocks for which porosity values have been determined by conventional core analysis, porosity analysis by gamma ray attenuation (Evans, 1965; Harms and Choquette, 1965), or other mass methods is helpful, particularly if the rocks are of the same facies as those being estimated. Suites of porosity reference standards for various rock types may permit direct visual comparison of measured porosity samples with "unknowns."

Porosity abundance normally is expressed as the percentage of the total sample volume occupied by pore space. In the notation of this classification (Fig. 2), abundance is stated in a percentage which follows the designation of the porosity type. If more than one basic type of porosity is present, the abundance of each may be useful. It can be noted as a percentage of each type or as a simple ratio of relative abundances of the two or more types, or as both a ratio and the total percent porosity abundance. Relative-abundance ratios are easy to make by quick visual inspection. Moreover, ratio estimates made in routine sample descriptions can be converted to absolute percentages once core-analysis data are available.

←

FIG. 9.—Examples of fenestral porosity.

A. Fenestral porosity in pisolitic wackestone dolomite. These fenestral pores, like those of examples **B-F**, are probably of secondary eogenetic origin. Pore sizes are dominantly in large-mesopore range. Classification: eogenetic, large mesofenestral porosity (SE-lmsFE). Tansill Formation (Permian), Guadalupe Mountains, New Mexico. Polished surface.

B. Fenestral porosity in irregularly laminated algal-mat dolomite. Mission Canyon Formation (Mississippian), North Dakota. Polished core surface.

C. Fenestral porosity in poorly laminated dolomitic packstone. Note large pore size in relation to grain size, and multigranular margins of pores (arrows), a distinctive and diagnostic feature of fenestral porosity. Phosphoria Formation (Permian), Cottonwood Creek field, Wyoming. Polished core surface.

D. More magnified thin-section view of rock shown in **C**, illustrating more clearly the multigranular roofs, walls, and floors of fenestral openings.

E. Completely filled former fenestral pores ("birdseyes") in laminated, pellet-intraclast limestone. Multigranular roofs characterized long former pores (arrow). Clear sparry calcite cement has filled pores. Classification: cement-filled fenestral porosity (cf-FE). To this characterization could be added descriptors noting pore size, inferred time of porosity origin, and even inferred time of cement filling. McLish Formation (Ordovician), Oklahoma. Polished surface. Specimen courtesy of W. E. Ham (see Ham, 1954).

F. Well-connected fenestral porosity in laminated pellet-intraclast dolomite packstone. Complex shapes of pores could be characterized as irregular (Ir) to equant (Eq). Tansill Formation (Permian), Guadalupe Mountains, New Mexico. Polished surface.

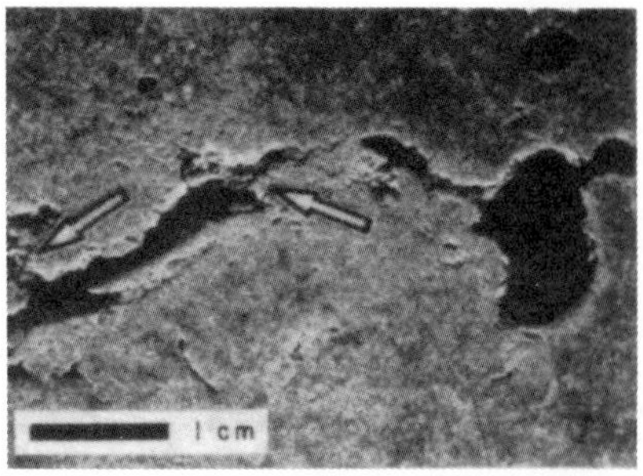

A

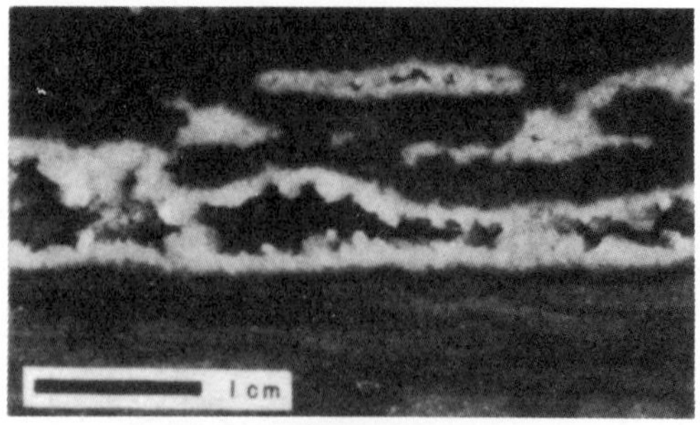

C

B

D

E

Porosity Descriptions and Code

A complete porosity description using the elements of this classification includes a designation of basic porosity type(s) and one or more accessory modifying terms relating to genesis, size (and shape), and abundance. The sequence in which these terms may be combined is shown in Figure 3, with examples illustrating various levels of descriptive detail. It is further illustrated in the captions of Figures 5 through 12.

Mnemonic letter symbols for the basic porosity types and most porosity modifiers, coupled with percentages and/or ratios for abundance, adapt this classification for brief porosity notation or coding for field or wellsite descriptions (Fig. 2). The symbols suggested in Figure 2 have proved easy to learn and useful.

The code symbols use upper case letters for the basic porosity types and the modifying terms "primary" (P) and "secondary" (S). Symbols of most modifying terms use lower case letters. The derivation of the symbols are apparent, except for the three porosity types, interparticle, intraparticle, and intercrystal, whose letter construction makes a direct mnemonic notation difficult. Code letters for these three basic types are the initial letters W for "within" (intra-) and B for "between" (inter-); thus the letter symbols are BP, WP, and BC, respectively. Vug is not abbreviated.

It can be useful to record pore-shape information directly, in place of or in addition to pore-size information, though in our experience shape modifiers are commonly unneeded. Pore shape can be expressed by the following symbols enclosed in parentheses: (Eq) for equant, (Tb) for tubular, (Pl) for platy, and (Ir) for irregular. In a code notation the shape modifier is placed just to the left of the pore-size modifier or basic porosity type symbol. Examples are (Eq)MO and (Tb)mgCH.

For some purposes it is desirable simply to record the size, abundance, or some genetic information about porosity without designating the basic porosity type. The symbol PO is used for porosity or pore, as in mcPO for micropores, S-PO for secondary pores, or PO(15%) for 15 percent porosity.

Compound and gradational porosity name and code designations involve the same basic construction as in Figure 3. To designate *compound* pore systems (Fig. 12), the individual porosity type and its modifying terms should be separated by the word "and." The most abundant porosity type should be listed last and followed by the porosity-abundance parameters; for example, "reduced moldic and reduced primary interparticle porosity (15%) (1:4)." For the porosity code, separate the individual porosity-type terms (and their modifier terms, if used) by a slash mark, so that the example just given would be represented as r-MO/rp-BP (15%) (1:4). *Gradational* porosity types (Fig. 11B) are separated by the word "to" if described in words, as in the description "solution-enlarged interparticle to channel porosity," and are separated by a long dash in code form, as in FR—CH.

Study of the illustrated examples of porosity and their code designations (Figs. 4–12), and some practice with actual rock specimens suffice to show the descriptive and interpretive leverage of the system, and the ease of learning it. For very detailed porosity characterizations, additional parameters can be added to

---

FIG. 10.—Examples of vug and channel porosity.

A. Channel pore system (CH) in dolomite. Large openings at left and right are interconnected in three dimensions. In places (see arrows) intercrystal porosity connects and is gradational with channel porosity. Leduc Formation (Devonian), Big Valley field, Alberta. Polished core surface.

B. Vug porosity in microcrystalline dolomite. A few vugs have been filled partly to completely by internal sediment (small arrows) prior to dolomitization. Vugs are mostly mesovugs (VUG). Leduc Formation (Devonian), Big Valley field, Alberta, Canada. Polished core surface.

C. Reduced channel porosity in dolomite. Channels of elongate to platy shapes that parallel lamination have been reduced or filled by cementation. Cement is coarsely crystalline dolomite. Classification code: cr-CH. Trenton Formation (Ordovician), Scipio field, Michigan. Polished core surface.

D. Irregular surface of non-fabric-selective megapore in dolomite. Distinction between large-scale channels and vugs in relatively small rock samples may not be possible. Devonian, Alberta. Two polished core surfaces.

E. Solution-developed megapores (cavern, solution-enlarged fractures, channels, and vugs) in bioclastic "lime" grainstone. Cavern is about 3½ m across. Solution development of cavern was selective to very permeable zone of primary interparticle porosity (interval shown by vertical bars) where fractures intersected this zone. Salem Limestone (Mississippian), quarry near Oolitic, Indiana.

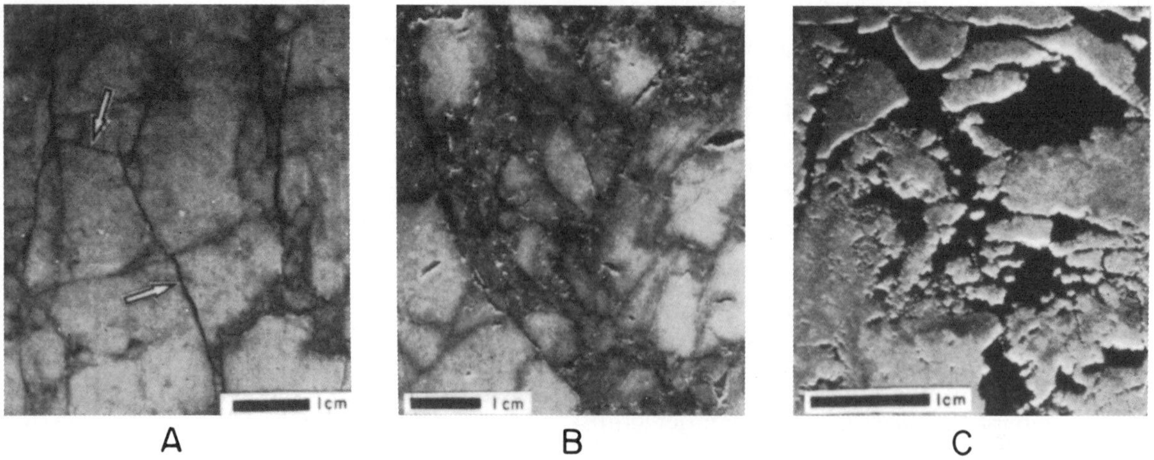

FIG. 11.—Examples of fracture and breccia porosity.

A. Fracture porosity in stylolitic "lime" mudstone. Dark gray patches are caused by oil stain. Madison Group (Mississippian), Oregon Basin field, Wyoming. Polished core surface.

B. Fracture porosity grading to breccia porosity (FR—BR) in microcrystalline dolomite. Porosity along microfractures is shown by oil stain (darker gray). Leduc Formation (Devonian), Big Valley field, Alberta. Polished core surface.

C. Solution-enlarged breccia porosity (sx-BR) in microcrystalline dolomite. Leduc Formation (Devonian), Big Valley field, Alberta. Polished core surface.

suit one's purpose. But for many uses, we find that simple combinations of only two or three of the parameters are satisfactory. The major advantage which the classification system provides is not, however, that of providing an easy method of characterizing porosity in sedimentary carbonates, but that of forcing more critical observations of the pores in relation to the enclosing rock. Use of the classification system should result in more accurate genetic interpretations.

CONCLUSIONS

1. The origin and modification of porosity are important for understanding sedimentary carbonates, and in exploring for and exploiting their fluids. A genetically oriented system of porosity nomenclature and classification helps to develop the requisite understanding.

2. Modifications in porosity are a major and commonly the predominant diagenetic process in most sedimentary carbonates. The vast reduction in porosity from the initial sediment to the negligible porosity of most ancient carbonates is accomplished largely by cementation. The volume of cement filling former pores approaches or exceeds the volume of the framework in many carbonate rocks.

3. Even though most porosity in limestones and dolomites can be related to primary features, many pores form after deposition (secondary).

4. Porosity in carbonate rocks is normally physically complex, genetically diverse, and distinct from that of other sedimentary rocks. Carbonate porosity generally differs significantly from that of sandstone (Table 1), with which it commonly is compared, in that the amount of pore space is ordinarily smaller; interparticle porosity is less important and intraparticle, intercrystal, moldic, and other types much more important; pore size and shape can be much more varied; and both the pre- and postdepositional periods are more important in forming and modifying porosity.

5. Pore space which reflects by its position and boundaries the depositional or diagenetic fabric elements of a sediment or rock is termed "fabric selective." Porosity formed early in diagenesis is commonly fabric selective, in contrast to much of the porosity formed later when unstable carbonate minerals and most or all former pore space has been eliminated. Much carbonate porosity is fabric selective.

6. The time of final deposition and burial provides a practical basis for subdividing the porosity history of sedimentary carbonates into three main stages: *predepositional, depositional,* and *postdepositional. Primary* porosity forms during the first two stages, and *secondary* porosity forms during the last one. The use of all these terms is independent of lithification.

7. Much postdepositional creation and modification of porosity occur either very early or

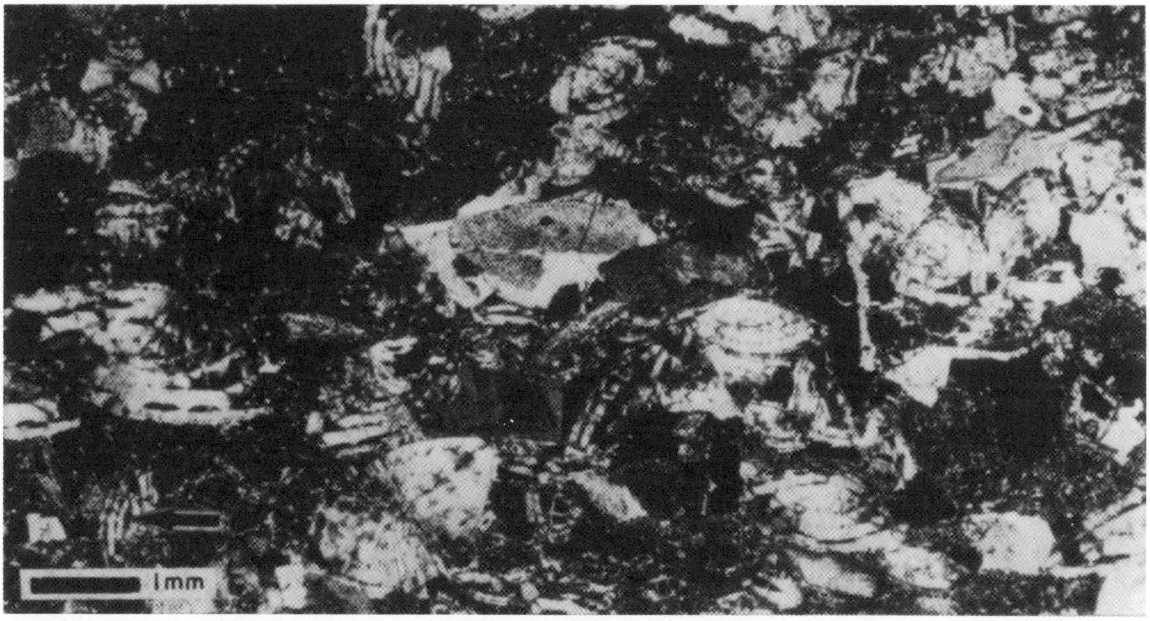

A

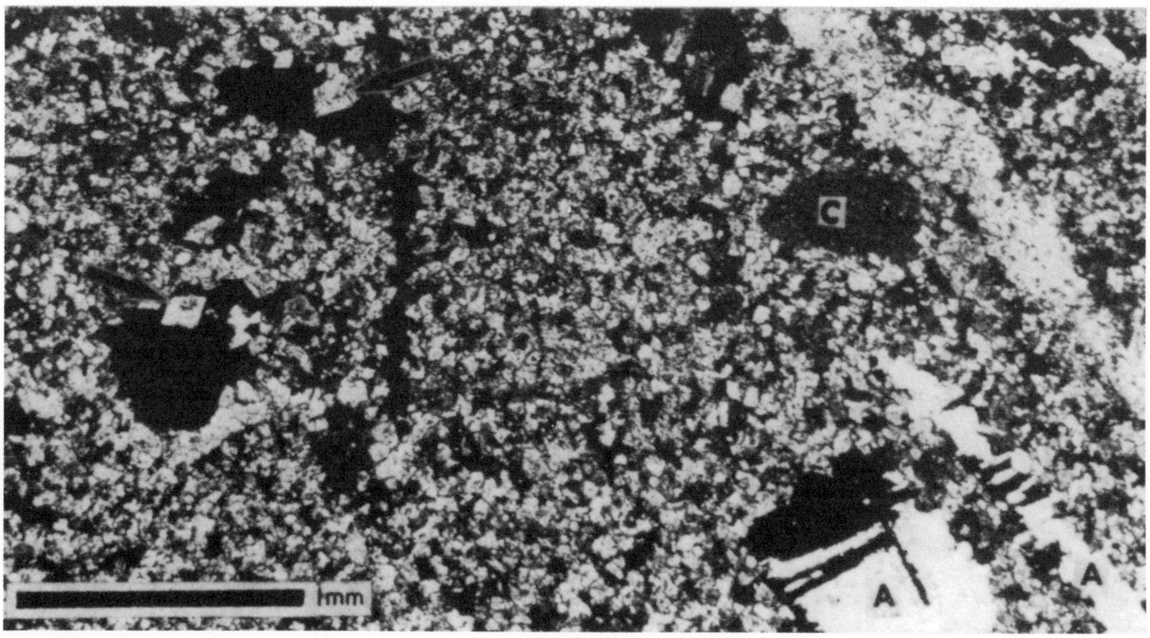

B

FIG. 12.—Examples of compound porosity types.

A. Mesopores (black) in skeletal packstone. Larger interparticle voids show evidence of solution enlargement. Some nummulitid forams contain small intraparticle mesopores (arrows). Notice partial pore fillings of calcite overgrowths on echinoderm debris, seen best in central part of photo. Classification code: smsWP/sx-lmsBP (1:10). Tertiary, Libya. Thin section, cross-polarized light.

B. Moldic and intercrystal porosity in sucrose dolomite. Molds (large black areas) have been filled in part by dolomite rhombs (arrows) and large anhydrite crystals, *A*. Several undissolved calcitic echinoderm fragments, *C*, are visible. Simple classification code representation would be cr-MO/BC; more complete designation would be cr-lmsMO/smsBC(1:12). Madison Group (Mississippian), Oregon Basin field, Wyoming. Thin section, cross-polarized light.

Table 2. Attributes Used to Define Basic Porosity Types
(Main attributes are indicated by "X" and attributes of lesser importance by "x." Detailed definitions are given in glossary)

Basic Porosity Type	*Size*	*Shape*	*Position in Fabric*	*Mode of Origin*	*Fabric Selective*	*Example (Fig. No.)*
Boring		x		X	Variable	6F, G
Burrow		x	X	X	Yes	
Breccia			X		Variable	11C
Cavern	X			x[1]	Uncommonly	10E
Channel		X		x[1]	No	10A, C
Fenestral	x[2]	x	X	x	Yes	9A–F
Fracture		x		X	Uncommonly(?)	11A, C
Growth framework			X	X	Yes	
Intercrystal			X	X[3]	Yes	7A–C, 12B
Interparticle			X		Yes	5A–E, 12A
Intraparticle			X		Yes	6D, E, 12A
Moldic				X	Yes	8A–H, 12B
Shelter	x[2]		X	X	Yes	6A, B
Shrinkage				X	Variable	
Vug		X		x[1]	No	10B, D, E

[1] Solution is the dominant process, but interpretation of process is not required for the definition.
[2] The size implication is that pore size is large in relation to the normal size of interparticle fabric elements.
[3] Intercrystal porosity applies largely to carbonate rocks composed of dolomite.

very late in burial history, when the sediment or rock is influenced significantly by surface-related processes. Therefore, it is useful to subdivide the postdepositional period into three main burial stages (Fig. 1): (a) the *eogenetic* stage, when newly deposited and/or recently buried deposits are subjected to processes operating from or related to a deposition surface or a surface of intraformational erosion; (b) the *telogenetic* stage, when long-buried rocks are affected by processes at or just below an erosion surface; and (c) the *mesogenetic* stage, or intermediate time of burial at depths below significant influence by surficial processes. These three terms also can be used to designate the porosity formed in each stage, the processes acting during each stage, or the corresponding burial zones.

8. Current porosity nomenclature can be improved by adding a few new terms and by sharpening or restricting the definitions of current terms. Key elements of the nomenclature we suggest are: (a) definition of primary and secondary and predepositional, depositional, and postdepositional as major porosity time terms; (b) recognition of the eogenetic, meso-

Table 3. Times[1] and Modes of Origin of Basic Porosity Types
(Letter symbols denote dominant, D; subordinate, s; and rare, r)

Basic Porosity Type	*Time of Origin Relative to Time of Final Deposition*			*Mode of Origin*			
	Before	*During*	*After*	*Framework Accretion*	*Sorting, Packing*	*Organic or Physical Disruption*	*Solution, Decomposition or Replacement*[1]
Boring	r[2]	s	D			D	
Breccia	r[2]	s	D		s	D	s
Burrow	r[2]		D			D	
Cavern		r	D			r	D
Channel	r[2]		D			r	D
Fenestral	r[2]	s	D	s?	s?	D	
Fracture	r[2]		D			D	
Growth framework	r[2]	D		D			
Intercrystal	r[2]	s	D[3]				D[3]
Interparticle	r[2]	D	s	s	D		s
Intraparticle	D	r	s	s	s	s	s
Moldic	r[2]		D				D
Shelter	r[2]	D			D		
Shrinkage	r[2]		D			D	
Vug	r[2]		D				D

[1] Exclusive of porosity of recycled extraformational rock fragments.
[2] This relates to porosity of individual particles, including intraformational clasts, that subsequently were moved to the site of final deposition.
[3] Intercrystal porosity of dolomites is of chief interest for purposes of this table.

genetic, and telogenetic time stages and corresponding burial zones; (c) restriction of "vug" and "channel" to pores of contrasting shape that are not fabric selective (see glossary); and (d) proposal of a size grade scale for porosity, the terms for which can be used as prefixes either to a porosity-type term (*e.g.*, micromold, mesovug) or to "pore" (*e.g.*, micropore, mesopore). A glossary with discussion of most porosity terms is appended to this article.

9. The geologic classification of porosity we propose incorporates most current nomenclature and the modifications cited, and is summarized in Figure 2. Its main elements are 15 basic porosity types defined by physical and/or genetic features. Of these types seven (interparticle, intraparticle, intercrystal, moldic, fenestral, vug, and fracture) are the dominant forms in sedimentary carbonates. Each basic type can be used independently or combined with modifying terms that give information about genesis, size, and abundance of porosity. Genetic modifiers pertain to the time of porosity origin, the process of porosity modification (solution, cementation, or internal sedimentation), and the direction or stage of porosity modification (enlarged, reduced, or filled). These genetic modifiers give the classification much of its interpretive value.

As a better understanding of porosity in sedimentary carbonates is developed, it undoubtedly will prove desirable to build more elaborate or different classifications of porosity, and an entirely new nomenclature may prove feasible for general and specialized purposes. Perhaps the system advocated here will speed these developments. In the interim we hope this article will help to focus more attention on the useful geologic information available from scrutinizing pores in relation to their carbonate host.

References Cited

Adams, J. E., 1953, Non-reef limestone reservoirs: Am. Assoc. Petroleum Geologists Bull., v. 37, p. 2566–2569.

——— and H. N. Frenzel, 1950, Capitan barrier reef, Texas and New Mexico: Jour. Geology, v. 58, p. 289–312.

——— and M. L. Rhodes, 1960, Dolomitization by seepage refluxion: Am. Assoc. Petroleum Geologists Bull., v. 44, p. 1912–1920.

American Geological Institute, 1960, Glossary of geology and related sciences with supplement: Washington, D.C., Am. Geol. Inst. Pub. 501, 397 p.

Archie, G. E., 1952, Classification of carbonate reservoir rocks and petrophysical considerations: Am. Assoc. Petroleum Geologists Bull., v. 36, p. 278–298.

Aschenbrenner, B. C., and G. V. Chilingar, 1960, Teodorovich's method for determining permeability from pore-space characters of carbonate rocks: Am. Assoc. Petroleum Geologists Bull., v. 44, p. 1421–1424.

Bathurst, R. G. C., 1959, The cavernous structure of some Mississippian *Stromatactis* reefs in Lancashire, England; Jour. Geology, v. 67, p. 506–521.

——— 1964, The replacement of aragonite by calcite in the molluscan shell wall, *in* J. Imbrie and N. Newell, eds., Approaches to paleoecology: New York, John Wiley & Sons, p. 357–376.

——— 1966, Boring algae, micrite envelopes and lithification of molluscan biosparites: Geol. Jour., v. 5, pt. 1, p. 15–32.

Beales, F. W., 1958, Ancient sediments of Bahaman type: Am. Assoc. Petroleum Geologists Bull., v. 42, p. 1845–1880.

——— 1965, Diagenesis in pelletted limestones, *in* L. C. Pray and R. C. Murray, eds., Dolomitization and limestone diagenesis—a symposium: Soc. Econ. Paleontologists and Mineralogists Spec. Pub. 13, p. 49–70.

Beales, F. W., and A. E. Oldershaw, 1969, Evaporite-solution brecciation and Devonian carbonate reservoir porosity in western Canada: Am. Assoc. Petroleum Geologists Bull., v. 53, p. 503–512.

Behrens, E. W., 1965, Environment reconstruction for a part of the Glen Rose Limestone, central Texas: Sedimentology, v. 4, p. 65–111.

Black, M., 1933, The algal sedimentation of Andros Island, Bahamas: Royal Soc. London Philos. Trans., ser. B, v. 222, p. 165–192.

Boekschoten, G. J., 1966, Shell borings of sessile epibiotic organisms as paleoecological guides (with examples from the Dutch coast): Palaeogeography, Palaeoclimatology, Palaeoecology, v. 2, p. 333–379.

Bonet, F., 1952, Urgonian facies of the middle Cretaceous in the region of Tampico: Asoc. Mexicana Geólogos Petróleros Bol., v. 8, p. 389–488.

Boyd, D. W., 1958, Observations on the Phosphoria reservoir rock, Cottonwood Creek field, Washakie County, Wyoming: Am. Assoc. Petroleum Geologists, Rocky Mountain Sec., Petroleum Inf., Geol. Rec., p. 45–53.

Bretz, J H., 1942, Vadose and phreatic features of limestone caverns: Jour. Geology, v. 50, p. 675–811.

Chilingar, G. V., 1957, A short note on types of porosity in carbonate rocks: Compass, v. 35, p. 69–74.

———H. J. Bissell, and K. H. Wolf, 1967, Diagenesis in carbonate rocks, *in* G. Larsen and G. V. Chilingar, eds., Diagenesis in sediments: Developments in Sedimentology, Amsterdam, Elsevier Pub. Co., v. 8, p. 179–322.

Choquette, P. W., and J. D. Traut, 1963, Pennsylvanian carbonate reservoirs, Ismay field, Utah and Colorado, *in* R. O. Bass, ed., Shelf carbonates of the Paradox basin: Four Corners Geol. Soc., p. 157–186.

Cloud, P. E., Jr., 1960, Gas as a sedimentary and diagenetic agent: Am. Jour. Sci. (Bradley vol.), v. 258-A, p. 34–45.

Davis, W. M., 1930, Origin of limestone caverns: Geol. Soc. America Bull., v. 41, p. 475–628.

Dunham, R. J., 1962, Classification of carbonate rocks according to depositional texture, *in* W. E. Ham, ed., Classification of carbonate rocks—a symposium: Am. Assoc. Petroleum Geologists Mem. 1, p. 108–121.

——— 1963, Early vadose silt in Townsend mound

(reef) in New Mexico (abs.): Am. Assoc. Petroleum Geologists Bull., v. 47, p. 356.
——— 1965, Vadose pisolite in the Capital Reef (abs.): Am. Assoc. Petroleum Geologists Bull., v. 49, p. 338.
Engelhardt, W. von, 1960, Der Porenraum der Sedimente: Berlin, Springer-Verlag, 207 p.
Etienne, J., 1963, Rock impregnation by colored resins for studying porosity in thin section: Inst. Français Pétrole Rev., v. 18, no. 4, p. 611–623.
Evamy, B. D., 1967, Dedolomitization and the development of rhombohedral pores in limestones: Jour. Sed. Petrology, v. 37, p. 1204–1215.
Evans, H. B., 1965, GRAPE—A device for continuous determination of material density and porosity: 6th Ann. Soc. Prof. Well Log Analysts Symposium Trans., Dallas, v. 2, p. B1-B25.
Fairbridge, R. W., 1967, Phases of diagenesis and authigenesis, *in* G. Larsen and G. V. Chilingar, eds., Diagenesis in sediments: Developments in Sedimentology, Amsterdam, Elsevier Pub. Co., v. 8, p. 179–322.
Fay, A. H., 1920, A glossary of the mining and mineral industry: U.S. Bur. Mines Bull. 95, 754 p.
Fischer, A. G., 1964, The Löfer cyclothems of the Alpine Triassic, *in* Symposium on cyclic sedimentation: Kansas Geol. Survey Bull. 169, p. 107–150.
——— and R. E. Garrison, 1967, Carbonate lithification on the sea floor: Jour. Geology, v. 75, p. 488–496.
Folk, R. L., 1959, Practical classification of limestones: Am. Assoc. Petroleum Geologists Bull., v. 43, p. 1–38.
Fraser, H. J., 1935, Experimental study of the porosity and permeability of clastic sediments: Jour. Geology, v. 43, p. 910–1010.
Friedman, G. M., 1965, Occurrence and stability relationships of aragonite, high-magnesian calcite, and low-magnesian calcite under deep-sea conditions: Geol. Soc. America Bull., v. 76, p. 1191–1196.
Gebelein, C. D., 1967, Origin and growth rates of subtidal algal stromatolites, Bermuda (abs.): Geol. Soc. America Prog. Ann. Mtg., New Orleans, p. 75; 1968, Geol. Soc. America Spec. Paper 115, p. 75.
Ginsburg, R. N., 1956, Environmental relationships of grain size of Florida carbonate sediments: Am. Assoc. Petroleum Geologists Bull., v. 40, p. 2384–2427.
——— ed., 1964, South Florida carbonate sediments: Geol. Soc. America Ann. Mtg. Field Trip Guidebook No. 1, Miami, Florida, 72 p.
——— E. A. Shinn, and J. H. Schroeder, 1967, Submarine cementation and internal sedimentation within Bermuda reefs (abs.): Geol. Soc. America Prog. Ann. Mtg., New Orleans, p. 78–79; 1968, Geol. Soc. America Spec. Paper 115, p. 78–79.
Ham, W. E., 1954, Algal origin of the "birdseye" limestone in the McLish Formation: Oklahoma Acad. Sci. Proc., v. 33, p. 200–203.
——— ed., 1962, Classification of carbonate rocks—a symposium: Am. Assoc. Petroleum Geologists Mem. 1, 279 p.
——— and L. C. Pray, 1962, Modern concepts and classifications of carbonate rocks, *in* W. E. Ham, ed., Classification of carbonate rocks—a symposium: Am. Assoc. Petroleum Geologists Mem. 1, p. 2–19.
Harbaugh, J. W., 1960, Petrology of marine bank limestones of Lansing Group (Pennsylvanian), southeast Kansas: Kansas Geol. Survey Bull. 142, pt. 5, p. 189–234.
——— 1967, Carbonate oil reservoir rocks, *in* G. V. Chilingar, H. J. Bissell, and R. W. Fairbridge, eds., Carbonate rocks, Pt. A, Origin, occurrence and classification: Developments in Sedimentology, Amsterdam, Elsevier Pub. Co., v. 9A, p. 349–398.
Harms, J. C., and P. W. Choquette, 1965, Geologic evaluation of a gamma-ray porosity device: 6th Ann. Soc. Prof. Well Log Analysts Symposium Trans., Dallas, v. 2, p. C1-C37.
Harris, W. H., and R. K. Matthews, 1967, Subaerial diagenesis of aragonitic carbonate sediments, Barbados, West Indies: Efficiency of the solution-reprecipitation process (abs.): Geol. Soc. America Prog. Ann. Mtg., New Orleans, p. 91–92; 1968, Geol. Soc. America Spec. Paper 115, p. 91–92.
Hohlt, R. B., 1948, The nature and origin of limestone porosity: Colorado School Mines Quart., v. 43, no. 4, 51 p.
Holmes, A., 1920, The nomenclature of petrology, with references to selected literature (1st ed.): London, Thomas Murby & Co., 284 p.
Howard, J. H., 1967, A classification of subsurface bodies of fragmented rocks: Am. Assoc. Petroleum Geologists Bull., v. 51, p. 945–951.
Howard, W. V., 1928, A classification of limestone reservoirs: Am. Assoc. Petroleum Geologists Bull., v. 12, p. 1153–1161.
——— and M. W. David, 1936, Development of porosity in limestones: Am. Assoc. Petroleum Geologists Bull., v. 20, p. 1389–1412.
Illing, L. V., 1954, Bahaman calcareous sands: Am. Assoc. Petroleum Geologists Bull., v. 50, p. 1–95.
——— 1959, Deposition and diagenesis of some upper Paleozoic carbonate sediments in Western Canada: 5th World Petroleum Cong. (New York) Proc., Sec. I, p. 23–52.
——— A. J. Wells, and J. C. M. Taylor, 1965, Penecontemporary dolomite in the Persian Gulf, *in* L. C. Pray and R. C. Murray, eds., Dolomitization and limestone diagenesis—a symposium: Soc. Econ. Paleontologists and Mineralogists Spec. Pub. 13, p. 89–111.
——— G. V. Wood, and J. G. C. M. Fuller, 1967, Reservoir rocks and stratigraphic traps in non-reef carbonates: 7th World Petroleum Cong. Proc., Mexico City, v. 2, p. 487–499.
Imbt, W. C., and S. P. Ellison, 1946, Porosity in limestone and dolomite petroleum reservoirs: Am. Petroleum Inst. Drilling and Production Practice, p. 364–372.
Ireland, H. A., 1951, Insoluble residues, *in* L. W. Leroy, ed., Subsurface geologic methods: Colorado School Mines, p. 140–156.
Jodry, R. L., 1966, Pore geometry of carbonate rocks (abs.): Am. Assoc. Petroleum Geologists Bull., v. 50, p. 619
Krynine, P. D., 1948, The megascopic study and field classification of sedimentary rocks: Jour. Geology, v. 56, p. 130–165.
Land, L. S., 1967, Diagenesis of skeletal carbonates: Jour. Sed. Petrology, v. 37, p. 914–930.
——— F. T. MacKenzie, and S. J. Gould, 1967, Pleistocene history of Bermuda: Geol. Soc. America Bull., v. 78, p. 993–1006.
Laporte, L. F., 1967, Carbonate deposition near mean sea-level and resultant facies mosaic: Manlius Formation (Lower Devonian) of New York State: Am. Assoc. Petroleum Geologists Bull., v. 51, p. 73–101.
Lees, A., 1964, The structure and origin of the Waulsortian (lower Carboniferous) "reefs" of west-central Eire: Royal Soc. London Trans., v. 247, p. 483–531.
Leighton, M. W., and C. Pendexter, 1962, Carbonate

rock types, *in* W. E. Ham, ed., Classification of carbonate rocks—a symposium: Am. Assoc. Petroleum Geologists Mem. 1, p. 33–61.
Levorsen, A. I., 1967, Geology of petroleum (2d ed.): San Francisco, W. H. Freeman & Co., 724 p.
Lindstrom, M., 1963, Sedimentary folds and the development of limestone in an Early Ordovician sea: Sedimentology, v. 2, p. 243–275.
Lucia, F. J., 1962, Diagenesis of a crinoidal sediment: Jour. Sed. Petrology, v. 32, p. 848–865.
——— and R. C. Murray, 1967, Origin and distribution of porosity in crinoidal rock: 7th World Petroleum Cong. (Mexico City) Proc., v. 2, p. 409–423.
Matthews, R. K., 1966, Genesis of recent lime mud in southern British Honduras: Jour. Sed. Petrology, v. 36, p. 428–454.
Milliman, J. D., 1966, Submarine lithification of carbonate sediments: Science, v. 153, p. 994–997.
Monty, C., 1965, Recent algal stromatolites in the windward lagoon, Andros Island, Bahamas: Soc. Géol. Belgique Annales, v. 88, Bull. no. 5–6, p. 269–276.
Murray, A. N., 1930, Limestone oil reservoirs of the northeastern United States and of Ontario, Canada: Econ. Geology, v. 25, p. 452–469.
Murray, R. C., 1960, Origin of porosity in carbonate rocks: Jour. Sed. Petrology, v. 30, p.. 59–84.
——— 1964, Preservation of primary structures and fabrics in dolomite, *in* J. Imbrie and N. D. Newell, eds., Approaches to paleoecology: New York, John Wiley & Sons, p. 388–403.
——— and F. J. Lucia, 1967, Cause and control of dolomite distribution by rock selectivity: Geol. Soc. America Bull., v. 78, p. 21–35.
——— and L. C. Pray, 1965, Dolomitization and limestone diagenesis, an introduction, *in* L. C. Pray and R. C. Murray, eds., Dolomitization and limestone diagenesis—a symposium: Soc. Econ. Paleontologists and Mineralogists Spec. Pub. 13, p. 1–2.
Muskat, Morris, 1949, Physical principles of oil production: New York, McGraw-Hill, 922 p.
Nuss, W. F., and R. L. Whiting, 1947, Technique for reproducing rock pore space: Am. Assoc. Petroleum Geologists Bull., v. 31, p. 2044–2049.
Pettijohn, F. J., 1957, Sedimentary rocks: New York, Harper & Brothers, 526 p.
Powers, R. W., 1962, Arabian Upper Jurassic carbonate reservoir rocks, *in* W. E. Ham, ed., Classification of carbonate rocks—a symposium: Am. Assoc. Petroleum Geologists Mem. 1, p. 122–192.
Pray, L. C., 1961, Compaction in calcilutites (abs.): Geol. Soc. America Bull., v. 71, p. 1946.
——— 1964, Limestone clastic dikes in Mississippian bioherms, New Mexico (abs.): Geol. Soc. America Spec. Paper 82, p. 154.
——— 1965, Limestone clastic dikes and marine cementation, Mississippian bioherms, southern New Mexico (abs.): Soc. Econ. Paleontologists and Mineralogists, Permian Basin Sec., Prog. Ann. Mtg., Midland, Texas, p. 21–22.
——— and P. W. Choquette, 1966, Genesis of carbonate reservoir facies (abs.): Am. Assoc. Petroleum Geologists Bull., v. 50, p. 632.
——— and J. L. Wray, 1963, Porous algal facies (Pennsylvanian) Honaker Trail, San Juan Canyon, Utah, *in* R. O. Bass, ed., Shelf carbonates of the Paradox basin: Four Corners Geol. Soc., p. 204–234.
Purdy, E. G., 1963, Recent calcium carbonate facies of the Great Bahama Bank. I. Petrography and reaction groups: Jour. Geology, v. 71, p. 334–355.
Rice, C. M., 1955, Dictionary of geological terms: Ann Arbor, Michigan, Edwards Brothers, 465 p.
Rittenhouse, G., 1959, There is a reason, *in* Reports and minutes of the 33rd Annual Meeting of the Society of Economic Paleontologists and Mineralogists: Jour. Sed. Petrology, v. 29, p. 289–291.
Robinson, R. B., 1966, Classification of reservoir rocks by surface texture: Am. Assoc. Petroleum Geologists Bull., v. 50, p. 547–559.
Roehl, P. O., 1967, Stony Mountain (Ordovician) and Interlake (Silurian) facies analogs of recent low-energy marine and subaerial carbonates, Bahamas: Am. Assoc. Petroleum Geologists Bull., v. 51, p. 1979–2032.
Rooney, L. F., 1966, Evidence of unconformity at top of Trenton Limestone in Indiana and adjacent states: Am. Assoc. Petroleum Geologists Bull., v. 50, p. 533–546.
Sander, N. J., 1967, Classification of carbonate rocks of marine origin: Am. Assoc. Petroleum Geologists Bull., v. 51, p. 325–336.
Schmidt, V., 1965, Facies, diagenesis, and related reservoir properties in the Gigas Beds (upper Jurassic), northwestern Germany, *in* L. C. Pray and R. C. Murray, eds., Dolomitization and limestone diagenesis—a symposium: Soc. Econ. Paleontologists and Mineralogists Spec. Pub. 13, p. 124–168.
Seilacher, A., 1964, Biogenic sedimentary structures, *in* J. Imbrie and N. D. Newell, eds., Approaches to paleoecology: New York, John Wiley & Sons, p. 296–316.
Shinn, E. A., 1964, Recent dolomite, Sugarloaf Key, *in* R. N. Ginsburg, ed., South Florida carbonate sediments: Geol. Soc. America Ann. Mtg. Field Trip Guidebook No. 1, Miami, Florida, p. 62–67.
——— 1968, Practical significance of birdseye structures in carbonate rocks: Jour. Sed. Petrology, v. 38, p. 221–224.
——— R. N. Ginsburg, and R. M. Lloyd, 1965, Recent supratidal dolomite from Andros Island, Bahamas, *in* L. C. Pray and R. C. Murray, eds., Dolomitization and limestone diagenesis—a symposium: Soc. Econ. Paleontologists and Mineralogists Spec. Pub. 13, p. 112–123.
Stanton, R. J., Jr., 1966, The solution brecciation process: Geol. Soc. America Bull., v. 77, p. 843–848.
Stout, J. L., 1964, Pore geometry as related to carbonate stratigraphic traps: Am. Assoc. Petroleum Geologists Bull., v. 48, p. 329–337.
Summerson, C. H., 1966, Crystal molds in dolomite: their origin and environmental interpretation: Jour. Sed. Petrology, v. 56, p. 221–224.
Taft, W. H., and J. W. Harbaugh, 1964, Modern carbonate sediments of southern Florida, Bahamas, and Espíritu Santo Island, Baja California—A comparison of their mineralogy and chemistry: Stanford Univ. Pubs. Geol. Sci., v. 8, no. 2, 133 p.
Tebbutt, G. E., C. D. Conley, and D. W. Boyd, 1965, Lithogenesis of a distinctive carbonate fabric: Wyoming Univ. Contr. Geology, v. 4, no. 1, 13 p.
Thomas, C. M., 1965, Origin of pisolites (abs.): Am. Assoc. Petroleum Geologists Bull., v. 49, p. 360.
Thomas, G. E., 1962, Grouping of carbonate rocks into textural and porosity units for mapping purposes, *in* W. E. Ham, ed., Classification of carbonate rocks—a symposium: Am. Assoc. Petroleum Geologists Mem. 1, p. 193–223.
Thrailkill, J., 1968, Chemical and hydrologic factors in the excavation of limestone caves: Geol. Soc. America Bull., v. 79, p. 19–46.

Todd, T. W., 1966, Petrogenetic classification of carbonate rocks: Jour. Sed. Petrology, v. 36, p. 317–349.

Waldschmidt, W. A., P. E. Fitzgerald, and C. L. Lunsford, 1956, Classification of porosity and fractures in reservoir rocks: Am. Assoc. Petroleum Geologists Bull., v. 40, p. 953–974.

Waring, W. W., and D. B. Layer, 1950, Devonian dolomitized reef, D-3 reservoir, Leduc field, Alberta, Canada: Am. Assoc. Petroleum Geologists Bull., v. 34, p. 295–312.

Appendix A

Glossary of Porosity Terms

In this Glossary most of the terms that have been used in the past few decades to characterize porosity in sedimentary carbonates are defined and/or discussed. The listing of terms is alphabetic. A usage is suggested for each term which either reflects prevailing usage as we understand it, or seems desirable in view of present knowledge about carbonate rocks. For some of the terms, the glossary gives the original definition and reviews significant subsequent usage. But for most terms, particularly the older ones and those which have evolved gradually and somewhat haphazardly from a nontechnical usage into more precise usage, details of the evolution of the term have little relevance.

The discussions of many terms not only consider definitions and usage, but briefly treat the geologic occurrence and/or the origin of the porosity features.

Birdseye, birdseye fabric, birdseye porosity.—In sedimentary carbonates, the term "birdseye" commonly is used for conspicuous, somewhat lens-shaped or globular masses of sparry carbonate cement a few millimeters to 1 cm or more in size. Although the term normally refers to either the sparry carbonate features themselves or to the carbonate rock containing them (Folk, 1959; Ham, 1954; Illing, 1959), it has also been applied to voids of like sizes and shapes; hence, the expression "birdseye porosity." Most birdseye features appear to be identical to what has been termed more recently "fenestral" (Tebbutt *et al.,* 1965). We recommend adoption of "fenestral" (*q.v.*) for the individual features whether open or infilled, and for the fabric of the rocks containing such features. The use of "fenestral" achieves more precision than "birdseye" and avoids possible confusion arising from the use of "birdseye" for lens-like or "augen" features of varied origins in nonsedimentary rocks.

Borings, boring porosity.—Openings created in relatively rigid constituents or rock by boring organisms. A rigid host is the feature which distinguishes borings from burrows; the latter form in unconsolidated sediment. Porosity created by boring organisms is not abundant in most ancient carbonate rocks, but borings constitute a distinctive and commonly genetically important minor type of porosity (Fig. 6F, G). Borings can be formed by a variety of organisms in a wide array of depositional or eogenetic environments and also can be formed in the telogenetic zone (Fig. 1). Recognition of borings (whether as porosity or as infilled openings) can be important in environmental and stratigraphic analysis. Discussions on borings in carbonate rocks and particles have been given by Ginsburg (1956), Behrens (1965), Bathurst (1964, 1966), Matthews (1966), and Boekschoten (1966).

Breccia porosity.—The type of interparticle porosity in a breccia. Breccias are rather common in many carbonate facies, but breccia porosity is only locally of quantitative importance. Carbonate breccias are of diverse origins (Howard, 1967). Some form by deposition of angular clasts. These depositional breccias may retain some primary porosity in the ancient geologic record if they were well sorted initially and were composed of relatively large particles; but typically, the more poorly sorted, matrix-rich depositional breccias, such as carbonate debris flows, retain negligible porosity. Postdepositional breccias form by fracturing of previously deposited sediment or rock. These can be termed "fracture breccia" and any associated porosity "fracture-breccia porosity." If the process responsible for fracturing is known, the fracture breccias can be identified more specifically as collapse breccias (Stanton, 1966), fault breccias, tectonic breccias, *etc.* Any associated breccia porosity can be designated similarly. Fracture-breccia porosity commonly intergrades with fracture porosity. We differentiate the two on the basis of the amount of displacement or chaos created by the fracturing process. If there has been rupture causing porosity by simple opening along a fracture, this is considered fracture porosity. But where there has been rupture and appreciable rotation or jumbling, resulting in chaotic rearrangement of the rock fragments, any resulting porosity is considered "fracture-breccia porosity." Illustrations of several rocks with breccia and fracture porosity are given in Figure 11. Beales and Oldershaw (1969) presented an excellent discussion of breccia-moldic porosity created by solution in Devonian oil reservoirs of western Canada.

Burrows, burrow porosity.—Features created by organic burrowing in relatively unconsolidated sediment, in contrast to borings, which formed in rigid sedimentary particles or rock. Porosity representing discrete burrows is relatively uncommon in ancient carbonate rocks, because most burrows collapse, become filled by sediment, or are backfilled by the burrow-forming organism itself. However, other types of porosity, such as interparticle porosity, may occur within burrow fillings in carbonate facies. In somewhat muddy facies, burrow fillings may have more permeability than the host sediment. Commonly these structures, perhaps because of higher permeability, appear to have localized dolomitization (Beales, 1958). In some carbonates the porosity is largely in dolomitized burrows. Features that aid in the recognition of burrows may be found in articles by Seilacher (1964) and Behrens (1965).

Cavern porosity.—A pore system characterized by large openings, or caverns. Although much cavernous porosity is of solution origin, the term is descriptive and not genetic. The term "cavern" or "cavernous" has been widely applied to porosity in carbonate rocks and reservoirs, and the only confusion in the use of the term is the size of opening large enough to warrant this designation. A practical lower size limit of "cavern" for outcrop studies is about the smallest opening an adult person can enter. Where the rock unit is known only from drilling, a practical lower size limit is that large enough to cause an easily recognizable drop of the drilling bit (a half meter or so). Cavern porosity is too large to be identified in subsurface cores of the usual diameters of only 7–12 cm.

Cellular porosity.—This term has diverse meanings. It appears to have been applied first in a technical sense to carbonate-rock porosity by Howard and David (1936, p. 1406), who used it for solution-formed molds and other generally equidimensional solution vugs, as opposed to more elongate channel-like openings. The term "cellular" also has been applied to organically created intraparticle openings within fossils, particularly chambered organisms such as corals or bryozoans. "Cellular" is little used currently to designate porosity. In view of its infrequent use, its diverse connotations, and the availability of more explicit terms, it seems best to abandon the term (see *intraparticle*).

Chalky, chalky porosity.—"Chalky" is a widely used surface-texture term denoting the distinctive dull and earthy character of many chalks, marls, and other minutely structured porous carbonates. We recommend continued use of the descriptive term "chalky" for such surface texture, as advocated by Ireland (1951, p. 146) and Archie (1952, p. 280). The term "chalky" also can be applied to the porosity of such very finely textured rocks. It is useful where a more specific size or porosity-type designation, such as "micropore," "microintraparticle," or "micromold," is not warranted. Chalky texture and chalky porosity are important in petroleum geology. Under low capillary pressures, chalky carbonates can have high water saturations and low oil saturations, and may form reservoir seals for petroleum. However, under extremely high capillary pressures, such as those in the upper part of an oil column several thousands of meters thick in the Asmari limestone of Iran, they can form productive reservoirs (Illing *et al.*, 1967, p. 490).

Channel, channel porosity.—A type of pore or pore system, here defined on the basis of its shape and origin. We propose that "channel" be used for markedly elongate pores or irregular openings with a marked elongation or continuity in one or two dimensions relative to a third dimension, and that it be applied only to such pores and openings which show by their boundaries or continuity that they have developed indiscriminately with respect to texture or fabric elements in the host rock. Thus, "channel" is a shape category of pore which does not display fabric selectivity. "Markedly elongate" is used to mean shapes whose lengths are 10 times or more their cross-section diameter or width. Channel pores or pore systems less than 1/16 mm in cross section or thickness are termed "microchannels." Most channels originate by indiscriminate solution along fracture systems or by lateral coalescence, through enlargement, of other types of pores.

The term "channel" normally does not connote the full range of elongate shapes or openings to which it is applied in the foregoing definition. But it has been used for similarly shaped pores in sedimentary carbonates. Howard and David (1936, p. 1403–1406) used "channel" specifically for solution-created porosity of an elongate or "continuous" nature in contrast to their "cellular" and "equisolution" porosity. Their discussions and illustrations indicate that some "channels" were tubelike or planar. Powers (1962, p. 140) used the term in a similar shape sense, noting that ". . . channels in some aphanitic limestones are commonly 0.5 mm wide and can be traced continuously along an irregular path for 10 to 20 mm." If a more specific shape connotation is desired than simply that of continuity, "channel" can be modified to denote more specifically various cross-sectional shapes along the direction(s) of elongation such as "tubular" or "platy." Examples of channel porosity are shown in Figure 10.

Channels constitute a basis type of porosity closely related to vugs by our definition (see *vug*). The two terms are contrasting shape end members. The more discrete, somewhat equant or equant-elongate voids are classed as vugs; the more markedly elongate pores or pore systems are channels. Channels, as defined heretofore, may grade into other porosity types. Gradations of channel and vug porosity are common. Other types of gradational porosity involving channel as one type are channel-to-fracture porosity, or combinations of solution-enlarged moldic or fenestral porosity with channel porosity. Although channels are classified as not showing fabric selectivity, indiscriminate solution may enlarge or cause coalescence of former pores and pore interconnections in a manner that creates large irregular pores of channel shape, but leaves at least part of the boundary showing fabric-selective relations to the host rock.

Compact.—A surface-texture term applied to rocks that break along smooth to conchoidal faces and generally have little or no porosity. The term was advocated in a classification of carbonate reservoir rocks by Archie (1952, p. 280) and currently has rather wide usage in this sense. "Compact" is useful as a gross indicator of low matrix porosity in a finely textured rock, and seems preferable to the somewhat synonymous term "dense," as it avoids connotations of mass.

Constructional void porosity.—See *growth-framework porosity.*

Continuous porosity.—A term proposed by A. N. Murray (1930) for systems of interconnected pores, in contradistinction to isolated or "discontinuous" porosity. The terms "continuous" and "discontinuous," though easily understandable, are little used now, and we see no reason to advocate their use. Other porosity terms also imply degree of continuity without needing a specific designation of "continuous" or "discontinuous." Somewhat related and more widely used terms are "effective" and "noneffective" porosity.

Dense.—See *compact,* the preferred term.

Discontinuous porosity.—Poorly connected or isolated pore chambers (Murray, 1930). See discussion of continuous porosity.

Disrupted porosity.—A general term proposed by Powers (1962, p. 140) for voids in some aphanitic limestones which are ". . . of irregular shape and result from diverse but unknown causes . . ." such as "burrowing organisms, entrapped gas, incomplete buckling and tearing of semiconsolidated mud, and slumping." Though little used by subsequent workers, it perhaps has utility as a "wastebasket" term. We prefer more specific porosity terms, such as "shrinkage" or "fenestral," where possible.

Earthy.—This is a widely used surface-texture term, with usage commonly like that of "chalky" (Archie, 1952, p. 280), denoting the dull, unreflective, finely textured appearance of many porous chalks, marls, carbonaceous micrites, and microcrystalline dolomites. The term "earthy" has been applied more restrictively by Thomas (1962, p. 194) to slightly argillaceous carbonates, and commonly implies porosity values as well as particle sizes similar to those of rocks with chalky textures. The term "earthy porosity" is useful if rock examination is made with the unaided eye or at low magnification and observable detail is insufficient to allow a more specific identification of porosity types.

Effective porosity.—The "intercommunicating void space of a rock" (Muskat, 1949, p. 114). As it is usually the effective and not the total porosity that is measured in standard core-analysis procedures, in petroleum engineering practice the term "porosity" normally means "effective porosity" (Muskat, 1949, p. 114).

Eogenetic stage, eogenetic porosity.—"Eogenetic stage" (Greek *eos:* dawn, early), a term herein proposed, applies to the time interval between final deposition and the time when the newly deposited sediment or rock is buried below the depth of significant modification by processes that either operate from the surface, or whose effectiveness is dependent upon proximity to the surface. Eogenetic porosity is that formed in the eogenetic stage. (See discussion in Part 2.)

Fenestra, fenestral fabric, fenestral porosity.—The term "fenestra" (pl. fenestrae) was proposed by Tebbutt *et al.* (1965, p. 4) for a ". . . primary or penecontemporaneous gap in rock framework, larger than grain-supported interstices." They specifically applied the term "fenestra" both to the void and to the feature resulting from partial to complete infilling of the original void. "Fenestra," "fenestral fabric," and "fenestral porosity" are broadly equivalent to the widely used terms "birdseye," "birdseye fabric," and "birdseye porosity" of most carbonate literature (*e.g.*, Folk, 1959; Ham, 1954; Illing, 1959). We prefer "fenestra(e)" and "fenestral" to "birdseye," partly to recognize the contribution made by Tebbutt *et al.* in defining the critical aspect of these types of features in carbonates—namely that the opening is larger than the grain-supported interstices—and partly because many geologists use "birdseye" in other connotations. The only modification we suggest in the Tebbutt *et al.* definition is application of "fenestra(e)" and "fenestral" to both mud-supported and grain-supported fabrics, provided the opening is larger than normal interparticle openings.

Fenestrae, fenestral fabrics, and fenestral porosity, illustrated in Figure 9, are important and distinctive features in sedimentary carbonates and warrant specific recognition. They run the entire textural gamut of clastic carbonates from carbonate mudstones to grainstones and some boundstones (textural rock classification terms of Dunham, 1962). Petroleum reservoirs formed of rocks containing fenestral porosity are a distinctive and important type of "specific reservoir facies" (Pray and Choquette, 1966) in strata of many ages. Many distinctive fabrics or pore systems in ancient carbonate facies interpreted as algal mat during the past decade are of fenestral nature.

Fenestrae occur as somewhat rounded features of spherical, lenticular, or more irregular shapes; their large size in comparison to normal interparticle openings and their multigranular roofs, floors, and other margins are key characteristics (Fig. 9D). Fenestrae are commonly somewhat flattened parallel with the laminae or stratigraphic planes of the rock. However, they may be round or very irregular, and some are elongate in a vertical dimension. Although isolated fenestrae occur in sedimentary carbonates, it is more common to find many in close association. They are most abundant along obscure partings or laminae in the rock. Such fabrics are termed "laminoid fenestral" by Tebbutt *et al.* (1965, p. 4). Rock with a well-developed fenestral porosity paralleling lamination normally has greater horizontal than vertical permeability.

Fenestral porosity is almost certainly polygenetic. A detailed discussion of its origin is given by Tebbutt *et al.* (1965, p. 11–13), who suggested numerous mechanisms of formation, chiefly involving decay of sediment-covered algal mats, shrinkage during drying, and accumulation of pockets of gas or water. Well-developed fenestral fabrics in Triassic backreef facies of Austria have been described by Fischer (1964), who emphasized a sediment-shrinkage origin and suggested the term "löferite" for rock fabrics "riddled with shrinkage pores" (fenestrae). Pores formed by gas bubbles in Holocene sediment cores described by Cloud (1960) closely resemble fenestrae. Features identical to many fenestrae of ancient rocks can be formed in carbonate sand and mud samples in the laboratory by acidizing the solution to generate small amounts of gas. We believe most fenestrae form by gas evolution and sediment distension shortly after deposition, an interpretation supported by recent observations by Shinn (1968).

Although fenestral porosity and fenestral fabrics need not be restricted to any particular environment, concentrations of fenestrae along laminae are usually interpreted, perhaps correctly, as indicators of high intertidal or supratidal algal-mat sedimentation. However, caution is needed in interpreting environments on the basis of these distinctive features of fabrics. Well-developed fenestral fabric is also prominent in many calcareous spring deposits (travertine), and certainly the generation of gas within a sediment need not be restricted to supratidal or intertidal environments. For a further sampling of geologic literature regarding fenestral fabrics of both Holocene and ancient deposits, see Black (1933), Boyd (1958), Shinn (1964), and Shinn *et al.* (1965).

Fracture porosity.—Porosity formed by fracturing. "Fracture porosity" generally is used for porosity occurring along breaks in a sediment or rock body where there has been little mutual displacement of the opposing blocks. Fracture porosity grades into breccia porosity with increasing dislocation or "chaos" (see Fig. 11; *breccia porosity*). In carbonate rocks, fractures and hence fracture porosity may originate in diverse ways, such as by collapse related to solution, slumping, or various kinds of tectonic deformation. A detailed geometric classification of fractures in reservoir rocks is given by Waldschmidt *et al.* (1956).

Growth-framework porosity.—Primary porosity created by the in-place growth of a carbonate rock framework. "Growth-framework porosity," as herein defined, is reserved specifically for the pore space of rock frameworks known or inferred to have grown in place as rigid or semirigid fabrics. It applies to "boundstones" rather than to "clastic carbonates" in Dunham's (1962) terminology. Growth-framework porosity can form by organic and/or inorganic processes. It may consist of a variety of geometric or genetic subtypes. The most important of these is "growth *intra*framework porosity," which designates such openings as the internal chambers of colonial framework organisms. "Growth *inter*framework porosity" is created where elements of the growth structure, such as the platy arms of the colonial coral *Acropora palmata* or irregular sheets of the alga *Lithoporella* or *Archaeolithothamnium* intergrow in such a manner as to isolate voids from sedimentation. Unusual types of growth-framework porosity occur in travertine and in pisolitic caliches (Dunham, 1965; Thomas, 1965).

Porosity occurring in "rigid or semirigid . . . sediment frameworks" was one of three general categories of primary porosity in carbonate rocks discussed by R. C. Murray (1960, p. 61), who used the term "constructional void porosity" for the "primary porosity of carbonate frameworks." Our definition of growth-framework porosity is drawn from Murray's concept of constructional void porosity, but is more restrictive in that we exclude intraparticle openings of any individual organisms or particles that were clastic components of the rock.

Although growth-framework porosity as applied to the primary porosity of in-place carbonate deposits may seem conceptually distinct from the primary intraparticle porosity of clastic limestones, the distinction may not be sharp, and differentiation in a specific carbonate rock can be arbitrary and troublesome. How large must the growth unit be to qualify as a growth framework or boundstone? Frameworks can range in scale from microscopic tests of individual organisms, through the size of oncolites or clams, to the massive organic growth frameworks of barrier reefs. The answer must be varied to fit the purpose of the classifier. We find it practical to restrict the use of "growth-framework porosity" to rock of hand-specimen or larger size in which the boundstone origin can be inferred.

Quantitatively, growth-framework porosity is a relatively minor type of primary porosity in ancient carbonate facies. This scarcity reflects not only the vast preponderance of clastic carbonates in the geologic record, but also the fact that even in facies interpreted as growth frameworks, such as true reefs, the porosity of intimately associated clastic carbonates commonly may exceed that of the framework itself.

Intercrystal porosity.—Porosity between crystals. Although this simple definition could apply in a strict sense to almost all porosity in carbonates, "intercrystal" (or "intercrystalline") normally is restricted to the porosity between individual crystals of somewhat equant and equal size, as in many porous dolomites (Fig. 7). Intercrystal porosity may be of either primary or secondary origin. The primary porosity of depositional fabrics composed of crystals such as halite and other evaporites that have grown in place is an intercrystal type.

Intergranular porosity.—Porosity between grains (*cf. interparticle porosity*). "Intergranular" is the most commonly used term for between-grain porosity in clastic sandstones and carbonates alike. Occasionally, but unfortunately, it has been considered synonymous with primary porosity; for example, the porosity classification of Fraser (1935), adopted for petroleum reservoir rocks by Levorsen (1967, p. 113), recognized two major categories, primary or intergranular porosity and secondary or intermediate porosity. It seems preferable to use "intergranular" to designate only the position of the porosity with respect to grains, independent of the time of its formation.

"Intergranular porosity" is used more commonly than the broader term, "interparticle porosity." If "intergranular" is used, we suggest it be restricted to between-grain porosity of carbonates whose particles are coarse enough to be considered "grains." The lower size limit of "grains" ranges from 0.004 to 0.06 mm in the major carbonate rock classifications (Ham, 1962).

Interparticle porosity.—Porosity between particles (*cf. intergranular porosity*). Interparticle porosity denotes position and not genesis. In clastic carbonates, interparticle porosity is generally of depositional (primary) origin, as illustrated in Figures 5A, B, and E, 6A-C, and 12A. But it also can form by several postdepositional processes of which the predominant mechanism is selective solution of finer textured matrix or micrite from between larger particles. Examples of solution interparticle porosity in carbonate packstones are shown in Figure 5C and D. The discrimination, if possible, between a primary depositional origin and a secondary origin is important and commonly difficult. Interparticle porosity is the dominant type of porosity in most carbonate sediments; it is important in some carbonate rocks but is unimportant in others.

We advocate the term "interparticle porosity" as the general name for between-particle porosity in sedimentary carbonates, in preference to the more widely used "intergranular porosity." "Interparticle" is being adopted increasingly following its use by R. C. Murray (1960). "Interparticle" is the broader of the two terms because "particle" has no lower size limit, unlike the term "grain" as defined in some modern carbonate rock classifications (Ham, 1962). Thus, "interparticle porosity" can be used for "lime muds" or micrites as well as for coarser clastic carbonates. In a strict sense, neither "grain" nor "particle" is satisfactory, as etymologically both imply small size. Yet current sedimentologic usage dating at least from Krynine (1948) recognizes no upper size limit to grains, and the same seems warranted for particles or clasts.

Many specialized geometric and genetic types of interparticle porosity can be recognized. Some related porosity types are fenestral and shelter porosity and some types of breccia and growth-framework porosity (*q.v.*).

Intracrystal porosity.—Porosity within individual crystals, pores in large crystals of echinoderms, and fluid inclusions form most of this category of porosity.

Intragranular porosity.—Porosity within individual grains. Despite wide usage of "intragranular," we prefer to use "intraparticle" as the general term.

Intraparticle porosity.—Porosity within individual particles or grains (see *interparticle porosity*). "Intraparticle porosity" as used here is a physical, positional, not a genetic porosity type. It is abundant in carbonate sediments and can be an important part of the preserved porosity in carbonate rocks. In spite of its importance, it has been neglected in comparison to other major types of carbonate porosity. Much intraparticle porosity in carbonates forms before final deposition of the sedimentary particle or grain (predepositional porosity); some forms during or after final deposition. Internal chambers or other openings within individual or colonial skeletal organisms are the most commonly recognized intraparticle pores. However, an appreciable amount of the primary intraparticle porosity in carbonate sediments consists of pore space within individual pellets, intraclasts, oöids, and other nonskeletal grains. It may be pertinent to designate specific types of intraparticle porosity as "intrabiotic," "intracoral," "intrapellet," *etc.* Intraparticle porosity of postdepositional origin forms chiefly by solution and borings. Examples of types of intraparticle porosity are shown in Figures 6 and 12A.

Matrix porosity.—The porosity of the matrix or finer portion of a carbonate sediment or rock, in contrast to porosity associated with the coarser particles or constituents; or the porosity of "blocks" of the rock in contrast to the porosity of the fractures.

Megapore, megaporosity.—Size terms herein proposed for large pores. Megapore is the largest of three pore-size classes (see pore-size discussion; *mesopore* and *micropore*). The "mega-" size designation is used for equant to equant-elongate pores whose average diameter is larger than 4 mm, and for tubular or platy pores whose average cross-sectional diameter or thickness, respectively, is larger than 4 mm. "Mega-" can be combined with "pore" as in "megapore," or with the basic porosity type as in "megamold" or "megamoldic porosity."

Mesogenetic.—A term herein proposed to designate the postdepositional stage between the time when newly buried deposits are acted upon chiefly by processes related to the depositional interface, and the time when long-buried deposits are acted upon primarily by processes operating from the erosional interface. The term also can be used to designate porosity formed in this stage, processes acting during this stage, or the burial zone in which they operate. (See discussion in Part 2.)

Mesopore, mesoporosity.—Size terms herein proposed for intermediate-size pores (see pore-size discussion; *megapore* and *micropore*). The "meso-" size designation is used for equant to equant-elongate pores whose average diameter is between 4 and 1/16 mm, and for tubular or platy pores whose average cross-sectional diameter or least diameter, respectively, is between 4 and 1/16 mm. "Meso-" can be combined with "pore" as in "mesopore" or with the basic porosity type, as in "mesomold" or "mesomoldic porosity."

Micropore, microporosity.—Size terms herein proposed for microscopic pores. Micropore is the smallest of three pore-size classes (see pore-size discussion; *mesopore* and *megapore*). The "micro-" designation is used for equant to equant-elongate pores whose average diameter is less than 1/16 mm and for tubular or platy pores whose average cross-sectional diameter or least diameter, respectively, is less than 1/16 mm. "Micro-" can be combined with "pore" as in "micropore" and "microporosity," or with the basic porosity type, as in "micromold" or "micromoldic porosity."

Mold, moldic porosity.—A mold is a pore formed by the selective removal, normally by solution, of a former individual constituent of the sediment or rock such as a shell or oölith. The resulting porosity is moldic. In general geologic usage, a mold is a "negative" of a former feature and two types, mold cavities and mold impressions, long have been recognized in sedimentary rocks. The terms "mold porosity" and "moldic porosity" have been used sporadically for many years by insoluble-residue workers (1946 Residue Conference reported *in* Ireland, 1951). But the use of "moldic" (or "mold") as a general name for a class of porosity in carbonate literature during the past decade has been meager. The term deserves wider usage. Molds are extremely abundant in many porous carbonate rocks. Many of the so-called "vugs" of petroleum reservoir facies are molds and should be so designated. Molds and moldic porosity are identified on the basis of shape, size, wall ornamentation, or relict features.

Most molds in sedimentary carbonates are created by the selective solution of various types of carbonate depositional particles. Especially common in limestones are molds of primary aragonitic constituents, notably oölites and molluscan shells. Molds in dolomite commonly have formed by selective solution of either aragonite or calcite primary constituents, and less commonly by solution of anhydrite (Murray, 1960, p. 80), gypsum (Summerson, 1966), or halite. A complex genesis of molds by dedolomitization has been documented by Evamy (1967). Decomposition of organic matter, such as that in the plant roots, stalks, or twigs, forms molds. Figure 13 portrays diagrammatically the origins of most molds in carbonate rocks, and examples of moldic porosity are shown in Figures 8 and 12B.

An unusual type of moldic porosity, apparently formed by organic decay and known in ancient rocks largely from features now completely filled by cement, is interpreted to have formed some of the types of structures loosely termed "Stromatactis" in the lower Carboniferous "reef knolls" of England and Ireland. The former cavities are believed (Bathurst, 1959; Lees, 1964) to record the presence of an unknown type of a somewhat rigid-bodied, noncalcareous organism that did not decay until at least partial cementation of the host sediment had occurred. Cavities like these may be more widespread in ancient carbonates than generally has been recognized. The oncolite biscuits described in Holocene carbonates of Florida (Ginsburg, 1964), the Bahamas (Monty, 1965), and Bermuda (Gebelein, 1967) could easily form a related type of moldic porosity.

If the identity of the mold is known, it can be added to the porosity term, a long-established practice with terms like "oömold" or "oömoldic" and "dolomold" or "dolomoldic" (Ireland, 1951, p. 146). A few other

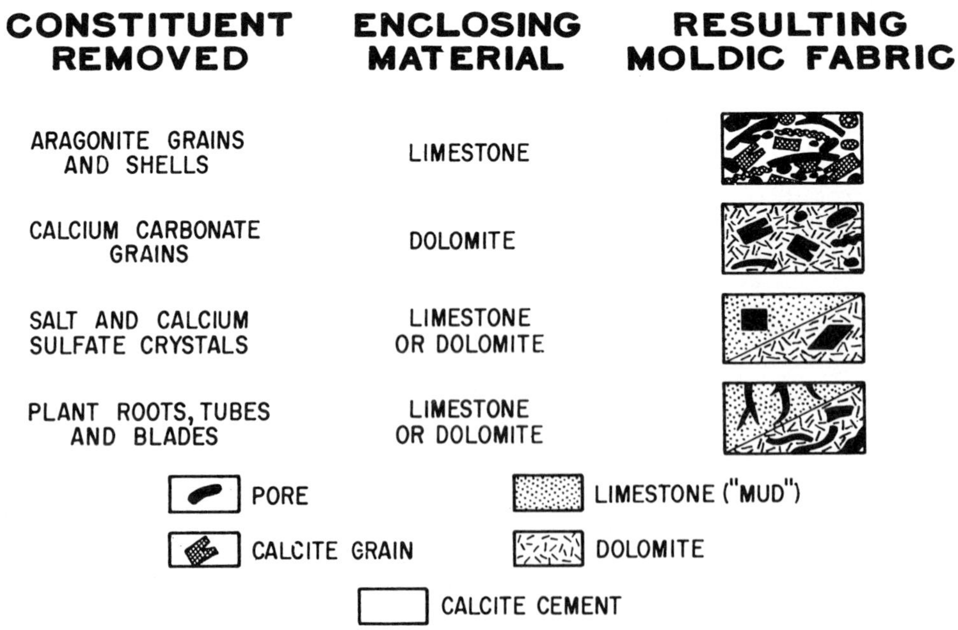

FIG. 13.—Common forms of moldic porosity in carbonate rocks.

examples that we have found useful are "pelmoldic" (pellets), "fusumoldic" (fusulinids), and "crimoldic" (crinoids). See breccia-moldic porosity of Beales and Oldershaw (1969).

Pinpoint porosity.—A wastebasket term for very small pores in carbonate rocks. This term has been used almost exclusively in the petroleum industry, usually in a semitechnical sense. It has no genetic implications. Pore sizes to which "pinpoint" has been applied range from a few hundredths of a millimeter to about a millimeter in diameter. Most commonly, "pinpoint" has been used to designate pores that are barely discernible to the unaided eye (about 0.1 mm in diameter). The term has been used most commonly for rocks with much less porosity than chalks or earthy dolomites.

The term "pinpoint" is most useful for simple, nontechnical, characterizations of pore size. For detailed characterization we prefer to note both size and type of porosity more specifically, either by giving dimensions and pore type directly or by using the size grade scale terms proposed in this article. Most pores of pinpoint sizes are micropores or small mesopores of moldic or intercrystal porosity types.

Primary porosity.—Porosity formed during final sedimentation or present within sedimentary particles at the time of deposition. The term "primary porosity" includes all predepositional and depositional porosity of a particle, sediment, or rock. It also applies to any remnant of the primary porosity (see Part 2).

Saccharoidal.—A textural term, essentially synonymous with "sucrose," referring to a textural resemblance to common table sugar (see *sucrose*).

Secondary porosity.—Porosity formed in a sediment or rock subsequent to final deposition. The term "postdepositional porosity" is essentially synonymous with "secondary porosity" as defined here (see Part 2).

Shelter porosity.—A type of primary interparticle porosity, proposed herein, created by the sheltering effect of relatively large sedimentary particles which prevent the infilling of pore space beneath them by finer clastic particles. Shelter porosity is a common and important type of primary porosity in many clastic limestones. It is found in many oölitic grainstones containing coarse shell fragments or intraclasts, and in carbonate packstones and wackestones and some carbonate mudstones containing platy skeletal particles. Excellent examples of shelter porosity are the "umbrella-like" structures of Harbaugh (1960), and other examples are shown by Dunham (1962). Shelter porosity also can be formed beneath marine grasses and other decomposable organic materials and may be preserved if rigidity was created by cementation prior to decomposition. Illustrations of shelter porosity are shown in Figure 6A-C.

As shelter pores are commonly larger than most of the associated interparticle primary pores, they tend to be sites of preserved primary porosity in many rocks whose finer interparticle pores have been filled by cementation. This tendency has been important in the favorable porosity and permeability of some petroleum reservoir facies, for example, those containing abundant phylloid algae in the Pennsylvanian of the Paradox basin (Choquette and Traut, 1963; Pray and Wray, 1963). Large shelter pores enhance rock permeability and may partly localize subsequent solution enlargement in the formation of vug and channel porosity.

Shrinkage porosity.—Porosity produced by sediment shrinkage. Drying commonly produces shrinkage porosity, but other processes can create contraction cracks (shrinkage porosity) in aqueous environments. Although most shrinkage porosity is a specialized type of fracture porosity, it can be formed by shrinkage of individual sedimentary particles (Schmidt, 1965).

Sucrose, sucrosic.—"Sucrose" and the common variant, "sucrosic," have been used in both a semitechnical and a technical sense to describe carbonate rocks that have appreciable intercrystal pore space and are composed dominantly of somewhat equant, uniformly sized, euhedral to subhedral crystals. Sucrose texture implies appreciable intercrystal pore volume, relatively uniform crystal size, and enough free planar surfaces on the crystals to create a host of reflecting surfaces. Many porous dolomites have a sucrose texture; aggregates of calcite crystals having sucrose texture are very uncommon. We prefer to use "sucrose" as a textural term and not as a porosity term (*cf.* Harbaugh, 1967). Examples of dolomites with sucrose texture are shown in Figure 7.

The terms "sucrose," "sucrosic," and "saccharoidal" are derived from the textural resemblance to common table sugar. The analogy to cube or lump sugar is particularly close. Some geologists follow the standard English-language dictionaries (Oxford, Webster's Unabridged, or Random House) and use "saccharoidal" as the textural term, reserving "sucrose" for the chemical disaccharid compound of common sugar. British geologists appear to have followed this usage more carefully than American geologists. Despite the dictionary authority, the word "sucrose" has had much more extensive usage than "saccharoidal" as a textural term for carbonate rocks (*e.g.*, Archie, 1952; Illing *et al.*, 1967; Levorsen, 1967; Murray, 1960). "Sucrose" seems too firmly entrenched to be eliminated, and as it is the simpler of the two terms we recommend it.

The crystal sizes of what have been termed "sucrose" dolomites range from that of common table sugar (normally 0.1–0.5 mm) to much smaller. Crystal sizes coarser than 0.5 mm are very unusual. Murray (1960, p. 67) states that the crystal size of sucrose dolomite ". . . ranges from less than 5 μ to more than 100 μ, with 25 to 50 μ being very common in reservoir rocks." Sucrose dolomites of interbedded carbonate-evaporite facies commonly have crystal sizes in the range cited by Murray. Most porous dolomites in which the individual crystals are so small as to be difficult to discern even with a hand lens are termed "chalky" or "earthy" by American geologists.

Although "sucrose" is useful as a textural term, the term "sucrose porosity" (*cf.* Harbaugh, 1967) has questionable merit, for although sucrose texture implies intercrystal porosity, many sucrose dolomites contain other types of porosity as well. Moldic porosity, in particular, is a common and important contributor to many sucrose dolomite reservoir rocks, and vug, channel, or fracture porosity also may be significant. We prefer to designate explicitly the porosity between crystals of sucrose carbonates as intercrystal, and to designate any associated porosity as moldic, fenestral, or other basic types.

Telogenetic stage, telogenetic porosity.—Telogenetic stage (Greek *telos:* end, completion), a term herein proposed, applies to the time interval during which long-buried carbonate rocks are influenced significantly by processes associated with weathering and subaerial and subaqueous erosion. Telogenetic porosity is that formed in the telogenetic stage. (See discussion in Part 2.)

Total porosity.—The designation for all the porosity of a rock, regardless of whether it is interconnected or isolated (*cf. effective porosity*).

Vug, vug porosity.—"Vug" and its derived forms, "vuggy" and "vugular," are probably the most commonly used carbonate porosity terms, and also have had the most widely varied definitions and usages. We define "vug" as a pore that (1) is somewhat equant, or not markedly elongate, (2) is large enough to be visible with the unaided eye (diameter greater than 1/16 mm), and (3) does not specifically conform in position, shape, or boundaries to particular fabric elements of the host rock (*i.e.*, is not fabric selective). Solution is the dominant process in formation of vugs as we use the term, but an interpretation of process is not essential to our definition. Most vugs may represent solution enlargement of fabric-selective pores to such an extent that the identity of the precursor cannot be discerned, and at a stage in diagenesis when solution is apparently indiscriminate of fabric elements. The evolution of vugs from molds, which undoubtedly are common precursors of vugs, is illustrated diagrammatically in Figure 4.

The definition of "vug" proposed here, involving size, shape, and the absence of fabric selectivity, is more restrictive than most earlier definitions and usages. It permits the recognition of an important type of porosity that was not specifically differentiated heretofore but was used in a broader sense by many geologists. The new definition also permits integration of vug with the other basic types of porosity in our classification (Part 3).

In our terminology, vugs and channels are similar in that neither is fabric selective. They differ in shape, however; "vug" is used for the more equidimensional pores and "channel" is used for pores more markedly elongate or continuous in one or more dimensions. We arbitrarily separate the two shape categories at a ratio of length to average cross-sectional diameter of about 10:1. Because of the complex shapes of pores in carbonate rocks that might qualify as vugs or channels, actual diameter ratios rarely are calculated; precision generally is not practical or necessary. The compound term "vug-channel" can be useful for pore systems displaying a range of shapes across this boundary. If practical, vug shapes that are somewhat elongate (diameter ratios 3:1 to 10:1) can be differentiated from the more equant forms.

In our usage, vugs, like other basic porosity types, can be designated more specifically as to size by terms such as "mesovug" (1/16–4 mm), "small mesovug" (1/16–½ mm), or "megavug" (4–256 mm). Those relatively few pores that are smaller than 1/16 mm but otherwise fit our definition of "vug" can be termed "microvugs." The practice of indiscriminately calling any visible pore a vug, whether or not it can be identified more exactly or informatively, can be abandoned in favor of combining a size and/or shape designation with the word "pore," as in "mesopore" or "platy megapore." If terminologic anarchy persists in the usage of "vug," a new term should be coined for the porosity here designated as "vug."

The widespread use of "vug," its importance in the classification system we propose, and its varied usage justify more discussion than has been accorded most of the other terms in this glossary. "Vug" seems to have been derived from the Cornish word, *vooga*, meaning underground chamber. Holmes (1920, p. 238) used "vug" or "vugh" as a mining term for ". . . an unfilled cavity . . . generally with a mineral lining of different composition from that of the immediately surrounding ore." Fay (1920) cited "a cavity in the rock, usually lined with a crystalline incrustation," and noted the alternate spellings, "voog," "vough," and "vugg." Rice (1955, p. 444) cited the definitions of both Holmes and Fay. Pettijohn (1957, p. 217) defined vugs differently as "solutional structures" that are "... irregular openings related in origin to the 'phreatic' passages produced by ground water action . . . common in carbonate rocks." The AGI Glossary (1960, p. 317) cited Holmes and Fay, but added the more specialized usage of "vug" by many petroleum geologists for " . . . any opening from the size of a small pea to the size of a boulder." Applied to carbonate porosity, "vug" usually connotes "large" size and a somewhat equant shape, as implied in the quoted AGI definition. Two points of difference among geologists are "what size?" and whether "vug" should be applied to any type of porosity of the requisite size and shape regardless of the nature of the porosity. Moreover, some geologists seem to designate all visible pore spaces as "vugs," making "vug" synonymous with "pore."

A few examples in recent literature further illustrate the use of "vug." Murray (1960, p. 61, 81, Figs. 17, 18) used the term in a sense essentially synonymous with "pore" and applied it to a variety of specific types of porosity, many of a fabric-selective type, as in "primary constructional vugs," "anhydrite mold vugs," and "intercrystal vugs." Likewise, Thomas (1962, Pls. 1–3) illustrated interparticle (intraparticle), intrabiotic, intercrystal, and birdseye (fenestral) types of vugs. Others differentiate vugs from interparticle and intercrystal pores (Illing *et al.*, 1967, p. 488; Waring and Layer, 1950, p. 310). Vugs are generally differentiated from more elongate or continuous openings such as channels (*e.g.*, Powers, 1962, p. 140–141; Rooney, 1966, p. 540). Though some use "vug" only for pores of a "pea to boulder size," others apply it to pores as small as ¼-1 mm in diameter or less (*e.g.*, R. C. Murray, 1960, Figs. 17, 18; Robinson, 1966, p. 550; Shinn *et al.*, 1965, Fig. 7a, b; Thomas, 1962, Pls. 1, 2; Waring and Layer, 1950, Fig. 13).

The examples cited are exceptions, however, to the more general practice of indiscriminately lumping porosity of many basic types, sizes, and shapes into a wastebasket category termed "vug." This practice clearly has resulted in the loss of information that might have helped in identifying and interpreting pores in carbonates. The definition of "vug" proposed here, involving size, shape, and restriction to pores that are not fabric selective, is suggested to encourage more critical observations and descriptions of porosity in carbonate facies, without abandoning the term itself.

First published in *Journal of Petroleum Technology,* V. 29,
(July 1977), pp. 873-885.

Distribution and Continuity of Carbonate Reservoirs

D. Jardine, Imperial Oil Ltd.
D.P. Andrews, SPE-AIME, Imperial Oil Ltd.
J. W. Wishart, Imperial Oil Ltd.
J. W. Young, Imperial Oil Ltd.

Introduction

Production experience in Canada includes carbonates ranging in age from Ordovician to Triassic. This experience encompasses the entire domain of carbonate reservoir types, from the small pinnacle reefs of Silurian and Middle Devonian age in Ontario and the Rainbow basin of northern Alberta, through the large Leduc and Slave Point reef complexes of central Alberta, to widespread shoals and carbonate banks of Devonian, Mississippian, and Triassic age extending over much of the Western Canada basin.

Carbonate reservoirs are characterized by extreme heterogeneity of porosity and permeability, often within a single pool. They range from massive, vuggy, and fractured reservoir types in the organic- reef facies to highly stratified, often vertically discontinuous reservoirs in the back-reef and shoal facies.

Depletion plans for these pools, most of which are on enhanced recovery operation, require detailed, integrated geological-engineering studies. Initially, reservoir description, consisting of lithofacies, correlation cross-sections, and fluid saturation studies, is used to develop reservoir engineering numerical models. Continued monitoring of operating performance is essential to assure that the geological model is valid.

The first part of this paper describes, in general terms, the geometry and internal porosity distribution of various types of carbonate bodies and the diagenetic influences that can modify the original textures. To relate this to real cases, four typical carbonate pools in Western Canada are discussed in the second part. Each pool has unique reservoir characteristics that require detailed integration of the geological-engineering disciplines for optimum development. In the third part, one of these pools, the Judy Creek Devonian reef reservoir, is described in detail to illustrate how integrated geological-engineering studies can change the operating practices within a pool.

Porosity of Carbonates

Primary Porosity

Carbonate rock types shown schematically in Fig. 1 and in the core photographs in Fig. 2 are found in a wide variety of depositional settings. These settings range from reefs covering less than 1 sq mile to carbonate banks that often extend over thousands of square miles. In Fig. 1, the types are arranged in order of decreasing energy of the depositional environment. There is a parallel decrease in average grain size, pore size, and permeability.

Biohermal reefs are characterized by their relatively small size, high degree of relief above the sea floor, steeply sloping sides, and a high percentage of skeletal material. An interior lagoonal facies of finer, more bedded carbonates is usually present. Porosity and permeability are highest in the skeletal-rich rim sediments (Fig. 2a). The interior is less porous and often has poor vertical permeability because of interbedding of porous and non-porous sediments. Biohermal reefs tend to develop when basin subsidence is relatively rapid.

Biostromal reefs are more sheet-like deposits that are associated with slowly subsiding basins. They can extend over hundreds of square miles. A high percentage of skeletal and coarse detrital material is usually present, as indicated in Fig. 2b. Highest porosity is found near the seaward edge but some degree of porosity is present throughout the deposit. Stratification is pronounced so that horizontal permeability is often higher than vertical permeability by at least an order of magnitude.

Carbonate banks are thick, tabular deposits of very large areal extent formed by buildup of carbonate sediments across a basin during extended periods of stillstand of sea level. Porosity is often confined to a narrow rim along the seaward edge or to local patches in the interior. Since bank

Carbonate reservoirs are characterized by extreme heterogeneity of porosity and permeability. This is related to the complexities of the original depositional environment and the diagenetic influences that can modify the original textures. Four pools are discussed to illustrate the relationship between internal reservoir geometry and pool performance.

carbonates are largely composed of lime mud (Fig. 2c), low porosity and permeability is the rule unless this has been enhanced by fracturing, dolomitization, or leaching.

Shoal carbonates are mound-like features containing a higher proportion of skeletal debris or other coarse-grained material than the enclosing fine-grained sediments (Fig. 2d). They are formed in various depths of water in a variety of geographic positions, including the interior portions of both reefs and banks as well as more basinal settings. Porosity is mainly of the intergranular type.

Nearshore deposits include beach, tidal flat, and supratidal deposits that often are subjected to minor fluctuations in sea level, resulting in cyclic sequences of evaporites and fine-grained carbonates. These sequences usually are thin, have a restricted areal distribution, and are strongly stratified (Fig. 2e). Although porosity in individual beds may be very high, permeability rarely exceeds a fraction of a millidarcy.

In summary, the predominant feature of carbonate porosity is heterogeneity, even within a single pool. Pore systems in sandstones are generally relatively simple, being mainly intergranular and controlled by the size of the grains and the amount of matrix and secondary cement. Carbonate pore systems, on the other hand, are usually more complex. The primary intergranular porosity is more variable because of the diversity of grain size and shape; in addition, the skeletal grains themselves often have cellular porosity characteristic of the original organism. Also, because of later diagenetic effects, the original pore systems often are modified to more complex pore networks.

Secondary Porosity

The five processes of porosity alteration that are noted in Fig. 1 and illustrated in Fig. 3 occur commonly in carbonates.

Leaching (Fig. 3a) generally has a favorable effect on reservoirs because it tends to improve porosity and markedly enhances permeability.

Dolomitization can play a dual role. Generally, it will improve a reservoir by increasing the pore size, as indicated in Fig. 3b; at other times, porosity may be destroyed when dolomitization creates a dense interlocking crystal fabric.

The influence of dolomitization on reservoir properties has been evaluated empirically by comparing undolomitized with dolomitized reefs in Western Canada. Limestone reefs have an average porosity of 8 percent and an average permeability of 68 md, whereas dolomitized reefs have an average porosity of 9 percent and a permeability of 800 md. These comparisons suggest that dolomitization does not significantly change porosity but that it does increase permeability. This increase can be attributed to better development of solution vugs in dolomites and to fracturing, which is more intense in dolomites because of their brittle nature.

Fracturing (Fig. 3c), which is most common in tectonically active areas, can create permeability in rocks where none existed previously and can form a pathway for leaching or cementing solutions. Fracturing often materially increases the producibility of carbonate reservoirs; but from an operational standpoint, it can also result in high water rates because of channeling effects.

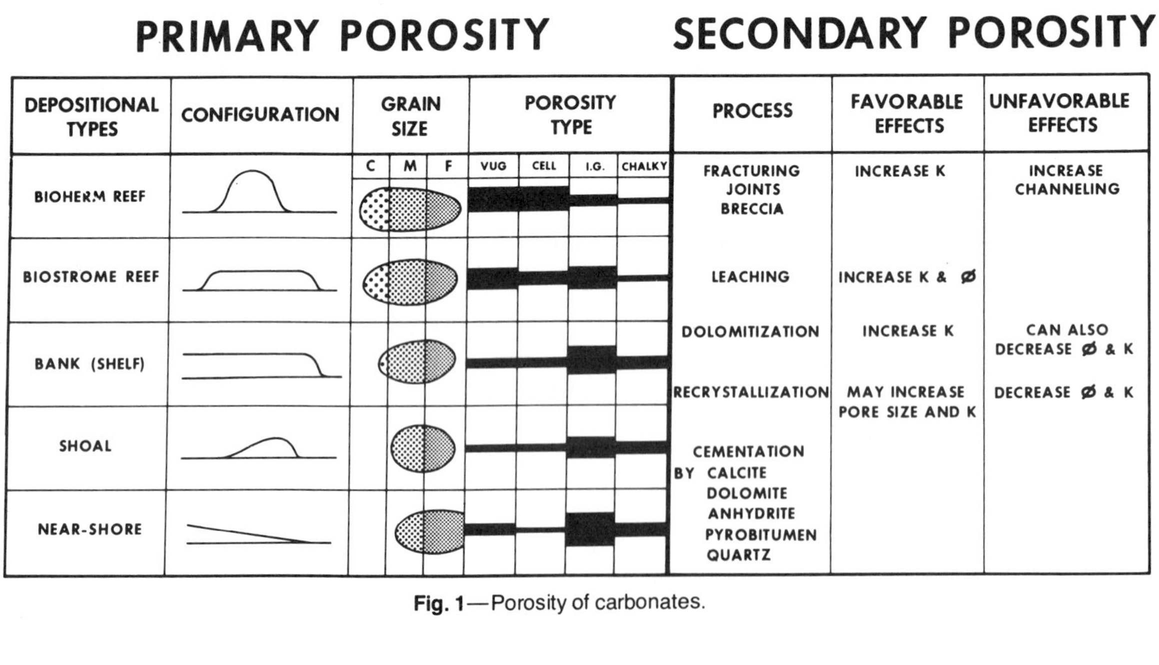

Fig. 1—Porosity of carbonates.

Recrystallization or micrite enlargement is generally a reservoir enhancing process that produces larger calcite crystals and a "chalky" texture (Fig. 3d). Carbonates with chalky textures can have high porosity that is evident on wireline logs, but very low permeability that is not so easy to determine without cores or production tests.

There are many different types of cements that can destroy porosity in carbonates. Even small amounts of certain cements such as pyrobitumen can seriously affect productivity because of the tendency to occupy the narrow connections between pores, thereby reducing the permeability. An example of cementation by calcite is shown in Fig. 3e.

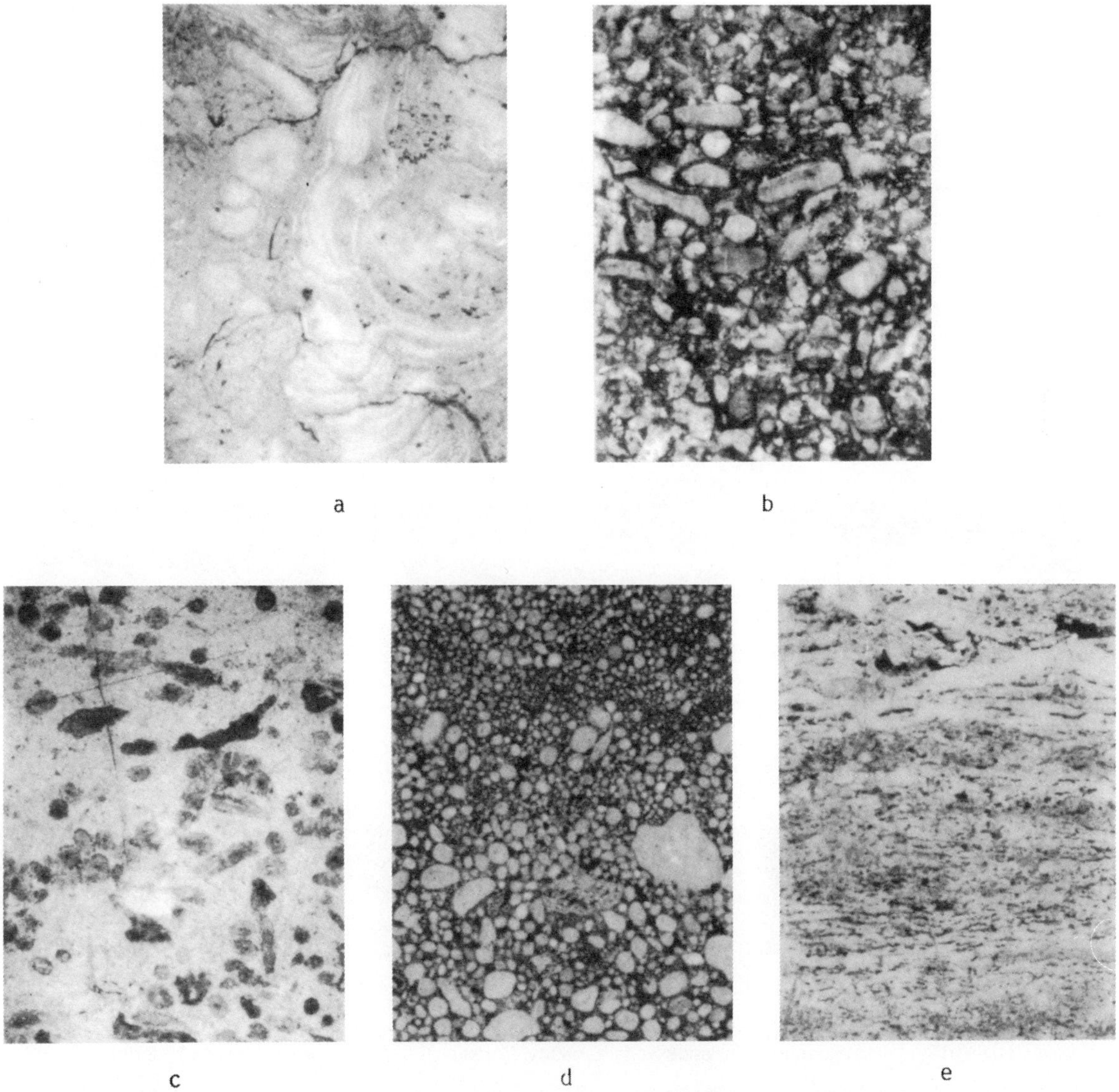

a—Bioherm reef. Stromatoporoid organic-reef framework typical of bioherm margins. Organic-cell porosity predominates. Porosity, 14 percent; permeability (k_{90}), 35 md.

b—Biostrome reef. Skeletal conglomerate composed of Stromatoporoid and Coral fragments. The original intergranular and organic-cell porosity have been destroyed by calcite cementation. Porosity, < 1 percent; permeability (k_{90}), < 0.001 md.

c—Bank (shelf). Micritic-skeletal (Amphiporid) limestone. Interior portions of banks characteristically have a high proportion of micrite (lime mud). Porosity in this sample is mainly of the organic-cell and chalky types. Porosity, 3 percent; permeability (k_{90}), 1 md.

d—Shoal. Oolitic-lump limestone with excellent intergranular porosity. Porosity, 18 percent; permeability (k_{90}), 700 md.

e—Nearshore. Algal laminite typical of the intertidal zone. The vuggy (birdseye) porosity is distinctly layered. Porosity, 9 percent; permeability (k_{90}), 200 md.

Fig. 2—Examples of deposition types.

The Geology and Pool Development of Four Carbonate Fields

The various depositional environments and diagenetic changes just described can strongly affect the pool performance of carbonate reservoirs and, therefore, the production practices used in their development and depletion. This cause-and-effect relationship is best illustrated by describing four oil and gas fields from Western Canada.

Boundary Lake

Boundary Lake field (Fig. 4), located in British Columbia, produces oil from Triassic nearshore carbonates. The field

c d e

a—Leached fossil casts in a skeletal micritic limestone. Some types of skeletal material are much more susceptible to leaching than others because of original differences in mineralogy. Porosity, 9 percent; permeability (k_{90}), 72 md.

b—Dolomitized skeletal-micritic limestone with leached fossil (Amphiporid) casts. Secondary vuggy and intercrystalline porosity predominate. Porosity, 16 percent; permeability (k_{90}), 300 md.

c—Fracture porosity in a dolomitized fore-reef breccia. The fractures have been partly infilled with white, coarsely crystalline dolomite cement. Two distinct fracture systems of different ages are evident. Porosity, 3 percent; permeability (k_{90}), 30 md.

d—Recrystallization in a skeletal-pellet-micritic limestone. The darker layers are relatively unaltered, nonporous micrite (lime mud). Recrystallization of the micrite to a coarser, porous, chalky texture (light patches) is confined to the skeletal (Amphiporid)-rich layers where some primary organic-cell porosity is still present. Porosity, 12 percent; permeability (k_{90}), 15 md.

e—Cementation in a skeletal (Stromatoporoid) conglomerate. All the originally intergranular porosity has been completely infilled with white, coarsely crystalline calcite. The primary organic-cell porosity in the Stromatoporoid fragments also has been destroyed as a result of calcite cementation. Porosity, < 1.5 percent; permeability (k_{90}), < 0.01 md.

Fig. 3—Examples of porosity types.

area is about 120 sq miles and the reservoir rarely exceeds 30 ft in gross section. This field is a combination stratigraphic-structural trap. Reservoir beds dip gently to the south and terminate updip at an unconformity where they are sealed above by tight siltstone beds. The base seal is an anhydrite that also terminates at the unconformity. Oil in place is 700 million bbl and oil production to the end of 1975 was 106 million bbl.

The reservoir occurs within a cyclic sequence of shallow-water evaporitic carbonates consisting of thinly interbedded lime muds and lime sands. The supratidal algal beds have been dolomitized selectively and form the best reservoirs. Up to five separate porous zones can be stacked within a net pay section of 28 ft. The thickness of the total unit at any location is determined by the amount of relief on the unconformity surface.

There is no water leg present and the pool is terminated downdip by the disappearance of effective porosity. A small gas cap is present in a few structurally high wells. Porosity averages 20 percent and permeability is 30 md.

Boundary Lake field has been on production since 1961 and was placed under waterflood operation in 1964. The period before waterflooding was characterized by high production rates, progressively increasing GOR's, and rapidly declining reservoir pressure. Continued production under natural depletion would have resulted in ultimate recovery of only 12 percent.

This type of reservoir, consisting of thin but continuous porous zones confined above and below by dense beds, lends itself to pattern waterflooding. Inverted nine-spot patterns as close as 80 acres have been implemented and will increase the ultimate recovery to more than 30 percent of the oil in place. Detailed correlation and mapping of the individual porous zones is a prerequisite for assuring that every zone is being waterflooded effectively.

Clarke Lake Field

The Clarke Lake field (Fig. 5), also located in British Columbia, produces gas from the Middle Devonian. The field covers an area of about 45 sq miles and has marketable gas reserves of 1.5 Tcf.

The cross-section and field outline on the map show that Clarke Lake field is located at the edge of a vast carbonate bank. The pool is sealed at the top by shale and at the base by tight dolomite. The lateral beds are basinal shales and tight limestone. Local dolomitization has occurred along fracture zones at the bank edge and along linear trends shelfward of the front. Porosity is limited to irregularly shaped dolomitized intervals and occurs either in the form of leached vuggy

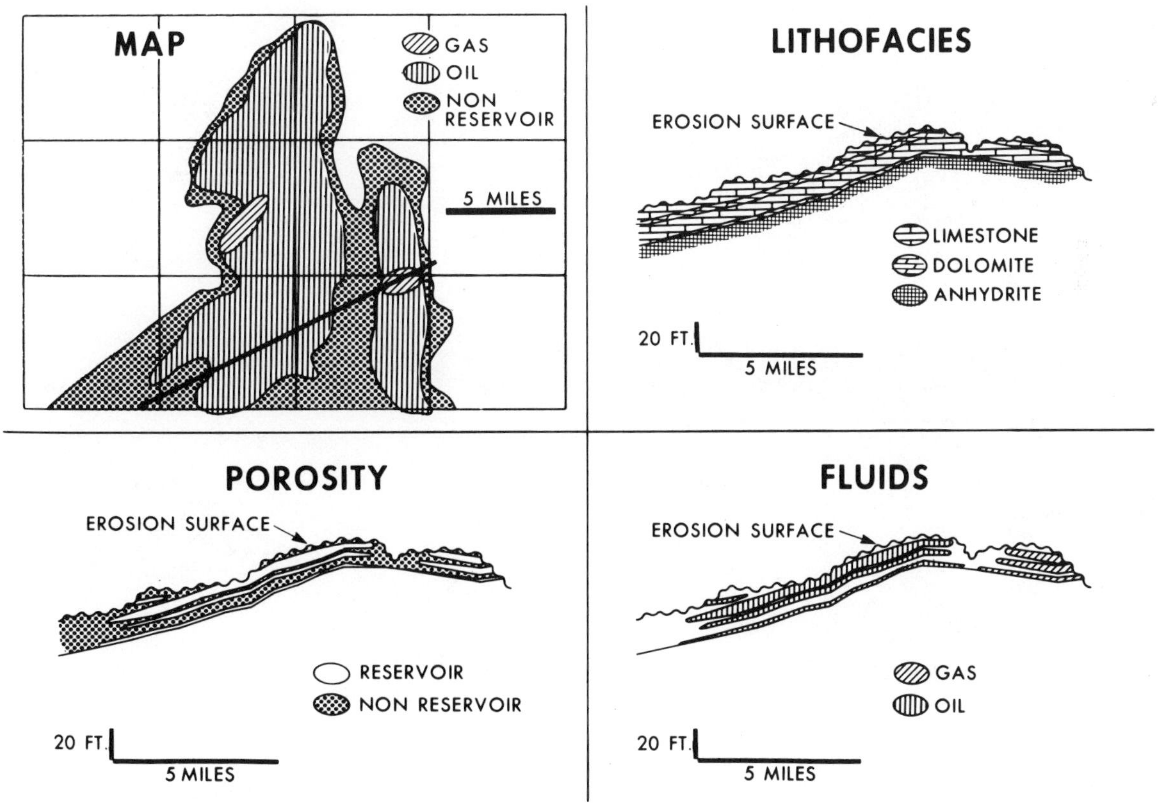

Fig. 4—Boundary Lake oil field. Map and cross-sections illustrating lithofacies, porosity, and fluid distribution.

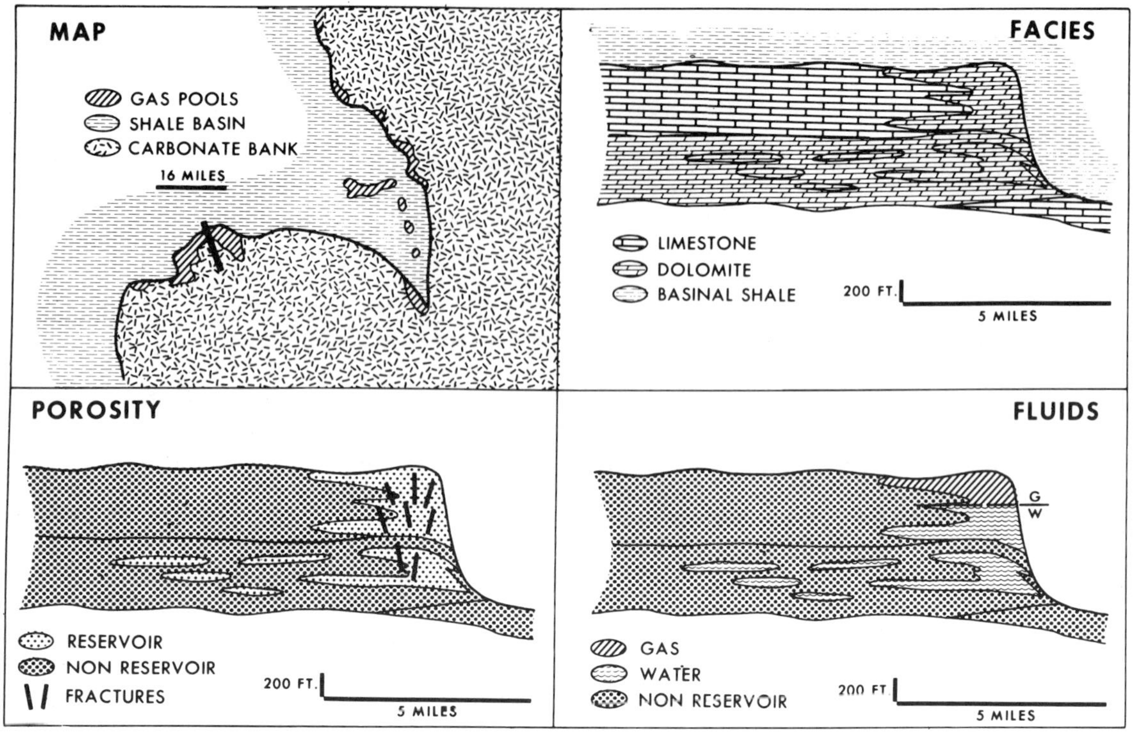

Fig. 5—Clarke Lake gas field. Map and cross-sections illustrating lithofacies, porosity, and fluid distribution.

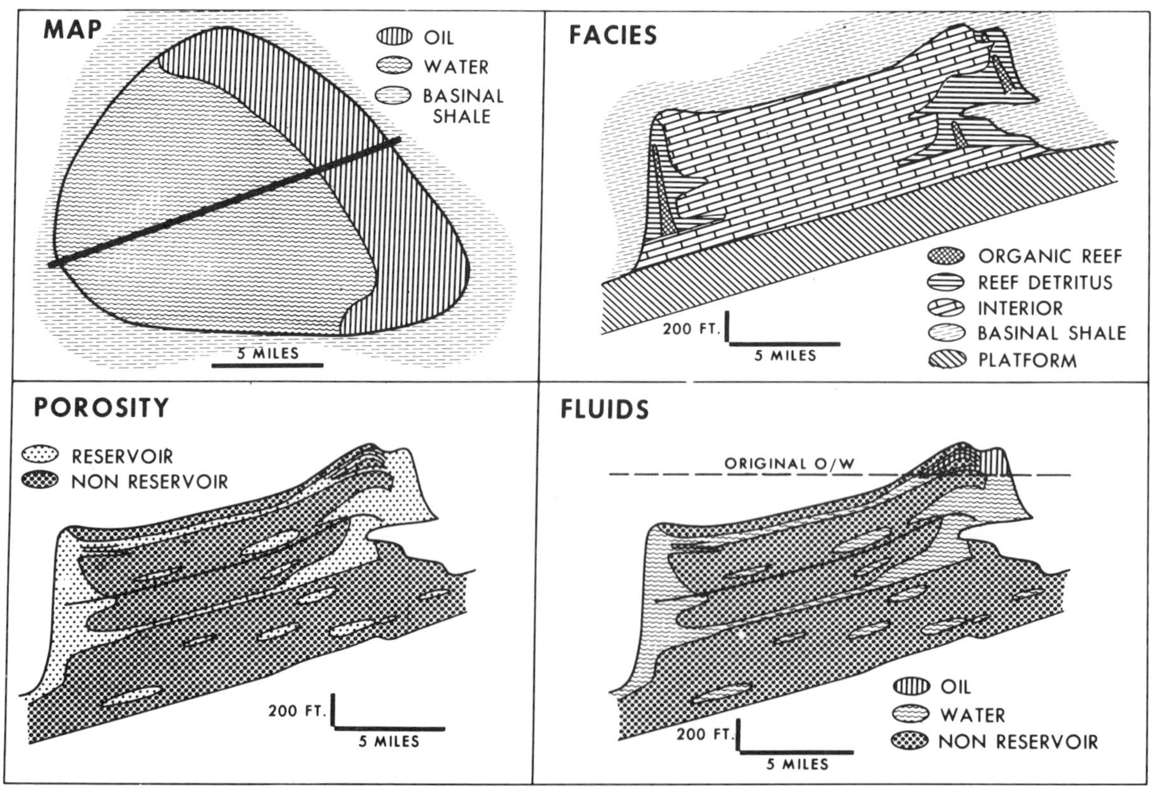

Fig. 6—Redwater oil field. Map and cross-sections illustrating lithofacies, porosity, and fluid distribution.

and intercrystalline porosity or as second-stage, coarsely crystalline dolomites along fractures. The dolomitized belts are narrow and difficult to map, resulting in a poor exploration and delineation drilling record. The ratio of successful wells to dry holes is about 1:1. Maximum pay is 350 ft and the gas is underlain by a local aquifer. Some of the better wells can deliver up to 30 MMcf/D. The field producing gas-water ratio is about 60 to 70 bbl/MMcf, and there has been a relatively slow pressure decline from the original 2,900 to 2,200 psig after 12 years and nearly 1 Tcf of gas production.

This pool can be visualized as having two distinct reservoir components. The host rock is composed of a very low-permeability reservoir; within this is a second reservoir type consisting of a complex of fractures and associated vugs with very high permeability.

The following production characteristics will be expected from this combination.

1. Very high initial production rates, mainly from the fractures. These will decline rapidly as the fracture gas is exhausted and the flow becomes dependent on the low matrix permeability of the host rock.

2. With an underlying aquifer, high water rates will result when it becomes as easy for the water to flow up the permeable channels as it is for the gas to escape from the tight matrix.

3. This reservoir probably will be abandoned at a relatively high pressure (about 1,500 psig) as productivity declines and water production exceeds the economic limit. Note that the conventional pressure/production plot will indicate substantially higher reserves than will actually result.

Redwater Field

The Redwater reef of central Alberta (Fig. 6) is a large bioherm resting on a carbonate platform but otherwise entirely enclosed in shale. It covers 200 sq miles and is 800 ft thick. In section, the reef has the appearance of an atoll with an exterior rim of organic reef with its associated skeletal detritus and an interior lagoonal and intertidal to supratidal facies of algal limestones and finer detritus. This reef is composed entirely of limestone.

From the point of view of volume and continuity, the best reservoir is in the organic facies of the rim. The fine-grained interior laminites are highly stratified and have lower and less continuous porosity. Most of the porosity in the rim has been enhanced by leaching.

Oil is trapped at the updip edge of the reef and is present in both the rim and interior facies. About 5 percent of the gross volume of the reef is occupied by the oil. Redwater field is

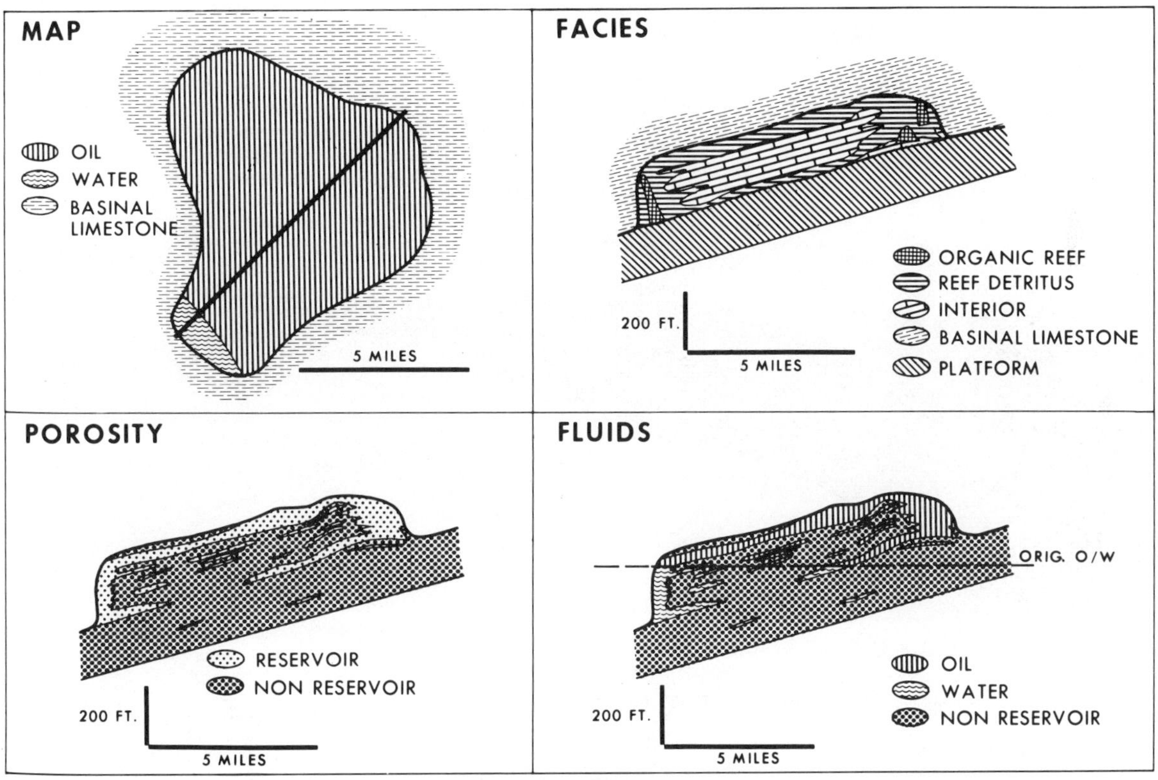

Fig. 7—Judy Creek oil field. Map and cross-sections illustrating lithofacies, porosity, and fluid distribution.

the second largest pool in Canada from the standpoint of recoverable reserves—850 million bbl or 65 percent of the original oil in place. From its discovery in 1949 to the end of 1975, it produced 570 million bbl or about two-thirds the ultimate recoverable reserves.

The geological description of the reef complex into its organic-reef, fore-reef, and back-reef systems is a key input into the definition of the flow patterns in both the aquifer and the oil zones. Production strategies, in terms of oilwell recompletions and location of water injectors, has depended greatly on the geological modeling of the barrier systems within the reef-rim and back- reef components.

Until recently, water disposal into the aquifer was limited to produced water, but there are now three water-injection wells supplementing the strong natural water drive associated with this pool. Supplemental water injection is aimed at raising the reservoir pressure to improve well productivity. Detailed stratigraphic studies were necessary to establish the optimum program for this water disposal.

Judy Creek Field

The Judy Creek field of central Alberta (Fig. 7) illustrates some interesting similarities as well as differences when compared with the Redwater field. The most obvious difference is size. Judy Creek field occupies only 47 sq miles compared with Redwater's 200 sq miles, and is 220 ft thick from reef top to platform. In cross-section, Judy Creek, like Redwater, is tilted gently to the southwest and consists of an organic reef rim with an interior of detrital carbonates.

Porosity at Judy Creek field is best developed in the organic reef and reef detritus of the perimeter. Excellent

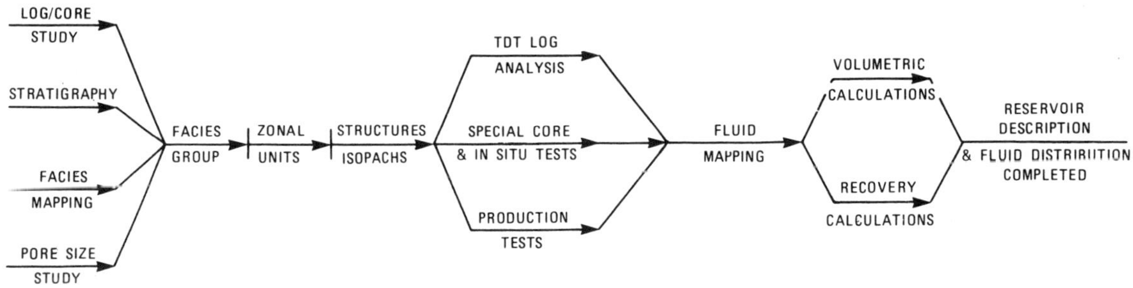

Fig. 8—Study sequence, descriptive phase, Judy Creek pool.

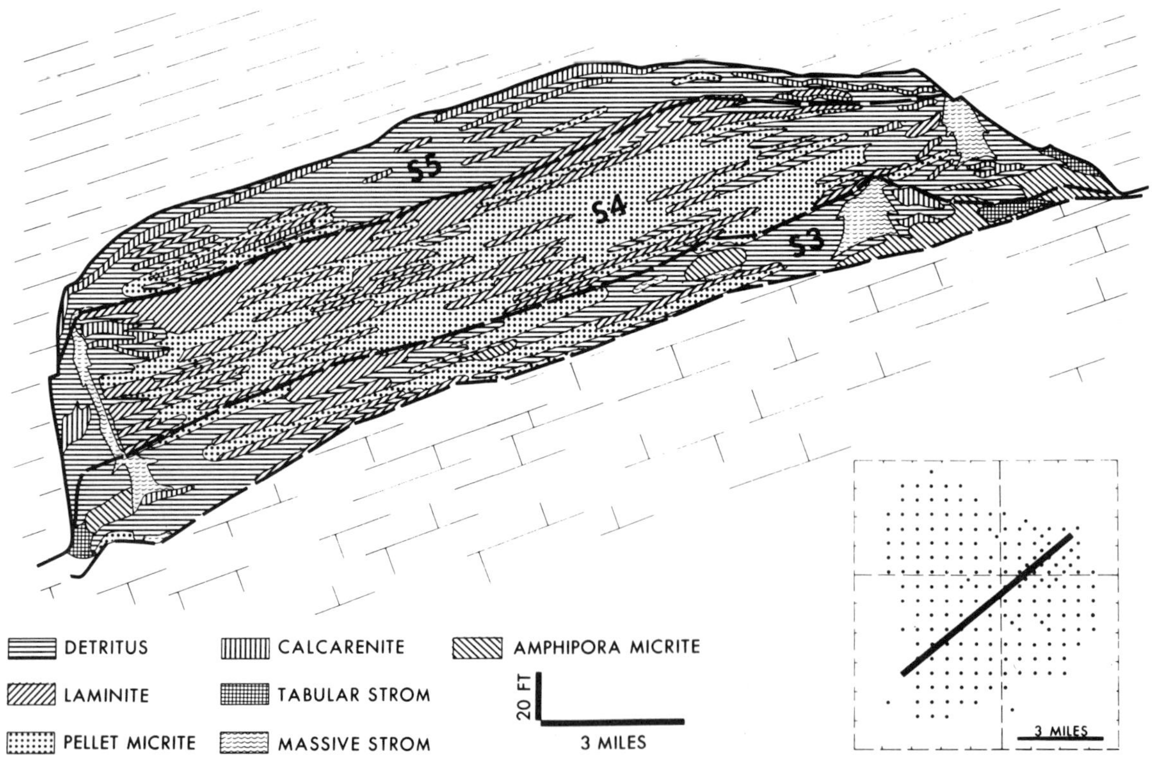

Fig. 9—Cross-section showing complexity of lithofacies, Judy Creek field.

porosity also occurs in a zone of reef detritus across the top of the reef. In the stratified reef interior, porosity is discontinuous and occurs as patches and lenses. About 75 percent of the reef is filled with oil. A total of 830 million bbl of oil in place compares with about 1.3 billion bbl at Redwater field. Unlike Redwater, the Judy Creek pool is not connected to a large aquifer. The recovery factor under pressure maintenance is expected to be about 47 percent, compared with 65 percent at Redwater.

The Judy Creek pool was discovered and put on production in 1959. By 1962, a waterflood had been installed to arrest the rapidly declining pressure. However, as pool production increased from about 30,000 B/D in the late 1960's to 70,000 B/D in the early 1970's, extreme pressure gradients developed. By early 1973, a gradient of 2,000 psi existed across the pool. At that time, reservoir pressure in the vicinity of the injection wells was about 1,000 psi above original reservoir pressure and some internal areas of the pool were below the bubble point, indicating the ineffectiveness of the peripheral waterflood. There obviously were permeability barriers that did not allow the pressure-maintenance scheme to function as expected. The remedial measures undertaken to correct this problem are described in the next section.

Judy Creek Geology and Depletion Plan

We have briefly reviewed four producing carbonate pools. Each pool has different geological parameters and internal reservoir heterogeneities (size, shape, stratification, porosity, permeability, etc.) that have resulted in performance problems requiring unique pool development techniques. This section deals with Judy Creek field in greater detail, and describes how geological and engineering studies were combined to achieve more effective depletion of the reservoir.

Fig. 8 shows the study plan for the Judy Creek pool. The study team included reservoir geologists, stratigraphers, log analysts, computer specialists, and reservoir and production engineers. On completion of Phase 1, the preparation of the reservoir description and fluid distribution, the study progressed to Phase 2, the operational and predictive aspects of the project. Phase 2 included the planning of well workovers, artificial lifting equipment, infill drilling, and design of field facilities. In the final phase, predictive studies were carried out using multidimensional, multiphase models based on the detailed reservoir description and fluid distribution compiled during the initial phase.

Detailed Lithofacies

Fig. 9, a west-east cross-section through the field, illustrates the complexity of facies that make up the Judy Creek bioherm. For simplicity, 11 distinct carbonate facies have been reduced to seven in this illustration. The reef rests on a platform of tight limestones designated as the S1 and S2 units. Above the platform, the reef buildup has been subdivided into three stratigraphic units representing three distinct periods of reef growth and designated as the S3, S4, and S5 units.

The S3 unit consists of a narrow, peripheral rim of organic reef with an interior lagoon of reef detritus and lime muds.

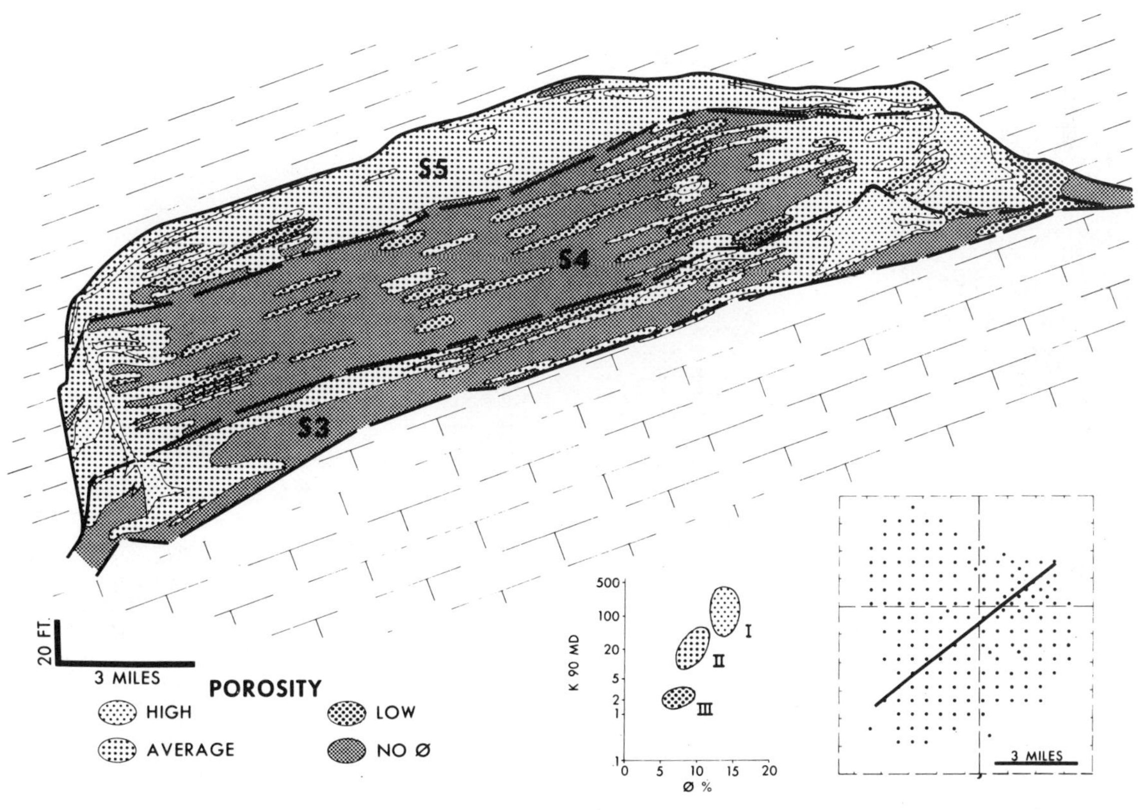

Fig. 10—Cross-section showing the distribution of three facies-controlled porosity groups, Judy Creek field.

The S4 unit has a thick organic-reef buildup and an extensive interior facies of interbedded, tight, and porous limestone. Immediately inside the rim is a zone of coarse detritus.

The uppermost S5 unit has no organic-reef framework. It consists predominately of coarse organic detritus and lime sand.

Even on this simplified cross-section, the distribution of facies exhibits an extremely complex stratification and lensing of rock types, particularly in the interior portion of the S4 unit.

Porosity Distribution

Fig. 10 illustrates the classification of porosity types within the Judy Creek reservoir. A cross-plot of horizontal permeability (k_{90}) vs porosity resulted in an even simpler subdivision of the environmental facies into three reservoir families, illustrated in the inset.

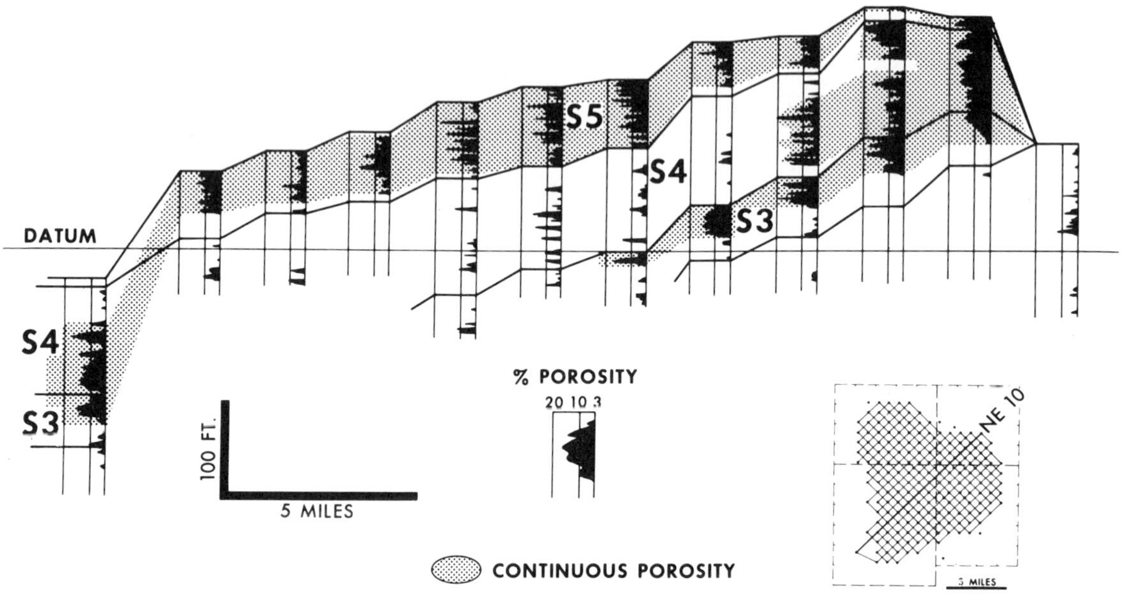

Fig. 11—Example of a correlation cross-section, Judy Creek field (after Flewitt[3]).

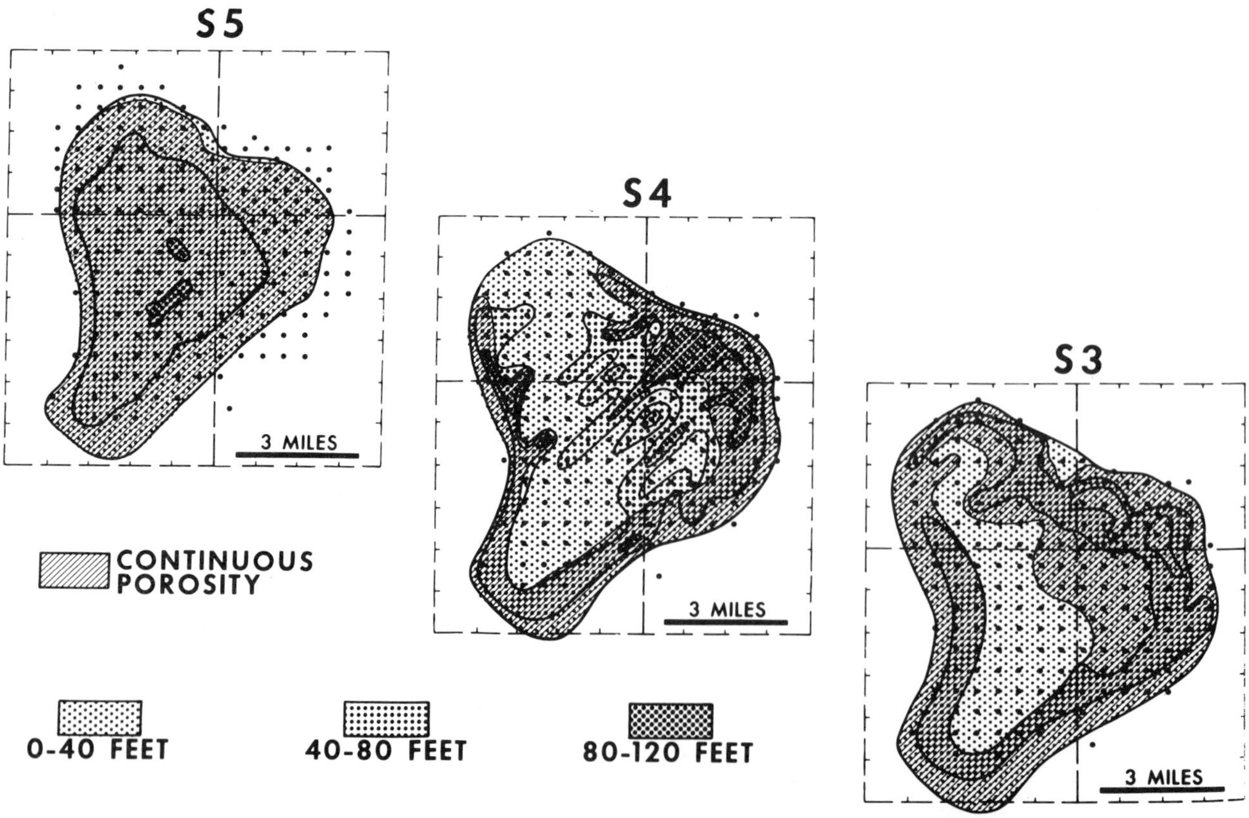

Fig. 12—Isopach of net porosity and zones of continuous porosity in the S3, S4, and S5 units, Judy Creek field (after Flewitt[3]).

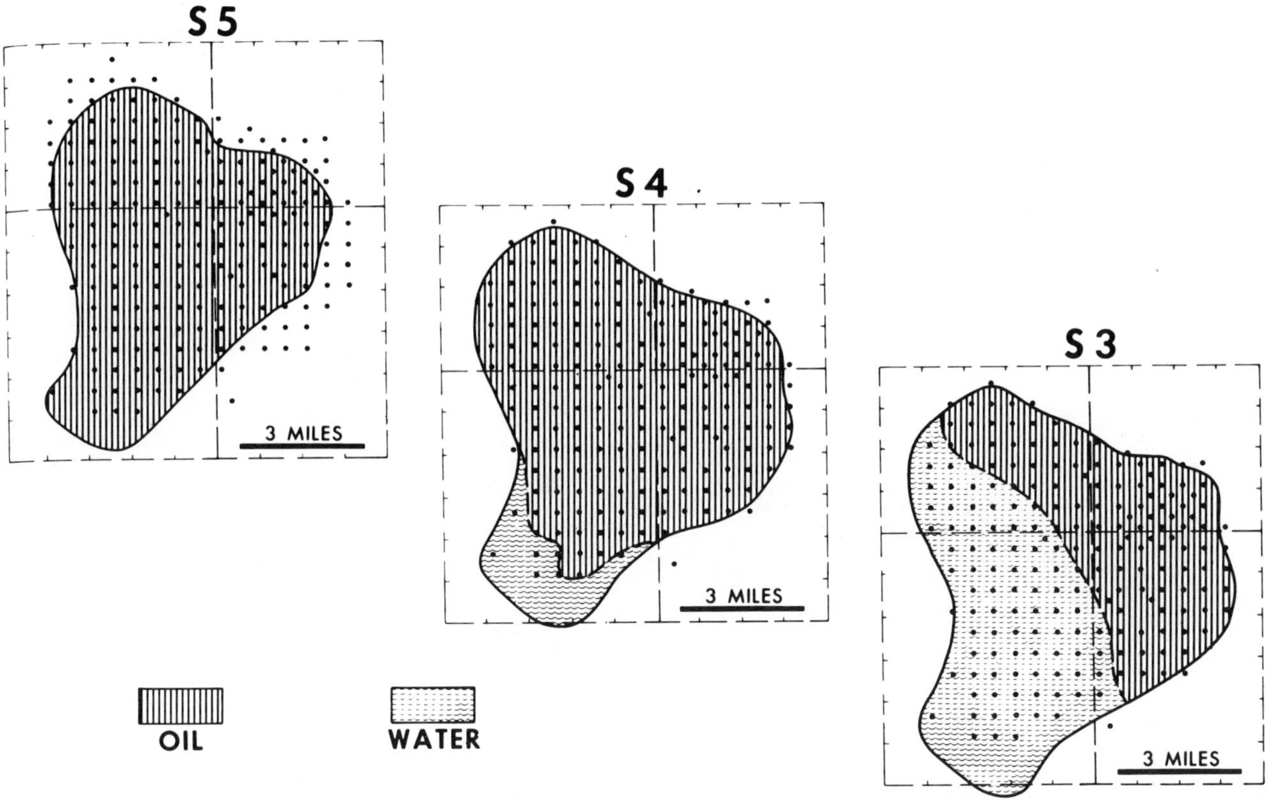

Fig. 13—Maps showing original distribution of oil and water in the S3, S4, and S5 units, Judy Creek field (after Flewitt[3]).

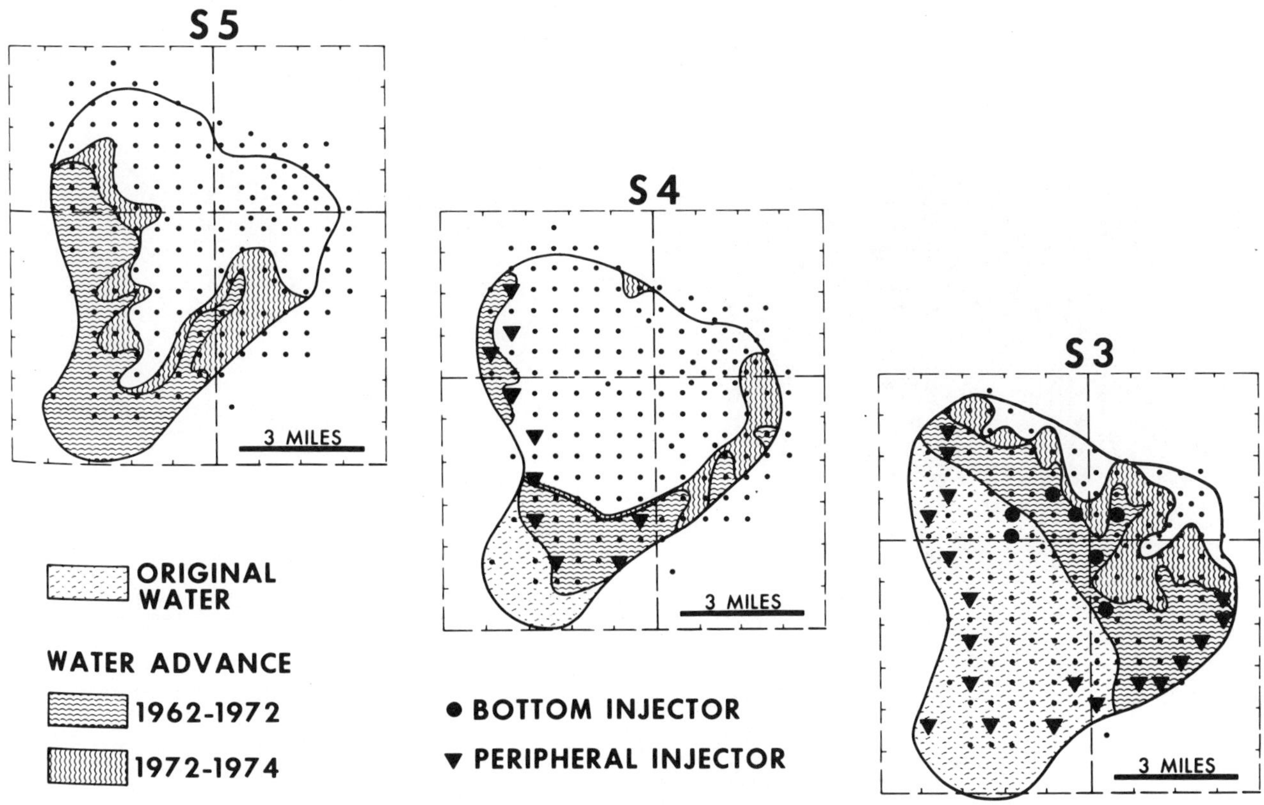

Fig. 14—Maps showing rate of water advance under peripheral and bottom waterflood in the S3, S4, and S5 units, Judy Creek field (after Flewitt[3]).

Group I reservoirs occur in the organic-reef and shallow shoal facies and consist of reef framework and associated reef detritus. These are the best reservoirs, having an average porosity of 12.5 percent and an average permeability (k_{90}) of 170 md.

Group II reservoirs are found in shoals in the fore and back reefs where reef detritus and algal laminite rocks were deposited. Porosity averages 9.5 percent and average permeability is 40 md.

Group III reservoirs occur in the deeper-water facies of the fore reef and in the interior lagoonal facies. Rock types consist of organic debris and pellets in a lime-mud matrix. Average porosity is 6.5 percent and average permeability is 3 md. Note that the major portion of the interior lagoonal facies of the S3 and S4 units is composed of tight lime muds.

Group I reservoirs provide vertically continuous porosity around the perimeter of the reef and laterally continuous porosity across the top. The interior of the reef is character-

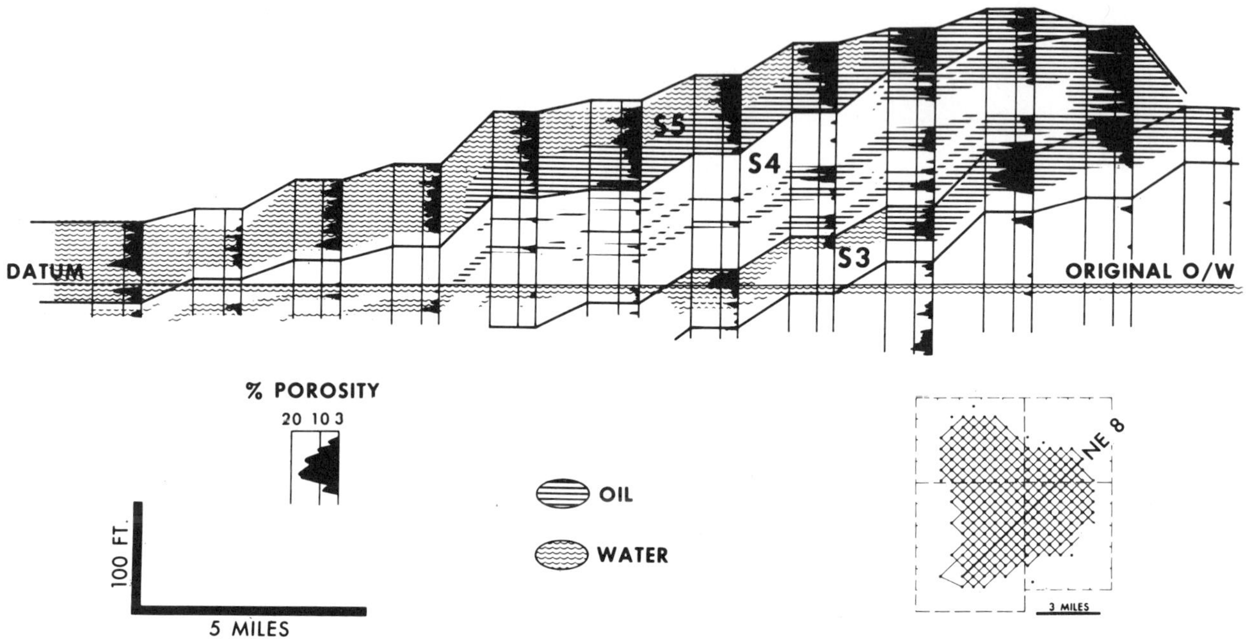

Fig. 15—Northeast-southwest structural cross-section, Judy Creek field, illustrating the differential advance of water injected downdip (after Flewitt[3]).

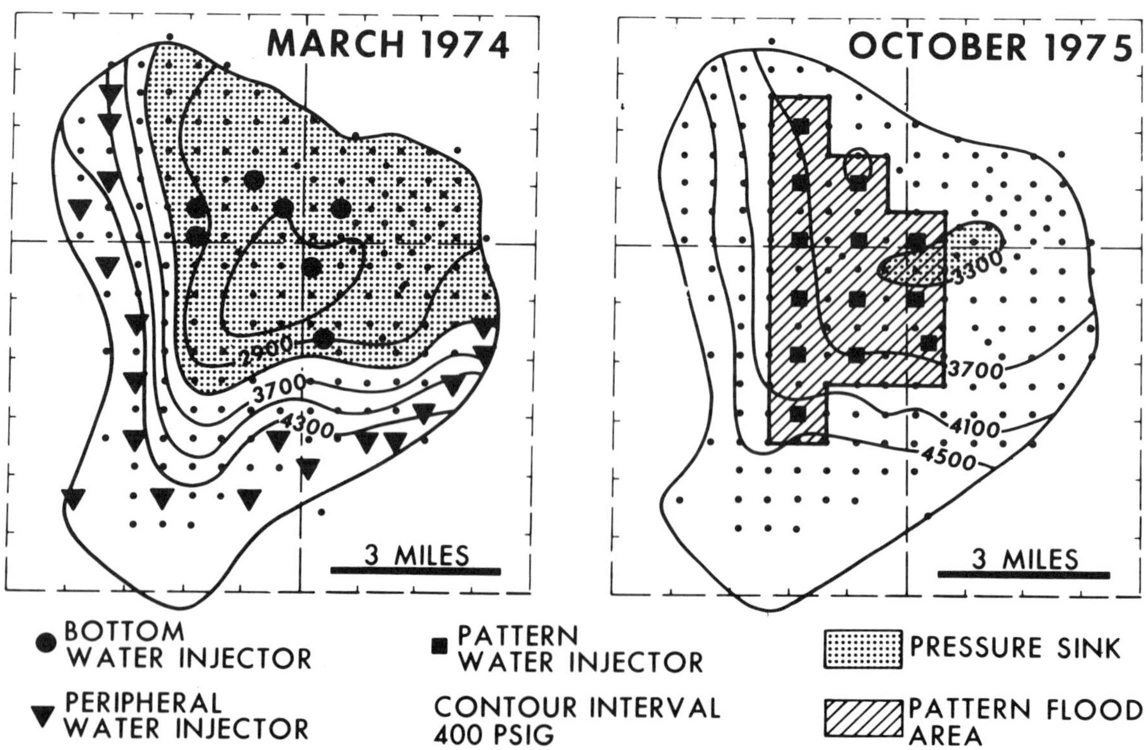

Fig. 16—Isobaric maps showing the result of conversion from peripheral to pattern flood, Judy Creek field (after Flewitt[3]).

ized by extreme stratification that results in thin porous zones that tend to be discontinuous both laterally and vertically. These zones consist mainly of Groups II and III type reservoirs.

A grid of 47 well-log cross-sections, encompassing every well drilled, was used to correlate porous units and tight bands throughout the pool. The section in Fig. 11 crosses the pool from the southwest to northeast and illustrates how the correlation was facilitated by using the facies descriptions and the simplified quantitative porosity profiles. Porous intervals, which are stippled in the figure, are interpreted to be in continuity either vertically or horizontally. These occur primarily in the reef perimeter and the S5 unit. Discontinuous porous beds are not stippled and occur in the interior deposits of the S3 and S4 units.

Net Porosity Maps

From the cross-section grid, net porosity maps were prepared for each stacked unit. Perhaps the most useful maps, in terms of understanding fluid flow, are those that delineate areas of continuous porosity. The cross-hatched area in Fig. 12 indicates that the uppermost unit, the S5, has continuous porosity across the top of the reef. The lower units, the S3 and S4, have zones of continuous porosity only around the periphery of the reef.

Fig. 13 shows the original fluid distribution within the S3, S4, and S5 units. The uppermost unit, the S5, was essentially water-free; the S4 had a limited downdip water leg; and the lowermost unit, the S3, was about half filled with water.

Pressure maintenance, initiated in 1962, concentrated water injection in the downdip periphery of the reef, principally into the S3 and S4 units. Fig. 14 shows the water movement within the reef over two time periods: the first period is from start of production to 1972, and the second period is from 1972 to 1974. Injected water tended to follow the highly permeable reef rim but the relatively tight lagoonal facies prevented fluid movement to the interior of the reef. In effect, the internal regions of the S4 and S5 units were not being swept by injection water.

The cross-section in Fig. 15 illustrates this discontinuity more strikingly. Water currently has advanced far updip in the S3 unit. Also apparent is the advance in the upper part of the S5, which contains the highest continuous porosity in the reef. Here, the water has overridden the lower part of the S5, which has somewhat lower permeability. Sandwiched between these two tongues of advancing water are the bypassed discontinuous porous beds in the S4, which still contain oil.

Fig. 16 shows pool isobaric maps for March 1974 and Oct. 1975. The most prolific producers, located in the northeastern area of the pool, were completed in the S5 unit and were experiencing severe pressure drawdowns. This area could communicate with the injectors only through the organic framework around the outside edge of the reef. Although several salt-water disposal wells were completed in the S3 unit in this area, injection was ineffectual in repressuring the S5 because of the intervening tight S4 sections. In addition to the poor pressure- maintenance performance of the overlying S5 section, there were many zones within the S4 unit that were not being drained through perforations or being pressure-maintained by injection in the S3 unit.

As a result of the reservoir model study and the use of reservoir simulators, it was established that (1) the best way to recover the oil in the S5 unit was through pattern flood, and (2) the discontinuous porous beds in the interior of the S4 unit were not being waterflooded effectively.

A revised program to optimize pool performance then was initiated and included the following steps.

1. Pattern waterflood injectors were to be completed in all porous intervals in the S3, S4, and S5 units. In addition, producing wells in the pattern waterflood area were to have all porous intervals perforated and high-volume pumps installed.

2. Wells now behind the flood front and previously deemed to be flooded out (although not all porous zones had been opened up) were to be recompleted and reactivated. Many reactivated wells are currently producing more than 100 BOPD and these are expected to recover more than 10 million bbl of oil. Reactivation of about 20 additional wells was planned for 1976 and 1977.

The pattern waterflood was placed in operation in 1974; Fig. 16 shows the dramatic improvement in reservoir pressure approximately 1 year later. Now that the reservoir pressure is under control, zonal injected and produced fluid balances will be maintained at approximately the original reservoir pressure of 3,500 psig. Water volumes to the peripheral injectors have been cut back to prevent overpressuring in this region.

Conclusions

1. Reservoir heterogeneity, in both a lateral and vertical sense, is commonly high in carbonate pools because of the wide variety of environmental facies present and their susceptibility to diagenetic alteration. The reservoir facies should be mapped and studied individually to define their geometry and the most likely patterns of fluid movement under various depletion programs.

2. Reservoir description is not static but must be upgraded continuously as depletion of the reservoir proceeds. Well performance must be monitored to follow fluid migration and pressure patterns and the geological model should be upgraded continually as new data become available.

3. In order for the reservoir engineer to develop an effective depletion plan, a multidisciplinary approach should be taken. The study team should include reservoir geologists and log analysts as well as production and completion engineers.

4. Accurate reservoir description is the key to constructing meaningful reservoir engineering models that are necessary for understanding the producing characteristics of a reservoir, and for developing the optimum depletion plan for the pool. This, in turn, translates into greater profitability of production operations and more efficient utilization of our valuable energy resources.

Acknowledgments

We thank Imperial Oil Ltd. for permission to publish this paper. Grateful thanks are also due R.P. Glaister and B.D. Stewart for critical reading of the paper.

Bibliography

1. Barss, D.L., Best, D.W., and Meyers, N.: *Geologic History of Western Canada*, Alta. Assn. of Pet. Geologists (1964) Chap. 9.
2. Fischbuch, N.R.: "Stratigraphy Devonian Swan Hills Reef Complex of Central Alberta," *Bull.*, Cdn. Pet. Geol. (Dec. 1968) 446.
3. Flewitt, W.E.: "Refined Reservoir Description to Maximize Oil Recovery," paper presented at SPWLA Sixth Annual Logging Symposium, New Orleans, June 1975.
4. Gray, F.F. and Kassube, J.R.: "Geology and Stratigraphy of Clarke Lake Gas Field, British Columbia," *Bull.*, AAPG (1963) 467.
5. "Submission Dated February 1973 to the Alberta Energy Resources Conservation Board in the Matter of the Judy Creek Beaverhill Lake 'A' Pool Reserves Evaluation Hearing," Application No. 6984, Imperial Oil Ltd., Calgary, Alta. (Sept. 1973).
6. "Submission Dated May 1974 to the Alberta Energy Resources Conservation Board in the Matter of the Judy Creek 'A' Pool Application for Waterflood, MER and Reserves Evaluation and Removal of Interwell Spacing Distance," Application No. 7779, Imperial Oil Ltd., Calgary, Alta. (1974).
7. "Application Dated June 1973 to the Alberta Energy Resources Conservation Board for Approval of Pressure Maintenance by Water Injection, Redwater D-3 Pool," Application No. 7181, Imperial Oil Ltd., Calgary, Alta. (1973).
8. Klovan, J.E.: "Facies Analysis of the Redwater Reef Complex, Alberta, Canada," *Bull.*, Cdn. Pet. Geology (1964) 1.
9. Klovan, J.E.: "Development of Western Canadian Devonian Reefs and Comparison With Holocene Analogues," *Bull.*, AAPG (May 1974) 787.
10. Palmer, C.E.: "Revised Flood Depletion Plan Judy Creek Beaverhill Lake 'A' Pool," paper CIM 75-29 presented at 26th Annual Meeting of Petroleum Society of CIM, Banff, June 11-13, 1975.
11. Roy, K.J.: "Stratigraphic Analysis and Environmental Reconstruction of the Boundary Member of the Charlie Lake Formation (Triassic), Northeastern British Columbia, Canada," PhD dissertation, Northwestern U., Evanston, Ill. (June 1968).
12. Willmon, G.J.: "A Study of Displacement Efficiency in the Redwater Field," *J. Pet. Tech.* (April 1967) 449-456.

Original manuscript received in Society of Petroleum Engineers office Aug. 12, 1976. Paper accepted for publication Jan. 26, 1977. Revised manuscript received Sept. 18, 1976. Paper (SPE 6139) was first presented at the SPE-AIME 51st Annual Fall Technical Conference and Exhibition, held in New Orleans, Oct. 3- 6, 1976.

The American Association of Petroleum Geologists Bulletin
V. 60, No. 2 (February 1976), P. 245-257, 14 Figs.

Pore Geometry of Carbonate Rocks as Revealed by Pore Casts and Capillary Pressure[1]

N. C. WARDLAW[2]
Calgary, Alberta, Canada

Abstract The three-dimensional aspects of pores in a selected group of dolomites show that pore connections are commonly sheet-like rather than tubular and that prevailing geometric styles of pores and connections can be described from examination of resin casts of pores with a scanning-electron microscope.

As dolomite crystals grow and porosity is reduced, polyhedral pores are transformed to tetrahedral pores and finally to interboundary-sheet pores. In this process of porosity reduction, pore structure becomes geometrically simpler and more regular and pore-size to throat-size ratios increase.

Mercury-ejection efficiency can be defined as the total volume of mercury ejected from a sample on reducing pressure from 100 kg/sq cm to 0.1 kg/sq cm expressed as a percentage of the total volume of mercury injected over the same pressure range. Ejection efficiency decreases linearly with decrease in porosity and probably results mainly from decrease in the size of throats relative to the size of pores.

Mercury-injection and ejection tests do not provide a close analogue for an oil-water system but may give a relative indication of the efficiency with which various rock types can expel a nonwetting phase when pressure is reduced. Because ejection efficiency correlates positively with porosity, it may be possible from porosity data to provide a factor that will indicate recovery efficiency, insofar as this is dependent on the pore geometry.

INTRODUCTION

Pore systems in carbonate rocks have been classified comprehensively with emphasis on their genetic aspects (Choquette and Pray, 1972), but few attempts have been made to describe the geometry, particularly in three dimensions, of pore systems.

It commonly is held that pore sizes, shapes, and connections are so complicated and variable as to defy description. Where it has been necessary to have information on the geometry of pore connections, as for example in the interpretation of capillary-pressure curves and for a variety of calculations on fluid movements, it has been common to assume that these connections are simple tubes or, alternatively, the types of spaces associated with various packings of spheres.

The objective of this paper is to describe the geometry of pores and pore connections or throats in a selected group of carbonate rocks. The term pore is used loosely to describe any space, but where it is necessary to recognize local enlargements in a pore system which are linked by smaller connecting spaces, the latter are referred to as pore connections or throats. Use of the term throat does not imply any particular shape. It will be shown that pore throats in dolomites commonly are sheet-like rather than tubular and that prevailing geometric styles of pores and throats can be described. Casts of pore systems were examined using optical and scanning-electron microscopes.

Mercury-injection capillary-pressure curves provide additional information on the size distribution of pore throats and this information is related to sizes of pores and throats, when measured directly from the pore casts. Additionally, mercury-ejection curves were plotted by determining the volume increments of mercury expelled from samples during pressure reduction from 100 to 0.1 kg/sq cm. In the carbonate rocks analyzed, the volume of mercury retained in the sample after pressure reduction (analogous to residual oil saturation), ranged from 90 to 40 percent of the injected mercury and this percentage increases linearly with porosity decrease. That is, low-porosity carbonate rocks eject mercury (nonwetting phase) less efficiently during pressure reduction than high-porosity rocks. This relation is explained in terms of changes in pore geometry which accompany changes of porosity.

Ejection of mercury from a small sample during pressure drop in a porosimeter does not provide a realistic analogue for the production of oil from a reservoir. There are large rheologic differences between mercury and the less viscous, pressure-dependent, natural pore fluids consisting of oil-water emulsions in the presence of dissolved gas. However, Dumoré and Schols (1974) investigated drainage capillary-pressure functions for the same rock with various fluid-fluid combinations and concluded that mercury-air functions

[1]Manuscript received, February 28, 1975; accepted, July 14, 1975.

[2]Department of Geology, University of Calgary.

The writer thanks Raymond Taylor for enthusiastic assistance in making the measurements of porosity and capillary pressure used in this paper and for many valuable discussions. Stephen Cheshire kindly undertook statistical analysis of some of the data presented. Alan Oldershaw is thanked for his assistance in operating the Cambridge Stereoscan 600 scanning-electron microscope. The project was financed by the National Research Council of Canada and this support is gratefully acknowledged.

can be transformed to predict the behavior of toluene and heptane in the presence of water. Comparisons were not extended to imbibition functions for the mercury-air system (mercury ejection or withdrawal). It is suggested here that pore systems with geometric attributes causing extensive retention of mercury are also likely to retain other nonwetting fluids in large proportions. The difficult experiments to confirm this should be performed. As porosity decreases, not only does storage capacity for a nonwetting phase decrease, but the tendency to retain that phase on pressure reduction increases. Such relations should receive increasing attention in a period of dwindling reserves when efficiency of production becomes more important.

The carbonate rocks studied are from Middle Devonian pinnacle reefs (A and B pools) of the upper Keg River Formation in the Rainbow Lake area, Alberta. Details of the stratigraphic setting can be found in Hriskevich (1966, 1967, 1970), Langton and Chin (1968), and Barss et al (1970).

The limestones sampled are mainly lagoonal mudstones now composed of microspar (5-10μ diameter) and grainstones from the upper third of the A pool reef. The lower two-thirds of this reef has been dolomitized completely and the limestone-dolomite contact is sharp and forms a nearly horizontal surface through the reef. A variety of dolomites was sampled from this pool and also from the B pool.

Mercury-Injection and Ejection Curves

Principles and Methods

Mercury, as a nonwetting liquid, must be forced under pressure into a porous solid and enters successively smaller spaces as pressure is increased. This provides the basis of a method for determining certain properties of a pore system (Purcell, 1949; Chilingar et al, 1972) and evidence has been presented that the method is not destructive (Burdine et al, 1950, p. 199; Dumoré and Schols, 1974, p. 440).

The volume of mercury entering an evacuated pore system during successive pressure increments is plotted as a pressure-volume curve, referred to as a mercury-injection capillary-pressure curve (Fig. 1). Following attainment of maximum pressure (100 kg/sq cm), pressure can be reduced incrementally and the volume of mercury ejected or expelled for each increment can be expressed as a mercury-ejection curve (Fig. 1).

The volume of mercury entering a sample between specific pressure limits is related to the fraction of the total pore volume that is connected by pore throats which are within specific size limits. Thus, a mercury-injection pressure curve

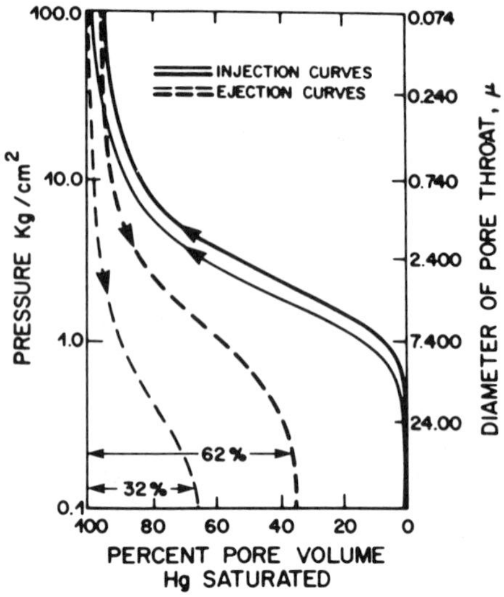

FIG. 1—Mercury-injection and ejection capillary-pressure curves for two Rainbow Lake (A pool) dolomite samples to illustrate that samples may have similar injection but differing ejection curves and therefore ejection efficiencies. Samples respectively ejected 62 and 32 percent of injected mercury. Both samples have porosities of about 20 percent.

provides information about the combined volumes of pores and throats which are connected by throats between specific size limits. However, there is a further complication. The sample is invaded by mercury from the exterior to the center and relatively large throats and pores in the interior may be shielded from invasion if they are connected to smaller throats within the system. Thus, a volume increment injected is an "apparent" volume accessible through throats within a defined size range, the size range being related to the pressure limits for the increment.

Volume of mercury injected is plotted against pressure and from this curve an apparent pore-throat size distribution is obtained. The expression normally used to derive size from pressure is:

$$P_c = \frac{4\sigma \cos\theta}{d} \quad (1)$$

where p_c = pressure, σ = surface tension for liquids, θ = contact angle for liquid on solid, and d = diameter of tube.

The preceding expression applies for tubular pore throats. In fact, pore throats in the dolomites and limestones studied are mainly sheet-like. Accordingly, the expression that applies for parallel plates is more appropriate and will, for a given pressure, give half the pore diameter obtained with the preceding expression. That is:

$$P_c = \frac{2\sigma \cos\theta}{d} \quad (2)$$

where d = distance separating two parallel surfaces.

Special problems exist for samples with large exterior pores (vugs) because mercury will invade these pores during measurement of bulk volume in a porosimeter. In this case low values of bulk volume, and therefore of pore volume, will be obtained. Also, in the mercury-injection procedure there are unnatural "boundary effects." Large pores on the exterior of the sample are intersected and therefore accessible by openings which are large compared with throats connecting similar pores in the interior of the sample. The form of the initial part of the capillary-pressure curve may be affected by large exterior pores and the volume of the unsaturated pores at the maximum pressure attained will be affected by inaccuracies of pore-volume measurements. However, detailed consideration of these factors will not be given here, for they do not affect the generalities presented.

Analyses were performed on cored cylinders of 2-cm diameter with volumes of about 8 cc that were treated with acetone in a Soxhlet's apparatus and dried in a vacuum oven. Pore volumes were obtained using a Ruska universal porosimeter and the precision of the method is discussed elsewhere (Taylor and Wardlaw, 1975). Mercury-injection and ejection pressure-volume curves were obtained using a Ruska porosimeter and contained mercury is expressed as a percentage of total pore volume. During the ejection procedure corrections must be made for "backlash" on the pump wheel; the method will be described in detail elsewhere.

Ejection Efficiency

Two samples may have closely similar mercury-injection curves yet differing ejection curves (Fig. 1). Because the form of the ejection curve also is dependent on pore structure, it is possible, although not common, for two samples to have significantly different pore structures yet similar injection curves. The injection and ejection curves should be considered in combination in interpretations of pore structure.

Ejection ratio or ejection efficiency can be defined as the total volume of mercury ejected from the sample, on reducing pressure from 100 kg/sq cm to 0.1 kg/sq cm, expressed as a percentage of the total volume of mercury injected over the same pressure range. Ejection efficiency ranged from less than 10 to more than 60 percent in the samples studied. There is a positive correlation between porosity and ejection efficiency (Fig. 2) and, from the relation, it appears that porosity decrease in carbonate rocks is accompanied by progressive changes not only in the volume of pore systems but in their geometric forms. These changes are evaluated by further examination of capillary-pressure curves and then by direct observations of pore systems.

Figure 3 illustrates typical mercury-injection and ejection curves for samples of 5 to 25-percent porosity. As porosity decreases two progressive changes in the forms of these curves are noticeable. Ejection efficiency falls and the median pore-throat size, as indicated by the mercury-injection curve, decreases. Lower porosity samples tend to have smaller average throat sizes (Fig. 4). Figure 5 illustrates that for dolomite samples pore size tends to increase with grain size, intergranular pores ranging in size up to that of the containing grains. There is no apparent correlation of pore size with porosity (Fig. 6), but pore-to-throat-size ratios increase with decrease in porosity (Fig. 7) and probably are of prime importance in controlling ejection efficiency. Ejection efficiency decreases as the throat diameters become smaller and the pore-to-throat-diameter ratio increases (Fig. 8). If large pores are connected by narrow throats, mercury "snap-off" occurs on pressure reduction at relatively high pressures and there is little tendency for mercury drainage from pores. Pores with relatively large throats

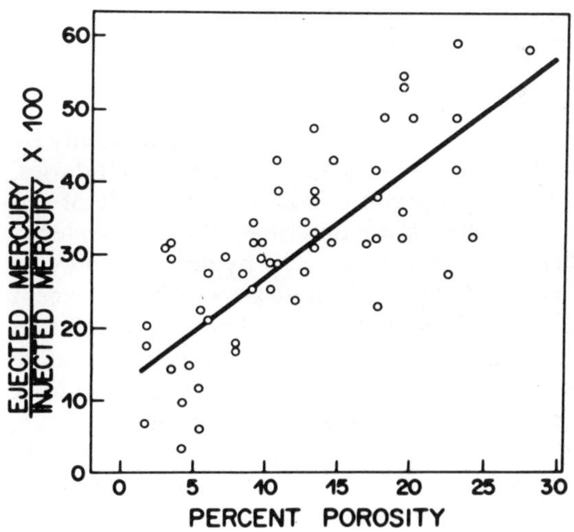

FIG. 2—Positive correlation between mercury-ejection ratio and porosity. Equation for best-fit regression line is: ejection ratio = $13 + 1.5\theta$. Dolomite samples from Rainbow Member at locations 2-32-109-8W6, 6-32-109-8W6, 8-32-109-8W6, 10-32-109-8W6, 12-3-109-8W6.

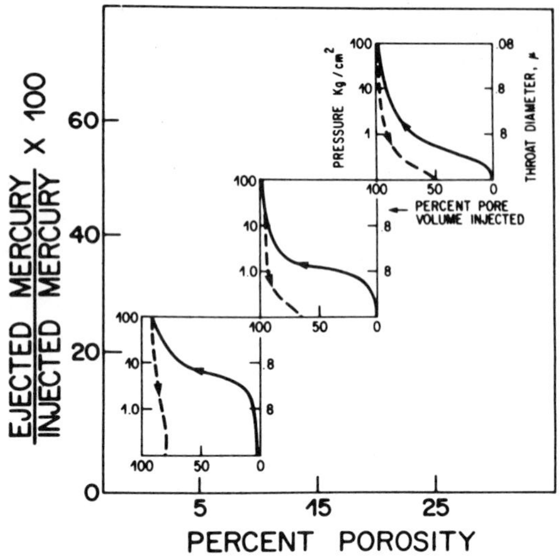

FIG. 3—Typical mercury-injection (solid lines) and ejection curves (broken lines) for samples of 5, 15, and 25-percent porosity. Ejection efficiency and median throat size both decrease as porosity decreases. Typical samples from locations identified in caption of Figure 2.

drain more efficiently. These relations can be demonstrated by simple experiments (Fig. 9). Additionally, throat-to-pore volume, number of throats per pore, and heterogeneity of pores and throats are all likely to affect ejection efficiency, but their effects have not been evaluated.

In conclusion, the reduction in ejection efficiency that accompanies decrease in porosity is thought to result largely from an increase of pore-to-throat-size ratio (Fig. 8).

MECHANISM OF DOLOMITE GROWTH

Pore spaces in dolomites are defined mainly by dolomite crystal faces and intergrain boundaries, and it is appropriate to consider the mode of dolomite growth before describing the pore systems.

The replacement of calcite by dolomite may be piecemeal. That is, dolomite grows while adjacent calcite dissolves, with only a thin film of solution separating the minerals. Alternatively, dolomite crystals may grow in larger pore spaces without the necessity of simultaneous solution of calcite. Evidence for this latter process is widespread in Rainbow limestones where the small proportion of dolomite is present preferentially as void fill in the larger pore spaces. The surfaces of dolomite grains in contact with pore walls are defined by the irregularities of the pore walls, whereas perfect crystal faces are developed into the adjoining pores (Fig. 10A). Even in rocks which are composed entirely of dolomite, void filling could have been the main mode of growth, provided that new spaces were produced periodically by solution of the host limestone.

Further support for such a mechanism is seen where dolomite has grown in a porous microspar and, apparently, included grains of microspar to give a poikilotopic texture (Fig. 10B, C). Similar forms of dolomite have been reported by Folk and Siedlecka (1974). Microspar grains within dolomite appear to be scattered irregularly and lack an obvious relation to crystallographic planes in the host dolomite, yet are similar in size and shape to surrounding microspar grains. From these observations, void-filling growth appears more probable than piecemeal replacement. Intracrystal porosity in the dolomite may have been created by subsequent partial solution of the microspar. Poikilotopic dolomite was observed only in rocks which are composed predominantly of microspar and not in rocks with a high percentage of dolomite.

During progressive growth of dolomite in limestone, whether by void-filling growth or by piecemeal replacement, a stage is reached at which dolomite grains touch and provide the supportive framework of the rock. At this stage little, if any, calcite remains in the system (Fig. 10D) and the rock may have as much as 25 percent porosity. However, growth of dolomite crystals may continue and a progression of pore forms is generated as porosity is reduced. These forms are observed conveniently by preparing pore casts.

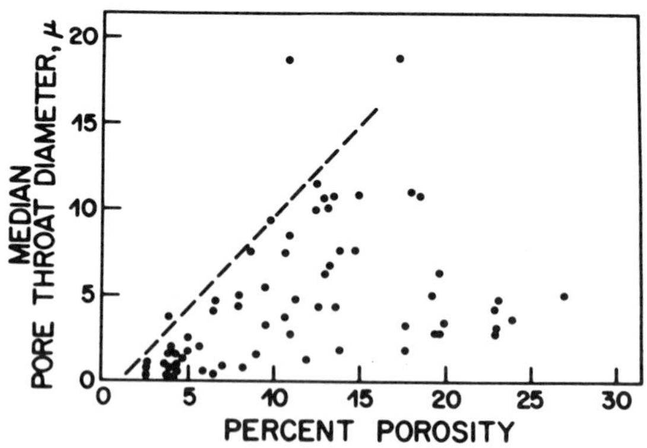

FIG. 4—Illustrates tendency for positive correlation between median pore-throat diameter (diameter of throat being injected when 50 percent of pore volume has been saturated) and porosity. Points are widely scattered but for given porosity there is maximum pore-throat diameter (broken line) that is not exceeded. This maximum increases with porosity increase. Samples are dolomites from locations identified in caption, Figure 2.

Pore Systems in Dolomites

Pore casts were obtained by impregnating samples with resin and then slowly etching with dilute acid to remove the carbonate host. A cast of the pore system then stands in relief and was examined using a binocular microscope and, after coating with gold-palladium, with a Cambridge Stereoscan 600 scanning-electron microscope.

Casts of fine pore structures down to 0.1μ were obtained by using a low-viscosity resin (Union Carbide resin ERL 4026 with Hooker Chemical het anhydride as the hardener), evacuating the sample before impregnation and applying a pressure of over 100 kg/sq cm during impregnation. The resin was stained red with Oracet Red B (Ciba-Geigy) to facilitate distinguishing it from residual mineral matter. The hardened resin is isotropic and has a refractive index of 1.534. Tests were made to establish that the resin does not react with carbonate rocks and is not affected by the acids used in leaching the host rock.

Dolomites consisting of a mosaic of rhombohedral crystals have high porosity (Fig. 10D). Individual pore spaces are defined by many randomly oriented, planar crystal faces which are described here as intercrystal *polyhedral pores.* Sizes of these pores are related to sizes of containing rhombohedra and pore connections or throats are large and numerous in relation to the pores (see Fig. 10F which is a pore cast of the rock illustrated in Fig. 10D).

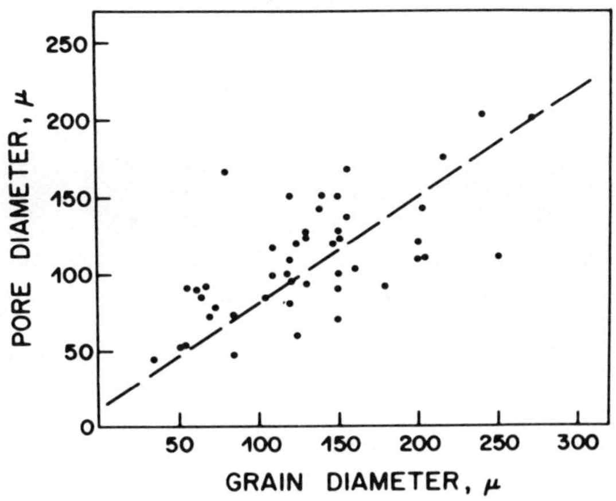

FIG. 5—Pore diameter increases with grain diameter, pore size tending to be less than grain size. Samples from locations identified in caption of Figure 2. Grains and pores were measured on random traverses of etched-pore casts and maximum diameters were recorded. Thirty measurements were made for grains and 30 for pores in each pore cast. Each point on graph is average for single pore cast.

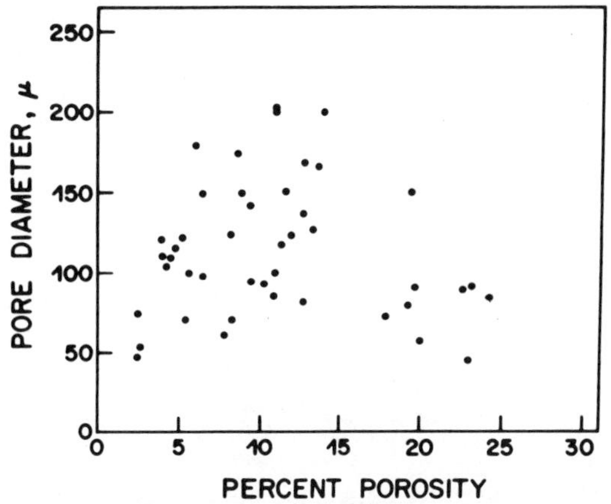

FIG. 6—Average pore diameters (measured as indicated in caption to Fig. 5) versus porosity. No correlation is apparent. Each point on graph represents average of 30 pore measurements on each pore cast. Samples from locations identified in caption of Fig. 2.

Not all dolomite rocks are composed predominantly of rhombohedral crystals. Where dolomite growth continued, planar compromise boundaries developed, rhombohedral form was lost, and the polyhedral pores were reduced in size (left side of Fig. 11A). The pores were reduced also in geometric complexity. Figure 12 illustrates a polyhedral pore which has been filled by equal growth of the adjacent crystal faces. With progressive growth the number of containing faces is reduced from five to three. What appears as a triangle in two dimensions is a tetrahedron in three dimensions, for it is improbable that all the faces would parallel a common axis. During the continued growth of dolomite, polyhedral pores tend to be reduced in size and simplified in form to tetrahedral pores. Tetrahedral pores are common in the lower porosity dolomites (Figs. 11A-E).

Further growth of dolomite crystals eliminates tetrahedral pore spaces so that grain boundaries meet at triple junctions. Whatever the complexity of the original polyhedral pore, if crystal growth continues, triple junctions form and the junction angles tend toward equality. This can be illustrated by considering three faces with interfacial angles of 20°, 70°, and 90° each growing at an equal rate. The triple-junction angles (100°, 125°, and 135°) are much closer to equality than the interfacial angles (Fig. 12). Consideration of several examples shows that this is a general result.

With the elimination of tetrahedral pores, the only pore spaces remaining are thin, sheet-like spaces at grain boundaries. *Interboundary-sheet pores* were observed in more than half of all the dolomite pore casts studied (Fig. 11A, B, D, E).

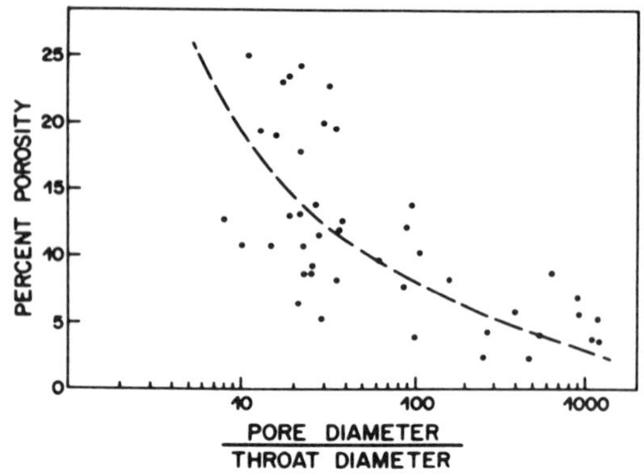

FIG. 7—Negative correlation between pore-to-throat-size ratio and porosity. Throat size was obtained from mercury-injection capillary-pressure curve and is equivalent radius of pore throat being injected when 50 percent of pore volume has been saturated. Average pore size was obtained from resin-pore casts by measuring maximum dimensions of randomly selected pores. Thirty pore sizes were averaged for each pore cast. Pore sizes were determined on same cylinders used in obtaining mercury capillary-pressure curves. Samples are from locations identified in caption of Figure 2.

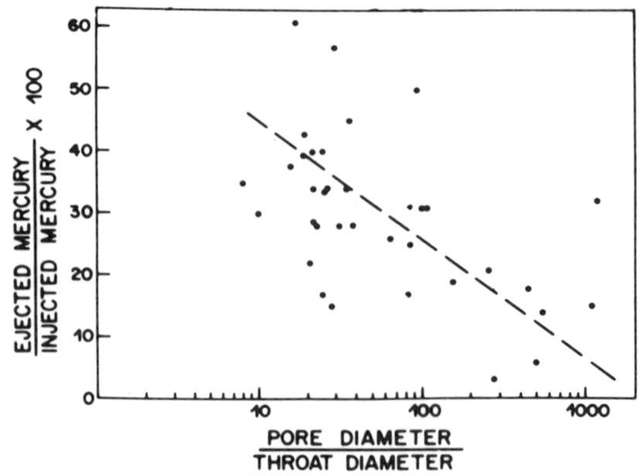

FIG. 8—Negative correlation between pore-to-throat-size ratio (obtained as indicated in caption to Fig. 7) and mercury-ejection efficiency. Samples are dolomites from locations identified in caption to Figure 2.

Forty measurements of sheet-pore thickness made from pore casts with a scanning-electron microscope ranged from 0.1 to 4μ and averaged 2μ. Sheet pores of less than 0.1μ no doubt are present and probably were not observed because this is the lower size limit for resin impregnation under the conditions used.

The capillary-pressure curves for dolomites have a steep initial part, where there is a relatively large increase of pressure for a small volume of mercury injected, and then a flattening of the curve where a considerable volume of mercury is injected for a relatively small pressure increment (Fig. 1). A "threshhold" injection pressure can be defined at this change in gradient and can be translated into an equivalent pore-throat size. Most of the pore volume is connected by throats which are smaller than this size. Dolomites with porosity of less than 5 percent have polyhedral and tetrahedral pores which are connected mainly by interboundary-sheet pores. The thickness of these pores, as estimated from capillary-pressure curves, ranges from 0.1 to 2μ and is in good agreement with direct measurements made from pore casts.

Initially it was suspected that the interboundary-sheet pores might be artifacts of depressuring, as the cores were relieved of 6,000 ft (1,818 m) of overburden, or perhaps an artifact of the resin-injection process. Apparently this is not the case, as some samples contain solid bitumen residues and these can be seen on etched surfaces as intergrain sheets of the same thickness and form as the sheets of resin (Fig. 11F). This provides direct evidence for the existence of interboundary-sheet pores in the ground during deposition of solid bitumen. Also, the relatively large volume occupied by interboundary-sheet pores in some dolomites provides indirect evidence that these pores are not the result of depressuring during core recovery. In a dolomite with 200μ average grain size and with interboundary-sheet pores as the only pore type, the measurable pores averaged 2μ in thickness. This can be computed as equivalent to about 3-percent porosity, which is the same as the porosimeter-measured value for the sample. This is greatly in excess of an expect-

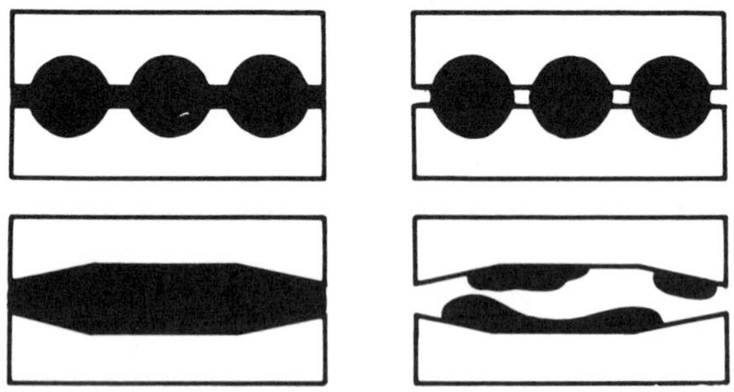

FIG. 9—Mercury (black) retention after pressure reduction is greater for pore systems with larger pore-to-throat-size ratios (upper figures) than for pore systems with smaller pore-to-throat-size ratios (lower figures). Figures on right illustrate condition after pressure reduction and mercury ejection. Pore systems are etched into glass plates and are transparent. They are evacuated and then saturated with mercury under pressure. As pressure is reduced mercury ejection can be viewed.

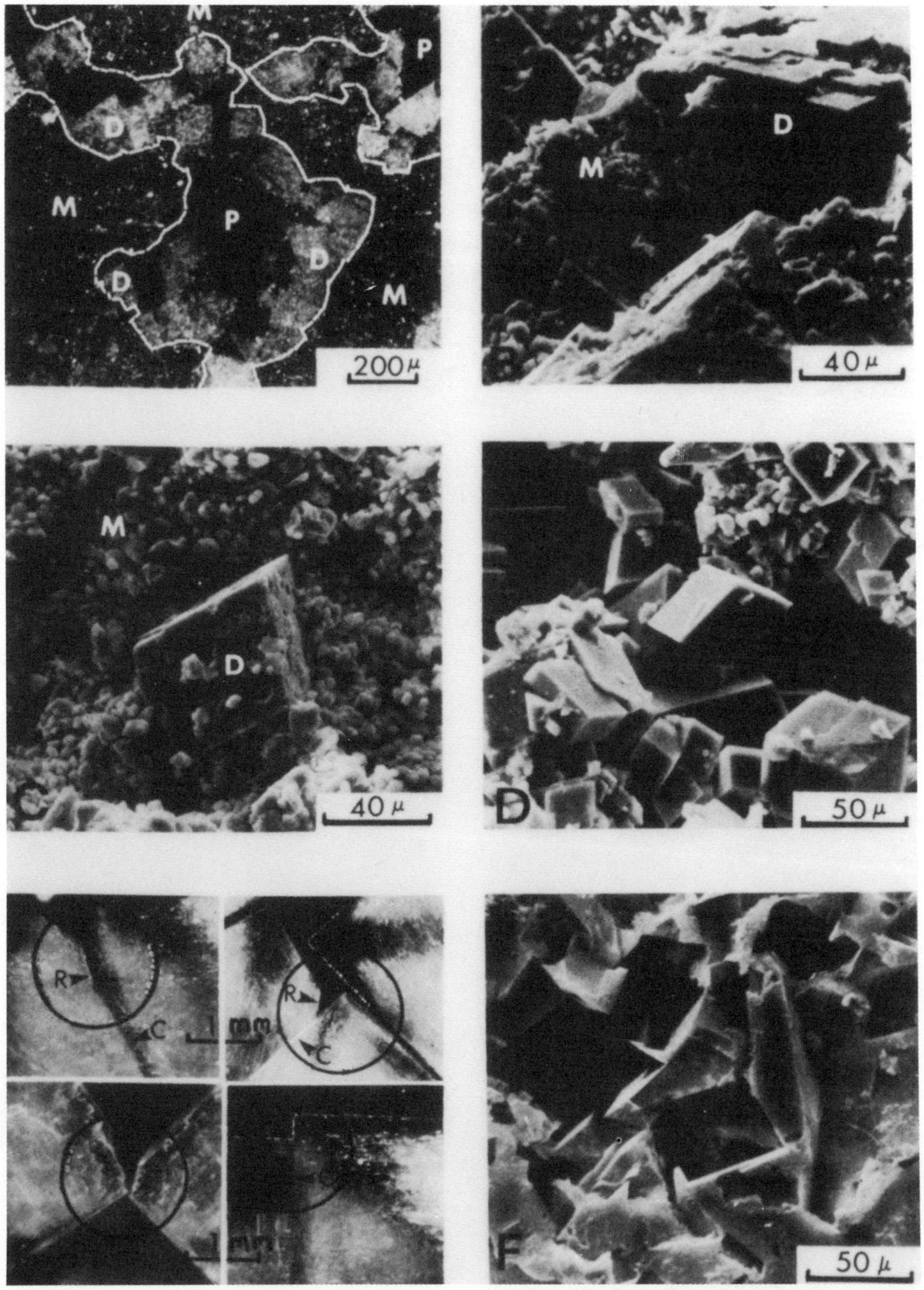

FIG. 10—**A.** Carbonate impregnated with dark epoxy and partly etched. Surface seen in reflected light. Large pores *P* in microspar *M* are outlined in white. Dolomite grains *D* are present preferentially as void fill in large pores. Core location 12-3-109-8W6 at 5,880 ft (1,782 m) subsurface.

B-C. Dolomite rhombohedra *D* have grown in microspar *M* and apparently include grains of microspar to give a poikilotopic texture. Photographed with scanning-electron microscope. Core location 12-3-109-8W6 at 5,879 ft (1,782 m).

D. Dolomite composed predominantly of rhombohedral grains. Photographed with scanning-electron microscope. Core location 12-3-109-8W6 at 5,880 ft (1,782 m) subsurface.

E. Halite crystallized in laboratory from cooling solution. At ends of compromise boundaries *C* (top figures) adjacent crystal faces are incompletely developed and small reentrants *R* are present. Two crystals (bottom left) with faces incompletely developed adjacent to points of contact. Where compromise boundary *C* formed between two crystals of similar orientation no reentrant is present (bottom right).

F. Pore cast of sample illustrated in **D** above. Dolomite host has been dissolved completely by acid so that all solid material in picture is resin cast of pore system. Complex pores defined by many planar surfaces are described as polyhedral. Note that pore throats are large in relation to pore size. Photographed with scanning-electron microscope.

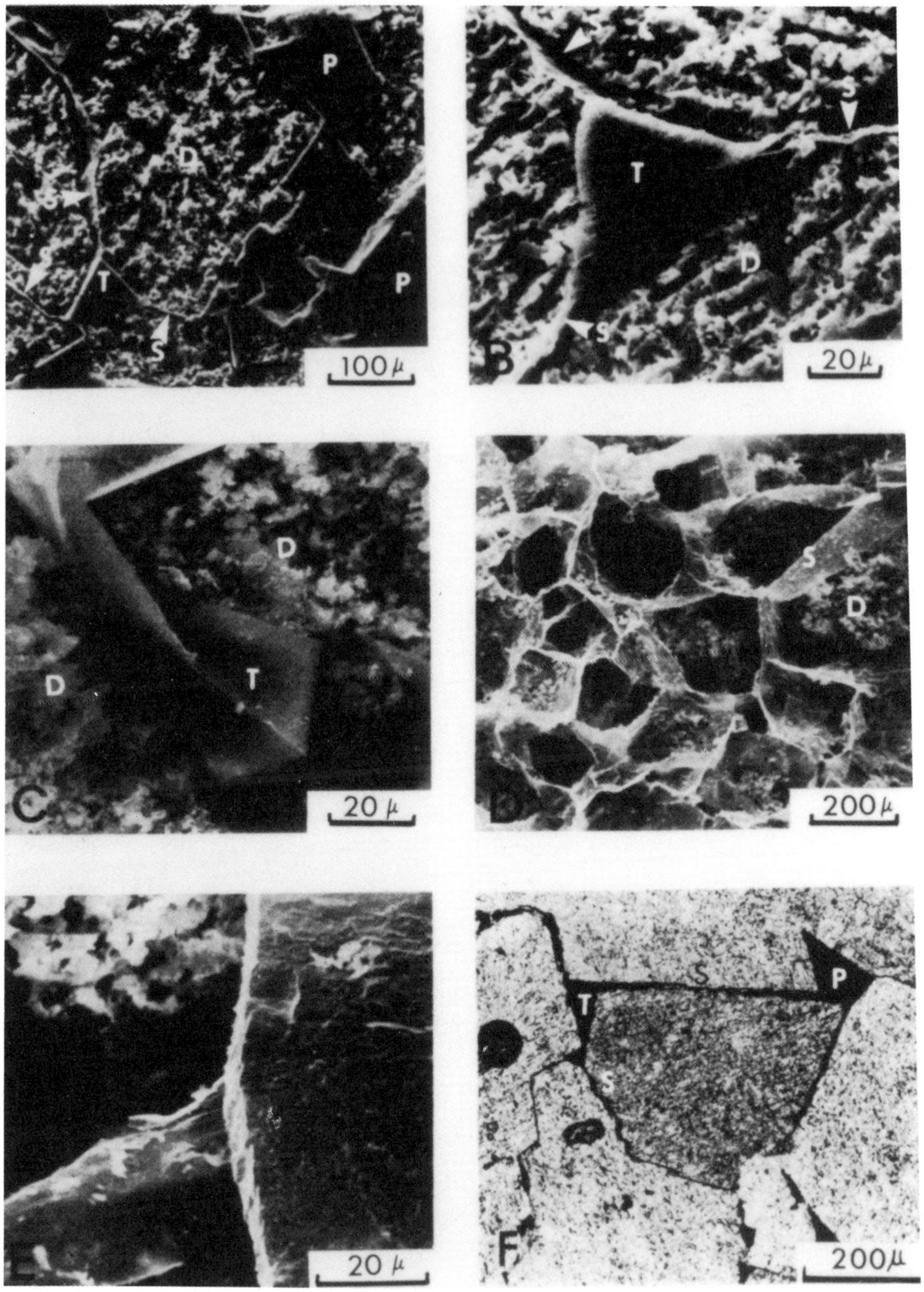
P
D
T
S
P
100 μ
T
S
D
S
20 μ
D
D
T
C
20 μ
S
D
200 μ
20 μ
S
P
T
S
F
200μ

ed expansion due to depressuring which, from data presented by Hall (1953) and Fatt (1958), should not exceed about 1 percent for the samples considered.

Where dolomite crystals have grown to form compromise boundaries, the pore throats are in the form of interboundary-sheet pores and should be considered as spaces between parallel plates rather than as tubes. As discussed previously, in translating capillary pressures to equivalent pore-throat sizes it is more appropriate to use equation (2) than equation (1).

Interboundary-sheet pores have a wider significance than for consideration of capillary pressure. Spaces at grain boundaries have been recognized by Thill et al (1973, p. 554) in Tennessee marble and are described in the following way: "the most prominent type of pore structure is intergranular microcracks that outlined individual grains in an almost continuous network." In this context, the significance of these intergranular spaces is seen as their effects on mechanical strength and seismic velocity. They are also likely to affect electrical properties.

In summary, growth of rhombohedral grains in a sucrosic dolomite reduces porosity and establishes compromise boundaries. These boundaries are the sites of interboundary-sheet pores up to only a few microns thick which act as pore connections and throats. Formation of sheet pores greatly increases the pore-to-throat-size ratio for the rock and it is this ratio which is most important in influencing the ejection efficiency for a nonwetting phase. As porosity is reduced, pore-to-throat-size ratio increases and ejection efficiency is reduced (Figs. 7, 8, 13). In the final stages of porosity reduction, when tetrahedral pores are reduced in size and boundary-sheet pores predominate, it can be anticipated that pore-to-throat-size ratio then will decrease but, at the low porosity (say <4 percent) which then prevails, the rock is of little interest as a reservoir rock.

Inhibition of Dolomite Growth

It is apparent from the preceding that the form and amount of pore space in a dolomite are inversely proportional to the ability of dolomite crystals to grow. In this section, some comments are offered on inhibition of the dolomite growth.

A sucrosic dolomite, such as illustrated in Figure 10D, is composed almost exclusively of rhombohedral crystals and therefore lacks compromise boundaries. Apparently, when two dolomite crystals touch, further growth to form a compromise boundary is inhibited, at least temporarily. If this were not so it would be a coincidence to find a rock with a predominance of rhombohedra.

A phenomenon which may be analogous is present during the growth of halite crystals from a saturated solution which is cooled slowly. Growth appears to be inhibited on a macroscopic scale at intercrystal boundaries. At the end of a compromise boundary adjacent crystal faces are developed incompletely and small reentrants are present (Fig. 10E). Such reentrants are not present where two faces of similar orientation have grown together. The cause of this phenomenon is not understood but it is of importance, because it favors the preservation of intercrystal pore space. During the growth of crystals there is a prevailing

---

← FIG. 11—All figures except **F**, photographed with scanning-electron microscope.

A. Etched-pore cast of dolomite rock. After resin impregnation, rock was etched with acid and residual dolomite grains *D* have spongy appearance and are recessed relative to resin. Illustrating resin-filled polyhedral pores *P*, tetrahedral pores *T* (triangular in cross section), and interboundary-sheet pores *S* along compromise boundaries. Core location 9-32-109-8W6 at 6,122 ft (1,855 m) subsurface.

B. Etched-pore cast of dolomite rock. Detail of resin-filled tetrahedral pore *T* connected by interboundary-sheet pores *S*. Dolomite host *D* is recessed slightly by etching which has selectively affected cleavage planes. Location as for **A**.

C. Etched-pore cast of dolomite rock. Detail of resin-filled tetrahedral pore *T* exposed by solution of dolomite grains *D* whose partly digested remains have spongy appearance. Core location 12-3-109-8W6 at 5,866 ft (1,778 m) subsurface.

D. Etched-pore cast of dolomite rock. Resin-filled interboundary-sheet pores *S* at dolomite-grain boundaries stand in relief. Some undigested remnants of dolomite grains *D* are visible. Core location 9-32-109-8W6 at 6,122 ft (1,855 m) subsurface.

E. Etched-pore cast of dolomite rock. Detail of resin-filled interboundary-sheet pores. Same location as **D**.

F. Polished dolomite surface viewed in reflected light. Solid bitumen occupies polyhedral *P*, tetrahedral *T*, and interboundary-sheet pores *S*. Provides evidence that interboundary-sheet pores were present in ground during deposition of solid bitumen. Bitumen occurs as intergranular films as thin as 2μ.

tendency for crystal faces to remain nearly flat, even where this means opposing constraints in the solution such as irregularities of diffusion gradients. Secondly, surfaces bounding crystals tend to be faces which are either parallel with, or simply related to, the edges of the unit cell. Because crystals generally are oriented irregularly, compromise boundaries will not be related simply to unit cells and will not be on a common plane with adjacent growing faces. Both of these factors may tend to retard the formation of compromise boundaries, although this explanation is far from adequate.

When the threshhold that appears to limit growth of touching rhombohedral dolomite grains is overcome, compromise boundaries form and polyhedral pores are reduced to tetrahedral pores which are connected by interboundary-sheet pores. From the abundance of tetrahedral pores in low-porosity dolomites, it appears that a second threshhold has been reached and that small spaces at triple junctions are difficult to eliminate by crystal growth from solution. This is perhaps not surprising because the free energy per atom along a triple junction is likely to be higher than that per atom in a grain boundary (McLean, 1957, p. 50) and presumably will increase as the space diminishes.

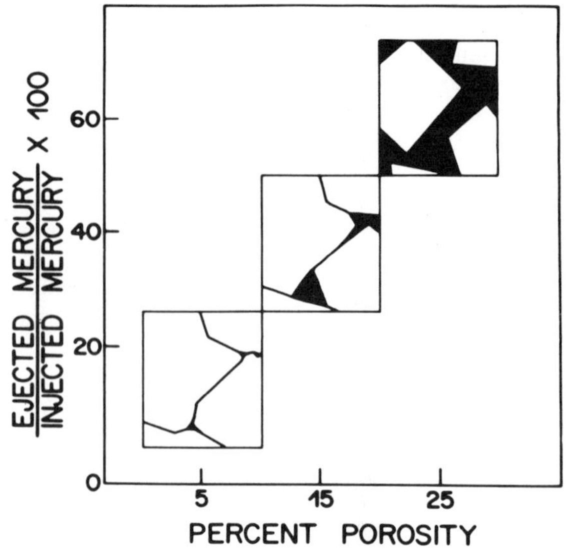

FIG. 13—Illustration of changes from polyhedral to tetrahedral to interboundary-sheet pores that accompany growth of dolomite crystals and porosity reduction. Porosity reduction is accompanied by reduction of mercury-ejection efficiency which is thought to be influenced largely by increase in pore-to-throat-size ratio.

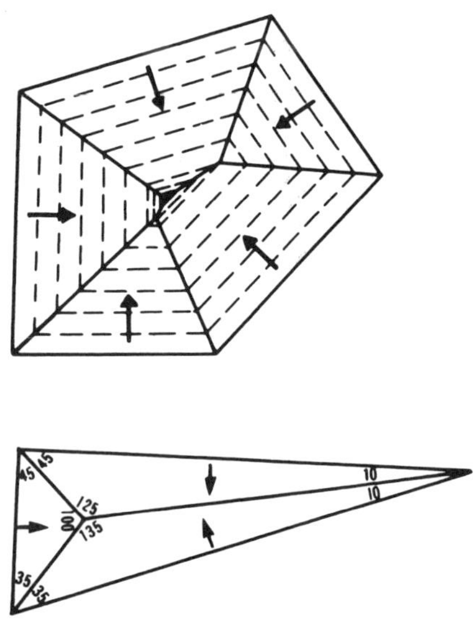

FIG. 12—Upper figure illustrates cross section of polyhedral pore. Number of containing faces is reduced from five to three as adjacent crystal faces (broken lines) extend themselves by equal growth to fill pore.

Lower figure illustrates cross section of pore space that has been filled by equal growth of three crystal faces to form compromise boundaries (arrows indicate growth directions for three crystal faces). Triple-junction angles (100°, 125°, and 135°) tend toward equality in spite of large inequalities in interfacial angles (20°, 70°, and 90°).

Interboundary-sheet pores are so widespread that apparently it is only with great difficulty that these residual intergrain spaces can become filled with mineral matter. Becker and Day (1916) and Correns (1949) have demonstrated that a growing crystal can lift an imposed weight. A solution film must persist between the solids and is not eliminated by growth of the crystal. Because the growth pressure may be greater than the hydrostatic pressure it seems that the water film must be capable of supporting a shear stress. Thin films of solution at crystal surfaces are also a requirement for the process of pressure solution. In this context Weyl (1959, p. 2005) has concluded that ". . . mineral grains are separated by a film of solvent which can support a shear stress but in which the ions of the solute can diffuse at a rate that is rapid compared with solid diffusion in the mineral grains." These examples attest to the persistence of solution films at intergrain surfaces although the mechanisms of such persistence remain obscure.

In conclusion, inhibition of crystal growth is of fundamental importance to the preservation of pore space in dolomites. Crystal growth seems to be arrested, at least temporarily, at three stages: when rhombohedra have grown sufficiently large to touch in three dimensions, when compromise

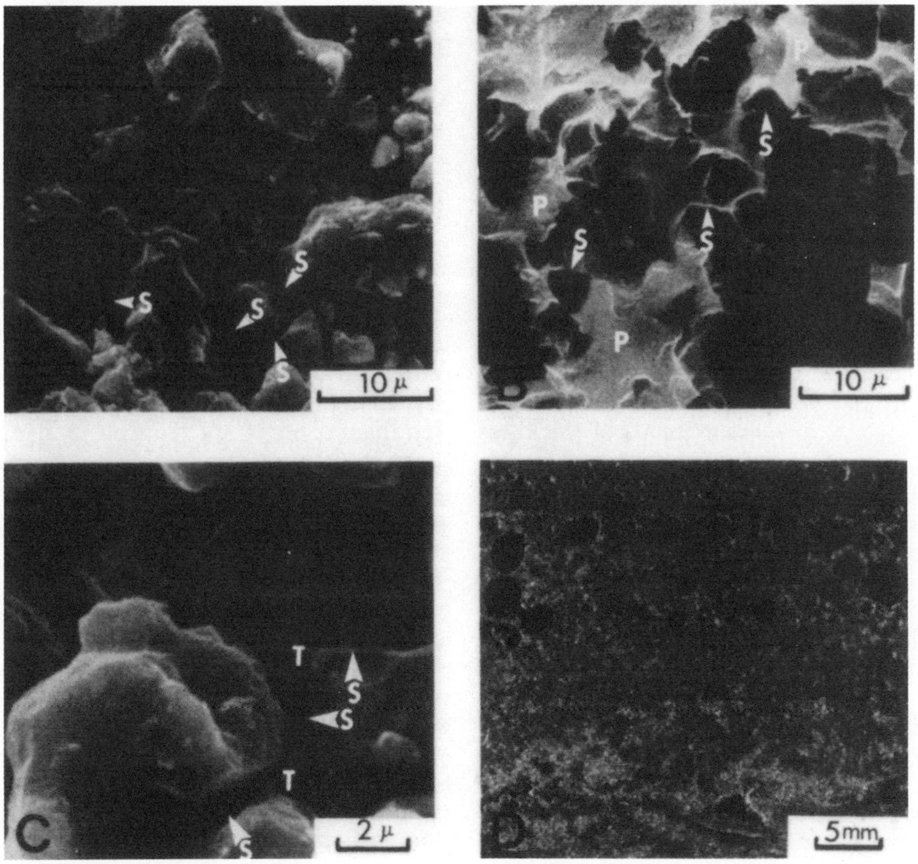

FIG. 14—**A-C** photographed with scanning electron microscope. **D** is hand specimen in reflected light.

A. Limestone rock. Microspar with interboundary-sheet pores *S* visible at grain boundaries. Core location 12-3-109-8W6 at 5,880 ft (1,782 m) subsurface.

B. Etched-pore cast of rock illustrated in **A.** All mineral matter removed by acid treatment. Resin-filled polyconcave pores *P* are connected by interboundary sheet pores *S*.

C. Limestone rock. Microspar with concave-tetrahedral pores *T* connected by interboundary-sheet pores *S*. Location as for **A.**

D. Limestone rock. Hand specimen in reflected light illustrating vesicular pores. Where these pores abut on upper surface of lamina they have preferentially flattened upper surfaces and appear like relict gas bubbles. Core location 2-32-109-8W6 at 5,992 ft (1,816 m) subsurface.

boundaries are sufficiently extensive to leave tetrahedral pores at triple junctions, and when interboundary-sheet pores are the only type of pore spaces remaining.

In addition to the intercrystalline pores described, many dolomites contain larger pores some of which are clearly molds after brachiopods, corals, stromatoporoids or gastropods, and others that are not obviously fabric selective and should be referred to as vugs (Choquette and Pray, 1972, p. 171). Additionally, channels associated with fine horizontal fractures are locally common. Where megapores of these types are not abundant enough to form connected networks, which appears to be the case in many of the rocks studied, communication must be made through the intercrystal pores which, for this reason, are of fundamental importance and have received most attention in this paper.

PORES IN LIMESTONE

The limestones studied are mainly fine grained or contain fine-grained matrices. Micrites (grains 1 to 4μ diameter) are present but are less common than microspars (grains 4 to 10μ diameter; Fig. 14A-C). The terminology is that of Folk (1959, p.

7-8; 1965, p. 37). The grains are mainly equidimensional but some are slightly elongate and they commonly are rounded with only local and imperfect expression of crystal faces. Pore casts typically have a spongy appearance and the pores differ from the polyhedral pores of the dolomites in being smaller and having concave surfaces, as the confining grains have rounded rather than planar surfaces. These pores are described here as polyconcave micropores (micropore, following the terminology of Choquette and Pray, 1972, p. 171) and constitute up to 20-percent porosity in some samples. As in the dolomites, where polyhedral pores do not intersect, connections are sheet-like (Fig 14B) and 20 measurements of thickness made from pore casts range from 0.1 to 0.5μ, averaging 0.3μ. Interboundary-sheet pores are also visible in the rock samples (Fig. 14A-C) but are less well displayed than in the pore casts.

Lower porosity microspars have interboundary sheet pores connected to concave-tetrahedral pores (Fig. 14C) analogous to the tetrahedral pores in dolomites. Cementation appears to be inhibited in these small spaces. Some limestones have larger pores (mesopores) some of which are within fossils and others which are fenestral. An additional type of mesopore is described as vesicular from its similarity to vesicles in volcanic rocks. These pores are present commonly in laminated microspars and are of up to 4-mm diameter, round to elliptical and not related to solution of components of similar shape in the rock. Where the vesicular pores abut on the upper surface of a lamina they have preferentially flattened upper surfaces and appear like relict gas bubbles. It is thought that they are of early diagenetic origin and their preservation testifies to early lithification of the deposit.

As with the dolomites, the intercrystal porosity in limestones is thought to be of fundamental importance because only locally are mesopores sufficiently abundant enough to intersect in three dimensions.

Conclusions

As dolomite crystals grow and porosity is reduced, there are progressive changes in pore geometry. Polyhedral pores are reduced to tetrahedral pores which may be eliminated to leave interboundary-sheet pores. All types may be present in the same sample. As pore size is reduced, pore structure tends to become simpler and more regular (Fig. 12). Pore connections are sheet-like rather than tubular and the expression

$$P_c = \frac{2\sigma \cos\theta}{d}$$

should be used to relate capillary pressure (P_c) to pore thickness (d).

Where polyhedral pores are volumetrically abundant they may form a connected network with relatively large pore throats. Where porosity has been reduced by further dolomite growth, polyhedral and tetrahedral pores are connected by interboundary-sheet pores and the ratio of pore size to throat size is greatly increased.

It is this increase in pore-to-throat-size ratio that is thought to be the most important factor contributing to the decrease in mercury-ejection efficiency which accompanies decrease of porosity. In the final stages of porosity reduction when tetrahedral pores are reduced in size and boundary-sheet pores predominate, it can be anticipated that pore-to-throat-size ratio then will decrease but, at the low porosity (say <4 percent) which then prevails, the rock is of little interest as a reservoir rock.

Mercury-injection and ejection tests do not provide a close analogue for an oil-water system, but mercury ejection-injection ratios may give a relative indication of the efficiency with which various rock types can expel a nonwetting phase when pressure is reduced. Because ejection efficiency correlates positively with porosity, it may be possible from porosity data to provide a factor that will indicate recovery efficiency insofar as this is dependent on pore geometry.

Inhibition of crystal growth is of fundamental importance to the preservation of pore space in dolomites. Crystal growth seems to be arrested, at least temporarily, in three sequential stages: (1) when rhombohedra have grown sufficiently large to touch in three dimensions; (2) when compromise boundaries are sufficiently extensive to leave tetrahedral pores at triple junctions; and (3) when interboundary-sheet pores are the only type of pore space remaining. The mechanisms by which growth is arrested at each of these three stages are not understood.

References Cited

Barss, D. L., A. B. Copland, and W. D. Ritchie, 1970, Geology of Middle Devonian reefs, Rainbow area, Alberta, Canada, *in* Geology of giant petroleum fields: AAPG Mem. 14, p. 19-49.

Becker, G. F., and A. L. Day, 1916, Note on the linear force of growing crystals: Jour. Geology, v. 24, p. 313-333.

Burdine, N. T., L. S. Gournay, and P. P. Reichertz, 1950, Pore-size distribution of petroleum reservoir rocks: AIME Petroleum Trans., v. 189, p. 195-204.

Chilingar, G. V., R. W. Mannon, and H. H. Rieke, 1972, Oil and gas production from carbonate rocks: New York, American Elsevier, 408 p.

Choquette, P. W., and L. C. Pray, 1972, Geologic nomenclature and classification of porosity in sedimentary carbonates, *in* Carbonate rocks II: porosity and classification of reservoir rocks: AAPG Repr. Ser. 5, p. 154-197.

Correns, C. W., 1949, Growth and dissolution of crystals under linear pressure: Faraday Soc. Discussions no. 5, p. 267-271.

Dumoré, J. M., and R. S. Schols, 1974, Drainage capillary-pressure functions and the influence of connate water: Soc. Petroleum Engineers Jour., p. 437-444.

Fatt, I., 1958, Pore volume compressibilities of sandstone reservoir rocks: AIME Petroleum Trans., v. 213, p. 362-364.

Folk, R. L., 1959, Practical petrographic classification of limestones: AAPG Bull., v. 43, p. 1-38.

——— 1965, Some aspects of recrystallization in ancient limestone, *in* Dolomitization and limestone diagenesis—a symposium: Soc. Econ. Paleontologists and Mineralogists Spec. Pub. 13, p. 14-48.

——— and A. Siedlecka, 1974, The "schizohaline" environment: its sedimentary and diagenetic fabrics as exemplified by late Paleozoic rocks of Bear Island, Svalbard: Sed. Geology, v. 11, p. 1-15.

Hall, H. N., 1953, Compressibility of reservoir rocks: AIME Petroleum Trans., v. 198, p. 309-311.

Hriskevich, M. E., 1966, Stratigraphy of Middle Devonian and older rocks of Banff Aquitaine Rainbow West 7-32 discovery well, Alberta: Bull. Canadian Petroleum Geology, v. 14, p. 241-265.

——— 1967, Middle Devonian reefs of the Rainbow region of northwestern Canada—exploration and exploitation: 7th World Petroleum Cong., Mexico City, Proc., v. 3, p. 733-763.

——— 1970, Middle Devonian reef production, Rainbow area, Alberta, Canada: AAPG Bull., v. 54, p. 2260-2281.

Langton, J. R., and G. E. Chin, 1968, Rainbow Member facies and related reservoir properties, Rainbow Lake, Alberta: Bull. Canadian Petroleum Geology, v. 16, p. 104-143.

McLean, D., 1957, Grain boundaries in metals: Oxford, England, Clarendon Press, 346 p.

Purcell, W. R., 1949, Capillary pressures—their measurement using mercury and the calculation of permeability therefrom: Jour. Petroleum Tech., v. 1, p. 39-48.

Taylor, R. P., and N. C. Wardlaw, 1975, Precision of porosity measurements using mercury pycnometer and Boyle's Law method: Jour. Canadian Petroleum Tech., v. 14, p. 33-37.

Thill, R. E., T. R. Bur, and R. C. Steckley, 1973, Velocity anisotropy in dry and saturated rock spheres and its relation to rock fabric: Internat. Jour. Rock Mech. Min. Sci., v. 10, p. 535-557.

Weyl, P. K., 1959, Pressure solution and the force of crystallization—a phenomenological theory: Jour. Geophys. Research, v. 64, p. 2001-2025.

The American Association of Petroleum Geologists Bulletin
V. 69, No. 5 (May 1985), P. 733-741, 14 Figs., 3 Tables

Selected Characteristics of Limestone and Dolomite Reservoirs in the United States[1]

JAMES W. SCHMOKER, KATHERINE B. KRYSTINIK, and ROBERT B. HALLEY[2]

ABSTRACT

Data from the United States Oil and Gas File (TOTL) developed by the University of Oklahoma, Norman, Oklahoma, are used to characterize the lithology, location (state and basin), geologic age, year of discovery, depth to top of pay, porosity, permeability, water saturation, volume of crude oil and nonassociated gas originally in place, and net-pay thickness of limestone and dolomite reservoirs in the United States. Distributions of these parameters, representing thousands of reservoirs, establish a framework to which individual carbonate reservoirs can be compared, and provide insights into geologic processes affecting reservoir characteristics.

Limestone reservoirs are more numerous in the United States than dolomite reservoirs (by a ratio of about 3 to 1) because limestones are more abundant than dolomites. However, in the eight states that account for over 90% of United States carbonate reservoirs, there is a statistical tendency for carbonate reservoirs to occur preferentially in dolomites. Dolomite reservoirs, on the average, are larger and deeper than those of limestone, yet they often have lower matrix porosities and permeabilities. This line of investigation offers supplemental evidence that dolomitization tends to improve the reservoir properties of a given formation, and that effective fracture systems at reservoir depths are more likely to occur in dolomites than in limestones.

INTRODUCTION

Traditions have evolved over the years with respect to differences between limestone and dolomite reservoirs. "We are all aware of the importance of dolomites as petroleum reservoirs" (Zenger et al, 1980, p. iii). "Because dolomitization generates a carbonate rock with improved capillary properties, the resulting rock is commonly an oil reservoir" (Murray and Pray, 1965, p. 2). "Ancient dolomite in the subsurface is much more commonly a porous rock than limestone. Indeed, approximately 80% of the oil and gas that will be recovered from carbonate rocks in North America will come from dolomite" (Blatt et al, 1972, p. 492). Statements regarding average or typical reservoir depths, porosities, and permeabilities are commonly included in introductory sections of reports. Such generalizations are usually presented without documentation.

In this paper, comparisons are made between thousands of limestone and dolomite reservoirs in the United States, using data from publicly available files. These data bases are by no means complete, but they are the largest compilation of reservoir information known to the authors, and they provide a basis for quantifying some of the traditions associated with the characteristics of limestone and dolomite reservoirs.

DATA SET

The principal source of data for this study is the United States Oil and Gas File (TOTL) developed by the University of Oklahoma Energy Resources Center (Norman, Oklahoma). TOTL includes over 100,000 reservoir and field records containing publicly available geological and engineering information. The data in TOTL are primarily from annual reports of state regulatory agencies, supplemented by sources that include hearing files, state geological surveys, and private information services.

Data were retrieved from TOTL (on magnetic tape) in April 1983, through the General Electric Information Services Company[3] computer network. The initial retrieval, of all reservoirs with limestone or dolomite in their lithologic description, was edited to remove those with clastic associations. Carbonate reservoirs with anhydrite or chert were retained. Each reservoir was classified as limestone or dolomite according to the more abundant component.

The various data subsets generated from the main file are listed by reservoir parameter in Table 1, together with the number of limestone and dolomite cases for each. Note that these data represent reservoirs—not fields; several carbonate reservoirs can be found in a given field. Based on the most recent entries for "year of discovery," the data are current through 1980.

The data file's large size and numerous contributors are an advantage in that random measurement errors and individual selection and measurement biases are not likely to significantly affect the overall descriptive statistics. However, a factor of uncertainty is introduced because, to a greater or lesser degree depending on the parameter, reservoir data are missing in TOTL.

Some of the distributions of reservoir parameters presented in following sections have been truncated by economic factors associated with hydrocarbon production that discriminate against less profitable reservoirs. The characteristics of these economic factors, although not quantitatively known, should be the same for both lime-

[1]Manuscript received, May 17, 1984; accepted, November 12, 1984.

[2]U.S. Geological Survey, Box 25046, Denver Federal Center, MS 921, Denver, Colorado 80225.

We thank Gwen Luttrell, U.S. Geological Survey, Reston, Virginia, for the retrieval from the Geonames data file.

[3]Use of corporate names in this report is for descriptive purposes only and does not imply endorsement by the U.S. Geological Survey.

Table 1. Summary of Data Set Listing Reservoir Parameters Investigated and Number of Limestone and Dolomite Cases for Each, U.S. Carbonate Reservoirs

	Number of Reservoirs		
Reservoir Parameter	Limestone	Dolomite	Total
Lithology	11,770	4,079	15,849
State	11,770	4,079	15,849
Basin	11,770	4,079	15,849
Year of discovery	10,936	3,928	14,864
Geologic period	10,965	3,850	14,815
Depth to top of pay (ft)	9,290	3,413	12,703
Average net-pay thickness (ft)	6,156	2,726	8,882
Crude oil originally in place (bbl)	2,947	1,319	4,266
Average porosity (%)	2,166	866	3,032
Average permeability (md)	1,352	510	1,862
Average water saturation (%)	982	499	1,481
Nonassociated gas originally in place (mcf)	828	306	1,134

stones and dolomites. Therefore, although distributions affected by economic truncation do not reflect true in-situ statistics, comparisons between limestone and dolomite histograms are valid.

RESULTS

Lithology, Location, and Age

About three-fourths of United States carbonate reservoirs are in limestone (Figure 1). However, even though limestone reservoirs are about three times more common than dolomite, the relative value of limestone and dolomite as reservoir rocks should not be judged on the basis of Figure 1 alone. Other factors are considered in subsequent sections.

Limestone reservoirs are reported from 25 states, and dolomite reservoirs from 24 states, but 93% and 92% of reported reservoirs, respectively, occur in the eight states of Texas, Oklahoma, Illinois, Kansas, Kentucky, Indiana, New Mexico, and Michigan (Figure 2). Texas alone accounts for about one-half of the limestone and dolomite reservoirs. Carbonate reservoirs are reported from 45 geologic provinces, but the Permian basin–Bend arch of Texas and New Mexico accounts for 40% (limestone) and 49% (dolomite) of the total (Table 2).

Distributions of limestone and dolomite reservoirs by geologic period (Figure 3) show that most United States carbonate reservoirs are of Paleozoic age. The principal exceptions are the Cretaceous and Jurassic limestone trends of the Gulf Coast, which constitute about 11% of

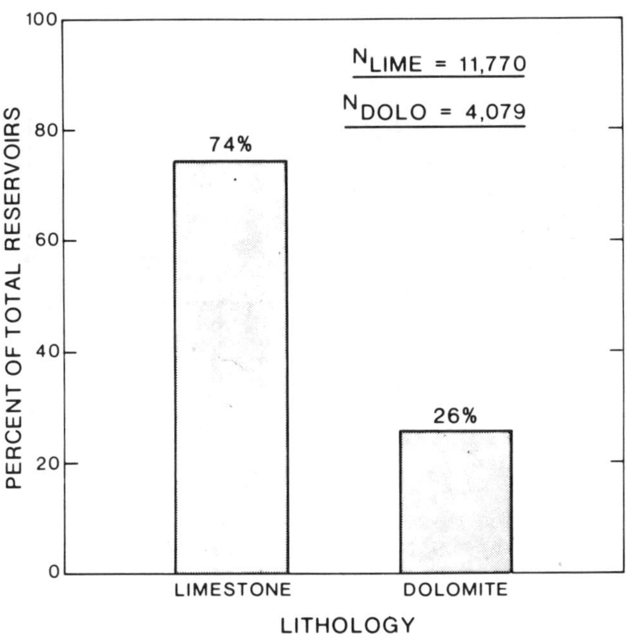

Figure 1—Lithology distribution, U.S. carbonate reservoirs.

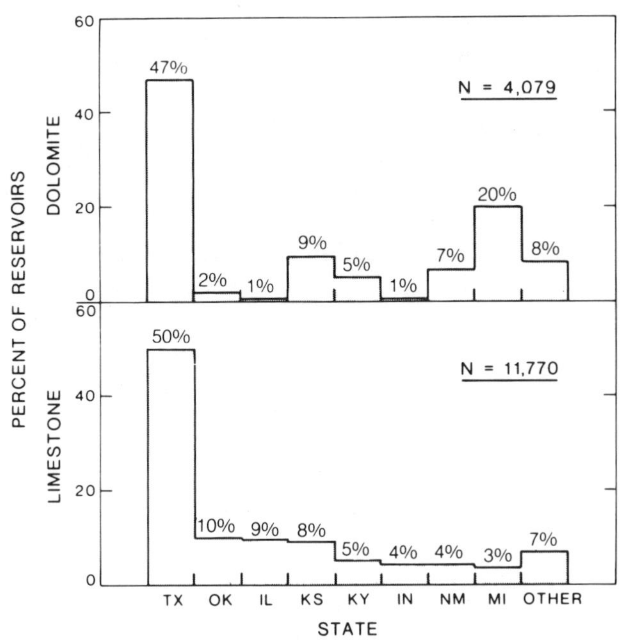

Figure 2—Location distributions by state, U.S. carbonate reservoirs.

Table 2. Location by Geologic Province, U.S. Carbonate Reservoirs

	Limestone		Dolomite	
Geologic Province	% of Reservoirs	Number of Reservoirs	% of Reservoirs	Number of Reservoirs
Permian basin–Bend arch	40	4,661	49	2,016
Illinois basin	14	1,589	1	39
Anadarko basin–Chautauqua platform	11	1,340	3	129
East Texas basin	4	504	1	21
Central Kansas uplift	4	497	7	304
Gulf Coast basin	4	436	0	3
Cincinnati arch	3	371	5	194
Michigan basin	3	330	20	825
Fort Worth syncline	3	318	2	67
Appalachian basin	2	265	3	137
Arkla basin	2	245	0	2
Williston basin	1	163	2	67
Other	9	1,051	7	275
Total	100%	11,770	100%	4,079

reported limestone reservoirs. A rough inverse correlation exists between the number of Paleozoic limestone and dolomite reservoirs as a function of geologic period (Figure 3).

Depth

Nearly 80% of reported carbonate reservoirs (79% of limestone and 77% of dolomite) are less than 7,000 ft (2,100 m) deep (Figure 4). The depth range of 3,000 to 4,000 ft (900 to 1,200 m) marks the peak for both distributions and includes 18% and 20% of the limestone and dolomite reservoirs, respectively. The main body of the dolomite histogram appears to be displaced about 1,000 to 1,500 ft (300 to 460 m) deeper, relative to that of limestone. The dolomite histogram is nearly constant at depths greater than 9,000 ft (2,700 m), but the limestone histogram decreases steadily (Figure 4).

Local geology in areas that dominate the data set (Figure 2, Table 2) might explain the fact that dolomite reservoirs tend to be deeper than those of limestone. On the other hand, Figure 4 can be taken as evidence of a fundamental tendency for dolomites to retain superior reservoir charac-

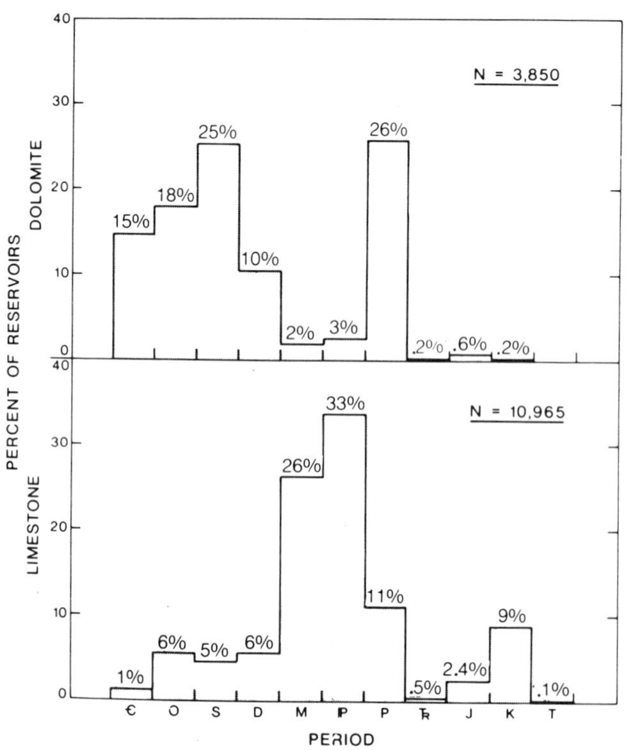

Figure 3—Age distributions by geologic period, U.S. carbonate reservoirs.

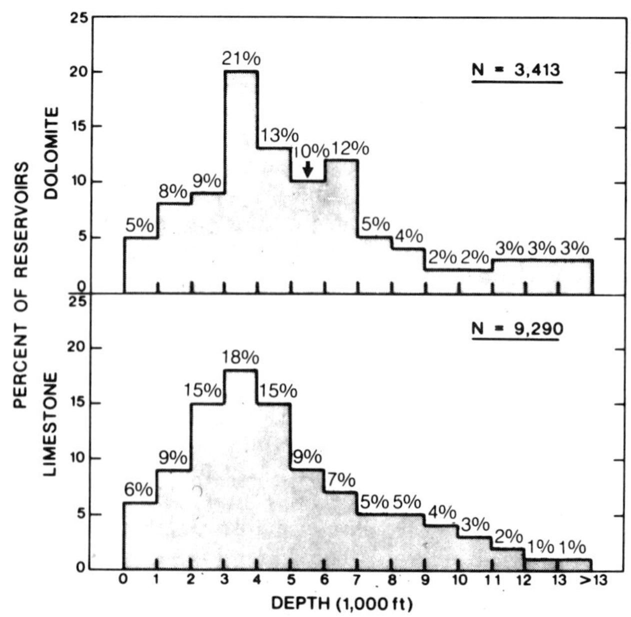

Figure 4—Distributions of depth to top of pay, U.S. carbonate reservoirs; 1,000 ft = 304.8 m.

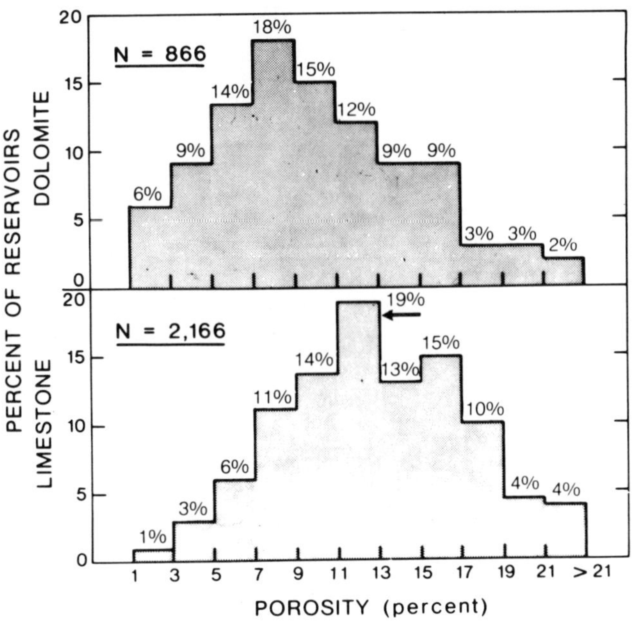

Figure 5—Average-porosity distributions, U.S. carbonate reservoirs.

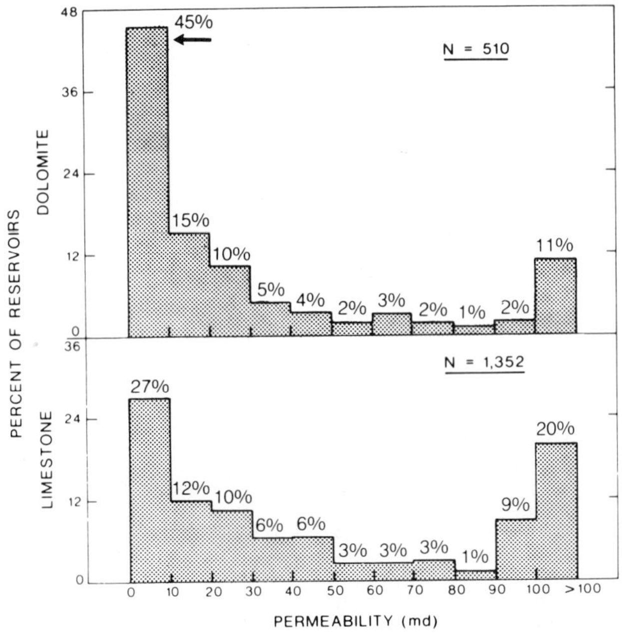

Figure 6—Average-permeability distributions, U.S. carbonate reservoirs.

teristics relative to limestones as depth increases. Arguments favoring the latter interpretation are presented in following sections.

Porosity and Permeability

Dolomite reservoirs tend to be less porous than those of limestone (Figure 5). The maximum of the distribution for limestone reservoirs occurs at 11-13 porosity units (with 19% of reported cases), and 61% of the data are in the 9-17 porosity-unit range. By comparison, the maximum distribution for dolomite reservoirs occurs at 7-9 porosity units (with 18% of reported cases), and 59% of the data are in the 5-13 porosity-unit range. Figure 5 gives no indication that dolomites have greater resistance to porosity loss by pressure solution than limestones, as suggested by data of Glover (1968) and of Schmoker and Halley (1982, their Figure 9).

A lower economic limit of about 6 porosity units is commonly chosen in mapping net feet of porosity (e.g., Kohm and Louden, 1978), and it seems doubtful that carbonate reservoirs with matrix porosities of 6 porosity units or less could be viable unless fractured. About 20% of the reservoirs of Figure 5 fall in this low matrix-porosity category, which suggests the overall importance of fractured reservoirs in carbonate production. The fact that only 10% of reported limestone reservoirs have porosities of 6 porosity units or less, compared to 29% of dolomite reservoirs (Figure 5), is interpreted as evidence that open fracture systems are more likely to occur in dolomites than in limestones.

The data of Figure 6 represent matrix permeability, rather than in-situ permeability, and the effects of fracture systems are not directly apparent. In the range of 10 to 90 md, the limestone and dolomite distributions are similar in form and include 44% of the limestone cases and 42% of the dolomite cases. The permeability range between 0 and 10 md includes 45% of the dolomite reservoirs but only 27% of the limestone reservoirs. The greater percentage of dolomite reservoirs with low matrix permeabilities is interpreted to mean that proportionately more dolomite reservoirs are enhanced by fracture permeability, which supports the argument that effective, open fracture systems are more likely to occur in dolomites than in limestones. This tendency—rather than better matrix porosity or permeability—gives dolomites apparently superior reservoir characteristics, relative to limestones, as depth increases (Figure 4).

Plots of porosity vs. permeability and of porosity vs. depth showed poor correlations from which little could be deduced. Most likely, significant correlations within reservoir populations of more or less homogenous character were masked by our grouping thousands of carbonate reservoirs with diverse lithologies, ages, depositional environments, and diagenetic and burial histories.

Water Saturation

The water-saturation distributions for limestone and dolomite reservoirs are quite similar (Figure 7). The maximums for both histograms occur at 27.5-32.5%, and both resemble normal distributions; 78% and 74% of reported limestone and dolomite reservoirs, respectively, have water saturations between 17.5 and 42.5%.

Reservoir Size

Dolomite reservoirs tend to have more oil and nonassociated gas originally in place than limestone reservoirs

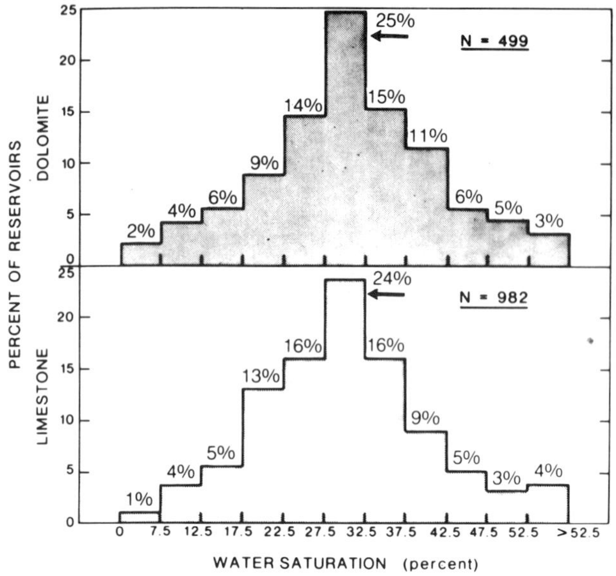

Figure 7—Average-water-saturation distributions, U.S. carbonate reservoirs.

(Figures 8, 9). Forty-eight percent of reported dolomite oil reservoirs and 44% of reported dolomite gas reservoirs exceed 1 million bbl (or equivalent), compared to only 29% and 20% of reported limestone oil and gas reservoirs, respectively. Also, the decline at large reservoir sizes is sharper for limestones than for dolomites. However, note that the geographic representation of Figures 8 and 9 is restricted because 91% of reported cases are from the Permian basin–Bend arch of Texas and New Mexico.

Figures 8 and 9 can be misleading because equal visual emphasis is given to both large and small reservoirs. A different perspective is obtained from Figure 10, which shows that the many small oil reservoirs dominating the appearance of Figure 8 contribute little to total reserves. Over 70% of the oil reported in each rock type is in giant reservoirs of 100 million bbl or more. Worldwide, about 85% of discovered hydrocarbons are in giant fields, which may include more than one reservoir (Fitzgerald, 1980).

The separation between the dolomite and limestone curves of Figure 10 illustrates again that dolomite reservoirs tend to be larger than limestone reservoirs. (The visual impact of the curve separation is reduced by the logarithmic reservoir-size scale.) Dolomites account for 31% of the reservoirs of Figures 8 and 10, but contain 57% of the oil originally in place.

Average net-pay thickness is another measure of reservoir size, and is more widely reported than oil and gas originally in place (Table 1). The net-pay thickness of dolomite reservoirs tends to be greater than that of limestone reservoirs (Figure 11). Sixty-six percent of reported dolomite reservoirs have net-pay thickness greater than 12 ft (3.7 m), compared to 42% of limestone reservoirs; 22% of dolomite reservoirs, but only 7% of limestone reservoirs, have net-pay thickness greater than 60 ft (18.3 m).

Three different sets of data—oil and nonassociated gas originally in place and average net-pay thickness—show that dolomite reservoirs are larger on the average than

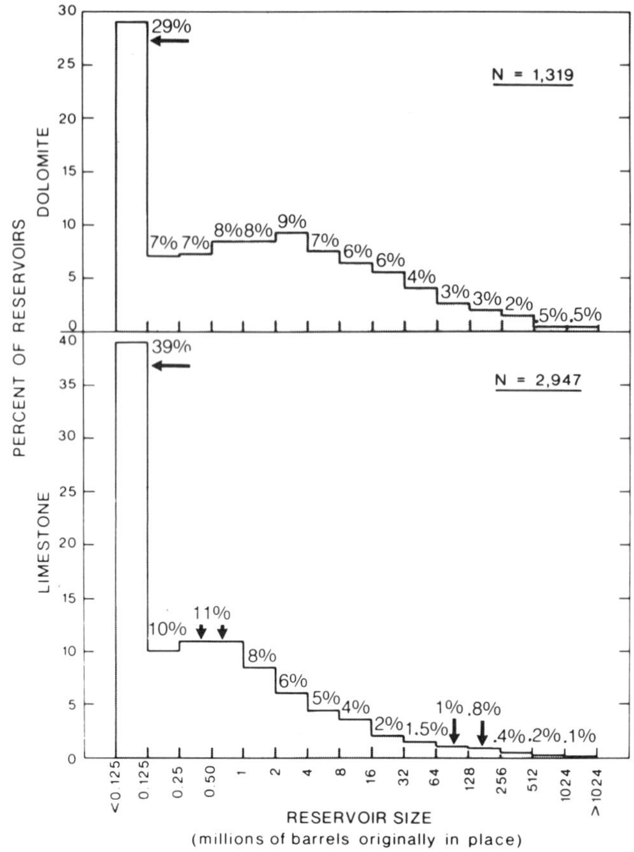

Figure 8—Distributions of reservoir size as measured by crude oil originally in place, U.S. carbonate reservoirs. Note logarithmic scale of abscissa.

limestone reservoirs (Figures 8-11). Based on the idea that a propensity toward fracture development would tend to increase effective reservoir size, this result is taken as additional evidence that open fracture systems are more likely to occur in dolomites than in limestones.

Discovery Rates

Discovery rates of United States carbonate reservoirs (the slopes of the curves of Figure 12) were approximately constant from 1910 to 1940, increased between 1940 and 1950, and remained high during the 1950s. The limestone and dolomite discovery-rate curves representing this 50-year period are nearly identical (Figure 12).

From 1960 to 1980, the discovery rate for limestone reservoirs gradually declined relative to that of the 1950s. The discovery rate for dolomite reservoirs initially declined, but then increased as a result of successful exploration in the Michigan basin.

The average sizes of newly discovered limestone and dolomite reservoirs, as measured by crude oil originally in place, have decreased by two orders of magnitude since 1930 (Figure 13). Throughout this period, new dolomite reservoirs were commonly larger than new limestone reservoirs. Although Figure 13 primarily represents the Per-

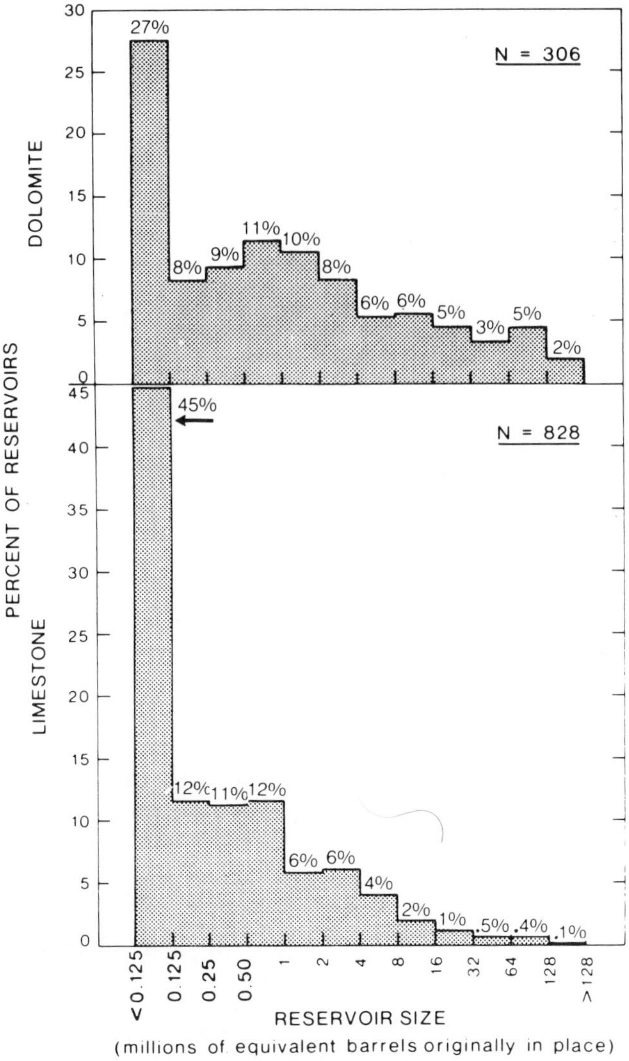

Figure 9—Distributions of reservoir size as measured by nonassociated gas originally in place, U.S. carbonate reservoirs. Note logarithmic scale of abscissa. Gas volume is expressed as energy-equivalent barrels of oil, with 1 bbl = 6,000 ft³.

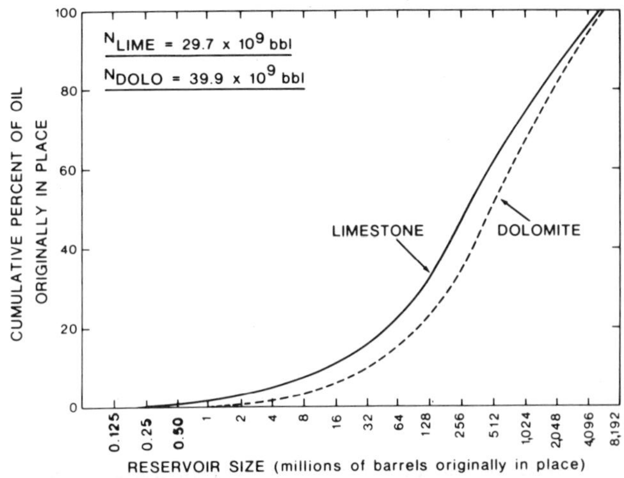

Figure 10—Cumulative curves of crude oil originally in place vs. reservoir size, U.S. carbonate reservoirs. Note logarithmic scale of abscissa.

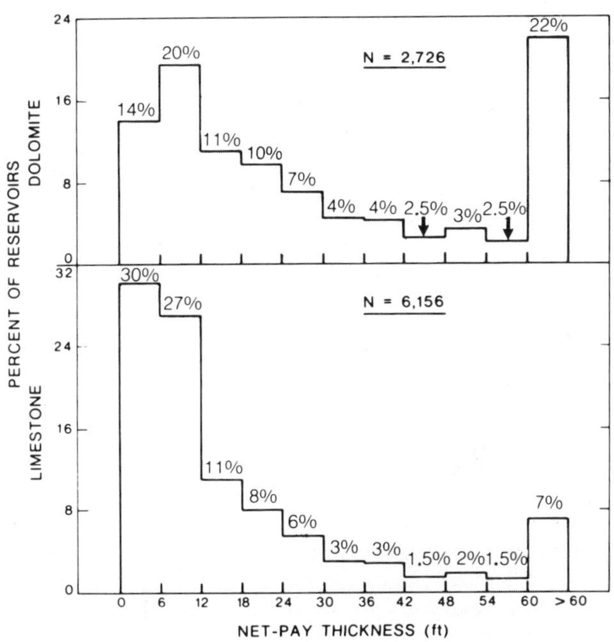

Figure 11—Distributions of average net-pay thickness, U.S. carbonate reservoirs; 1 ft = 0.3048 m.

mian basin-Bend arch of Texas and New Mexico, it is broadly typical of other mature petroleum provinces (e.g., Drew et al, 1982, p. 3).

Since 1960, the discovery rate for United States carbonate reservoirs (with the exception of dolomite reservoirs in the Michigan basin) has been declining (Figure 12); the size of newly discovered carbonate reservoirs also has been declining (Figure 13). It is notable that these declines occurred during a period of concentrated research effort and significant technological improvement in all aspects of hydrocarbon exploration and development.

SUMMARY AND DISCUSSION

Limestone/Dolomite Ratios for United States

In this section, the limestone/dolomite ratio of reservoirs in the United States (Figure 1) is compared to the overall United States volumetric ratio of limestone and dolomite rocks of Paleozoic age and younger. The latter ratio is estimated here in two ways:

1. Formally named formations with primary lithologies of limestone and of dolomite are counted, and the number of limestone and dolomite formations is assumed proportional to the relative volumes of the two rock types.
2. Type-section thicknesses of formally named formations with primary lithologies of limestone and of dolomite are summed, and the total thickness of limestone and dolomite type sections is assumed proportional to the relative volumes of the two rock types.

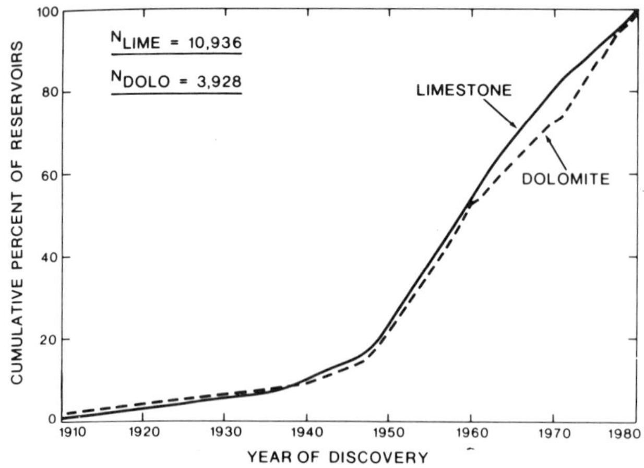

Figure 12—Cumulative curves of percentage of total reservoirs vs. year of discovery, U.S. carbonate reservoirs.

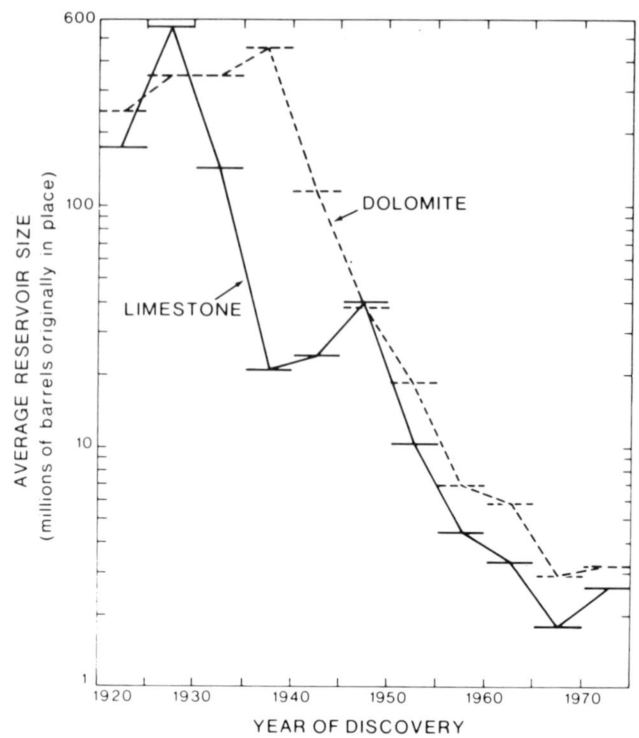

Figure 13—Reservoir size as measured by crude oil originally in place vs. year of discovery, U.S. carbonate reservoirs. Data are averaged over 5-year periods.

Errors tend to balance out using these impersonal, statistical approaches, and results are considered accurate enough to be useful.

Formation names, lithologies, geologic ages, and type-section thicknesses were obtained from the Geonames data file maintained by the U.S. Geological Survey in Reston, Virginia. At the time of this study, Geonames was updated through 1980 and contained information on about 17,000 United States geologic names. The 993 limestone formations and 297 dolomite formations retrieved from Geonames have type-section thicknesses totaling 558,000 ft (170,000 m) and 153,000 ft (47,000 m), respectively.

Dolomites constitute about 22-23% of all carbonate rocks in the United States younger than Precambrian age (Table 3), and contain 26% of United States carbonate reservoirs (Figure 1). However, in Texas, Oklahoma, Illinois, Kansas, Kentucky, Indiana, New Mexico, and Michigan, dolomites constitute only 15-17% of the carbonate rocks but contain 26% of the carbonate reservoirs. Thus, carbonate reservoirs occur preferentially in dolomites in the significant hydrocarbon provinces that account for over 90% of United States carbonate reservoirs (Figure 2). Limestone reservoirs are more numerous than dolomite reservoirs only because limestones are much more common than dolomites.

One of the traditions alluded to in the introduction is that dolomitization tends to improve reservoir properties relative to those otherwise expected. Such observations are supported by the data of Table 3, which confirm a tendency for dolomites to be favored reservoir rocks.

United States Limestone/Dolomite Ratio Through Time

Because the Geonames data file includes geologic age, limestone/dolomite ratios can be calculated as a function of time. The limestone/dolomite ratio for both carbonate

Table 3. Relative Proportions of Limestone and Dolomite (Paleozoic Age and Younger)

	Limestone (%)	Dolomite (%)	Limestone/Dolomite Ratio
Conterminous United States and Alaska			
Determined by formation names	77	23	3.3
Determined by type-section thicknesses	78	22	3.5
States of Texas, Oklahoma, Illinois, Kansas, Kentucky, Indiana, New Mexico, and Michigan (Figure 2)			
Determined by formation names	83	17	4.9
Determined by type-section thicknesses	85	15	5.7
U.S. carbonate reservoirs (Figure 1)	74	26	2.8

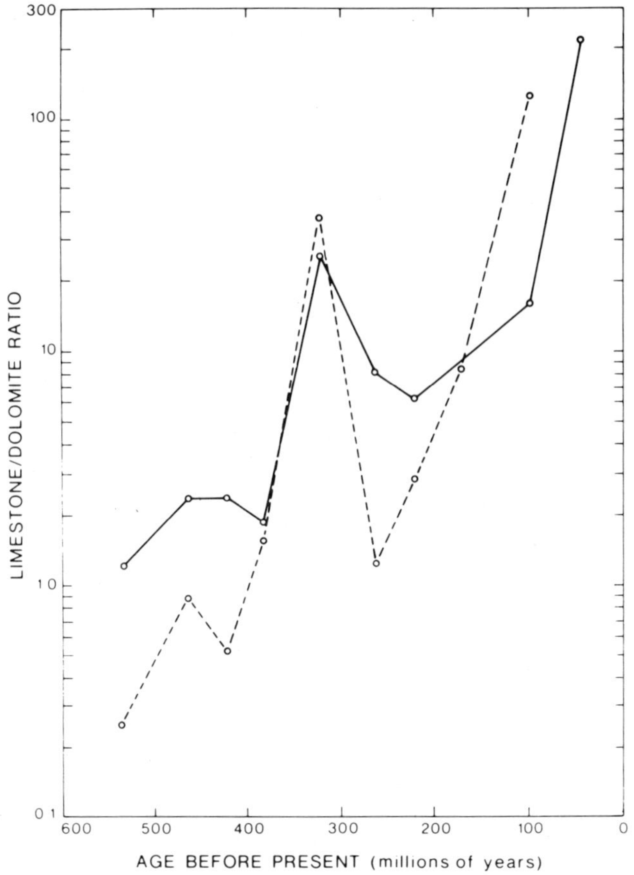

Figure 14—Limestone/dolomite ratios vs. age before present. Dashed line = U.S. carbonate reservoirs (Figure 3). Solid line = all U.S. carbonate rocks, based on type-section thicknesses from Geonames data file. See text for explanation.

reservoirs and all carbonate rocks in the United States increases by two orders of magnitude between the beginning of the Paleozoic and the end of the Mesozoic Eras (Figure 14). Data scatter and periodic fluctuations superpose irregularities upon the general increase that mask possible inflection points of geochemical significance. The data of Figure 14 are in basic agreement with laboratory analyses demonstrating an increase from past to present in the average calcium/magnesium ratio of carbonate rocks (Daly, 1909; Chilingar, 1956).

The origin of dolomite is an active research topic (e.g., Ricketts, 1982; Tucker, 1982; Zenger, 1982; Kastner, 1983; Saller, 1984). The data of Figure 14, representing extremely large rock volumes, add to the limited number of published limestone/dolomite ratio measurements, and bear on the global geochemical implications inherent in models of dolomite formation.

Dolomites as Superior Reservoir Rocks

Dolomites are commonly regarded as better reservoir rocks than limestones, and the results of our study generally support this idea. The tendency documented here for carbonate reservoirs to occur selectively in dolomites is statistical evidence that dolomitization can improve the reservoir characteristics of a given formation relative to values otherwise expected. The apparent paradox that matrix porosities and permeabilities of dolomite reservoirs are lower on the average than those of limestone reservoirs (Figures 5, 6) can be explained by assuming that dolomitization improves reservoir properties primarily in a local sense, relative to adjacent limestones that would otherwise make poor reservoirs.

The data showing that dolomite reservoirs on the average are larger and deeper, yet less porous and permeable, than limestone reservoirs suggest that effective fracture systems at reservoir depths are more likely to occur in dolomites than in limestones. All else being equal, an open fracture system connecting formation and well bore would act to increase a reservoir's effective size and would tend to make a reservoir viable at otherwise poor conditions of depth, porosity, and permeability.

Laboratory experiments and field observations (Handin et al, 1963; Stearns, 1967; Stearns and Friedman, 1972; Hugman and Friedman, 1979) show that, under similar conditions, dolomite has greater ultimate strength and is less ductile than limestone. Therefore, dolomite is more likely than limestone to yield by fracturing than by intragranular flow. In addition, fractures in dolomite may be more effective permeability channels than those in limestone because of the greater mechanical strength of asperities that act to keep fractures propped open (Hugman and Friedman, 1979). These differences in mechanical properties explain why effective fracture systems at reservoir depths could occur more frequently in dolomites than in limestones.

REFERENCES CITED

Blatt, H., G. Middleton, and R. Murray, 1972, Origin of sedimentary rocks: Englewood Cliffs, New Jersey, Prentice-Hall, 634 p.

Chilingar, G. V., 1956, Relationship between Ca/Mg ratio and geologic age: AAPG Bulletin, v. 40, p. 2256-2266.

Daly, R. A., 1909, First calcareous fossils and the evolution of the limestones: GSA Bulletin, v. 20, p. 153-170.

Drew, L. J., J. H. Schuenemeyer, and W. J. Bawiec, 1982, Estimation of the future rates of oil and gas discoveries in the western Gulf of Mexico: USGS Professional Paper 1252, 26 p.

Fitzgerald, T. A., 1980, Giant field discoveries 1968-1978: an overview, *in* Giant oil and gas fields of the decade 1968-1978: AAPG Memoir 30, p. 1-5.

Glover, J. E., 1968, Significance of stylolites in dolomitic limestones: Nature, v. 217 (March 2), p. 835-836.

Handin, J., R. V. Hager, M. Friedman, and J. N. Feather, 1963, Experimental deformation of sedimentary rocks under confining pressure: pore pressure tests: AAPG Bulletin, v. 47, p. 717-755.

Hugman, R. H. H., III, and M. Friedman, 1979, Effects of texture and composition on mechanical behavior of experimentally deformed carbonate rocks: AAPG Bulletin, v. 63, p. 1478-1489.

Kastner, M. 1983, Origin of dolomite and its spatial and chronological distribution—a new insight (abs.): AAPG Bulletin, v. 67, p. 2156.

Kohm, J. A., and R. O. Louden, 1978, Ordovician Red River of eastern Montana and western North Dakota: relationships between lithofacies and production, *in* The economic geology of the Williston basin: Montana Geological Society 24th Annual Conference, 1978 Williston Basin Symposium, p. 99-117.

Murray, R. C., and L. C. Pray, 1965, Dolomitization and limestone diagenesis—an introduction, *in* Dolomitization and limestone diagenesis: SEPM Special Publication 13, p. 1-2.

Ricketts, B. D., 1982, Comment on "Precambrian dolomites: petrographic and isotopic evidence that they differ from Phanerozoic dolomites": Geology, v. 10, p. 663.
Saller, A. H., 1984, Petrologic and geochemical constraints on the origin of subsurface dolomite, Encwetak atoll: an example of dolomitization by normal seawater: Geology, v. 12, p. 217-220.
Schmoker, J. W., and R. B. Halley, 1982, Carbonate porosity versus depth: a predictable relation for South Florida: AAPG Bulletin, v. 66, p. 2561-2570.
Stearns, D. W., 1967, Certain aspects of fracture in naturally deformed rocks, *in* R. E. Riecker, ed., NSF advanced science seminar in rock mechanics: Bedford, Massachusetts, Air Force Cambridge Research Laboratory Special Report, p. 97-118.
——— and M. Friedman, 1972, Reservoirs in fractured rock, *in* Stratigraphic oil and gas fields—classification, exploration methods, and case histories: AAPG Memoir 16, p. 82-106.
Tucker, M. E., 1982, Precambrian dolomites: petrographic and isotopic evidence that they differ from Phanerozoic dolomites: Geology, v. 10, p. 7-12.
Zenger, D. H., 1982, Comment on "Precambrian dolomites: petrographic and isotopic evidence that they differ from Phanerozoic dolomites": Geology, v. 10, p. 662.
——— J. B. Dunham, and R. L. Ethington, 1980, Concepts and models of dolomitization: SEPM Special Publication 28, 320 p.

THE AMERICAN ASSOCIATION OF PETROLEUM GEOLOGISTS BULLETIN
V. 51, NO. 10 (OCTOBER, 1967), P. 1979-2032, 48 FIGS.

STONY MOUNTAIN (ORDOVICIAN) AND INTERLAKE (SILURIAN) FACIES ANALOGS OF RECENT LOW-ENERGY MARINE AND SUBAERIAL CARBONATES, BAHAMAS[1]

PERRY O. ROEHL[2]
Brea, California

ABSTRACT

During Late Ordovician and Silurian times, epeiric carbonate sediments were deposited over a negative metastable lineament of the western Williston basin. These sediments are now the Stony Mountain Formation and Interlake Formation dolomites. The original sedimentary and early diagenetic fabrics are still preserved and, as a group, characterize a complete cycle of low-energy carbonate sedimentation. They also determine the distribution and internal geometry of interparticle and intercrystalline porosity. Finally, these fabrics are analogous with modern platform infratidal (shallow subtidal), intertidal, supratidal, and subaerial diagenetic terrane deposits of the Bahama Islands.

Minor transgressive deposits punctuate the sequence but are important mainly in that they reinitiated offlap deposition. Ultimately the basin became fully emergent and the sediments underwent the last but most widespread penecontemporaneous dolomitization, solution alteration, and evaporite deposition similar to the present-day *sebkha-wadi* complex (supratidal) of the Trucial Coast, Persian Gulf. An extensive period of subaerial exposure ensued, linking the positive attitude of the Williston basin with that of the continent until Middle Devonian time. During emergence a microkarst developed over minor positive structural elements. As a consequence some reservoir rock of the uppermost Interlake was destroyed by superposed stream erosion.

A generalized working model of low-energy facies relations is presented showing the proposed environment of offlap deposition and some of the fabric types in which producible oil is found. These fabrics are pelleted, laminated, and burrowed "lime mud" and silt, algal mats, and stromatolites, flat-pebble breccia, epigenetic (non-tectonic) fragmental and solution breccia, and a few cut-and-fill structures. Leaching of fossils and anhydrite in certain places has accentuated and improved pore structures.

Infratidal and intertidal facies deposits, including their storm and spring high-tide equivalents onshore, are the most widespread and consistent, but least diagnostic, of facies environments. Supratidal and subaerial diagenetic terrane remnants on the other hand are most diagnostic, having several distinctive microfacies fabrics. Because these fabrics commonly grade from one to the other in short distances, they are difficult to evaluate as prospective major oil-bearing units.

INTRODUCTION

During the past few years, geologists and engineers have learned a great deal about carbonate sediments and their peculiarities as reservoir rock. This has been a result of intensive concurrent study by researchers in modern and ancient carbonate sediments and by applied specialists investigating subsurface carbonates in operational areas of interest. Until recently, one of the most perplexing problems confronting each group has been an understanding of "low-energy" (largely synonymous with "tidal flat") sediments, their relative importance in the sedimentary record, and the petrophysical attributes which permit these unpromising sediments to develop into widespread potential hydrocarbon reservoirs.

In order to understand all of the fabrics observed in a sedimentary record, geologists usually look for similar fabrics produced by modern processes of sedimentation and organic activity. Two fundamental methods of study have been used to good advantage. The first is a study of the lithic and organic content, textures, sedimentary fabrics, and their diagenetic alterations. The second is the development of an appropriate model of

[1] Read before the Rocky Mountain Section of the Association at Denver, Colorado, October 26, 1966. An early abstract of this paper was submitted to the 1962 San Francisco annual meeting of the Association and was published in the *Bulletin,* v. 46, no. 2, February, 1962, p. 276. Manuscript received and accepted, July 1, 1967.

Modified from EP Report 33795 (1961) Shell Oil Company of Canada, Ltd., Calgary, Alberta, Canada, and Shell Development Company (A Division of Shell Oil Company), Exploration and Production Research Division, Houston, Texas.

[2] Union Research Center, Union Oil Company of California. The writer is pleased to acknowledge the contributions of many persons in Shell Oil Company, Denver Area Operations, particularly C. N. Tinker, J. E. Tuttle, D. C. Jones, D. W. Solmonson, F. P. Van West, A. D. Mitchell, and J. A. Pederson, each of whom shared directly in the effort. R. N. Ginsburg and M. M. Ball of the Coral Gables Office, Shell Development Company, gave freely of their time in collaborating with the writer. Many of the Recent sample descriptions, photos of Recent sediments, and opinions appearing superficially as the writer's are those of Messrs. Ginsburg and Ball. Finally, the writer is grateful to P. E. Cloud, Jr., for permission to use the underwater photo of Figure 7.

the topographic, hydrologic, and climatic scheme against which the individual elements of deposition can be portrayed. Usually the model is a hybrid form adapted to several constituent fabric and facies relations modified by successions of biota which are more or less unique to a particular situation. This degree of uniqueness places important constraints on the general application of any model. Therefore, formulation of an analog model must rely not only on comparison properties but also on the observer's inductive capacity. The following outline indicates the major interpretive methods of fabric analysis.

Methods of Fabric Interpretation

A. *Comparison*
  1. Direct analogy by content and geometry with Recent sediments and diagenetic products
  2. Indirect comparison
     a. *Vertical* position and fabric succession
     b. *Horizontal* association with other fabrics or facies

B. *Induction*
  1. Reconstruction of essential paleoenvironmental factors, many of them ephemeral, such as hydraulic regimen, organic activity, and diagenetic processes

The importance of a sedimentary fabric to carbonate-reservoir development depends on the subsequent diagenesis. This is especially true where weathering alteration and early dolomitization are involved. If the proper analog has been chosen, it should provide a suitable solution for these diagenetic effects. The common association of dolomite with low-energy deposits, especially evaporites, implies a predilection of each for the other. Recent work by Wells (1962), Deffeyes *et al.* (1965), Shinn (1964), Shinn *et al.* (1965), Illing *et al.* (1965), and Kinsman (1965) has confirmed this predilection. Hence, a great deal of the mystery has been removed from this diagenetic process. As an early diagenetic event, then, the process of dolomitization can be linked more confidently to patterns of sedimentation.

To the extent that dolomitization alters porosity and pore geometry by modifying particle size and shape, or crystal size and euhedralism, it cannot be ignored from the reservoir viewpoint. It is well known that dolomitization very commonly enhances the petrophysical development of porosity in fine-grained carbonates. Because the latter are most widespread in low-energy environments, they commonly are dolomitized and any discussion of their reservoir capacity must be concerned with potential petrophysical changes related to the process.

The modern Bahamas tidal flat and adjacent supratidal and subaerial terrane zones have excellent oil-bearing counterparts in both fabric and facies character in ancient epeiric tidal flats and shoreline deposits. The extensive carbonate-evaporite plain (*sebkha-wadi* complex) of the Trucial Coast, Persian Gulf, also contains important correlative fabrics resulting from severe subaerial exposure. Some ancient analogs are here considered in detail, including those showing the effect of post-depositional alterations which influenced the character and distribution of some reservoir quality fabrics and facies. The case history cited is for the Upper Ordovician and Silurian formations of the Williston basin. Other workers have cited equally impressive occurrences of low-energy carbonates from all parts of the geologic record. The writer would particularly like to emphasize here the identification, association, distribution, and alteration of some important, *oil-bearing* types of low-energy-environment rocks.

Regional Lithology and Stratigraphy

Carbonate rocks predominate in the lower Paleozoic sequence of the Williston basin. Most of the detailed outcrop descriptions and nomenclature have been carried into the southwestern subsurface from exposures in southern and west-central Manitoba and east-central Saskatchewan (Fig. 1). The overall conformity of sedimentary sequences in this vast area, in conjunction with problems of fossil-invertebrate age criteria, has led to a systemic boundary controversy. The problem has carried over into the subsurface of the Cedar Creek anticline area where the arbitrary top of the Ordovician chosen for field development has been a radioactive log marker "t" identified above the highest dark shale in the Stony Mountain sequence and below the lowest buff, clean dolomite characteristic of the Interlake Formation (Figs. 1, 2).

Stony Mountain Formation.—The Stony Mountain Formation was described by Dowling (1900), and assigned a Richmond age by Whiteaves (1906), Foerste (1929), Okulitch (1943), and later workers. In the western Dakotas-Montana area the formation overlies the Red River Formation dolomite and limestone of Late Ordovician age. It is composed, from base to top, of (1), the Stony Mountain "Shale" Member, which

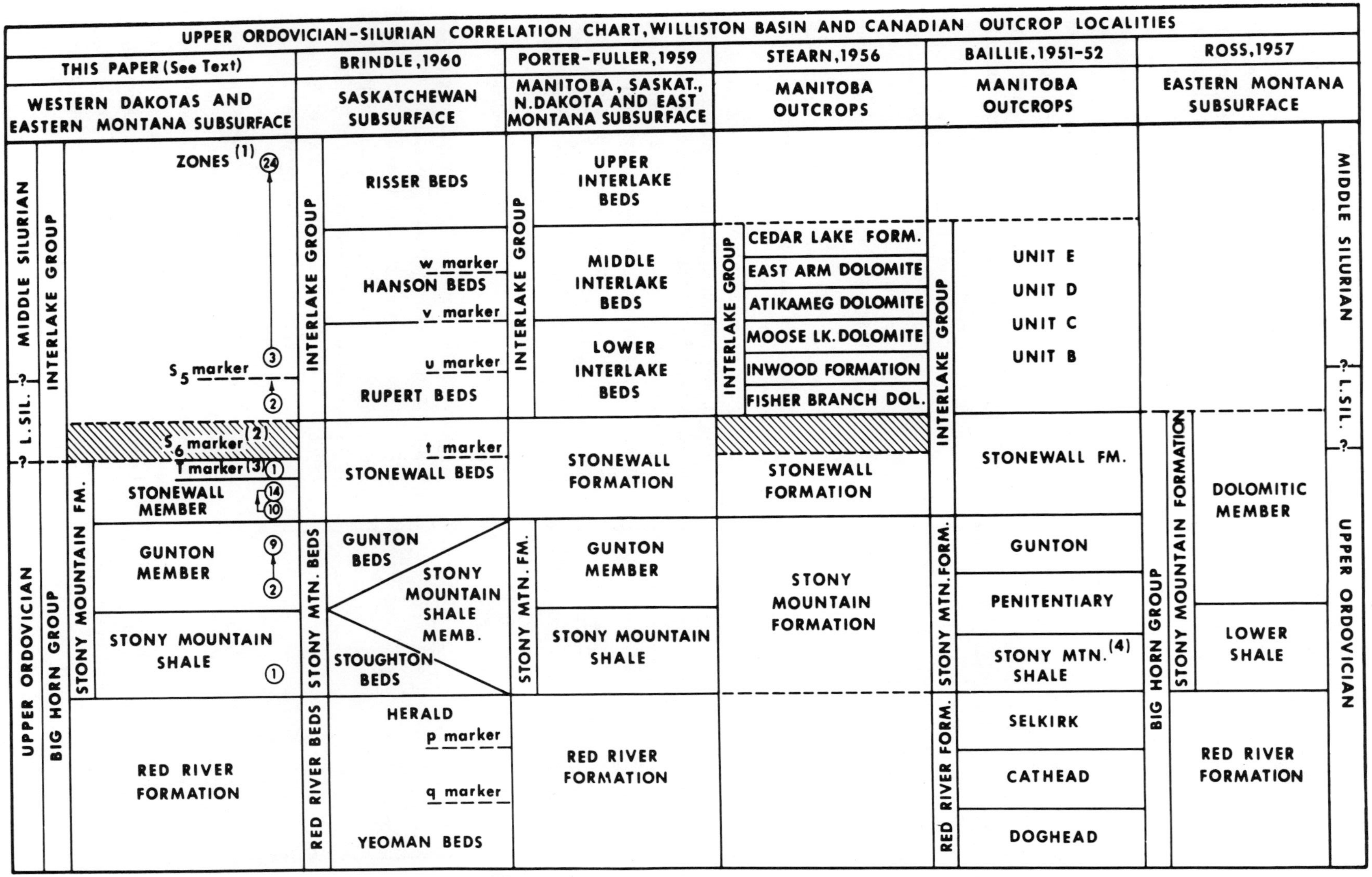

(1) Petrophysical Zones.

(2) Log Marker, Top of Tripolitic Zone (Top Ordovician?)

(3) Engineer's Log Top, Ordovician.

(4) Approximately = Gunn Member (Sinclair and Leith, 1958)

FIG. 1.—Upper Ordovician and Silurian correlation chart, Williston basin of United States and Canada, and Canadian outcrop localities.

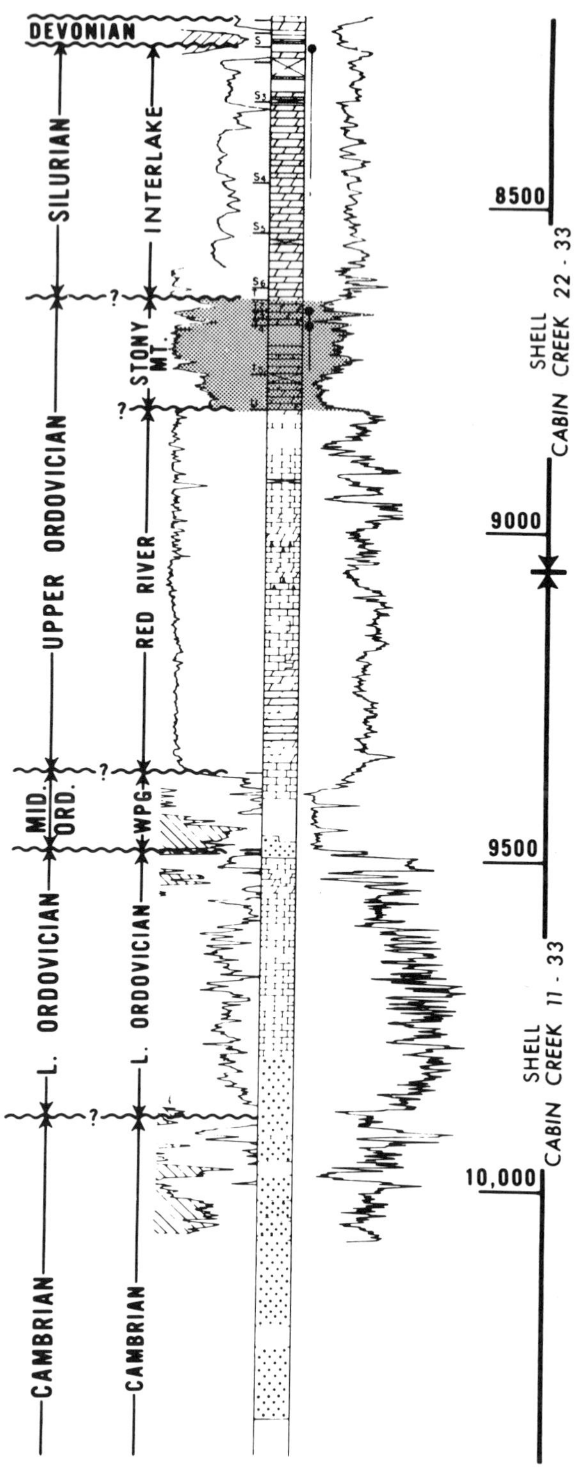

Fig. 2.—Type radioactivity log of lower Paleozoic formations, Cedar Creek anticline, Montana. Letters left of lithologic log are engineer log markers; at right, productive intervals are shown by solid circles and thickness lines.

in fact is a very argillaceous, calcareous, fossil-bearing dolomite approximately 50 ft (16 m) thick, (2) the Gunton Member, which contains about 70 ft (23 m) of brown, illitic, mottled and laminated dolomite, and (3) the Stonewall Member, about 40–50 ft (13–16 m) of argillaceous, arenaceous, and leached fossil dolomite, probably equivalent to the lower Stonewall of Porter and Fuller (1959).

Stearn (1953, 1956), on the basis of meager fossil evidence, argued that there is an important sedimentation gap between the Ordovician and Silurian. Porter and Fuller (1959) and Brindle (1960) did not agree. Brindle (1960) consequently proposed that the Lower Silurian may include the sequence between the Stonewall "t" marker and the "u" marker of mid-Rupert beds in the Saskatchewan subsurface. The latter marker appears to be the top of petrophysical zone "2" (the "S_5" marker) of the Cedar Creek anticline (Fig. 1).

As is discussed in a subsequent section, a progressive change in environment of deposition near the end of Ordovician time bears on this problem. It can be noted here however, that another engineer's log marker in the anticline area, "S_6," is a highly weathered tripolitic surface of coarsely crystalline tabulate corals and, from a time viewpoint, may represent the hiatus referred to by Stearn and be equivalent to Brindle's "t" marker. Although the "S_6" marker is not as pronounced as the shaly marker intervals below it, this silicified surface may be more important than any other marker with regard to the Siluro-Ordovician boundary problem.

Interlake Formation.—The Interlake Formation was proposed originally by Baillie (1952) for the sequence of biostratigraphic carbonate units between the Stony Mountain Formation and the Ashern Formation (Silurian or Early Devonian) in the Interlake outcrop area of Manitoba, Canada. Subsequent studies by Rader (1952), Porter and Fuller (1959), Fuller (1961), and others have made possible extensive subsurface correlations to the southern limits of the Williston basin. The top of the Interlake (and Silurian) in the western Williston basin is selected below the first thick rubble and/or reddish shale beds called "Ashern" in Manitoba. There are additional shale breaks, as well as other gamma ray-neutron log breaks, both below and above this contact. This is related in part to the channel and karst origin of the Ashern which is discussed subsequently.

In the area of the Cedar Creek anticline, and through most of the United States part of the Williston basin, the Silurian is principally dolomite, with varying but minor amounts of limestone, illitic shale, quartz siltstone and sandstone stringers, and very disseminated lenses and nodules of anhydrite. Chert is important at the base and top of the section. Meager fossil materials (*e.g.*, tabulate corals, brachiopods, and blastoid ossicles) occur in the section but diagenesis has rendered most of these generically unidentifiable.

Both the Stony Mountain and Interlake Formations in the western Williston basin contain sedimentary fabrics which indicate a shallow-water, intertidal, and subaerial-terrane environment of deposition and diagenesis. This is demonstrated by the occurrence of burrows, stromatolites, algal mats, pelletoidal mud, and bioclastic- and flat-pebble breccia. These are extraordinarily similar to present sediment types along the western margin of Andros Island, Bahamas. Though there are few lithologic breaks, the fact that the entire sequence of fabrics in the Stony Mountain has been preserved permits reconstruction of the typical depositional cycle which matches the regional cycle development discussed by other workers.

Although many of the fabrics of the overlying Silurian Interlake are well preserved and very diagnostic, the zonation of the formation has been based primarily on petrophysical properties using modified Archie rock types (Archie, 1952) because of (1) the severity of diagenesis, (2) the increased number of probable cycles, and (3) the much lower reliability for correlation of many of the fabrics. Nevertheless, many of the surviving Interlake fabrics are very distinctive, correlative, and oil-productive and the details of their formation and implication are considered here. The use, convenience, and reservoir importance of the petrophysical zonation is the subject of a later paper.

Depositional Analogs

The importance of carbonate sediments deposited in the low-energy environment as potential hydrocarbon reservoirs has been long suspected but only recently has it been fully appreciated. By low-energy environment the writer means a marine shallow-water and tidal-flat environment of deposition where the wave fetch extends across many miles of shallow water or where a remote barrier effectively reduces the hydrokinetic energy below that expected in open-marine conditions. These sediments possess fabrics which result from combined organic and hydraulic processes, and which frequently are modified by periods of subaerial exposure resulting from the offlap nature of their deposition. Analogous fabrics are common in the epeiric deposits of Paleozoic and Mesozoic carbonates. Where preserved under favorable structural conditions, they usually determine the extent and configuration of oil entrapment because of direct influence on pore-size distribution. This is commonly true even where complete dolomitization by replacement has occurred.

The purpose of this paper is to show the correspondence of oil-bearing fabrics preserved in the dolomites of the Upper Ordovician Stony Mountain and Silurian Interlake Formations of the Williston basin (Fig. 1) to modern examples from the Bahama platform. Illustrations of the modern fabrics are taken from samples collected or localities examined on western Andros Island and northern Eleuthera Island (Fig. 3). Figure 4 shows the position of the Cedar Creek anticline and several of the oil fields in eastern Montana from which samples have been taken to compare with those of the Bahamas.

Cedar Creek anticline is an ancient structure that came into being in the early Paleozoic, apparently as a result of basement faulting. Three known periods of uplift have been largely responsible for the continued existence of the structure: pre-Mississippian–post-Silurian, Mississippian–Triassic, and post-Paleocene. The present closure is due to moderate folding of a few hundred feet, accentuated by important northwest-southeast linear faulting. A result of this faulting has been the isolation of a sliver horst along the Cabin Creek part of the anticlinal axis. Several productive fabric types are discussed which were taken from well cores obtained in this area.

REGIONAL ENVIRONMENT OF SILURO-ORDOVICIAN DEPOSITION

During Late Ordovician and Silurian times, much of the interior North American continent

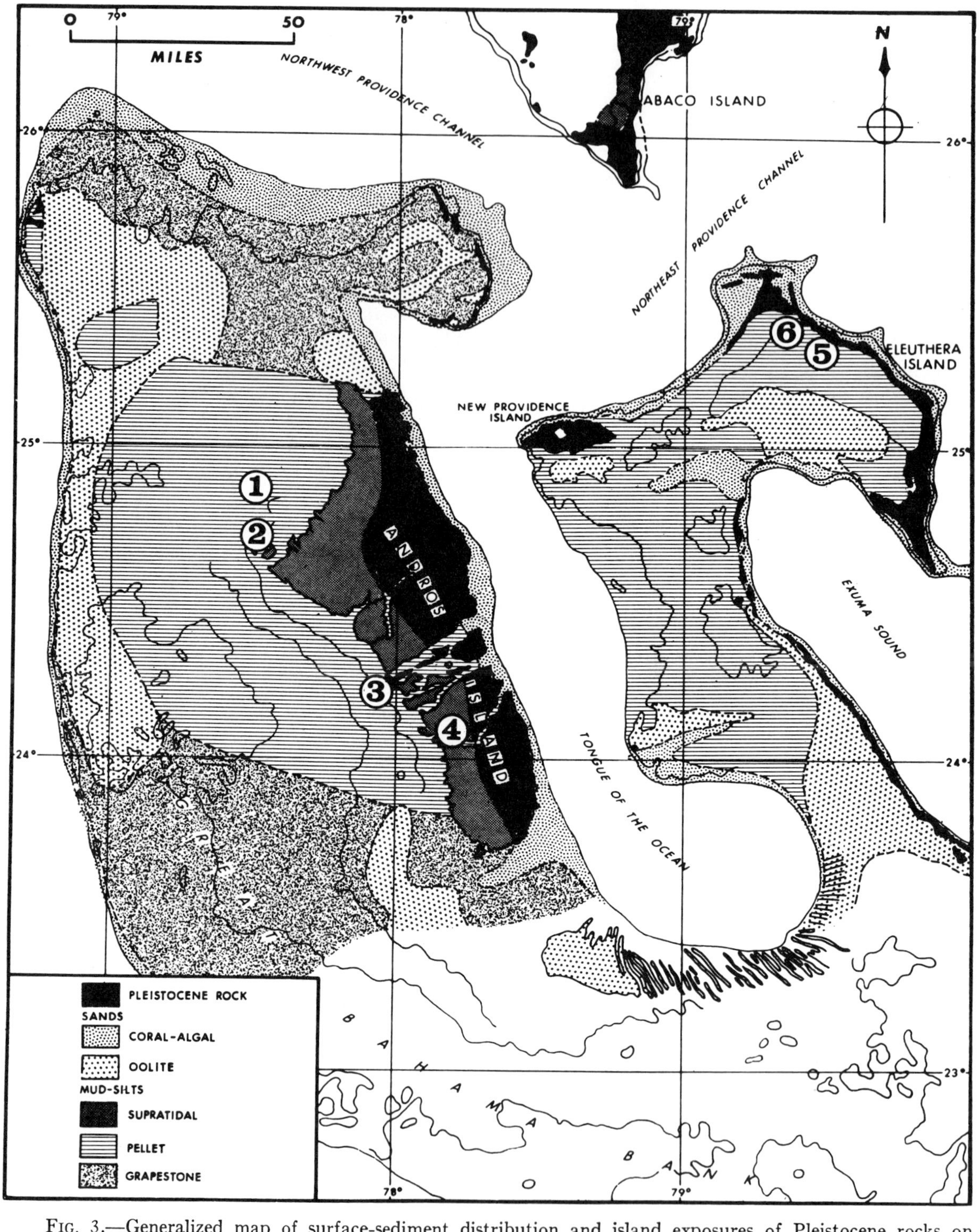

FIG. 3.—Generalized map of surface-sediment distribution and island exposures of Pleistocene rocks on Bahamas platform. Facies boundaries are only approximate. Data compiled largely from Newell and Rigby (1957), Field (1931), Shinn *et al.* (1965), Purdy (1963), and Purdy and Imbrie (1964). Localities discussed: 1—U.S.G.S. Station G4; 2—Williams Island; 3—Little Loggerhead Creek; 4—South Bight-west; 5—Shores Estate Cut; and 6—Glass Window.

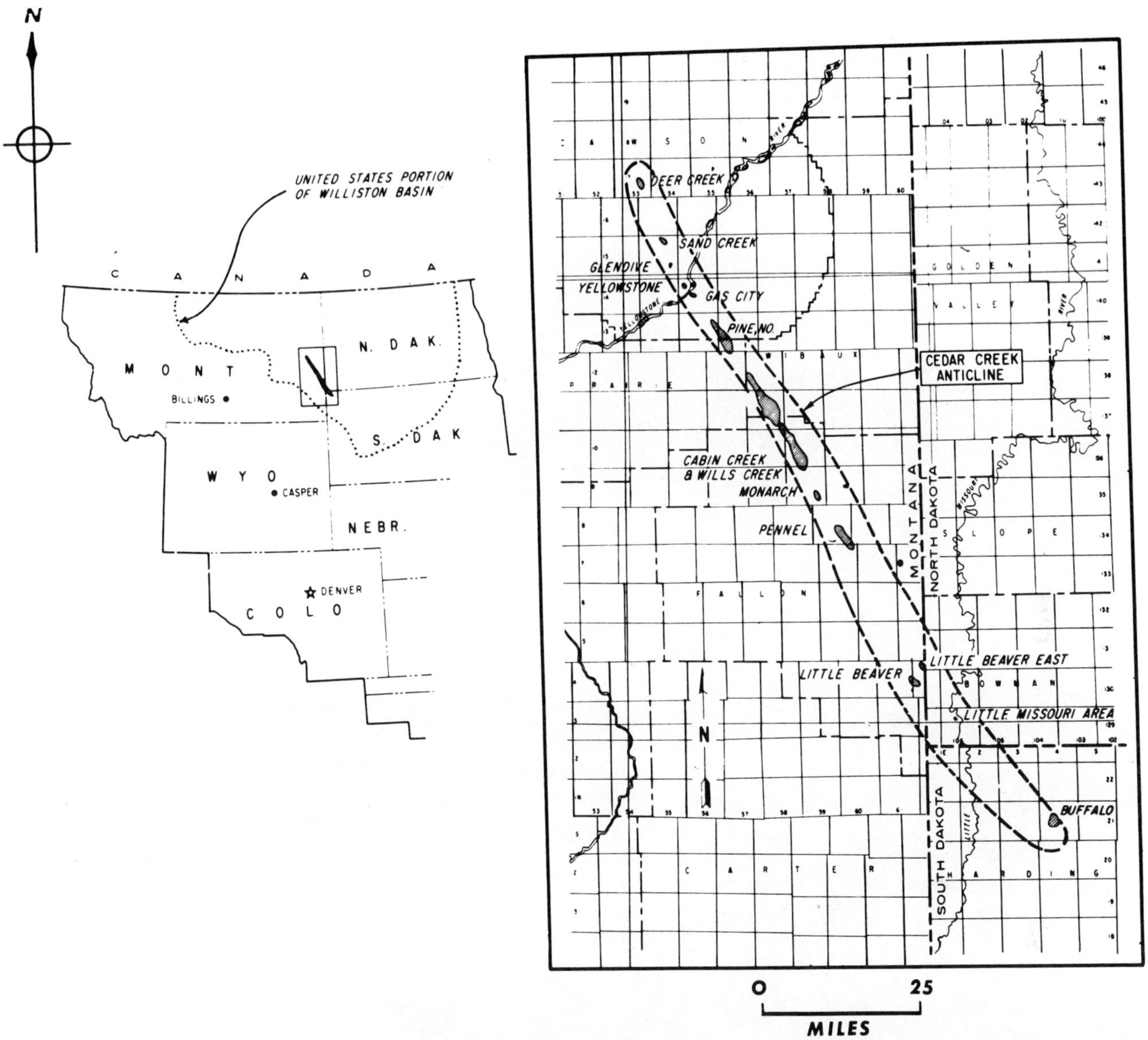

FIG. 4.—Index map of eastern Montana, western North Dakota, and western South Dakota showing location of Cedar Creek anticline and some fields which produce from Ordovician and Silurian reservoirs.

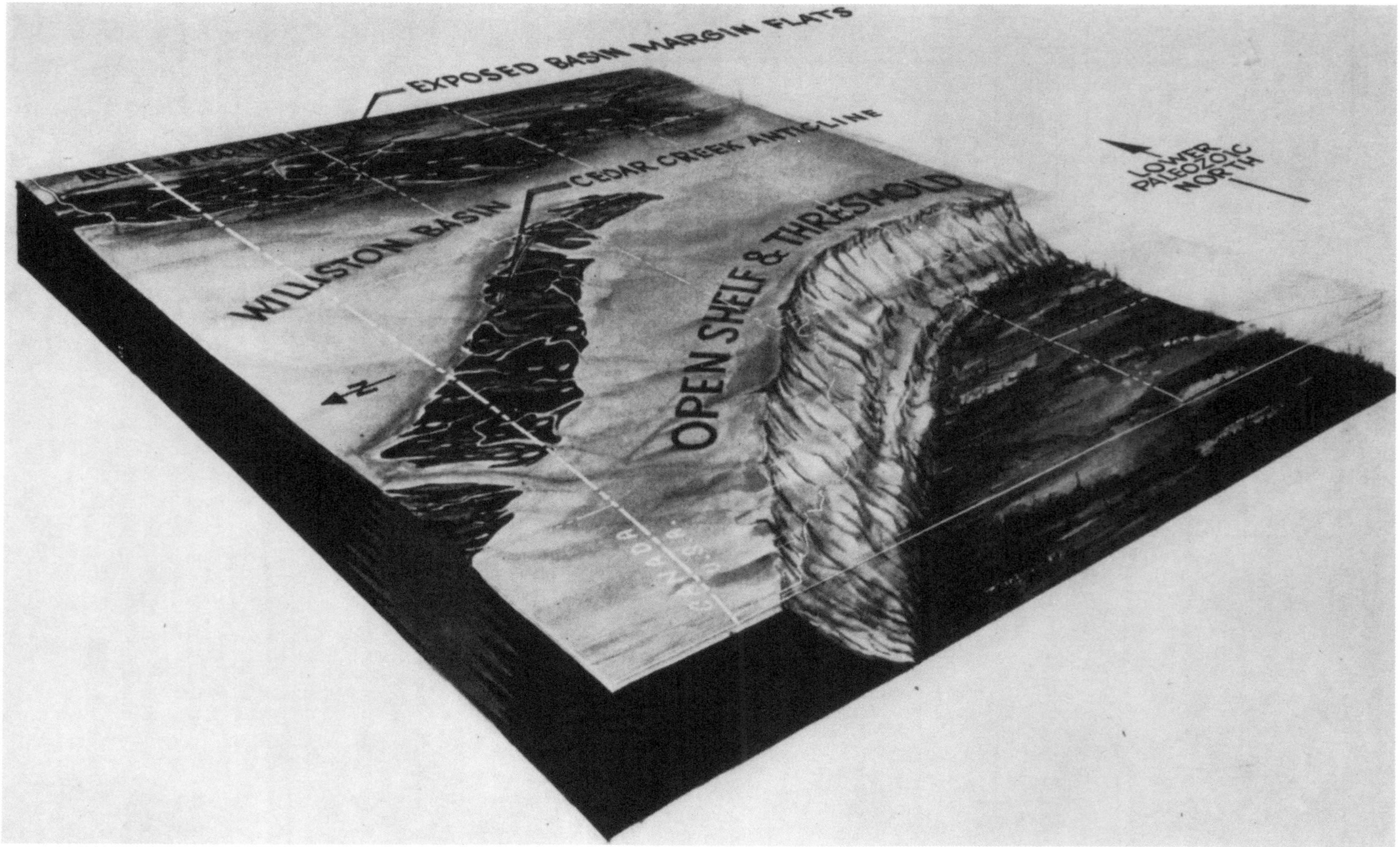

FIG. 5.—Block diagram of Williston basin region illustrating proposed regional depositional setting during period of emergence of western metastable lineament and simultaneous offlap deposition of infratidal, intertidal, and supratidal sediments. With exception of brief transgressive episodes, this depositional pattern persisted until basin became filled and emergent, probably near end of Silurian. Present-day north is toward lower left. Direction of lower Paleozoic north is only approximate (see text). Vertical and horizontal scales unequal.

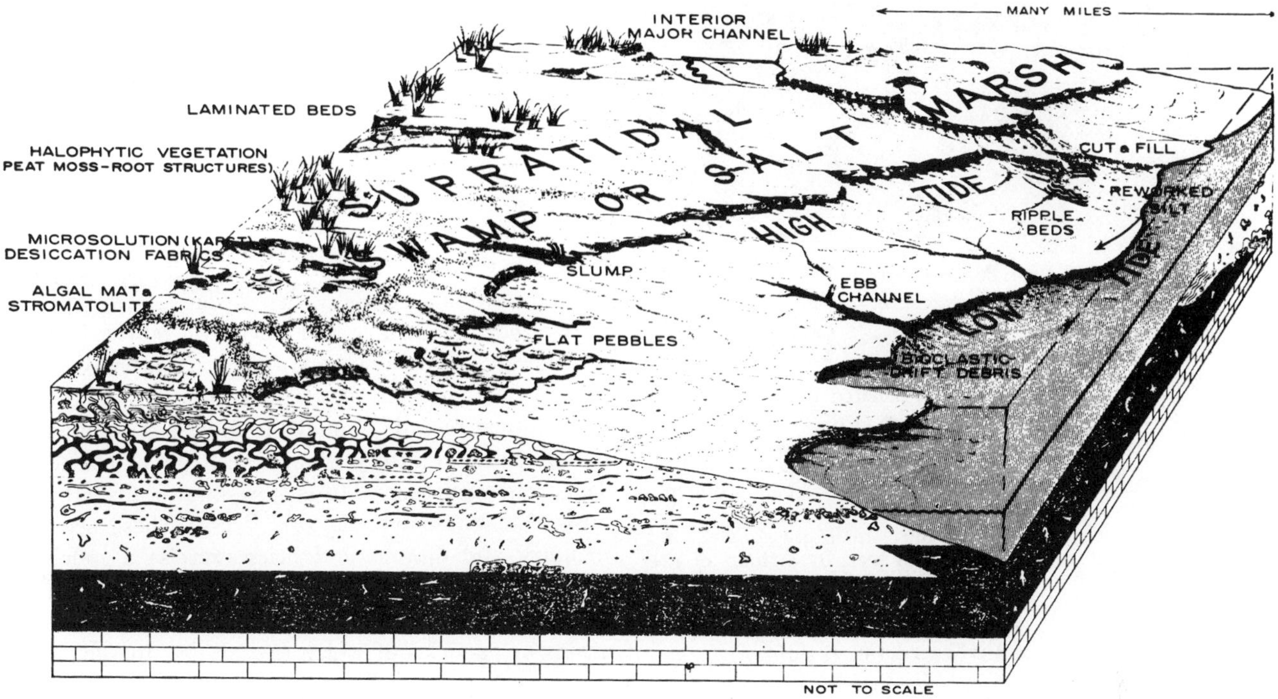

FIG. 6.—Conceptual scheme of epeiric tidal flat and supratidal environment of low-energy carbonate deposition. Succession of offlap microfacies and their fabrics are presented to conform with those described for Stony Mountain and parts of Interlake Formations. Vertical and horizontal scales unequal.

was covered with a veneer of shallow-marine water from the Arctic to Texas. If, as some workers postulate, the Ordovician poles were situated near the Philippines and in Africa (Creer *et al.*, 1957), a faunal belt of equatorial marine invertebrates probably extended roughly south to north as a result of near-equatorial conditions then existing at a high angle to the present-day equator. There is an impressive uniformity of fauna in both Red River and Stony Mountain Formations and their respective equivalents extending from New Mexico to Greenland (Flower, 1946; Ma, 1956; Nelson, 1959). The fact that this fauna (corals and brachiopods) is so unlike that of equivalent strata in eastern North America has been attributed to an intervening barrier, a change in facies, or a transition to temperate conditions. The faunal distribution patterns of Nelson and the paleomagnetic data of Creer *et al.* favor the choice of tropical to temperate adjustments of the fauna.

Within the western regional depositional setting (Fig. 5), the Cedar Creek anticline area was part of a major metastable lineament featuring mud banks along the western margin of the extensive, shallow epeiric Williston basin. The term "basin" is used here loosely to describe a geomorphologically negative area under the control of epicontinental tectonics. The banks lay between the basin and a broad oceanic threshold. Depending on the position of epeiric sea level, these shallow-water mud banks (or group of low islands) are interpreted to have been a large semi-restricting feature to the Williston basin whose lateral extent is as yet unknown and whose width varied according to the degree of basin restriction. The evidence supporting this regional reconstruction is based primarily on the published works of several authors (Flower, 1946; Ma, 1956; Creer *et al.*, 1957; Pye, 1958; Nelson, 1959; Porter and Fuller, 1959; Andrichuk, 1959; Kent, 1960; Fuller, 1961; Churkin, 1962; Macauley, 1964) and unpublished data, part of which appear here for the first time.

If undisturbed by eustatic and tectonic forces, or if sedimentation keeps pace with relative sea-level changes, such mud-bank deposits, because they are constructional, prograde seaward, and especially basinward, in a net regressive manner (Fig. 6). The location and stability of the major

FIG. 7.—"Lime mud" and pelleted bottom sediment in about 3 fm water a few miles northwest of Williams Island at U.S.G.S. Station G4. Photograph was taken by P. E. Cloud, Jr., U.S. Geological Survey, and he provides the following description: "The sediments here are mainly aragonite mud at the surface. The grasslike plant is the so-called turtle grass *Thalassia;* the fan-shaped alga *Udotea* is seen by a 'loaf' of the sponge *Hircinia,* with the algae *Laurencia* and *Microdictyon* clustered around the sponge. The mound at the upper right, with a hole down the middle of it is probably that of a white nocturnal shrimp."

platform edge (oceanic threshold; Fig. 5) have an important regional effect on such regressive carbonate deposition. In addition, the water regime is under the influence of local structures. Therefore, regional facies distribution reflects the combined continental and structural orientations and the latter influence circulation patterns.

Similar controls operate today on the Bahamas platform where high-energy sedimentary facies parallel the open ocean foredeeps along platform margins which are structurally positioned. These margins are identifiable by geophysical anomalies (Talwani, 1960). Trends of the sheltered "low-energy" deposits within the segmented platform also parallel these structurally controlled margins, but have more complicated patterns because of climatic differences, local geomorphology, and organic activity (Fig. 3).

If regressive sedimentation proceeds without interruption for some time, the area of original generation may be expected to develop supratidal features and low topography, which, in effect, bar the basin and establish conditions necessary for the sedimentation of evaporites, the processes of dolomitization, and eventually subaerial diagenesis which terminates the depositional cycle. Such regressive sedimentation predominates in both the Stony Mountain and Interlake Formations and indicates long periods of relative stability except for minor subsidence. The width of each facies belt on the basin side of the mud banks may range from a few hundred feet to many miles. In contrast, marine transgressions seem to have taken place abruptly in this period of Williston basin history. In many examples the transgression was so abrupt that certain facies which should be present in the sequence are absent. The transgressions are therefore important mainly in that they begin a new offlap cycle of deposition.

Are the sedimentary and diagenetic fabrics in a low-energy regressive environment distinctive? Would they normally be preserved as part of the sedimentary record? Would the fabrics determine or influence the distribution of primary or secondary void structures? It is now known that the answer to these questions is "yes." These sedimentary and diagenetic fabrics exercise a much stronger control on porosity, permeability, and capillarity properties of the rock than do the grains or particles of which they are composed.

WESTERN ANDROS ISLAND SEDIMENTARY ENVIRONMENTS

The regional depositional setting in the Bahamas and the details of the local Andros Island depositional environments have already been well described by several authors (*e.g.,* Newell and Rigby, 1957; Cloud, 1962; Purdy, 1963). Therefore, the writer points out only a few of the more important features.

The water depth just offshore is less than 2 fm and throughout the leeward part of the bank it exceeds 4 fm in very few places. The bottom (Fig. 7) is composed of soft "lime mud," rich in aragonite needles, and both soft and hard fecal pellets, the latter reportedly much more common toward the platform edge (Cloud, 1962). Both Cloud and Ginsburg (personal commun.) refer to polychaete annelid worms as the major pellet producer but many other organisms, especially crustaceans, form preservable pellets in this region. On the muddy surface is a thin growth of the "turtle grass" *Thalassia testudinum* and sparse bottom fauna of mollusks, crustaceans, green algae, and sponges. Hypersaline waters, organism population, and sediment type ("lime mud" and silt) are interacting products of the long fetch length, water temperature, and shallow depth of water. Except under storm or hurricane conditions the diurnal tidal currents diminish considerably toward the western shoreline. These tidal currents usually result in water-level variations of only 18 in. (0.5 m) to about 3 ft (1 m) along the low-lying western shoreline of Andros. Their velocity near the shore is markedly reduced because of the sheltering effect of the emergent low island, the major current drift, long fetch length, and the baffling action of shallow patches of turtle grass and faunal obstructions (Fig. 7).

The shoreline itself is a mildly sinuous, low-gradient mud flat with a few pronounced promontories and reentrants having differential relief up to 4 ft (Figs. 8A, B). The wind-driven current refracts off the mud flat but the reduced velocity provides no surf during periods of calm. As a result, the mud flat grades almost imperceptibly outward into the submerged nearshore area (Fig. 9). The water depth usually is less than 2 ft and the light-colored "lime mud" is quite soft (Fig. 8B). Because the mud is returned easily to suspension, it is constantly being reworked by the stronger tides. Because of the offshore volume and ease of transport of this sediment, it is a dominant contributor to all major zones of deposition in the area. The reentrants usually are floored with coarse, well-sorted bioclastic detritus, dominantly small, high-spired gastropods (*Cerithium*), and rounded bioclasts and lithoclasts (Fig. 9). The landed portion beyond the reach of normal high tides (commonly called "swash zone" but here called "supratidal") is low and comprised of palmetto hammocks (*Thrinax* and *Pseudophoenix*), mangrove swamps, tidal pools of highly variable salinity, algal mats and stromatolites, salt bush (*Avicennia*), and buttonwood (*Conocarpus*). This zone is breached by sluggish and abandoned channel distributaries and major ac-

FIG. 8A.—Aerial view, 300 ft elevation looking east over shoreline of western Andros Island near Little Loggerhead Creek. Shoreface has relief of about 4 ft. Although waves are refracting onshore there is no surf during periods of calm.

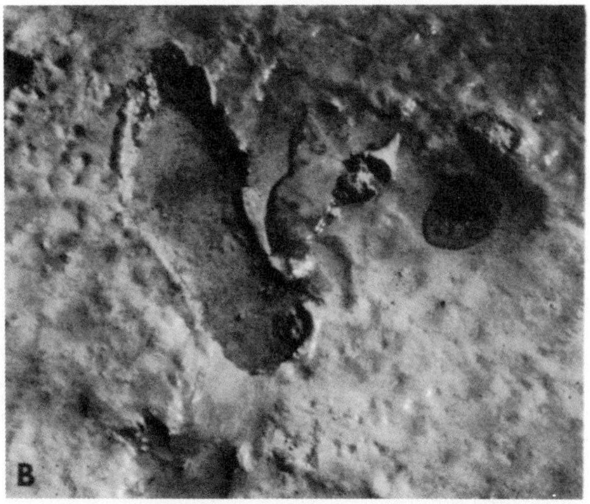

FIG. 8B.—Intertidal "lime mud" and pellets showing soft, paste-like nature. Location same as Figure 8A.

FIG. 9.—Shoreline reentrants floored with coarse bioclastic detritus. View is toward northwest near Little Loggerhead Creek. Shallow, almost flat gradient of nearshore submerged platform is shown by long stretch of light-colored "lime mud" underlying less than 2 ft (0.7 m) of water.

FIG. 10A.—Aerial view of supratidal or "swash" zone of western Andros Island. Note major channels and distributaries that show meander patterns and individual levee development. More pronounced inner banks or "point bars" commonly are covered by heavy growth of black mangrove (arrow).

tive tidal channels complete with natural levees (Figs. 10A, B).

The mangroves are thickest in two distinctive locations in the supratidal zone. The first is along the inner channel bank or "point bar" (arrow, Fig. 10A). The second is the perimeter of palmetto hammocks and levees of abandoned water courses where the recent dolomite "crusts" occur (Shinn *et al.*, 1965). Finally, elements of the older-beach or tidal-flat rocks crop out irregularly along the western Andros supratidal zone (Fig. 36) and pass abruptly into the low eolian dunes of Pleistocene age on the extreme eastern side of the Andros platform. Because of earlier sea-level changes, much of this surface is karsted and mantled by a poor laterite or *terra rossa* soil which in many places is lithified. Whether these soils constitute a true laterite is open to question. According to the latest discussions on laterites (Maignien, 1966), the Bahama soils would qualify from a profile, color, structure, and induration point of view (Figs. 37, 38). Because detailed studies of these soils on Andros are incomplete, the writer uses the term *"terra rossa"* as defined by the AGI glossary (1957): "Residual red clay mantling limestone bed rock."

In this setting the sediment source for fabric construction is substantially the same for all local environments; *i.e.*, the transport of mud, pellets, and bioclastic material is in the shoreward direction and so is shared by all of the sites of deposition and trapping, even though the sharing usually is unequal (Fig. 11). To provide the energy to pick up the accumulated offshore fine sediment and other debris for delivery into the supratidal zone, one or a combination of storm, hurricane, or spring high tides is required. The normal tide range does not exceed about 3 ft (1 m) and, except for seasonal or wind-generated patterns, it usually conforms with existing bathymetric conditions. During major storms, however, the bottom profiles on the shallow platform are disturbed and tides of several feet discharge a large sediment load far inland from the intertidal zone. The storm also disrupts the unprotected or weakly lithified surfaces in the supratidal zone.

Although all of the sand-size and smaller sediment is easily suspended for transport during spring-tide or storm-tide movement, the gradual reduction of sheet flow and current velocity onshore into the supratidal zone results in deposition of graded beds. In general, the finest or lightest material is carried farthest inland. Much of this is a "lime-mud" fraction. Some of it also is channeled back toward the shoreline under reduced energy conditions. This provides an additional fine sediment layer which accentuates the gradation of the recently deposited flood layer beneath. The sediment profile is thus somewhat unique in that it is the product of two opposed current systems.

This seemingly paradoxical situation is expectable from the geological point of view. In spite of the general day-to-day low-energy state of the tidal-flat and shoreline environment, the require-

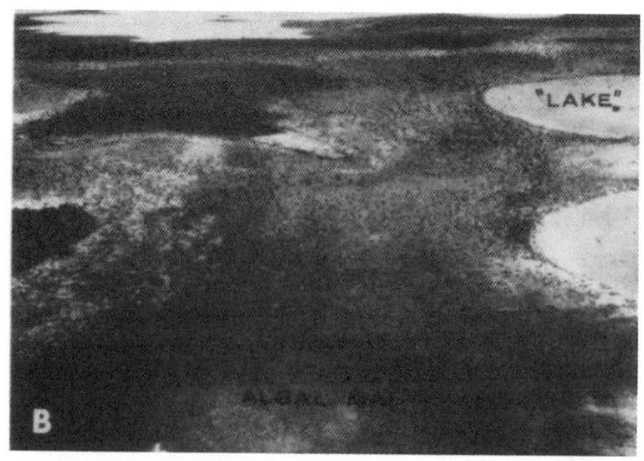

FIG. 10B.—Aerial view of supratidal or "swash" zone showing palmetto hammocks of *Thrinax* and *Pseudophoenix*, algal mats, and bypassed pools and lakes of brackish and saline waters, western Andros, south of South Bight.

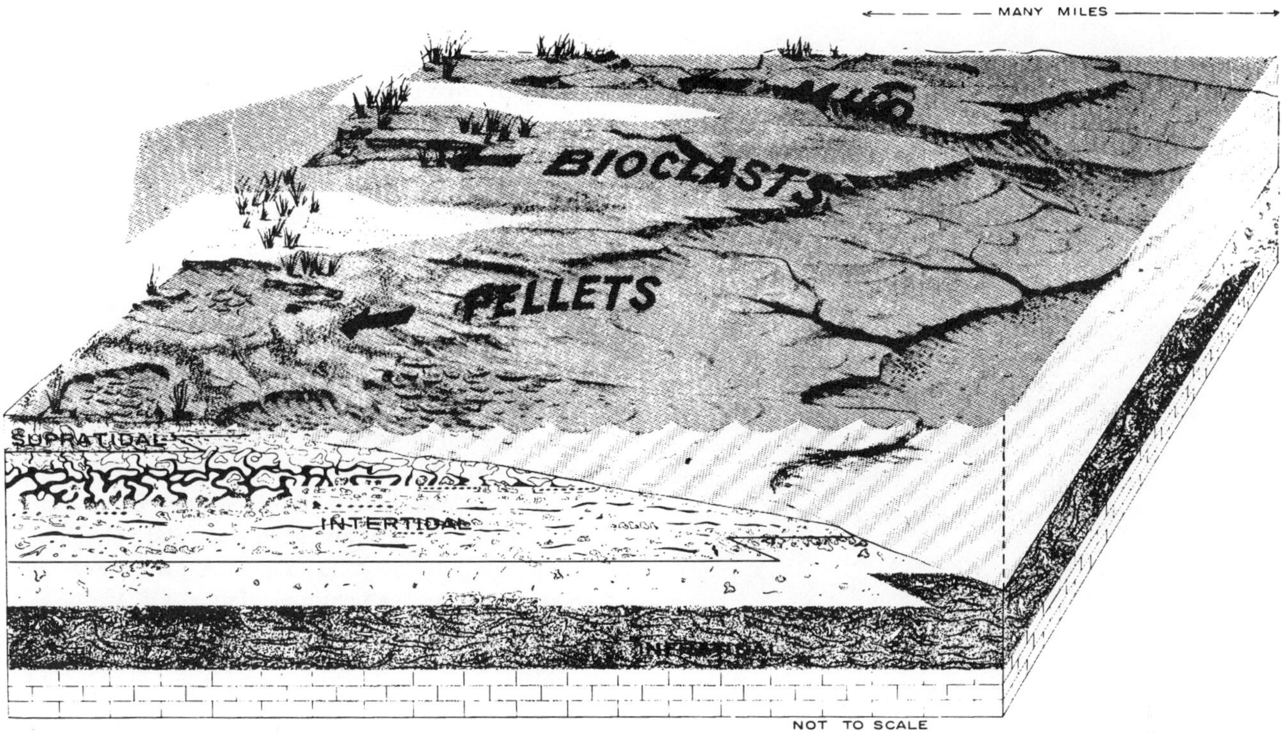

FIG. 11.—Onshore distribution of common sediment types from which low-energy environment fabrics originate. Vertical and horizontal scales are unequal.

ment for renewed sediment distribution, processing, and later offlap deposition is for repeated high-energy hydraulic and climatic events. These uncommon daily occurrences are nevertheless common in the context of geologic time.

The four specific subenvironments which are considered here are:

1. The *infratidal* zone, which is below mean low tide, including most of the shallow offshore marine.

2. The *intertidal* zone, which is between mean low and high tides.

3. The *supratidal* zone, which is above mean high tide and is only subject to spring- and storm-tide floodings.

4. The *subaerial diagenetic terrane,* or island zone, above tidal influence and subject to early diagenetic alteration or destruction by subaerial or vadose processes.

Andros or any other such model will not be complete in the content and distribution of possible sedimentary and diagenetic structures. The setting of Andros is not an interior continental platform and it certainly undergoes many different adjustments because of geographic aspects peculiar to it and humid climate conditions. However, the several sedimentary mechanisms, organic processes, diagenetic alterations, and general geometry (including facies progression) are adequate for the analogy.

Analog Facies Fabrics

The carbonate part of the Stony Mountain and Interlake reservoir rocks is almost 100 percent dolomite and almost all of the reservoirs have substantial porosity. The relative reservoir quality depends secondarily on the changes brought about by dolomitic replacement. It depends primarily on the original pore size and distribution which is controlled by the depositional fabric except where diagenetic alterations are major.

There was perceptive offlap development throughout Late Ordovician and Silurian times as is indicated by the predominantly shaly offshore sediments and shaly burrowed intertidal sediments in the Stony Mountain Formation which grade laterally into intertidal and supratidal sediments in the Interlake Formation. Much of the Interlake subsequently has been transformed by early diagenesis culminating in a long period of subaerial exposure in Late Silurian time and continuing into Early Devonian time.

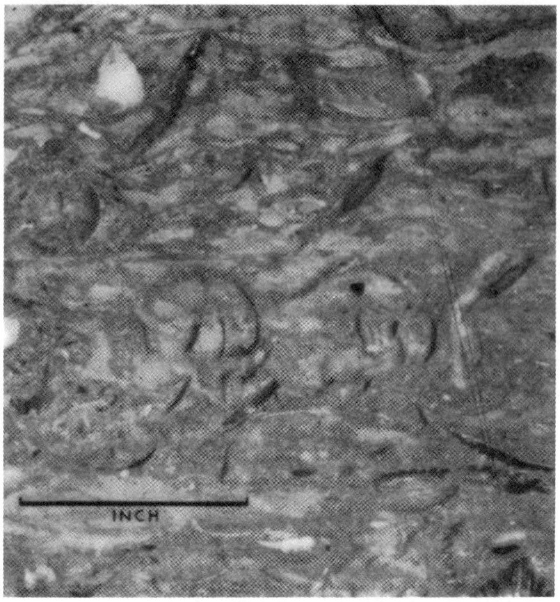

FIG. 12.—Polished section of very argillaceous dolosiltite of Stony Mountain "Shale" composed of about 30 percent calcareous articulate brachiopod shells and trepostome bryozoan fragments. Fabric represents "basin" facies of epeiric framework and corresponds to lowest or infratidal zone of Figure 6. This rock is nonproductive. Sample loc.: Pine field, well 12–15A, depth 9,863 ft.

BASIN FACIES

Before considering analogous facies between the Andros Island area and the Ordovician-Silurian epeiric tidal flats of the Williston basin, brief mention should be made of the basin facies of the Williston basin which does not have an exact analogy at Andros because of the lack of a clastic sediment source near Andros. This is the so-called Stony Mountain "Shale" Member, or Gunn Member, of Sinclair and Leith (1958).

In all barred basins or extensive epeiric seas, one is likely to find an offshore fine sediment rich in clays and admixed with variable amounts of bioclastic debris. Figure 12 is an illustration of such a deposit from the Stony Mountain "Shale" Member. It is not shale but a dolomite composed of a very shaly (illite) matrix (30 percent) and randomly oriented shells of articulate brachiopods and fragments of trepostome bryozoans, both of which are still composed of calcium carbonate. All such samples of the Stony Mountain "Shale" throughout the western Williston basin have similar compositions. The disorientation of the shell matter, the types of organisms, and the mineral content suggest deposition in an extensive shallow infratidal site with early post-depositional reworking by organisms.

This rock type everywhere has a porosity greater than 10 percent but, because of high clay content, cannot be a potential reservoir rock. The source of the clay has been the subject of speculation by several geologists. Although its origin is unclear, it is an important element in the differentiation of good reservoir rock in the overlying facies of the Gunton Member.

INFRATIDAL ZONE

The infratidal deposits are perhaps the least diagnostic of fabric types. Extensive reworking by organisms leaves little fabric except burrow traces composed of silt- and sand-size fecal pellets in a suspension of fine mud (Fig. 13A), disoriented shell debris, and a benthonic fauna. Off western Andros, Cloud (1962) has confirmed the overwhelming abundance and diversity of fecal matter and aragonitic mud below the low-tide level of the shallow platform (Fig. 7). Some variation in development and extent of preservation should occur in the direction of the low-tide position. In the ancient record it fortunately is possible to find *in situ* accumulations of the benthonic forms, to verify the vertical-facies sequence position, and to check lateral correlation and uniformity.

In the absence of diagnostic fabrics or facies associations, however, this type of sediment can-

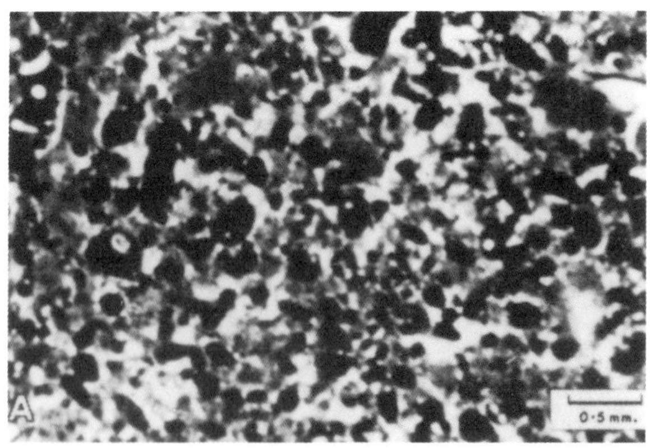

FIG. 13A.—Photomicrograph of muddy pelletal calcarenite from core about 1 ft below sediment-water interface 4 mi (6 km) west of Andros Island. Darker pellets are slightly indurated. "The fine-grained matrix is initially in the form of pellets but these are soft and become distorted upon very slight compaction." (Ginsburg, pers. commun.)

not be referred with certainty to any facies. This is true because the sediment is the major source material for movement by storms and tides into the intertidal and supratidal zones. Figure 13B shows a Silurian example which, by association with other widespread supratidal fabrics, clearly does not represent the same environment as that of Figure 13A, even though the fabrics are substantially the same.

Because of the high "lime-mud" content and extensive breakdown of soft matrix pellets, such sediment probably will not become reservoir rock unless extensively dolomitized. In the example of Figure 13B the Silurian pelletoidal dolomite is one of the most widespread and uniform of potential reservoir rocks. If the constituent grains and intercrystalline pore spaces are sufficiently large after dolomitic replacement, a volumetrically important reservoir may be formed. The crystals and pellets of the Silurian example are 20–40 microns in size, and the rock is a good reservoir. The quality of these porous rock types, however, is related to capillarity, and for many such rocks the entry pressure necessary for oil to overcome the connate water cannot be attained. The importance of these uniform, fine-grained rocks is discussed further in the section on supratidal deposits.

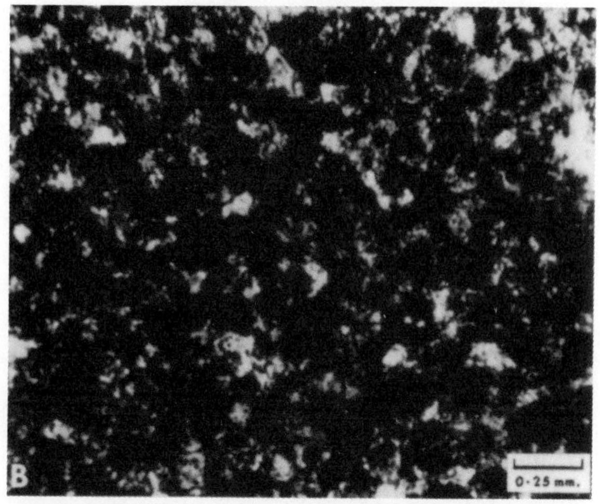

Fig. 13B.—Photomicrograph of pelletoidal dolosiltite from Interlake Formation similar to Figure 13A although pellets are smaller and outlines indistinct. Matrix also appears to be mashed pellet form. This rock type is a potential reservoir. Sample loc.: Cabin Creek field, well 21–34, depth 8,706 ft.

INTERTIDAL ZONE

There is much conjecture concerning what structures and fabrics are likely to be preserved in deposits of tidal or near-tidal origin. In contrast to the offshore-marine and infratidal environment, the sediments within the intertidal zone are subject to many diverse adjustments, both hydraulic and organic. An important distinction to be made between terrigenous and calcareous mud and silt in this environment is the degree of cementation or cohesiveness of the sediment immediately after burial. The textures and structures of carbonate and noncarbonate deposits may be similar superficially because of comparable hydraulic environments, but the diagenetic action immediately following burial will likely be dissimilar, except for certain kinds of biologic activity. Carbonates commonly attain early competence through cementation in the aerated zone (Ginsburg, 1957), whereas noncarbonate deposits in tidal flats are subject to frequent relocation, reprocessing, and removal before ultimate stabilizing (van Straaten, 1951, 1954a, b, 1959). The very early diagenetic history is therefore commonly critical in determining what kind of preservation takes place.

Although considerably more has been learned about noncarbonate than carbonate tidal-mud flats, the actual depositional environment and details of diagenetic processes may be sufficiently similar to permit some valid comparisons between the two, independent of their likelihood of survival. Van Straaten noted that Waddensee high-tide flats are extensively burrowed, the vertical burrows being preserved and the other, essentially horizontal, types probably being compacted beyond recognition due to sediment incompetency. The low-tidal flat also is burrowed, but gullies responding to tides which range up to 8 ft (2.6 m) reprocess the sediment and thus destroy a feature which might otherwise be widely preserved. Relicts of both types of burrowing apparently are widespread in the Stony Mountain and Interlake carbonates. There are really only two dominant ways in which such fabrics can be destroyed or altered without changing the general depositional environment, *viz.*, differential compaction and lateral removal by channels and gullies. Early cementation in carbonates would reduce the effect of both substantially.

Figure 14 is an example from the intertidal zone of western Andros Island. Close inspection

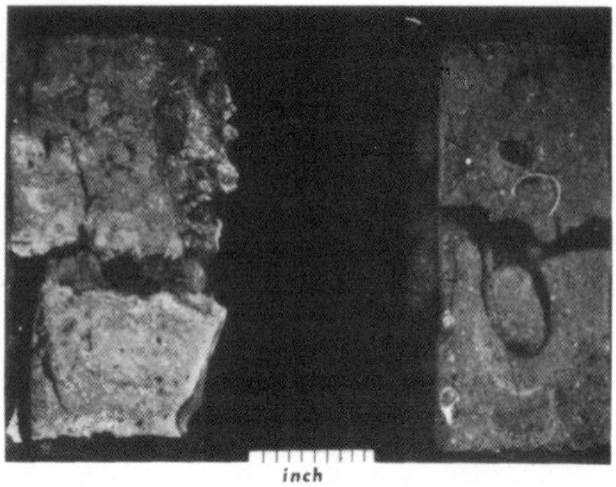

Fig. 14.—Burrowed tidal fabrics in Recent sediment cores from western Andros Island. Burrowing is revealed in particle intermixing and color (tone) mottling.

of this sediment reveals many faint traces of a burrow network which is common in the zone. In comparable material from the Gunton Member of the Stony Mountain Formation there is a succession of burrowed, churned, and otherwise reworked sediment. The character of such fabrics in the Stony Mountain Formation has been enhanced because the sediment-processing organisms preferentially segregated small amounts of illitic clay and other noncarbonate minerals. The fine, delicate, and, in places, bifurcating habit of small, numerous, horizontal burrows should be noted, together with some larger, more convolute masses in Figure 15A.

This kind of burrowing may have occurred near and across the juncture with the infratidal zone and was the most "near-normal" marine environment as far as organic activity is concerned, *i.e.*, the shallowest, most easily aerated, phototropic, and near normal salinity. Numerous mud and bottom dwellers probably interacted at and just below the water/sediment interface from the uppermost reaches of the infratidal through the lower intertidal zone. The latter is the position of most continually shifting water, which should satisfy most benthonic organisms. Local variations are noted where ripple beds, cut-and-fill structures, and water-drift accumulations of bioclastic debris result (Fig. 6). These probably relate to atypical water movements, climatic processes, or erratic action of ebb gullies having fixed lateral excursions and reaching down through the intertidal zone to interchange with the open-shelf waters.

Commonly these fabrics receive breccia fragments which are the result of ebb-flow channel downcutting of a shoreward facies. The fragments can be composed therefore of many different deposits. In the Stony Mountain the most common type of redeposited fragment type in the infratidal and lower intertidal zone is the lithified larger burrows occurring just shoreward in the upper intertidal facies. These are everywhere very angular. If brecciated, the large burrow of Figure 15B (lower left) would be a typical example. It is significant that these are apparently lithified (diagenetically?) very early in their history and that they are limited in lateral extent even though locally abundant. They are most common in zone 5 (lower intertidal, Fig. 45).

Moving higher in the inferred intertidal zone of the Stony Mountain Formation, Figures 15B and C show the larger burrows and reworked

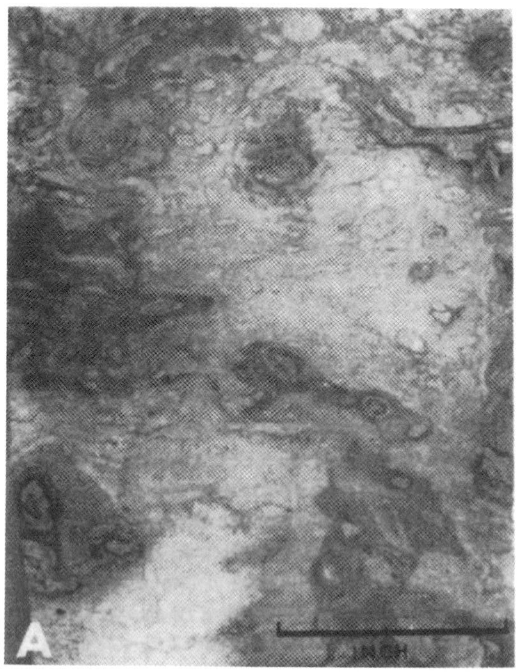

Fig. 15A.—Fine-burrowed low-intertidal fabric from dolomitized Gunton Member, Stony Mountain Formation, showing pronounced horizontal linearity, rare bifurcating habit, and crude circular cross sections. It is intermixed in some places with larger convolute burrows. Fabric is accentuated by illitic clay and pyrite "dust." Sample loc.: Pine field, well 14–29, depth 8,789 ft.

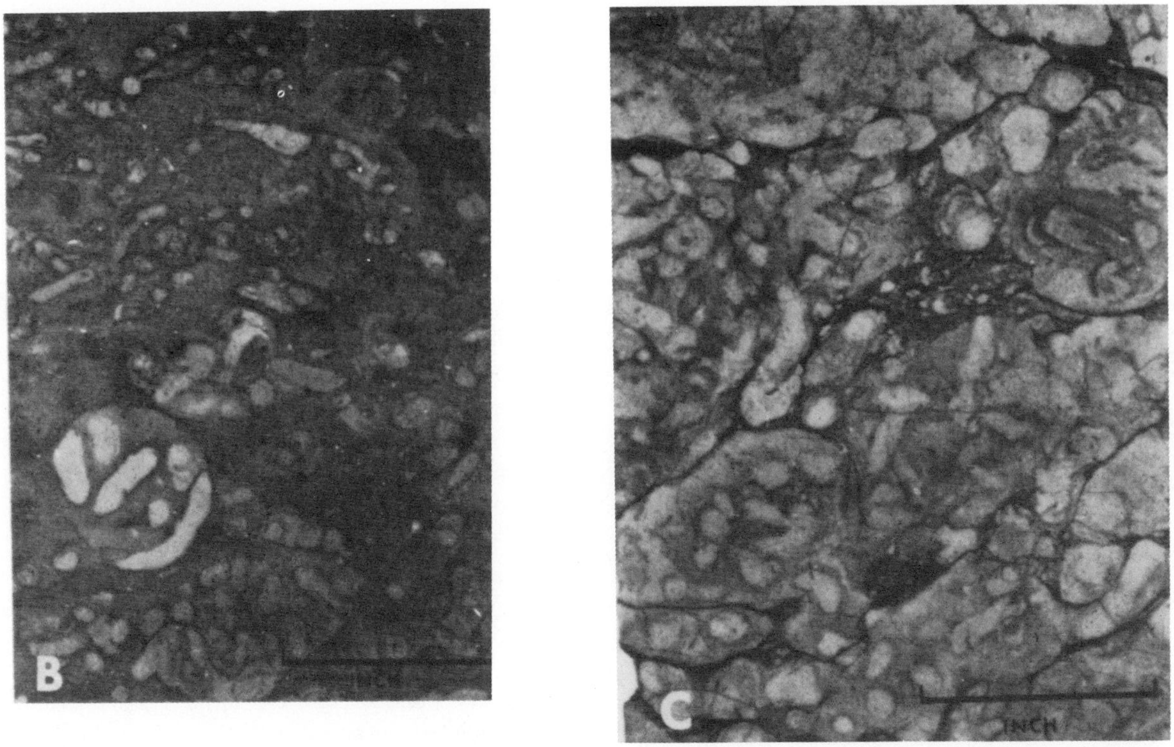

Figs. 15B, C.—Two examples of thick-burrowed intertidal fabrics which retain internal detail of organism processing. Small clay content (dark) is restricted to interburrows. Samples are of Gunton Member, Stony Mountain Formation. Sample locs.: 15B: Pine field, well 14–29, depth 8,791 ft; 15C: Pine field, well 12–15A, depth 8,834 ft.

masses which predominate. They retain much internal detail and traces of previous burrows. Figure 15D illustrates a near maximum of reworking, with interference between soft masses of sediment and gradual disappearance of internal structure. The illitic clay content is here reduced to a minimum for the zone ($>$ 5 percent) and is most effectively eliminated from the large reworked masses. This material initially may also have been finely burrowed but thorough reworking of the sediment has obscured any earlier fabrics.

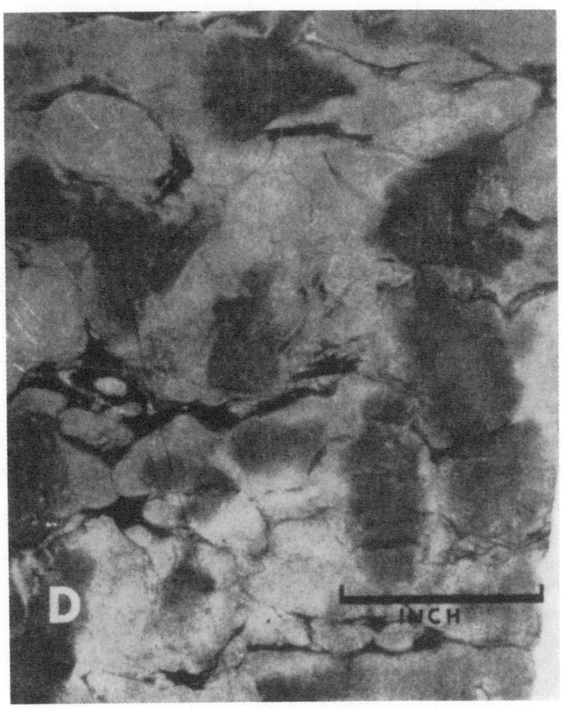

Fig. 15D.—Terminal phase of burrowed and reworked fabrics in dolomitized Gunton Member, Stony Mountain Formation. This fabric inferred to be near top of intertidal zone. Very large reworked masses gradually lose the internal excreta structures shown in Figure 15C. Illite is well-confined between masses. Specimen is completely oil-stained. Sample loc.: Pine field, well 12–15A, depth 8,823 ft.

FIG. 16A.—Diagenetically altered fabric at top of oil-stained intertidal facies of the Gunton Member, Stony Mountain Formation. Laminations, *in situ* fracturing of "crust"-like layer, and horsetail solution stylolites suggest subaerial exposure during and immediately after deposition. Sample loc.: Pine field, well 12–15A, depth 8,810 feet.

The intertidal zone is capped by thin laminae of fine-grained dolomite, some of algal-mat origin (Figs. 16A, B). These laminae are not part of the previous fabric suite in the sense of having been organically reworked. They are part of an important transitional fabric which is related to the supratidal zone. Evidence of early diagenetic alteration as related to the first stages of subaerial exposure is the *in situ* fracturing of "algal" crusts and some leaching effects demonstrated by a filagree of "horsetail" stylolites shown in Figure 16A. This early diagenesis has aided in the development of favorable petrophysical properties of the rock.

The unaltered, well-laminated dolomite type illustrated by Figure 16B is in this example non-oil-bearing. Because it is extremely widespread and can be an important reservoir, as in the Interlake Formation, it cannot be dismissed easily. This rock type is discussed fully for the Interlake Formation. It is important here mainly because it terminates an important reservoir facies in the Stony Mountain Formation.

There is a progressive increase in reservoir quality with increased burrowing and reworking. The dolomite matrix of the fine-burrowed fabric is composed of 10–25 micron anhedral crystals whereas loose 25–40 micron euhedral crystals characterize the thickly reworked fabric. The improvement is reflected in both porosity and capillary properties. The thickly reworked upper intertidal fabric has twice the porosity (15 *versus* 7 percent), a much lower entry pressure (49 *versus* 123 psia), and superior permeability of (2.5 *versus* 0.1 md) compared with the fine-burrowed fabric. The log-log capillary pressure curves show that the thickly reworked fabrics have a superior pore-size distribution because of their much lower pore geometrical factor value of 0.17 *versus* 0.39. This value describes the curve shape in terms of the hyperbolic notation *G*. It is related to two asymptotic end members: (1) the extrapolated entry pressure, and (2) the extrapolated bulk volume occupied at infinite pressure. The mathematical relation is given by Thomeer (1960) as follows (Figs. 17A, B):

$$(\log P_c - \log P_d)(\log BV_f - \log BV_{p\infty}) = -C^2$$

where

$$C^2 \ln 10 = G, \text{ and}$$

P_c, capillary pressure, air/Hg (psia);
P_d, extrapolated entry pressure;
BV_f, fractional bulk volume at P_c;
$BV_{p\infty}$, extrapolated bulk volume occupied at infinite pressure; and
G, pore "geometrical" factor.

Dolomitization has enhanced these petrophysical properties of the burrowed sediments. Based on recent findings concerning the origin of modern dolomite, the most reasonable hypothesis for

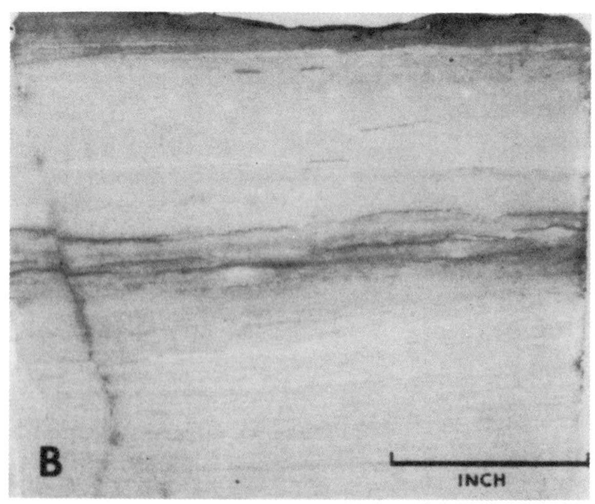

FIG. 16B.—Thinly laminated, dense, transitional supratidal dolomite. This rock type overlies that of Figure 16A. Sample loc.: Pine field: well 21–10A, depth 8,952 ft.

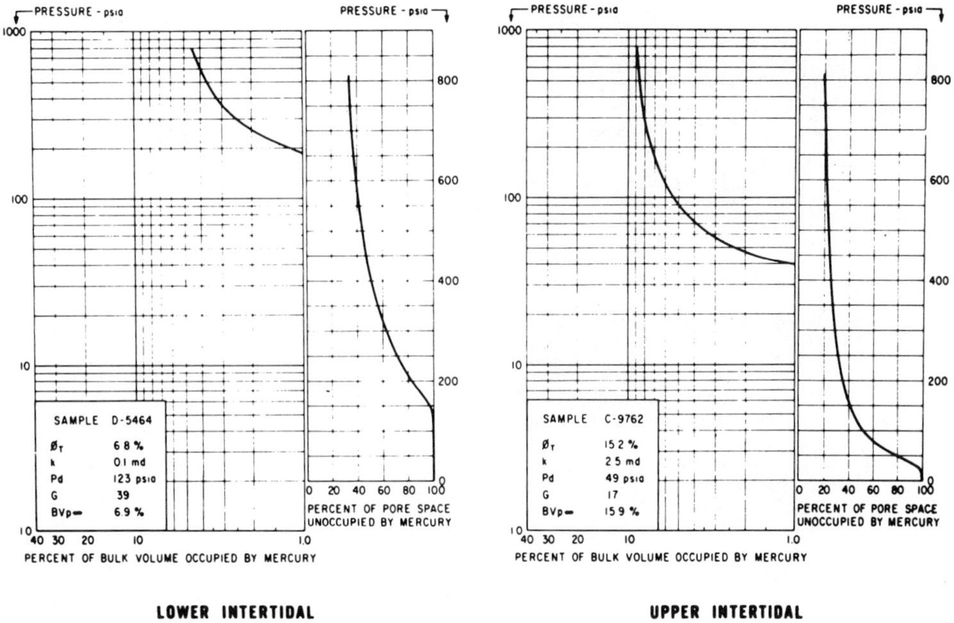

FIG. 17A.—Mercury-air capillary pressure curves of burrowed and reworked fabrics, Gunton Member, Stony Mountain Formation.

the formation of dolomite in the Stony Mountain Formation is one of early epigenetic replacement. This is discussed in a subsequent section. Whatever the origin, the dolomitization of the Stony Mountain permitted retention of the initial sedimentary fabric, inhibited compaction, and retained the pre-existing density of susceptible nucleation sites. The volumetric importance of these oil-bearing fabrics in the Stony Mountain Formation is borne out by their known occurrences throughout the basin. (For example, compare Fig. 15D with the "nodular" limestone, Pl. 1, Fig. 3, of Porter and Fuller, 1959, and with Kupsch, 1953; these workers also report its common occurrence in the Gunton Member.)

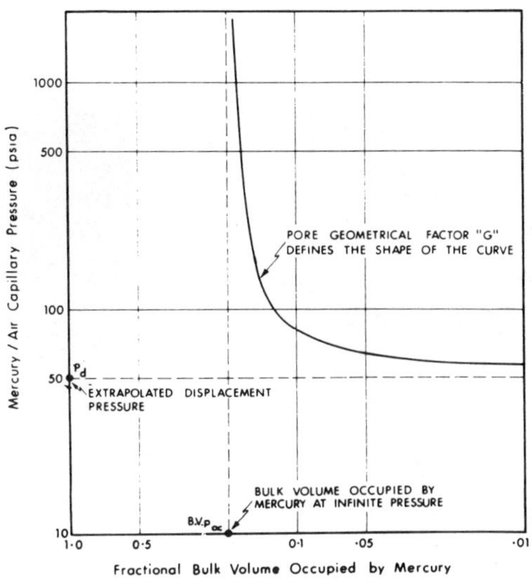

FIG. 17B.—Hyperbolic notation of typical log-log capillary pressure curve as defined by Thomeer (1960).

Burrowed and reworked fabrics cannot be uniquely positioned in the tidal environment solely on the basis of their attributes. However, their relation to each other as well as to fabrics of different facies in the vertical succession indicates distinct locations in the intertidal zone. Once these intertidal fabrics have been related directly to the larger sequence, additional interpretations of events, resulting from lateral correlation and change, commonly present themselves. Because the intertidal zone is areally extensive and shifts laterally in time, its sediments may be a major part of the record and comprise a volumetrically important potential reservoir.

A detailed description of the Gunton Member intertidal deposits is given in the Appendix and is graphically portrayed in Figure 45.

Overlying the Gunton Member in the western Williston basin is a thin series of interrupted cycles which are very highly leached and dolomitized, and separated by very argillaceous beds. These argillaceous beds correspond reasonably well to the cyclic units described by Porter and

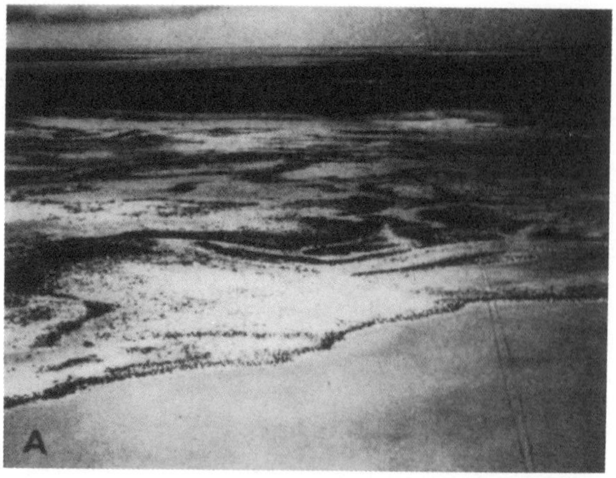

FIG. 18A.—Northeasterly view over western shoreline of Andros Island, South Bight. Vast stretches of unvegetated, supratidal layers of calcisiltite and calcilutite are common where not overgrown with algal mats, mangroves, palmettos, salt bush (*Avicennia*), and buttonwood (*Conocarpus*). Such layers result from combined storm and spring high tides. Similar deposits in geologic record usually are closely associated with other, more distinctive supratidal deposits.

Fuller (1959). Although no Stony Mountain anhydrite beds appear to extend across the Cedar Creek anticline, the argillaceous zones occur at approximately the positions where the evaporites would be expected (Cummings *et al.*, 1959; Kent, 1960). In addition, the disseminated anhydrite content exceeds 5 percent only in these zones. It cannot be determined with certainty whether this anhydrite is a residual relict of the upslope edge of the evaporite facies (phase 3 of Porter). Its presence does strongly suggest, however, that the intertidal units in the Gunton Member, and presumably those of the more frequently interrupted deposits of the overlying Stonewall, were progressively built up until the basin became barred. A strong case can therefore be made, based on platform emergence and lithologic similarity, for a supratidal origin of disseminated anhydrite similar to that described by Kerr and Thomson (1963) for the Permian of Texas and New Mexico and by Kinsman (1965) for recent *sebkha* deposits of the Trucial Coast, Persian Gulf. The comparison is discussed again under the subject of subaerial diagenetic terranes.

SUPRATIDAL ZONE

Supratidal sediments may not at first seem volumetrically important, either as preserved sediments or as potential reservoirs. Such deposits are, however, surprisingly well preserved in the epeiric sedimentary record. In fact, these deposits are predominant in the Silurian Interlake Formation on the Cedar Creek anticline. No one sedimentary process or fabric is exclusively diagnostic but fortunately there are many diverse fabrics which as a group indicate the distinctive aspect of the environment. The quality of any one of these as a potential reservoir rock depends on continuity of a porous matrix. At any one time and locality supratidal sediments of one fabric type may be fairly continuous but more commonly they are interrupted by the formation of other impinging supratidal fabrics, by meandering channels, and by irregular low-profile topography (Figure 18A).

Laminated calcilutite and calcisiltite.—Perhaps the most important, volumetrically, are the laminated deposits which result from numerous periods of flood and runoff sedimentation (Figure 18B). These sediments are widespread along Andros Island and the separate laminae are variably composed of calcilutite and calcisiltite. The oil-stained example in Figure 18B shows a similar fabric in the Interlake Formation and emphasizes the importance of minor changes in size and accumulation of sediment which may be preserved. Obviously the reservoir quality of this type of fabric changes greatly and is dependent on the size range of the original sediment and the type and degree of subsequent diagenesis. Sediments of this type along the western Andros Island shoreline are subjected to some burrowing and disruption, not only by animals but also by plants

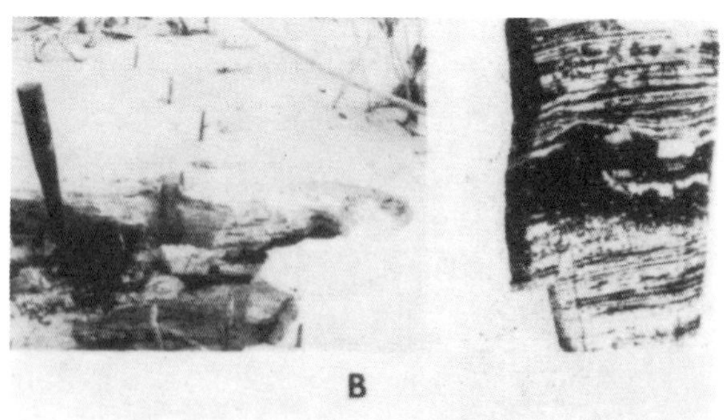

FIG. 18B.—Comparison of ancient and modern laminated carbonates. Left: laminated calcilutite and calcisiltite from the supratidal zone of Williams Island, Bahamas. Right: linear porosity and oil staining in inferred supratidal laminated fabric from dolomitized Interlake Formation, Cabin Creek field, well 21–34, depth 8,630 ft.

such as the breather tubes and roots of mangroves (Figure 19). Except for algae and perhaps some low-order halophytic vegetation, plants do not seem to figure importantly in Ordovician or Silurian depositional history.

Whether or not the organic activity in the supratidal zone permits some homogenization, especially by burrowers, is related to the harshness of the environment. On humid, rainy, and frequently flood-inundated Andros Island, there is some burrowing in the supratidal zone (Shinn *et al.*, 1965). However, none has been reported so far from the Trucial Coast except as burrowed relicts of earlier offlap sediments stratigraphically beneath the supratidal layer (Illing *et al.*, 1965). Because of the severity of the environment the writer doubts that any should be expected except perhaps at the intertidal juncture.

Several inches of "lime mud" were deposited over the Florida Bay Islands during Hurricane *Donna* (Ball *et al.*, 1963). Deposits of this thickness in the supratidal zone should be sufficient to be preserved and too thick and widespread for thorough burrowing even if the burrowing organisms are available for a short time after the initial flood.

On Andros Island, the supratidal mud connects the small "islands" building their own supratidal outliers offshore (Purdy and Imbrie, 1964). Where not overgrown by large plants, the sediment is covered by algal mats and stromatolites. Thus the depositional association of laminated deposits is with algal laminae and the rip-up clasts from their surfaces which comprise flat-pebble breccias. This association is what is observed in ancient deposits, including the Interlake Formation.

The Interlake laminated deposits are uniform and widespread. They are similar to the Stony Mountain example (Fig. 16B) but the original, purer carbonate has been dolomitized with the result that the laminae are very difficult to see (Fig. 20A) except where the original pellets (Fig. 13B), ripple structures, or thin graded bedding are observable in thin sections. Although the laminated deposits occur with normal-marine fabrics rich in corals (*Paleofavosites* sp.), they are most prevalent as interbeds with algal mats and flat-pebble breccias. No doubt this same very fine-grained sediment served as the original material for the later, more distinctive supratidal fabrics.

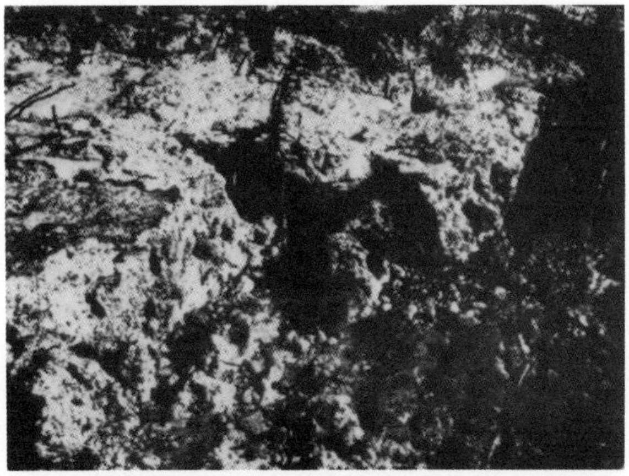

FIG. 19.—Erosion profile of Recent laminated calcilutite and pelletoidal calcisiltite from shore of Williams Island, Bahamas. Sediment has moderate competence, having been extensively penetrated by roots and breather tubes of mangroves, and excavations of burrowing crabs.

The Interlake laminated deposits are thoroughly dolomitized and porous, the crystal sizes ranging generally from 5 to 50 microns. This range is the petrophysical Type II of Archie (1952). Because there is a minimum average crystal size below which the dolomite is non-oil-bearing, an additional distinction is introduced here. Type II dolomites are confined to the size range between 5 and 20 microns, and are non-oil-bearing (Fig. 20B); Type II–III (micro-saccharoidal) dolomites are here introduced and defined as ranging in size from 20 to 50 microns, and are oil-bearing in the Interlake Formation (Fig. 21B).

The porosity of Type II commonly is high, as much as 16 percent. The effective porosity is proportionately good and the pore sizes and connections are fairly uniform, yielding a well-defined capillary pressure plateau of low G value (Thomeer, 1960). This value ranges up to 0.4 (Fig. 20C). However, it normally has a low oil saturation in the Silurian reservoirs because of its high extrapolated entry pressure (P_d).

In contrast, the Type II–III dolomites have porosity values which in a few samples reach 24 percent. The effective porosity and pore geometrical factors (G) of Type II–III are as good or better than those of Type II. Petrographically, the differences are accentuated further, because Type II–III crystals tend to be euhedral except in crystal "clusters," whereas Type II crystals usually are anhedral to subhedral with a trace of "sparry" cementation. The range of crystal sizes

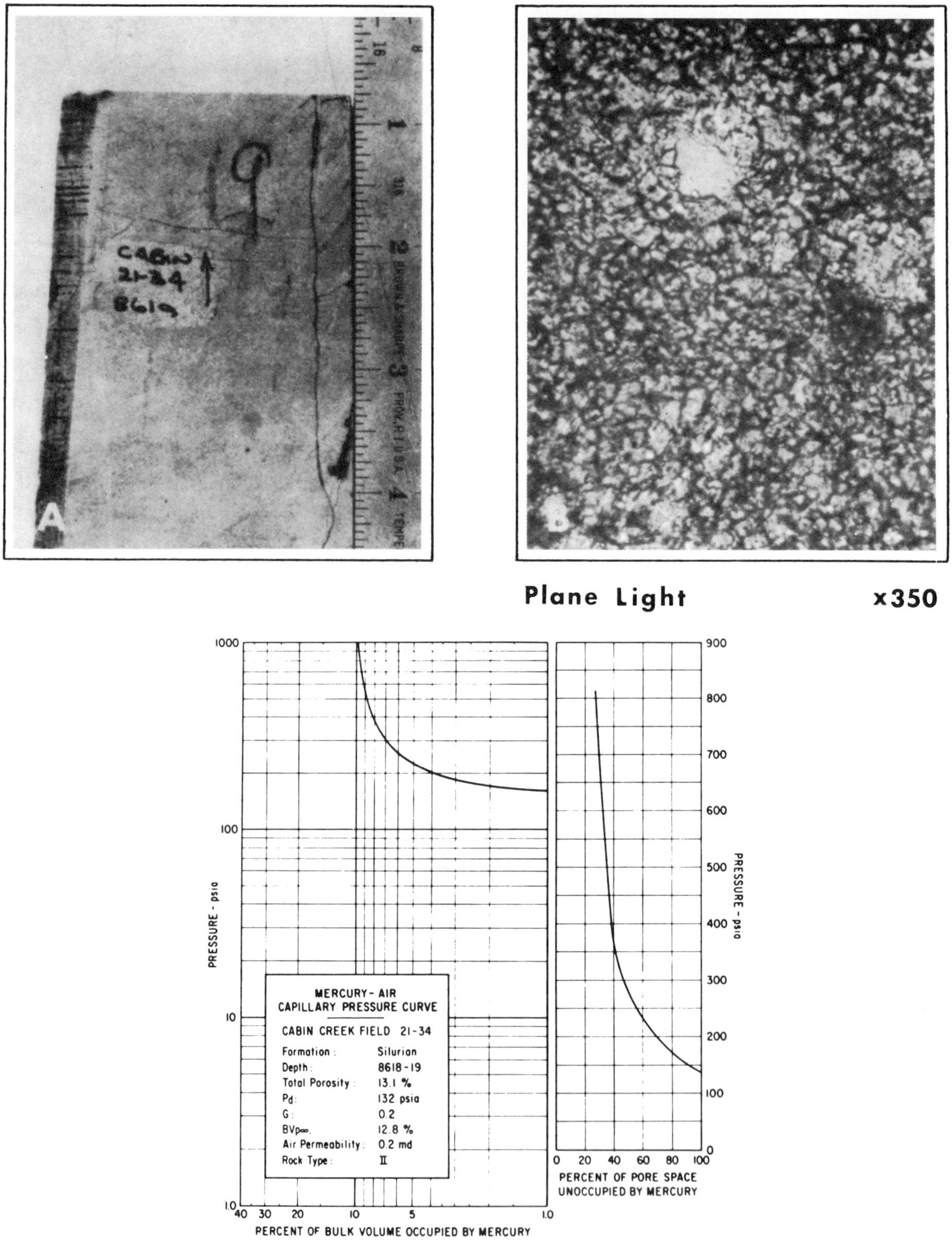

C

FIG. 20A.—Laminated dolosiltite similar to Figures 16B and 18B, except it is uniformly cream-colored, extra-fine-grained, chalky dolomite.

FIG. 20B.—Photomicrograph of rock in Figure 20A showing anhedral to subhedral, 5-20-micron dolomite crystals with uniform to sutured contacts.

FIG. 20C.—Capillary-pressure curve of laminated dolosiltite using Thomeer notation. Shape factor ($G = 0.2$) indicates good pore-size distribution and interconnection; however, relatively high entry pressure ($P_d = 132$ psia) reflects very small size of intercrystalline pores.

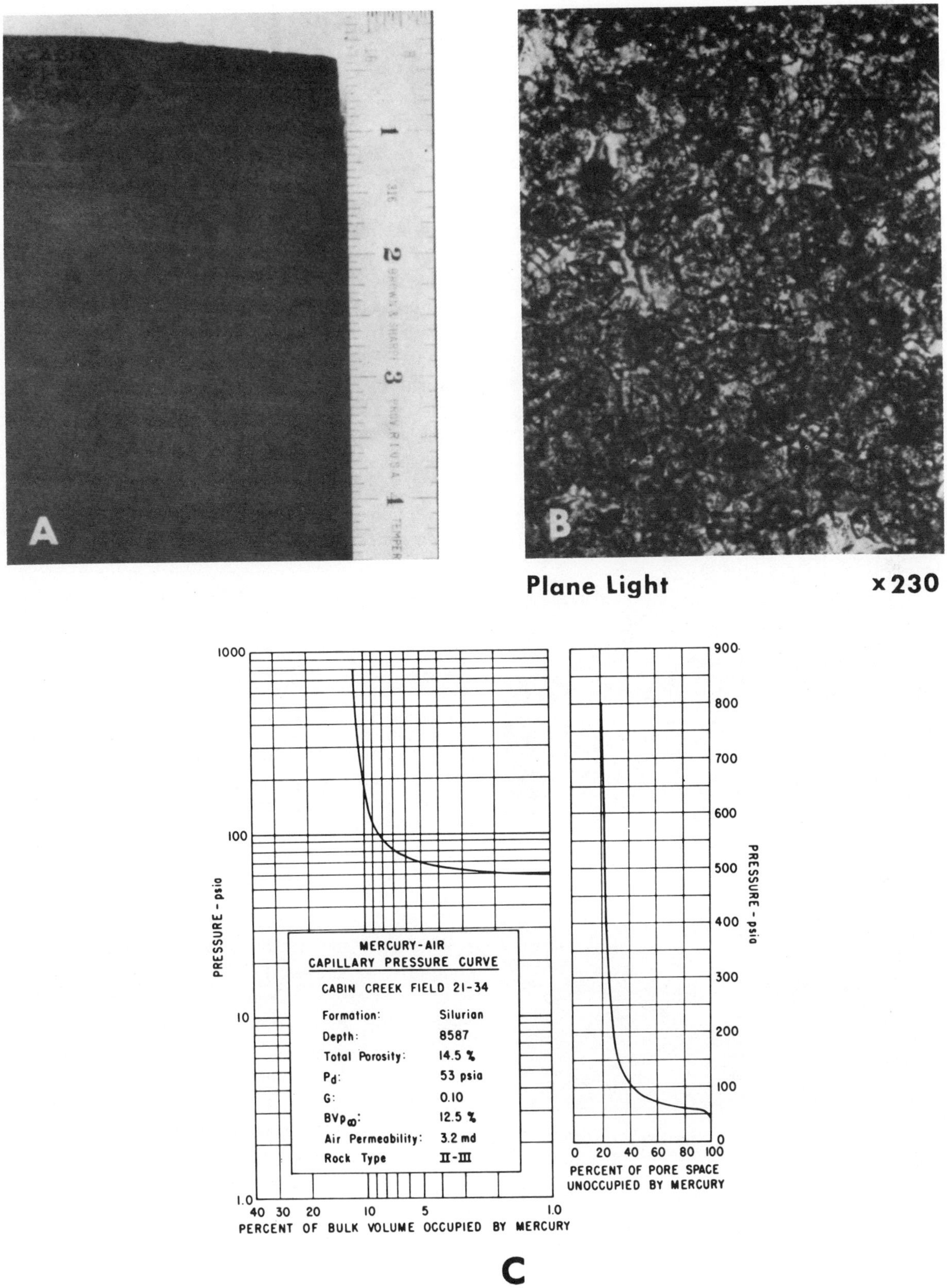

FIG. 21A.—Laminated dolosiltite similar to Figure 18B and Figure 20 except it is uniformly micro-saccharoidal and is oil-stained.

FIG. 21B.—Photomicrograph of rock in Figure 21 showing euhedral to subhedral; 15-35-micron dolomite crystals which have incomplete mutual contacts.

FIG. 21C.—Capillary-pressure curve of laminated dolosiltite using Thomeer notation. Shape factor (G = 0.1) and displacement pressure (P_d = 53 pisa) indicate superiority of this rock compared with Figure 20 rock. Pores are larger and better interconnected.

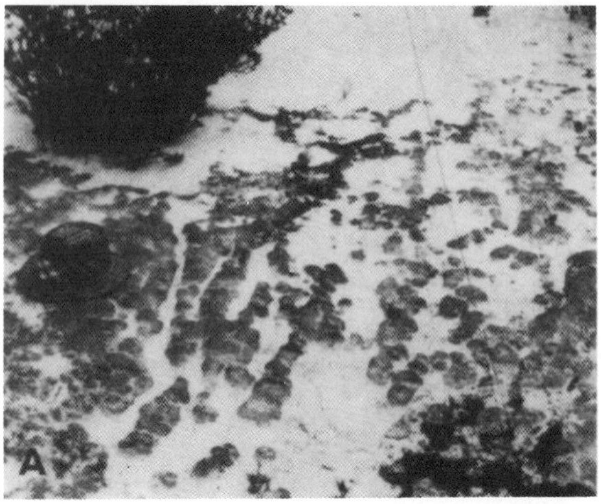

FIG. 22A.—Linear algal stromatolites in supratidal zone of western Andros Island. Depressions are floored with "lime mud."

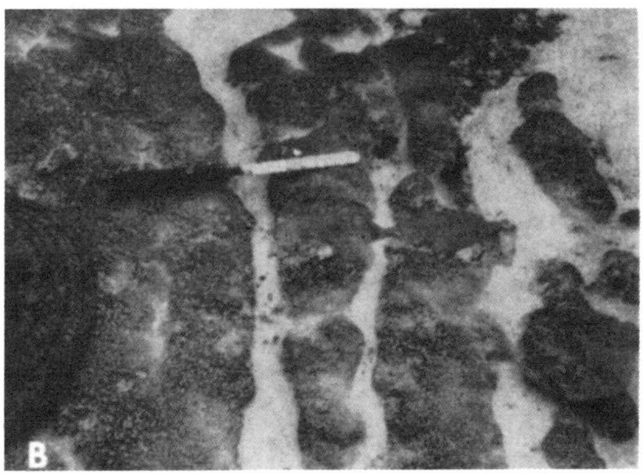

FIG. 22B.—Close view of Figure 22A.

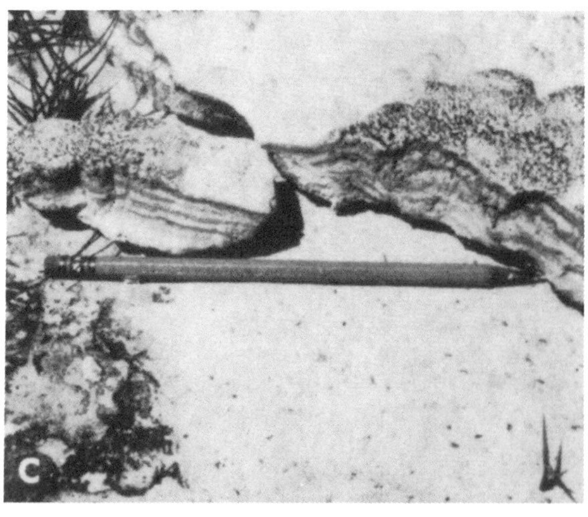

FIG. 22C.—Algal mat and LLH stromatolite from supratidal zone, Little Loggerhead Creek, western Andros Island.

within each rock type differs with the result that packing is rather close, thus masking invisible intercrystalline porosity. In either rock, but especially Type II–III, the packing of crystals is less close where crystal sizes are more uniform and where euhedralism is more complete. The larger pores commonly show horizontal linearity, a property derived from their original deposition and a condition which probably also prevails in the invisible intercrystalline porosity zones.

Algal mats and stromatolites.—One of the more important fabric types in the ancient record is the filamentous blue-green algal mat and stromatolite. Beginning with Black's work in 1933 and continuing with that of Young (1935), Ginsburg *et al.* (1954), Ginsburg (1957, 1960), Rezak (1957), Logan (1961), Logan *et al.* ((1964), and many more recent workers, an understanding is rapidly being acquired of the sediment-binding and fabric-contributing activities of filamentous and unicellular algae at Andros Island and other sites. On Andros the filamentous blue-green algae form both linear stromatolitic (Figs. 22A, B) and wavy continuous mat (Fig. 22C). Such algae are important for at least three reasons:

1. They hold vast areas of fine sediment, both along the shore zone and, most important, large interior areas of restricted circulation (Figs. 10B, 22).

2. They protect primary porosity from subsequent destruction by firmly binding the incoherent material (Fig. 23A).

3. They may themselves contribute to porosity by the subsequent decay of their organic constituents (Fig. 23B).

The algal mat-stromatolite of Figure 23C is composed of silt-size pellets and biogenic debris. Because of external factors, such as variable sediment influx, desiccation, drainage linearity, and swelling and bulging of the mat, such fabrics may assume complex geometric attitudes (Figs. 24A, B). The algae are capable of conforming with any previous surface configuration, even vertical surfaces, as shown in Figure 24C. These complex surfaces may be easily perpetuated once started. Where they persist through time and are buried, they show columnal, bifurcating, and more complex forms commonly described in the older literature as "form-genera" (Figs. 25A–C).

Figure 25A is a very common solitary growth form from the Interlake Formation known as "gymnosolen" in other occurrences. It is obviously built up by successive layers of algal mat composed of a single raised hemispheroid which persisted with deposition. Presumably it is connected with others like it (Figs. 25B, C) and would be described geometrically as LLH-S (Logan *et al.*, 1964). Depositionally it is possible to observe the close similarity between the dense, non-oil-stained layers of the Interlake sample lying in trough positions, and the nearly identical Recent Andros example (shown filled with very fine "lime mud") in Figures 22A and B. This mud periodically settles out in the quiet, low-energy, ebb-flow waters and apparently inhibits algal growth.

As time passes these stromatolites bifurcate and coalesce both vertically (Fig. 25B) and laterally (Fig. 22A). During dry periods they generally desiccate, exfoliate, and crack open into the well-known polygons and other complex geometric forms. These become re-established during flood tides unless deeply buried beneath a storm layer which serves as a phototropic barrier. Where

FIG. 23A.—Algal filaments binding silt-size carbonate pellets from supratidal zone as shown by Figure 23B.

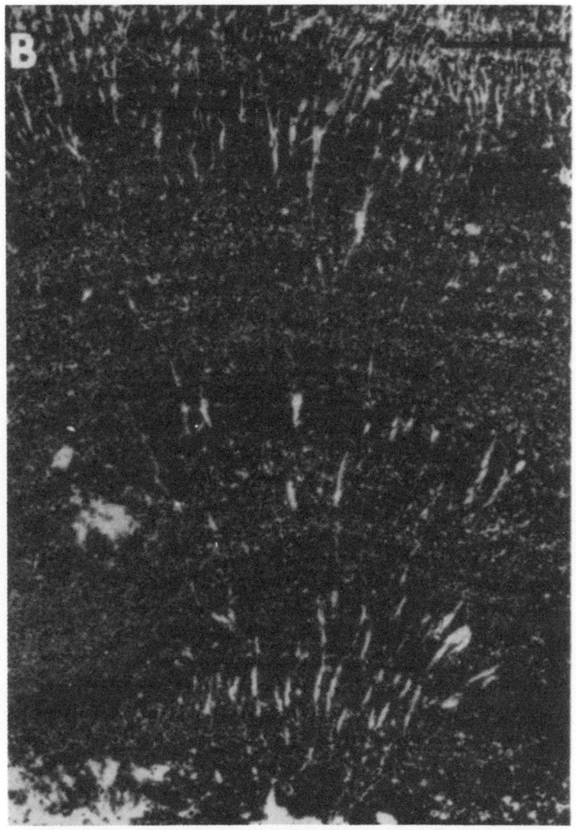

FIG. 23B.—Laminated array of calcareous pellets in Figure 23A transected vertically by radiating filament traces. Filaments were dissolved out to emphasize their control on "antigravity" nature of fabric and their potential contribution to porosity. Transmitted plane-light photograph.

uniform drainage direction (ebb flow) exists, a linear habit approximately perpendicular to the shoreline may dominate the algal growth form for several square yards. This condition is common along the open shorelines of western Andros and Williams Islands (Figs. 22A, B). In the Interlake Formation reservoir this feature is commonly preserved (Fig. 25C). The linear stromatolite shown in the figure extends through 4 ft of core. Because of extremely slow depositional rates, this sequence represents an extraordinarily uniform period of directed flood and ebb flow and hemispheroidal organic-growth succession.

In examples from the Interlake Formation attention is drawn to the localization of oil staining, and thus better porosity development, which for the most part is within the succession of hemispheroidal "crests." Although dolomitization has had some influence here, the primary difference in void structure between crests and

FIG. 23C.—Interlake Formation algal stromatolite showing filament traces which are now vertically enlongated pores. Sample loc.: Shell Northern Pacific, well 12–3, depth 8,940 ft, T. 10 N., R. 58 E.

troughs is one of initial variation in entrapment rate, conversion of organic matter to porosity, and the dominance of carbonate mud in the "troughs" where no algae existed to entrap larger size sediment during tidal floods. The facts that algal filaments have held and preserved initial porosity in the loose sediment of the Silurian stromatolites and have themselves contributed importantly to this porosity are well illustrated by Figure 23C.

The succession and lateral interplay of the various fabrics commonly are seen on a microfacies scale. These are represented by the compound fabrics of Figures 26 and 27. In Figure 26, a

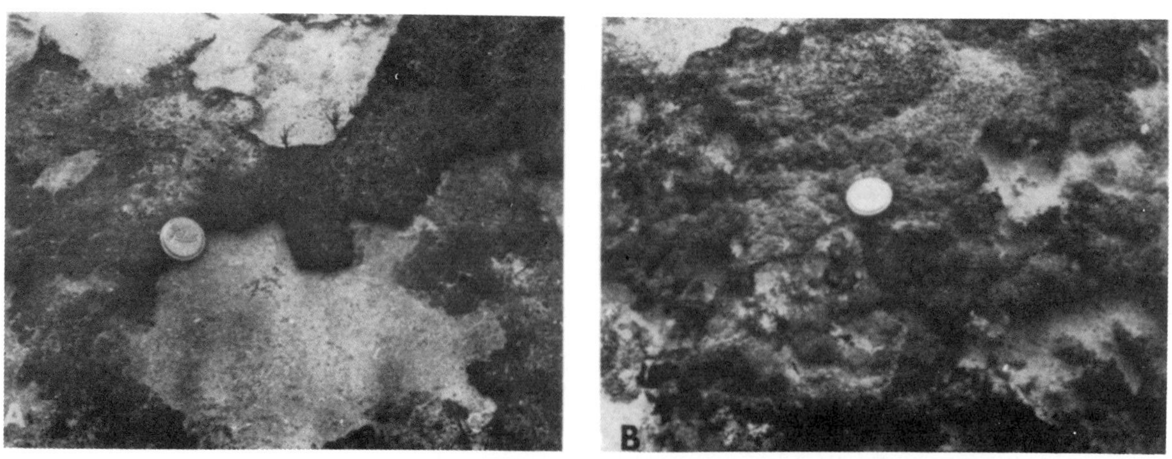

FIGS. 24A, B.—Irregular surfaces bound by algal mat near Little Loggerhead Creek, western Andros Island. Many surfaces are initially irregular due to erosion and infilling. Continuing algal growth could then develop complex stromatolitic intergrowths which are seen in stratigraphic record. Lens cap for scale.

FIG. 24C.—Leathery algal mat covering vertical surface left of lens cap. Large holes are animal burrows. Loc.: Little Loggerhead Creek, western Andros Island.

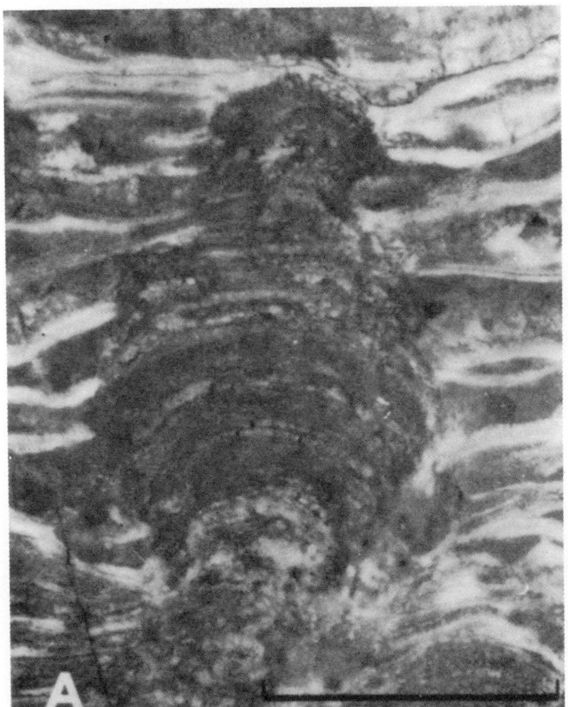

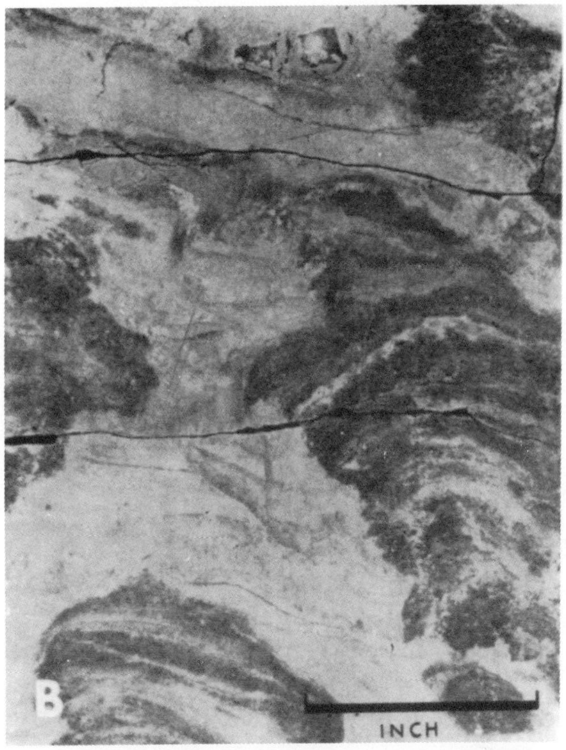

FIG. 25.—Stromatolites from Interlake Formation showing fabrics determined by differential growth habits resembling form-genera. Porosity and oil-staining predominant in hemispheroidal "crests." (A) Solitary stromatolitic growth form resembling "gymnosolen." Sample loc.: Cabin Creek field, well 21–34, depth 8,734 ft. (B) Example of branching and coalescing growth habit. Dense, light areas are interpreted as troughs filled with "lime mud" similar to Figure 22B. Sample loc.: Shell Northern Pacific 12–3, depth 8,940 ft, T. 10 N., R. 58 E. (C) Linear growth fabric which extends more than 4 ft in core. Troughs are dense. Compare linearity with Figure 22B. Sample loc.: Pine field, well 34–25A, depth 8,708 ft.

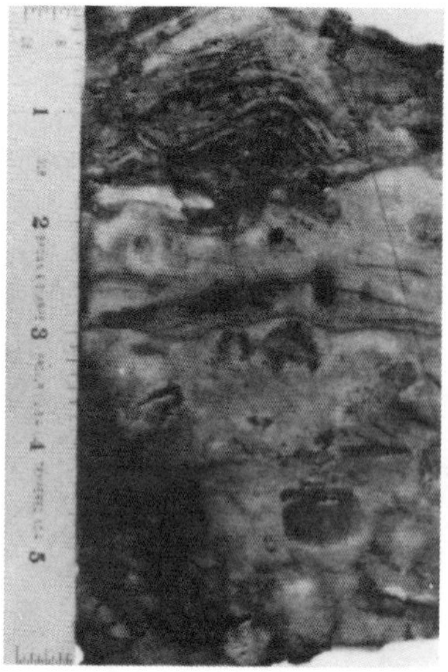

FIG. 26.—Combination of algal stromatolite succeeding burrowed fabric in upper intertidal or transitional supratidal zone. Note porous oil-bearing development in burrows and hemispheroidal crests. Such dual fabric successions aid environmental analysis and reservoir evaluation of tidal deposits. Sample loc.: Pine field, well 24–15A, depth 8,560 ft.

stromatolitic buildup has succeeded a burrowed fabric. Both types have reservoir properties, though unpredictable even on the scale of several inches. Such a relation necessarily leads to complex pore-size distribution. The example of Figure 27 shows small lithoclasts, flat pebbles, and other bioclastic particles bearing secondary overgrowths. These are loosely packed with little or no finer matrix sediment and provide a primary microvugular porosity which has survived dolomitization. The capping fabric is a dense algal "crust" which abruptly terminates but also protects the porous phase. Although the lower, clastic fabric in the figure probably differs from place to place, it is probable that the algal crusts persist laterally for several yards to several hundred feet. Both examples demonstrate that complex fabric interrelations can be used to define the intermixing of microfacies within a larger depositional environment.

Flat-pebble breccias.—Another fabric diagnostic of the supratidal zone is the flat-pebble breccia or conglomerate. This fabric is very com-

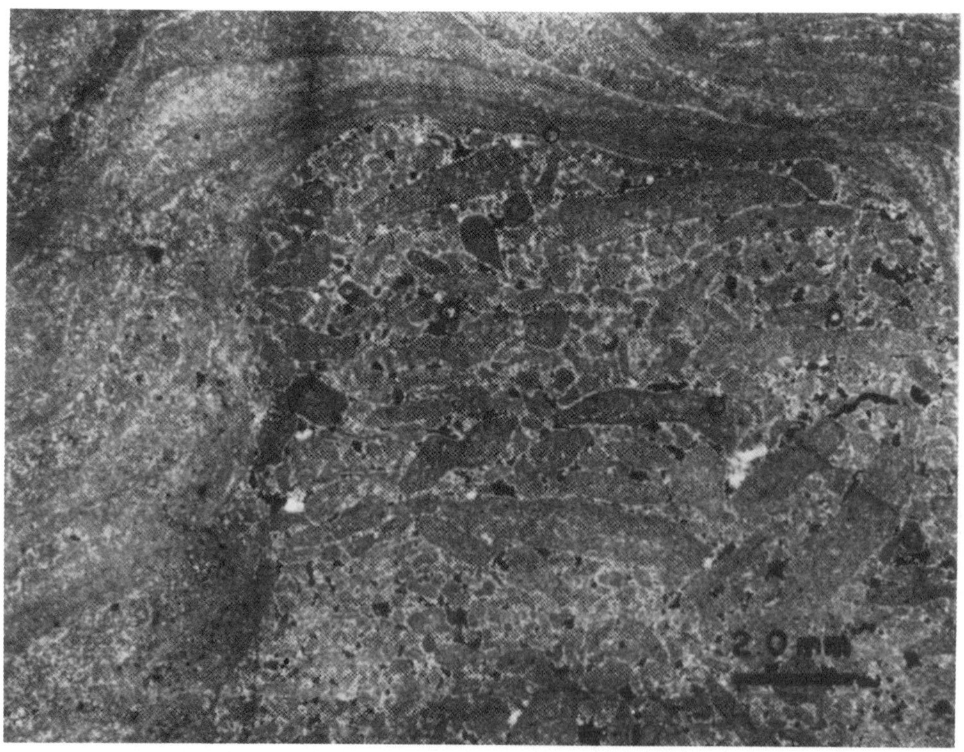

FIG. 27.—Photomicrograph of algal mat-stromatolitic crust which binds small mound of flat chip and other coarse detritus. Note interparticle porosity (dark) preserved within druse-coated loose detritus. Petrophysical values are: $\emptyset = 8.9$, $G = 0.7$, and $P_d = 20$ psia. Sample loc.: Interlake Formation, Cabin Creek field, well 21-34, depth 8,689 ft.

mon in the sedimentary record and forms some interesting reservoir rock. Flat-pebble breccias are commonly associated with the other supratidal deposits, particularly the algal mats and stromatolites (Fig. 31C), because many of them are composed primarily of small slabs and chips of such mats (Fig. 28). These blue-green algal surfaces show several stages of induration and desiccation in the Bahamas depending on frequency of flooding (duration of subaerial exposure), abrasion or profile erosion, and the thickness (organic richness) of the mat. In Figure 28 the discolored mat surface is an indurated crust which is hard and is gradually being broken into large chips. The underlying laminated "lime mud" is weakly consolidated and weathers back to its primary-size sediment for redeposition with the next flood.

Brecciation also can result from the forceful growth of mangrove pneumatophores (aerating roots) which penetrate the overlying laminated calcilutite and break up the resistant layers. On Williams Island these become strewn about the surface (Fig. 29) and form part of an abrupt facies change on the supratidal surface which reflects sharply contrasting sediment-load capacities of adjacent flood waters moving through the mangrove surface growth and between low-profile erosion remnants of earlier sediment. The smaller chips (Fig. 30) on the Andros supratidal surface are concentrated and probably have a diverse ori-

FIG. 29.—Flat-pebble breccia in supratidal sediment of Williams Island, Bahamas. Note abrupt local transitions to finer chip sediment facies (compare with Fig. 31) and low-profile erosion remnant on right.

FIG. 28.—Contemporary fragmentation of calcilutite layer in supratidal zone at Little Loggerhead Creek, western Andros Island. Surface is discolored and well indurated whereas underlayer is incompletely lithified. Chips of this material may provide sediment for flanking deposits and resemble flat-pebble breccia in older rocks.

FIG. 30.—Smaller chips of laminated calcisiltite from Little Loggerhead Creek area, western Andros Island. Similar material retains good interparticle porosity in Ordovician and Silurian supratidal fabrics, especially where bound by filamentous algal mat (see Fig. 27).

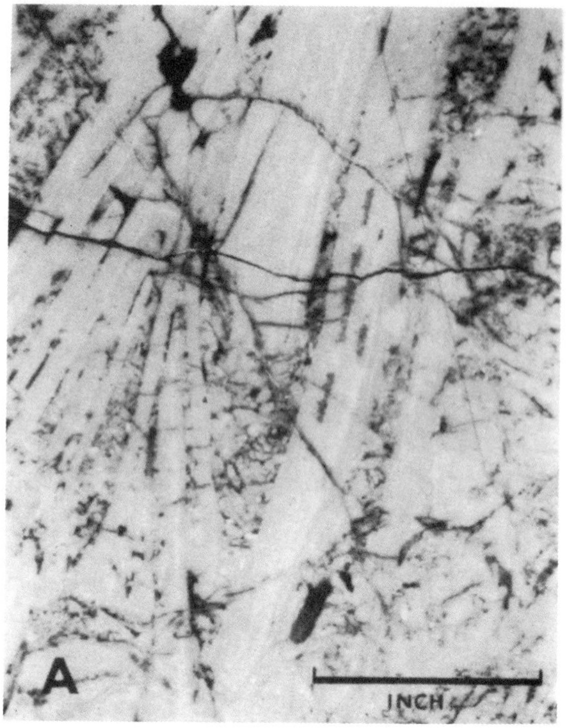

FIG. 31A.—Flat-pebble breccia with interparticle porosity similar to Figure 31B except for greater prevalence of dense larger chips. Note high angle of repose of longest chips. See also Figure 47. Sample loc.: Interlake Formation, Cabin Creek field, well 21–34, depth 8,689 ft.

gin, in part bioclastic. These resemble the Interlake forms protected by algal laminae (Fig. 27).

The paucity of surface vegetation (except algae) and the widespread effects of desiccation in the early Paleozoic suggests that flat "pebbles" may have been even more common than on Andros Island. The Interlake Formation is replete with dolomitized flat-pebble-breccia deposits, most of which retain fabrics that required at least a brief storm or local high-energy tidal-flow regime for their formation. From a reservoir viewpoint, the most important characteristics of a flat-pebble breccia are the porosity and continuity of the host-matrix rock. Although the presence of the flat pebbles implies subaerial exposure in the supratidal zone, subsequent transport ultimately may cause their relocation in other environments. In the Stony Mountain and Interlake Formations, the local abundance and size, together with subjacent and superjacent fabrics, suggest, however, that in most places in this area they were redeposited within the supratidal zone. The flat slabs shown in Figure 31A are quite dense. The steep angle of repose should be noted, as well as the indurated, wafer-like nature of the individual pieces. There is no doubt that these fragments are brecciation products of an early, diagenetically indurated layer of fine-grained algal mat. The underlying beds represent a highly leached solution breccia on an old diagenetic terrane (Fig. 47), and the overlying beds are flat layers of algal mat. The writer's interpretation is that these fragments were dumped suddenly by a high-energy event together with a somewhat sorted host matrix comprised of large, loose pellets and particles. Consequently, this is an important potential reservoir rock.

A second breccia, similar to the previous example, and probably also a higher energy event. is shown in Figure 31B. Here likewise a single depositional process has provided a graded bed of large and small breccia fragments. These fragments similarly are dense and have been derived from algal stromatolites as shown by Figure 31C. In each example, the fragments are suspended in a more porous oil-saturated matrix than the first example (Fig. 31A).

In contrast to the three examples with good

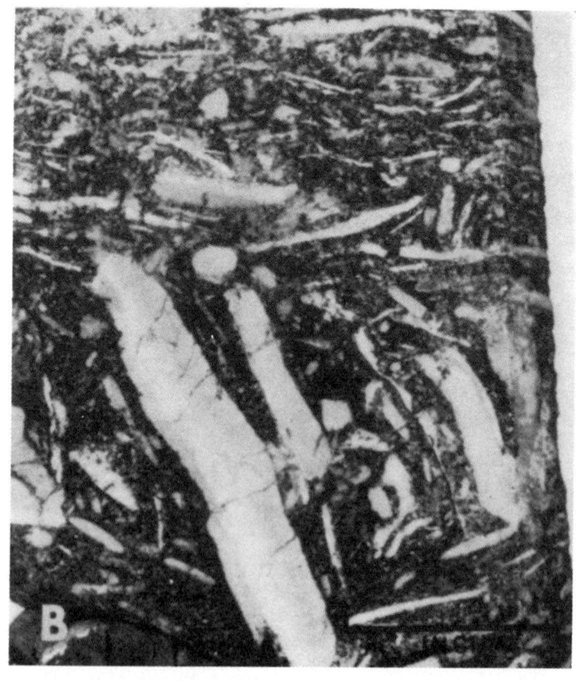

FIG. 31B.—Dense, unsorted, flat-pebble breccia in finer grained, heavily oil-stained matrix. Crude size grading of chips and their orientation suggest this deposit was single, brief, high-energy, supratidal event. Sample loc.: Interlake Formation, Pine field, well 34–25A, depth 8,711 ft.

matrix porosity is the flat-pebble breccia of Figure 31D. Although the size and abundance of chips do not differ much from the other examples, this example may not represent a high-energy event for three reasons. First, the larger chips or slabs are unsorted or ungraded. Second, these large pieces exhibit horizontality, which implies a lack of high-energy turbulence. Third, the original "lime mud" is preserved as a matrix. Consequently, these large pieces probably were floated in by quiet water, being easily moved because of their large surface-to-volume ratio, low bulk density (due to earlier internal porosity development), and convenient shape for flotation. The internal porosity of chips and slabs is retained, revealing in one example an oil-stained graded bed within the largest slab (it is a discontinuous slab in the third dimension). Because these slabs and chips were floated and admixed with a dense mud matrix, the reservoir properties of this rock are doubtful.

FIG. 31D.—Porous, oil-bearing algal mat chips suspended in dense matrix. In contrast to Figures 31B and 31C, this porosity is sharply localized. Note internal graded character of upper large chip. Sample loc.: Cabin Creek field, well 32–7, depth 8,447 ft.

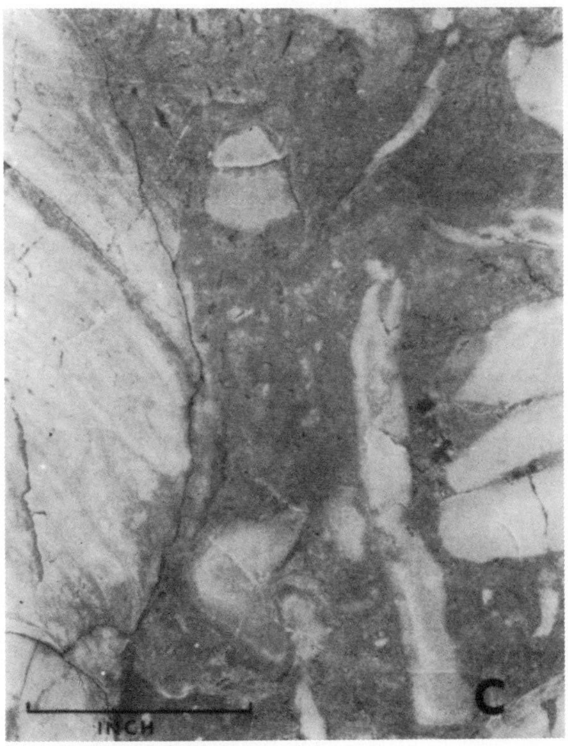

FIG. 31C.—Flat pebbles suspended in finer, porous matrix associated with large algal stromatolite (left). Occurrence of flat-pebble breccias and stromatolites together is common because stromatolites commonly supply material for breccias. Sample loc.: Interlake Formation, Pine field, well 34–25A; depth 8,712 ft.

The initial energy state of the depositing waters seems to be important in that it determines the nature of the matrix and therefore the reservoir quality of the total deposit. Thus, the *edgewise breccia* of Figure 31A is supported by a higher energy calcarenite matrix. The *graded breccia* of Figure 31B is suspended in a somewhat lower energy matrix of calcisiltite particles. In both examples the dolomitization has not altered the primary grain size. However, this may not be true of other ancient examples. Finally the *float breccia* shown in Figure 31D is composed of suspended, unsorted slabs and chips in a fine-grained, nonporous dololutite. The first two are good reservoir rocks but the last is not. Thus, the depositional geometry of the chips and slabs, and the matrix grain-sizes, not only provide information about the initial energy of deposition but also predetermine the potential reservoir prospects of the breccias.

Channels and gullies.—Channel and gully deposits, especially those which breach all three

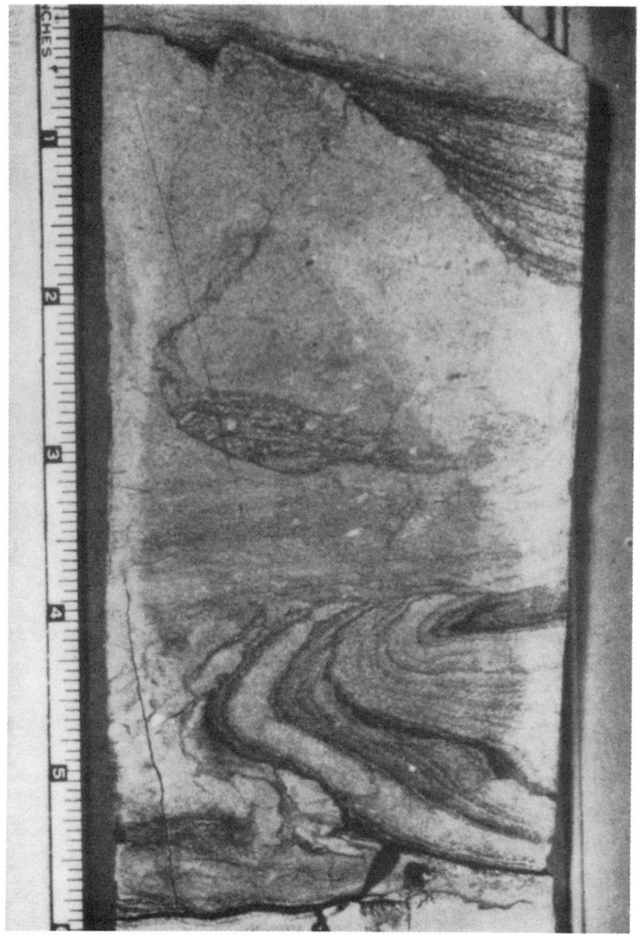

FIG. 32.—Slump, and cut-and-fill structures of probable channel or gully origin in Stony Mountain Formation. Quantitative importance of this fabric type cannot be ascertained. Rare occurrence of this type suggests that there are fewer such deposits compared with Recent sediments because of climatic or other factors affecting preservation in geologic record. Sample loc.: Pine field, well 21–10A, depth 8,957 ft.

environments, are potentially important because sorted and perhaps cross-bedded skeletal and lithoclastic particles may be preserved and would make good reservoir rock. Even if this were not true, such deposits would be important because their occurrence would probably indicate the termination of many other porous facies, notably those in the supratidal environment. At present, evidence is insufficient to predict whether or not such deposits would be widely preserved. If they are preserved, as is seen in a few cores of the Interlake Formation (Fig. 32), their volume and importance depend on climate and regional geography, both of which probably are unpredictable and different for each ancient sedimentary locale.

In arid climates such as the southwestern coast of the Persian Gulf, the tidal channels, although in some places several miles long, seem to be confined generally to the intertidal zone below the level of the *sebkha* (Illing *et al.*, 1965; Evans *et al.*, 1964). Inspection of their air-photograph mosaics reveals only a few short tributaries to the channels, little or no meandering courses, an apparent shallow subsurface drainage shoreward of the channel, and gully heads which reach far inland to surface wadis and saline playas. In the lower Paleozoic, a similar general aridity may explain fewer recognizable channel deposits associated with intertidal and supratidal zones. In the extreme case, channels confined to the intertidal zone may be reworked by the tide and burrowing organisms, thus eliminating a clear record of their existence.

In contrast, the Bahamas are influenced by a

FIG. 33A.—Position of imminent gully (small channel) capture on migrating gully loops, western Andros Island.

wet climate with many large tidal channels and supratidal water courses which migrate laterally (Figs. 10A, 18A). This migration results from persistent water movement generated by both storms and tidal flux. As a consequence meander loops, cut-off meanders, and capture are common (Fig. 33A). This appears to be related to the ease of bank erosion as much as the movement of water down the channel and scour by the sediment load. Many features of the intertidal and supratidal deposits also apply to channel and levee deposition so that, in the study of the ancient equivalents, the sedimentary fabrics probably are insufficient, and detailed stratigraphic study through short intervals is required to determine the preservation and extent of channel cutouts, levee construction, and channel fill.

Figure 33B shows a cut-bank levee surface on the western part of Andros Island which stands 1–2 ft above normal high tide. These levees usually are not wider than a few tens of yards (Fig. 10A). They are commonly bare, especially after a new surface of fresh mud has been put down. In these cases, the surface may be perfectly smooth or show a kind of striation resulting from sheet flow "shadows." A depression shadow develops leeward from stems and pneumatophores of levee vegetation, and ridges rise behind short obstructions such as pebble lithoclasts. Older surfaces, and especially abandoned levees, are veneered with dolomitic crusts (Shinn *et al.*, 1965). The example of Figure 33C shows such a surface and, in contrast to the previous examples cited where black mangrove pneumatophores caused loosening of the surface crusts, here the areas between the pneumatophores have been stripped of the latest surface.

The levee flood layers are, in fact, no different than the broad intervening low supratidal flats

FIG. 33B.—Cut-bank levee surface of gully in supratidal zone of western Andros Island. Surface is characterized by build-ups and depressions of "lime mud" leeward of plant stems and pneumatophores. These form during sheet flow in "shadow" of obstructions. Flow direction toward right.

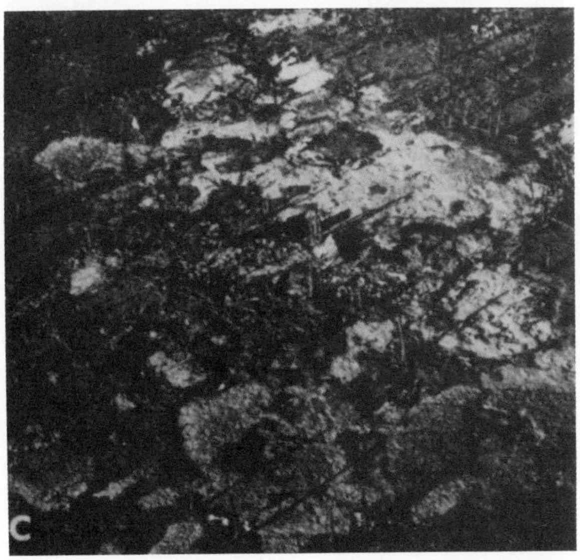

FIG. 33C.—Levee surface on cut-bank side of gully in supratidal zone of western Andros Island. Dark layer contains Recent dolomite (Shinn *et al.*, 1965). Note that surface fragmentation and removal have occurred in spots free of mangrove pneumatophores.

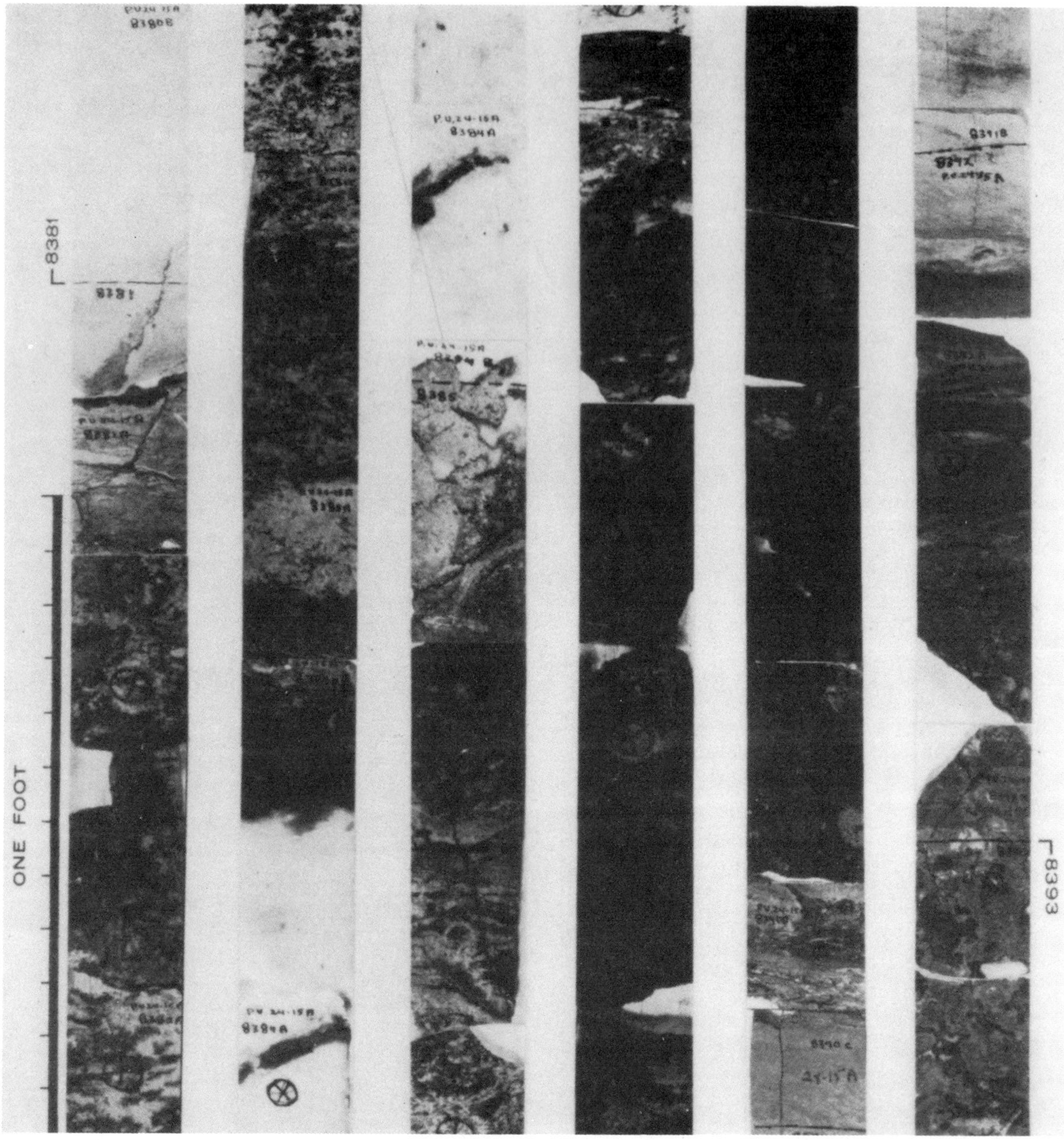

FIG. 34.—Photograph of slabbed core from near top of Interlake Formation of Pine field, well 24-15A. Core interval is continuous from left to right and covers about 13 ft (4.3m). One-ft white, coarsely crystalline, anhydrite layer occurs at base of left row 2 and top of left row 3 at depth of 8,384 ft. Lithologic associations and interbedding with dolomitized supratidal fabrics necessitate similar supratidal origin for anhydrite.

except that perhaps they are thicker and show less algal growth layering. Otherwise, if preserved, they may be expected to show little difference in the ancient record from almost any other laminated deposit (Figs. 16B; 18B, right; 20; 22).

Dolomite.—A widespread association of gypsum, anhydrite, and dolomite has been established recently in modern supratidal studies (Wells, 1962; Curtis *et al.*, 1963; Illing *et al.*, 1965; Shinn *et al.*, 1965; Kinsman, 1965). Anhydrite-dolomite phases also have been described from the cycles of Ordovician-Silurian deposits of the Williston basin (Porter and Fuller, 1959). The present evidence that anhydrite beds occur as lateral and intermixed equivalents of supratidal facies (Figs. 34, 47) across the western metastable lineament of the Williston basin suggests that dolomitization began and was completed during periods of platform emergence. This implies an early diagenetic development for Stony Mountain and Interlake dolomites. However, it does not reveal whether dolomitization was syngenetic or epigenetic.

The entire Stony Mountain and Interlake sequence in the western Williston basin is dolomite. Because modern primary dolomite is not yet discovered and defined, and because the modern depositional environment is replete with nondolomitic carbonates, there is no reason to expect that the many organic fabrics described here were composed originally of anything but aragonite and calcite. Moreover, there is a segregation of dolomite crystal sizes and euhedralism in many of the burrowed rocks, the smaller anhedral crystals and clay being confined to the interburrow areas (Fig. 15). Certainly the animals did not (and do not today) process sediment of loose dolomite crystals either for shape or size. These kinds of observation and reasoning are insufficient to eliminate the possibility that primary dolomite deposition was important within the sequence. A possible solution to this problem is to build a case for epigenetic dolomitization based on some facts and circumstantial evidence related to modern processes.

Normal sea water is deficient in the Mg ion concentration necessary for dolomitization. Many equivalent pore volumes of water are therefore required to convert calcite sediments to dolomite, or pore water will have to be enriched considerably to make the process more efficient. The first process requires paleohydrodynamic flow and the second requires paleoclimatic aridity. In the first process a good and persistent gradient and/or brine gravity instability is necessary. In the second, a high evaporation rate within the surface sediments and some periodic replenishment by hypersaline waters are necessary.

The existence of the evaporitic cycles of the Williston basin already suggests that there were long periods of widespread aridity and an impoverishment of Ca with respect to Mg in the waters through deposition of $CaSO_4$. The dependence of dolomitization on removal of calcium as sulfate should mean that at least a sheath of dolomite should be present below and above the evaporite layers where deposited throughout the basin above the older limestones. Such sheaths are observed in association with the "G" and "S" anhydrites of the Stony Mountain Formation cycles (Porter and Fuller, 1959). The deeper limestones probably escaped dolomitization because they were laterally remote from any possible refluxing gravity head and were too far beneath the periodic surface brines trapped in very small intergranular capillaries under the control of climate and tidal rhythm.

In the Silurian, however, the general lack of limestone requires that a similar mechanism be continually in operation. Because the evaporative-capillary mechanism operates at or near the surface, it must depend on supratidal processes. Such processes became predominant during the time of Interlake deposition and are reflected in the extremely low rate of preservable deposition.

Harland *et al.* (1964) give two useful radioactive dates: (1) base of Late Ordovician $= 445 \pm 10 \times 10^6$ yr; and (Ludlow) Late Silurian $= 400 \pm 10 \times 10^6$ yr. Therefore the maximum and minimum time intervals and their respective rates of preservable deposition for a selected maximum of 2,500 ft of combined Stony Mountain and Interlake carbonates would be:

Time interval		*Time necessary to deposit 1 ft of preservable sediment*
Maximum	65×10^6 yr	26,000 yr
Minimum	25×10^6 yr	10,000 yr

The chosen thickness of 2,500 ft is *twice* the observed thickness measured beneath the Siluro-Devonian unconformity in the central part of the basin where the thickest section is preserved. Nevertheless, even this assumed liberal rate of preservable deposition indicates that nondeposition or reworking prevailed most of the time. Such sedimentary conditions would most closely resemble the daily regimen of the intertidal and especially the supratidal zones as they are observed today. The alternative would be to assume sudden removals of large deposits by erosion. Except for exposures already noted, there is no evidence for such removal.

Pray and Choquette (1966) have measured modern carbonate-mud porosity values in excess of 50 percent. Supratidal muds, although variable, commonly exceed 30 percent. Undistorted preservation of the many diverse fine-grained fabrics reviewed here suggests that compaction is not important in the reduction of pore space, and early epigenetic dolomitization would seem to favor retention of initial high porosity values. Yet the Interlake, Stony Mountain, and most other ancient dolomitized "lime muds" have very low porosity values, usually less than 20 percent. Pray and Choquette (1966) favor early calcite cementation, a phenomenon which is certainly common. On the other hand, many fine-grained rocks show no positive evidence for either widespread dolomitization of such cement or for leaching of the cement after dolomitization of a primary matrix as suggested by Murray (1960).

Kinsman (1965) and Illing *et al.* (1965) record extensive dolomitization along the Trucial Coast and Qatar Peninsula, respectively, occurring in unlithified sediment in the supratidal and subjacent intertidal offlap deposits, apparently by preferential replacement of aragonite. Illing *et al.* also record leaching of calcite particles in the same sediment. There is no evidence, however, that any earlier calcite cement is being leached. Extrapolation of these processes to the ancient deposits would favor (1) a process of dolomite-rhomb nucleation governed by aragonite occurrence and (2) penecontemporary leaching of calcite grains, perhaps for use of Ca in additional dolomite-rhomb growth. Simple replacement, however, is insufficient to explain the larger crystal sizes of dolomite occurring in the ancient carbonate mudstones.

Although not completely satisfactory, the most reasonable explanation at this time for dolomitization is that it takes place in conjunction with precipitation of anhydrite in the supratidal zone. The dolomitic replacement process probably was induced and sustained by the combined action of interstitial evaporation and capillary-brine replenishment.

Anhydrite.—Preserved in the center of the Williston basin are a series of cycle-culminating evaporites, some several feet thick. Along the western flank of the basin the lateral stratigraphic equivalents are argillaceous dolomite beds with higher than normal anhydrite content. This content usually is less than 30 percent (Fig. 35D) except for clear, white, massive occurrences that are a few inches to 2 ft thick. The massive anhydrite is everywhere associated with subjacent scattered nodules of similar anhydrite and enclosed stratigraphically in disturbed or laminated supratidal facies. Figure 34 shows such an example occurring near the top of the Interlake Formation. It should be noted in this figure that, in left row 2 at the bottom, and in left row 3 at the top, a 1-ft massive, slightly argillaceous anhydrite is present. It is interbedded with supratidal dolomite. At its base is a disturbed transition layer characteristic of nodular growth in modern supratidal deposits (compare with Fig. 5 of Kinsman, 1965, p. 310). The same type of relationship occurs in Figure 47 (well 24–15A, Pine field) where the ancient surface deposits, composed of anhydrites and solution diagenetic fabrics, abruptly give way to renewed algal-mat formation (at reference level 1, right-hand column of Fig. 47).

Is this ancient cyclic anhydrite, even though marginally discontinuous at times, the result of hypersaline precipitation in a large barred basin or in a series of shallow saline playas interconnected by wadis and tidal channels across a supratidal (*sebkha*) evaporite plain? All the anhydrite cores which the writer has examined from the underlying Red River, Stony Mountain, and Interlake Formations, and those figured in the literature, have been crudely interlayered in nodular form with siltstone or claystone, or interlaminated with deposits thought to be of supratidal origin. The formation of such deposits probably is similar to that described by Kerr and Thomson (1963); *i.e.*, the known occurrences of evaporites in the Ordovician and Silurian of the Williston basin cyclic sequences can be explained as more or less densely packed anhydrite masses which grew in a supratidal location as small, large, and even coalesced crystals and thin beds (Kinsman, 1965). The facts that these beds may be fairly continuous and are a few feet thick do not preclude a supratidal saline-playa origin. Throughout the countries of the Near East the strong desert *shamal* winds deflate much of the supratidal surface, including areas remote from the current strandline (Illing *et al.*, 1965; Evans *et al.*, 1964). These depressions are salt encrusted, and are studded above and below the surface with large gypsum and/or anhydrite crystals and rosettes. The salts commonly are predominant, as in the large Sahara desert playas where as much as 12 ft of sulfate evaporite is present today.

SUBAERIAL DIAGENETIC TERRANE

Toward the end of Silurian time, protracted subaerial exposure ensued (Fig. 43). This was part of much wider emergent conditions which affected the continent at that time and which persisted into the Devonian. Because of their elevation, large parts of this terrane zone are behind and above the supratidal zone, although many of the sedimentational and diagenetic features pass imperceptibly from one zone to the other. Minor changes in sea level would convert one zone immediately into the other because of the expected low-profile characteristics of epeiric shorelines and supratidal interiors. In a modern example, interior Andros Island rises only a few feet above the supratidal zone except for the Pleistocene rock exposures. Also, Illing *et al.* (1965) report that the Qatar Peninsula is, with few exceptions, elevated less than 50 ft.

The writer believes that many of the earlier Stony Mountain and Interlake cycles also reflect brief episodes of terrane exposure which resulted in significant diagenetic alteration. Within such a terrane, saline waters would occur under both subterranean tidal and capillary influence. In addition, vadose waters of highly variable salinity would occur from place to place, depending on surface geomorphology, local rainfall, and the locations of saline playas and shallow wadis cut off from the sea. Sediments and older rocks in this environment would be subject to strong diagenetic alteration. In contrast to the previously described zones, diagenetic processes occurring at or just below the subaerial terrane usually cause significant modification of potential reservoir properties. These changes, however, generally are incomplete except where near-total erosion or leaching occurs.

The climatic factors which so strongly influence the supratidal processes are even more important for the subaerial terrane. Major synoptic changes affect all of the surface-water bodies in size and salinity, the behavior and vigor of vadose water movements, and the deflation, desiccation, and eolian infilling of dry plains and recesses. These processes can interact in complex ways. During long periods of time, such interaction invariably leads to the complete destruction of the rock. The common end product is a ragged karst surface supporting a poor soil and regolith profile. The distinctive red- and brown-colored sesquioxides of iron and alumina concentrated in these residues are observed to overlie all warm-climate carbonate exposures today, both recent and ancient. Subsurface profiles commonly show similar preservation (Fig. 37).

In general, and for discussion purposes, two kinds of diagenetic fabrics can be distinguished. *Epigenetic fabrics* are those which result from diagenesis in the shallow subsurface. They may be either early or late with respect to the time lapse since the sediments were originally deposited. Epigenesis in this case is regarded as encompassing supratidal dolomitization, on the one hand, and vadose alterations, on the other. The second type is the classic *karst fabric* which develops at the expense of the exposed carbonate surfaces through erosion and soil-forming processes. Karsting also is obviously related to epigenesis because it is an early phase of vadose alterations.

Epigenetic fabrics.—Four examples of epigenetic fabric complexity are shown on Figure 35. In Figure 35A the Interlake supratidal fabric is mottled with oil-staining and characterized by laminae of variable porosity. This normal supratidal fabric has been diagenetically altered as shown by a strong upper-surface desiccation, infilling, slight internal deformation, and minor alterations. The sample underlies about 5 in. of similar non-oil-stained, fine dolomite whose fractures are filled with anhydrite. The next several feet of rock overlying both samples is shown in Figure 47. The relation between the lower shaly unit and the figured sample is not clear, but it is probable that a long lapse in carbonate sedimentation occurred here while the area of deposition was subaerially exposed. The major reasons for believing this are: (1) the interval is "bracketed" above and below by supratidal fabrics; (2) the sequence is common to at least three locations separated by 10 mi; and (3) the event is correlatable throughout the western Williston basin. The writer believes that the diagenetic effects are characteristic of the arid conditions common in early Paleozoic sedimentary environments.

The example in Figure 35B displays internally crushed debris of admixed porous and less porous layers exemplified by variable oil staining. Commonly, after such an *in situ* adjustment, leaching

FIG. 35.—Solution-epigenetic breccias from inferred supratidal zone fabrics of the Interlake Formation. Original character of sediment obscured by early diagenetic alteration. Porosity, and thus oil-bearing capacity of rock, may be slightly enhanced (desiccation and solution-Fig. A; epigenetic fragmentation-Fig. B), substantially enhanced (solution alteration-Fig. C), or completely destroyed (solution, fragmentation, and resealing by anhydrite-Fig. D). Sample locs.: (A) Cabin Creek field, well 32–18, depth 8,528 ft; (B) Cabin Creek field, well 32–7, depth 8,410 ft; (C) Cabin Creek field, well 32–18, depth 8,518 ft; and (D) Pine field, well 34–25A, depth 8,467 ft.

occurs, as well as, perhaps, sealing with gypsum or anhydrite. Leaching results in the progressive obliteration of all of the previous fabrics, both depositional and diagenetic. This is clearly evident in the example of Figure 35C. It has been thoroughly shattered and subsequently frittered by leaching. The dolomitization process could be contributory here, but in view of the excellent state of associated fabric retention shown, it seems more likely that severe changes accompanying a long subaerial exposure are responsible, either prior to or contemporaneous with dolomitization.

If not healed with evaporite minerals, as are the crushed blocks of dolomite in Figure 35D, the severely leached rocks can be very good reservoirs. This is demonstrated by Figure 47. The interval just below the marginal number 1 shows that all three cores are composed of very porous solution breccia (SB), although that of Pine Unit well 24–15A is not oil-bearing. It should be noted that each of the breccias has been re-covered by algal mats (AM) and that the flat-pebble breccia of Figure 31A is between the solution breccia and the flat-algal mat in the Cabin Creek well, core 21–34. These flat pebbles must have been derived from flanking algal mats and dumped into a depression on an old solution-breccia surface.

In contrast to the other supratidal fabrics, these, although largely secondarily developed, may be widespread because the severity of forces leading to their formation may permit a similar conversion of many different primary fabrics.

Karst fabrics.—The importance of karst formation in the removal of carbonates and the modification of carbonate surfaces is well known. Such removal has been common throughout geological history and commonly the thin soil profiles developed on karst surfaces have been used as the only reliable correlation markers in many ancient carbonate sequences. Karst formation is in progress today on both ancient limestone exposures and very recently lithified supratidal deposits such as those on western Andros Island (Fig. 36).

The removal and/or displacement of carbonate rock is accomplished by a complex combination of forces and processes. In general, these are: (1) leaching by percolating vadose waters; (2) cavitation due to gravity collapse of carbonate blocks which have been excavated and undermined by erosion; and (3) organic processes which supply leaching solutions and growth pressure to the rock, while maintaining and augment-

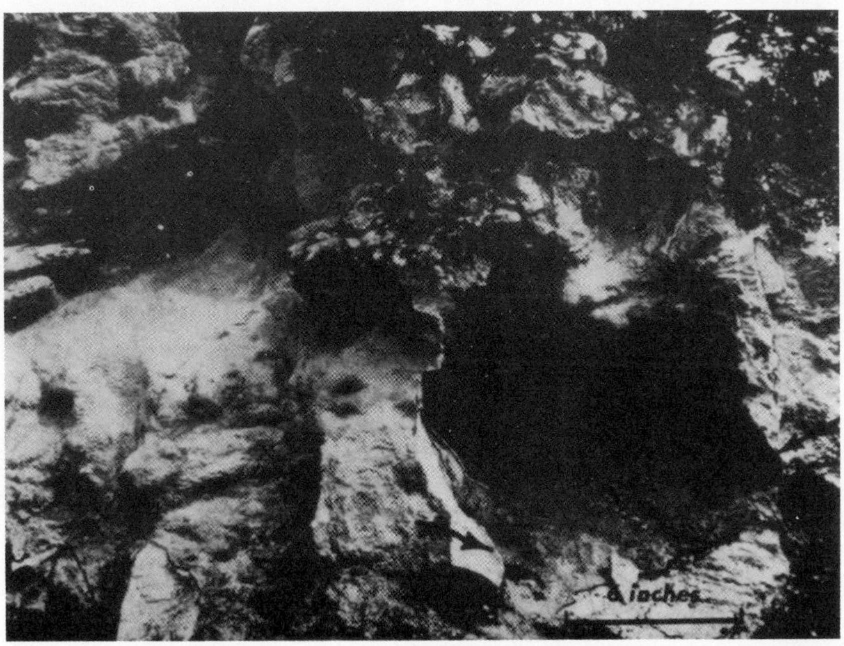

Fig. 36.—Craggy, corroded, cliff-face rock at Loggerhead Creek, western Andros Island. This lithified buildup of laminated and algal mat carbonate silt is being simultaneously leached by unsaturated vadose waters, and case-hardened by calcite precipitation over exposed surface area. Note case-hardened rind (crust layer) outlining recently exposed vadose cavity between two arrows which point to broken rind edges.

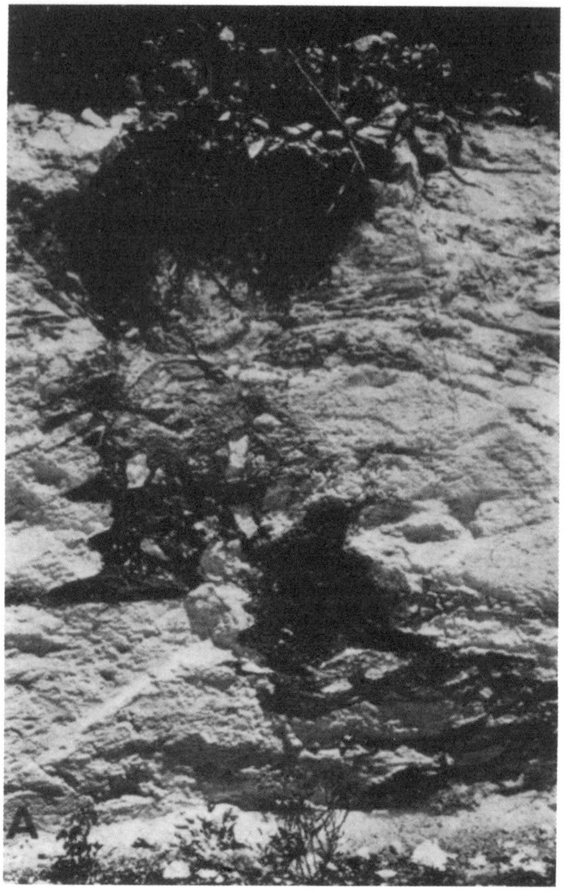

FIG. 37A.—Red laterite (*terra rossa*) forming at expense of eolian carbonate rock outcrop on Shores Road, Eleuthera Island. Penetration below karstlike sink hole is approximately 9 ft. Note detached and rotated slabs of carbonate bypassed during solution alteration.

ing a soil cover. Reprecipitation of carbonate cement by supersaturated waters at the surface tends to slow down the overall process. The most widespread result of this slowing process is well-known case-hardened layer which covers exposed surfaces where wind, rain, or surf spray alternate with dry periods (Fig. 36). Unless the rock is quickly buried, however, it will be subject to the same forces and processes as the original rock which it has in part replaced.

Any question whether a widespread karst surface developed in the Williston basin and, if so, whether it is recognizable, can be answered affirmatively by comparing the fabrics in Figures 37 and 38. The two Recent examples are from Eleuthera Island. The soft red soil penetrating deeply into the eolian dune surface rock occurs in a road cut serving Shores Estates (Figs. 37A, C). The indurated, deeply altered *terra rossas* of Figures 37D and 38A are exposed near the narrowest part of the island called "Glass Window," and are developed as an alteration of a poorly sorted calcarenite, probably also of eolian origin. Such soils are common throughout the Caribbean. They are well-developed on dunes at Salvador Point, Andros, and in numerous cuts through the lithified dune ridges of New Providence. They occur also on the surfaces of the Key Largo reefal limestones on Windley quarry, Key Largo, Florida, and on the Miami Oölite of mainland Florida. It is reported (but not published) also in certain localities along the Trucial Coast.

There is marked similarity between the detached-block and red soil fabric of Figure 37C and the Silurian example of Figure 37B. In the latter, most of the blocks are still horizontal and slightly separated. Nevertheless, the red-brown

FIG. 37B.—Detached dolomite slabs at the top of the eroded Interlake Formation. Infilling material is mostly iron oxide with suspension of dolomite euhedra. Sample loc.: Cabin Creek field, well 21-34, depth 8,559 ft.

"soil" pervading the interstices has caused the loosened blocks to fragment. Somewhat more rotation and suspension of the blocks have occurred in the Recent example. Examination of Figure 37A indicates that a core barrel passing through a similar profile in the subsurface might intersect only a thin surface cover of the soil, or a major part of the soil pocket. It is clear that plants have settled in the modern surface lows and have enhanced the downward penetration of alteration fluids. No such plant occurrences and activity can be cited for the Ordovician and Silurian periods; nevertheless, poor "soils" did develop.

Whether these ancient "soils" are laterites cannot be argued here. They have the profile, color, structure, and induration described for laterites (Maignien, 1966) but perhaps retain too much of the host carbonate. The more nondescript term *terra rossa* probably is better for use here. The example of Figure 38A was selected for study from the Glass Window outcrop area of Eleuthera. It shows a striking similarity to the Silurian example of Figure 38B. Both examples show a host carbonate which has been pervaded by leaching vadose solutions. In both cases, complex residues have been left behind. In the light-colored areas on the right of each figure, behind the alteration front, large carbonate particles have been leached. In each case this area is depleted of Fe and Al sequioxides whereas the adjacent red and red-brown areas retain high concentrations of both. Furthermore, each specimen shows a thin frontal halo indicative of the movement of the diagenetic fluids. An incomplete analysis of the halo and dark red zone of the Figure 38A sample reveals only a trace of clay (so far unidentified) and the following preliminary elemental analysis by mass spectrometry after leaching:

Sample Region	*10+%*	*1.0–10%*	*0.1–1.0%*	*Trace*
Halo		Al	Si, Fe, Ca	Cr, Cu, Sr, Ti, Ag
Red zone	Si, Al	Fe	Mg, Ti, Ca	Cr, Ni, V, Mn, Cu, Na, Sn, K, Sr

The silica occurred in the form of a few silt grains and cannot be explained at present. The high aluminum and iron contents are compatible with the expectable oxides referred to by soil specialists for these soil-regolith types. No analysis is yet available for the Interlake sample.

The development of karst features and the erosion which it implies are of more than academic interest because they bear directly on the occurrence and distribution of petroleum in the Interlake Formation. This is true because of a combination of circumstances. The emergence of the metastable lineament in Late Silurian time, coupled with the persistent constructional nature of the deposition, ended with the elevation of an area having karst-like features. This elevated area was breached by erosion in the manner of superimposed drainage or depression form, controlled by local subjacent highs in the region (Figs. 39,

Fig. 37D.—Lithified *terra rossa* outcrop at Glass Window, Eleuthera Island. Dark band is a bright red zone of iron oxide enrichment. Note pervasive character of lower interface where fingers of solution alteration are in process of detaching additional slabs and chunks of original rock.

Fig. 37C.—Close view of bottom part of Figure 37A. Arrow in photo indicates diagenetically relithified soil and rock.

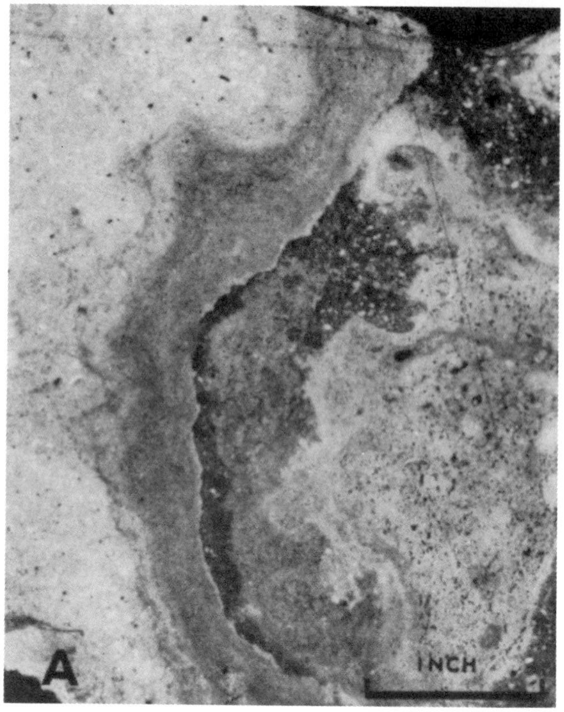

FIG. 38A.—Transition boundary between host carbonate (left) and diagenetic *terra rossa* which is complex mixture of chalky zones from which all carbonate particles have been leached (center right) and red zones (dark) rich in unsorted carbonate particles cemented by iron-oxide-bearing carbonate. Sample loc.: Glass Window, Eleuthera Island.

40). This alone might be more or less inconsequential if it were not for the fact that much of the oil in the Interlake Formation occupies porous supratidal facies high in the stratigraphic section. The unfortunate result is that many of these high-stratigraphic potential reservoir facies were removed by erosion (Fig. 41).

Another important result is the lack of reliability of *present* structural position as a guide to hydrocarbon productivity. Without understanding the reservoir development of the supratidal facies, its lateral variations, and its susceptibility to karst erosion as related to paleostructural highs, both wildcat and development drilling are hazardous. Seismic highs can be stratigraphic thins!

CONCLUSIONS

Regional epeiric model.—What has been learned here about the general nature of the Stony Mountain and Interlake reservoirs by examining their fabric detail and by comparing them with Recent carbonate depositional and diagenetic settings? A series of three block diagrams is presented to summarize their history. The first stage of development during the Late Ordovician was the deposition of infratidal and intertidal deposits across a shallow submerged mud bank (Fig. 42). These deposits were the Stony Mountain "Shale," Gunton, and Stonewall Members which reached a shoaling stage and developed a perceptive offlap recognizable in a detailed sequence of burrowed and reworked deposits. These were succeeded through short cycles by increasing subaerial exposures, resulting in the widespread deposition of the Interlake supratidal facies and multiple complex fabrics, in part organic and in part mechanical (Fig. 5). This progressive offlap depositional process combined with a regional emergence to culminate in the development of a vast carbonate desert, probably resembling the *sebkha-wadi* features of the Trucial Coast and the Interior Quarter of the Persian Gulf region (Fig. 43). The metastability of the western Williston basin led to the elevation of an area of

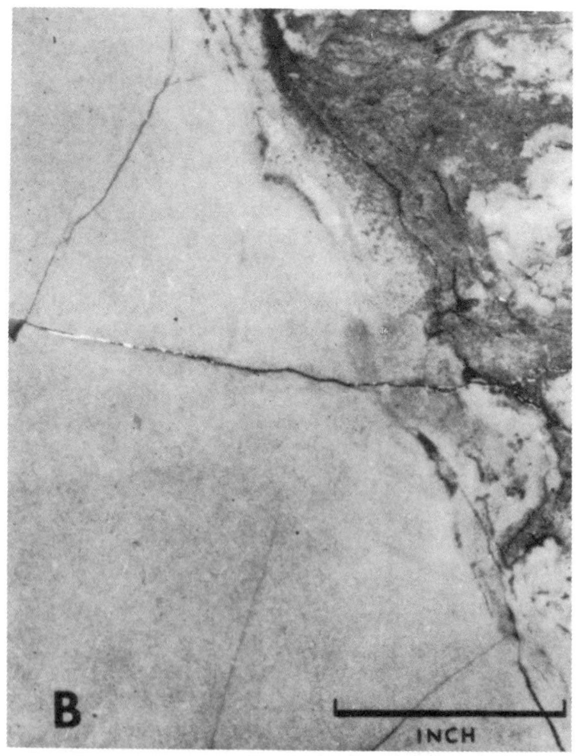

FIG. 38B.—Upper Interlake dolomite showing *terra rossa* diagenesis similar to Figure 38A. Note faint pervasive halo preceding iron-oxide-rich zone (upper right). Sample loc.: Cabin Creek field, well 32–18, depth 8,423 ft.

karstlike features which were breached by erosion in the form of superposed drainages that were concentrated over the subjacent anticlinal highs. The karst and desert regions are grouped under the term "subaerial diagenetic terrane" and, together with the supratidal zone, are used to explain the coexistence of supratidal fabrics, evaporites, dolomitization, and complex vadose alteration products.

Low-energy facies models.—As an example of the use of depositional fabrics in the study of epeiric–tidal-flat sediments and the localization of oil accumulations, a working model of the environment of deposition of the Gunton Member (Stony Mountain Formation) is presented (Fig. 44). It portrays a reconstructed sequence of development postulated for the vertical succession of fabrics and other features observed in the rocks. Although the depositional model is hypothetical, it is thought to approximate closely the actual mode of occurrence. Other similar examples may well exhibit differences from this one because of different local conditions, *i.e.*, the shore-to-sea gradient, transgressions, degree of carbonate purity, types of sediment-ingesting organisms, degree (rate) of interplay between sedimentation and the organic population, and diverse climatic factors.

The Gunton sediments and included pore space are controlled by fabrics. These fabrics are in turn limited within a particular facies, based on the extent of influence and preservation of a particular sedimentary part of the paleoenvironment. Knowledge of fabric detail for the entire succession of strata permits some intelligent guesswork concerning where each fabric fits into its own subenvironment. The reservoir parts of the Gunton sequence occur in and are preferentially controlled by (1) local burrows in the infratidal-zone deposits; (2) local solution fabrics, breccias, and

FIG. 39.—Isopachous map of thickness interval between Devonian datum and unconformity at top of Interlake Formation, Cabin Creek field. Positive axes in Interlake Formation are overlain by erosional and soil-filled karst depressions. Figure 40 gives interpretation.

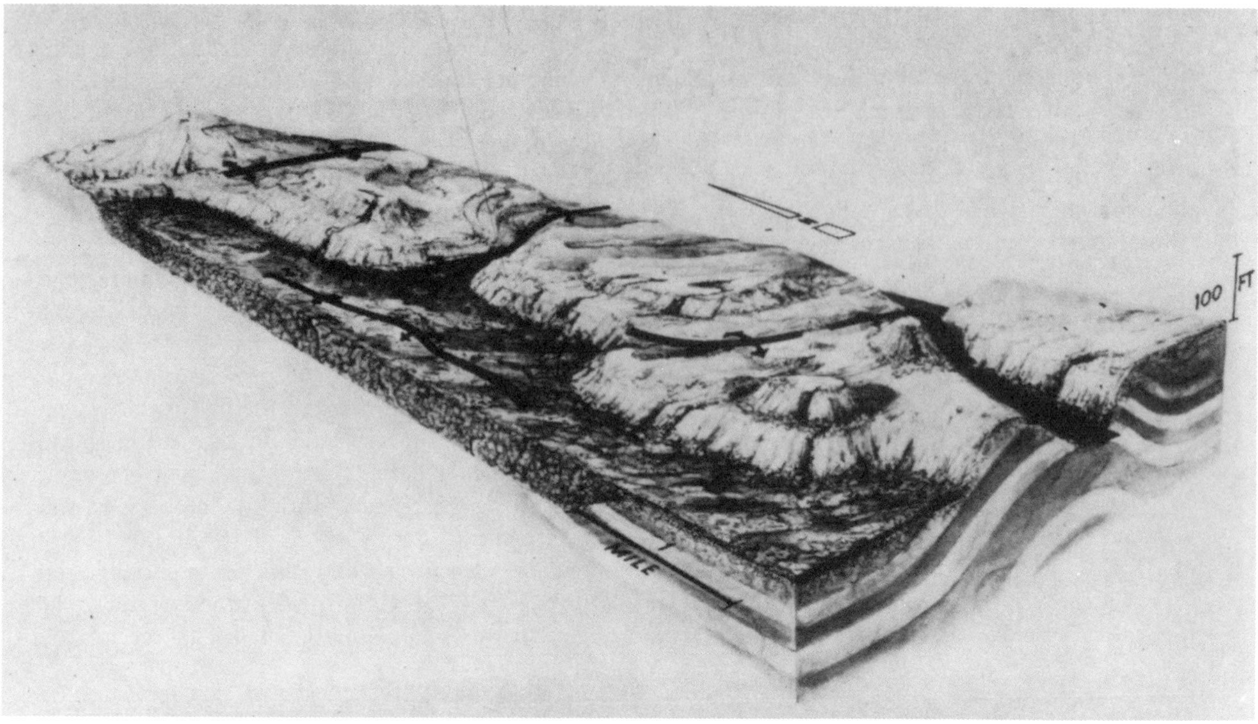

FIG. 40.—Block diagram of the Interlake unconformity of Figure 39 portrayed as karst terrane. The drainage configuration was predetermined by local underlying anticlinal axes of pre-Devonian datum age.

thick burrows in the lower intertidal zone; (3) extensive churned fabrics of the upper intertidal zone; and (4) laminated and stromatolitic fabrics of the supratidal zone.

Figure 45 recapitulates the Gunton facies and petrophysical properties. The zones are numbered for ease of use in engineering logging. It should be noted that interpretation of the gamma ray-neutron log is good only for subzone 7c. The porosity and clay contents of these clayey carbonates appear to be related. Porosity increases progressively higher in the section and decreases only in the transition zone to supratidal deposits.

Local thick burrows in the lower intertidal zone are almost identical with those of the upper intertidal zone. However, there are important petrophysical differences between them as illustrated by the histograms of pore-size frequency distribution based on composite capillary-pressure data (Fig. 46). The upper intertidal fabric possesses effective pores whose radial dimensions are less than 2 microns. These rocks are nevertheless superior reservoirs because of the uniformity of the pore-size distributions and a much lower percentage of pore space in the range below 0.2 micron. It is not, therefore, either the porosity or entry pressure but the distribution of available pore space which determines the reservoir quality of the rock. In addition to the above features, there is a lack of lateral continuity of the lower intertidal fabric due to general facies heterogeneity. Replacement dolomitization has influenced these intercrystalline relations but it is thought that ultimate control is exercised by the original fabric through readjustment of individual matrix elements.

Although the Silurian fabric types are now known in detail, lateral predictability is poor, especially where thin detailed intervals are involved. This is true for two reasons: (1) many different supratidal fabrics have abrupt lateral transition zones with each other and, (2) the common diagenetic influence on many of these fabrics (a) reduces the ease with which they may be identified and (b) produces important petrophysical changes. Because petrophysical changes are measured easily in core material and can be correlated with radioactive and other in-hole logs, the Silurian Interlake has been subdivided on this basis (Fig. 41) with the additional control of cer-

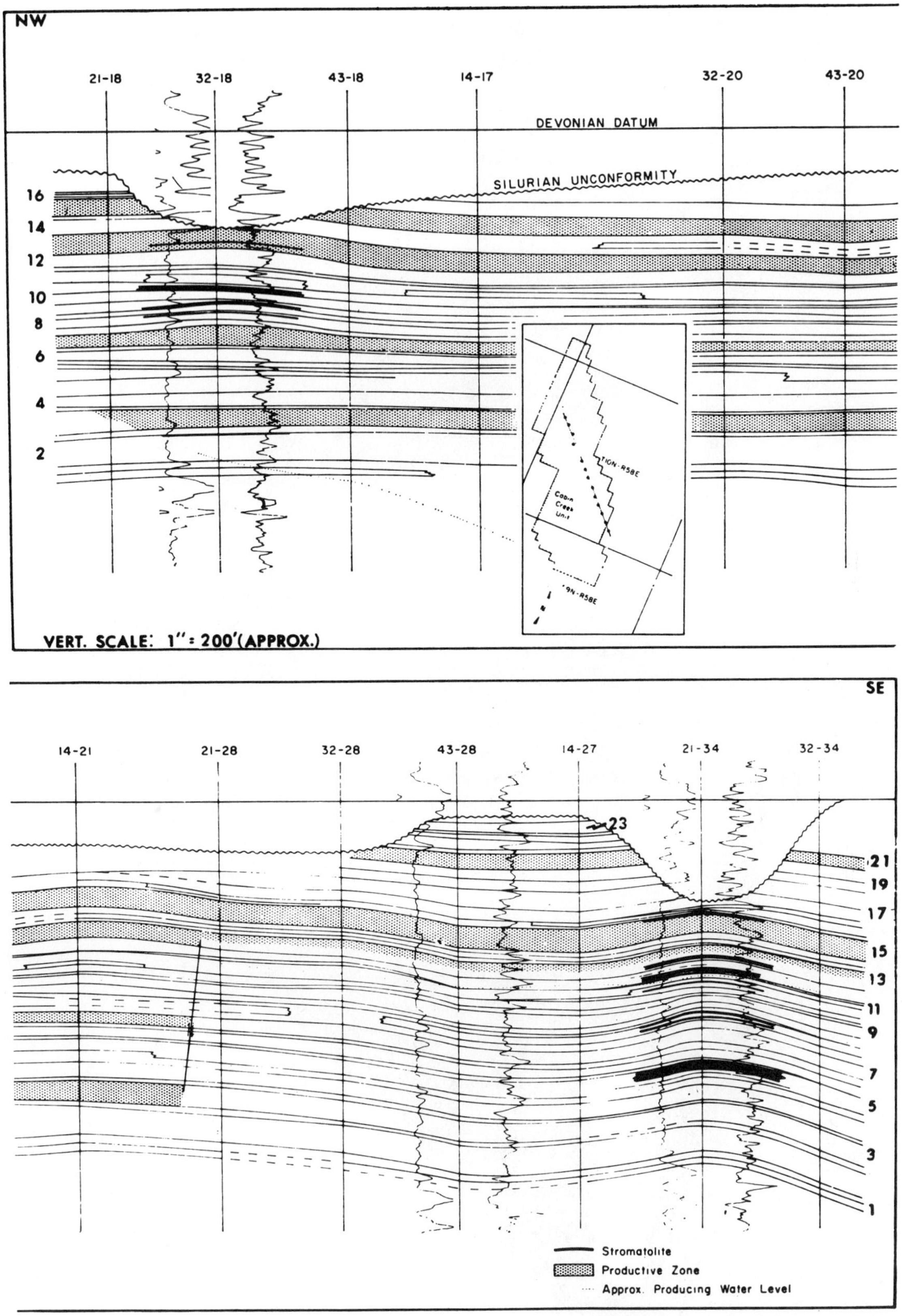

FIG. 41.—Northwest-southeast cross section through Cabin and Wills Creek fields showing position of two pre-Middle Devonian positive axes which underlie channels cut into Interlake Formation. At least one productive zone has been eroded from upper Interlake of well 21-34 and at least two zones eroded from well 32-18.

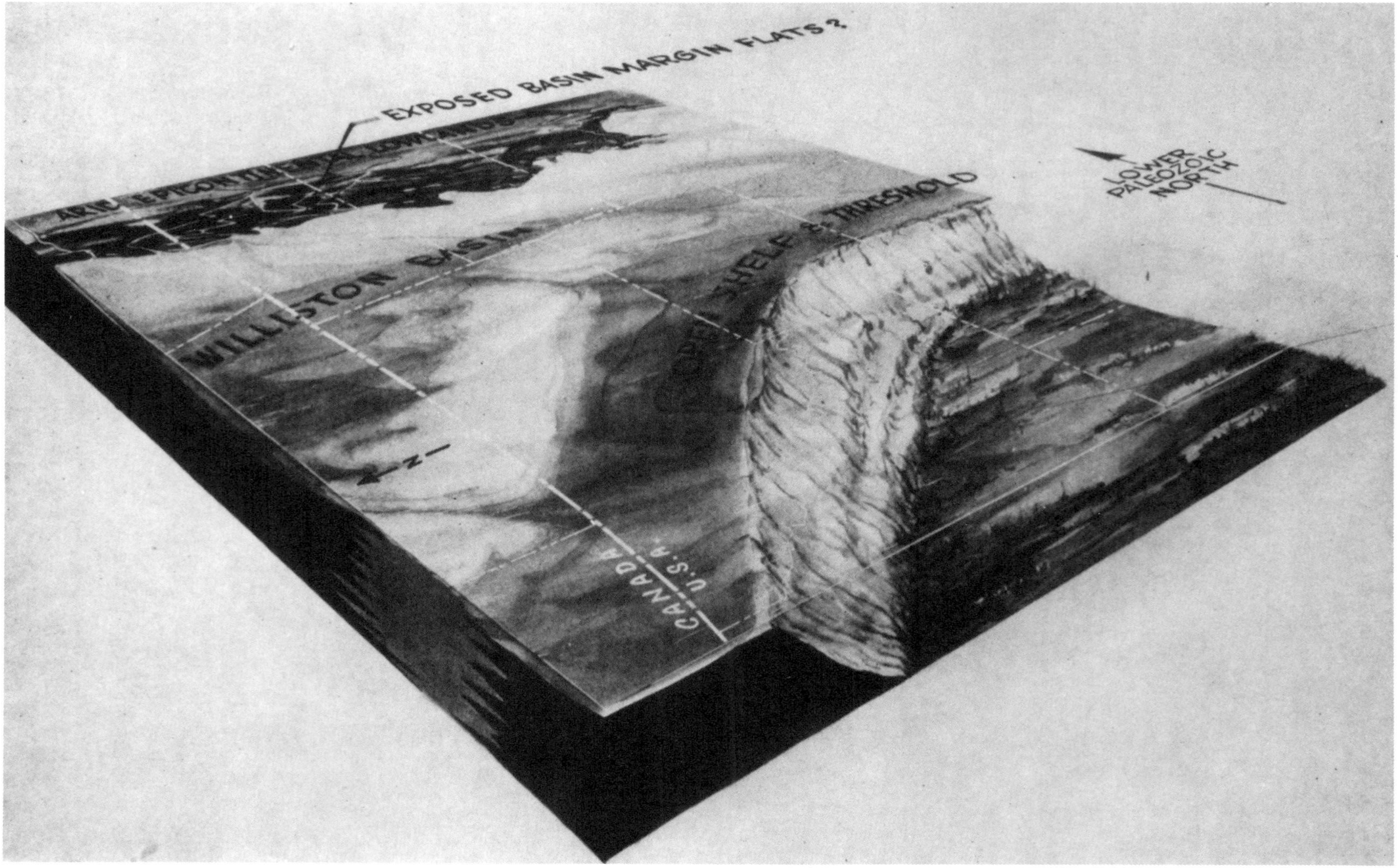

FIG. 42.—Block diagram of proposed initial shoaling stage of deposition over western metastable lineament of the Williston basin. Negative epeirogenesis probably reinitiated regional transgression during Late Ordovician time following intertidal and supratidal deposition of underlying Red River Formation. Note that present-day north is toward lower left.

FIG. 43.—Block diagram suggesting fully emergent stage of Williston basin region in Late Silurian and Early Devonian times. Because of early differential uplift along metastable lineament, microkarst topography developed on overlying surface and probably separated large interior evaporite plain from western shelf margin.

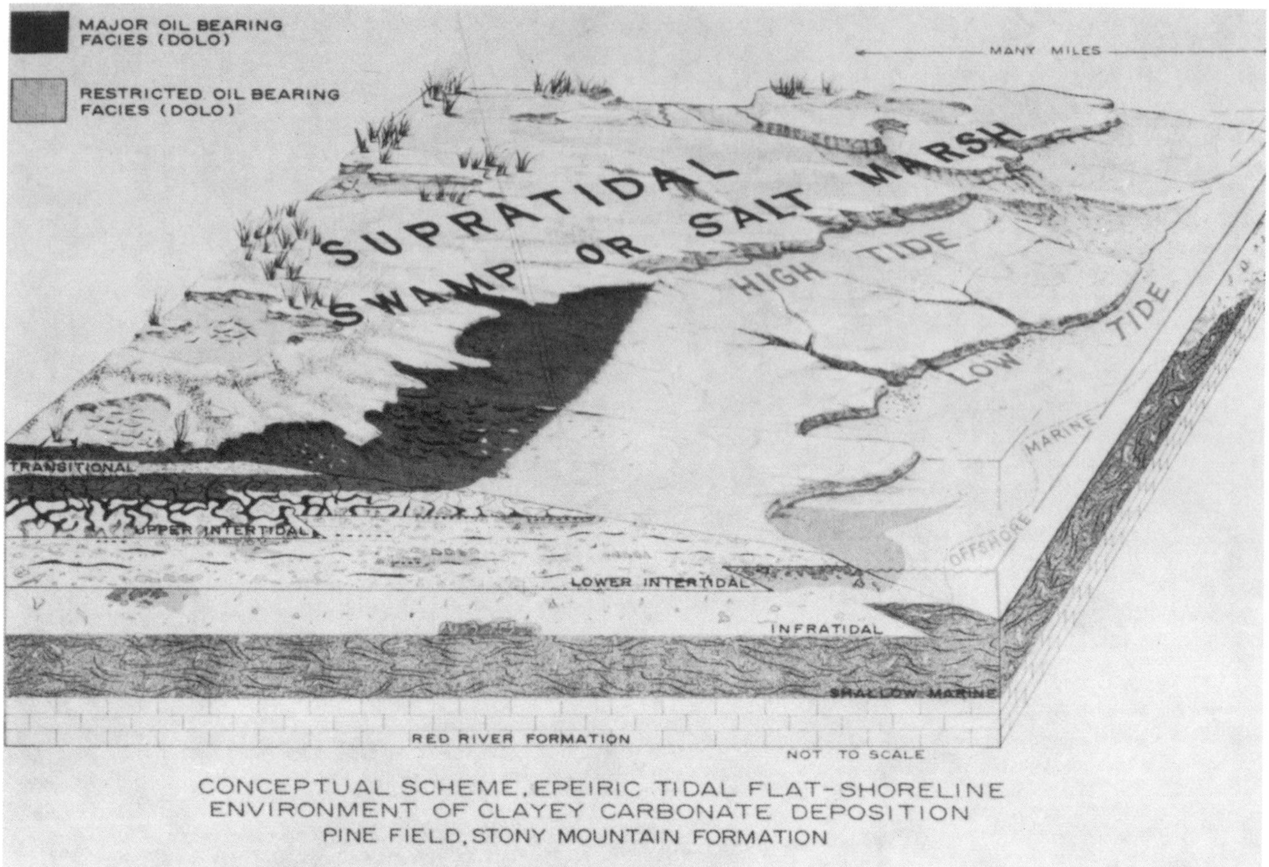

Fig. 44.—Conceptual scheme of epeiric tidal flat-shoreline environment of carbonate deposition for Gunton Member, Stony Mountain Formation, Pine field, Montana. Restricted oil-bearing fabrics occur in inferred infratidal and lower intertidal facies; however major producing zone occurs in upper intertidal facies and contiguous transitional zone of supratidal facies.

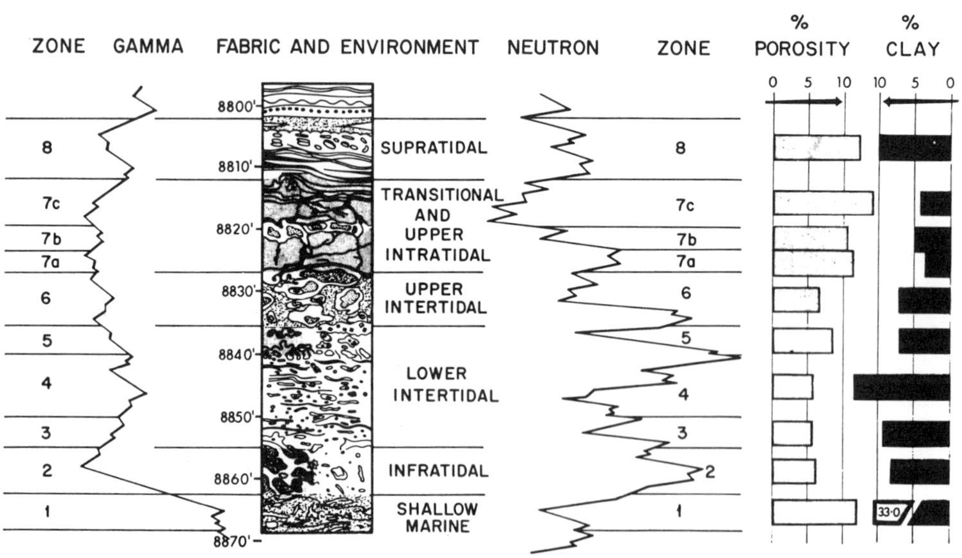

Fig. 45.—Environment of deposition and productive zonation sequence established for the Gunton Member, Stony Mountain Formation, of Pine field. Gamma ray-neutron curves are taken from well 12–15A. Oil-**productive fabrics are stippled.**

tain widespread supratidal rock types (Figs. 18, 20, 21, 25).

Prediction within some intervals is more reliable than in others, and where prediction is possible, the need for detailed step-out stratigraphy is not so great. One example is shown in Figure 47, in which cores from three wells are illustrated. Each of these wells is separated from the next by about 5 mi (8 km) along the line of depositional and structural strike. The interval selected is petrophysical zone 8, which is between two very good shaly markers and an overlying supratidal deposit. The total interval is 23 ft (7 m). There is virtually no gap in the core record. The open spaces are caused by a choice of a larger scale necessitated by the greater initial magnification of core 24–15A and by a very slight thickness reduction of the total interval in core 21–34.

Above the two shale intervals (SH) the carbonate is pure, and is a uniformly dense Archie type I dolomite. At each level, there is an abrupt termination reflected by the return to recognizable organic or diagenetic products overlying a basal terrane breccia (BTB). An anhydrite mass (A) caps the lower sequence in well 24–15A and suggests that a brief period of dolomitization and solution brecciation (SB) occurred which affected all three well locations, *i.e.*, a regional event.

Several supratidal correlations are possible here because of good core control and log correlations. The most persistent and reliable are the four labelled levels of algal mat (AM) and algal stromatolite (AS). These are commonly, but not invariably, flanked and underlain by flat-pebble breccia (FPB).

Summary

Examples have been presented of distinctive reservoir fabrics which, despite replacement dolomitization, appear to have originated under conditions similar to the sedimentary and diagenetic environments of western Andros Island, Bahamas, and the Trucial Coast, Persian Gulf. Both ancient and modern sediment examples reflect the importance of organic processes in fabric development. Fabrics control the configuration, distribution, and quality of reservoir rock. There are four major subenvironments (Fig. 48). Three acquire mud and particulate sediment from a common offshore source and the fourth shares these sediments and modifies them:

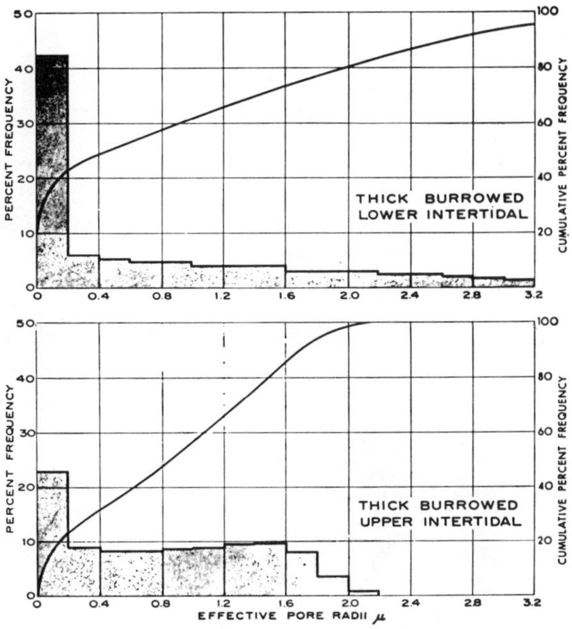

Fig. 46.—Histograms of computed effective pore-size distributions based on composite capillary-pressure data for similar-appearing thick-burrowed fabrics from inferred lower and upper intertidal facies in Stony Mountain Formation.

1. The *infratidal* fabrics are least diagnostic and commonly absent. The continuity of sediments is quite uniform and widespread except next to the mean low-tide line.

2. The *intertidal* fabrics are most diagnostic because of mutual associations and their vertical-sequence position. Their lateral uniformity may be either good or extremely erratic.

3. The *supratidal* zone possesses the greatest number of diagnostic fabrics as a group, but individual fabrics are the least uniform laterally. These sediments are volumetrically more important in the sedimentary record than previously supposed.

4. The *subaerial diagenetic terrane* develops distinctive fabrics by altering previous ones and shares in several supratidal fabrics related to evaporite deposition and dolomitization. Karst surfaces are usually reliable for correlation. Correlations may be enhanced by a thin soil profile. Prolonged karsting may destroy some subjacent potential reservoirs.

In contrast to criteria commonly used to interpret open-shelf or deep-water carbonates, it is not the type, size, or frequency of particles which necessarily defines a microfacies in the low-energy environment. Instead, the tendency of most

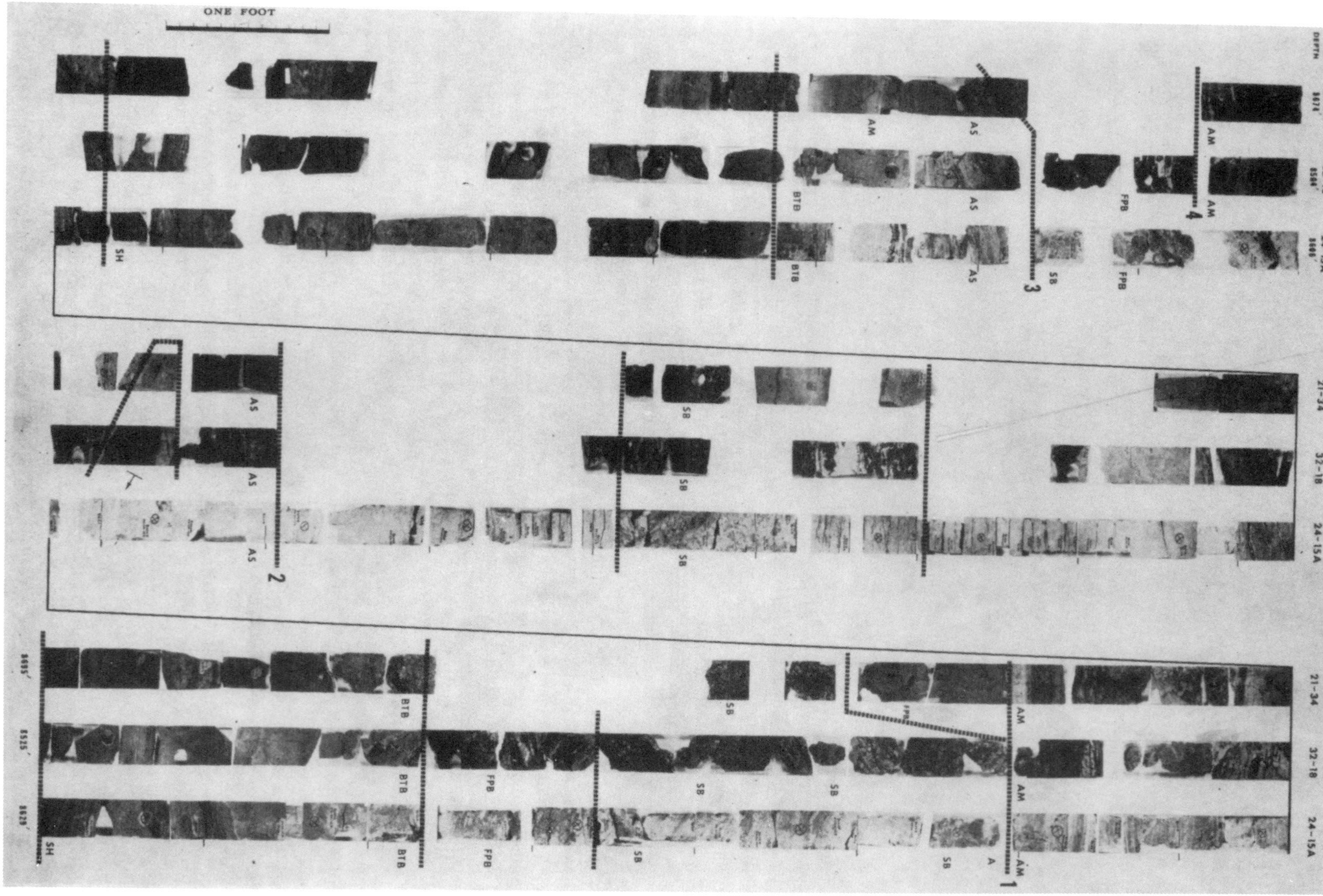

FIG. 47.—Fabric correlations of Interlake cores from three wells: Cabin Creek field 21-34, 32-18, and Pine field 24-15A. The 23 ft (7.6 m) interval includes petrophysical zone 8 (Fig. 42), which lies between two prominent shale markers (SH), and the first overlying supratidal sequence. Other symbols are: BTB, basal terrane breccia; FPB, flat-pebble breccia; SB, solution breccia; AM, algal mta; A, anhydrite; and AS, algal stromatolite.

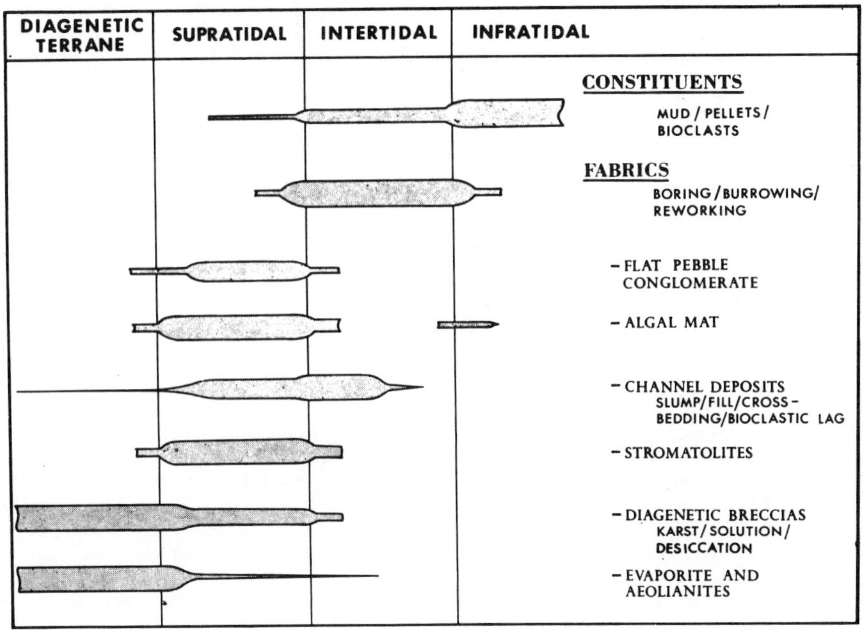

FIG. 48.—Estimate of relative frequency of occurrence of potential carbonate fabrics in epeiric-tidal flat and subaerial terranes. With exception of evaporites, none of these fabrics should be overlooked as possible reservoir rocks in ancient carbonates.

facies to possess common matrix elements derived from the nearby shallow-marine waters makes it necessary that fabrics which result from combined organic and hydraulic forces or diagenetic processes be chosen to identify facies development and history. In conclusion, it appears certain that further knowledge of modern ecological relations and diagenetic processes pertaining to low-energy environments such as those on the western part of Andros Island and the Trucial Coast will aid greatly in the interpretation of the ancient epeiric–tidal-flat deposits which are shown here to be important hydrocarbon reservoirs.

APPENDIX

SYNOPSIS OF PETROLOGY AND PETROPHYSICS GUNTON MEMBER, STONY MOUNTAIN FORMATION

The following sequential description summarizes the character of the rock fabrics of the Gunton Member of the Stony Mountain Formation at Pine field, Montana. For stratigraphic reference the zones are keyed to Figure 45. The dominant lithology of each zone is now dolomite. Two porosity values (total ($\emptyset_t$) and effective at 800 psia ($\emptyset_e$)) are averaged for each zone and given together with representative or best Thomeer (1960) values for capillary properties. G is the pore geometrical factor and P_d is the extrapolated entry pressure. Rock types are given in Archie notation except for Type II–III which refers to rocks composed of crystals ranging in size from 20 to 50 microns and Type II rocks which are restricted to a crystal-size range of 5–20 microns.

DESCRIPTION

Zone 1.—This zone is about ⅓ disseminated illite and ⅔ calcareous invertebrate shell debris in a very fine dolomitic matrix. At this locality it is the thinned equivalent of the silty, dolomitic Stony Mountain shale. Effective porosities are nil except for two which measured 8.6%. The zone is therefore essentially impermeable (Fig. 12).

Rock type: I (shaly)

$$\bar{\phi}_t = 12\%, \bar{\phi}_e = nil$$

Zone 2.—This is a 7-ft dolomite composed of finer dolomitized shell debris and much less clay than Zone 1. Many shells are whole and all are suspended in clayey dolomite matrix. The zone is laterally variable, with local thickly burrowed fabrics and some minor influx of argillaceous lithoclasts. The burrowing reflects changes in crystal sizes which affect porosity distribution. The potential reservoir type ($G = 0.45$) has loose 50–400 micron crystals in hiatal association with fine crystal aggregates. Thus porosity values will not have great significance because of sharp diversity of fabric.

Rock type: III/I, I

$$\bar{\phi}_t = 6.2\%, \bar{\phi}_e = 3.7\%$$

$$G = .45 - 1.4, P_d \geq 40 \text{ psia}$$

Zone 3.—Six feet of fine-grained illitic dolomite comprises the zone which contains extensive burrowed fabric disposed horizontally. Burrows have a bifurcating habit with crude circular cross sections. In a few places they intermix with larger burrows and are accentuated by pyrite dust. These rocks are of uniformly finer grain size than those of Zone 2, and the areas of variable granularity are more isolated. The latter have definable but high-pressure capillary properties where clay content has localized into irregular laminae (Fig. 15A).

Rock type: II-I, I

$\bar{\phi}_t = 5.8\%, \bar{\phi}_e = 3.4\%$

$G =$ variable, $P_d \geq 100$ psia

Zone 4.—This zone consists of 10 ft of rock genetically similar to Zone 3 but with less effective porosity and higher illite content ($\pm 13\%$), partly aggregated as stringers or "flocks." Capillary response is extremely erratic. The occurrence of leached bioclastic vugs reduces some P_d values below 100 psia.

Rock type: I (shaly)

$\bar{\phi}_t = 5.8\%, \bar{\phi}_e = 1.7\%$

$G =$ variable, $P_d \geq 100$ psia

Zone 5.—This is a terminating phase of Zone 3 and 4, about 5 ft thick. It is characterized by pronounced contrasts in compound burrowing fabrics and fecal distribution. Clay content is reduced from Zone 4 and localized between burrows. These interburrow areas are comprised of 30-micron subhedral crystals with a few relict fossil phenocrysts and leached phenocryst vugs up to 125 microns. Float breccia and brecciated pieces of contemporary lithified burrow fillings occur locally. These interburrows have medium to high *G*-factor values and are locally oil-bearing. Crystal sizes are smaller ($\sim$ 15 microns), nonvuggy, and more uniform in the burrowed part (Fig. 15B).

Rock type: III/I (variable G, P_d)

$\bar{\phi}_t = 8.4\%, \bar{\phi}_e = 4.8\%$

$G = .71,\ P_d = 17$ psia

Zone 6.—This 8-ft zone is characterized by the predominance of thickly burrowed fabric first developed in Zone 5. Variations of crystal size occur abruptly at burrow margins and between fecal fillings within burrows. Uniform dense, anhedral crystals of about 15 microns predominate in the burrow-fill dolomite. Other crystals range in size from 10 to 30 microns (Fig. 15C).

Rock type: II-I, II

$\bar{\phi}_t = 6.9\%, \bar{\phi}_e = 2.8\%$

$G = .7,\ P_d = 116$ psia

Zone 7.—This zone is subdivided into three parts: A, B, and C. Overall it is a 15-ft thick, highly reworked sequence. Zones A and C are nearly identical. Zone B is a denser 2–4-ft interval. Zone C is capped by transitional pelleted and laminated dolomites.

7A.—This is a modified Archie II–III dolomite about 4 ft thick with a small clay content concentrated in border areas and pockets between organically processed original fabric. Crystals in clay-free regions are euhedral and very fine-grained (35 micron average). There are numerous incipient fractures associated with the periphery of fabric element boundaries (Fig. 15D).

Rock type: II–III

$\bar{\phi}_t = 11.4\%, \bar{\phi}_e = 8.8\%$

$G = .41,\ P_d = 31$ psia

7B.—This thin 2–4–ft subzone occurs almost in the middle of the zone and has wide lateral persistence. It does not differ greatly from the major part of the zone but, because of uniform finer grain size (25 microns), the slightly higher illite content (5%) renders the rock non-oil-bearing under present saturation-pressure conditions. Oil-stained fractures are common.

Rock type: II, and II–III (shaly)

$\bar{\phi}_t = 10.7\%, \bar{\phi}_e = 9.1\%$

$G = .47,\ P_d = 103$ psia

7C.—In some respects this 7-ft subzone is the most predictable and uniformly favorable zone in the reservoir. It is generally similar in fabric and crystallinity to Zone 7A except for gradation at the top into laminated and convoluted (slump) microstructures. At the base of the interval, crystals are euhedral and average 35 microns. They are commonly in loose aggregates or in seriate contrast at margins of the larger fabric boundaries, ranging in size from 5 to 50 microns. Overlying the euhedral dolomite layer is approximately 1 ft of uniform "agglutinated" masses of fecal pellets with an irregularly connected interpelletal porous network. The pellets are composed of microgranular dolomite crystals 1–5 microns in size.

Capping the subzone is 2–4 ft of texturally banded dololutite and dolosiltite with traces of an older pelletoidal fabric; it is contorted and slumped in places. Some contortions and crenulations appear to be algal. Porosity is rarely microvugular, commonly intercrystalline (Fig. 17A).

Rock type: II–III (w/some II)

$\bar{\phi}_t = 14.6\%, \bar{\phi}_e = 12.2\%$

$G = .17,\ P_d = 49$ psia (Zone 7C)

Zone 8.—This is a continuation of (algal) crenulated and laminated dolomites (Fig. 16) that appeared in Zone 7C, to which has been added thin shale stringers, containing *boudinage*-like dolomite structures, additional flat-pebble breccias, and particulate masses. Although these rocks have fairly high porosity, many are impermeable. Porosity values of those that are permeable are only 50–70% effective because of clay and micrograin sizes of matrix. These latter explain the relatively high *G*-factors and P_d values. In places the zone is microvugular. At its top, in conjunction with transition to Zone 9, a marked secondary anhydrite content occurs.

Rock type: II, I (with shale laminae)

$\bar{\phi}_t = 12.4\%, \bar{\phi}_e = 8.0\%$

$G = .33,\ P_d = 154$ psia

References Cited

American Geological Institute, 1957, Glossary of geology and related sciences: Am. Geol. Inst., J. V. Howell, chm., Natl. Acad. Sci.-Natl. Res. Council Publ. 501, 325 p.

Andrichuck, J. M., 1959, Ordovician and Silurian stratigraphy in southwestern Manitoba, Canada: Am. Assoc. Petroleum Geologists Bull., v. 43, no. 10, p. 2333–2398.

Archie, G. E., 1952, Classification of carbonate reservoir rocks and petrophysical considerations: Am.

Assoc. Petroleum Geologists Bull., v. 36, no. 2, p. 278–298.
Baillie, A. D., 1952, Ordovician geology of Lake Winnipeg and adjacent areas, Manitoba: Manitoba Dept. Mines Nat. Resources Mines Br. Pub. 51-6, 64 p.
Ball, M. M., E. A. Shinn, and K. W. Stockman, 1963, Geologic effects of hurricane Donna (abs.): Am. Assoc. Petroleum Geologists Bull., v. 47, no. 2, p. 349.
Black, Maurice, 1953, The algal sediments of Andros Island, Bahamas: Roy. Soc. London Phil. Trans., ser. B, v. 222, p. 165–192.
Brindle, J. E., 1960, The faunas of the lower Paleozoic carbonate rocks in the subsurface of Saskatchewan: Sask. Dept. Min. Resources Rept. 54, 45 p.
Churkin, M., Jr., 1962, Facies across Paleozoic miogeosynclinal margin of central Idaho: Am. Assoc. Petroleum Geologists Bull., v. 46, no. 5, p. 569–591.
Cloud, P. E., Jr., 1962, Environment of calcium carbonate deposition west of Andros Island, Bahamas: U. S. Geol. Survey Prof. Paper 350, 138 p.
Cole, L. H., 1913, Gypsum in Canada: Canada Mines Br., Rept. 245, 256 p.
Creer, K. M., E. Irving, and S. K. Runcorn, 1957, Palaeomagnetic investigations in Great Britain, VI. Geophysical interpretation of palaeomagnetic directions from Great Britain: Roy. Soc. London Phil. Trans., v. 250, ser. A, p. 144–156.
Cumming, A. D., J. G. C. M. Fuller, and J. W. Porter, 1959, Separation of strata—Paleozoic limestones of the Williston basin *in* Symposium on concepts of stratigraphic classification and correlation: Am. Jour. Science, v. 257, p. 722–733.
Curtis, R., G. Evans, D. J. J. Kinsman, and D. J. Shearman, 1963, Association of dolomite and anhydrite in the Recent sediments of the Persian Gulf: Nature, v. 197, no. 4868, p. 679–680.
Deffeyes, K. S., F. J. Lucia, and P. K. Weyl, 1965, Dolomitization of Recent and Plio-Pleistocene sediments by marine evaporite waters on Bonaire, Netherlands Antilles, *in* Dolomitization and limestone diagenesis: Soc. Econ. Paleontologists and Mineralogists Spec. Publ. 13, p. 71–88.
Dowling, D. B., 1900, Report on the geology of the west shore and islands of Lake Winnipeg: Geol. Survey Canada Ann. Rept. 11, pt. F., 100 p.
Evans, G., D. J. J. Kinsman, and D. J. Shearman, 1964, A reconnaissance survey of the environment of Recent carbonate sedimentation along the Trucial Coast, Persian Gulf, *in* Developments in sedimentology, v. 1, Deltaic and shallow marine deposits: Amsterdam, Elsevier Publ. Co., p. 129–135.
Field, R. M., 1931, Geology of the Bahamas: Geol. Soc. America Bull., v. 42, p. 759–784.
Flower, R. H., 1946, Ordovician cephalopods of the Cincinnati region: Bull. Am. Paleontology, v. 29, no. 116, 656 p.
Foerste, A. F., 1929, The cephalopods of the Red River Formation of southern Manitoba: Denison Univ. Bull., Jour. Sci. Lab., v. 24, p. 129–235.
Fuller, J. G. C. M., 1961, Ordovician and contiguous formations in North Dakota, South Dakota, Montana, and adjoining areas of Canada and United States: Am. Assoc. Petroleum Geologists Bull., v. 45, no. 8, p. 1334–1363.
Ginsburg, R. N., 1957, Early diagenesis and lithification of shallow-water carbonate sediments in south Florida, *in* Regional aspects of carbonate deposition: Soc. Econ. Paleontologists and Mineralogists Spec. Publ. 5, p. 80–100.
——— 1960, Ancient analogues of Recent stromatolites: Internatl. Geol. Congr., 21st sess., Copenhagen, pt. 22, p. 26–35.
——— L. B. Isham, S. J. Bein, and J. Kuperburg, 1954, Laminated algal sediments of south Florida and their recognition in the fossil record: unpubl. rept. no. 54-21, Univ. Miami Marine Lab., Coral Gables, Fla., 33 p.
Harland, W. B., A. G. Smith, and B. Wilcock, eds., 1964, The Phanerozoic time-scale: a symposium: Geol. Soc. London, v. 120s, suppl. to Quart. Jour. Geol. Soc. London, 458 p.
Illing, L. V., A. J. Wells, and J. C. M. Taylor, 1965, Penecontemporary dolomite in the Persian Gulf, *in* Dolomitization and limestone diagenesis: Soc. Econ. Paleontologists and Mineralogists Spec. Publ. 13, p. 89–111.
Kent, D. M., 1960, The evaporites of the Upper Ordovician strata in the northern part of the Williston basin: Saskatchewan Dept. Min. Resources Rept. 46, 46 p.
Kerr, S. D., Jr., and A. Thomson, 1963, Origin of nodular and bedded anhydrite in Permian shelf sediments, Texas and New Mexico: Am. Assoc. Petroleum Geologists Bull., v. 47, no. 9, p. 1726–1732.
Kinsman, D. J. J., 1965, Gypsum and anhydrite of Recent age, Trucial Coast, Persian Gulf, *in* 2d Symposium on salt, v. 1: No. Ohio Geol. Soc., p. 302–326.
Kupsch, W. O., 1953, Ordovician and Silurian stratigraphy of east central Saskatchewan: Saskatchewan Geol. Survey Dept. Nat. Resources Rept. 10, 62 p.
Logan, B. W., 1961, Cryptozoon and associate stromatolites from the Recent Shark Bay, Western Australia: Jour. Geology, v. 69, no. 5, p. 517–533.
——— R. Rezak, and R. N. Ginsburg, 1964, Classification and environmental significance of stromatolites: Jour. Geography, v. 72, no. 1, p. 68–83.
Ma, T. Y. H., 1956, Climate and the relative positions of the continents during the Ordovician; research on the past climate and continental drift: Taiwan, China, Geol. Soc. China Bull., v. 11, p. 1–29.
Macauley, G., 1964, Regional framework of Paleozoic sedimentation in western Canada and northwestern United States, *in* Third international Williston basin symposium: Billings Geol. Soc., N. Dak. Geol. Soc., and Sask. Geol. Soc., p. 7–36.
Maignien, R., 1966, Review of research on laterites, *in* Natural resources research, IV: United Nations Educational, Scientific and Cultural Organization (UNESCO), 148 p.
Murray, R. C., 1960, Origin of porosity in carbonate rocks: Jour. Sed. Petrology, v. 30, no. 1, p. 59–84.
Nelson, S. J., 1959, Arctic Ordovician fauna: an equatorial assemblage: Alberta Soc. Petroleum Geologists Jour., v. 7, no. 3, p. 45.
Newell, N. D., and J. K. Rigby, 1957, Geological studies on the Great Bahama Bank, *in* Regional aspects of carbonate deposition: Soc. Econ. Paleontologists and Mineralogists Spec. Publ. 5, p. 15–79.
North Dakota Geological Society, Devonian Committee, 1961, Devonian system, *in* Stratigraphy of the Williston basin: N. Dak. Geol. Soc., 50 p.
Okulitch, V. J., 1943, The Stony Mountain Formation of Manitoba: Roy. Soc. Canada Trans., 3d ser., v. 37, sec. 4, p. 59–74.
Porter, J. W., and J. G. C. M. Fuller, 1959, Lower Paleozoic rocks of the Williston basin: Am. Assoc.

Petroleum Geologists Bull., v. 43, no. 1, p. 124–188.
Pray, L. C., and P. W. Choquette, 1966, Genesis of carbonate reservoir facies (abs.): Am. Assoc. Petroleum Geologists Bull., v. 50, no. 9, p. 2034.
Purdy, E. G., 1963, Recent calcium carbonate facies of the Great Bahama Bank, 2. Sedimentary facies: Jour. Geology, v. 71, p. 472–497.
——— and J. Imbrie, 1964, Carbonate sediments, Great Bahama Bank, *in* Guidebook for field trip no. 2: Miami, Fla., Geol. Soc. America Convention.
Pye, W. D., 1953, Structural deformation in the Williston Basin (abs.): Geol. Soc. America Bull., v. 64, no. 12, p. 1552.
——— 1958, Habitat of oil in northern Great Plains and Rocky Mountains, *in* Habitat of oil: Am. Assoc. Petroleum Geologists, p. 178–224.
Rader, M. T., Jr., 1952, Ordovician, Silurian and Devonian stratigraphy of a portion of northern Montana: Billings Geol. Soc., 4th Ann. Field Conf. Guidebook, p. 64–66.
Rezak, R., 1957, Stromatolites of the Belt series in Glacier National Park and vicinity, Montana: U. S. Geol. Survey Prof. Paper 294-D, p. 127–154.
Roehl, P. O., 1962, Oil-bearing analogs of modern calcareous sedimentary fabrics in the low-energy environment (abs.): Am. Assoc. Petroleum Geologists Bull., v. 46, no. 2, p. 276.
Ross, R. J., Jr., 1957, Ordovician fossils from wells in the Williston basin, eastern Montana: U. S. Geol. Survey Bull. 1021M, p. 439–510.
Sinclair, G. W., and E. I. Leith, 1958, New name for Ordovician shale in Manitoba: Jour. Paleontology, v. 32, p. 243–244.
Shinn, E. A., 1964, Recent dolomite, Sugarloaf Key, *in* Guidebook for field trip no. 1, South Florida carbonate sediments: Miami, Fla., Geol. Soc. America Convention, p. 62–67.
——— R. N. Ginsburg, and R. M. Lloyd, 1965, Recent supratidal dolomite from Andros Island, Bahamas, *in* Dolomitization and limestone diagenesis: Soc. Econ. Paleontologists and Mineralogists Spec. Publ. 13, p. 112–123.
Stearn, C. W., 1953, Ordovician-Silurian boundary in Manitoba (abs.): Geol. Soc. America Bull., v. 64, p. 1477–1478.
——— 1956, Stratigraphy and paleontology of the Interlake Group and Stonewell Formation of southern Manitoba: Geol. Survey Canada Mem. 281, 162 p.
Talwani, M., 1960, Gravity anomalies of the Bahamas and their interpretations: unpubl. Ph.D. dissert., Columbia Univ.
Thomeer, J. H. M., 1960, Introduction of a pore geometrical factor defined by the capillary pressure curve: Jour. Petroleum Technology Tech. Note 2057.
van Straaten, L. M. J. U., 1951, Texture and genesis of Dutch Wadden Sea sediments: Groningen-Wageningen, Netherlands, 3d Internatl. Congr. Sedimentology Proc., p. 225–244.
——— 1954a, Sedimentology of Recent tidal flat deposits and the psammites du Condroz (Devonian): Geol. en Mijn., n. ser., jaarg. 16, no. 2, p. 25–47.
——— 1954b, Composition and structure of Recent marine sediments in The Netherlands: Leidse Geol. Mededel., v. 19, p. 1–110.
——— 1959, Minor structures of some Recent littoral and neritic sediments: Geol. en Mijn., n. ser., jaarg. 21, no. 3, p. 197–216.
Wells, A., 1962, Primary dolomitization in Persian Gulf: Nature, v. 194, no. 4825, p. 274–275.
Whiteaves, J. F., 1906, Fossils of the Silurian rocks of Keewatin, Manitoba, the northeastern shore of Lake Winnipegosis, and the lower Saskatchewan River: Geol. Survey Canada, Paleozoic fossils, v. 3, pt. 4, p. 243–298.
Young, R. B., 1935, A comparison of certain stromatolitic rocks in the dolomite series of South Africa with marine algal sediments in the Bahamas: Geol. Soc. South Africa Trans., v. 37, p. 153–162.

CARBONATE DEPOSITIONAL ENVIRONMENTS

American Association of Petroleum Geologists Memoir 1,
Classification of Carbonate Rocks, edited by William E.
Ham, copyright 1962, pp. 108-123.

CLASSIFICATION OF CARBONATE ROCKS ACCORDING TO DEPOSITIONAL TEXTURE[1]

ROBERT J. DUNHAM[2]
Houston, Texas

ABSTRACT

Three textural features seem especially useful in classifying those carbonate rocks that retain their depositional texture (1) Presence or absence of carbonate mud, which differentiates muddy carbonate from *grainstone;* (2) abundance of grains, which allows muddy carbonates to be subdivided into ***mudstone, wackestone,*** and ***packstone;*** and (3) presence of signs of binding during deposition, which characterizes ***boundstone.*** The distinction between grain-support and mud-support differentiates packstone from wackestone—packstone is full of its particular mixture of grains, wackestone is not. Rocks retaining too little of their depositional texture to be classified are set aside as ***crystalline carbonates.***

INTRODUCTION

The increasing use of thin sections and oiled slabs has shown that most limestone and much dolomite retain their depositional texture, in a more or less ghostly fashion, despite diagenesis. A descriptive classification based on depositional texture is thus a generally helpful adjunct to other classifications, particularly to those based on genetic kind of particles and to those based on mineralogic composition. It will not substitute for further description and classification, nor will it produce ready-made interpretations, but it will focus attention on whichever few textural properties are chosen as particularly significant for interpretation of depositional environment, and it will provide the convenience of class names based only on depositional texture.

Ideally, one could confidently divide unaltered carbonate rocks into two main groups, which Grabau (1904) termed clastic and biogenic. According to the ideal, one group is controlled mainly by hydraulic conditions. Origin of particles is relatively insignificant, and subdivision is according to particle size. Calcirudite thus is distinguished from calcarenite. The other group is controlled mainly by the biologic or biochemical processes responsible for producing carbonate. Here, particle size is relatively insignificant, and subdivision is made according to origin of particles. Crinoidal limestone thus is distinguished from coral limestone. As it turns out, most samples are neither clearly in one group nor clearly in the other. Instead, they fall in the middle ground where hydraulic conditions and biologic or biochemical processes are in joint control.

Focusing attention on whether or not the larger particles are transported (fragmented or disarticulated) does not re-establish the sharp boundary between the two groups. Too often one is forced to conclude that a carbonate sediment was transported a little but not far, or that some components were transported and others were not,

[1] Part of a symposium arranged by the Research Committee, and presented at Denver, Colorado, April 27, 1961, under joint auspices of the Association and the Society of Economic Paleontologists and Mineralogists. Manuscript received, April 23, 1961.

[2] Shell Development Company, Exploration and Production Division. I am deeply indebted to R. L. Folk, who shared his knowledge of carbonates and classification prior to publication, and repeatedly contributed to the evolution of the present concepts; to R. N. Ginsburg, whose studies of Recent carbonates clarified problems in ancient carbonates; and to D. L. Amsbury, K. J. Hsu, S. D. Kerr, O. P. Majewske, P. F. Moore, R. C. Murray, J. M. Parks, B. F. Perkins, G. Rittenhouse, C. I. Smith, B. W. Wilson, and J. L. Wilson, who gave encouragement and stimulating criticism.

Explanation of Plate I

Looser packing would require mud-support. Samples were impregnated on the beach in such a way as to preserve natural packing and were then reimpregnated in the laboratory and sliced. All samples were from Cayo Centro, Cayos Arcas, Campeche Banks, Mexico.

a. Coral lime gravel, ×1, reflected light. South side of island, AC-64.
b. Coral-red algae lime sand, ×25, transmitted cross-polarized light. Northeast side of island, AC-47.
c. Red algae lime sand, ×5, transmitted cross-polarized light. Southeast side of island, AC-83.

PLATE I.—Grain-support in beach deposits.

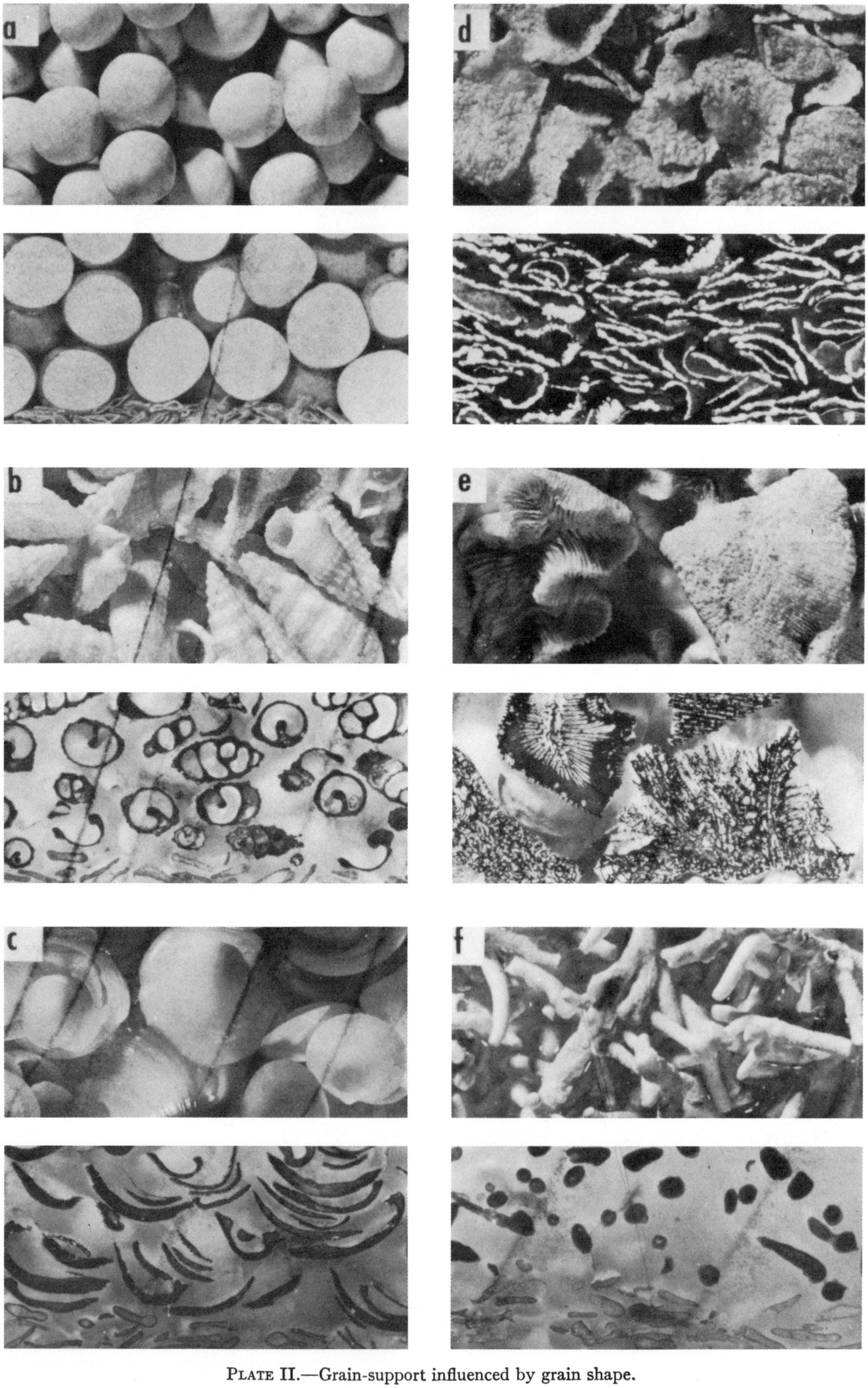

PLATE II.—Grain-support influenced by grain shape.

or that degree of transportation is indeterminate. Indeed, the outstanding differences between land-derived sediment and carbonate sediment stem from the simple fact that, as a rule, land-derived sediment is produced by destruction far from the site of deposition, whereas carbonate sediment is produced by destruction and construction at or near the site of deposition. The recognized usefulness of classifications based on origin of particles attests to the idea that carbonate grains are unlike land-derived sediment in distance of transport; if grains did not generally remain near where they were produced, such classifications would lack the environmental significance they are known to have. Possibly, the ease with which carbonate grains are destroyed, or cemented, accounts for their relatively short transport.

SIGNIFICANCE OF MUD

The distinction between sediment deposited in calm water and sediment deposited in agitated water is fundamental. Evidence bearing on this problem thus deserves to be incorporated in class names. This can be accomplished in several ways. One is to focus attention on average or predominant size, which erroneously assumes that all sizes in a sample are equally significant hydraulically. Another is to focus attention on the size, abundance, and condition of the coarse material brought to the site of deposition. This emphasis on what might be called currents of delivery has long been successful in dealing with land-derived sediments, but does not work well in lime sediment because of the local origin of many coarse grains. A third way is to focus attention on the fine material that was able to *remain* at the site of deposition. This emphasis on what might be called currents of removal seems advisable if we wish to characterize carbonate sediment systematically in terms of hydraulic environment.

Inasmuch as calm water is characterized by mud being able to settle to the bottom and remain there, it seems that the muddy rocks deserve to be contrasted with mud-free rocks, regardless of the

---

←

EXPLANATION OF PLATE II

Looser packing would require mud-support. Aggregates were made by sedimenting wet grains in quiet water onto a bed of previously deposited grains (*Halimeda* plates visible at base of aggregate), drying, impregnating with clear plastic, slicing, then staining. Plate II-*d* was sedimented dry.

Per cent grains is expressed in two ways, both values being rounded to the nearest 5 per cent. Per cent grain-solid, which is measured by water displacement, refers to the volume of solid matter in the grains divided by the volume of the aggregate. Per cent grain-bulk, which is measured by point-counting, differs in that openings and deep indentations in grains are conventionally counted as grain instead of as pore.

a. Well sorted, well rounded, highly spherical porcelain balls, ×1, 65 per cent grain-bulk, 60 per cent grain-solid.

b. High-spired snails, ×1½, 45 per cent grain-bulk, 25 per cent grain-solid.

c. Small clams, ×1½, 30 per cent grain-bulk, 25 per cent grain-solid.

d. Corn flakes, ×1, 30 per cent grain-bulk.

e. Rose corals, ×¾, 70 per cent grain-bulk, 20 per cent grain-solid.

f. Branching red algae broken to ¾-inch lengths, ×1¾, 20 per cent grain-bulk, 15 per cent grain-solid.

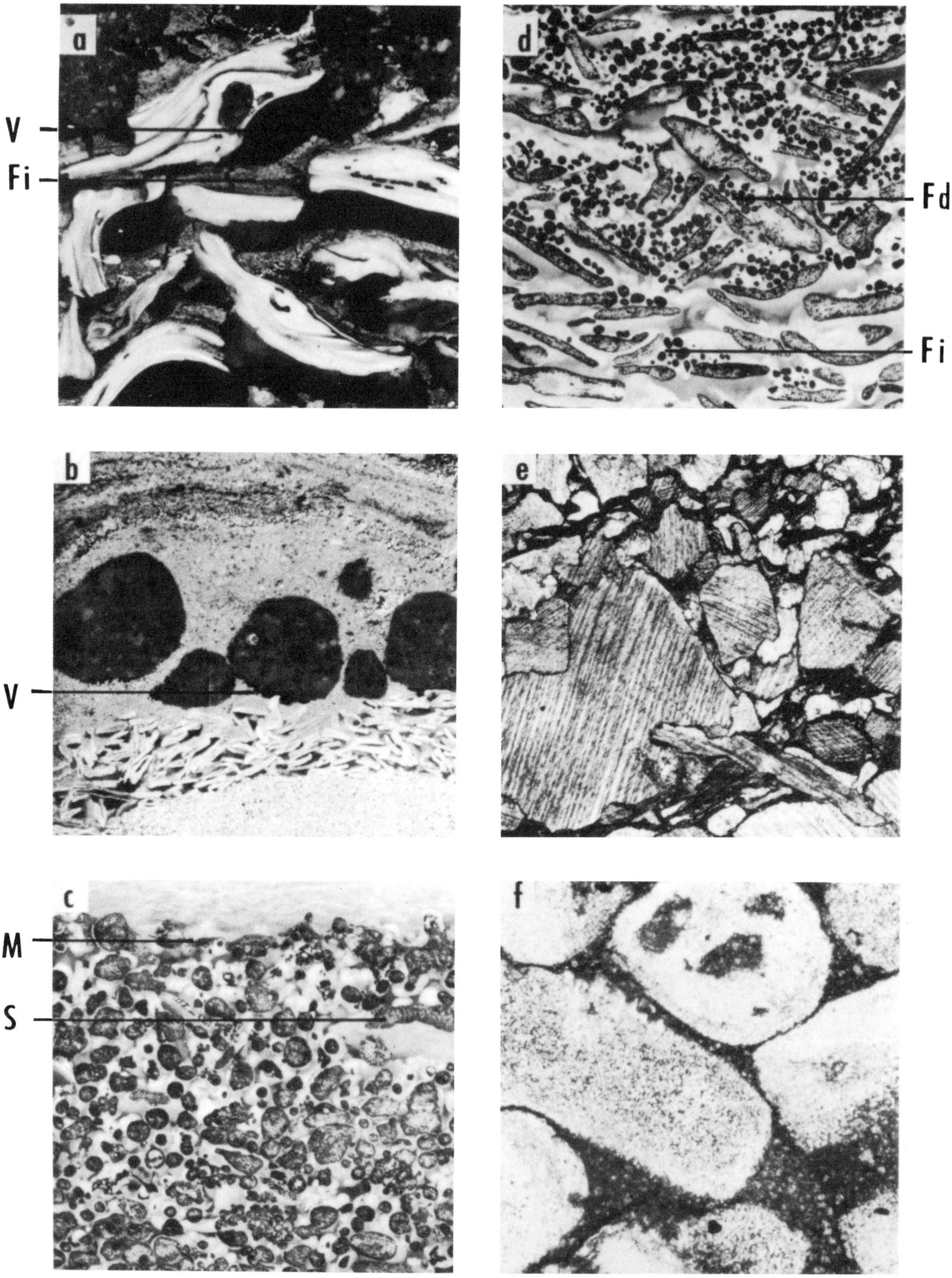

PLATE III.—Indications of grain-support.

amount and size of included coarse material. Folk (1959) and Bramkamp and Powers (1958) in their new classifications of carbonate rocks have much the same viewpoint. The terms mud and grains are used variously by different authors. Usage here is based on particle size, grains being larger than 20 microns and mud being smaller than 20 microns. The distinction thus parallels the distinction between matrix and grains in sandstone (Pettijohn, 1957, p. 284). Freedom from mud is taken to mean virtual absence—less than 1 per cent.

GRAIN-SUPPORT AND MUD-SUPPORT

Rocks bearing carbonate mud constitute the bulk of many carbonate sequences and therefore require subdivision. The most useful textural subdivision of such rocks seems to be on the basis of abundance of grains. Such subdivision allows mapping of gradients in rate of production of grains relative to rate of accumulation of mud.

Three degrees of abundance can be recognized. In the most common case, grains are abundant enough to be prominent, say more than 10 per cent, but are not so abundant as to support one another. In such a texture, the grains are sometimes said to be "floating." Here they will be called "mud-supported." Rocks in which grains are less abundant than 10 per cent constitute a second category. Rocks in which grains are so abundant as to support one another, just as they do in mud-free rocks, which are necessarily "grain-supported," are a third category.

The distinction between mud-support and grain-support seems to be more meaningful than

---

←

EXPLANATION OF PLATE III

Floored interstices, shelter effects, embayed contacts, and overly close packing indicate grain-support.

a–d—Floored interstices are produced by fine sediment filtering into coarser sediment, and by concurrent deposition of fine and coarse sediment. Marine infiltering reaches deeply only if the size difference is large, and even then does not fill interstices to the roof.

a. Oyster shells infiltered by lime mud in the laboratory, ×1, reflected light. Mud floors (*Fi*) are layered, and drape over irregularities. Shrinkage during drying reduced the amount of mud in the interstices, but voids (*V*) were beneath the shells before drying.

b. Gravel, *Halimeda*, and oöliths infiltered by very fine sand in the laboratory, ×1, reflected light. Sand largely fills interstices in the gravel layer, reaches part way through the *Halimeda* layer, and does not enter the oölith layer. The voids (*V*) beneath the pebbles are original shelter effects—the sediment did not shrink during drying.

c. Beach sand infiltered by lime mud in the laboratory, ×5, reflected light. Infiltering (*M*) is negligible except in the top layer of grains. Note the shelter effects at the large grain (*S*).

d. Floored interstices (*Fd*) produced by concurrent deposition of oöliths and *Halimeda* in the laboratory, in volume ratio of 3:1, ×3, reflected light. Oöliths making floors (*Fi*) in lower quarter of picture filtered into interstices in previously deposited bed of pure *Halimeda*.

e. Embayed contacts, ×20, transmitted unpolarized light. Lake Valley limestone, Mississippian, Sacramento Mountains, New Mexico, AGN.

f. Overly close packing, ×50, transmitted unpolarized light. Mission Canyon formation, Mississippian, Shell Richey NP-1, Richey field, Dawson County, Montana, NP 1-7312.

any alternative percentage boundary. A grain-supported rock is full of its particular mixture of grains, whereas a mud-supported rock is not. In land-derived sediments, the figures 65 per cent grains, 35 per cent porosity or interstitial debris, have been used to mark the boundary between grain-support (framework) and mud-support (disrupted framework). If all carbonate grains were as equidimensional as oöliths, a percentage boundary might equally well be substituted for visual distinction between mud-support and grain-support in carbonate rocks; but carbonate grains commonly are shaped like potato chips and twigs instead of like marbles. Because of this, a grain-supported rock whose grains are *Ivanovia*, a platy alga shaped rather like a cornflake, or *Halimeda*, would contain a far smaller percentage of grains than would a grain-supported rock whose grains are crinoid ossicles. A 65 per cent boundary, or a 50 per cent boundary, would put the *Ivanovia* rock in one group and the crinoid rock in another group; yet both rocks would contain as much of their particular mixture of grains as their volume allows them to hold.

Objection may arise that differentiating mud-supported from grain-supported rocks is impossibly subjective because of the need to envision a three-dimensional arrangement of irregular shapes by looking at a two-dimensional view. The difficulty is real but not so bad as it at first seems. (The errors are not much greater than those encountered in attempting to measure or estimate per cent grains in rocks having hollow or indented grains.) Experience gained in examining mud-free carbonates, which are necessarily grain-supported is an aid in determining what the kind of support is in muddy carbonates. Other aids are the floored interstices, embayed grains, overly close packing, and sheltering effects seen in grain-supported rocks and not in mud-supported rocks. Illustrations of these and of grain-supported natural and artificial aggregates are shown in Plates I–III.

The phenomenon of grain-support has the added importance of bearing on the postdepositional history of the sediment. Soluble grains that are grain-supported are in contact with each other, and thus have a chance of forming a connected network of molds (or incomplete molds) unlike the relatively disconnected molds in mud-supported rocks. Furthermore, compaction affects grain-supported sediment differently than it does mud-supported sediment. The grains in grain-supported sediment carry the weight of the overburden. This tends to cause weak grains to brecciate. It also tends to protect from compaction

---

Explanation of Plate IV

(×5, transmitted unpolarized light)

a. Coral lime *boundstone*. Toronto limestone, Pennsylvanian Virgil, Greenwood County, Kansas, L-38-8b.

b. Laminated lime *boundstone* (stromatolite). Cabbage-head structure in West Spring Creek limestone, Ordovician, Arbuckle Mountains, ADN-1.

c. Floored lime *boundstone*. Floored openings (*F*) constructed by intergrown complex (*C*) of encrusting Foraminifera, algae, and hydrozoans? Permian Wolfcamp. Shell State ETA-5, Townsend field, Lea County, New Mexico, ETA5-10368.2.

PLATE IV.—Binding in boundstone.

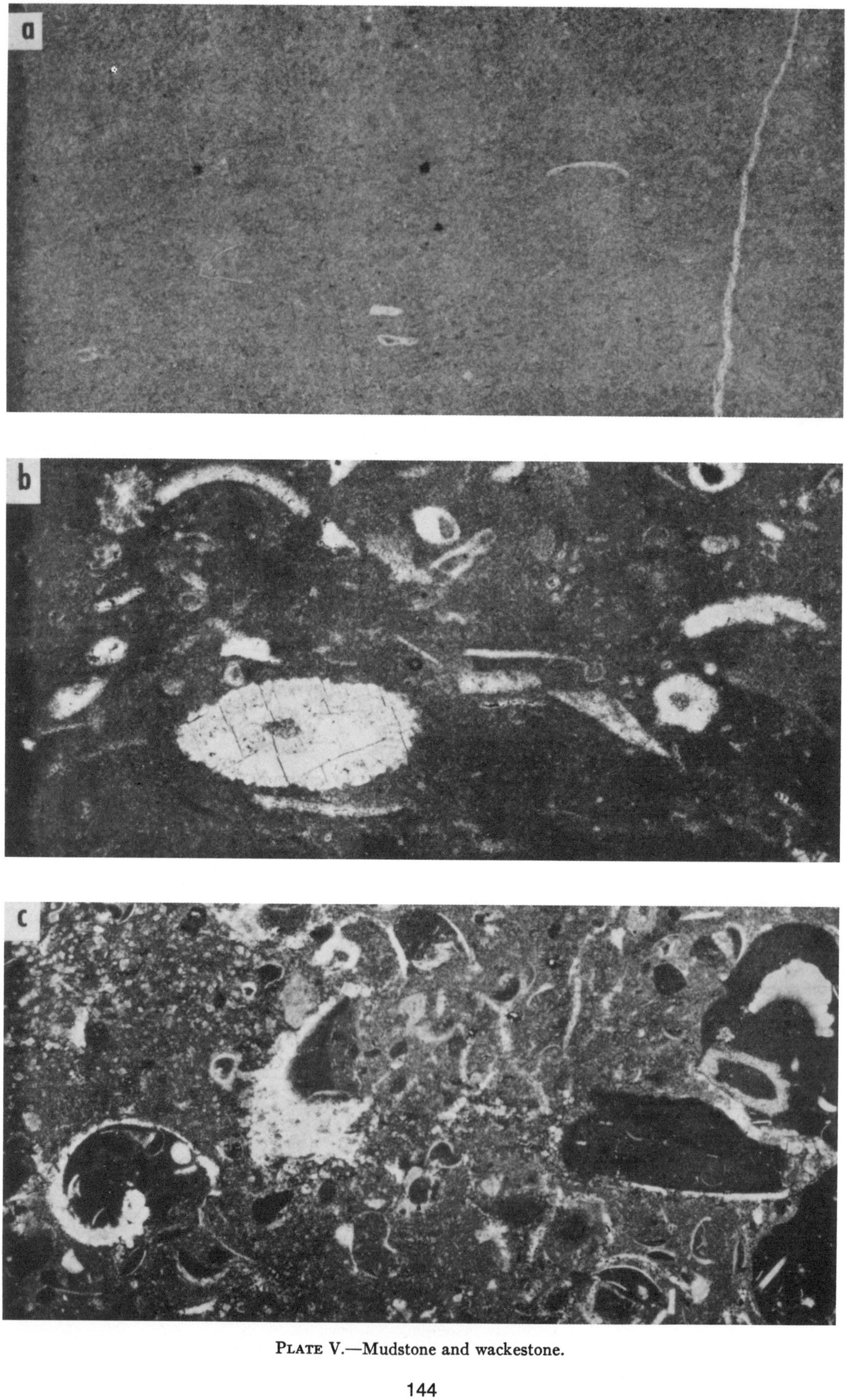

PLATE V.—Mudstone and wackestone.

Table I. Classification of Carbonate Rocks According to Depositional Texture

DEPOSITIONAL TEXTURE RECOGNIZABLE					DEPOSITIONAL TEXTURE NOT RECOGNIZABLE
Original Components Not Bound Together During Deposition				Original components were bound together during deposition... as shown by intergrown skeletal matter, lamination contrary to gravity, or sediment-floored cavities that are roofed over by organic or questionably organic matter and are too large to be interstices.	
Contains mud (particles of clay and fine silt size)			Lacks mud and is grain-supported		
Mud-supported		Grain-supported			
Less than 10 percent grains	More than 10 percent grains				(Subdivide according to classifications designed to bear on physical texture or diagenesis.)
Mudstone	Wackestone	Packstone	Grainstone	Boundstone	Crystalline Carbonate

any sheltered mud beneath strong grains, perhaps making the mud more susceptible to leaching. Collapse brecciation improved permeability in muddy grain-supported limestone composed mostly of *Ivanovia* in the Paradox formation of the Desert Creek field, Utah (Murray, 1960, p. 66). Leaching of mud from interstices in muddy grain-supported limestone composed mostly of crinoids caused the porosity in some of the Devonian reservoirs in Andrews County, Texas (F. J. Lucia, personal communication).

BINDING

Most carbonate rocks retaining their depositional texture are lithified sediment made of clearly discrete and originally loose particles. The rocks that differ by showing signs of being bound during deposition are scarce, but are worth special attention. Three signs of binding during deposition are recognized (Pl. IV). One is interconnected skeletal matter, such as occurs where colonial corals or encrusting Foraminifera grow one on the other. Another is lamination contrary to gravity, such as occurs in the crinkly lamination of stromatolites. The third is sediment-floored cavities that are too large to be interstices and are roofed over by organic or questionably organic matter, such as the large and small tunnels and grottoes in coral reefs.

CLASSES AND NAMES

The concepts outlined above allow five textural classes to be recognized (Table I). The names tentatively attached to these classes are fairly

---

Explanation of Plate V

(×25, transmitted unpolarized light)

a. Ostracod lime *mudstone*, less than 1 per cent grain-bulk. Duperow formation, Late Devonian, Brown's Gulch section, Blaine County, Montana, BG-17.

b. Mixed-fossil dolomite *wackestone*, 20 per cent grain-bulk. The rock contains no calcite. San Andreas Permian, Shell D. Roberts No. 15, 5,124 feet, Wasson field, Yoakum County, Texas, I-300.

c. Ostracod-lithiclast lime *wackestone*, slightly dolomitized, 40 per cent grain-bulk. Duperow formation, Late Devonian, Shell Richey NP-1, 8,900 feet, Richey field, McCone County, Montana, 1.562.

short, meaningful English nouns that apply only to texture. They can be combined with names for grain-kind classes and mineralogic classes, as shown in the figure captions, and they also can stand alone where mineralogy and grain-kind are not at issue.

Mudstone.—Muddy carbonate rocks containing less than 10 per cent grains (10 per cent grain-bulk as defined in Pl. II) are termed mudstone Pl. V). The name mudstone is synonymous with calcilutite, except that it does not specify mineralogic composition and thus avoids such ambiguities as dolomite calcilutite, and it does not specify that the mud is of clastic origin. The significance of mudstone, aside from the implication of calm water, is the apparent inhibition of grain-producing organisms.

Wackestone.—Mud-supported carbonate rocks containing more than 10 per cent grains (10 per cent grain-bulk) are termed wackestone (Pl. V). The name has much against it, but it does have the advantage of calling to mind a mixture of mud and grains similar to that seen in some sandstones, and it is less awkward than expressions such as calcarenitic calcilutite or calcarenitic limestone.

Packstone.—Grain-supported muddy carbonate rocks are termed packstone (Pl. VI). Grain-support is generally a property of rocks deposited in agitated water, and muddiness is generally a property of rocks deposited in quiet water. A rock exhibiting both properties is peculiar, and it is well to have it isolated for further study. It may record simple compaction of wackestone, as is suggested where interstices are completely filled with mud. It may record early or late infiltering of previously deposited mud-free sediment, or prolific production of grains in calm water, as is suggested where interstices are floored with mud. It may record mixing by burrowers or incomplete winnowing or partial leaching of mud, as is suggested by patchily distributed mud.

Grainstone.—Mud-free carbonate rocks, which are necessarily grain-supported, are termed grainstone (Pl. VII). Grainstones are not all of the same hydraulic significance. Some are current laid; some are the product of mud being bypassed while locally produced grains accumulate, or of mud being winnowed from previously deposited muddy sediment; and some, conceivably, are the product of locally produced grains accumulating too rapidly to be contaminated by mud. Commonly, origin cannot be definitely known from

---

»»→

EXPLANATION OF PLATE VI

(×5, transmitted unpolarized light)

a. Crinoid lime *packstone* overlying lens of lime mudstone. Overly close packing (75 per cent grain-bulk) suggests that grain-support was acquired during compaction. Mission Canyon formation, Mississippian, Shell Richey NP-1, Richey field, Dawson County, Montana, NP 1-7312.

b. Hydrozoan? lime *packstone.* Floored interstices (*F*) indicate that grain-support is original and suggests infiltering. Permian Wolfcamp, Shell Hilburn No. 1, Townsend field, Lea County, New Mexico, SH-1-10,556.

c. Coated-grain lime *packstone.* Abraded nuclei of grains indicates turbulent water; presence of mud indicates quiet water; patchy distribution of mud suggests that burrowers mixed originally interbedded sand and mud together. Toronto limestone, Pennsylvanian Virgil, Kansas, Bed A.

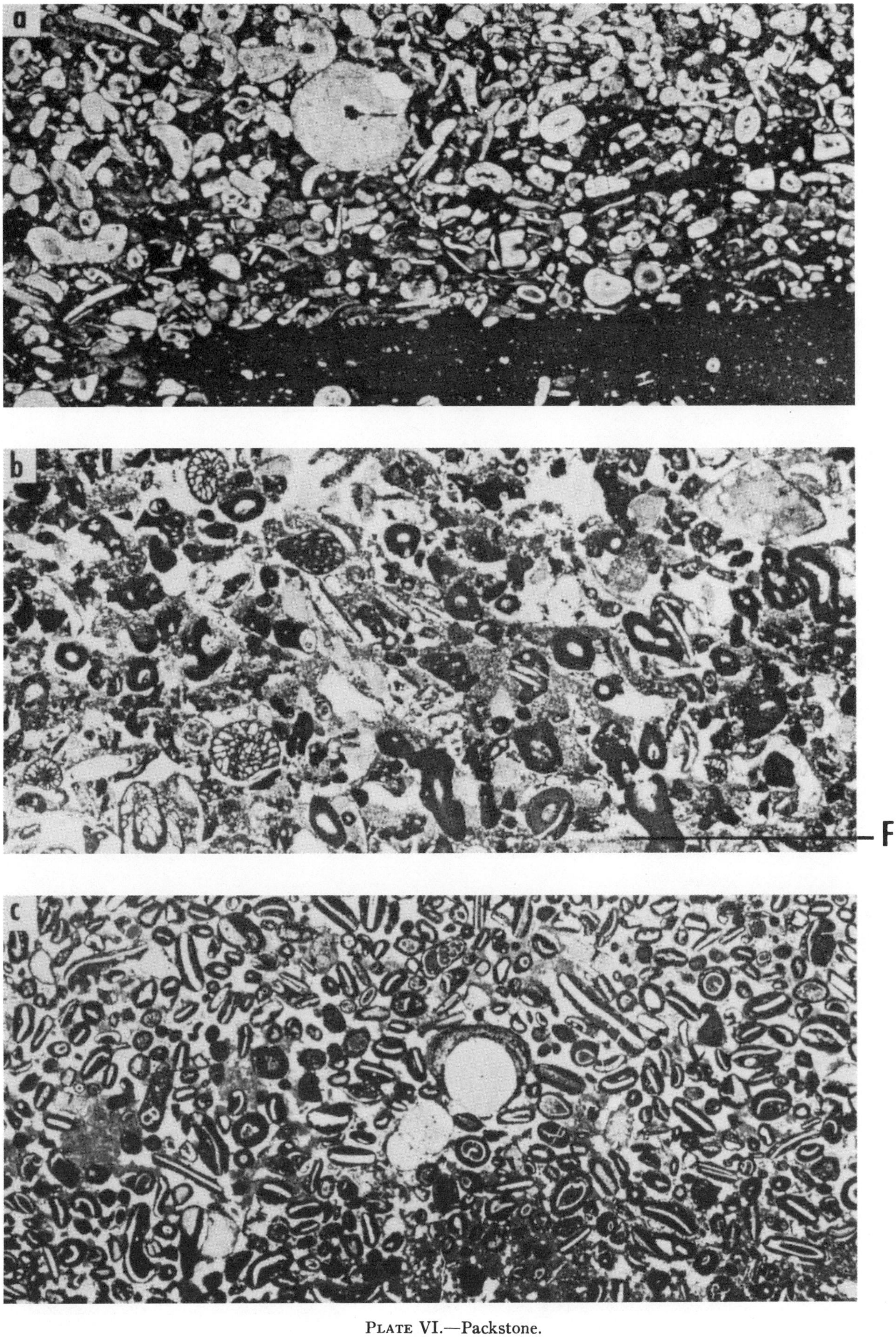

PLATE VI.—Packstone.

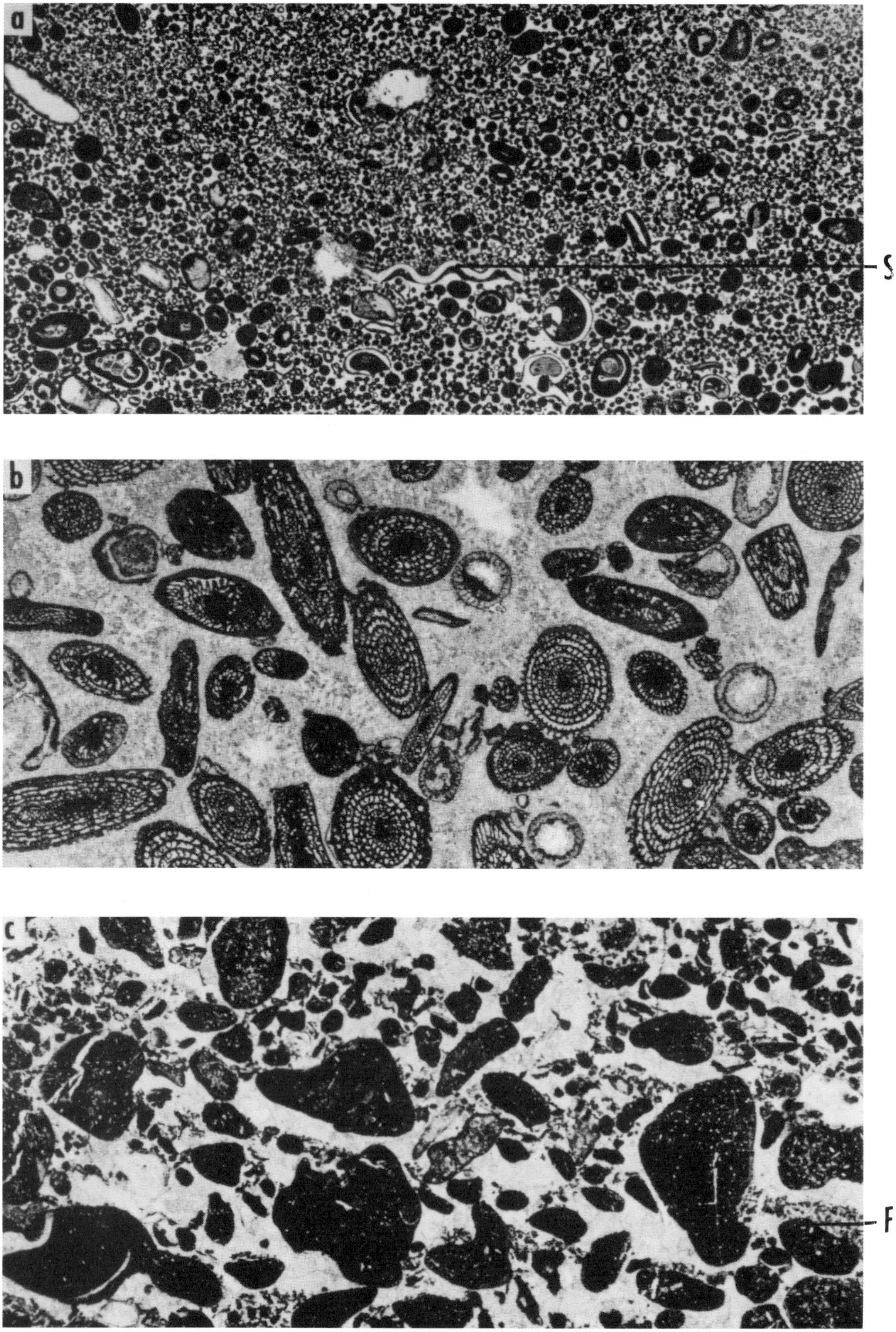

PLATE VII.—Grainstone.

the single sample that is being classified. The class name thus denotes merely the absence of mud, and, of course, the corollary that the grains are supported by each other. If the grains were not self-supporting, the former presence of now-recrystallized mud would be indicated. Subdividing grainstone, so as to include more evidence in class names, is a temptation. Many possible subdivisions suggest themselves. Grain size is, of course, one possibility. Sorting is another, for the well-sorted grainstones are scarce enough to be noted. Wear is another. Names such as *calcirudite*, *sortedstone*, and *wornstone* are useful in studying some suites of rocks, but, at present, none of these properties seems of wide enough application to warrant being incorporated in a general classification.

Boundstone.—Carbonate rocks showing signs of being bound during deposition are termed boundstone (Pl. IV). The signs of binding are specific, and they occur within the sample being classified. Except for that, the concept of boundstone plays the same role as the reefoid carbon te, biohermal carbonate, constructed biogenic carbonate, klintite, and biolithite of other classifications.

Crystalline carbonates.—Inasmuch as this classification is concerned with depositional environments, the rocks retaining too little of their depositional texture to be classified must be set aside. Such rocks are here termed crystalline carbonates (specifically, crystalline dolomite, crystalline limestone). Although depositional texture is lacking, relics or ghosts of grains commonly allow classification according to origin of grains; for example, crinoid-bearing crystalline dolomite.

REFERENCES

Bramkamp, R. A., and Powers, R. W., 1958, Classification of Arabian carbonate rocks: Geol. Soc. America Bull., v. 69, no. 10, p. 1305–1318.

Folk, R. L., 1959, Practical petrographic classification of limestones: Am. Assoc. Petroleum Geologists Bull., v. 43, no. 1, p. 1–38.

Grabau, A. W., 1904, On the classification of sedimentary rocks: Am. Geologist, v. 33, p. 228–247.

Murray, R. C., 1960, Origin of porosity in carbonate rocks: Jour. Sed. Petrology, v. 30, no. 1, p. 59–84.

Pettijohn, F. P., 1957, Sedimentary rocks, 2d ed.: New York, Harper and Bros.

EXPLANATION OF PLATE VII

(×5, transmitted unpolarized light)

a. Oölith lime *grainstone* (oölite). Note the sheltering done by the coated brachiopod shell (*S*) near the center —grains are smaller and more numerous above the shell than below the shell; compare Plate III-*c*. Wapanucka limestone, Pennsylvanian Morrow, Bromide, Oklahoma, W-23.

b. Fusulinid lime *grainstone*, druse-cemented. Seemingly loose packing is due to fusulinids being unusually elongate. Uniform orientation (long axis perpendicular to page) and breakage indicate that the grains are current-laid. Back-reef limestone of Capitan reef, Permian Guadalupe, Pinery Canyon, Guadalupe Mountains, Texas, PN-6-A.

c. Lithiclast lime *grainstone*. Note the sand-floored interstices (*F*); compare Plate III-*d*. El Paso limestone, Ordovician, Franklin Mountains, Texas, AGE.

Reprinted by permission of the Canadian Society of Petroleum Geologists. Published in *Facts and Principles of World Petroleum Occurrence,* A.D. Maill, editor, CSPG Memoir No. 6 (January 1980), pp. 95-115.

A REVIEW OF CARBONATE RESERVOIRS

JAMES LEE WILSON

ABSTRACT

Porosity and permeability in carbonate strata result from interactions between certain favored types of original sediment and a variety of diagenetic processes. Alteration may be induced by several generations of hydrographic conditions controlled in turn by the total geologic history of the reservoir area. Reservoir development is therefore highly complex. Carbonate sediment is produced and deposited with high porosity. This is normally reduced from 50% to less than 10% during diagenesis but permeability may be increased by some processes.

In an analysis of hydrocarbon reservoirs, the following combinations of sediment type, stratigraphic relations and favorable diagenetic overprint persistently recur in the geologic record:

1. Lime sand bars and sheets on shelves with histories of incomplete cementation (Chesteran, Oklahoma; Murban on Trucial Coast; Arab D, Saudi Arabia; Smackover of Louisiana).

2. Carbonate organic buildups whose original high porosity and permeability along margins of banks and platforms have not been completely occluded, or whose fabric has been dolomitized and fractured (Devonian buildups in Alberta and the Golden Lane atoll of eastern Mexico).

3. Downslope talus accumulation below major carbonate bank of platform margins (Poza Rica and Reforma pools of Mexico).

4. A stratigraphic trap formed by subtidal grainstone and intertidal dolomitized mudstone reservoir which change facies updip to a sabkha anhydrite seal (Permian San Andres of West Texas and the Mississippian strata of Williston Basin).

5. Leaching and dolomitization at regional unconformities, processes aided by an original grainy texture and/or fracturing (Natih and Fahud Cretaceous fields of Oman; Late Paleozoic phylloid algal limestone of southwestern U.S.A.; basin margin pinnacle reefs of Devonian Zama-Rainbow; and Silurian of Michigan basin).

6. Chalk textured pure carbonate muds which have not been "solution welded" by burial below 1 000 m and have been protected from Mg poor solutions (North Sea Danian; Austin Chalk).

RESUME

Dans les sédiments carbonatés, la porosité et la perméabilité sont le résultat de réactions internes entre certaines caractéristiques privilégiées du sédiment original et des réactions diagénétiques variées. Une altération peut se produire en plusieurs épisodes de conditions hydrographiques, contrôlées à tour de rôle par l'histoire géologique totale de la région où se trouve le réservoir. L'évolution du réservoir est donc vraiment complexe. Le sédiment carbonaté se forme et se dépose avec une porosité élevée. Cellci se réduit normalement de 50% à moins de 10% pendant la diagenèse mais la perméabilité peut s'acroître par quelques réactions.

Lors de l'analyse des réservoirs à hydrocarbures, il ya a un renouvellement persistant dans l'histoire géologique des combinations suivantes du sédiment type, de ses rélations avec la stratigraphie et d'une surimpression diagénétique favorable:

1. Feuillets et barres de sables calcaires sur les plateaux continentaux accompagnés de cimentation incomplète (Chesteran en Oklahoma, Murban sur la côte des Pirates, Arab D en Arabie Séoudite—Smackover en Louisianne).

2. Constructions carbonatés faites d'organismes dont la porosité et la perméabilité élevées, en bordure de talus et de plateforme, n'ont pas été complétement bouchées, ou dont la texture a été dolomitisée et fracturée (récifs Devoniens en Alberta et atoll de "Golden Lane" dans le Mexique oriental).

[1]Department of Geology and Mineralogy, The University of Michigan, Ann Arbor, Michigan 48109.

3. Sédimentation de bas de pente à flanc de talus, en contrebas des bordures de talus et de plateforme carbonatés de grande extension (réservoirs de Poza Rica et Reforma au Mexique).

4. Piège stratigraphique formé de réservoirs calcaires "grainstones subcotidaux" et "mudstones dolomitisés interidaux, qui passent en amont à des lits anhydritiques de sebkha. (Permien San Andres del'Ouest Texas et Mississipien du bassin de Williston).

5. Lessivage et dolomitisation au niveau des discordances régionales, réactions facilitées par une texture granulaire d'origine et/ou par le fracturation (champs crétacés de Natih et Fahud en Oman—calcaire alguaire de Paléozoique Terminal du S.W. des Etats-Unis—récifs biohermes de la bordure du bassin devonien de Zama-Rainbow et dans le Silurien du bassin de Michigan).

6. Boues calcaires pures à texture crayeuse n'ayant pas subi de diagenèse par cimentation/solution dûe à un enfouissement supérieur à 1 000 m et ayant été protégés de solutions pauvres en Mg (Mer du Nord danoise Craie d'Austin).

INTRODUCTION

It is a truism that carbonate rocks are important reservoirs. Two-thirds of the major free world reserves in the Middle East reside in carbonates. Despite the predominance of the Gulf Coast terrigenous Tertiary province, about one-third of the major North American oil and gas fields are in carbonates and they produce about half of the total oil produced from the North American giant fields (Halbouty, 1970). Most carbonate reservoir rock is a product of sedimentation in shallow tropical water. The sediment is deposited in an established spectrum of nine or so facies from offshore low energy areas through a shelf marginal belt of high energy to a shelf interior with low energy. Each facies has its own special biota and texture and individual bodies of sediment. The carbonate facies spectrum will form by outward accretion around gentle positive areas, by building out from edge to center of shallow cratonic basins or by developing progradational cycles across shelves.

Carbonate sediment is deposited with initially high porosity (reefs with more than 50%, sands with 40-80%, muds with 40-50%). Most of this porosity is occluded after deposition. Limestone in the geologic record bears porosity from less than 5% to a maximum of about 25% in chalk and some reefs.

Carbonate diagenesis is of over-riding importance in determining whether a sediment will have porosity and permeability. The processes are as follows: early organic rotting, decay, and coating of particles; cementation; neomorphic alteration of calcium carbonate minerals; preferential solution of less stable aragonite; replacement by silica, dolomite, calcium sulphate, phosphate, and manganese oxide; solution-compaction; mechanical compaction; pressure solution to form stylolites; fracturing, vein-filling; and collapse-brecciation. The major factors responsible for these processes are: amount and rate of fluid flow, salinity and chemical composition (pH) of meteoric and connate waters, organic content in buried sediment, overburden and tectonic pressure. These factors are controlled by the environments of diagenesis created within the surficial and buried sediments along the facies spectrum developed across a typical carbonate platform (Choquette and Pray, 1970). These include (1) the varied *eogenetic* (pre- and early depositional) environments of deep water, shallow tropical, splash zone, and inshore brackish to hypersaline water, and early influence of meteoric water; (2) the *telogenetic* environment where meteoric water has affected later lithified sediment; and (3) the *mesogenetic* environment (deep burial, subsurface, or connate water).

The sequence in which these various processes act and react on the sediment is important. Certain early processes control later ones. For example, initially high porosity may lead to leaching if the grains are coarse enough to permit fluid flow. But some early cementation will inhibit later fluid flow, protecting the grainstone from dolomitization (a process which normally increases porosity-permeability) or protecting from solution-compaction-cementation (a process which decreases porosity-permeability). The presence of some original interstitial lime mud will control the extent of later diagenesis, inhibiting cementation and encouraging dolomitization (Lucia, 1962). The early development of coatings or decay rinds on certain particles such as echinoderm plates may inhibit later overgrowth cementation and prevent porosity occlusion.

TYPES OF RESERVOIRS

The large number of diagenetic processes and environments, the varied factors responsible for them and the long periods of geologic time permitting opportunity for variations in sequence result in very complex diagenetic relationships to porosity and permeability. Much study of the cause of porosity preservation in specific reservoirs has been accomplished but many examples are still being worked out through laborious case history studies. The approach of the present review is therefore empirical: What are the major known types of carbonate reservoirs? This paper describes them very briefly and characterizes them in terms of the dominant diagenetic processes imprinting the original sediment types. Six important categories are mentioned below but the list is not exhaustive and some major reservoirs are actually combinations of the categories.

1. Sheets and linear bodies of grainstone with preserved primary porosity and both leaching and early cementation modifying the porosity slightly. Such reservoir rock is commonly found in shelf cycles.

2. Reefy buildups at shelf margins with some of the high initial porosity preserved. Even though partly cemented, such fabric may have porosity enhanced by subsequent leaching and dolomitization.

3. Downslope debris off carbonate shelf margins with some primary porosity preserved in coarse rubble; permeability is commonly enhanced by dolomitization and fracturing.

4. Stratigraphic traps naturally occurring in the carbonate facies spectrum when it is developed in shelf cycles. The reservoir consists of porous grainstone grading to intertidal dolomite. Key to the trap is an updip and commonly overlying seal of sabkha anhydrite.

5. Solution, fracturing, and dolomitization beneath a regional unconformity.

6. Chalky-textured reservoir with special conditions to preserve the fine-grained fabric from solution compaction welding, to permit permeability, or to permit fluid flow through normally impermeable strata.

1. Grainstones with preserved primary porosity

Sedimentary geometry Grainstones may form in widespread prograded sheets such as occur in the Arab zone cycles or occur as wide fringes around basin edges (Smackover of the Gulf of Mexico coast; *see* Bishop, 1968; Wilson, 1975). Such sediment is, of course, the product of high water energy and in detailed configuration it often forms linear bars of lime sand. Belts of such bars may parallel shelf edges, but individual bars may be elongate parallel to currents running across shelves, or barchan-shaped, or the sands may be piled in ridges against headlands within the sheif.

Diagenesis and porosity Oolite is common but round and slightly coated grains are equally so; the initial porosity of such sand is about 40 percent but with increased content of coarse and irregular bioclastic grains it may be much higher, about 75 percent. There is a tendency for early fringing cement to protect the fabric from solution-compaction, but coating and micritization of the particles may cause inhibition of crystal overgrowth. Echinoderm grainstones almost always "cement-up" unless they are coated early in diagenetic history. Dolomitization is not common in grainstone but is known to occur in interbedded wackestones. Later generations of blocky anhedral calcite cement may occlude remaining porosity; this is often seen in outcrops, causing grainstones to commonly form resistant ledges. In searching for grainstone reservoirs one should look to causes of lack of cementation: areas which lack fluid flow during geologic time, strata which were saturated early in water depleted in calcium, early fringes of cement which might have inhibited solution-compaction-welding, areas which lacked burial deep enough for overburden to effect the above process, or areas where early hydrocarbon migration inhibited cementation.

Occurrence Such grainstone reservoirs are rare in the North American Paleozoic but are more common in North African and Middle East Tertiary and Mesozoic strata. The largest current carbonate production in the world comes from the Jurassic Arab C-D zones which are poorly cemented bioclastic-oolitic lime sand with porosity of 20-25% and very good permeability. Initial production potential from these wells on the Saudi Arabian mainland is on the order of 950-2 860 m^3 (6,000-18,000 barrels) of oil per day (Fig. 1 and 2). Other examples include the Smackover of Gulf Coast (Fig. 3), Minagish (Lower Cretaceous of Neutral Zone), the Metloui nummulitic grainstone of offshore Tunisia (Bishop, 1974) and the Chester oolite of the Anadarko basin (Lucia and Murray, 1966).

2. Organic carbonate buildups at shelf margins

Sedimentary geometry Size varies from very large linear, composite masses at the platform margins, with thicknesses of hundreds of meters, to low fringes and individual loaf-shaped mounds with a relief of a few tens of meters. Such masses have complex composition; reef framework, debris piles, flank beds, trapped finer material. Trends of these organic reefs and banks of bioclastic sediment debris piled at shelf margins are normally parallel to the shelf edge. En echelon multiple linear trends are known which cut across basins along rejuvenated fault blocks. Swarms of pinnacle reefs occur downslope from shelf margins.

Diagenesis and porosity Organic reefs are recognized by the relative abundance of sessil benthonic colonial animals and plants. Originally high porosity is caused by microporous interior of coral fasciculate structure, by internal cavities in all biotic framework, by rapid growth enclosing and encrusting over large cavities, and by reef growth in large scale pillars. It is common for original porosity to be so great that it cannot be completely filled in by the normal accumulation of internal sediments, bioclastic debris, and lime mud, or by organic and inorganic marine cement. Additional high porosity-permeability is caused by rotting and boring in reefs. High porosity-permeability is also common in adjacent coarse reef rubble derived from buildups in high energy environment.

The tropical waters in which organic buildups grow induce dominant, rapid precipitation of porous aragonite frameworks (modern Mesozoic corals, hydrozoans, stromatoporoids, green algae, tropical mollusks) and high magnesian calcite (red algae). These forms of calcium carbonate are notably unstable outside of the marine environment. The original porosity and the appreciable remaining voids are therefore easily subjected to meteoric vadose and phreatic diagenesis. Any sea level lowering exposes first the build-up reefy facies and hence for a longer period of time permits preferential leaching of aragonite. At the same time low Mg calcite cement may be precipitated in phreatic zones established lower in the reefs. Such processes are well documented by Mathews (1968) and Land (1973) for uplifted Pleistocene reefs on Caribbean islands. Stacked thin reservoirs with multiple oil-water contacts present in Pennsylvanian phylloid algal mounds in West Texas probably result from such zones of vadose-phreatic diagenesis. (For further discussion see Category 5.)

The massive construction of carbonate buildups commonly occurs surrounded by thinner beds with argillaceous partings or interbeds. In folding or even simple compaction the massive reef rock often fractures rather than responding plastically, and joints and faults at sides and within buildup masses are common. The process of jointing-fracturing may begin early in buildup history and continue over eons of geologic time. This may enhance later leaching by increasing permeability. The notable porosity and permeability of buildups is probably responsible for their considerable amount of dolomite. The dolomitization process in some places enhances matrix permeability, adding appreciably to good reservoir quality.

Occurrence Giant reef fields are numerous. Some of these are at edges of major carbonate platforms (Kirkuk in Iraq) and many are associated with large offshore banks bordered by reefy buildups (Golden Lane rudist trend, see Figs. 7, 8); Scurry County Horseshoe Atoll (algal and other bioclastic debris). Or reefs may extend linearly across basins (Devonian and Alberta, Figs. 4-6). Smaller but numerous and important buildups occur within basins as clusters on marginal ramps (Upper Pennsylvanian on the eastern shelf of the Midland basin); Devonian pinnacle reefs of Zama and Rainbow of northern Alberta, and the Silurian pinnacles of the Michigan basin. In some instances these basinal buildups are made reservoirs by dolomitization, particularly when encased in evaporitic strata. The basinal location of such buildups places reservoir in proximity to source beds and so commonly forms hydrocarbon accumulations (see also discussion below under Category 5).

3. Downslope carbonate debris with preserved primary void space and some enhanced secondary porosity.

Sedimentary geometry Important carbonate marine debris flows have been described as irregular patches with linear trends below major shelf margins and around perimeters of offshore banks. Outcrop studies of these sediments have educed much information about them in recent years. Rigby (1958) and Jacka *et al.* (1972) on the Permian Reef Complex; Hopkins (1977) and Cook *et al* (1972) on the Devonian of Alberta; Carrasco and Enos (1977) on the Mexican Cretaceous; and Biddle (1978) on the Triassic of the Dolomites, Italy. They occur as sheet-like masses tens of meters thick or spread in patches many kilometers out into basins beneath the slopes of carbonate shelf margins. Their sediments may be coarse blocks up to many meters across, surrounded by much bioclastic debris. Recent sedimentary models for such have been described below the Holocene reef rim at Belize (Ginsburg and James, 1973) and on the east side of the Bahama Banks (Lynts *et al.*, 1973; Schlager *et al.*, 1976). About 20,000 years ago the drastic Wisconsin sea level lowering and the following rapid rise induced much sedimentary debris in the seaward and sloping edges of many carbonate platforms. This is found downslope, below 100 m under the present sea level. The processes responsible appear to be those of terrigenous clastics: mass flow induced by gravity, major slumping and turbiditic density flows.

Diagenesis and porosity The microfacies typically contain a jumbled variety of normal marine shelly organisms. The sediments are characteristically very fossiliferous. Angular lithoclasts of slope sediments and blocks of cemented reefy boundstone may be admixed. Some slope beds consist of abundant particles of loosely articulated organisms living on the upper slopes or as niche dwellers in the boundstone reef framework. For example, the modern calcareous alga *Halimeda*, crinoid debris in Ordovician to Mississippian beds, and phylloid platy algae in the later Paleozoic.

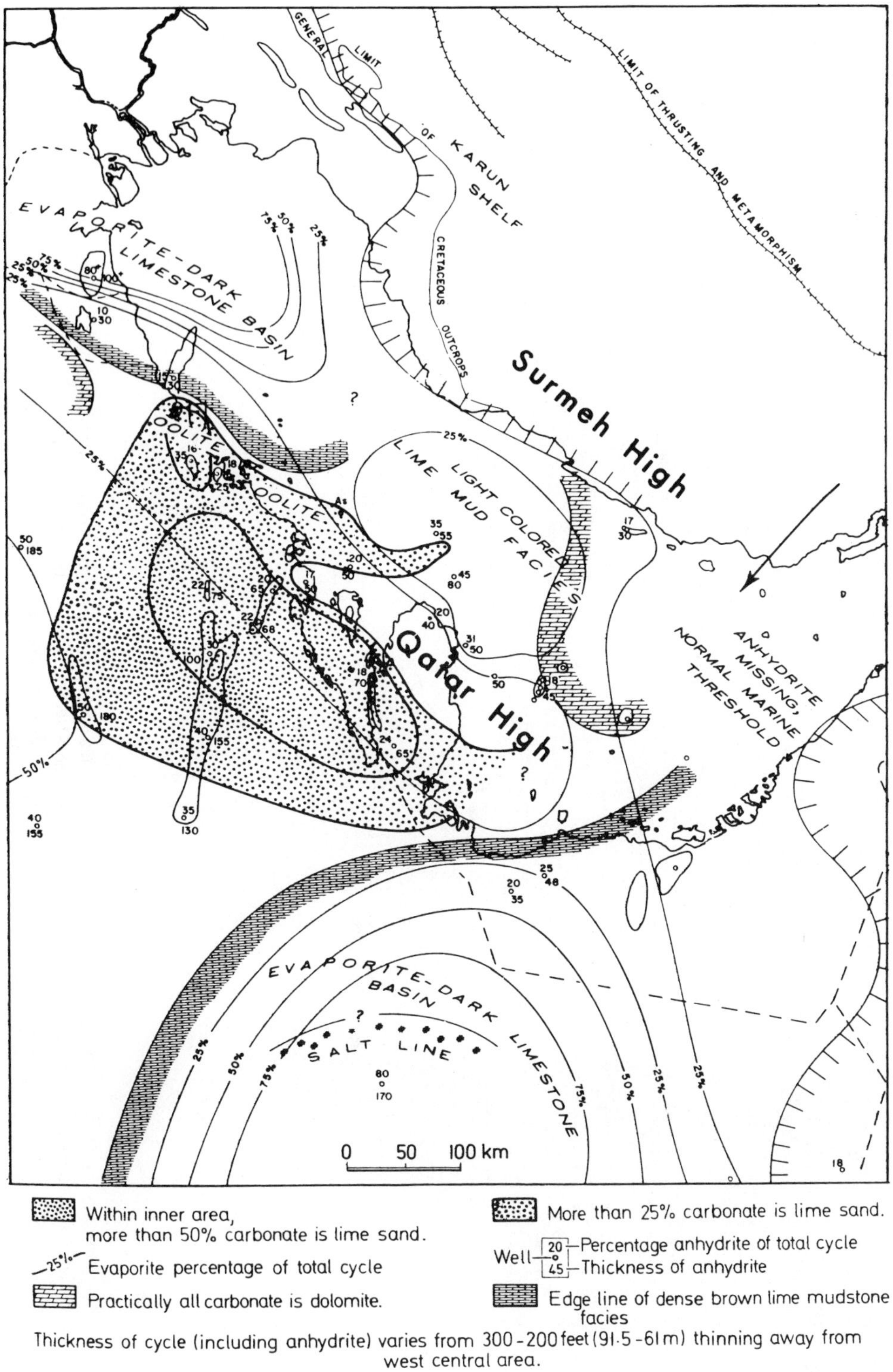

Fig. 1 Primary Grainstone Porosity. Facies map of Arab D cycle, percentage lines indicating evaporite and dark shaly limestone in basins north and south of Hasa coast of Saudi Arabia. Grainstone buildup is in shelf area between the basins and between the Arabian Shield to the west and the Qatar-Surmeh high on the east (from Wilson, 1975).

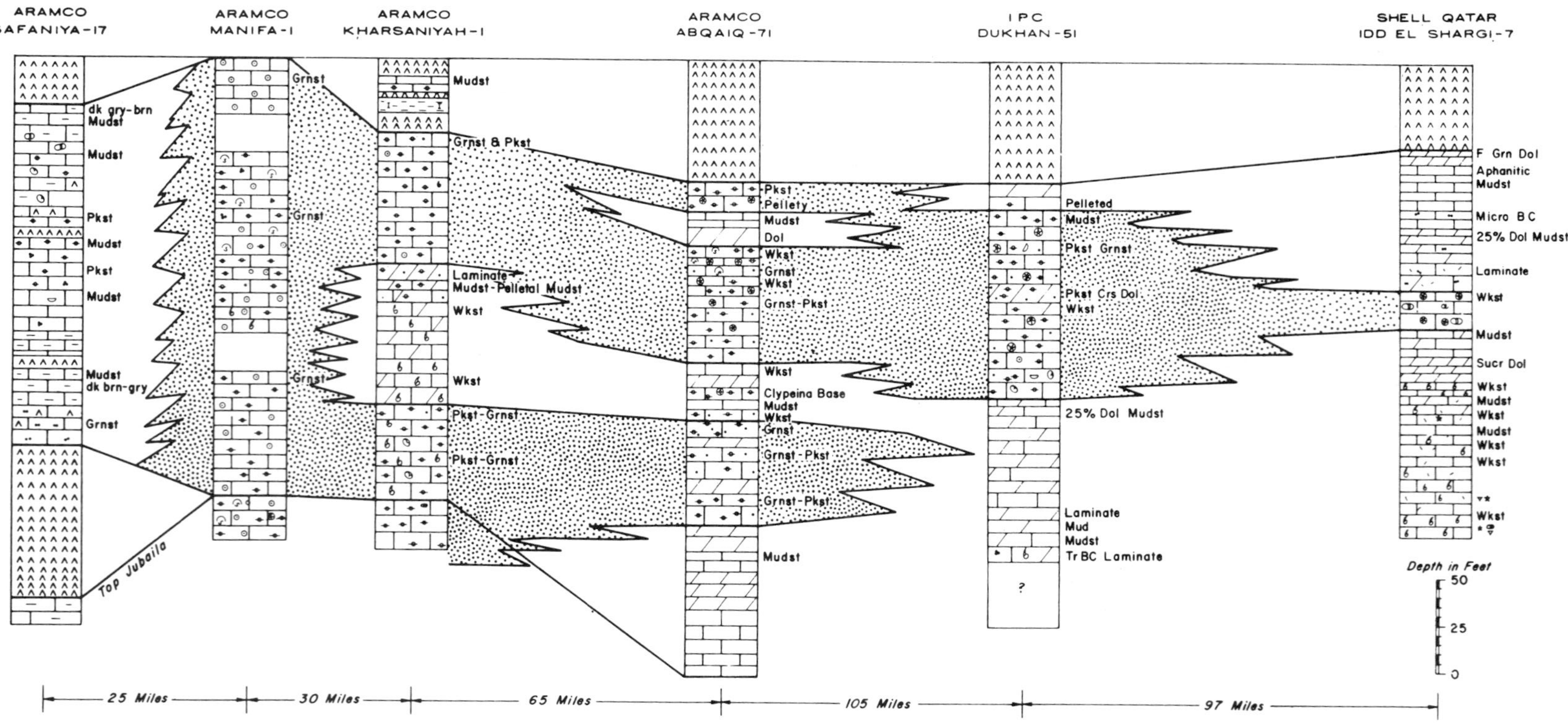

Fig. 2 Primary Grainstone Porosity, Arab D facies in cross section down Hasa coast from Neutral zone on north (left) to Qatar and east into Persian Gulf—about 550 km. Shows double body of 20-35% porous lime sand in Arab D. Note northern sand buildup before change in Basrah "basin" and the eastern disappearance of grainstone into tidal flat dolomite at Id el Shargi (from Wilson, 1975).

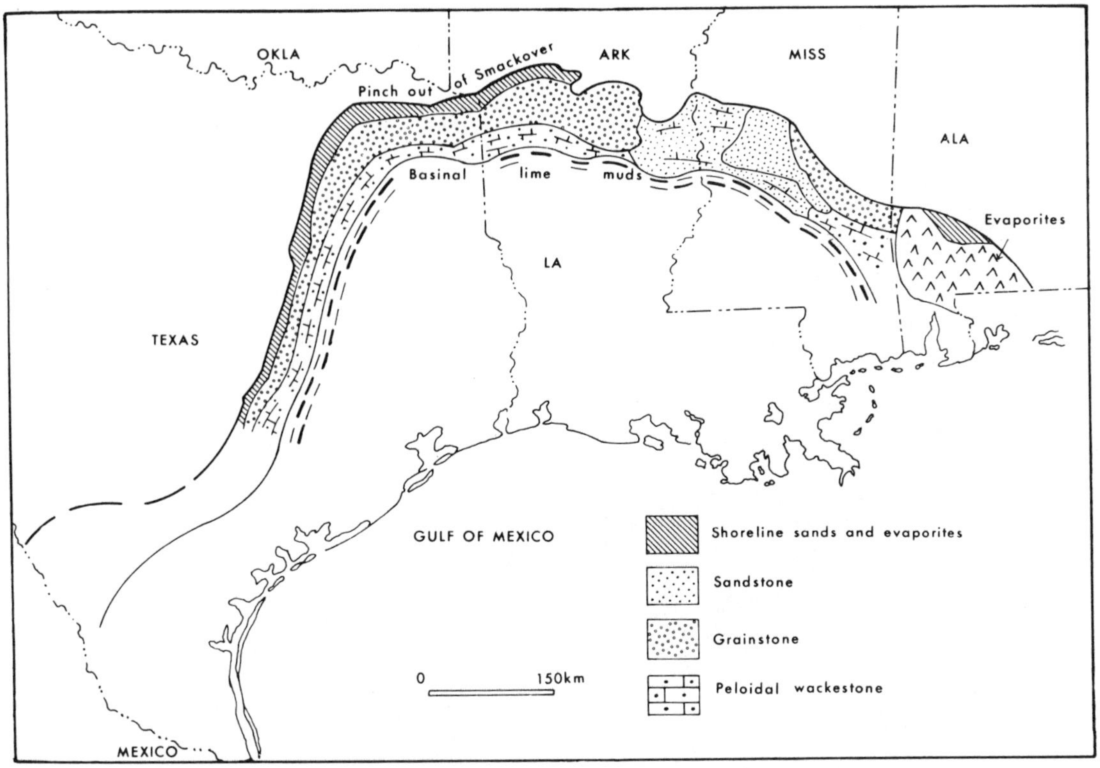

Fig. 3 Primary Grainstone Porosity. Upper Smackover facies, northern Gulf of Mexico, after Bishop (1968), showing the peripheral grainstone buildup a few 100 m thick and about 75 km wide around the Gulf grading into very thick dark lime wackestone and basinal lime mudstone which reaches more than 1 000 m. Primary porosity in southern Arkansas and northern Louisiana gives way in northeast Texas to secondary porosity caused by dolomitization and leaching (from Wilson, 1975).

Diagenetic characterization of such sediment is still proceeding. Its coarseness results in unusually high initial porosity. In most instances burial in the downslope areas has resulted in protection from subsequent dolomitization but the giant Poza Rica field below the Golden Lane of Mexico is a notable exception to this rule. The role of subsurface lenses of meteoric water in cementing, leaching and stabilizing this debris is important.

Occurrence Despite the detailed outcrop studies on this type of carbonate sediment the only reservoirs known to date are those associated with the Mexican rudist reef trends of the Middle Cretaceous along the western Gulf of Mexico and possible the forereef shoals area of Oligo-Miocene Main limestone of Kirkuk (Henson, 1950). Production also exists in the lower part of Kirkuk (Iraq) below the Lower Cretaceous rudist fringe and in forereef slope deposits of Bu Hasa in the United Arab Emirates of the Arabian-Persian Gulf (Harris *et al.*, 1968). Some of these are major reservoirs. The Poza Rica accumulation alone is at least $2.4 \times 10^8 m^3$ (1.5 billion barrels) (Fig. 7) and an appreciable part of the many fields in the Reforma trend of south Mexico is in foreslope deposits, in some of which fracturing plays a role (Guzman, 1978). As more offshore exploration proceeds along carbonates developed over passive margins of cratons, this type of downslope reservoir should become more prominent.

4. Stratigraphic traps in shelf cycle

Sedimentary geometry The depositional spectrum of carbonate facies across shelves has built into it a stratigraphic trap for hydrocarbons. Commonly shelf strata are cyclic. The facies progress from offshore, argillaceous, deeper and quieter water deposits (affording source beds for petroleum) to upslope shelf marginal strata with naturally porous organic boundstones and lime sands. Shelfward sediments may include fine-grained carbonates deposited in lagoons or bays of moderate depth, some of which had restricted circulation and euxinic conditions not too different from offshore source beds. The shelf interior in the carbonate regime commonly ends in a tidal flat environment and, in an arid climate, in suptratidal sabkha evaporites. This climatic factor has two results. 1. The evaporites enhance the possibility of dolomitization of the interti-

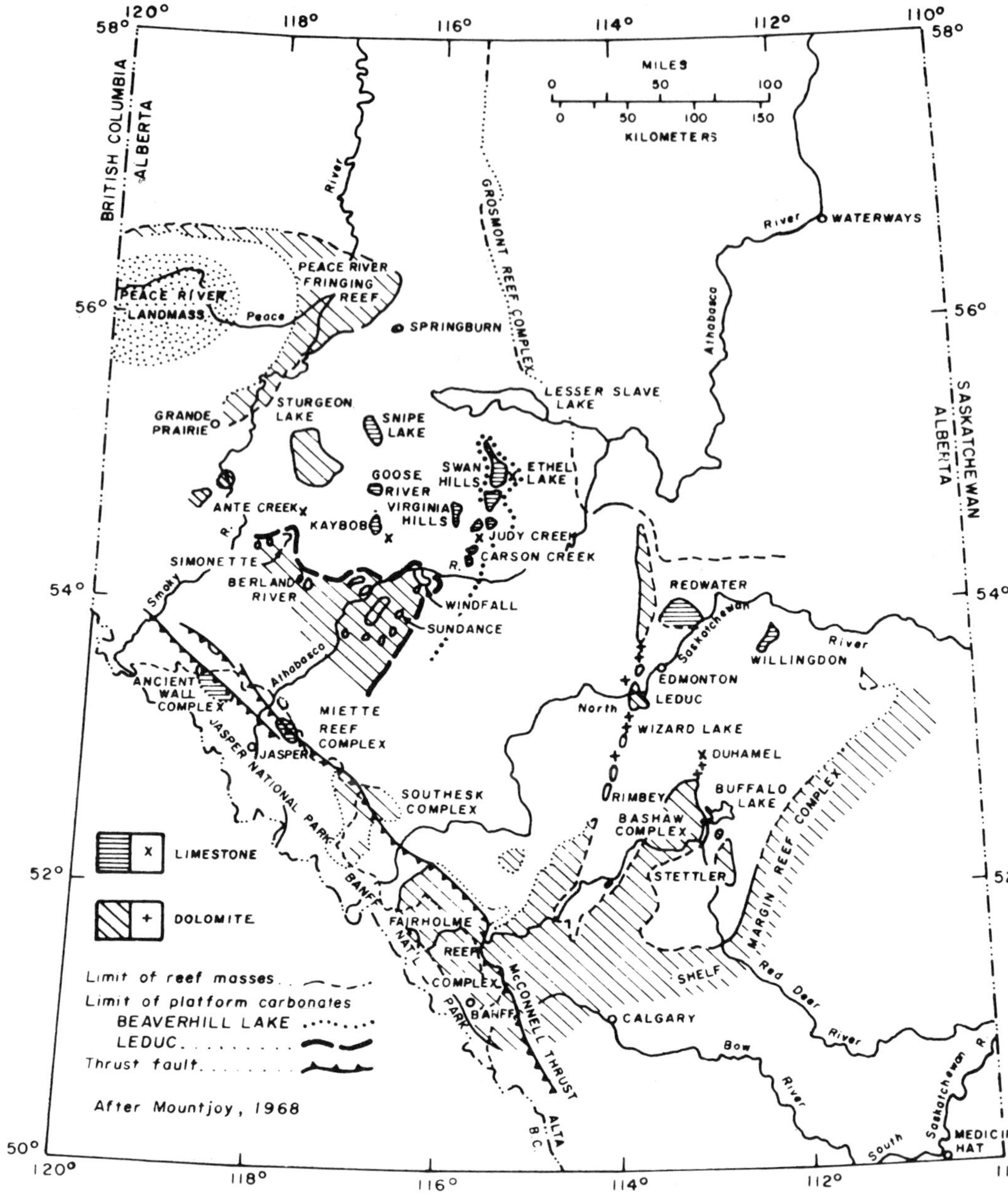

Fig. 4 Reefy Shelf Margin Porosity. Linear trends of Late Devonian platforms across Alberta basin. All the reservoirs in these great buried platforms are dolomite except for those isolated ones to the northeast far out in the basin (Toomey et al., 1970, Fig. 1).

dal and subtidal lagoonal carbonates rendering impermeable lime muds to more permeable finely sucrose-textured dolomite and thus developing a second updip reservoir; 2. The sabkha evaporites (anhydrite or gypsum) form an updip seal. A seal is also common above the cycle as the base of the next cycle may be a dense argillaceous limestone farther out in the basin. In many examples the whole series of sedimentary cycles is progradational, each cycle becoming more evaporitic higher in the section and stepping out basinward. In several such sequences the basinward outbuilding is so great that the updip sabkha anhydrite of the higher cycles directly overlies the grainstone or intertidal dolomite reservoir of the next underlying cycle. Hydrocarbons are thus trapped and cannot escape either updip (laterally) or vertically.

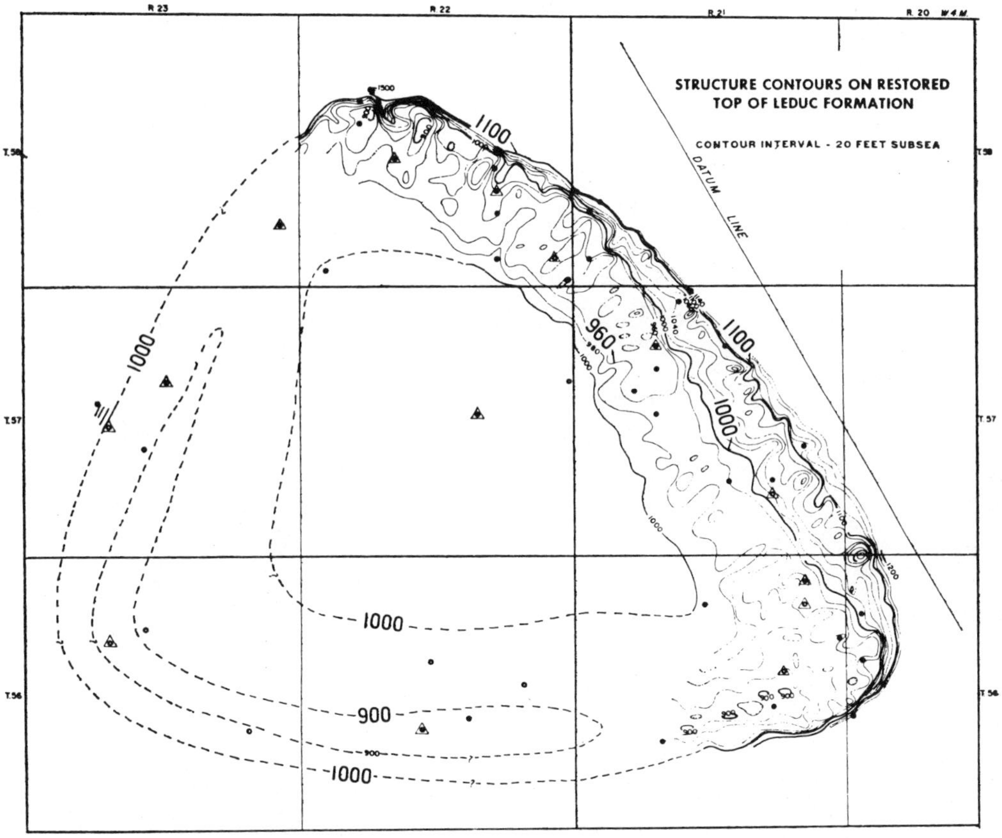

Fig. 5 Reefy Shelf Margin Porosity. Detail of surface of Redwater offshore carbonate bank, Alberta basin, Canada. Contours are in feet (Klovan, 1964, Fig. 5).

Diagenesis and porosity The two common reservoirs in these cycles include: 1. pelloidal and oolitic lime sands commonly affected strongly by splash zone cementation which develops vadose pisolite. This sediment was reworked somewhat as the shoals were placed in and out of water in an arid environment with rare periodic incursions of fresh water; 2. bioclastic lime wackestones were commonly dolomitized and contain high porosity caused by solution of the scattered but abundant shelly debris. This dolomite may be coarse because the rhombs grew more slowly further downdip from the source of the magnesium-rich fluid.

The true intertidal algal laminites and swamp deposits are also commonly dolomitized but the fine grain, lack of bioclastic debris, increased dolomite growth, and contamination with replacement anhydrite and perhaps silt and clay, may result in less satisfactory reservoir rock.

Occurrence Documented sequences of this character with dolomite reservoirs best developed in the subtidal environment are those studied by Lucia (1972) in the West Texas Middle Permian San Andres (Fig. 9) and the Ordovician-Silurian studied by Roehl (1967) in the Cedar Creek anticline of Williston Basin. A well studied model for this facies trap is the Mississippian section of the Williston Basin (Irwin, 1965; Edie, 1958; and more recently, Heck 1978). Numerous small fields in the Mission Canyon-Charles section of the east side of the Willison basin are developed in both lime sand and dolomitized mudstone (Murray, 1969) (Figs. 10, 11).

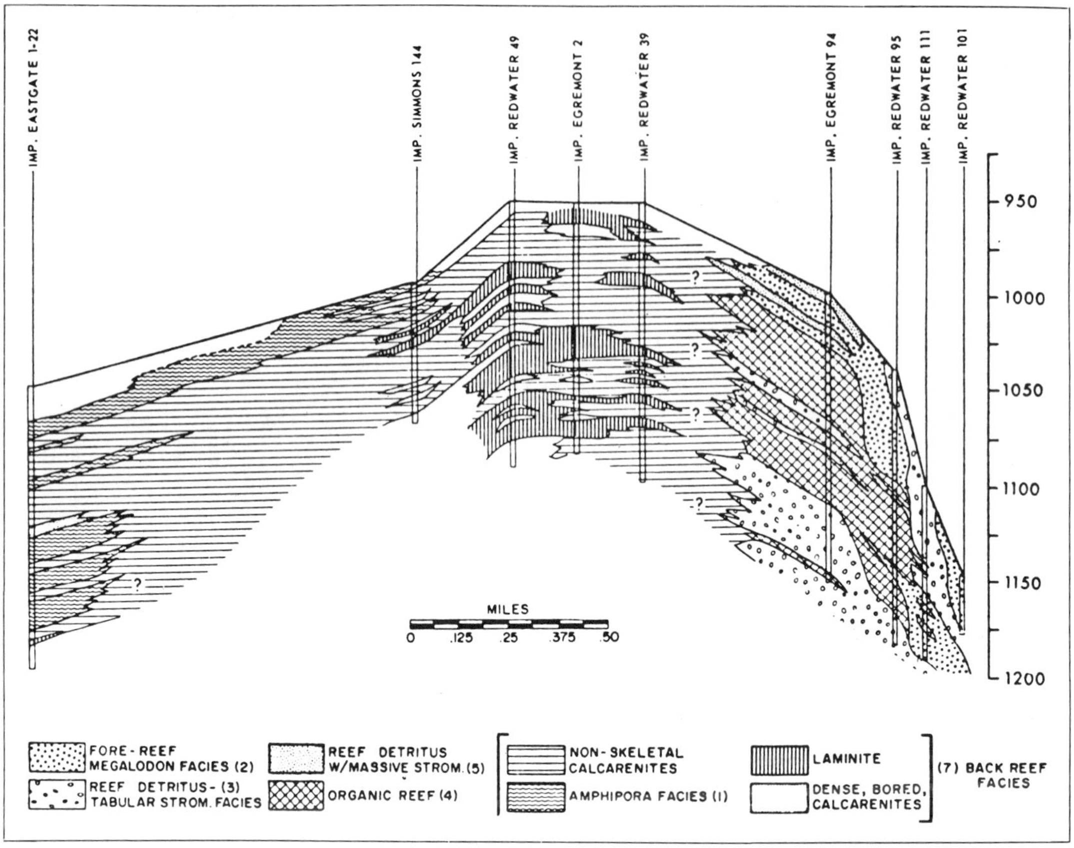

Fig. 6 Reefy Shelf Margin Porosity. Composite cross-section of upper Redwater reef complex showing facies across eastern edge of the large offshore bank in the Alberta basin, Canada. Porous reefy and grainstone facies on the eastern margin owe their origin to primary voids in the boundstone and bioclastic debris plus some secondary leaching (Klovan, 1964, Fig. 11).

5. Porosity-permeability developed beneath a regional unconformity through leaching and dolomitization

Sedimentary geometry Areas of such porosity are developed on surfaces of subaerial exposure beneath widespread unconformities. These surfaces may be irregular, with hydrocarbon occurrence controlled by facies variation in the reservoir, paleomorphologic (karstic) relief on the surface, and some degree of fracturing and faulting of the reservoir beneath the unconformity. In addition initial relief created by upbuilding of carbonate masses, whose crests are subjected to later subaerial exposure, may be important in controlling distribution and intensity of leaching and dolomitization. Edges of carbonate buildups theoretically should have more opportunity for diagenesis than central interiors because of more fluid flow through the strata on the bank edges. Sea level drops would naturally increase hydraulic head at edges owing to proximity to steep slopes off the bank. Some major fracturing is known to control diagenetic patterns beneath an unconformity. For example, the Albion-Scippio trend in the top of the Trenton Formation of southern Michigan. Another setting of diagenesis beneath an unconformity is known in certain basins: evaporitic draw-down following formation of intrabasinal pinnacle reefs exposes them to leaching and dolomitization. After this event these carbonate pinnacles may be surrounded and overlain by salt and anhydrite formed during the post-reef phase (Mesolella *et al.*, 1974, Silurian of Michigan basin).

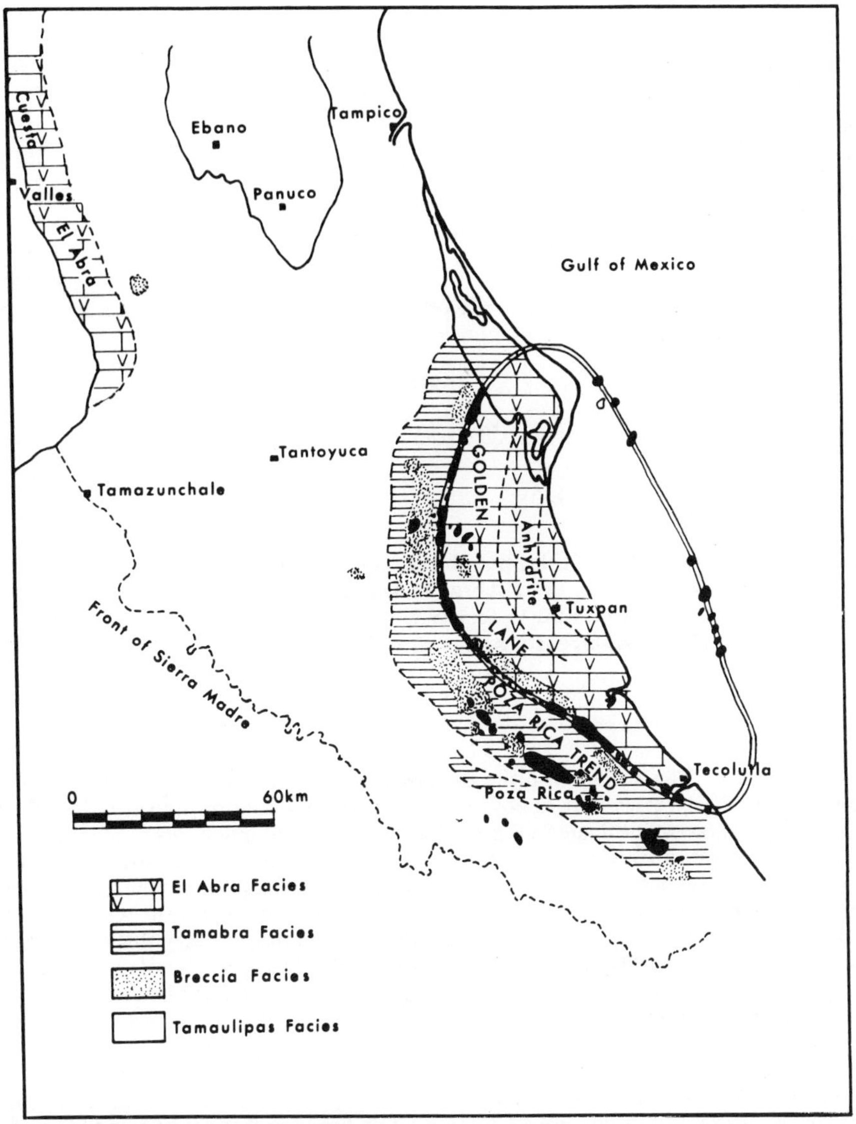

Fig. 7 Reefy Shelf Margin Porosity and Slope Debris Porosity. Oil fields and facies of Middle Cretaceous Golden Lane and Poza Rica trends of the Gulf coast of Mexico. The dotted pattern includes porous bioclastic-lithoclastic breccia of Cretaceous age along the Poza Rica trend. This is in debris flow deposits deposited in pelagic lime muds below the platform edge (Enos, 1977). The Golden Lane lies in the subsurface. The eastern margin of the outcropping Valles platform is shown at the Cuesta de El Abra in the northwest corner of the map (after Coogan et al., 1972, Fig. 2; from Wilson, 1975).

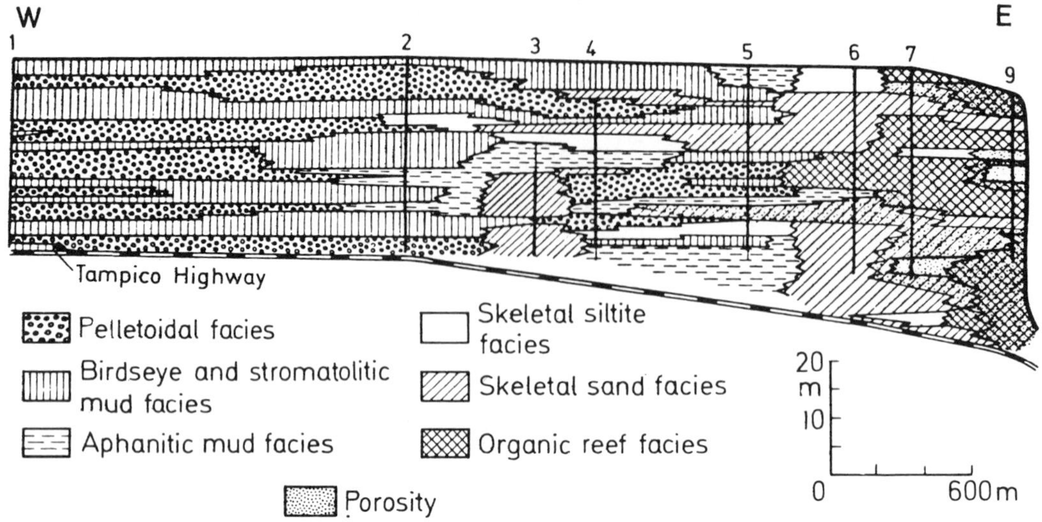

Fig. 8 Reefy Shelf Margin Porosity. Facies distribution of El Abra limestone as determined by factor analysis, in a traverse across the Cuesta de El Abra, 15 km east of Valles, San Luis Potosi, Mexico. The superimposed fine dot pattern indicates that porosity is best developed in the bioclastic sand facies and rudist accumulations at the shelf margin, eastern edge of the Valles platform (from Griffith et al., 1969, Fig. 8).

Diagenesis and porosity The dominant processes creating porosity and permeability are leaching and dolomitization, but other processes may follow them (collapse-brecciation and dedolomitization). The chances for fluid flow and favorable diagenesis are enhanced by lack of much early cementation, but previously lithified carbonate strata are also susceptible. Leaching and immediate and nearby precipitation in the vadose zone can be recognized petrographically by meniscus and pendent cement patterns (Dunham, 1971; Ward, 1970) and the blocky mosaic calcite of phreatic zone cementation below the water table is now well documented (Land, 1970). Generally subsurface fresh water lenses may be very thick even under a water table which is thin. Aragonite bioclasts may be leached thoroughly even though micrite matrix, pellets, and calcitic fossils have been preserved, but it may be difficult to determine in many instances whether the solution was caused by fresh water lenses or subsurface connate water. This is still an unsolved problem in carbonate geology.

Dolomitization beneath regional unconformities is commonly found in patterns superimposed on more regular, earlier, and stratigraphically controlled dolomite. Old broad arches are known to have dolomitized strata beneath unconformities; e.g., Wisconsin Arch (Badiozamani, 1973), the San Marcos Arch (Rose, 1972) and the Wyoming shelf (Andrichuk, 1955), among many others. The fluid regime downdip from the unconformable surface may include evaporitic brine or fresh water depending on climate at the time. Possibly both types of fluid are necessary for dolomitization beneath such unconformities: brines to concentrate magnesium ions and migrating lenses of fresh water from updip to force a mixing zone through strata saturated with high magnesium brines or just normal sea water. If this is so, a climatic change from evaporite to more humid conditions would favor dolomitization.

Petrography shows that late dolomitization tends to be coarse-grained, more permeable, and commonly vein-filling as well as replacement in habit. Dedolomitization is a common late diagenetic phenomenon beneath regional unconformities, particularly when evaporitic sulfates in updip strata may have been leached to provide abundant calcium sulphate in the meteoric water affecting strata beneath the unconformity.

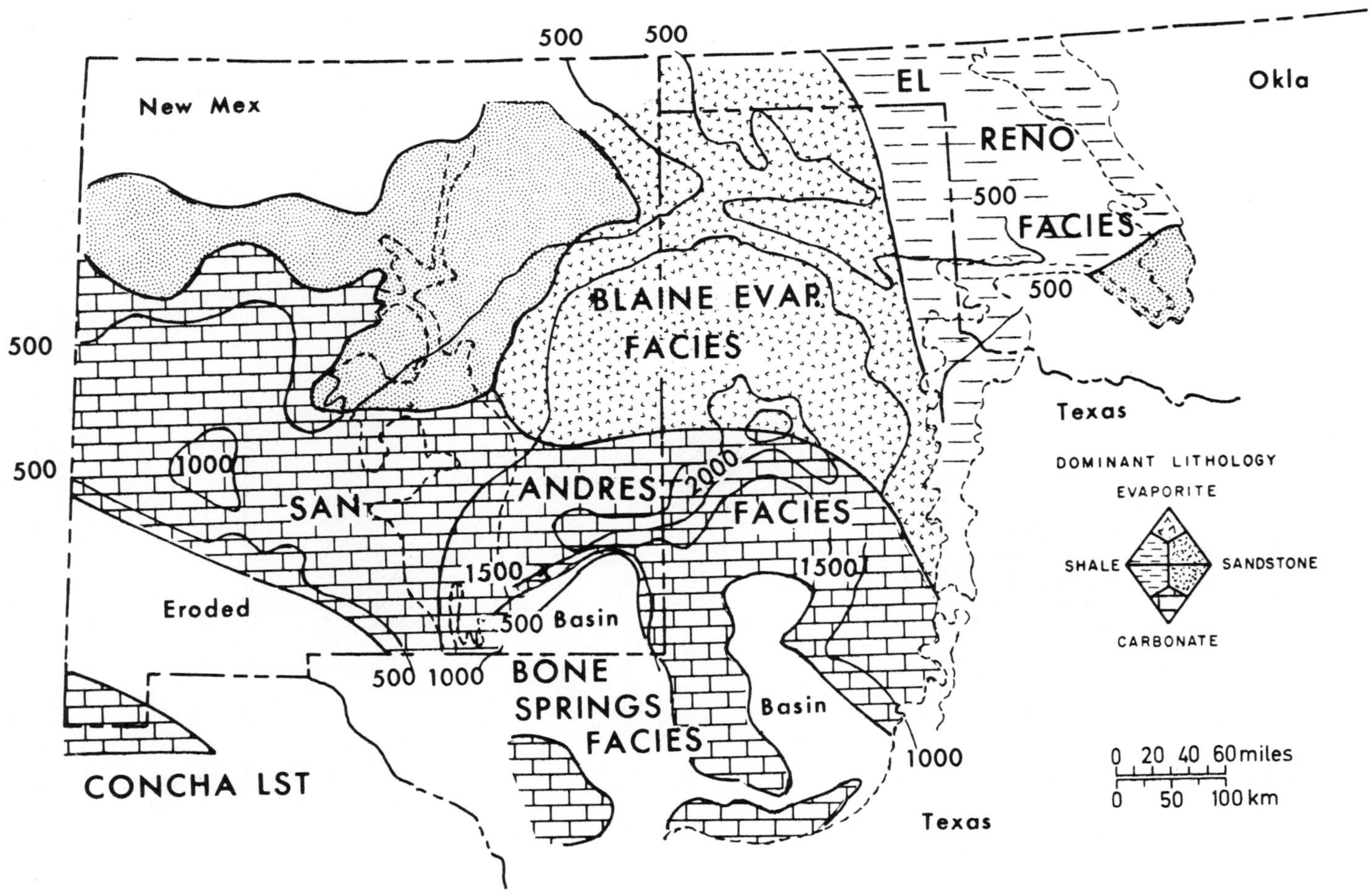

Fig. 9 Regional Facies as Stratigraphic Trap. Lithofacies and isopachous map of the Middle Permian San Andres Formation. The El Reno facies is red silt and shale; the sandstone (dotted pattern) is the Glorieta of New Mexico. The Blaine is terrigenous sediment and extensive gypsum. It constitutes a regional updip seal for the hydrocarbons which accumulated in porous dolomite of the San Andres and Concha formations. These are cyclic shelf carbonates with a lime sand rim at the northern edge of the basins. The basinal Bone Springs facies is black laminated limestone (from Meissner, 1972, Fig. 14).

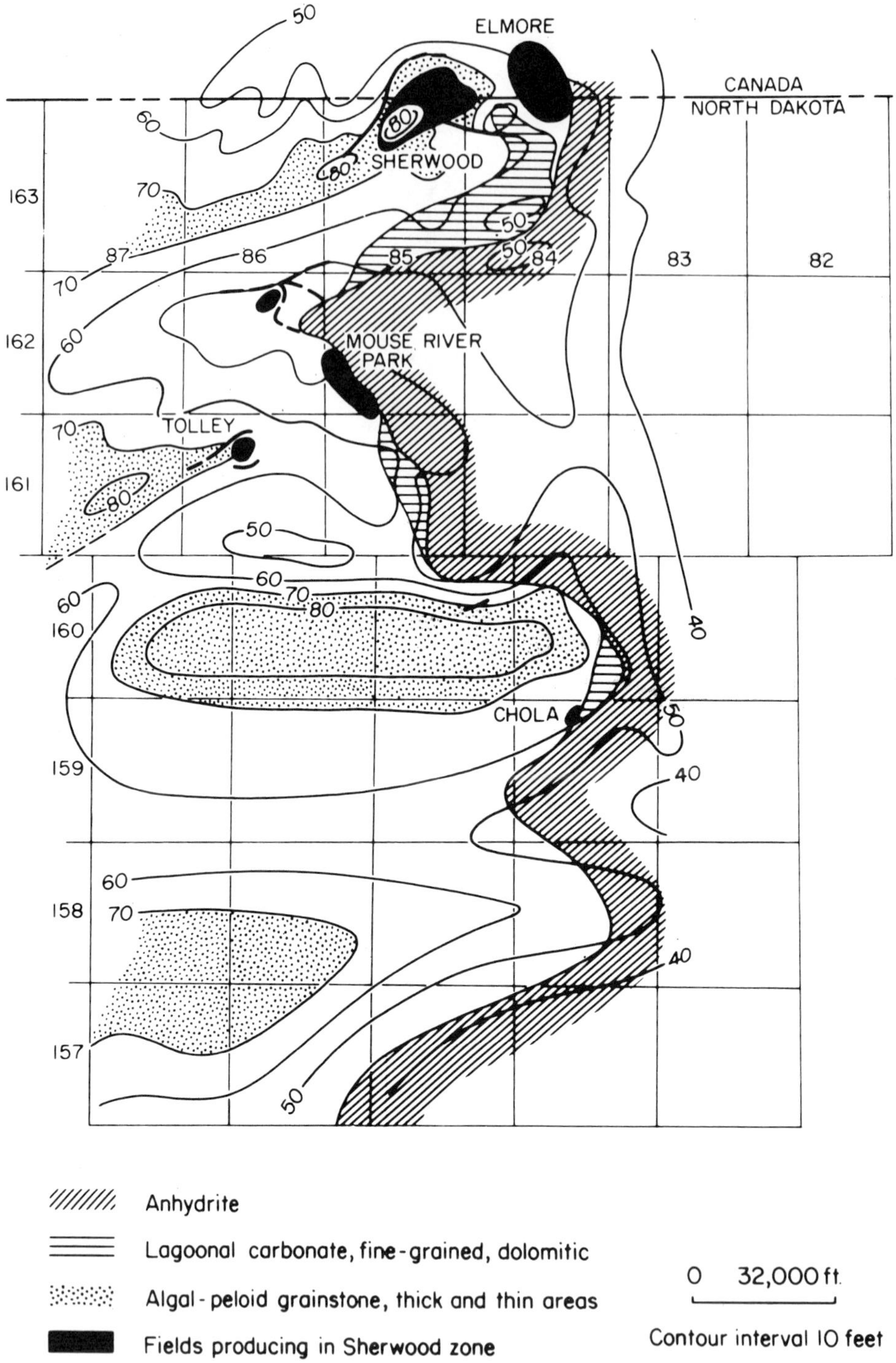

Fig. 10 Regional Facies as Stratigraphic Trap. Facies map of the Sherwood zone in the northeastern side of the Williston basin. This time-stratigraphic unit contains updip anhydrite with a thin irregular area of tidal flat and lagoonal dolomite to the west. The thickness lines show linear shoals (flattened lenses) of grainstone 6-12 m (20-40 ft) thicker than normal and oriented as 24-32 km (15-20 mile) long "bars" perpendicular to the north-south facies strike. The oil fields are concentrated where the grainstone and dolomite facies grade eastward to the sabkha anhydrite.

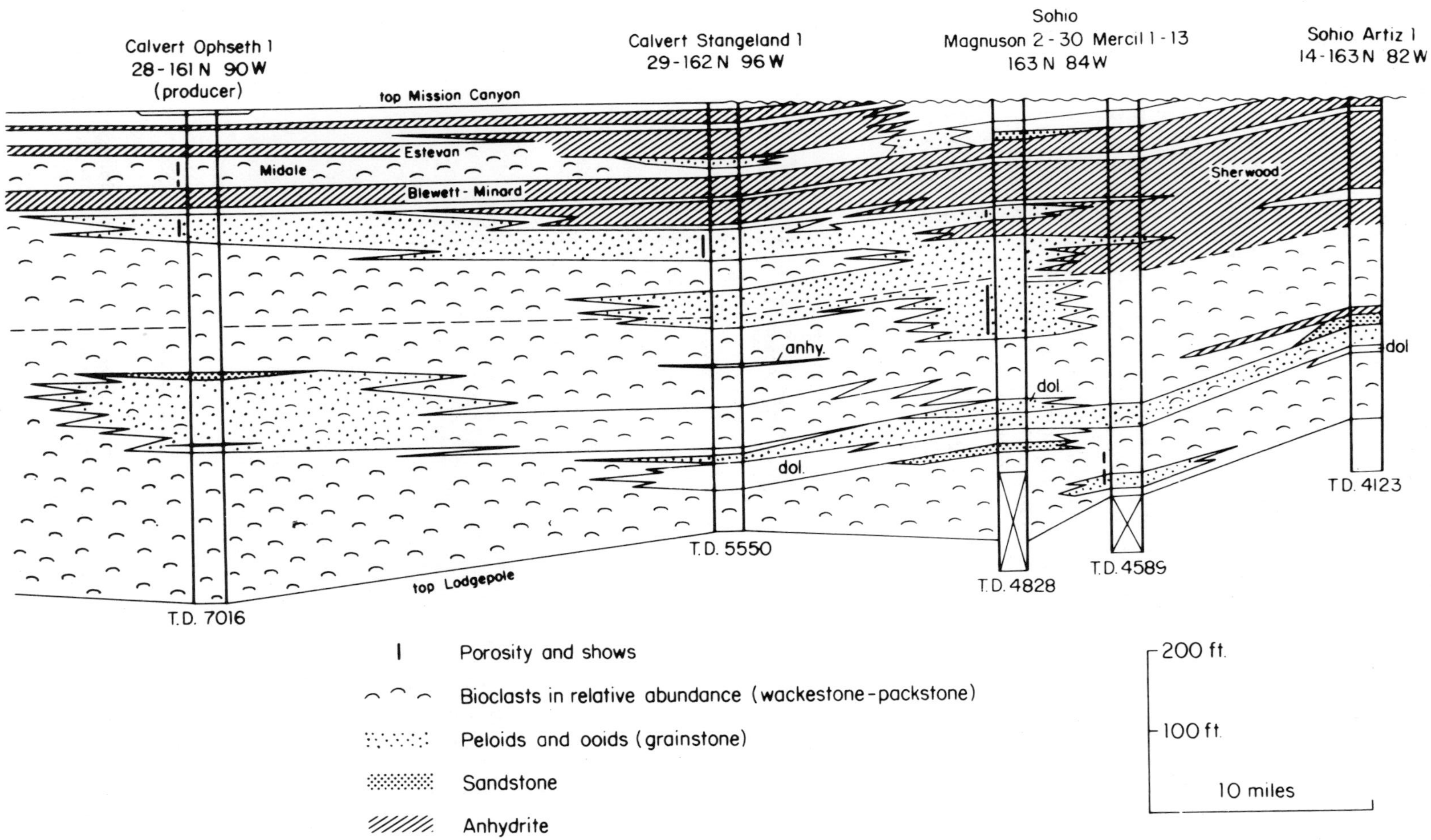

Fig. 11 Regional Facies as Stratigraphic Trap. Cross section of Mission Canyon strata (150-180 m; 500-600 ft thick) about 88 km (55 miles) long along the U.S.A.-Canadian boundary in north central North Dakota (Burke and Renville counties). Shows of hydrocarbon and some production occur generally where the grainstone facies is overlain and replaced easterly by the evaporites.

Occurrence Several settings may be taken as models for hydrocarbon accumulation associated with unconformities. Regional unconformities around edges of large basins commonly induced the diagenesis necessary for reservoir development. Certainly paleomorphology aids in entrapment, but the whole unconformity surface is generally worth exploration. For example, the northern shelf of the Anadarko basin has three such regional unconformities which are productive: 1. the top surface of the Arbuckle, 2. the top of the Hunton with selective downdip dolomitization beneath the Woodford black shale cover, and 3. the top of the "Mississippi Lime and Chat" a residual chert breccia and collapse zone under the Pennsylvanian stretching north over the central Kansas uplift. Widespread production is known at the post-Edwards unconformity over the San Marcos Arch of Texas (Luling field) (Fig. 12, 13). The Cretaceous also produces in mixed grainy and muddy limestones leached beneath an unconformity in the Wasia Formation of Oman in the Natih and Fahud structures. One of the oldest fields in the U.S.A. (Lima, Indiana-Ohio) owes its origin to unconformity dolomitization in the Trenton Limestone on the edge of the Findlay Arch.

A further model for production enhanced by unconformity is provided by the Golden Lane trend of central Mexico (Fig. 7) and from the top Shuaiba unconformity of Bu Hasa in the United Arab Emirates sheikdoms of eastern Arabia. In these giant Cretaceous fields the leaching and karsting occurs at the top of rudist reefs where the initial relief and high original porosity contributed to the effective diagenetic processes.

In addition, fracturing and faulting at the time of unconformity formation may contribute to porosity and permeability trends and control the type of solution. Linear breccia zones and pipes are known at unconformable surfaces in the Albion-Scippio Trenton of southern Michigan and the zinc-lead ore emplacement areas of eastern and central Tennessee.

A fourth model for unconformity production is given above in the brief mention of the effect of evaporitic drawdown in basins. Examples of this type of production in leached and dolomitized pinnacle reefs include not only the Michigan Basin Niagaran but also the Middle Devonian Winnipegosis of the Williston Basin and the major Zama-Rainbow trend of northern Alberta.

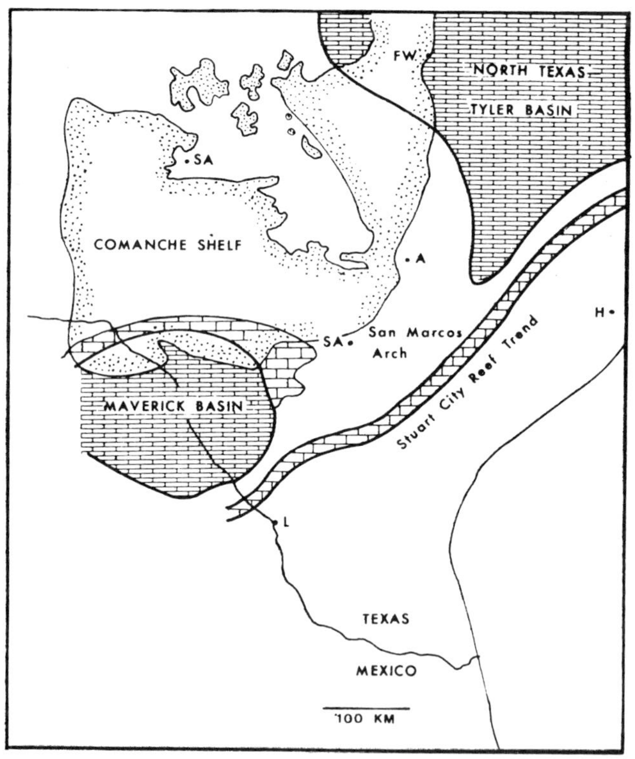

Fig. 12 Shelf Exposed Beneath an Unconformity. Central Texas paleogeography during Albian time showing position of the San Marcos arch. This area was subjected to subaerial exposure and leaching during Albian time before being inundated by the late Albian "Georgetown" transgression (from Wilson, 1975).

Lastly, leaching of phylloid algal rich carbonate strata in the Pennsylvanian of the southwestern U.S.A. forms a large number of reservoirs including the Scurry County Mass (Vest, 1970). This diagenesis seems clearly associated with repetitive sea level lowering during the Pennsylvanian in which mounds formed by the algae and associated encrusting foraminifera were exposed, leached, brecciated, and fractured (Toomey *et al.*, 1977) (Fig. 14).

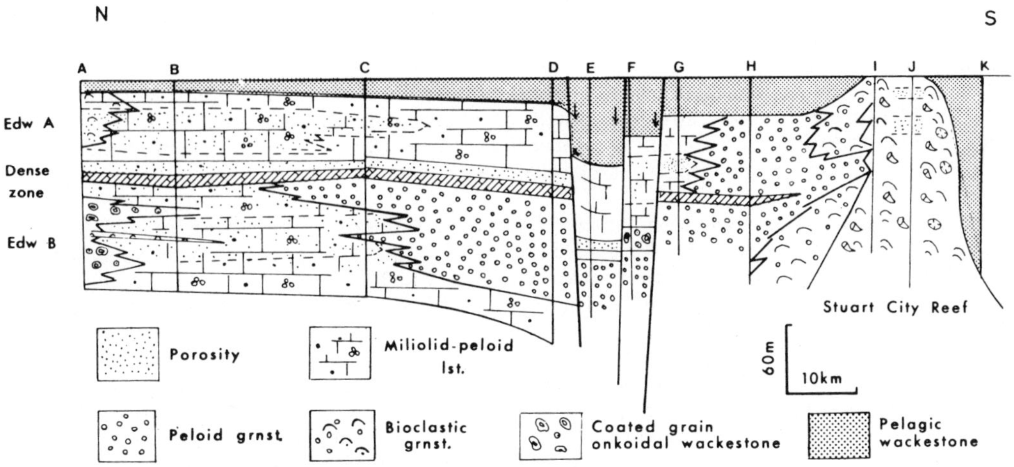

Fig. 13 Shelf Exposed Beneath an Unconformity. Cross section from north to south across the San Marcos arch (after Griffith et al., 1969) of about 200 m of section over a distance of 120 km. Note that the Stuart City reef contains little porosity, the near back reef grainstones practically none, and that porosity is best developed in the dolomitic tidal flat and lagoonal facies. It was caused by leaching by meteoric water over the San Marcos arch. Accumulations are governed by later faulting (from Wilson, 1975).

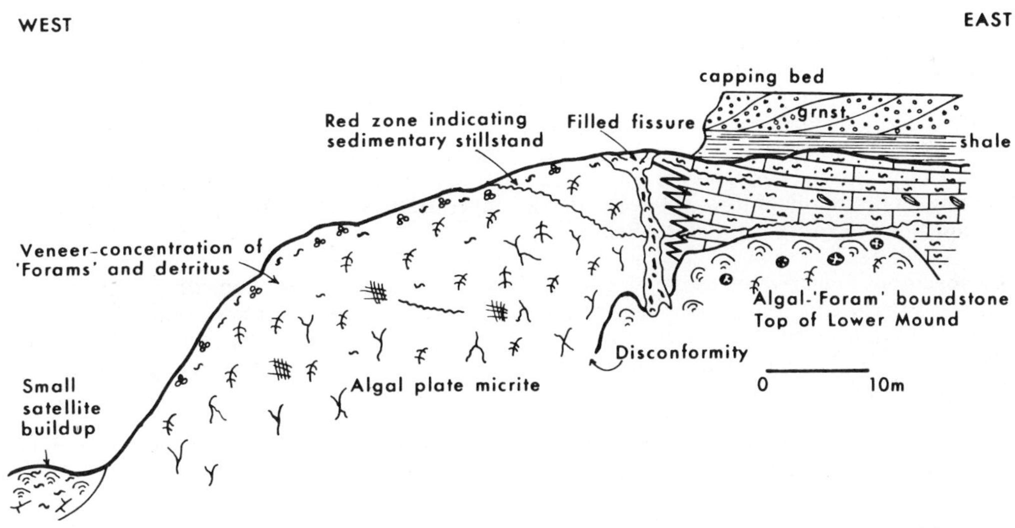

Fig. 14 Mound Beneath an Unconformity. Sequence of facies in typical Late Pennsylvanian mound, ± Yucca Mound in Sacramento Mountains, New Mexico (see Toomey et al., 1977). Note numerous disconformities within complex buildup indicated by alteration of lower mound boundstone, and higher red zones, veneer concentration of foraminifera detritus, and filled fissure. The brecciated fabric of micrite and phylloid algae of the main mass is filled with secondary cement in this outcrop example but commonly leached and porous in subsurface buildups (from Wilson, 1975).

6. Chalky-textured reservoirs

Chalky-textured carbonate is defined as that having well-disseminated microscopic pore space in a micritic matrix. The North Sea Chalk of Late Cretaceous age typifies this texture as well as much oceanic carbonate ooze. In these examples, the matrix consists of low magnesium calcite coccolith fragments (from 2 to 20 μm) in various stages of preservation from perfect disc-shaped forms to partially dissolved plates encompassed by tiny rhombic calcite. The grain size of the secondary rhombs remains exceedingly small, measured in only a few micrometers. Coccophores are pelagic microscopic algae and typically thrive in the offshore photic zone of the open sea.

In addition, other chalky-textured carbonate is widely distributed in Cretaceous and Tertiary beds in North Africa and the Middle East among strata deposited in shallow subtidal to tidal flat environments. Such beds have little or no coccolith remains and were probably originally dominated by aragonitic shelly microscopic debris. The well-disseminated microporosity in such strata must result from special circumstances which inhibited the cementation so characteristic of carbonate sediment.

Sedimentary geometry Reservoirs from pelagic chalks form widespread sheets with no known discrete sedimentary bodies controlling porosity. A possible exception to this is the presence of beds and lenses of bryozoan-rich detrital sediment in the type Danian outcrops, which may represent a facies approaching shoreline. In general the facies is so uniform because much chalk is an offshore pelagic sediment deposited in marine basins of at least a hundred meters.

Diagenesis and porosity Initial porosity of the microbioclastic debris comprising chalky micrite is very high. Studies of cores taken in the modern ocean by the Deep Sea Drilling Project indicate up to 80% porosity in the original sediment. It has been estimated that early physical compaction might reduce this initial porosity by 30-40% to about 50%. Lime mud settled in shallow modern lagoons retains almost 50% porosity from about 30 cm below the sediment-water interface to at least 10 m depth. But even after burial and further diagenesis, porosity in chalk remains very high; 20 to 35% is common today. Permeability is a different matter. It is low because of the fine grain size.

Controls on the preservation of the original high porosity have been studied by Scholle (1977) for the North Sea Cretaceous as well as by Neugebauer (1974). In the first place, the original mineralogy of pelagic chalk is dominated by stable low magnesian calcite so that the sediment is not as reactive as shallow water carbonate. Yet burial beneath 1 000 to 2 000 m promotes extensive solution-compaction welding of the tiny low magnesian calcite grains, particularly when interstitial waters low in magnesium are present. Temperature of burial obtained by 0^{18} to 0^{16} ratios correlates well with depth of burial and porosity reduction and petrographic changes observable with the scanning electron microscope. Effects of pressure solution such as small stylolites and microscopic wispy seams, which truncate burrows and larger clasts, also indicate burial diagenesis. That this happened under reducing conditions of burial is indicated by the presence of late generation ferroan calcite cement filling interiors of fossil grains within the chalky matrix. Thus shallow burial seems most favorable for preservation of chalky-textured porosity. Another factor of importance is the absence of argillaceous material which would encourage solution-compaction. In addition, early oil migration and subsequent seal and burial may protect the fabric from solution-compaction-welding. After burial, overpressuring results and fluid flow will overcome the impermeability caused by fine grain size (as in the Danian production at Ekofisk in the North Sea).

In a homogeneous matrix with rigidity effected by a certain amount of solution-compaction-welding, fracturing may be induced just by overburden pressure (e.g., production in certain parts of the Austin Chalk in the Texas Gulf Coast which has up to 20% porosity but permeability of only .3 mD). Obviously fracturing may be very important in improving permeability in such rocks. It has been proposed that sediment with grains scattered in a micrite matrix may be more susceptible to fracturing than pure micrite.

Orientation of micrite grains may be important in determining and predicting permeability. This and the effect of tectonic compressive stress on chalk fabrics is just now being investigated.

In pelagic chalks dolomitization is very rare, but in chalky-textured shallow water carbonates it is important and is discussed below.

Occurrence The major and almost unique occurrence of oil in the Danian pelagic chalk of the North Sea has obscured the fact that major reservoirs exist in chalky-textured (presumably little cemented) shallow water carbonates of Mesozoic and Tertiary age in the Persian-Arabian Gulf and Libya's Sirte Basin. Production is found in numerous horizons of chalky packstone in the Trucial Coast Thamama (Lower Cretaceous) of the Murban field. These are separated by dense, somewhat argillaceous, heavily stylolitized zones. The Wasia Formation (Middle Cretaceous) of Fahud and Natih fields in Oman produce from similarly grainy-textured rock. Sufficient grain content provides permeability in a chalky textured matrix wherein considerable porosity (20-30%) is retained. The effect of an overlying disconformity in enhancing permeability of such chalky strata cannot be overlooked. A combination of microscopic fracturing and some leaching at the unconformity probably improves the permeability of chalk.

Partial dolomitization and interbedded fine grain dolomites in intertidal chalky-textured sediments provide excellent reservoirs in the Jurassic Arab zones of Id-el-Shargi (offshore Qatar in the Persian-Arabian Gulf) and in the Paleocene of Libya (Defa field). Porosity and permeability of such fine-grain (15 μm) sucrose dolomite was described by Murray (1960) from the Mississippian Midale Field of Saskatchewan.

CONCLUSIONS

In summary: A carbonate reservoir is formed in an original body of sediment which has often undergone a long and important diagenetic history. To understand the reservoir the petroleum geologists must study all events which took place to form and alter the sediment over many millions of years. This double study (depositional and diagenetic) usually leads to demonstration of complex geologic history. The following conclusions are derived from a variety of such studies found in the literature.

1. Carbonate reservoirs may develop from sediments formed in many varied environments: Shelf grainstones, shelf tidal flat muds, shelf margin reefy and grainy sediments, debris flows, and calcareous turbidites on slopes, and offshore pelagic chalks.

2. Despite the widely varied original environments and the importance of the diagenetic overlay, the depositional details are worth considerable study because primary porosity enhances leaching, cementation, dolomitization and other diagenetic processes. In addition, carbonate particles and structures may give specific information about the type, shape, and trend of sedimentary bodies which develop the reservoir. E.g., well-formed multicoated ooids in grainstone usually indicate belts of tidal bars (Ball, 1967). Furthermore, mounds composed of phylloid algae are commonly (not always) oriented parallel to shelf margins. Experience in the geologic record shows that both types of sedimentary bodies are of preferred and predictable dimensions.

3. The importance of diagenesis is overriding. Trace element and isotopic studies leading to understanding the past history of water movements are adding significantly to modern petrographic studies distinguishing environments of cementation. Negative thinking about diagenesis is important. Carbonate sediment is made with high porosity, most of which is destroyed in time. The wise petroleum geologist searches for areas and past conditions where sediment has been protected from carbonate cementation, solution-compaction, and certain types of replacement.

4. Two diagenetic processes are very important, however, in creating permeability and porosity: Dolomitization and leaching at unconformities. Probably close to half of all carbonate reservoirs are in dolomitized strata—far more than half in Paleozoic rocks. Leaching at early stages of diagenesis during deposition, as well as later, beneath regional unconformities constitutes the other major reservoir creating process.

5. Stratigraphic relations of strata associated with carbonate reservoirs must be carefully worked out. Some shelf reservoirs are the direct result of entrapment due to updip pinchout of porous facies. These facies are a part of the standard pattern of carbonate sediments deposited across shallow shelves (Wilson, 1975). Regional map relations of unconformities and petrographic indications of unconformities in cores and samples must be carefully considered when searching for reservoirs. Lastly, the relationships of carbonate reservoirs to evaporites are important. Evaporites serve as updip and overlying seals, as porosity plugs; also, leaching of secondary anhydrite leads to enhancement of porosity in dolomites or to porous and permeable collapse breccias.

REFERENCES

Andrichuk, J., 1955, Mississippian Madison group stratigraphy and sedimentation in Wyoming and southern Montana: Am. Assoc. Petrol. Geol. Bull., v. 39, p. 2170-2210.

Badiozamani, K., 1973, The Dorag dolomitization model—application to the Middle Ordovician of Wisconsin: J. Sediment. Petrol., v. 43, p. 965-984.

Ball, M. M., 1967, Carbonate sand bodies of Florida and the Bahamas: J. Sediment. Petrol. v. 37, p. 556-591.

Biddle, K., 1978, Submarine cementation in basinal exotic boulders, and the nature of Triassic platform margins, Dolomite Alps, Italy: Proc. Tenth Congr. Internat. Assoc. Sediment., Jerusalem, Abs.

Bishop, W. F. 1968, Petroleum of upper Smackover limestone in North Haynesville field, Claiborne Parish, Louisiana: Am. Assoc. Petrol. Geol. Bull., v. 52, p. 92-128.

——— , 1974, Petroleum geology of Tunisia and adjacent parts of Algeria and Libya (abs.): Am. Assoc. Petrol. Geol. Ann. Mtg., p. 8.

Carrasco, V. B., 1977, Albian sedimentation of submarine autochthonous and allochthonous carbonates, east edge of the Valles-San Luis Potosi platform, Mexico: *in* H. E. Cook and P. Enos, *eds.,* Deep water carbonate environments; Soc. Econ. Paleont. Mineral. Spec. Pub. 25, p. 263-272.

Choquette, P. W., and Pray, L. C., 1970, Geologic nomenclature and classification of porosity in sedimentary carbonates: Am. Assoc. Petrol. Geol. Bull., v. 54, p. 207-250.

Coogan, A. H., Bebout, D. G., and Maggio, C., 1972, Depositional environments and geologic history of Golden Lane and Poza Rica trend, Mexico, an alternative view: Am. Assoc. Petrol. Geol. Bull., v. 56, p. 1419-1447.

Cook, H. E., McDaniel, P. N., Mountjoy, E. W., and Pray, L. C., 1972, Allochthonous carbonate debris flows at Devonian bank ("reef") margins, Alberta, Canada: Bull. Can. Petrol. Geol., v. 20, p. 439-497.

Dunham, R. J., 1971, Meniscus cement: *in* O. P. Bricker, Carbonate cements; The Johns Hopkins University studies in geology, Johns Hopkins Press, Baltimore, Maryland, v. 19, p. 297-300.

Edie, R., 1958, Mississippian sedimentation and oil fields in southeastern Saskatchewan: Am. Assoc. Petrol. Geol. Bull., v. 42, p. 94-126.

Enos, P., 1977, Tamabra limestone of the Poza Rica trend, Cretaceous, Mexico: *in* H. E. Cook and P. Enos, *eds.,* Deep water carbonate environments; Soc. Econ. Paleont. Mineral. Spec. Pub. 25, p. 273-314.

Ginsburg, R. N., and James, N. P., 1973, British Honduras by submarine: Geotimes, v. 18, no. 5, p. 23-24.

Griffith, L. S., Pitcher, M. G., Rice, G. W., 1969, Quantitative environmental analysis of a lower Cretaceous reef complex: *in* G. M. Friedman, *ed.*, Depositional environments in carbonate rocks: Soc. Econ. Paleont. Mineral. Spec. Pub. 14, p. 120-138.

Guzman, E. J., 1978, Petroleum geology of the Reforma area, southeastern Mexico (abs.): Houston Geol. Soc. Bull., September, p. 2.

Halbouty, M., *ed.,* 1970, Geology of giant petroleum fields: Am. Assoc. Petrol. Geol. Mem. 14, 575 p.

Harris, T. J., Hay, J. T. C., Twombley, B. N., 1968, Contrasting limestone reservoirs in the Murban field, Abu Dhabi: *in* Second regional technical symposium; Soc. Petrol. Eng., A.I.M.E., Dhahran, Saudi Arabia, p. 149-182.

Heck, T., 1978, Depositional environments of Bottineau interval, Lodgepole Formation in North Dakota: *in* Williston basin symposium; 24th ann. conf., Montana Geol. Soc., Billings, Montana, p. 191-199.

Henson, F. R. S., 1950, Cretaceous and Tertiary reef formations and associated sediments in the Middle East: Am. Assoc. Petrol. Geol. Bull., v. 34, p. 215-238.

Hopkins, J. C., 1977, Production of foreslope breccia by differential submarine cementation and downslope displacement of carbonate sands. Miette and Ancient Wall buildups, Devonian, Canada: *in* H. E. Cook and P. Enos, *eds.,* Deep water carbonate environments: Soc. Econ. Paleont. Mineral. Spec. Pub. 25, p. 155-170.

Irwin, M. L., 1965, General theory of epeiric clear water sedimentation: Am. Assoc. Petrol. Geol. Bull., v. 49, p. 445-459.

Jacka, A. D., Thomas, C. M., Beck, R. H., Williams, K. W., and Harrison, S. C., 1972, Guadalupian depositional cycles of the Delaware basin and northwest shelf: *in* J. C. Elam and S. Chuber, *eds.*, Cyclic sedimentation in the Permian basin, 2nd ed., West Texas Geol. Soc. Pub., p. 151-195.

Klovan, J. E., 1964, Facies analysis of the Redwater reef complex, Alberta, Canada: Alta. Soc. Petrol. Geol. Bull., v. 12, p. 1-100.

Land, L. S., 1970, Phreatic versus vadose meteoric diagenesis of limestones: evidence from a fossil water table: Sedimentology, v. 14, p. 175-185.

——— , 1973, Holocene meteoric dolomitization of Pleistocene limestones, North Jamaica: Sedimentology, v. 20, p. 411-424.

Lucia, F. J., 1962, Diagenesis of a crinoidal sediment: J. Sediment. Petrol., v. 32, p. 848-865.
——, 1972, Recognition of evaporite-carbonate shoreline sedimentation: *in* J. K. Rigby and W. K. Hamblin, *eds.*, Recognition of ancient sedimentary environments; Soc. Econ. Paleont. Mineral. Spec. Pub. 16, p. 160-191.
——, and Murray, R. C., 1966, Origin and distribution of porosity in crinoidal rock: Proc. 7th World Petrol. Cong., p. 409-423.
Lynts, G. W., Judd, J. B., and Stehman, C. F., 1973, Late Pleistocene history of the Tongue of the Ocean, Bahamas: Geol. Soc. Am. Bull., v. 84, p. 2665-2684.
Mathews, R. K., 1968, Carbonate diagenesis: equilibration of sedimentary mineralogy to the subaerial environment; coral cap of Barbados, West Indies: J. Sediment. Petrol., v. 38, p. 1110-1119.
Meissner, F. F., 1972, Cyclic sedimentation in middle Permian strata of the Permian Basin, West Texas and New Mexico: *in* J. C. Elam and S. Chuber, *eds.*, Cyclic sedimentation in the Permian Basin, 2nd ed., West Texas Geol. Soc., Midland, Texas, p. 203-232.
Mesolella, K. J., Robinson, J. D., and Ormiston, A. R., 1974, Cyclic deposition of Silurian carbonates and evaporites in Michigan basin: Am. Assoc. Petrol. Geol. Bull., v. 58, p. 34-62.
Murray, R. C., 1960, Origin of porosity in carbonate rocks: J. Sediment. Petrol., v. 39, p. 59-84.
Neugebauer, J., 1974, Some aspects of cementation in chalk: *in* K. Hsu and H. C. Jenkyns, *eds.*, Pelagic sediments: on land and under the sea: Internat. Assoc. Sediment. Spec. Pub. 1, p. 149-176.
Rigby, J. K., 1958, Mass movement in Permian rocks of Trans-Pecos, Texas: J. Sediment. Petrol., v. 28, p. 298-315.
Roehl, P. O., 1967, Stony Mountain (Ordovician) and Interlake (Silurian) facies analogs of recent low-energy marine and subaerial carbonates, Bahamas: Am. Assoc. Petrol. Geol. Bull., v. 51, p. 1979-2032.
Rose, P. R., 1972, Edwards group, surface and subsurface, central Texas: Bur. Econ. Geol., Univ. Texas, Austin, Rep. Invest. 74, 198 p.
Schlager, W., Hooke, R. L., and James, N. P., 1976, Episodic erosion and deposition in the Tongue of the Ocean (Bahamas): Geol. Soc. Am. Bull., v. 87, p. 1115-1118.
Scholle, P. A., 1977, Chalk diagenesis and its relation to petroleum exploration: Oil from chalks, a modern miracle?: Am. Assoc. Petrol. Geol. Bull., v. 61, p. 982-1009.
Toomey, D. F., Mountjoy, E. W., and Mackenzie, W. S., 1970, Upper Devonian (Frasnian) algae and foraminifera from the Ancient Wall carbonate complex, Jasper National Park, Alberta, Canada: Can. J. Earth Sci., v. 7, p. 946-981.
——, Wilson, J. L., and Rezak, R., 1977, Evolution of Yucca Mound complex, Late Pennsylvanian phylloid-algal buildup, Sacramento Mountains, New Mexico: Am. Assoc. Petrol. Geol. Bull., v. 61, p. 2115-2133.
Vest, E. L., 1970, Oil fields of Pennsylvanian-Permian Horseshoe Atoll, West Texas: *in* M. Halbouty, *ed.*, Geology of giant petroleum fields; Am. Assoc. Petrol. Geol. Mem. 14, p. 185-203.
Ward, W. C., 1970, Diagenesis of quaternary eolianites of northeast Quintana Roo, Mexico: Unpub. Ph.D. dissertation, Rice Univ., Houston, Texas, 207 p.
Wilson, J. L., 1975, Carbonate facies in geologic history: Springer-Verlag, Heidelberg, New York City, 417 p.

The American Association of Petroleum Geologists Bulletin
V. 69, No. 1 (January 1985), P. 1-21, 10 Figs.

Carbonate Platform Facies Models[1]

J. FRED READ[2]

ABSTRACT

Various types of carbonate platforms are characterized by distinctive profiles, facies, and evolutionary sequences. Ramps may be homoclinal or distally steepened, and may have fringing or barrier shoal-water complexes of ooid-pellet sands or skeletal banks. Homoclinal ramps pass seaward into deeper water without major break in slope, and lack deep-water breccias. Distally steepened ramps may be low energy, and characterized by widespread, shallow, subwave-base mud blankets, or high energy with coastal beach/dune complexes and widespread skeletal sand blankets. Slope facies may contain abundant breccias of slope-derived clasts.

Rimmed shelves have relatively flat tops, and marked break in slopes at the high energy, shallow-shelf edge where they pass into deep water. Such shelves may be aggraded with peritidal facies extending over much of the shelf, or incipiently drowned, depending on magnitude of sea level fluctuations. They may be accretionary, or bypass types that include gullied slope, escarpment, and high-relief erosional forms. Intrashelf basins occur on some shelves, controlling distribution of reservoir and source beds. Isolated platforms are surrounded by deeper water and may be located on rifted continental margins, or on submarine volcanoes. Most have high-relief rimmed margins.

Platforms that have been subjected to rapid sea level rise may be incipiently drowned, and characterized by raised rims, elevated patch or pinnacle reefs, and widespread subwave-base carbonate or fine clastic blankets. Completely drowned shelves develop where the shelf is submerged to subphotic depths, terminating shallow water deposition, and commonly resulting in blanketing of the shelf by deeper water facies. Some margins show extensive down-to-basin faulting that is contemporaneous with carbonate deposition, or associated with thick prograding clastic sequences.

The various types of platforms change in response to variations in sedimentation, subsidence or sea level rise, and may form distinctive evolutionary sequences. The relatively few models presented appear to accommodate most geological examples, some of which contain major reservoir facies.

INTRODUCTION

Carbonate platform models are important aids in understanding distribution of carbonate facies and to a lesser extent, primary porosity distribution, preservation of which largely is a function of diagenetic history. Many of the terms that are commonly used to describe the different platforms have various meanings to geologists. This lack of uniformity of usage has hampered the geologic application of platform facies models and has inhibited our understanding of the different facies sequences. This paper outlines major types of carbonate platforms, their facies distribution and criteria for their recognition, and examines influence of sea level and tectonics on platform evolution. The models outlined here are end members of a spectrum of carbonate-platform types, and are useful because relatively few types accommodate most geologic examples. However, real examples should not be forced to fit the model, because it is commonly the difference between the real example and the model that provides insight into platform evolution.

The classification of platform margins outlined here is based on that used by Ahr (1973) who recognized differences between rimmed shelves and ramps; Ginsburg and James (1974), who outlined characteristics of rimmed and drowned shelves; and Wilson (1975) who provided the first comprehensive model of platform margins. The classification outlined in Read (1982) uses the terms platform (a general term), ramp, rimmed shelf, isolated platform, and drowned platform to describe geomorphic, two-dimensional features (Figure 1).

The following facies are briefly described to avoid later repetition.

Tidal-flat complex.—Facies are generally arranged in cyclic, upward shallowing units 1-10 m (3-33 ft) thick. Sequences in humid zones are mainly subtidal-intertidal burrowed limestone with supratidal cryptalgal laminites, and may have inland freshwater algal marsh deposits, coal, or siliciclastics. Sequences in arid zones have burrowed to nonburrowed lagoonal limestone and cryptalgal heads, overlain by abundant intertidal sheetlike cryptalgal laminites, supratidal evaporites, or eolian-fluvial clastics.

Lagoonal facies (present behind barrier complexes).—These are mainly bedded pellet limestone or lime mudstone, or cherty, burrowed skeletal packstone to mudstone, with local biostromes of colonial metazoans. Minor, thin interbeds of peritidal fenestral or cryptalgal carbonates reflecting periods of shallowing of lagoon to tide levels.

Shoal-water complex of banks, reefs, and ooid/pellet shoals.—These may occur as shallow-ramp skeletal banks

[1]Manuscript received, October 27, 1983; accepted, May 7, 1984.

[2]Department of Geological Sciences, Virginia Polytechnic Institute and State University, Blacksburg, Virginia 24061.

This paper is an outgrowth of notes prepared for AAPG Fall Education Conference short courses. Many people kindly provided reprints and preprints, discussion and ideas, including W. M. Ahr, M. J. Brady, L. B. Collins, H. Cook, P. Crevello, R. N. Ginsburg, N. P. James, C. G. St. C. Kendall, B. W. Logan, H. T. Mullins, C. Neumann, W. Schlager, J. Wendte, J. L. Wilson, and my students, past and present. I thank Donna Williams (typing) and Martin Eiss (drafting). The paper was prepared during tenure of grants EAR 7911213 and 8108577 from the National Science Foundation.

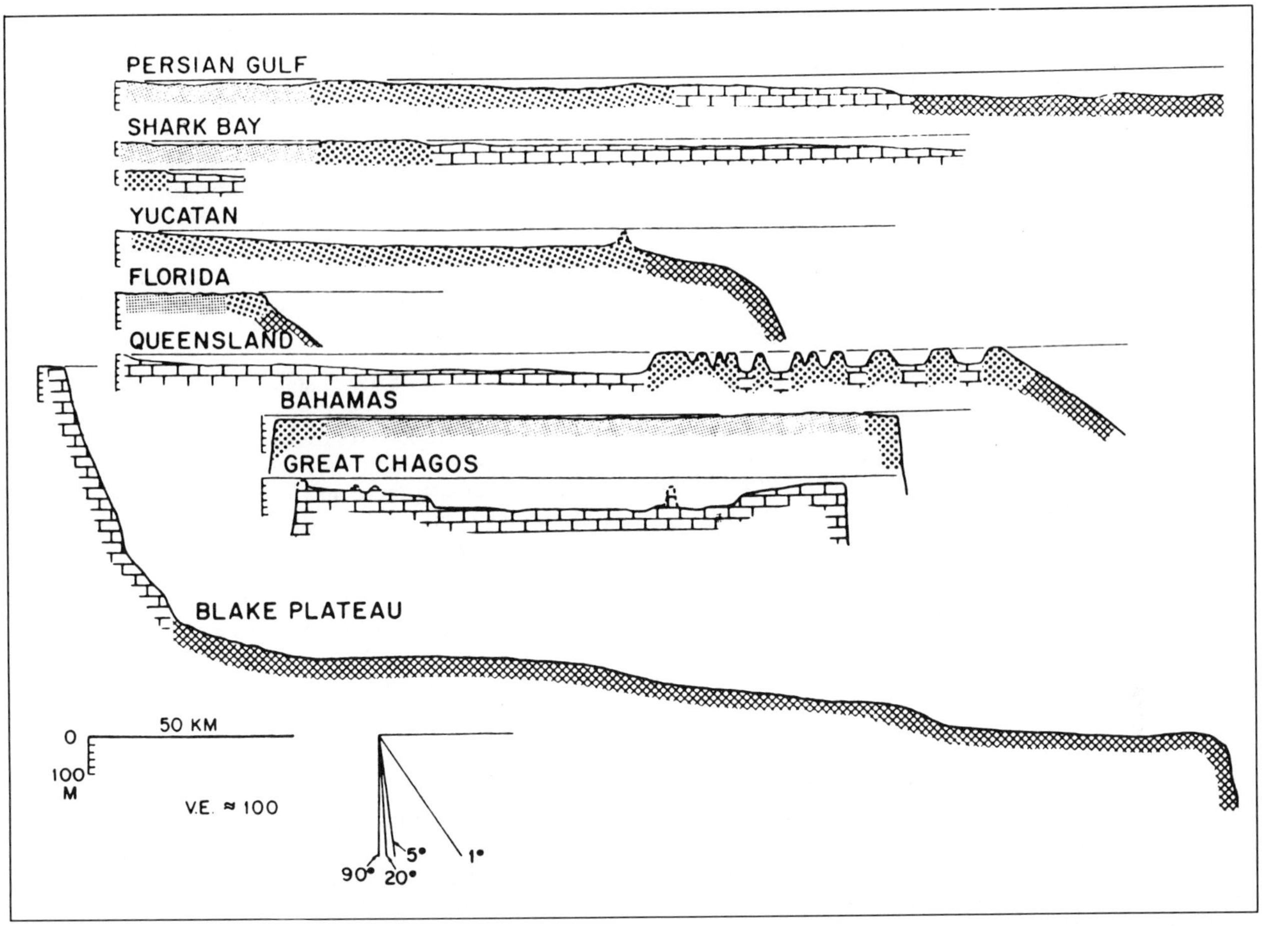

Figure 1—Profiles of carbonate ramps, rimmed shelves, and drowned platforms plotted at same scale. Persian Gulf and Shark Bay are homoclinal ramps. Yucatan is distally steepened ramp with local buildups on outer platform. Florida and Queensland are rimmed shelves, and the Bahamas and Great Chagos are isolated platforms. Note Queensland and Great Chagos also reflect incipient drowning. Queensland transect oblique to shelf trend. Blake Plateau is a drowned shelf.

or lime-sand shoals, or shelf-edge skeletal reefs and lime sands, to be described in detail later. These may pass gradually downslope into deep-ramp facies. On steeply sloping shelf edges, they pass downslope into foreslope and slope deposits marginal to deep shelf or basin.

Deep shelf and ramp facies.—These consist of cherty, modular bedded, skeletal packstone or wackestone, with abundant whole fossils and diverse open-marine biota. They may have upward-fining, storm-generated beds. Water depths range from 10 to 40 m (33 to 130 ft). The lithotope is largely below fair-weather wave base, but it may be influenced by storm waves.

Slope and basin facies.—Adjacent to steeply sloping platforms, foreslope and slope deposits have abundant breccias and turbidites interbedded with periplatform lime and terrigenous muds. Adjacent to most ramps, slope and basin deposits are thin-bedded, periplatform lime and terrigenous muds that generally have few sediment-gravity flow deposits. Basinal deposits in Paleozoic rocks commonly are shale, with carbonate content increasing toward the platform. Basinal deposits in Mesozoic and Cenozoic rocks may be shale or pelagic limestone. Slope and basin floor may be anoxic and lacking benthic organisms; thus, deposits will be laminated and nonburrowed. Where slope and basin waters are oxic, deposits may be burrowed and fossiliferous.

CARBONATE RAMPS

Carbonate ramps (Figures 1, 2) have gentle slopes (generally less than 1°) on which shallow wave-agitated facies of the nearshore zone pass downslope (without marked break in slope) into deeper water, low-energy deposits (Ahr, 1973). They differ from rimmed shelves in that continuous reef trends generally are absent, high-energy lime sands are located near the shoreline and deeper water breccias (if present) generally lack clasts of shallow shelf-edge facies. Ramps may be subdivided on the basis of profile into homoclinal ramps and distally steepened ramps (Figure 1).

Homoclinal Ramps

Homoclinal ramps have relatively uniform, gentle slopes (1 to a few meters/km or a fraction of a degree) into the basin (Figures 1, 2C). Facies include:

1. Tidal flat and lagoonal facies.
2. Shoal-water complex of banks or ooid-peloid sand shoals.
3. Deeper ramp argillaceous lime wackestone/mudstone, containing open marine, diverse biota, whole fossils, nodular bedding, upward-fining storm sequences, and burrows; may also have downslope buildups that are heavily marine cemented.
4. Slope and basin lime muds and interbedded shale; breccias and turbidites rare.

Modern ramps such as the Persian Gulf (Purser, 1973) and Shark Bay (Logan et al, 1974) are homoclinal. Ancient homoclinal ramps include the Middle Ordovician, Virginia (Read, 1980), and the Devonian, New York (Laporte, 1969).

Distally Steepened Ramps

Distally steepened ramps have some characteristics of ramps (agitated shoal-water to subwave-base facies transition occurs well back on platform) and some characteristics of rimmed shelves (slope facies contain abundant slumps, breccias, and allochthonous lime sands) (Figures 1, 2F). However, they differ from rimmed shelves in that the major break in slope does not occur at the seaward margin of a high-energy rim but many kilometers seaward of high-energy shoals (cf. Figures 2F, 4A). Because the shoal-water facies occur some distance back from the break in slope, deep-water breccias lack clasts of shallow-platform sands or reefs, but instead contain clasts of deep-ramp or slope facies.

Low-energy, distally steepened ramps have widespread deep-ramp mud blankets seaward of the shoal-water complex (Figure 2F) whereas high-energy, distally steepened ramps have broad lime-sand blankets over much of the deep ramp, with muds (and slope breccias and turbidites) being restricted to the slope and basin margin (Figure 2G). Facies belts seaward of the shoal-water complex (facies 1 and 2 above) of distally steepened ramps include:

3. Deep ramp (cf. above), skeletal wackestone/mudstone, argillaceous, nodular, burrowed, with open marine biota; may also have slumps, breccias, and turbidites along basin margin.
4. Slope and basin margin (cf. above), even-bedded, gray to black lime mudstone and lesser wackestone; may be argillaceous or shaly, laminated, unburrowed, abundant intraformational truncation surfaces, slumps and breccias with clasts of slope facies (units up to 10 m or 33 ft thick); clasts of shallow water facies rare; breccias commonly channel form or sheetlike; some interbedded allochthonous lime-sand beds (turbidites and contourites). Facies reflect relatively high (several degrees) slopes into the basin.

Examples include the Upper Cambrian-Lower Ordovician sequences of the western United States, described by Cook and Taylor (1977), H. E. Cook (1979), and Brady and Rowell (1976). This ramp type might be developed where a shelf is drowned to form a ramp (Holocene Yucatan platform, Ahr, 1973; Middle Cambrian Marjum-Pole Canyon sequence, western United States, Brady and Rowell, 1976). The earlier rimmed shelf surface is submerged below wave base and muddy carbonates mantle the slope at angles promoting downslope sediment-gravity transport. The ramp-type also might develop where the deeper ramp is developed over a zone of flexuring and rapid downwarping.

Shoal-Water Complexes on Ramps

Shoal-water complexes on homoclinal and low-energy, distally steepened ramps include skeletal banks or ooid-pellet sand shoals; these may be either fringing or barrier complexes. High-energy, distally steepened ramps have wide beach-dune complexes and extensive shelf sand blankets.

Ramps with fringing banks.—These are characterized by skeletal banks that pass landward into a tidal/supratidal complex without intervening lagoonal facies (Figure 2A). Holocene examples include fringing seagrass banks, Shark Bay (Davies, 1970; Hagan and Logan, 1974).

Facies belts include:

1. Tidal/supratidal complex.
2. Sublittoral sand sheet; composed of quartzose or skeletal/pellet sands with abundant micritized grains; rippled, plane bedded, and cross-bedded. Skeletal grains are from sand-flat biota and eroded bank sediments. Sand sheet overlies.
3. Fringing bank of skeletal carbonate; linear accumulations that parallel the shoreline; wedge shaped in cross section, thickening seaward. Relief on seaward face of bank may range from a few meters to tens of meters. Slopes on bank tops are extremely low; slopes on seaward margins are gentle (several degrees) to 20°-30° where stabilized by organic baffles or marine cementation. Fringing banks may be of (a) skeletal packstone/grainstone with local wackestone/mudstone bioherms, or (b) wackestone-mudstone grading up into skeletal grainstone cap. Fringing banks may be cut by channels up to 10 m (33 ft) deep. Others, especially narrow, fringing banks in areas of low tide range may lack channels. Channels on fringing banks may have cross-bedded lime-sand fills that contain rip-ups of bank sediments, and variable amounts of quartz sand transported from intertidal to shallow subtidal sand sheets, prograding terrestrial sand sheets, or fluvial systems.
4. Deep ramp/slope facies.

Ramps with barrier-bank complexes.—Ramps characterized by barrier banks of skeletal carbonate (Figure 2B) occur in the Holocene in Shark Bay (Read, 1974; Hagan and Logan, 1974). Barrier banks are separated from tidal flat and deltaic facies by lagoonal carbonates or prodelta shale.

Facies belts include:

1. Tidal-supratidal complex.

2. Lagoonal carbonates.

3. Barrier bank complex: typically flat-topped bank (water depths 2 m or 6.5 ft or less), 2-20 km (1.2-6 mi) wide (measured in a downdip direction), composed of biostromal thickenings arranged parallel to strike and elongate downdip, and separated by broad tidal exchange channels (100 m or 330 ft to several kilometers wide, up to 10 m or 33 ft or more deep) with terminal fans. Slopes on bank margins may be 1 to 15° or more; syndepositional relief may be low (10 m or 33 ft). Banks are of skeletal packstone or wackestone with a thin cap of skeletal grainstone. Bank sediments may be burrowed and structureless, flat bedded, or have large scale, gently inclined accretion bedding parallel to margins. Channel fills include cross-bedded skeletal sands with reworked clasts of bank facies, and wedge to lenticular units of accretion-bedded and cross-bedded lime sand that extend out into lagoonal and deep-ramp muds. Banks may be located on local or regional highs.

4. Deep-ramp/slope carbonates.

Ancient examples of barrier banks on ramps include the Devonian Helderberg Group, New York (Laporte, 1969) and the Wardell-Wassum sequence, Middle Ordovician, Virginia (Read, 1980). Heckel (1972) describes platy-algal mounds with numerous channels filled with calcarenite and sandstone in the Pennsylvanian, Kansas. Platy algal barrier banks also include the Pennsylvanian, north Texas (Brown, 1972).

Ramps with isolated shallow ramp buildups and downslope buildups.—The distinguishing feature of this ramp type is that buildups rarely form continuous linear barriers, but form isolated buildups on both the shallow ramp and on the deep ramp and basin slope (Figure 2C); other ramps may have isolated, downslope buildups located seaward of barrier banks. Examples include the Holocene, Persian Gulf (Purser, 1973) and the Middle Ordovician Rockdell and Effna Limestones, Virginia (Read, 1980).

Facies belts include:

1. Tidal-supratidal complex.

2. Lagoonal facies.

3. Shallow ramp banks and local patch reefs; separated laterally by intermound fine carbonate. Elongate parallel to trend of ramp. Low syndepositional relief (up to 10 m or 33 ft), biostromal geometry with widths from 1 km (0.6 mi) to tens of kilometers. Mainly skeletal banks of lime wackestone/mudstone (as massive cores or small pods) and skeletal sand (commonly as basal, flanking, and capping facies) may be differentiated with respect to weather aspect, with reefal rims on windward side. With continued growth, shallow ramp banks may coalesce laterally to form barrier-bank complex.

4. Deep ramp and basin slope with isolated downslope mounds. Mounds are less than 1 km (0.6 mi) to 10 km (6 mi) or more wide, are generally circular, lack any differentiation with respect to prevailing wind/wave approach, and have high syndepositional relief (up to 50 m or 160 ft or more), with gentle to steep marginal slopes (tens of degrees). Mounds may be wackestone/mudstone bioherms, some of which have flanking skeletal sands. Others are dominantly skeletal-sand buildups with local, crestal reefs, or scattered wackestone/mudstone pods. Deep flank beds commonly are shaly, packstone/wackestone that intertongue with basin facies containing detrital carbonates shed from the mounds. Downslope mounds commonly have abundant marine cement filling stromatatoid cavities and intergranular voids; this acts to stabilize high depositional slopes on buildup flanks.

Ramps with fringing ooid-shoal complex.—Fringing shoals of ooid-pellet sand occur along some coastlines of ramps (Figure 2D). Holocene examples include oolitic sublittoral platforms in Hamelin Pool, Shark Bay (Hagan and Logan 1974) and the Persian Gulf (Loreau and Purser, 1973).

Facies include:

1. Tidal-supratidal complex, passing seaward into:

2. Fringing shallow subtidal sand flat (depths up to 2-3 m or 6-10 ft), widths 0.5 km-more than 5 km (0.3-3 mi). Rippled and megarippled ooid sands may completely cover shoal, be confined to landward part, or restricted to seaward parts of promontories; slightly deeper parts of shoals may have intraclast sands reworked from hardgrounds on skeletal, ooid, or quartz sand; reefs (with relatively restricted biota) may occur at seaward edge. Quartz and skeletal sands with restricted sand-flat biota may occur landward and leeward of intraclastic and ooid sands. Seaward margin is steep, reaching angle of repose of sand. Passes seaward into:

3. Skeletal packstone and wackestone (locally oncolitic near shoal) forming in several meters to 10 m (33 ft) of water on shallow, lower energy ramp floor.

Ramps with barrier ooid pellet shoal complex.—Barrier shoals of ooid/pellet sand occur on some ramps (Figure 2E), for example the Holocene Trucial Coast, Persian Gulf (Purser and Evans, 1973).

Facies belts include:

1. Tidal/supratidal complex.

2. Lagoonal carbonates.

3. Ooid-pellet barrier complex: shore-parallel complex of beach ridge/dune barriers and subtidal shoals cut by broad tidal channels (up to 1 km—0.6 mi—or more wide, by 10 m or 33 ft deep) with oolitic tidal deltas. Shoal consists of megarippled and rippled, cross-bedded ooid-pellet sands that may have large-scale foreset bedding. Ooid complex may be located on paleohighs. Small patch reefs may occur in channels and between tidal channel deltas in front of the shoal. Larger reefs may occur seaward of the shoal, where they may be localized on salt domes, buried topography, or structural highs.

4. Deep ramp facies of skeletal packstone/wackestone, locally oncolitic near shoal.

Ancient examples include the Jurassic Smackover Formation, U.S. Gulf Coast (Bishop, 1969; Baria et al, 1982), the Middle Cambrian, Virginia-Tennessee (Erwin, 1981;

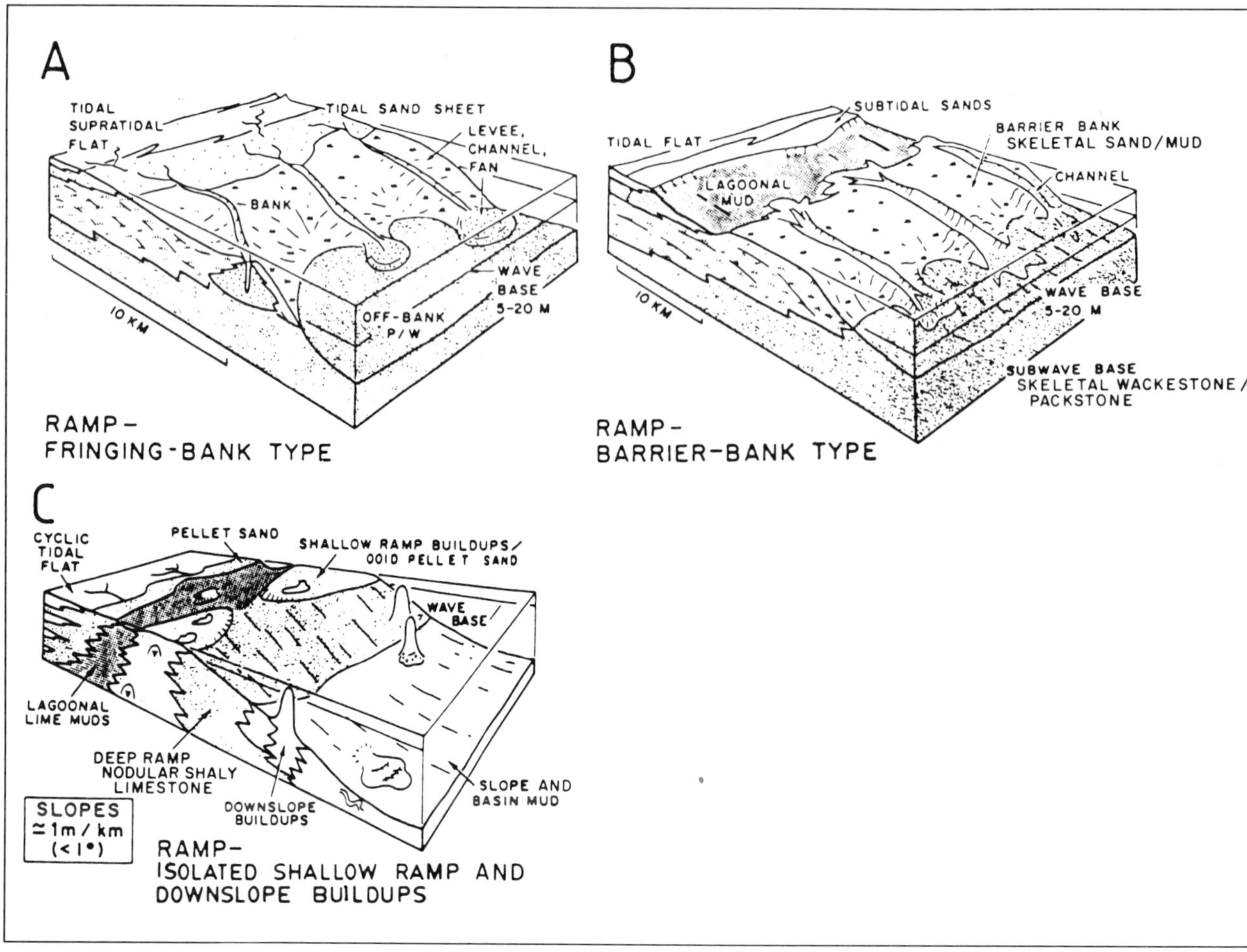

Figure 2—Carbonate ramps. A. Fringing bank complex on carbonate ramp. B. Barrier bank complex on carbonate ramp. C. Shallow ramp and downslope buildups on carbonate ramp. Ramp is homoclinal. D. Fringing ooid shoals on carbonate ramp. E. Barrier ooid shoals on carbonate ramp. F. Distally steepened ramp formed under low-energy conditions. G. Distally steepened ramp formed under high-energy, swell-dominated conditions.

Markello and Read, 1981), Mississippian Lodgepole oolitic cycles, Montana (Edie, 1958; Wilson, 1975), and the Mississippian Ste. Genevieve Limestone, Illinois basin (Choquette and Steinen, 1980).

High-energy ramps with coastal beach/dune complexes.—Carbonate ramps that are developed on mature continental shelves that lack reefal rims may be subjected to high-energy conditions due to ocean swell and hurricanes.

These ramps are characterized by high-energy wide coastal beach/dune complexes and extensive shelf sand blankets (Figure 2G). Holocene examples include the southwest Australian continental shelf (Collins, 1981), northeast Yucatan (Ward and Brady, 1979), and the Pleistocene Tamala Eolianite, Shark Bay (Logan et al, 1970). Such ramps are typically distally steepened (having developed following drowning of the continental shelf).

Facies belts include:

1. Coastal complex of dunes, beach ridges and beach deposits (few meters to 250 m or 800 ft thick). Sediments are lime sands and mature quartz sands, large scale eolian cross-bedding in dunes, swash laminated sands in beach deposits, grading seaward into festoon cross-bedded shelly sands, skeletal gravels, and small bioherms. Sequences cyclic and capped by unconformities and caliches that may extend over much of the shelf where large sea level (and water-table) fluctuations have occurred. Grade seaward into:

2. Inner ramp blanket (tens of kilometers wide) of skeletal or lithoclast sand; may be quartzose adjacent to major rivers. Common cross-bedding, plane lamination, upward fining storm deposits and ripples. Clean lime sands (grainstone) and local reefs grade seaward into fine lime packstone. The grainstones may extend tens of kilometers across the shelf. They contain abraded fragments from shallow water biota. Grade seaward (below depths of several tens of meters) into:

3. Outer ramp muddy lime sands (skeletal packstone), consisting of whole and fragmented, angular gravel to silt-sized skeletal material. Facies likely to contain storm-generated sequences. Grades seaward into:

4. Slope facies of highly bioturbated, fine skeletal wackestone.

An ancient example is the Tertiary of the Nullabor Plain, Australia (Lowry, 1970; L. B. Collins, personal

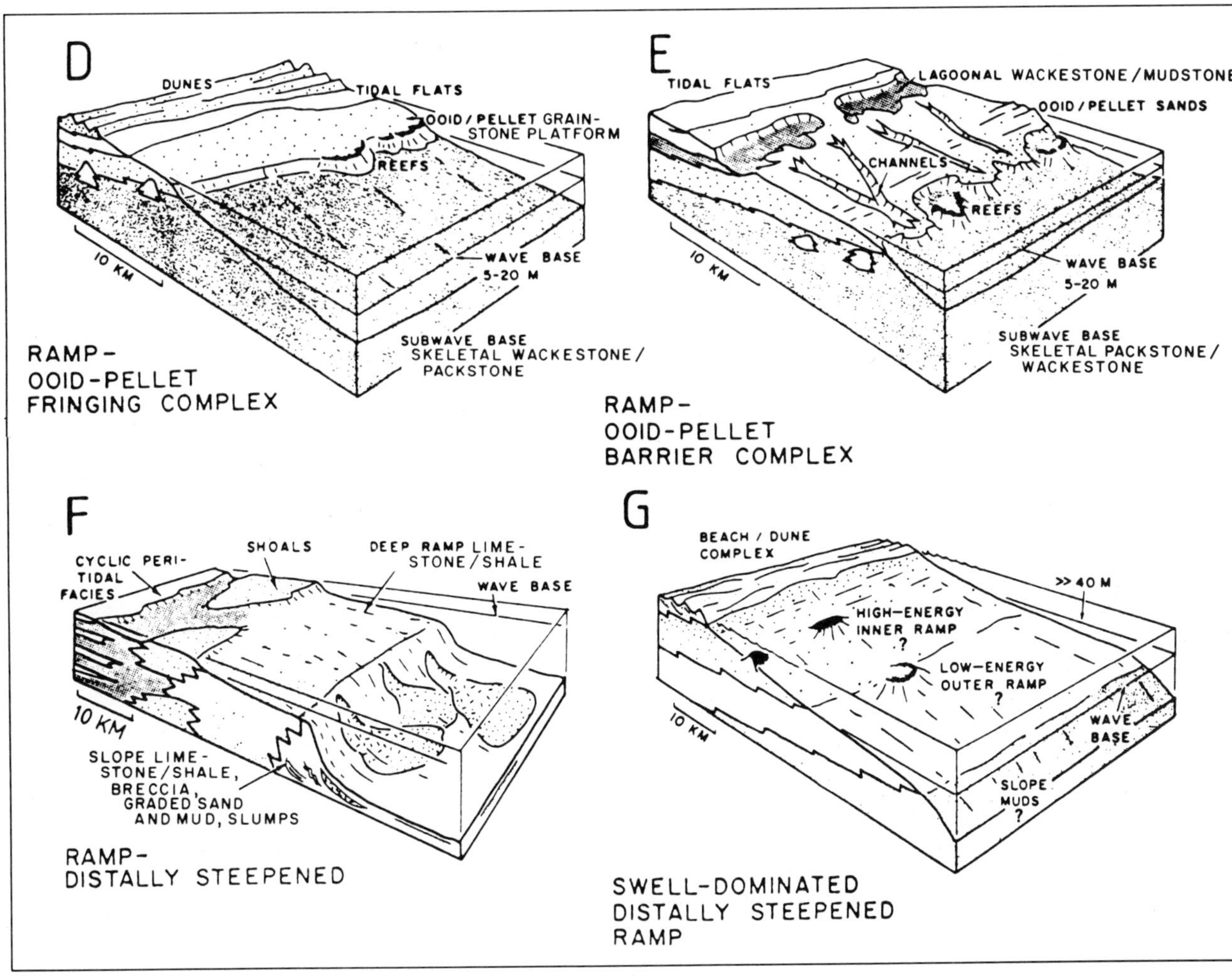

Figure 2—Continued.

communication, 1981).

Relations Between Ramp Types

On ramps, resident communities function mainly as sediment producers and as bafflers, trappers, and binders; thus ramps may be more likely to develop in areas, or at times, of tectonic or climatic crises in which reef formers were poorly represented (James, 1979). Homoclinal ramps develop on gentle regional paleoslopes. Such low slopes occur where ramps are located well landward of the continent–ocean crust boundary on continental margins, on underthrusting continental crust in foreland basins, or in continental interiors (Figure 10).

Ramps with fringing banks may develop on paleoslopes with higher gradients than those with barrier banks, enabling bank-building biota to colonize nearshore regions. Similarly, fringing ooid shoals may develop on higher gradients than barrier shoals, the shoals being localized in the zone of maximum tidal and wave energy. Both banks and ooid shoals commonly accrete around pre-existing highs. Skeletal banks on ramps may be more likely to develop in humid climatic settings, which inhibit development of hypersalinity. In contrast, arid settings that promote hypersalinity (and carbonate precipitation) may favor ooid complexes and associated evaporites. Fringing complexes seem likely to evolve into barrier complexes as platforms prograde, and may pass through an intermediate stage characterized by isolated shallow ramp and downslope buildups (Figures 3A-C). Ramps with isolated shallow ramp and downslope buildups also are common where ramps are undergoing rapid relative sea level rise, which prevents lateral outgrowth of the banks but favors upbuilding. Consequently, these ramps will have transgressive sequences and may culminate in drowning (Figure 3E). Barrier complexes will develop with decreasing subsidence/sea level rise that promotes basin filling and shallowing. Given time, ramps may evolve into rimmed shelves (Figure 3D). As margins steepen to a few degrees and skeletal banks prograde seaward, ramps may become transitional into rimmed shelves, with accumulations of much allochthonous skeletal sand beds in the slope facies.

Distally steepened ramps may develop where faulting or flexuring steepens the outer part of the ramp. More com-

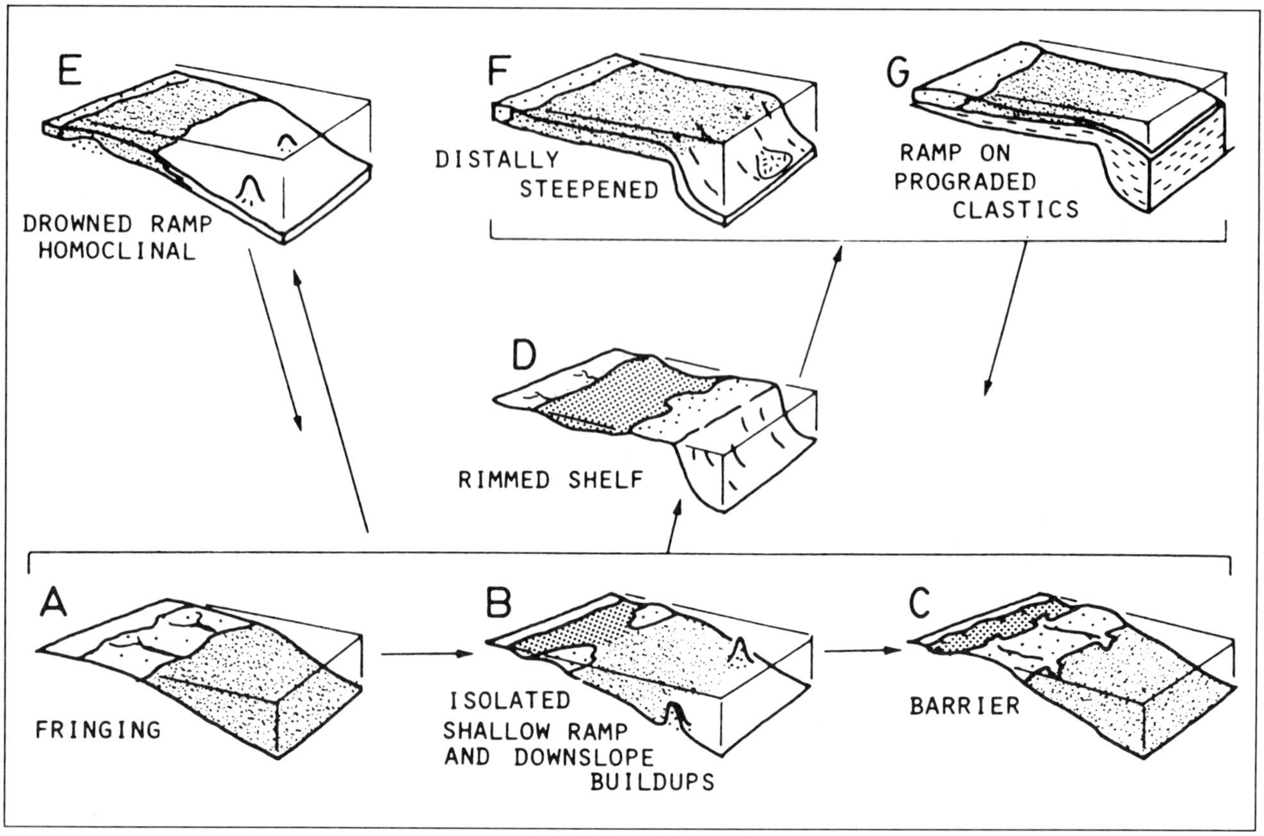

Figure 3—Ramp evolution. Ramps may start with fringing shallow water complexes (A) that change with time into barrier complexes (C), possibly by way of coalescence of shallow ramp buildups (B). These ramps may evolve into rimmed shelves (D) or into drowned homoclinal ramps (E). Where the rimmed shelves are drowned, these form distally steepened ramps (F). Where clastics bury the rimmed shelf, ramps (G) will be developed if carbonate sedimentation resumes.

monly, distally steepened ramps develop where earlier rimmed shelves undergo widespread drowning (Figure 3F). Ramps also develop on rimmed shelves that are prograded by clastics prior to renewed carbonate deposition (Figure 3G). High-energy distally steepened ramps with widespread, blanket skeletal sands probably are most likely to develop on continental shelves adjacent to large ocean basins in temperate latitudes, where reef builders are inhibited and rims are unlikely to develop. These ramps are likely to be subjected to oceanic swells that rework bottom sediments over large areas of the shelf, favoring accumulation of large amounts of shoreline sands in beach and dune complexes, especially where the shelf is subjected to transgressive and regressive events. Low-energy, distally steepened ramps might develop adjacent to foredeeps, small marginal basins, or on west sides of continents in low latitudes where winds are predominately offshore.

Examples of Reservoirs on Ramps

Reservoirs on ramps include shallow water banks, such as the Pennsylvanian shallow phylloid algal mounds (Baars and Stevenson, 1982) that formed isolated banks on a ramp subjected to repeated transgression and regression. Downslope buildup reservoirs on ramps include the Mississippian of Texas (Ahr and Ross, 1982) and the Devonian of New York (Kissling and Polasek, 1982). Oolitic ramp reservoirs include the Jurassic Smackover ooid sands (Bishop, 1969; Ahr, 1973), and reefs seaward of the ooid shoals (Baria et al, 1982), the Jurassic Arab A to D oolitic reservoirs, Middle East (the world's richest single oil habitat; Murris, 1980), and the Mississippian oolitic barriers, Mission Canyon and Charles Formations, Williston basin (Edie, 1958). In some ramps such as the Permian Grayburg (Longacre, 1980) and the Mississippian Mission Canyon (Lindsay and Kendall, 1980), the reservoirs occur seaward of ooid sands, in dolomitized muddy carbonates and downdip dolomitized skeletal packstone/mudstone. Seals on ramps may be provided by regressive evaporites, peritidal carbonates or fine clastics, or by transgressive deep-ramp, slope, and basin facies.

RIMMED CARBONATE SHELVES

Rimmed carbonate shelves (Ginsburg and James, 1974) are shallow platforms whose outer wave-agitated edge is marked by a pronounced increase in slope (commonly a few degrees to 60° or more) into deep water (Figures 1, 4). They have a semicontinuous to continuous rim or barrier

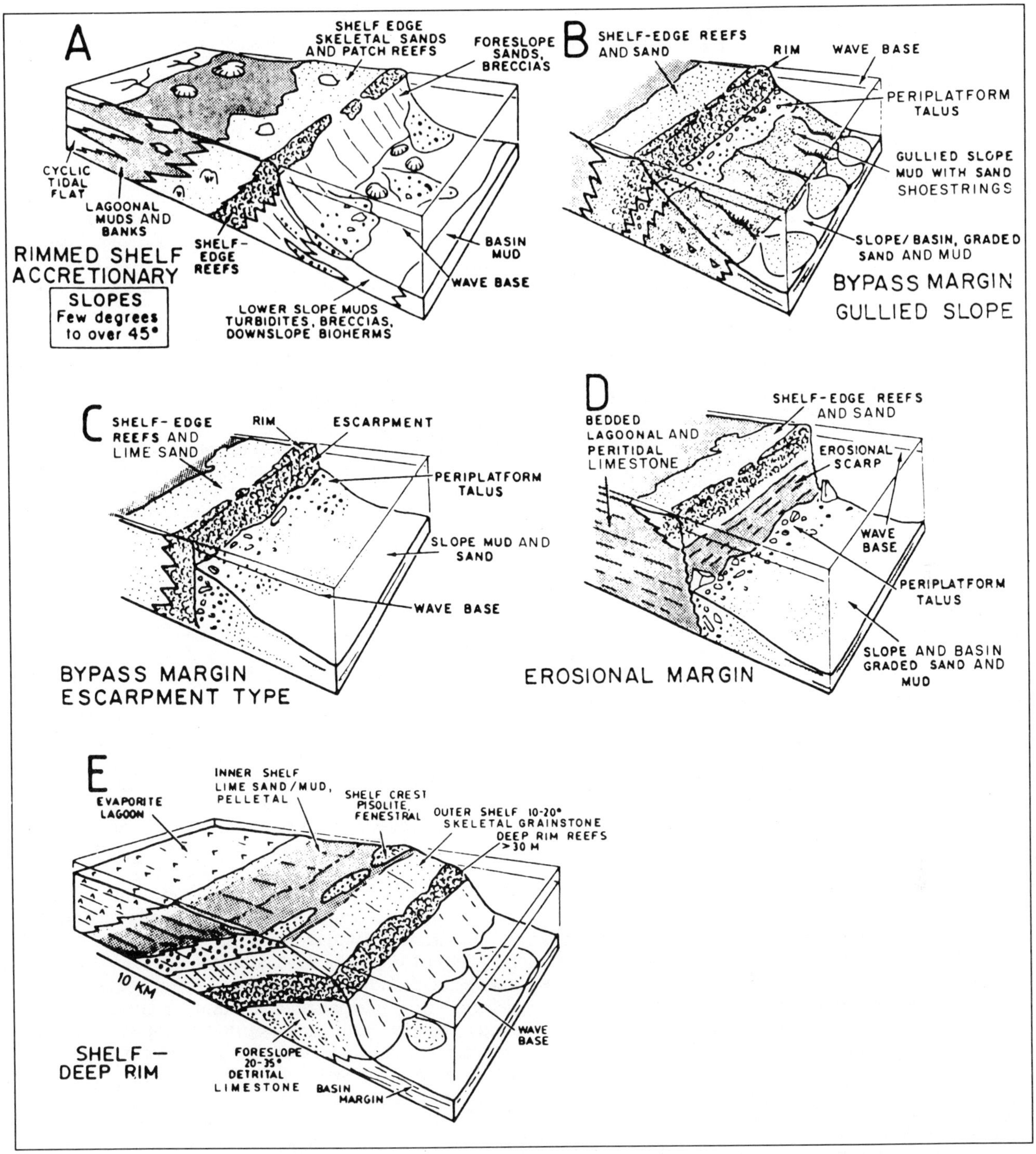

Figure 4—Rimmed shelves. A. Accretionary rimmed shelf. Reflects sedimentation exceeding relative sea level rise, causing shelf to prograde as well as build upward. B. Rimmed shelf with gullied bypass slope. C. Rimmed shelf with escarpment that functions as bypass slope. D. Rimmed shelf with erosional margin that exposes bedded platform-interior facies on escarpment. E. Rimmed shelf with deep reefal rim. Note that rim stays relatively deeply submerged throughout its growth and does not grow to sea level.

along the shelf margin which restricts circulation and wave action to form a low-energy lagoon to landward (Ginsburg and James, 1974). Rims may consist of barrier reefs, skeletal and ooid sands, or islands (eolianite or reefs) from an earlier depositional phase. Holocene rimmed shelves include the Great Barrier Reef, Australia (Maxwell, 1968), the south Florida Shelf (Enos and Perkins, 1977), and the Belize Shelf (Ginsburg and James, 1974).

Rimmed shelf margins and their slopes may be divided into (1) depositional or accretionary, (2) bypass, and (3) erosional margins. They may have reef-dominated or lime-sand–dominated rims (McIlreath and James, 1979).

Depositional or Accretionary Margins

Depositional or accretionary margins show both up-building and out-building; they generally lack high marginal escarpments; and shelf edge and foreslope/slope facies may intertongue (rather than abut) (Figure 4A). Major facies belts include:

1. Cyclic tidal-flat and lagoonal wackestone/mudstone and local patch reefs or banks widespread on flat-topped aggraded shelves. Tidal flats periodically may cover the whole shelf and extend to within a few kilometers of the rim. On some shelves, tidal flats also may extend from the rim into the lagoon or intrashelf basin. Where shelf is incipiently drowned, lagoon is relatively deep (tens of meters), and floored by fine-grained lime muds, siliciclastics, and high-relief patch reefs.
2. Shelf-edge skeletal or oolitic sands, cross-bedded; patch reefs and reef-fringed banks; lime sands muddier to landward.
3. Shelf-edge reefal carbonates, skeletal sands, and reef-derived rudites; abundant synsedimentary marine cement; reefs commonly zoned with respect to depth (James, 1979), with robust branching or encrusting frame-builders in high-energy zone, passing downslope into domal forms and then into sheetlike, encrusting, and fine branching forms.
4. Periplatform or foreslope lime sands, breccias, and some hemipelagic lime-mud beds. Typically clinoform bedded. Lime sands become muddier with water depth. Breccias have abundant clasts of reef and cemented lime sand of platform-margin/foreslope. Exotic blocks, slumps, pull-aparts, downslope mounds common.
5. Lower slope/basin margin lime turbidites, shale, sheet-form and channel-form breccias (sediment-gravity flows). Polymictic breccias have clasts of reef limestone, cemented shelf-edge and foreslope sands, and clasts of dark, fine-grained slope facies; oligomictic breccias mainly composed of slope facies.
6. Deep-water pelagic and hemipelagic lime muds, distal turbidites and shale.

Accretionary rimmed shelves commonly will have prograding (offlap) relations between reef, foreslope, slope, and basin facies (Figure 4A). Examples include Mesozoic Baltimore Canyon Trough and Scotian basin (Jansa, 1981), and Devonian (Fammenian) margins, Canning basin, Western Australia (Playford, 1980). The Cretaceous Stuart City Shelf, U.S. Gulf Coast, may be a rimmed shelf with relatively gentle slopes (2°) (Bebout and Loucks, 1974; Wilson, 1975). The shelf morphology is indicated by the marked break in slope at the shelf edge, and linear trends of reefs and shelf-margin, skeletal sands. The Silurian of the Michigan basin also is a reef-rimmed shelf that has a gentle basinward slope (10 m/km; 53 ft/mi) with numerous downslope pinnacle reefs (Mesolella et al, 1974; Sears and Lucia, 1979); the low slope seems to be important in development of such downslope buildups.

Bypass Margins

Bypass margins occur in areas of rapid upbuilding where shallow water sedimentation keeps pace with sea level rise. Bypassing may be associated with a marginal escarpment (Figure 4C) and/or a gullied bypass slope (Figure 4B) (McIlreath and James, 1979; Schlager and Chermak, 1979). Facies belts along the platform margin include:

1. Reefal carbonates and lime sands and gravels of the rim.
2. Escarpment (may be 200 m—660 ft—or more high), represents a zone of sediment bypassing from rim to slope.
3. Periplatform talus (sands, breccias, some mud interbeds). Where rim is reef-dominated, talus will have abundant reefal blocks; where rim is lime sand-dominated, periplatform sands will be abundant, together with clasts of cemented lime sand. Abuts base of escarpment; fines out into:
4. Gullied bypass slope of lime mud (commonly nodular bedded) with shoestring sand and gravel gully fills. If bypass slope is absent, periplatform talus (3) would fine out into:
5. Lower slope proximal graded turbidites, breccias, and lime mud; some massive sands and slumps, fining out into:
6. Basinal distal turbidites and lime mud or shale.

The Proterozoic Rocknest Formation of the Wopmay orogene, Canada (Hoffman, 1973; Grotzinger and Hoffman, 1983), is a rimmed continental shelf that may be of the escarpment-bypass type. McIlreath (1977) describes an example of a reef-dominated bypass margin in the Cambrian Cathedral Formation, western Canada. Here the bypass margin is vertical escarpment 200 m (650 ft) high, composed of calcareous algal reefs. The periplatform talus is dominated by clasts of reef. An example of a lime-sand–dominated bypass margin is the Cambrian "boundary limestone," a wedge (100 m or 330 ft thick, 3 km or 2 mi wide) of skeletal-peloid grainstone, wackestone, and hemipelagic mud, grading out into a distal wedge of hemipelagic lime mudstone and rare thin allochthonous lime sands; the unit accumulated at the foot of the Cathedral escarpment following cessation of reef growth (McIlreath, 1977). Bypassing associated with an escarpment also occurs in the Upper Devonian, Canning basin, Australia, where periplatform beds abut a platform margin unconformity (Playford, 1980). The Mesozoic of northwest Africa is a high-relief margin (over 2.5 km or 8,200 ft relief) (Todd and Mitchum, 1977) that may be a bypass or erosional margin.

Erosional Margins

These commonly are characterized by high, steep escarpments up to 4 km (13,000 ft) relief (Figure 4D). Reefal carbonates rim the platform, and are exposed on the upper few hundred meters of the upper escarpment. Downslope, due to erosional retreat of the escarpment by mechanical defacement, the escarpment exposes bedded, cyclic lagoonal, and peritidal beds. Facies belts include:

1. Reefal carbonates and lime sands/gravels of the rim.
2. Escarpment; in lower part exposes peritidal beds, and is abutted by periplatform talus.
3. Periplatform talus, lime sand, and mud, fining outward. Distinguishing features are clasts of fenestral, stromatolitic, and lagoonal carbonates in breccia beds, indicating large-scale retreat of margin. These are mixed with clasts of reefal carbonate and cemented lime sand.

Other erosional margins are characterized by large-scale erosion of outer shelf and slope deposits by bottom currents (Jansa, 1981). Examples of erosional margins are the Blake Bahama Escarpment (Ryan, 1980; Freeman-Lynde et al, 1981) and subsurface Mesozoic carbonates of eastern North America (Ryan and Miller, 1981; Jansa, 1981).

Shelves with Deep Rims

The Permian Capitan reef complex appears to be a rimmed shelf, in which the reefal rim, rather than being near sea level, remained relatively deeply submerged (over 30-m or 100-ft depth) throughout much of its growth (Yurewicz, 1977). Note that incipiently drowned reefs are excluded from this group because they have the potential to build to sea level, and thus would be unlikely to remain deeply submerged for long, geologically. The Capitan reef complex (Hileman and Mazzullo, 1977) consists of:

1. Subaqueous cyclic evaporites, carbonates, and clastics that pass seaward into:
2. Lagoonal (inner shelf) pellet-skeletal muds (Sarg, 1977), and:
3. Fenestral carbonates and vadose marine pisolite (shelf crest); these pass seaward into:
4. Outer shelf skeletal lithoclast sands with low (10°-20°) seaward dips.
5. Massive, marine-cemented, skeletal boundstone, and detrital carbonate of the deep rim (dips 20°-35°, water depths 30 m or 100 ft extending to 200 m or 650 ft depth), grading out into:
6. Foreslope talus, sands, and muds fining out into basinal facies (depths 300 to 600 m or 1,000-2,000 ft).

Such deep rims probably are uncommon in the geologic record, because shelf-edge biota have the greatest growth potential compared to the rest of the shelf; thus, deep rims would tend to build to sea level, unless there were unusual controls inhibiting such upbuilding. Capitan shelf-edge biota had high growth rates, indicated by the highly prograded margin. Perhaps the rim may have been unable to build to sea level because of the depth preferences of the reefal biota, or because of hypersaline shallow waters on the shelf. Lack of a shallow rim on these shelves probably would favor development of an emergent shelf crest, tidal flats, or islands a few kilometers landward of the rim as a result of wave-induced shoreward transport of shelf sands, causing these to be heaped into emergent barriers. Sea level fluctuations which allowed clastics to bypass the shelf and reach the basin also appear to complicate the evolution of the Capitan shelf edge.

Relations Between Rimmed Shelves and Other Platforms

Rimmed shelves typically develop from ramps with high carbonate production localized along the incipient shelf edge (Figures 5C, D). High rate of upbuilding commonly associated with reefal biota, coupled with sediment starving in the basin, increases relief and steepens the margin. Coupled with upbuilding, margins may change with time from accretionary and prograding type to bypass (gullied slope and escarpment type) to erosional margins (Figures 5D-G). Where transgression occurs across an unconformity, initial rims may be fringing, later evolving into barriers with progradation (Figures 5A, B). Fringing reefs may also form in areas with high paleoslopes undergoing periodic rapid uplift (Darwin, 1842).

During growth, margins may interchange between ooid dominated, where the platform is flat topped and shallow and keeps pace with sea level rise, to reef dominated, where relative sea level rise results in a deep lagoon and raised rim (Figure 5I), and there is much flushing between lagoon and open sea (Schlager and Ginsburg, 1980). Increased rate of relative sea level rise or down-to-basin faulting may cause platformward migration of the reefal margin (Playford, 1980). This may occur by abrupt backstepping, where the margin reestablishes some distance back from the original margin, reflecting rapid relative sea level rise (Figure 5J). Drowning may also convert the rimmed shelf into a distally steepened ramp (Figure 5H). Where drowning is gradual, reefal facies may onlap onto back-reef sands and lagoonal muds. Where sedimentation equals relative sea level rise, facies boundaries of the reef complex will be vertical. Margins may become downfaulted during growth causing drowning of seaward blocks, or faulting may occur after growth, triggered by later sediment loading.

Rimmed shelves are most likely to develop on continental shelves in low-latitude areas (Figure 10B), where reefal biota are abundant. Such rimmed shelves also are common in the tropics bordering volcanic/sedimentary arcs in areas of plate convergence, as in New Caledonia and New Guinea (Chevalier, 1973), where they may form fringing and barrier reef complexes hundreds of kilometers long (Figure 10E).

These complexes may become covered by pelagites, terrigenous or volcaniclastics, or pyroclastics, and may accrete onto continental margins as carbonate bodies in exotic or "suspect" terranes. Rimmed shelves seem to be relatively uncommon in foreland basins (Figures 10G, H), where waters may become restricted owing to tectonic uplands associated with plate convergence, and where fine terrigenous influx causes high turbidity.

Rimmed shelves develop peripherally to some interior or intracratonic basins (e.g., the Michigan basin; Mesolella et al, 1974), fault-bounded basins opening into continental shelves (e.g., Canning basin; Playford, 1980), or on regional platforms following drowning (e.g., western Canadian reefs; Klovan, 1974).

Rimmed shelves are likely to be absent from higher latitudes (temperate to cold-water shelves) where ramps would be expected. It would seem that rimmed shelves would be more likely to develop during periods when reef

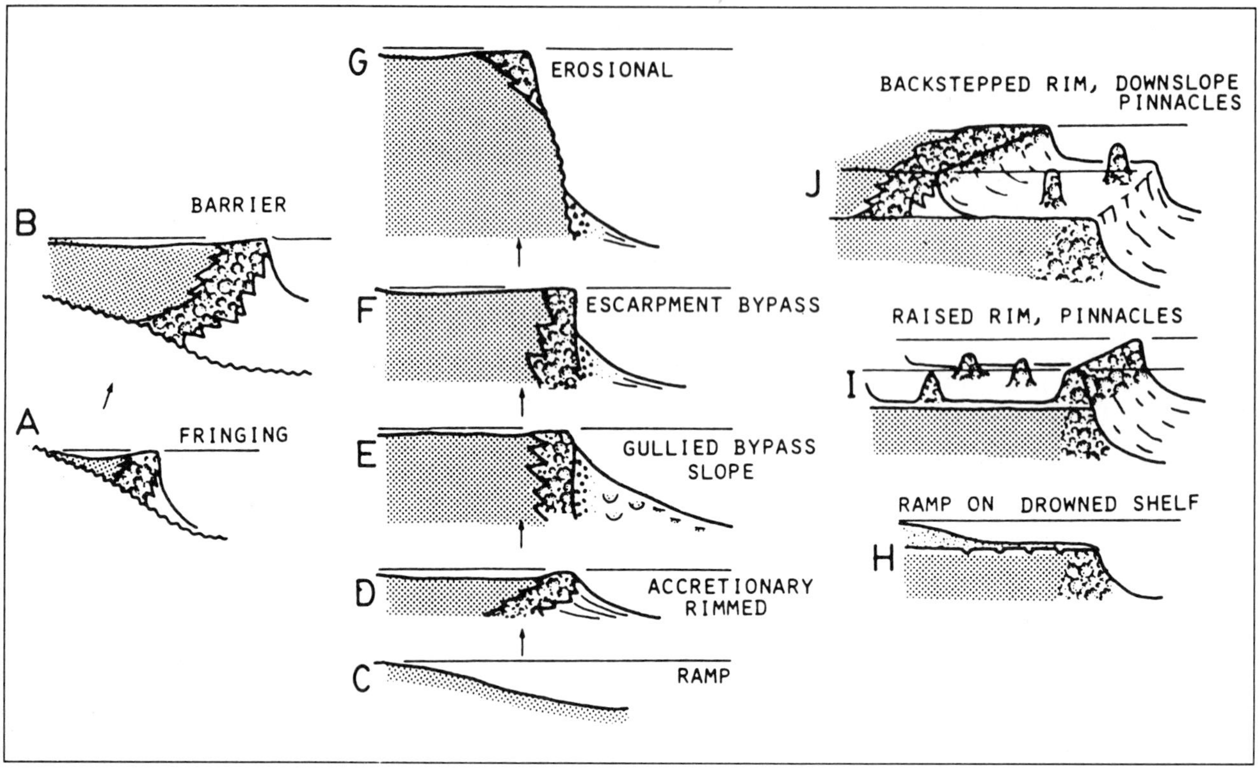

Figure 5—Evolution of rimmed shelves. A. Fringing-reef complex developed following transgression of high-relief surface. This later evolves into barrier reef complex (B). Many rimmed shelves develop from earlier ramps (C) into rimmed shelves, passing through accretionary (D), gullied bypass slope (E), escarpment bypass (F) to erosional rimmed margins (G). With drowning, rimmed shelves may develop into ramps (H) or into incipiently drowned shelves (I) with raised rim and high-relief reefs in the deep lagoon, or into drowned shelves (J) with backstepped rim with pinnacle reefs on deep shelf seaward of rim.

builders (skeletal metazoa capable of secreting large, robust, branching, hemispherical or tabular skeletons) were abundant. These were present during the Middle Ordovician (but capable of producing only relatively small reefs), the Silurian-Devonian, Late Triassic, Late Jurassic, Cretaceous, Oligocene, Miocene (?), and Plio-Pleistocene (Heckel, 1974; James, 1979). During the Precambrian and Cambrian, blue-green and skeletal algae appear to have been able to construct reefal rims (Hoffman, 1973; McIlreath, 1977).

Examples of Reservoirs on Rimmed Shelves

Reservoirs in the shelf-edge complex of rimmed shelves include the Cretaceous Stuart City trend, Gulf Coast (Griffith, et al, 1969; T. D. Cook, 1979), and the Permian Townsend-Kemnitz field (Malek-Aslani, 1970). Pinnacle reefs located on gentle slopes seaward of the rim are common reservoirs in the Silurian of the Michigan basin (Mesolella et al, 1974) and the giant Intisar D field, Paleocene, Libya (Brady, et al, 1980). Reservoirs associated with peritidal dolomites that pinch out updip into evaporites and dolomite seals form reservoirs on some rimmed shelves, as in the Permian basin (Meissner, 1974).

INTRASHELF BASINS ON RIMMED SHELVES AND RAMPS

Many rimmed shelves have inshore basins lying behind the shallow carbonate rim. The basins commonly pass landward into coastal siliciclastics. Seaward and along depositional strike, they may pass onto the shallow carbonate rim by way of a gently sloping ramp that may be skeletal or ooid-dominated.

Basins have water depths of a few tens of meters, and the basin floors may lie below fair-weather wave base but above storm wave base. Sediments filling intrashelf basins are shale with thin beds of quartz sand and lime silt, intraformational conglomerate, glauconite, and radial-ooid packstone arranged in storm-generated, upward-coarsening, and upward-fining sequences (Eliuk, 1978; Markello and Read, 1981). Where basin floors lie below storm wave base, basin fills may be euxinic to dysaerobic limestone and shale (Murris, 1980).

Intrashelf basins commonly develop during relative sea level rise, which allows rapid upbuilding of the carbonate rim while the basin floor lags behind because of slower rates of sedimentation. In arid settings, evaporite deposition occurs in the basins (Wilson, 1975, p. 326; Murris, 1980), possibly during regressive phases.

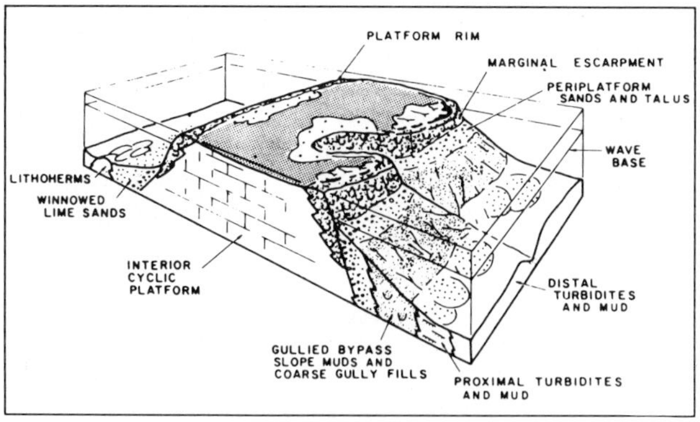

Figure 6—Block diagram of isolated platform aggraded to sea level.

Examples of intrashelf basins include the Cambrian of western Canada (Aitken, 1978), Cambrian of the southern Appalachians (Markello and Read, 1981), Mesozoic of eastern Canada (Eliuk, 1978), the Middle East (Murris, 1980), and the Gulf Coast (Wilson, 1975; p. 326, Bay, 1977), and perhaps deeper, siliciclastic portions of Holocene shelves such as the Sahul, Queensland, and Belize shelves (Ginsburg and James, 1974), although the last two are not bathymetrically closed.

Reservoirs Associated with Intrashelf Basins

Intrashelf basins may be associated with giant fields, because they contain source beds that are in close proximity to reservoir carbonates that include oolitic ramps and rudistid buildups prograding out across the intrashelf basins, as, for example, the Arab A to D fields and Shuaiba field, Middle East (Murris, 1980), and the Bombay High field, offshore India (Rao and Talukdar, 1980).

ISOLATED PLATFORMS (BAHAMA TYPE) AND OCEANIC ATOLLS

Pericratonic Isolated Platforms

Isolated or detached shallow water platforms offshore from continental shelves commonly are tens to hundreds of kilometers wide, located above rifted continental or transitional crust and surrounded by deep water—commonly several hundred meters and even exceeding 4 km (13,000 ft) (Figure 6). Some of these platforms have been termed atolls, especially where they have a deeper lagoon and elevated reefal rim, but they differ from true oceanic atolls which rise from volcanic foundations on oceanic crust. Isolated platforms include the Bahamas (Enos, 1974b) and adjacent platforms, platforms rising from the Coral Sea plateau (Orme, 1977), and Glovers Reef and Lighthouse Reef off the Belize shelf (James and Ginsburg, 1979).

Interiors of reef-rimmed platforms may be dominated by skeletal limestone, where interiors are relatively deep (up to 20 m or 65 ft). In contrast, where platforms are shallow and flat-topped, interior facies may be dominated by cyclic nonskeletal peloidal sands and muds, and platform margins are shoals and eolian islands of ooid grainstone with subordinate reefs (Beach and Ginsburg, 1980).

One of the major differences between isolated platforms and other types is that margins may be windward or leeward (Mullins and Neumann, 1979). Windward margins that are open (lack energy barriers) are sediment barren (except for skeletal sands in lee of local reefs). Some windward margins also have islands that augment offshelf sand transport and inhibit bankward transportation (Hine, Wilber, and Neuman, 1981; Hine, Wilber, Bane et al, 1981). Leeward margins that are open have widespread peloidal sands that are being transported offbank; islands on leeward margins are energy barriers which inhibit offbank transport. Tide-dominated margins have broad ooid sand lobes migrating onto the bank. Margin facies of isolated platforms also reflect whether they face an ocean, a protected seaway, or a narrow basin; some also are influenced by deep oceanic currents, and have winnowed sands, hardgrounds, and lithoherms (Mullins and Neuman, 1979).

Rare isolated platforms may have gently sloping margins (ramp-like profiles) (Matti and McKee, 1977; Purser, 1972, in Wilson, 1975, p. 285-288). Facies belts of low-relief, gently sloping, isolated platforms resemble those of ramps, discussed previously (cf. Matti and McKee, 1977).

More commonly, margins are steeply sloping, resembling those of rimmed shelves (Mullins and Neumann, 1979). Those with steeper profiles may have a marginal escarpment (up to 60° or more, and few hundred meters to 4 km or 13,000 ft high), grading down into a more gently sloping (1°-15°) deep-water sediment wedge that passes out into relatively flat-lying basin-plain deposits (slopes less than 1 m/km or 5 ft/mi) (Mullins and Neumann, 1979). Schlager and Ginsburg (1980) recognize accretionary margins, bypass margins and erosional margins (cf. Figures 4A-D) the sequence reflecting evolution of the margin with upbuilding and steepening of slopes through time.

Facies belts of isolated platforms (Figure 6) with steeper profiles (Mullins and Neumann, 1979; Schlager and Chermak, 1979) are:

1. Platform and platform rim: reefal carbonates, skeletal and oolitic sands, cemented islands; the platform may be covered by bedded, cyclic, pelletal sands and muds (locally peritidal) and evaporites; or by skeletal sands. Siliciclastics are absent.

2. Marginal escarpment: variably developed; upper parts expose back-reef, reef, and fore-reef sediments; lower parts of deeper escarpments (below 1 km or 3,300 ft) expose bedded lagoonal and tidal-flat carbonates, probably as a result of mechanical defacement (Ryan, 1980; Freeman-Lynde et al, 1981).

3. Talus slope or periplatform sands: muddy lime sands (mixed shallow water sediment and pelagics) and talus blocks. Commonly 1 to 3 km (0.62-2 mi) wide. The talus-slope deposits may pass downslope into basinal facies on high-relief, erosional, windward margins, or into:

4A. Slump and gravity flow deposits (accretionary margins), or:

4B. Winnowed slope (current swept slopes of accretionary margins): sands composed of planktonics, rock fragments, and lesser shallow water sediment, abundant hardgrounds. May prograde out onto and unconformably overlie basin facies, or pass downslope into lower slope lithoherms (5B), or into lower slope, nodular, pelagic limestone (cf. 4C), or:

4C. Gullied bypass slope (gullied bypass margins): composed of pelagic lime mud and shoestring lime sands and rubble (gully fills). May be nodular bedded (reflecting patchy submarine cementation and reworking). Hardgrounds and erosional cliffs are present.

5A. Lower slope or basin margin: alternating proximal, graded turbidites, and carbonate ooze. Some massive lime sands, debris flows, and slumps, or:

5B. Lithoherm belt: individual mounds up to 70 m (230 ft) thick; hardgrounds, sand waves (reflect presence of deep ocean currents; probably atypical).

6. Basin or basin interior: alternating graded distal turbidites and carbonate ooze.

Slope facies of accretionary margins (cf. Figure 4A) may be dominated by periplatform talus, grading out into lime sands and downslope lithoherms, or slumps and gravity flows. Slope facies of gullied-slope bypass margins (cf. Figure 4B) may be periplatform sands and talus, passing out into gullied-slope muds and shoestring sands, then into proximal turbidites. Commonly, marginal escarpments (cf. Figure 4C) and bypassing of gullied upper slopes are associated. Material accumulates at the foot of the escarpment by rock fall, sliding, or creep. The coarse debris may then be carried as erosive sediment gravity flows down gullies, bypassing the muddy upper slope, to accumulate on the lower slope and basin margin as turbidites and minor debris flows, together with pelagic and hemipelagic limestone (cf. Schlager and Chermak, 1979). Slope facies of high relief, erosional escarpments (cf. Figure 4D) appear to be dominated by periplatform talus. Other erosional margins show erosional truncation of slope facies due to turbidity-current erosion of the slope, which steepens as deep-water canyons incise basin facies (Schlager and Ginsburg, 1980).

Deep, elongate troughs between isolated platforms may be traversed by headwardly eroding, axial valleys cut by turbidity currents (Hooke and Schlager, 1980). These currents increase in strength with time as platforms are built upward, and valleys are down-cut. Channels are eroded into relatively flat-bottomed basins, to form V-shaped valleys that ultimately tap gullied bypass slopes, which then merge with flanks of valleys (and associated tributary system). These result in erosional slopes on deep margins of platforms.

Besides the Bahamas (Mullins and Neumann, 1979; Schlager and Chermak, 1979), examples of isolated platforms include the Cretaceous Golden Lane and El Doctor platforms, Mexico (Enos, 1974a; 1977), probably some Triassic platforms in the Dolomites, Italy (Bosellini and Rossi, 1974; Wilson, 1975), and Jurassic platforms in the Venetian Alps. Bosellini et al (1981) describe filling of a narrow, elongate basin of Jurassic age in the Venetian Alps by deep-sea fans of oolite (Vajont Limestone, up to 1,000 m or 3,300 ft thick) derived from the adjacent windward margin of the Friuli platform (Bosellini et al, 1981). Mesozoic platforms of Sicily (Catalano and D'Argenio, 1981), the late Paleozoic Horseshoe Atoll, Texas (Vest, 1970; Schatzinger, 1983), the Devonian Tor limestone, western United States (Matti and McKee, 1977), and parts of the Jurassic Paris basin (Purser, 1972, in Wilson, 1975, p. 285-288) are also isolated platforms.

Oceanic Atolls

Oceanic atolls are circular to elliptical platforms (1-km—0.62-18 mi—and rarely to 130 km—80 mi—diameter), with raised reefal rims and deep lagoons, and are developed above oceanic volcanoes that commonly rise from depths of 3 to 5 km (10,000–16,000 ft). Slopes to depths of a few hundred meters are about 40°, flattening with depth until they merge with the deep ocean floor. They are characterized by deep lagoons, commonly 30-90 m (100-300 ft) depth, that are floored by carbonate muds and sand, and dotted by numerous high-relief reef knolls (submerged) and patch reefs. These pass toward the rim into a complex of back-reef sands with small patch reefs and isolated large coral heads, cemented reef rubble, islands, and beach rock. Reefal boundstones dominated by coral form the rim, with coralline algal boundstones forming a topographically high rim; the reefal boundstones extend a few tens of meters downslope, commonly with spur and groove structures. Downslope, these pass into coral and algal sands fining into skeletal sands and silts and scattered reefal blocks. These grade out below 4 km (13,000 ft) into red clays. High slopes on the margins promote much slumping and debris slides.

Evolution of Isolated Platforms

Most isolated platforms on passive continental margins appear to develop on faulted, rapidly subsiding continental or transitional crust (Mullins and Lynts, 1977), commonly during early phases of opening of ocean basins (Figure 10). Many are underlain by regional shallow water carbonates, and are localized over horsts, the adjacent grabens becoming sites of deeper water sedimentation. Others may be localized over linear submarine ridges (Matti and McKee, 1977), and some are developed over structural highs in continental interiors during periods of high sea level. Initially, some isolated platforms may have ramp-like slopes, but these would develop into high-relief rimmed margins with time (Figure 7A, B). Subsidence, coupled with upbuilding of isolated platforms, creates large bank-to-basin relief, and the margins may progress through ramps to depositional rimmed platforms to bypass margins to erosional margins.

With rapid sea level rise, isolated platforms may become covered by extensive reefal carbonates and skeletal sands (where they are able to stay near sea level, Figure 7C) or they may develop raised rims and skeletal sands in the deep lagoon (Figure 7D), or they may become completely drowned and covered by basinal facies or periplatform detritus shed from extant shallow platforms (Figure 7E, F, G).

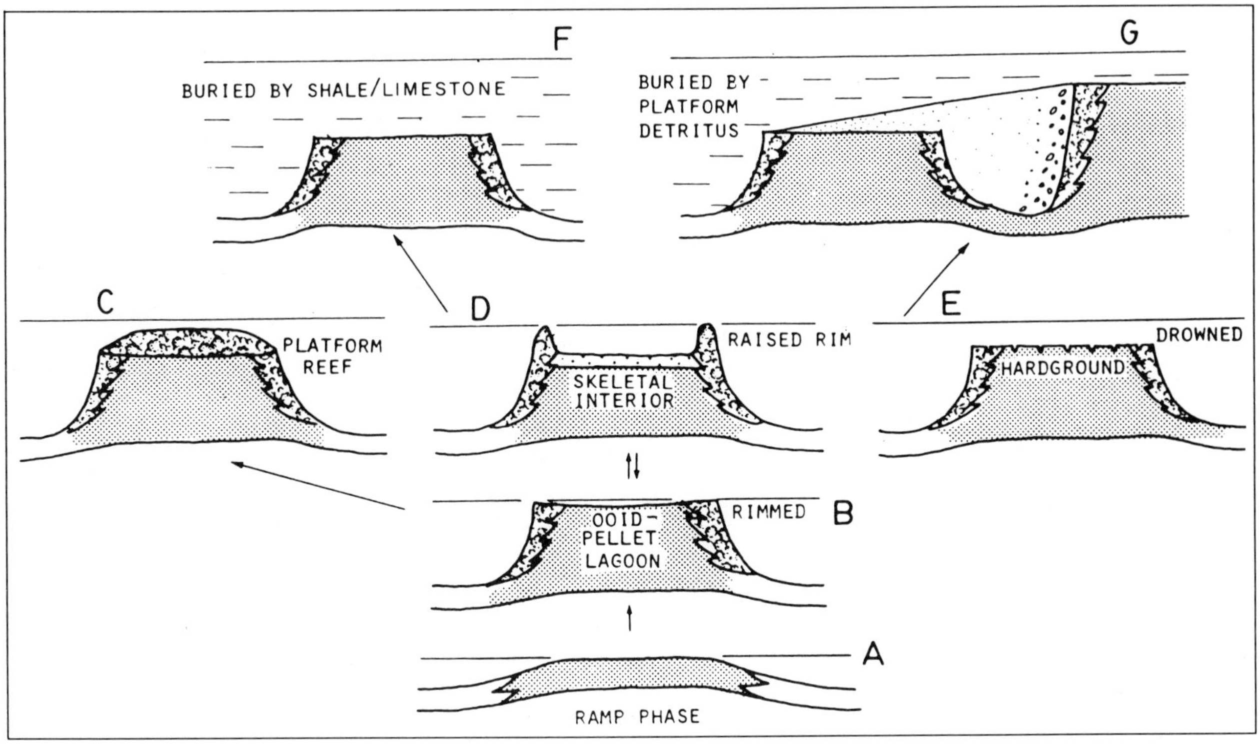

Figure 7—Evolution of isolated platforms. A. Isolated platform with initial ramp phase, evolving into (B) high relief rimmed platform with ooid-pellet interior aggraded to sea level. With sea level rise or subsidence, platform may become covered by extensive reefal carbonates and skeletal sands (C), or develop a raised rim with skeletal sands flooring a deep lagoon (D), or become drowned and surfaced by hardgrounds (E). Ultimately these drowned platforms become covered with basinal facies or periplatform detritus shed from adjacent platforms (F, G).

Most oceanic atolls (Darwin, 1842; Emery, et al, 1954; Ladd, 1973; Stoddart, 1973) develop on subsiding oceanic volcanoes. Many may start as fringing reefs that progress into barrier reefs, and finally into atolls as the volcanic edifice subsides below sea level (Darwin, 1842). Development may also be influenced by marine erosion and by subaerial erosion during time of lowered sea level (Steers and Stoddart, 1977). Modern atolls have been greatly influenced by the rapid postglacial transgressions, which have favored upbuilding of the rapidly growing rims and formation of deep lagoons. When sea levels were more stable, it seems likely that the atolls would develop into flat-topped platforms, given the low subsidence rates (few centimeters per thousand years) compared to sedimentation rates that may be 10 or 100 times higher.

In the geologic record, oceanic atolls should overlie oceanic volcanics, and they may become covered by deep-water pelagites, red clays, or pyroclastics. Oceanic atolls and their oceanic volcanic foundations may accrete onto margins of continental blocks during closure of ocean basins, to become discontinuous carbonate masses within exotic or suspect terranes.

Examples of Reservoirs in Isolated Platforms

Giant fields in isolated platforms include the upper Paleozoic Horseshoe atoll, Texas, where the reservoirs occur in the raised rim (Vest, 1970), and the Cretaceous Golden Lane atoll, where the reservoirs are the Golden Lane shelf-edge carbonates and the Tamabra periplatform debris of the Poza Rica trend (Enos, 1977).

DROWNED PLATFORMS

Where subsidence or sea level rise exceeds upbuilding, ramps, rimmed shelves, and isolated platforms may undergo incipient or complete drowning (Kendall and Schlager, 1981; Schlager, 1981) (Figures 1, 8).

Drowning occurs where rate of relative sea level rise exceeds vertical accumulation rate and the platform is submerged below the euphotic zone, terminating rapid production and accumulation of carbonate by photosynthetic organisms (Kendall and Schlager, 1981). The euphotic zone in the open ocean may extend down to 100 m (330 ft), but may be as little as 30 m (100 ft) in basins where fine-grained carbonate or clastics are abundant. Following drowning, platforms may become surfaced by hardgrounds, by deep-water, nodular, argillaceous limestone, by pelagic carbonates, or by periplatform talus shed from adjacent shallow parts of platforms. Condensed sequences with numerous hardgrounds may develop, or in areas of nondeposition, submarine unconformities or chemical sediments (iron, manganese, phosphorite, or sulfide crusts) may develop.

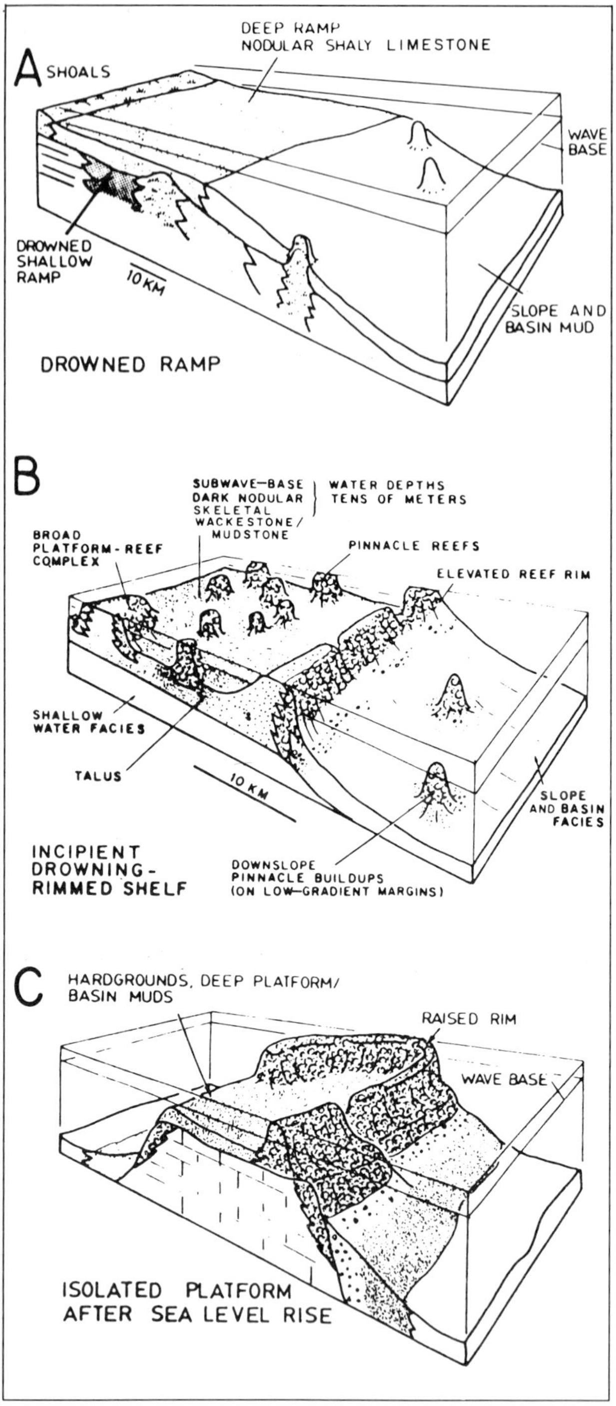

Figure 8—Drowned platforms. A. Ramp after rapid sea level rise, showing onlap of basinal and deep ramp facies onto shallow ramp carbonates. B. Rimmed shelf after rapid sea level rise showing development of raised rim and of pinnacle reefs in deep lagoon and downslope. C. Isolated platform after sea level rise, showing development of raised rim and deep interior.

Incipient drowning (Kendall and Schlager, 1981) occurs where relative sea level rise exceeds rate of carbonate upbuilding, but the platform surface stays within the euphotic zone. Consequently, the system is able to recover because rate of sea level rise decreases relative to rate of sediment deposition. In some incipiently drowned sequences, deepening may push the platform below the euphotic zone, but deeper water benthonic assemblages are able to build up into the photic zone, assisted by accumulation of lime and siliciclastic muds carried in from shallow platform areas. Facies typical of incipiently drowned platforms are nodular and thin-bedded argillaceous limestones (whole fossil wackestone/mudstone with some lime-sand layers) tens to hundreds of meters thick that overlie shallow platform facies from which rise scattered large buildups. Storm-generated, upward-fining beds may be common in the deeper water facies, along with hardgrounds.

Aggraded vs. Incipiently Drowned Platforms

Many ancient rimmed shelves are flat topped and aggraded to sea level for most of their development. They consist of many upward-shallowing sequences or cycles of peritidal carbonates that cover most of the platform, extending to within a few kilometers of the rim, where they merge with back-reef sands. Peritidal facies may even prograde landward from the rim, gradually deepening into shallow lagoon facies. Such aggraded shelves probably are more typical of the past, considering that carbonate sedimentation generally far exceeds long-term subsidence or sea level rise (Schlager, 1981). Further, the cyclic, aggraded platform sequences also reflect small-scale (few meter) sea level oscillations (periods of 20,000 to over 100,000 yr superimposed on long-term subsidence). Consequently, the record reflects short-term transgressive pulses of a few meters amplitude followed by shallowing to sea level and widespread peritidal deposition. Sea level falls, if present, are minor and rarely leave a record such as karst surfaces, regolithic breccias, or soils on tops of cycles.

In contrast, most modern platforms reflect incipient drowning by rapid post-glacial sea level rise. Thus they have relatively deep lagoons, elevated pinnacle and patch reefs, raised rims, and tidal-flat facies are far removed from the rim. Shallowing-upward sequences are disconformable, with well-developed karst features, regolith, and soils or caliches. Surfaces of transgression have considerable relief inherited from earlier, high-relief buildups on the shelf or from erosion or dune formation during the previous low sea level stand. These sequences reflect large-scale (over 100 m or 330 ft) glacio-eustatic sea level fluctuations (frequency 20,000 to over 100,000 yr) superimposed on long-term gradual subsidence. Because of the magnitude of sea level fluctuations, the shelves are rarely built to sea level, and tidal-flat facies rarely cap cycles but are restricted to coastal locations. Thus, modern platforms generally are poor analogs of ancient aggraded platforms. The modern platforms more closely resemble ancient incipiently drowned platforms or platforms that developed during periods of continental glaciation, as in the Carboniferous-Permian when large-scale glacio-eustatic fluctuations occurred.

Drowning of Ramps, Rimmed Shelves, and Isolated Platforms

Drowning causes a major, landward shift in the shallow-platform facies. New belts of platform-margin facies associated with ramp, shelf, and isolated platforms may be developed adjacent to positive elements (e.g., fault blocks, eolianite dunes, arches, and cratonic shorelines) at considerable distance from the earlier platform margin.

Where ramps are drowned (Figure 8A) resulting facies will have gradually transgressive or onlapping facies relationships. Drowning is diachronous, and youngest to landward. On shallower parts of incipiently drowned ramps, shallow-ramp facies may be overlain regionally by deep-ramp carbonates; whereas downslope, complete drowning occurs, and ramp facies and downslope buildups are overlain by chemical sediments and slope/basin pelagic or hemipelagic facies.

Where flat-topped rimmed shelves (Figure 8B) or isolated platforms (Figure 8C) are drowned, the drowning may be synchronous over large areas. During relative sea level rise, the rim may backstep, leaving in front of the rim a deeply submerged shelf (Figure 5J). Where relative sea level rise is slower, the rim may retreat gradually, and reefal facies will onlap back-reef beds (Playford, 1980). On rimmed shelves, drowning may initially cause upbuilding of the rim above the adjacent deepening lagoon (Figure 5I, 8B). Selective upbuilding of rims of platforms may result in "atoll"-like or elevated rim morphologies (Figure 8B, C); where such drowning is incipient, pelletal/ooid facies of earlier shallow platforms interiors may be succeeded by skeletal facies (Schlager and Ginsburg, 1980).

During drowning of ramps and rimmed shelves, numerous thick, isolated buildups may develop on the deeply submerged platform (Figure 8A, B). These may range from narrow pinnacle reefs to broad reef-rimmed banks or "shelf atolls," and downslope banks (Klovan, 1974; Kendall and Schlager, 1981; Read, 1982). These may develop seaward of any newly established rim on the carbonate shelf (Figure 5J), or they may develop in deeply submerged lagoons landward of the reefal rim (Klovan, 1974).

After rapid relative sea level rise, buildups and rims commonly show three phases of accumulation: a lag phase, during which the buildup lags behind sea level, and deeper water biotas may be established; a catch-up phase, when the upward growth exceeds sea level rise, and a shallowing-upward sequence may develop; and a keep-up or tracking phase, when the buildup keeps pace with relative sea level rise (Kendall and Schalger, 1981; Schlager, 1981).

Vertical transition from shallow platform to deeper water facies of the drowned phase may be abrupt or gradational. It may be marked by basal lime sands and gravels that result from migration of a high-energy transgressive environment over the low-energy platform interior. If drowning followed a period of sea level lowering, transgressive sands may rest unconformably on limestone with soil fabrics, caliches, or vadose features, as in many Pleistocene-Holocene sequences. If drowning followed shallowing to tide levels, basal lime sands and gravels may rest on tidal-flat carbonates with only minor evidence of subaerial weathering. Following drowning, progradation and upbuilding may tend to return the shelf or isolated platform to the original rimmed state; or in a ramp, it may result in progradation of shallow ramp facies over deep ramp and basinal facies. On isolated platforms, filling of the deep lagoon will result in pellet/ooid facies succeeding skeletal facies.

The problem of the drowning of a carbonate platform, where upbuilding potential is generally greater than tectonic subsidence or sea level rise, is discussed in detail in Schlager (1981) and Kendall and Schlager (1981). Carbonate platforms (and especially reefs) grow at 1-10 m/1,000 yr (3-33 ft/1,000 yr), maximum. Long-term tectonic subsidence of platforms commonly is 1-10 cm/1000 yr (0.4-4 in./1,000 yr) on passive margins and over 50 cm/1,000 yr (20 in./1,000 yr) in foredeeps. These rates generally are exceeded by reef and bank biotas. Eustatic sea level rise may reach several meters/1,000 yr, which may be matched by upbuilding of reefs. Thus drowning generally requires pulses of subsidence or sea level rise much greater than average, or stressing of resident communities by environmental or climatic changes.

Incipient to completely drowned ramps include the Middle Ordovician, Virginia (Read, 1980), and the Devonian Helderberg Group, New York (Laporte, 1969).

Incipiently to completely drowned distally steepened ramps or open shelves (Ginsburg and James, 1974) are common in the Holocene, reflecting the rapid postglacial rise in sea level, and include the Yucatan, West Florida, and Sahul Shelves.

Incipiently drowned rimmed shelves include the Holocene Queensland Shelf (Great Barrier Reef) (Maxwell, 1968), the southern Belize Shelf (Ginsburg and James, 1979), and the western Canada Devonian Rainbow-Zama reefs and underlying shelf (McCamis and Griffiths, 1968). Completely drowned rimmed shelves include the Cretaceous Blake Plateau (Sheridan, 1974) and the Edwards–Stuart City to Georgetown sequence of Texas (Griffith et al, 1969; Cook, 1979).

Examples of incipiently drowned isolated platforms include many oceanic platforms or atolls, and deeply submerged "banks," such as Great Chagos Bank (Stoddart, 1973) and Cay Sal Bank (Hine and Steinmentz, 1983). Drowned isolated platforms include Cretaceous platforms, Mexico (Enos, 1974a), and Mesozoic platforms in the Mediterranean (Bernouli and Jenkyns, 1974; Bosellini et al, 1981; Winterer and Bosellini, 1981; Catalano and D'Argenio, 1981).

Buildups developed on drowned platforms include some buildups of western Canada (Klovan, 1974), Ordovician buildups, Virginia (Read, 1980), and upper Paleozoic downslope buildups, Montana (Smith, 1977). An extensive listing of platforms and buildups associated with drowning is given in Kendall and Schlager (1981).

Examples of Reservoirs in Drowned Platforms

Reservoirs on drowned ramps include Devonian Onondaga bioherms, New York (Kissling and Polasek, 1982). Reservoirs associated with buildups on drowned platforms seaward of shallow shelves include the Paleocene giant Intisar D field, Libya, in which the reef is filled to spillpoint (Brady, et al, 1980); western Canada Upper

Devonian Swan Hills–Judy Creek fields (Wendte and Stoakes, 1982; Viau, 1983), where the reservoir facies commonly are platform-margin and slope-deposits; and Upper Devonian Nisku reefs or coral mudmounds where the porosity is primary or in coral-moldic and vuggy dolomites (Pounder et al, 1980; Machel, 1983). Longman (1980) describes a reservoir in highly fractured forereef talus adjacent to a lower Miocene atoll on a drowned continental shelf, Philippines. The Middle Devonian Rainbow-Zama fields are examples of reservoirs in buildups that formed landward of the rim following rapid deepening (Schmidt et al, 1980). Reservoirs may also occur in fringing-reef complexes that develop around basement highs (e.g., granite knobs or volcanic highs) undergoing rapid submergence as in the Upper Cretaceous Elaine field, Texas (Luttrell, 1977), and the Devonian Slave Point field, Alberta (Dunham, et al, 1983).

The giant fields of the upper Paleozoic Horseshoe atoll, west Texas, are examples of reservoirs associated with an isolated platform undergoing rapid deepening. The raised rim which forms the reservoir developed during rapid deepening which drowned much of the central and northern parts of the platform (Vest, 1970). Much of the porosity is related to leaching during emergence prior to drowning (Schatzinger, 1983). Many of the seals in the above reservoirs are either deeper water fine-grained clastics or regressive basin-filling evaporites.

FAULTED MARGINS

Faulted margins (Figure 9) of carbonate platforms are evident on some seismic profiles and also have been recognized in outcrop. Faulting may be contemporaneous with the carbonate platform and may cause drowning of seaward edges of the platform. Downfaulting may occur as a single downdropped block or as series of downstepped blocks (down-to-basin faults). Downdropped blocks may be overlain by downslope buildups, by deep-water facies, hardgrounds, or unconformities. Faulting that postdates the carbonates commonly is associated with progradation of thick clastic sequences. Reactivation of faults which earlier controlled location of the platform margin may cause intense shearing of shelf-edge facies, especially during periods of wrench faulting. Examples of faulted margins include the Mesozoic, eastern Canada (Jansa, 1981), Mesozoic, southern Alps, Italy (Winterer and Bosellini, 1981; Bosellini et al, 1981), and the lower Paleozoic, Greenland (Hurst and Surlyk, 1983).

CARBONATE PLATFORMS, TECTONICS AND EUSTACY

On passive margins, carbonate platforms commonly develop over basal rift volcanics, immature clastics and evaporites, or more mature shelf clastics. Initially, ramps develop typically on the gently sloping surface of rift or shelf clastics (Figure 10A). Later, they evolve into rimmed carbonate shelves (Figure 10B, right), as a result of high carbonate production on the developing shelf edge and sediment starving of off-shelf environments. Initial ramps or rimmed shelves also may evolve into isolated platforms during crustal extension (Figure 10B, left). The earlier, extensive platform carbonates are faulted to form horsts and grabens and undergo rapid submergence, with carbonate upbuilding being localized on the highs and the grabens become sites of drowning and deep-water sedimentation.

The reverse development (rimmed shelf into a ramp) commonly occurs where the earlier rimmed shelf is drowned (Figure 10C). A shallow-to-deep ramp transition forms some distance back from the drowned shelf edge, and this ramp may then evolve into a rimmed shelf, bordered to seaward by a broad, deep (drowned) shelf. Where rimmed shelves are prograded by clastics, subsequent carbonate platforms generally will be ramps (Figure 10D). Isolated oceanic platforms may be developed on submarine volcanoes on oceanic crust, and fringing- and barrier-reef complexes may be developed peripheral to volcanic arcs (Figure 10E).

During arc-continent or continent-continent collision, rimmed shelves may evolve into ramps where platforms are prograding out into filling foredeep basins or where the basin margin is undergoing uplift (Figure 10F). Also, carbonates on oceanic volcanoes may become incorporated into the subduction complex (Figure 10F). During convergence, rimmed shelves of passive margins commonly are unconformably overlain by carbonate ramp sequences extending into the developing foredeeps or foreland basins (Figure 10G). These ramps may be overlain by deep-water shales and graywacke turbidites (flysch). As convergence becomes advanced, large-scale overthrusting causes slope reversal by filling of foreland basins with shallow-marine to continental clastics. Ramps now deepen (slope) onto the craton (Figure 10H). Carbonates of the miogeocline and foreland basin will be preserved within thrust sheets transported toward the craton. Carbonates associated with oceanic volcanoes and arcs may be preserved in exotic or suspect terranes following collision (Figure 10H).

On stable continental interiors during high sea level stands, ramps are widely developed extending out from positive areas and down regional paleoslopes. Adjacent to basins, sea level rise, increased slope, and sediment starving within the basin may convert intracratonic ramps into rimmed shelves or high relief bank structures.

Periodic sea level oscillations ranging from 20,000 to 100,000 yr or more (4th and 5th order cycles; Vail et al, 1977) appear to have been superimposed on long-term subsidence of platforms, and controlled sequences developed. Small-scale fluctuations of a few meters appear to be associated with cyclic, upward-shallowing sequences on aggraded flat-topped platforms. In contrast, large scale (greater than 100 m or 330 ft) fluctuations appear to be associated with incipiently drowned platforms with numerous high-relief buildups and disconformity-capped cycles.

One to 10-million-year transgressive–regressive events (3rd order cycles) result in formation of carbonate sequences tens to hundreds of meters thick. Some of these consist of a single upward-deepening–upward-shallowing succession. Regressive events may be marked by unconformity development or offlap of clastics onto the carbonates. Finally, these smaller scale cycles are superimposed on 2nd order (10 to 80 million years or more) relative sea

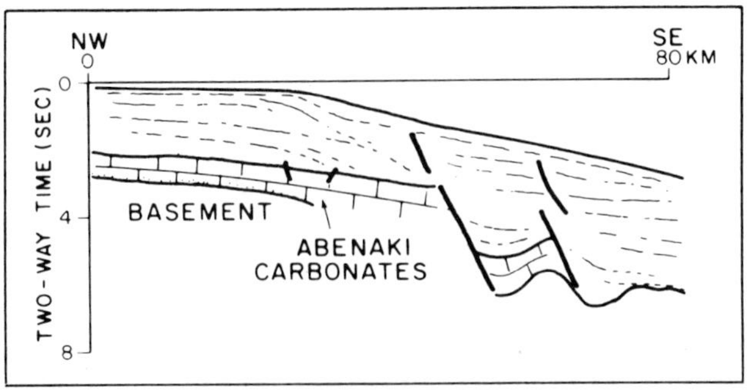

Figure 9—Sketch of seismic cross section (after Jansa, 1981) showing faulted margin. Mesozoic, eastern Canada. Faults drop down the seaward edge of the shelf. Limestone shown by brick patterns. Clastics above limestones are blank.

level cycles during which carbonate platforms 1-4 km (3,300-13,000 ft) thick, may be developed.

Incipient to complete drowning of platforms occurs where submergence (related to subsidence or sea level rise) outstrips upbuilding by biological production of carbonate sediment. Ramps or rimmed shelves may develop on landward parts of drowned shelves, and pass seaward into deeply submerged shelves with many large downslope buildups. Drowning is important in that it may result in large buildups on platforms, which become encased in deep-water potential source beds.

SELECTED REFERENCES

Ahr, W. M., 1973, The carbonate ramp—an alternative to the shelf model: Gulf Coast Association of Geological Societies Transactions, v. 23, p. 221-225.

——— and S. L. Ross, 1982, Chappel (Mississippian) biohermal reservoirs in the Hardeman basin, Texas (abs.): Gulf Coast Association of Geological Societies Transactions, v. 32, p. 185-193.

Aitken, J. D., 1978, Revised models for depositional grand cycles, Cambrian of the southern Rocky Mountains, Canada: Bulletin of Canadian Petroleum Geology, v. 26, p. 515-542.

Baars, D. L., and G. M. Stevenson, 1982, Subtle stratigraphic traps in Paleozoic rocks of Paradox basin, *in* The deliberate search for the subtle trap: AAPG Memoir 32, p. 131-158.

Baria, L. R., O. L. Stoudt, P. M. Harris, and P. D. Crerello, 1982, Upper Jurassic reefs of Smackover Formation, United States Gulf Coast: AAPG Bulletin, v. 66, p. 1449-1482.

Bay, T. A., Jr., 1977, Lower Cretaceous stratigraphic models from Texas and Mexico, *in* D. G. Bebout and R. G. Loucks eds., Cretaceous carbonates of Texas and Mexico, applications to subsurface exploration: University of Texas Bureau of Economic Geology Report of Investigations 89, p. 12-30.

Beach, D. K., and R. N. Ginsburg, 1980, Facies successions of Pliocene-Pleistocene carbonates, northwestern Great Bahama Bank: AAPG Bulletin, v. 64, p. 1634-1642.

Bebout, D. G., and R. G. Loucks, 1974. Stuart City trend, Lower Cretaceous, south Texas, a carbonate shelf-margin model for hydrocarbon exploration: University of Texas Bureau of Economic Geology Report of Investigations 78, 80 p.

Bernoulli, D., and H. C. Jenkyns, 1974, Alpine, Mediterranean, and central Atlantic Mesozoic facies in relation to the early evolution of the Tethys, *in* Modern and ancient geosynclinal sedimentation: SEPM Special Publication 19, p. 129-160.

Bishop, W. F., 1969, Environmental control of porosity in the upper Smackover Limestone, North Haynesville field, Claiborne Parish, Louisiana: Gulf Coast Association of Geological Societies Transactions, v. 19, p. 155-169.

Bosellini, A., D. Masetti, and M. Sarti, 1981, A Jurassic "Tongue of the Ocean" infilled with oolitic sands: the Belluno Trough, Venetian Alps, Italy: Marine Geology, v. 44, p. 59-95.

——— and D. Rossi, 1974. Triassic carbonate buildups of the Dolomites, northern Italy, *in* Reefs in time and space: SEPM Special Publication 18, p. 209-233.

Brady, M. J., and R. B. Koepnick, 1979. A Middle Cambrian platform-to-basin transition, House Range, west central Utah: Brigham Young University Geology Studies, v. 26, p. 1-8.

——— and A. J. Rowell, 1976, An Upper Cambrian subtidal blanket carbonate, eastern Great Basin: Brigham Young University Geology Studies, v. 23, p. 153-163.

Brady, T. J., N. D. J. Campbell, and C. E. Maher, 1980, Intisar 'D' Oil field, Libya, *in* Giant oil and gas fields of the decade, 1968-1978: AAPG Memoir 30, p. 543-564.

Brown, L. F., Jr., 1972, Virgil-lower Wolfcamp repetitive environments and the depositional model north-central Texas, *in* J. C. Elam and S. Chuber, eds., Cyclic sedimentation in the Permian basin, 2nd edition: West Texas Geological Society, p. 41-54.

Catalano, R., and B. D'Argenio, 1981, Paleogeographic evolution of a continental margin in Sicily: Guidebook of the field trip in western Sicily, GSA Penrose Conference on Controls of Carbonate Platform Evolution, 142 p.

Chevalier, J. P., 1973, Coral reefs of New Caledonia, *in* O. A. Jones and R. Endean, eds., Biology and geology of coral reefs, v. 1: Geology 1: New York, Academic Press, p. 143-167.

Choquette, P. W., and R. P. Steinen, 1980, Mississippian non-supratidal dolomite, Ste. Genevieve Limestone, Illinois basin—evidence for mixed-water dolomitization, *in* Concepts and models of dolomitization: SEPM Special Publication 28, p. 163-196.

Collins, L. B., 1981, Post-glacial non-tropical shelf carbonate sedimentation of the Rottnest Shelf, Western Australia (abs.): Perth, 5th Australian Geological Convention, Geological Society of Australia, p. 54.

Cook, H. E., 1979, Ancient continental slope sequences and their value in understanding modern slope development, *in* L. J. Doyle and O. H. Pilkey eds., Geology of continental slopes: SEPM Special Publication, 27, p. 287-305.

——— and M. E. Taylor, 1977, Comparison on continental slope and shelf environments in the Upper Cambrian and lowest Ordovician of Nevada, *in* Deep-water carbonate environments: SEPM Special Publication 25, p. 51-81.

Cook, T. D., 1979, Exploration history of south Texas Lower Cretaceous carbonate platform: AAPG Bulletin, v. 63, p. 32-49.

Darwin, C. R., 1842, On the structure and distribution of coral reefs: London, Smith Elder and Co.; reprinted by University of California Press, Berkeley and Los Angeles, 1962, 214 p.

Davies, G. R., 1970, Carbonate bank sedimentation, eastern Shark Bay, Western Australia, *in* Carbonate sedimentation and environments, Shark Bay, Western Australia: AAPG Memoir 13, p. 85-168.

Dunham, J. B., G. A. Crawford, and W. Panasiuk, 1983, Sedimentology of the Slave Point Formation (Devonian) at Slave field, Lubicon Lake, Alberta, *in* Carbonate buildups—a core workshop: SEPM Core Workshop No. 4, p. 73-111.

Edie, R. W., 1958, Mississippian sedimentation and oil fields in southeastern Saskatchewan: AAPG Bulletin, v. 42, p. 94-126.

Eliuk, L. S., 1978, The Abenaki Formation, Nova Scotia shelf, Canada—A depositional and diagenetic model for a Mesozoic carbonate platform: Bulletin of Canadian Petroleum Geology, v. 26, p. 424-514.

Emery, K. O. J. I. Tracy, Jr., and H. S. Ladd, 1954, Geology of Bikini and nearby atolls: U.S. Geological Survey Professional Paper 260-A, 265 p.

Enos, P., 1974a, Reefs, platforms, and basins in middle Cretaceous in northeast Mexico: AAPG Bulletin, v. 58, p. 800-809.

——— 1974b, Surface sediment facies map of the Florida-Bahamas plateau: GSA Map and Chart, MC 5, scale 1:3,168,000.

——— 1977, Tamabra Limestone of the Poza Rica trend, Cretaceous, Mexico, *in* Deep-water carbonate environments: SEPM Special Publication 25, p. 273-314.

——— and R. D. Perkins, 1977, Quaternary sedimentation in south Florida: GSA Memoir 147, 198 p.

Erwin, P. N., Jr., 1981, Stratigraphy, depositional environments, and dolomitization of the Maryville and upper Honaker Formations

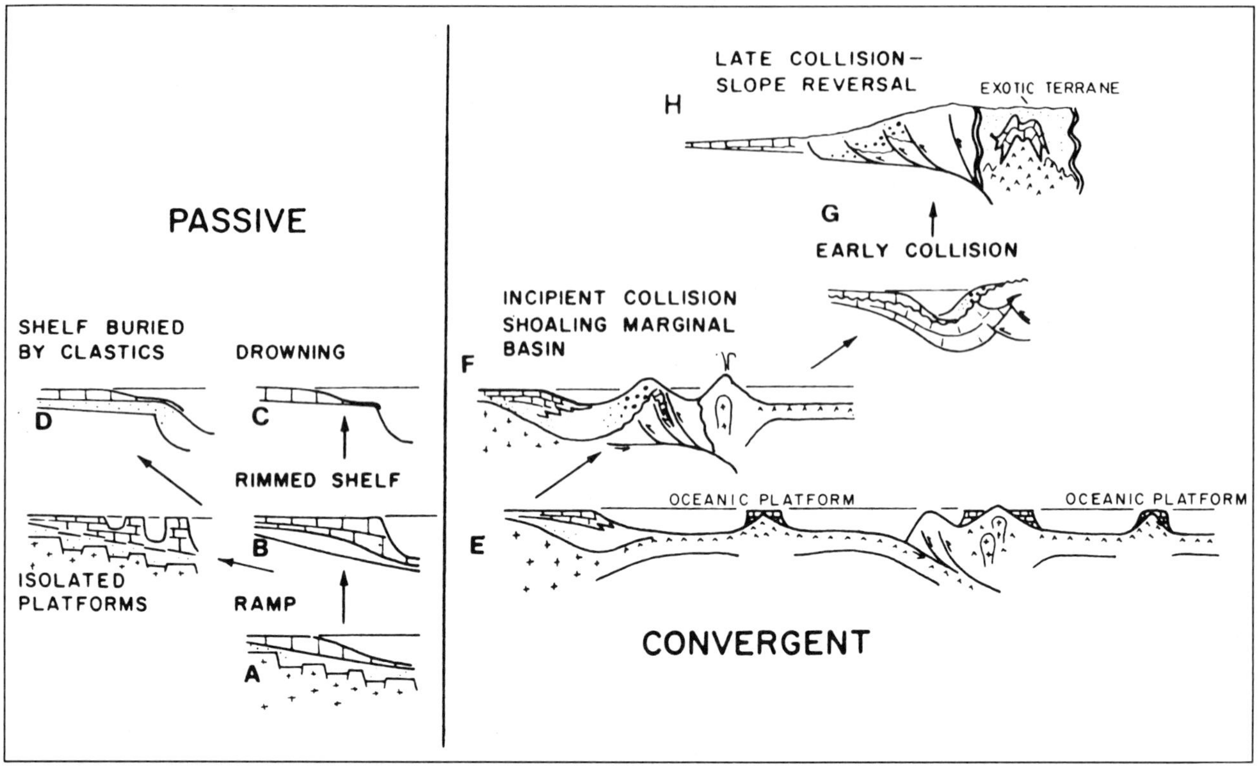

Figure 10—Evolution of ramps, rimmed shelved, drowned shelves, and isolated platforms in passive to convergent margin settings. A. Ramp develops on continental or marine shelf clastics of passive margin, and evolves into rimmed shelf (B, right) or isolated platform (B, left). Shelf may be drowned to form ramp (C), or shelf may be prograded by clastics and ramp established following cessation of clastic deposition (D).

In ocean basins, isolated platforms may form on oceanic volcanoes, and carbonate platforms (fringing and barrier-reef complexes) may develop around volcanic arcs (E). With convergence, earlier rimmed shelves may be converted to ramps during filling of basin or tectonic uplift of basin margin (F). Commonly, convergence (G) is accompanied by regional uplift of shelf and unconformity development, followed by foundering of continental shelf and widespread development of carbonate ramp extending into foredeep. With late convergence (H) and large-scale overthrusting, foredeep may be filled, and slope reversal results in ramp deepening out onto craton. Carbonate platforms associated with oceanic volcanoes and island arcs may accrete onto continental margin as exotic terranes (H).

(Cambrian), Tennessee and Virginia: M.S. thesis, Duke University, Durham, North Carolina, 232 p.

Freeman-Lynde, R. P., M. B. Cita, F. Jadoul, E. L. Miller, and W. B. F. Ryan, 1981, Marine geology of the Bahama Escarpment: Marine Geology, v. 44, p. 119-156.

Ginsburg, R. N., and N. P. James, 1974, Holocene carbonate sediments of continental shelves, *in* C. A. Burk and C. L. Drake, eds., The geology of continental margins: New York, Springer-Verlag, p. 137-155.

Griffith, L. S., M. G. Pitcher, and G. W. Rice, 1969, Quantitative environmental analysis of a Lower Cretaceous reef complex, *in* Depositional environments in carbonate rocks: SEPM Special Publication 14, p. 120-138.

Grotzinger, J. P., and P. Hoffman, 1983, Quantitative paleobathymetry of early Proterozoic (1.9 b.y.) continental slope, Rocknest Formation, Wopmay Orogen, N.W.T., Canada (abs.): AAPG Bulletin, v. 67, p. 425.

Hagan, G. M., and B. W. Logan, 1974, Development of carbonate banks and hypersaline basins, Shark Bay, Western Australia, *in* Evolution and diagenesis of Quaternary carbonate sequences, Shark Bay, Western Australia: AAPG Memoir 22, p. 61-139.

Heckel, P. H., 1972, Pennsylvanian stratigraphic reefs in Kansas, some modern comparisons and implications: Geologische Rundschau, v. 61, p. 584-598.

——— 1974, Carbonate buildups in the geologic record—a review, *in* Reefs in time and space: SEPM Special Publication 18, p. 90-154.

Hileman, M. E., and S. J. Mazzullo, eds., 1977, Upper Guadalupian facies, Permian reef complex, Guadalupe Mountains, New Mexico and west Texas: Permian Basin Section, SEPM Publication 77-16, p. 45-92.

Hine, A. C., B. J. Wilber, J. M. Bane, A. C. Neumann, and K. R. Lorenson, 1981, Offbank transport of carbonate sand along open, leeward bank margins, northern Bahamas: Marine Geology, v. 42, p. 327-348.

——— R. J. Wilber, and A. C. Neumann, 1981, Carbonate sand bodies along contrasting shallow bank margins facing open seaways in northern Bahamas: AAPG Bulletin, v. 65, p. 261-290.

——— and J. C. Steinmetz, 1983, Cay Sal Bank, Bahamas—A partially drowned carbonate platform (abs.): AAPG Bulletin, v. 67, p. 484.

Hoffman, P., 1973, Evolution of an early Proterozoic continental margin: the Coronation geosyncline and associated aulacogens of the northwestern Canadian shield: Philosophical Transactions of the Royal Society of London, series A, v. 273, p. 547-581.

Hooke, R. LeB., and W. Schlager, 1980, Geomorphic evolution of the Tongue of the Ocean and the Providence Channels, Bahamas: Marine Geology, v. 44, p. 343-366.

Hurst, J. M., and F. Surlyk, 1983, Initiation, evolution, and destruction of an early Paleozoic carbonate shelf, eastern North Greenland: Journal of Geology, v. 91, p. 671-691.

James, N. P., 1979, Facies models 11. Reefs, *in* R. G. Walker ed., Facies models: Geoscience Canada Reprint Series 1, p. 121-132.

——— and R. N. Ginsburg, 1979, The seaward margin of Belize barrier and atoll reefs: International Association of Sedimentologists Special Publication 3, 201 p.

Jansa, L., 1981, Mesozoic carbonate platforms and banks of the eastern North American margin: Marine Geology, v. 44, p. 97-117.

Kendall, C. G. St. C., and W. Schlager, 1981, Carbonates and relative changes in sea level: Marine Geology, v. 44, p. 181-212.

Kissling, D. L., and J. F. Polasek, Jr., 1982, Exploration and reservoir characteristics of Onondaga bioherms, (abs.): West Virginia Geological and Economic Survey Circular C-26, p. 34-35.

Klovan, J. E., 1974, Development of western Canadian Devonian reefs and comparison with Holocene analogues: AAPG Bulletin, v. 58, p. 787-799.

Ladd, H. S., 1973, Bikini and Eniwetok Atolls, Marshall Islands, *in* O. A. Jones and R. Endean, eds., Biology and geology of coral reefs, v. 1: Geology 1: New York Academic Press, p. 93-111.

Laporte, L. F., 1969, Recognition of a transgressive carbonate sequence within an epeiric sea—Helderberg Group (Lower Devonian) of New York State, *in* Depositional environments in carbonate rocks: SEPM Special Publication 14, p. 98-119.

Lindsay, R. F., and C. G. St. C. Kendall, 1980, Depositional facies, diagenesis and reservoir character of the Mission Canyon Formation (Mississippian) of the Williston basin at Little Knife field, North Dakota, *in* Carbonate reservoir rocks: Notes for SEPM Core Workshop No. 1, p. 79-104.

Logan, B. W., J. F. Read, and G. R. Davies, 1970, History of carbonate sedimentation, Quaternary Epoch, Shark Bay, Western Australia, *in* Carbonate sedimentation and environments, Shark Bay, Western Australia: AAPG Memoir 13, p. 38-84.

——— ——— G. M. Hagan, P. Hoffman, R. G. Brown, P. J. Woods, and C. D. Gebelein, eds., 1974, Evolution and diagenesis of Quaternary carbonate sequences, Shark Bay, Western Australia: AAPG Memoir 22, 358 p.

Longacre, S. A., 1980, Dolomite reservoirs from Permian biomicrites, *in* Carbonate reservoir rocks: Notes for SEPM Core Workshop No. 1, p. 105-117.

Longman, M. W., 1980, Carbonate petrology of the Nido B-3A core, offshore Palawan, Philippines, *in* Carbonate reservoir rocks: Notes for SEPM Core Workshop No. 1, p. 161-183.

Loreau, J. P., and B. H. Purser, 1973, Distribution and ultrastructure of Holocene ooids in the Persian Gulf, *in* B. H. Purser, ed., The Persian Gulf: New York, Springer-Verlag, p. 279-328.

Lowry, W. D., 1970, Geology of the Western Australian part of the Eucla Basin: Western Australia Geological Survey Bulletin 122, 199 p.

Luttrell, P. E., 1977, Carbonate facies distribution and diagenesis associated with volcanic cones—Anacacho Limestone (Upper Cretaceous), Elaine field, Dimmit County, Texas, *in* D. G. Bebout and R. G. Loucks, eds., Cretaceous carbonates of Texas and Mexico—Applications to subsurface exploration: University of Texas Bureau of Economic Geology Report of Investigations 89, p. 260-285.

Machel, H. G., 1983, Facies and diagenesis of some Nisku buildups and associated strata, Upper Devonian, Alberta, Canada, *in* Carbonate buildups—a core workshop: SEPM Core Workshop No. 4, p. 144-181.

Malek-Aslani, M., 1970, Lower Wolfcampian reef in Kemnitz field, Lea County, New Mexico: AAPG Bulletin, v. 54, p. 2317-2335.

Markello, J. R., and J. F. Read, 1981, Carbonate ramp-to-deeper shale shelf transitions of an Upper Cambrian intrashelf basin, Nolichucky Formation, southwest Virginia Appalachians: Sedimentology, v. 28, p. 573-597.

Matti, J. C., and E. H. McKee, 1977, Silurian and Lower Devonian paleogeography of the outer continental shelf of the Cordilleran miogeocline, central Nevada, *in* Paleozoic paleogeography of the western United States: SEPM Pacific Section, p. 181-215.

Maxwell, W. G. H., 1968, Atlas of the Great Barrier Reef: New York, Elsevier Publishing Co., 258 p.

McCamis, J. G., and L. S. Griffith, 1968, Middle Devonian facies relations, Zama area, Alberta: AAPG Bulletin, v. 52, p. 1899-1924.

McIlreath, I. A., 1977, Accumulation of a Middle Cambrian, deep-water limestone debris apron adjacent to a vertical submarine carbonate escarpment, southern Rocky Mountains, Canada, *in* Deep-water carbonate environments: SEPM Special Publication 25, p. 113-124.

——— and N. P. James, 1979, Facies models 12. Carbonate slopes, *in* R. G. Walker, ed., Facies models: Geoscience Canada Reprint Series 1, p. 133-143.

Meissner, F. F., 1974, Hydrocarbon accumulation in San Andreas formation of Permian basin, southeast New Mexico and west Texas (abs.): AAPG Bulletin, v. 58, p. 909-910.

Mesolella, K. J., J. D. Robinson, L. M. McCormick, and A. R. Ormiston, 1974, Cyclic deposition of Silurian carbonates and evaporites in Michigan basin: AAPG Bulletin, v. 58, p. 34-62.

Mullins, H. T., and G. W. Lynts, 1977, Origin of the northwestern Bahama Platform—Review and reinterpretation: GSA Bulletin, v. 88, p. 1447-1461.

——— and A. C. Neumann, 1979, Deep carbonate bank margin, *in* Geology of continental slopes, structure and sedimentation in the northern Bahamas: SEPM Special Publication 27, p. 165-192.

Murris, R. J., 1980, Middle East—stratigraphic evolution and oil habitat: AAPG Bulletin, v. 64, p. 597-618.

Orme, G. R., 1977, The Coral Sea Plateau—a major reef province, *in* O. A. Jones and R. Endean, eds., Biology and geology of coral reefs, v. 4, Geology 2, New York, Academic Press, p. 267-306.

Pfeil, R. W., and J. F. Read, 1980, Cambrian carbonates platform margin facies, Shady Dolomite, southwestern Virginia, U.S.A.: Journal of Sedimentary Petrology, v. 50, p. 91-116.

Playford, P. E., 1980, Devonian "Great Barrier Reef" of Canning basin, Western Australia: AAPG Bulletin, v. 64, p. 814-840.

Pounder, D. A., E. R. Venour, and L. Tremblay, 1980, Pinnacle reef reservoirs, Zeta Lake Member, Nisku Formation (Upper Devonian), West Pembina area, Alberta, Canada, *in* Carbonate reservoir rocks: Notes for SEPM Core Workshop No. 1, p. 64-78.

Purser, B. H., 1973, The Persian Gulf: New York, Springer-Verlag, 471 p.

——— and G. Evans, 1973, Regional [carbonate] sedimentation along the Trucial Coast, SE Persian Gulf, *in* B. H. Purser, ed., The Persian Gulf: New York, Springer-Verlag, p. 211-231.

Rao, R. P., and S. N. Talukdar, 1980, Petroleum geology of Bombay High field, India, *in* Giant oil and gas fields of the Decade 1968-1978: AAPG Memoir 30, p. 487-506.

Read, J. F., 1974, Carbonate bank and wave-built platform sedimentation, Edel Province, Shark Bay, Western Australia, *in* Evolution and diagenesis of Quaternary carbonate sequences, Shark Bay, Western Australia: AAPG Memoir 22, p. 1-60.

——— 1980, Carbonate ramp-to-basin transitions and foreland basin evolution, Middle Ordovician, Virginia Appalachians: AAPG Bulletin, v. 64, p. 1575-1612.

——— 1982, Carbonate platforms of passive (extensional) continental margins—types, characteristics and evolution: Tectonophysics, v. 81, p. 195-212.

Ryan, W. B. F., 1980, Tectonic significance of Atlantic carbonate platform escarpments (abs.): Congres Geologique Internationale, 26e, Resumes (Paris, France), p. 539.

——— and E. L. Miller, 1981, Evidence of a carbonate platform beneath Georges Bank: Marine Geology, v. 44, p. 213-228.

Sarg, J. F., 1977, Sedimentology of the carbonate-evaporite-facies transition of the Seven Rivers Formation (Guadalupian, Permian) in southeast New Mexico, *in* Upper Guadalupian facies, Permian reef complex, Guadalupe Mountains, New Mexico and west Texas: Permian Base Section, SEPM Publication 77-16, p. 45-92.

Schatzinger, R. A., 1983, Phylloid algal and sponge-bryozoan mound-to-basin transition—a late Paleozoic facies tract from the Kelly-Snyder field, west Texas, *in* Carbonate buildups—a core workshop: SEPM Core Workshop No. 4, p. 244-303.

Schlager, W., 1981, The paradox of drowned reefs and carbonate platforms: GSA Bulletin, v. 92, p. 197-211.

——— and A. Chermak, 1979, Sediment facies of platform-basin transition, Tongue of the Ocean, Bahamas: SEPM Special Publication 27, p. 193-208.

——— and R. N. Ginsburg, 1980, Bahama platforms and troughs present and past (abs.): Congress Geologique Internationale, 26e, Resumes (Paris, France), p. 540.

Schmidt, V., D. A. McDonald, and I. A. McIlreath, 1980, Growth and diagenesis of Middle Devonian Keg River cementation reefs, Rainbow field, Alberta, *in* Carbonate reservoir rocks: Notes for SEPM Core Workshop No. 1, p. 43-63.

Sears, S. O., and F. J. Lucia, 1979, Reef-growth model for Silurian pinnacle reefs, northern Michigan reef trend: Geology, v. 7, p. 299-302.

Sheridan, R. E., 1974, Atlantic continental margin of North America, *in* C. L. Burk and C. Drake, eds., The geology of continental margins: New York, Springer-Verlag, p. 391-407.

Smith, D. L., 1977, Transition from deep- to shallow-water carbonates, Paine Member, Lodgepole Formation, central Montana, *in* Deep-water carbonate environments: SEPM Special Publication 25, p. 187-201.

Steers, J. A., and D. R. Stoddart, 1977, The origin of fringing reefs, barrier reefs, and atolls, *in* O. A. Jones and R. Endean, eds., Biology and geology of coral reefs, v. 4: Geology 2: New York, Academic Press, p. 21-57.

Stoddart, D. R., 1973, Coral reefs of the Indian Ocean, *in* O. A. Jones and R. Endean, eds., Biology and geology of coral reefs, v. 1, Geology 1: New York, Academic Press, p. 51-92.

Todd, R. G., and R. M. Mitchum, Jr., 1977, Seismic stratigraphy and global changes in sea level, Part 8—Identification of Upper Triassic, Jurassic, and Lower Cretaceous seismic sequences in Gulf of Mexico and offshore west Africa, *in* Seismic stratigraphy—applications to hydrocarbon exploration: AAPG Memoir 26, p. 146-163.

Vail, P. R., R. M. Mitchum, Jr., and S. Thompson, III, 1977, Seismic stratigraphy and global changes of sea level, Part 4—Global cycles of relative changes of sea level, *in* Seismic stratigraphy—applications to hydrocarbon exploration: AAPG Memoir 26, p. 83-97.

Vest, E. L., Jr., 1970, Oil fields of Pennsylvanian-Permian Horseshoe atoll, west Texas, *in* Geology of giant petroleum fields: AAPG Memoir 14, p. 185-203.

Viau, C., 1983, Depositional sequences, facies and evolution of the Upper Devonian Swan Hills reef buildup, central Alberta, Canada, *in* Carbonate buildups—core workshop: SEPM Core Workshop No. 4, p. 112-143.

Ward, W. C., and M. J. Brady, 1979, Strandline sedimentation of carbonate grainstones, upper Pleistocene, Yucatan peninsula, Mexico: AAPG Bulletin, v. 63, p. 362-369.

Wendte, J. C., and F. A. Stoakes, 1982, Evolution and corresponding porosity of the Judy Creek complex, Upper Devonian, central Alberta, *in* W. G. Cutler, ed., Canada's giant hydrocarbon reservoirs: Canadian Society of Petroleum Geologists 1982 Core Conference, p. 63-81.

Winterer, E. L., and A. Bosellini, 1981, Subsidence and sedimentation on Jurassic passive continental margin, southern Alps, Italy: AAPG Bulletin, v. 65, p. 394-421.

Wilson, J. L., 1975, Carbonate facies in geologic history: New York, Springer-Verlag, 470 p.

Yurewicz, D. A., 1977, Origin of massive facies of the lower and middle Capitan Limestone (Permian), Guadalupe Mountains, New Mexico and west Texas, *in* Upper Guadalupian facies, Permian reef complex, Guadalupe Mountains, New Mexico and west Texas: Permian Basin Section, SEPM Publication 77-16, p. 45-92.

Reproduced with the permission of the Geological Association of Canada. Published in *Facies Models,* R.G. Walker, editor, *Geoscience Canada,* Reprint Series No. 1, second edition, 1984, pp. 213-228.

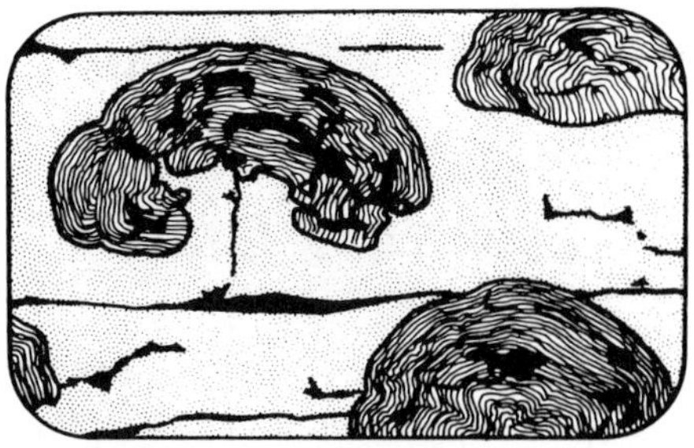

Shallowing-Upward Sequences in Carbonates

NOEL P. JAMES
Department of Earth Sciences
Memorial University of Newfoundland
St. John's, Newfoundland A1B 3X5

INTRODUCTION

Perhaps the most commonly encountered carbonates are laterally persistant, evenly bedded limestones and dolomites of apparent shallow water origin, as demonstrated by abundant fossil mud cracks and stromatolites. These deposits, which are usually found on the continents and in relatively undeformed portions of mountain belts, are not only important sources of paleontological and sedimentological information, but are also common host rocks for hydrocarbons and metallic ores (particularly lead and zinc). As such, it is critical that we be able to determine, as precisely as possible, the environment in which each of the interbedded sediments was deposited.

A quantum jump in our understanding of these deposits occured when modern carbonate tidal flats were examined in detail, notably by Robert Ginsburg and his colleagues in Florida and the Bahamas about 25 years ago. It was quickly realized that there was a host of sedimentary structures and textures on these flats that would allow a much more precise definition of environments of deposition than was possible before: these findings were quickly applied to fossil sequences (Fischer, 1964; Laporte, 1967; Aitken, 1966; Roehl, 1967). This application in turn generated two different lines of investigation: 1) description of other areas of modern tidal flat deposition, in particular the southern shore of the Persian Gulf where evaporites are common, and Shark Bay, Western Australia, where a great variety of modern stromatolites are forming; and 2) documentation of different styles of tidal flat deposits in the geologic record.

THE MODEL

Carbonate sediments characteristically accumulate at rates much greater than the rate of subsidence of the shelf or platform upon which they are deposited (Schlager, 1981). This is because carbonate sediments are produced mainly in the environment of deposition – especially in shallow water where conditions for the biological and physicochemical fixation of carbonate are optimum. As a result, carbonate accumulations repeatedly build up to sea level and above, resulting in a characteristic sequence of deposits in which each unit is deposited in progressively shallower water. This *shallowing-upward sequence* commonly is repeated many times in a succession of shallow water deposits (Fig. 1).

Readers will recognize that such a shallowing upward sequence also may be termed a 'regressive sequence'. This term has led to much confusion in the past, because it has been used to describe deposits associated with a high rate of sediment production and accumulation under relatively static sea level - sea bottom conditions. I have, therefore, abandoned the term 'regressive' altogether in favor of a rock-descriptive term, albeit interpretive; the *shallowing-upward sequence.*

1) The Model as a Norm. The ideal carbonate shallowing-upward sequence comprises four units illustrated in Figure 2. The basal unit, which is generally thin, records the initial transgression over pre-existing deposits and so is commonly a high energy deposit. The bulk of the sequence, which may be of diverse lithologies, consists of normal marine carbonate, as discussed below. The upper part of the sequence consists of two units: the intertidal unit within the normal range of tides; the other a supratidal unit deposited in the area covered only by abnormal, windblown or storm tides. Each of these units exhibits the characteristic criteria of subaerial exposure.

Figure 1
Bedded carbonates ranging in age from Middle to Late Cambrian near Fortress Lake, B.C. Arrows mark the top of large-scale shallowing-upward sequences (L - Lyell Fm., S - Sullivan Fm., W - Waterfowl Fm., A - Arctomys Fm., E - Eldon and Pica Fms.). Striping of the Waterfowl Fm. is caused by repetitive smaller scale shallowing-upward sequences between subtidal-intertidal limestones (dark) and supratidal dolomites (light). Photo courtesy J.D. Aitken.

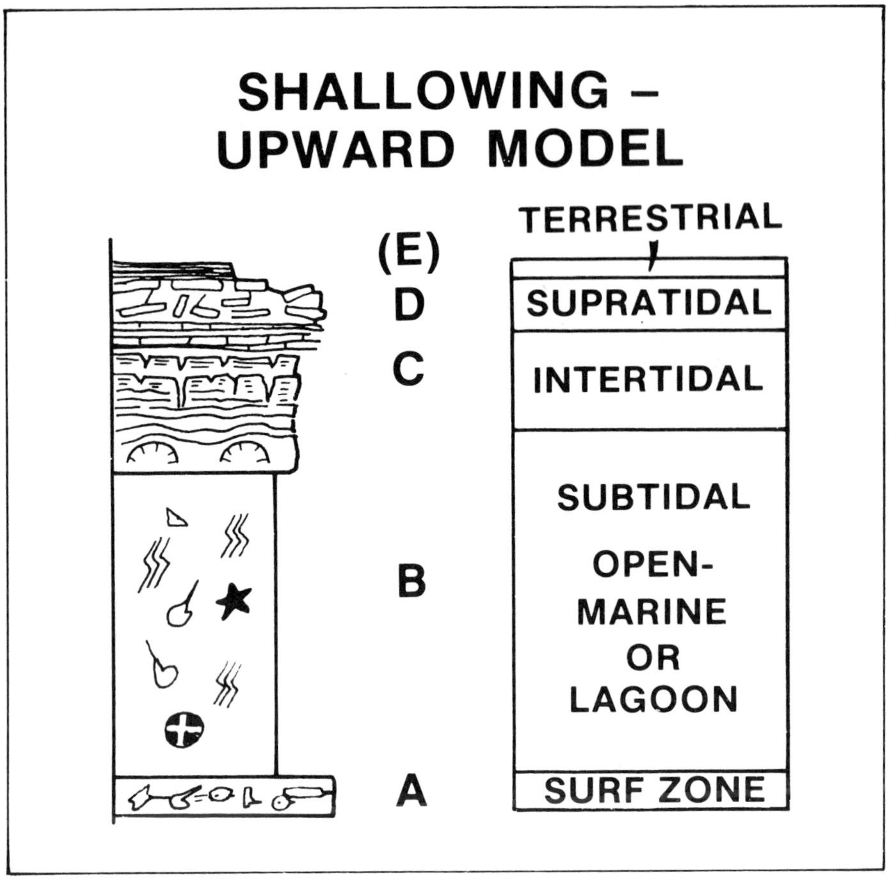

Figure 2
Five divisions of the shallowing-upward model for carbonates: A) lithoclast rich lime conglomerate or sand. B) fossiliferous limestone. C) stromatolitic, mud-cracked cryptalgal limestone or dolomite. D) well laminated dolomite or limestone, flat-pebble breccia. E) shale or calcrete, bracketed to emphasize that the unit is often missing - see text. Symbols used throughout are from Ginsburg (1975).

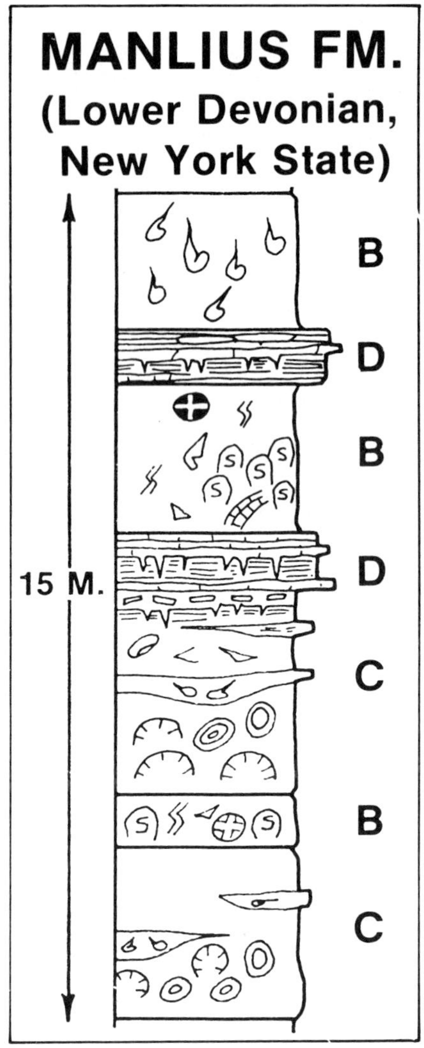

Figure 3
Actual sequence of several shallowing-upward sequences from the Manlius Fm., New York State (From Laporte, 1975).

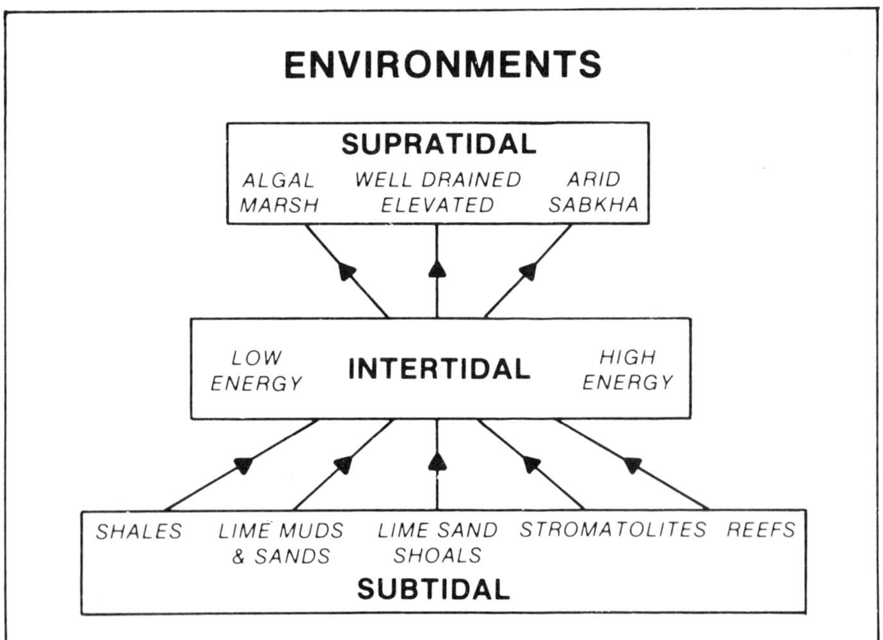

Figure 4
A flow diagram indicating the various possible environmental transitions present in a carbonate shallowing-upward sequence.

2) The Model as a Predictor. The thread that binds all such sequences together is the presence of the distinctive intertidal unit, which, once recognized, allows one to interpret the surrounding lithologies in some kind of logical sequence (Fig. 3), and thus predict what lithologies should occur in the rest of the succession under investigation.

First-order variation on the basic model revolves around the two main types of intertidal environment: 1) quiet, low-energy situations, commonly referred to as tidal flats, and 2) agitated, high-energy situations, or quite simply, beaches. Second-order variation involves the kind of subtidal units below and supratidal units above: the subtidal reflects the type of marine environment adjacent to the tidal flat and supratidal reflects the adjacent terrestrial environment, in particular the climate (Fig. 4).

For purposes of discussion I will begin with those sequences that contain low-energy intertidal units (tidal-flats) because they exhibit the greatest variety of distinctive features and consequently are well documented, both in modern and ancient settings. To place the observed features in context we should first examine modern carbonate tidal flats.

SEQUENCES WITH A LOW-ENERGY INTERTIDAL UNIT

Modern Tidal Flats

The main elements of a modern carbonate tidal flat system as exemplified by the narrow shelf and embayments of Shark Bay, Western Australia, the southern coast of the Persian Gulf, and wide platform of the Bahama Banks are shown in Figure 5. The sedimentary features of these tidal flats are beautifully illustrated by Shinn (1983). A characteristic of most modern examples is that they occur in protected locations: protected that is from the open ocean waves and swells, yet still affected by tides and severe storms. This unique setting is commonly afforded by the presence of a semi-protective barrier composed of lime sand shoals, locally associated with reefs and/or islands. The barrier commonly is dissected by tidal channels through which flow high velocity tidal currents. A shallow muddy lagoon lies in the lee of this barrier. The lagoon may be enormous as in the case of the Bahamas, relatively narrow and elongate as in the Persian Gulf, or very small as in the pocket embayments of Shark Bay. In such an arrangement, tidal flats are present as: 1) small flats atop and on the lee side of the emergent sand shoals of the barrier, and 2) large flats along the shoreline of the shallow lagoon (Fig. 5). Thus tidal flats occur in association with two separate carbonate accumulations, high energy sand bodies and low energy lime muds. A third type, which is less common in modern situations, is the association with reefs, especially the interior of large reef complexes.

Intertidal Environments. The intertidal zone, especially along rocky coasts and beaches is commonly a gradual transition from sea to land without much noticeable variation. On wide, gradually sloping tidal flats this zone can be the familiar gradual transition or a complex area of many subenvironments. At one end of the spectrum the flats have few, very shallow, short tidal creeks (Fig. 6). At the other end of the spectrum the flats are dissected by many tidal creeks flanked by levees. Slight depressions between the creeks are occupied by tidal ponds (which fill and partially empty during each rise and fall of the tide) and the whole complex is fronted by small beach ridges or erosional steps (Fig. 6). Perhaps in this case it would be better to refer to the whole zone as the "pond and creek belt" because some of the areas are dry most of the time (levees and beaches) whereas others are continuously submerged (ponds and creeks). These complications have led some workers (e.g., Ginsburg and Hardie, 1975) to despair of conventional terms and instead to relate different zones to the per cent of time that they are exposed rather than to their position.

On some tidal flats where there are many tidal creeks and noticeable relief between levee and tidal pond (about 1 m), as in the Bahamas, the true intertidal zone which lies between the levee and pond may comprise only 60 to 70% of the intertidal environment. In other areas such as the Persian Gulf, where there are fewer creeks and less relief, almost the whole flat is truly intertidal. The most important point to grasp is that numerous environments may exist in very close proximity not only perpendicular to the shoreline but parallel to it as well, so that in the geologic record rapid, local lithological variations are to be expected, both vertically and laterally, rather than a smooth succession of progressively shallower environments.

The tidal flat wedge is built up of fine grained sediments brought onto the flats from the adjacent offshore marine zone by storms rather than by daily tides. Large storms such as hurricanes which flood the flat with sheets of water, white with suspended sediment, are particularly effective. Shinn *et al.* (1969) have suggested that the tidal flat is a river delta turned wrong-side out, with the sea as the "river" supplying sedi-

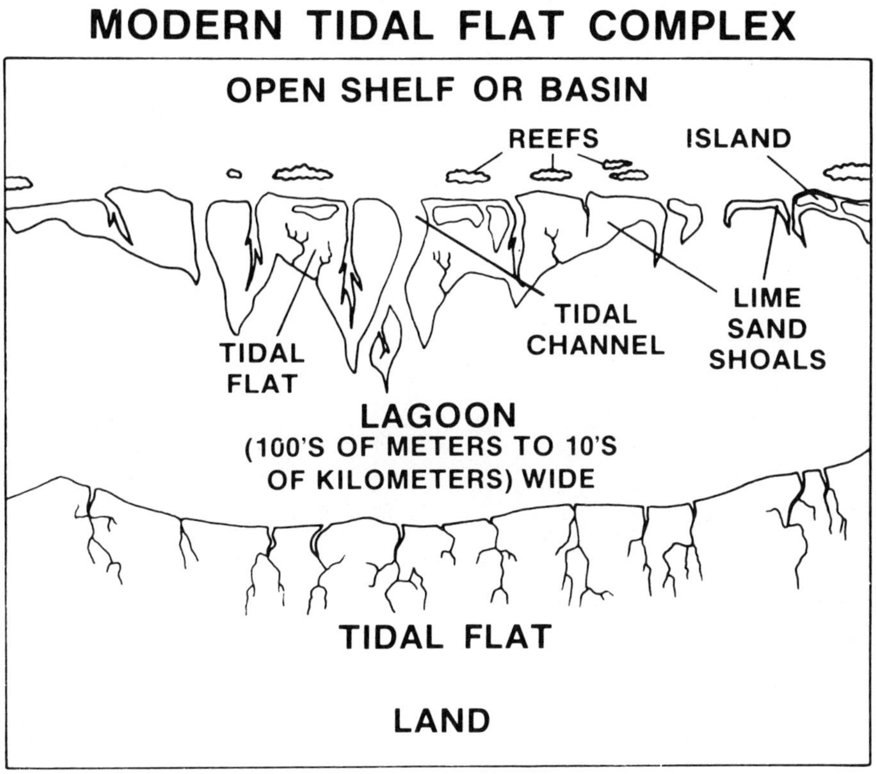

Figure 5
Plan view of the geometry of a modern tidal flat complex. Note that tidal flats can be present both adjacent to the land or in the lee of lime sand shoals.

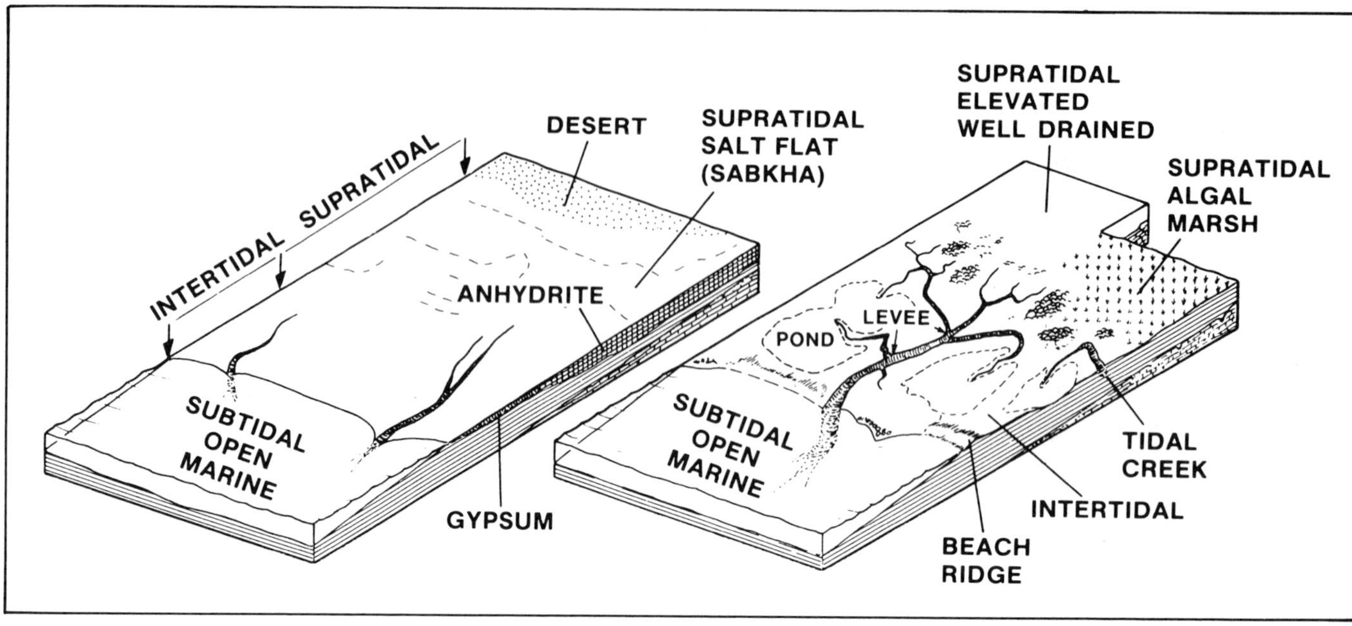

Figure 6
Block diagrams showing the major morphological elements of a tidal flat; left – a hypersaline tidal flat with few channels and bordering a very arid desert (similar to the modern Persian Gulf), right – a normal marine tidal flat with many channels and ponds and bordering an elevated well-drained area of low swamp algal marsh in a humid climate (similar to the modern Bahamas).

ment to the channeled flats as the "delta".

Sediments of the intertidal zone are characterized by three distinctive features, not found elsewhere: 1) algal mats, 2) irregular to even laminations (cryptalgal laminites), with fenestral porosity, and 3) desiccation features.

The algal mats are gelatinous to leathery sheets of blue-green algae which grow on top of the sediment surface. They are widely regarded as the signature of intertidal deposits. These mats are constructed solely or primarily by blue-green algae, which although photo-synthetic like other algal and higher plants, are prokaryotic and have much stronger affinities with bacteria than other eukaryotic forms. They are more correctly called cyanobacteria. It is probably more correct to refer to these mats as cyanobacterial mats or microbial mats (Bauld, 1981) but since the term algal mats is so entrenched in the literature I shall, for the time being, continue to use it in this paper. The mats may occur throughout the intertidal zone but their precise distribution is controlled by climate and the presence or absence of other organisms. The upper limit is controlled by climate; in arid areas they cannot grow above the high intertidal into the suprtidal zone, whereas in areas of high rainfall where the supratidal zone is moist or flooded for days at a time, mats are prolific. The lower limit is more variable and controlled by several factors. Garrett (1970) indicates that the main culprits are gastropods that eat algae. In areas of normal salinity, mats are prevented from developing below the middle intertidal zone because they are browsed by gastropods; in areas of hypersalinity (deadly for gastropods) mats grow down into the subtidal zone. In addition, algal mats will colonize only a temporarily or permanently stable bottom, and will not grow on shifting sand. Pratt (1982) points out that stromatolites are more common in post-Lower Ordovician rocks than generally realized, and that grazing by gastropods may be overemphasized as a limiting factor in their distribution. Rather it may be that substrate competition from various metazoans, together with increased rates of sedimentation during the Phanerozoic are more important controls.

Although the algal mats may themselves vanish with time, evidence of their presence during deposition remains because of the peculiar pores that they help to create, generally referred to as 'laminoid fenestrae'. These are irregular, elongate to mostly sub-horizontal sheet-like cavities (loferites or birds-eyes of some workers) with no obvious support and much larger than can be explained by grain packing. They are simply due to the fact that the mats are covered with sediment. The mats eventually rot away during burial, leaving voids as well as holes due to entrapped gas and mat shrinkage. Caution should be used when interpreting these structures, however, as similar features can be produced by submarine cementation of pellets, ooids and aggregate grains (Shinn, 1983).

Other sediments recording the presence of blue-green algal mats are the finely laminated carbonates (Fig. 7) ranging from stratiform and lightly crenulated to the familiar arched domes of stromatolites. These have been called cryptalgal (hidden, algal) laminations by Aitken (1967) in reference to the fact that the influence of algae in the rock-forming process is more commonly inferred than observed.

Lower Intertidal Zone. Much of the subtidal character remains evident in sediments from this part of the environment, and the deposits are commonly well burrowed and bioturbated. In hypersaline areas, however, the surface of the sediment is veneered with a thick algal mat, frequently broken into desiccation polygons. Beneath the mat, grains are

Figure 7
Cryptalgal laminites that have been mud-cracked. The intertidal unit of a shallowing-upward sequence in the Petit Jardin Fm. (Upper Cambrian) on the south shore of the Port-au-Port Peninsula, Nfld. (Photo courtesy R. Levesque).

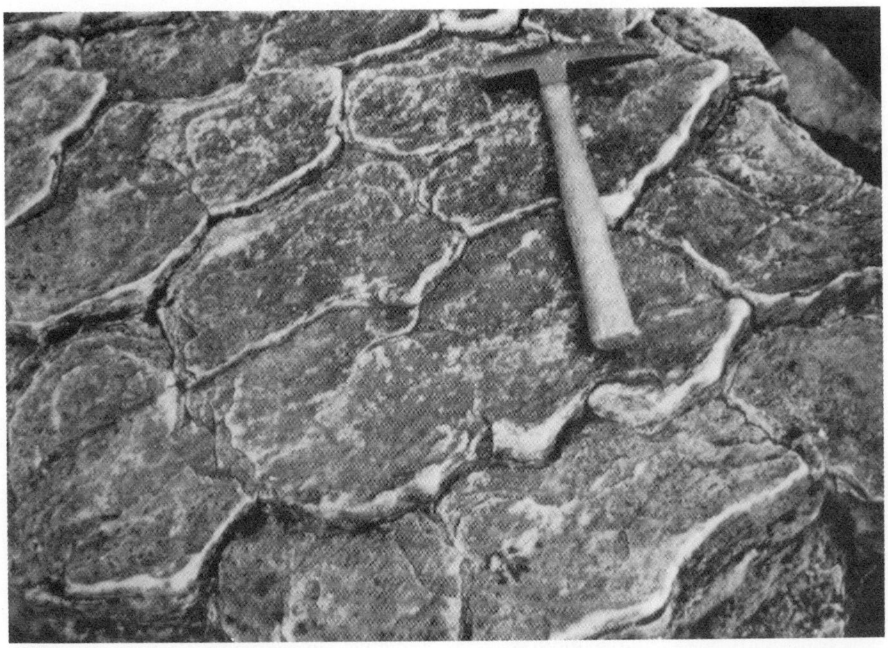

Figure 8
A bedding plane of mud-cracked polygons with the edges of each polygon curled up, probably because the algal mats in the polygons shrivelled upon exposure and drying out. Near the top of a shallowing-upward sequence in the East Arm Fm., (Upper Cambrian), Bonne Bay, Nfld.

blackened due to reducing conditions and altered by boring algae to peloids of lime mud.

Tidal ponds and the creeks that drain them on hypersaline tidal flats support the most prolific growth of algal mats anywhere on the flat. The algal mat flourishes in water depths greater than those in the immediate offshore area because of relatively elevated salinities in the ponds. On tidal flats where the salinity is closer to normal, marine tidal ponds are populated by a restricted but prolific fauna of foraminifers and gastropods and the gastropods prevent the growth of algal mats. Similarly, if tidal creeks are common in such areas, the channels are devoid of mats but do contain concentrations of the pond fauna that are washed out during total exchange and which may accumulate as bars of skeletal lime sand. As the channels migrate these skeletal sands commonly form a basal lag deposit.

Middle and Upper Intertidal Zone. Sediments here are commonly light-grey to light-brown (oxidizing conditions), have good fenestral porosity (the variable growth of algal mats), are graded (episodic storm deposition) and are broken into desiccation polygons (prolonged exposure). There is generally good growth of algal mats throughout, and in the lower parts thick leathery mats are separated into desiccation polygons a few centimetres to a metre in diameter with cracks filled by lime mud in the lower parts (Fig. 8). In the central parts, thinner leathery mats have surfaces that are puffed up into blisters and convoluted into crenulated forms. In the upper parts, shriveled, crinkled and split mats are found. Bedding generally is irregular, especially in the upper zones, with mats alternating with graded storm layers.

In some settings, sediment in the upper intertidal zone dries out to form chips of lime mud while in others the sediment below the mats is lithified to a depth of as much as 10 cm.

Although sediments commonly are laminated throughout the intertidal environment, they are also riddled with small-scale tubules produced by insects and worms, and with larger tubes produced by crabs and other crustaceans.

Sediments also may be penetrated by the prolific shallow roots systems of salt tolerant plants.

Supratidal Environment. In all situations (including channel levees) this area is characterized by long periods of exposure. This is reflected by the lithification of storm deposited sediments in the form of surface crusts several centimetres thick, and which in turn are fractured into irregular polygons. These polygons may be pushed up by the force of crystallization (or by plant roots) to form 'teepees', or dislodged completely to form pavements of flat-pebble breccia. Clasts are commonly cemented on modern tidal flats by cryptocrystalline aragonite or calcite, and characteristically contain considerable (25 to 50%) fine crystalline dolomite.

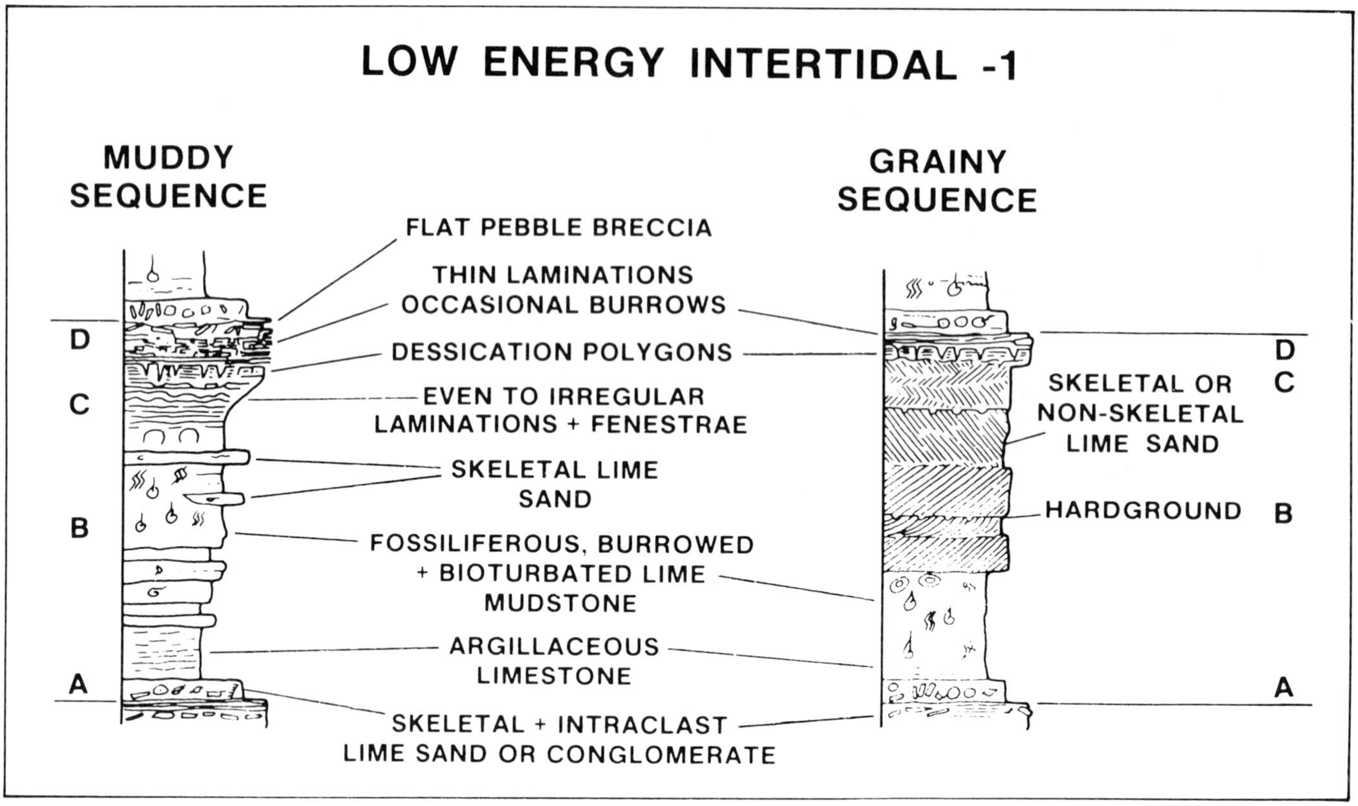

Figure 9
Two hypothetical sequences with a low energy tidal flat unit developed on a low energy subtidal unit (left) and a high energy lime sand unit (right).

If the creek levees in the intertidal zone have built up above normal high tide level, they consist of hard, finely to very finely laminated sediment, extremely regular and composed of alternating layers of sediment and thin algal mats with excellent fenestral porosity.

The landward parts of the supratidal zone may grade into various terrestrial environments, the end members of which are: 1) areas of elevated, pre-existing bedrock and no sedimentation in which the surface of the rock is characterized by intensive subaerial diagenesis, and the development of caliche (calcrete crusts); 2) areas of contemporaneous sedimentation which grade between: a) low-lying environments in regions of high rainfall occupied by algal marshes, b) low-lying environments in arid, desert regions, characterized by evaporite formation, and c) well-drained zones, often slightly elevated and with little deposition.

Algal marshes, flooded by fresh water during the rainy season, are an ideal environment for the growth of algal mats and these mats are periodically buried by layers of sediment swept in during particularly intense storms. The preserved record is therefore one of thick algal mats alternating with storm layers. With progressive aridity the supratidal zone dries out. If the chlorinity of the groundwaters remains constantly above 39 ‰ cementation, particularly by aragonite, is common. Cementation is most common if there is minor but consistent input of fresh water from inland to dilute the hypersaline groundwaters somewhat. If the chlorinity of the groundwaters remains constantly about 65 ‰ then authigenic evaporites precipitate within the sediment below ground level. In this setting (called a supratidal sabkha, or salt flat in the Middle East; see "Continental and Supratidal (Sabkha) Evaporites", this volume) dolomitizaiton is also common in the subsurface, saline brine pools occur at the surface, and terrigenous wind-blown sand is common in the sediment.

In relatively well-drained zones the supratidal environment is a deflation surface occasionally cut by the upper reaches of tidal creeks, sometimes damp from rising capillary waters and covered by a thin film of algal mat. Scoured and rippled sediment is common and clasts are sometimes encrusted with algae to form oncolites.

COMMON SEQUENCES WITH A LOW—ENERGY INTERTIDAL UNIT

Muddy and Grainy Sequences. These sequences developed either by progradation of the wide continental tidal flat or by shoaling the lime sand bodies that formed the barrier offshore (Fig. 9). The climate in the region of deposition was generally too wet or the ground-water table too low or diluted by fresh water to permit precipitation of evaporites.

The muddy sequences, those in which skeletal lime muds or muddy lime sands are the main subtidal unit, are well developed today in well-drained areas of Shark Bay where salinities are too high to permit development of a normal marine fauna as well as browsing of the algal mats by gastropods. Muddy sequences are also well developed in the tidal creek and pond belt of the Bahamas. These sequences are generally regarded as the 'classic' tidal flat sequences. The basal unit, if present, records the initial incursion of the sea onto land and as such is commonly coarse-grained, composed of clasts: all diagnostic of surf-zone deposition. The

Figure 10
Shallowing-upward sequences comprising lower intertidal-subtidal limestones (L) overlain by supratidal dolomites (D - Cryptalgal laminites, sandy in part) in the Lyell Fm. at Takakkaw Falls, Yoho National Park, B.C. (Photo courtesy J.D. Aitken).

Figure 11
Numerous shallowing-upward sequences comprising thick subtidal oolite lime sands and thin intertidal-supratidal cryptalgal laminites with fenestrate porosity: Petite Jardin Fm., Port-au-Port Peninsula, Nfld.

subtidal unit is characteristically a bioturbated lime wackestone to packstone with a normal and diverse marine fauna, commonly containing stromatolites in deposits older than middle Paleozoic. In Precambrian and lower Paleozoic deposits the characteristic tidal flat features such as desiccation polygons, well-laminated sediments and fenestrae will occur at the base of the intertidal zones (Figs. 10 and 11). In deposits younger than middle Paleozoic, the prolific browsing and burrowing activity in the lower intertidal zone (unless the water mass was hypersaline) has homogenized the sediment, so that the signature of intertidal deposition is recorded only within the mid and upper intertidal sediments.

If the tidal flat was extensively channeled, the migration of channels back and forth may also have destroyed some of the subtidal character, forming instead a partial fining-upward sequence (much like that a river), with a basal skeletal lime sand.

Where fenestrae are present they show a zonation: horizontal to laminated in the lower intertidal environments (smooth mat), irregular and, in some cases, vertical in the middle and upper intertidal environments (pustular, shriveled and crinkled mats).

Desiccation polygons are most common near the top, apparently coincident with cementation. The supratidal zone is characterized by very evenly laminated deposits or flat pebble breccias.

Readers interested in the finer details of such sequences are referred to studies by Laporte (1967) and Fischer (1964), the latter outlining and documenting a similar facies sequence but in reverse order, forming a deepening-upward sequence.

A common early Paleozoic subtidal lithology in these sequences is alternating thin-bedded limestone and shale, forming ribbon to parted limestones. Sepkowski (1981) notes that flat-pebble conglomerates are conspicuous in these Cambrian and Ordovician sequences and suggests that they may be formed by early seafloor lithification of true carbonate followed by erosion and redeposition as storm deposits. Thus the presence of flat-pebble conglomerates alone need not indicate tidal flat deposition. Expansion of infauna in middle Ordovician time led to greater burrowing in the subtidal zone and so reduced the opportunity for early lithification in younger sediments. Shoaling sequences also may be present off-shore from the low-energy tidal flat, on the lime-sand shoals. Here low energy tidal flats developed in the lee of the leading edge of the shoal once beach ridges were developed or currents had swept sand together to form islands. This will be reflected in the sequence as a sudden change from obvious high energy deposition to low energy intertidal deposition. The subtidal unit is generally well-sorted, oolitic, pelletoidal or skeletal lime sand (pelmatozoans are particularly common in the Paleozoic), with a few containing oncolites. Bedding is characteristically planar, with herringbone cross-laminations, large at the base and becoming smaller upwards, and individual bedding planes commonly covered with small-scale ripples. Early cementation is characteristic, and so deposits contain many intraclasts of cemented lime sand, and bored surfaces. Once the shoals, or parts of the shoals are inactive they may be burrowed and much of the original cross-bedding may be destroyed.

The intertidal to supratidal units are similar to those described above but are generally relatively thin. If the shoal is exposed for a long time caliche and soil profiles commonly develop, reflected by brown irregular laminations, breccias, and thin shale zones.

An excellent description of muddy and grainy sequences can be found in Demicco and Mitchell (1982).

Stromatolite and Reef Sequences. One common variation on the model is the development of shoaling-upward sequences in association with abundant stromatolites in the lower Paleozoic/-Precambrian and with reefs in the Phanerozoic in general.

In Shark Bay, Western Australia, where all environments are hypersaline and so stromatolites abound, the interrelationship between stromatolite morphology and environment has only recently been documented (Hoffman, 1976). In the intertidal zone columnar to club-shaped forms up to one metre high are found rimming headlands. In relatively high energy, exposed environments the relief of the columns is proportional to the intensity of wave action. The columnar forms grade laterally away from the headlands to the lower energy bights, where the stromatolites are more prolate and elongate oriented normal to the shoreline. In tidal pools digitate columnar structures abound.

These growth forms are the result of active sediment movement; algal mats only grow on stablized substrate, thus columns are nucleated upon pieces of rock, or cemented sediment; growth is localized there and does not occur on the surrounding shifting sands. Early lithification of the numerous superimposed layers of mat and sediment turns the structure into resistant limestone. Moving sand continuously scours the bases of the stromatolites. The mounds or pillars are largest in subtidal or lower intertidal environments and decrease in synoptic relief upwards, finally merging with stratiform mats in upper intertidal zones, above the zone of active sediment movement.

The resulting model sequence, summarized in Figure 12, is integrated from the Shark Bay example and the summary sequence of 200 or more shoaling sequences present in the Rocknest Formation of middle Precambrian age near Great Slave Lake (Hoffman, 1976). In the intertidal zone deposits reflect higher energy than normal, indicating a more exposed shoreline. These sediments underlie and surround the domal (Fig. 13) to columnar stromatolites, which in turn grade up into more stratiform stromatolites, and finally into very evenly bedded structures. The supratidal unit of this sequence will be characterized by both desiccation polygons and flat-pebble breccias as well as occurrences of delicate branching stromatolites (Fig. 14), formed in supratidal ponds. Care should be taken in delineating this sequence because stromatolites that are similar to those in the intertidal zone also occur in the subtidal (Playford and Cockbain, 1976).

Shallowing-upward sequences are also common as the last stage of sedimentation in large bioherms, as numerous successions within the large back-reef or lagoonal areas of reef complexes, and as 'caps' on widespread biostromes. In this type of sequence the shoaling upward is first reflected in the subtidal unit itself, generally as a transition from large massive hemispherical colonial metazoans of the reef facies, to the more delicate, stick-like forms that are common in the shallow protected

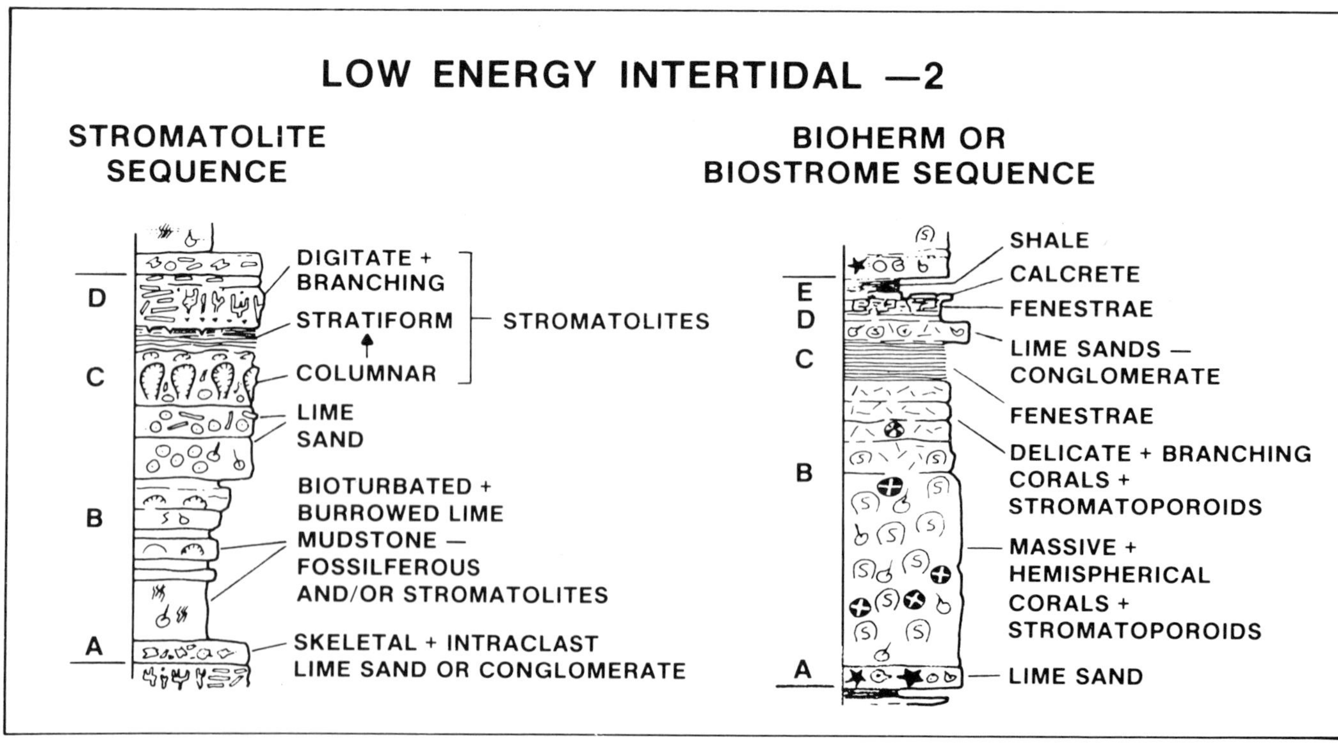

Figure 12
Two hypothetical sequences with a low energy intertidal unit developed in conjunction with stromatolites (left) and on top of a skeletal metazoan bioherm or biostrome (right).

Figure 13
A columnar to club-shaped stromatolite of Late Cambrian age from the Petite Jardin Formation, Western Newfoundland.

Figure 14
Digitate stromatolites from a shallowing-upward sequence of Late Cambrian age Western Newfoundland.

locations. These stick-like skeletons may be swept together on beaches at the edge of the tidal flat. As a result, the intertidal unit commonly contains a conglomerate within it, or at the base. The upper part of the sequence is otherwise similar to the others described. For a more detailed description of "reefy" sequences see studies by Havard and Oldershaw (1976), Read (1973), and Wong and Oldershaw (1980).

Carbonate-Evaporite Sequences (also see Kendall's paper in this volume). The other major variation on the model proposed at the beginning of this article is at the opposite end of the environmental spectrum, in the supratidal zone, in this case emergent in a very arid environment and flushed by hypersaline groundwaters. The hypersalinity of the groundwaters and attendant high evaporation results in the formation of authigenetic evaporites. This in turn raises the Mg^{++}/Ca^{++} ratio of the groundwaters and induces dolomitization of the sediment. The processes occur within the sediment, above the water table in the intertidal zone, and both above as well as below the water table in the supratidal zone. If the water compositions are barely within the field of gypsum precipitation, and there are fluctuations due to brackish flow of groundwater from the mainland, evaporites will occur in the form of isolated masses or crystals in the upper part of the sequence. If the groundwater compositions are continously well within the field of gypsum precipitation, growth of evaporite minerals takes place: 1) as a mush of gypsum crystals in the intertidal zone or as layers of anhydrite nodules, 2) as complex masses with a characteristic chickenwire texture, and 3) as layers contorted into enterolithic (intestine-like) shapes (Fig. 15). The important point, which is often ignored, is the growth of the evaporites within the sediment, as a diagenetic overprint on depositional facies of various environments. As evaporite growth is porphyroblastic, the host sediment commonly is displaced to intercrystalline areas and earlier fabrics are destroyed. Accompanying dolomitization is commonly intense with sediments of the intertidal and much of the subtidal zones affected.

Evaporites, however, are very soluble when exposed to percolating meteoric waters of low salinity and have a tendency to vanish from the record. Dissolution of the evaporites affects the sequences in several ways, but the most important is the formation of collapse breccias (Fig. 16). This collapse occurs when the evaporites dissolve leaving no support for the overlying sediments which subside into the void created by evaporite removal. Thus the top of the sequence is a breccia of marine limestone from the overlying sequence with a mixture of terrigenous sand, if a terri--genous facies capped the original sequence (Fig. 15). Isolated anhydrite crystals in lower parts of the sequence may be leached out, forming vugs which may be subsequently filled with quartz or chalcedony (usually length-slow). The dolomite, at least in the

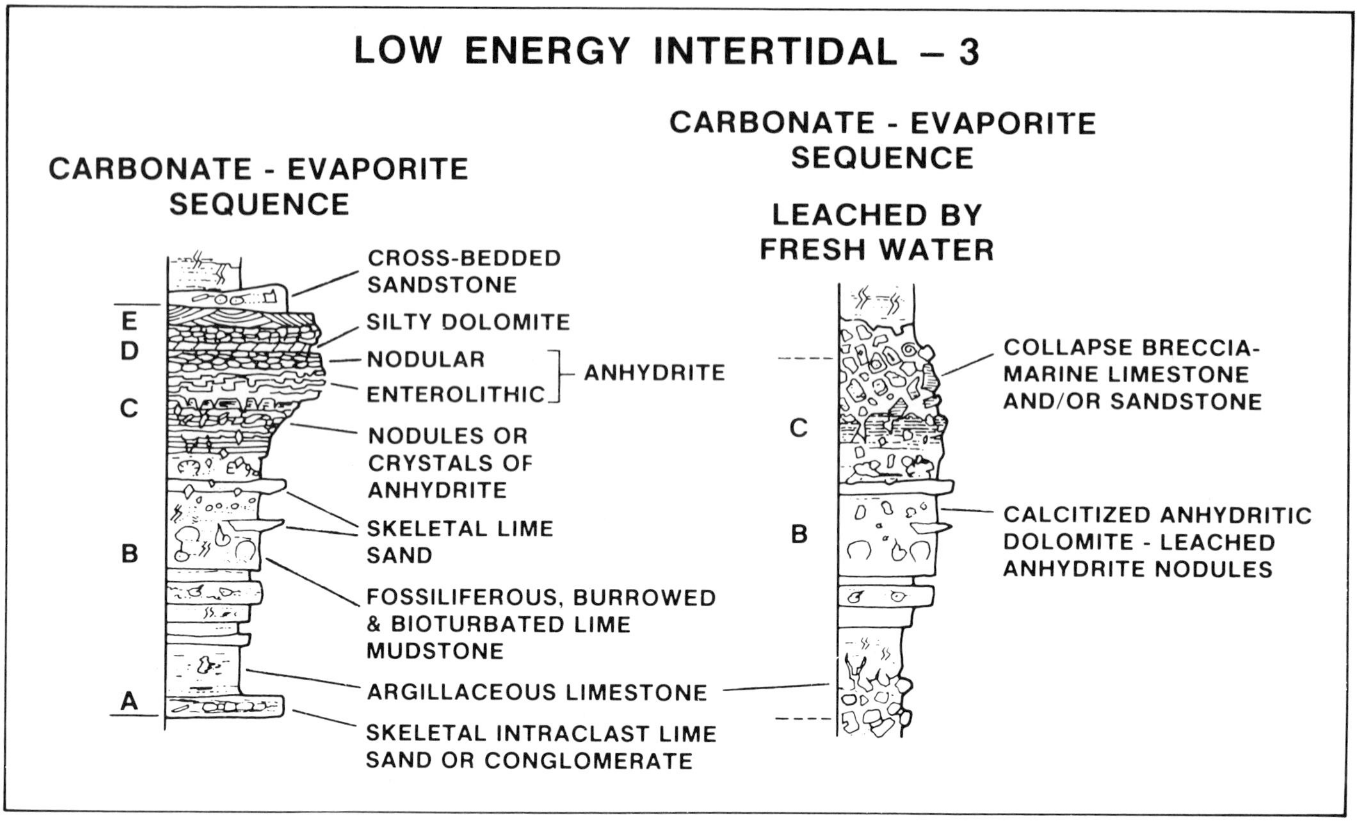

Figure 15
Two hypothetical sequences with a low-energy intertidal unit and a supratidal unit developed under arid conditions; on the right the evaporties have been dissolved by percolating fresh waters.

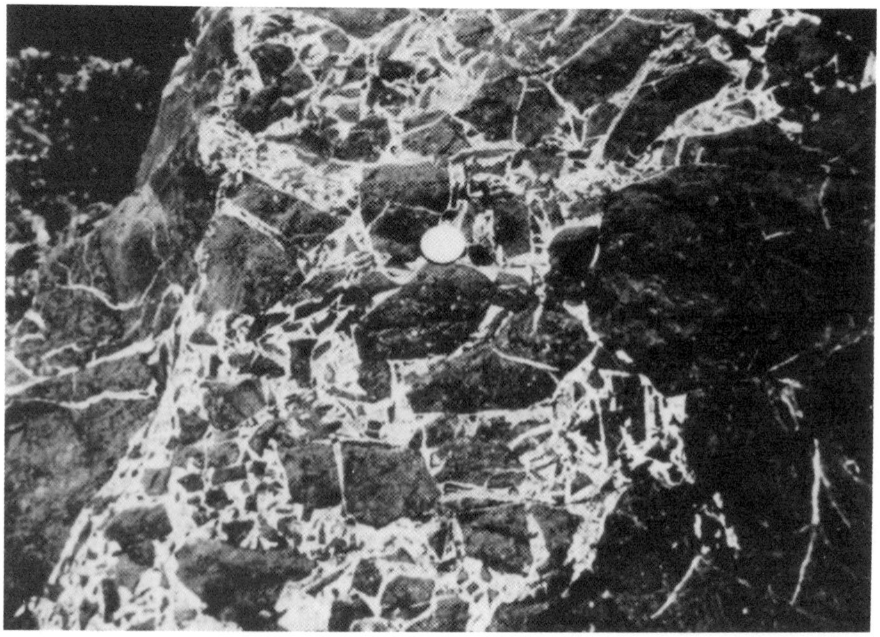

Figure 16
A collapse breccia of subtidal lime mudstone clasts in white calcite: caused by the solution of anhydrite at the top of a shallowing-upward sequence in the Shunda Fm. (Mississippian) at Cadomin, Alberta (Photo courtesy R.W. Macqueen).

upper part, is commoly altered to calcite, in the reverse of the dolomitization process (so-called "dedolomitization").

SEQUENCES WITH A HIGH ENERGY INTERTIDAL UNIT

In contrast to the low-energy intertidal (the tidal flat) the higher energy beach zone is not commonly recognized in the rock record. This may be partly because it resembles many subtidal grainstone deposits and hence is not obviously distinctive. Also, it is relatively narrow compared to the tidal flat, and has a lower preservation potential. Finally, the beach deposits lack the distinctive sedimentary features of the tidal flat. These very reasons illustrate the value of the concept of a shoaling-upward sequence as a guide. Once the potential for such a sequence is recognized in the geologic record, then one can concentrate on the search for subtle features that characterize beach deposition, which otherwise might go unnoticed.

Modern Carbonate Beaches

The sedimentology of carbonate beaches is nicely illustrated by Inden and Moore (1983). The beach is characterized by two zones: 1) the lower foreshore, usually below the zone of wave swash, and 2) the upper foreshore, the zone of wave swash. Sediments of the lower foreshore are coarse grained, poorly sorted, have a matrix of lime mud (if it is available), and are characterized by small and large-scale festoon cross-bedding, oriented parallel to the shoreline and generally attributed to longshore drift. The upper foreshore comprises thick-bedded, internally laminated, very well-sorted lime sands and gravels in planar cross-bedded accretionary beds that dip gently seaward (generally less than 15°). Sediments in the upper foreshore zone may have many open-space structures, the equivalent of the fenestrae of muddy intertidal sediment called keystone vugs (Dunham, 1969) or microcaverns (Purser, 1972). These are due to gas escape, and in the geological record are partly to completely filled with cement.

As on the tidal flat, periodic exposure of beach deposits leads to cementation and partial subaerial diagenesis. The textures thus created are difficult to recognize in the field but are important keys to recognizing the beach environment. The two most important of these diagenetic phenomena are beachrock and calcrete.

Shallowing-Upward Sequence With a High-Energy Intertidal Unit

The lower two units of this type of sequence are similar to those described in the preceding sections on sequences with low-energy intertidal units (Fig. 17). In this sequence, however, characteristic subtidal carbonates grade upward into coarse-grained lime sands with all the characteristics of the lower and upper foreshore described above (Fig. 9). The supratidal unit may be present in the form of a thin shale (soil), but more commonly the supratidal environment is represented not by a deposit but by intensive diagenesis of the upper unit (cementation, dissolution, calcrete formation and microkarst). This is in many ways similar to the diagenetic overprint of other facies by supratidal evaporite formation.

Beachrock is composed of seaward-dipping beds of lime sand and gravel that are generally cross-laminated and occur in the lower intertidal to middle intertidal environment. It is formed by the precipitation of carbonate cement out of seawater or mixed seawater and rainwater. The beds of limestone may be up to one metre thick, are commonly jointed at right angles to the beach, and

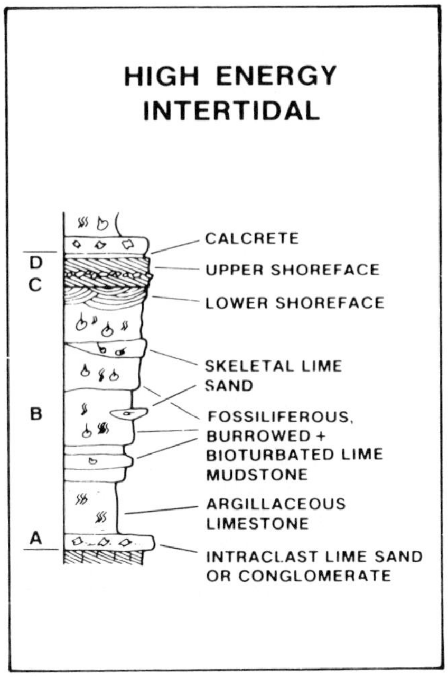

Figure 17
A hypothetical sequence with a high-energy intertidal unit: a beach, developed, in this case adjacent to a low energy subtidal environment.

are encrusted and/or bored by numerous intertidal organisms. Lithification disappears seaward and rarely extends higher than the intertidal zone. The partly cemented beds may be broken up and redeposited as conglomerates, made up of cemented sand clasts. In the upper parts of the intertidal zone cementation takes place in intergranular voids partly filled with air: the cements, as a result, are often stalactitic (more extensively developed on the undersides of grains).

If exposed for long periods of time and if located in an environment where there is at least periodic rainfall, the lime sands will begin to undergo subaerial diagenesis (see Bathurst, 1975, for an extended discussion of subaerial diagenesis). In addition the upper metre or so of such subaerially exposed deposits develop calcrete or caliche horizons which have many features that closely resemble those produced by laminar to laterally-linked stromatolites and oncolites. These features are discussed in detail by James (1972) and Read (1976).

The supratidal unit in these sequences may be any of the ones described above, although calcrete (caliche) is very common. Beaches may act as small barriers protecting supratidal ponds and flats so that the cap in such sequences will be thin beds of lime mud (often dolomitized) with all of the associated supratidal features. One variation not found elsewhere occurs where the high energy surf zone of the overlying sequence erodes the top of the sequence down to the cemented portions, resulting in truncation layers or hardgrounds that separate sequences.

CYCLICITY

The shallowing-upward facies model is demonstrably one of the most useful concepts a sedimentologist can have when working with platform carbonates. In the rock record these sequences occur on two scales; *small-scale,* or those less than a metre to a few 10s of metres in thickness at most, and *large-scale* or those many tens to hundreds of metres in thickness, and often likened to the Grand Cycles of Aitken (1967). In spite of the model's obvious utility, the precise mechanisms by which numerous sequences are generated remains obscure. Since the first edition of this paper (James, 1979) attention has shifted from the details of the model itself to the possible mechanisms of this accretion.

On the surface the explanation seems simple enough. Because the rate of carbonate deposition exceeds the rate of platform subsidence or sea level rise, sediments will rapidly accrete to sea level. This is fine for a single shallowing-upward sequence, but how are repeated sequences formed and how do they exist over vast carbonate platforms?

Small-Scale Sequences

Current thinking on how repeated small-scale sequences are formed is succinctly summarized by Wilkinson (1982), who points out that there are currently two end-member models, both of which generate virtually identical sequences (Fig. 18).

The *eustatic model* is one in which the rate of carbonate sedimentation is constant but the rate of subsidence or the absolute position of sea level are not constant and change in a non-uniform or periodic fashion. During periods of stability or slowly rising sea level (Figs. 18 and 2) the whole sequence progrades out across the shelf yielding a typical shallowing-upward or regressive sequence. This pattern is interrupted by a sudden and rapid sea level rise, flooding the platform (Fig. 18-3) and resulting in a short period of arrested- or

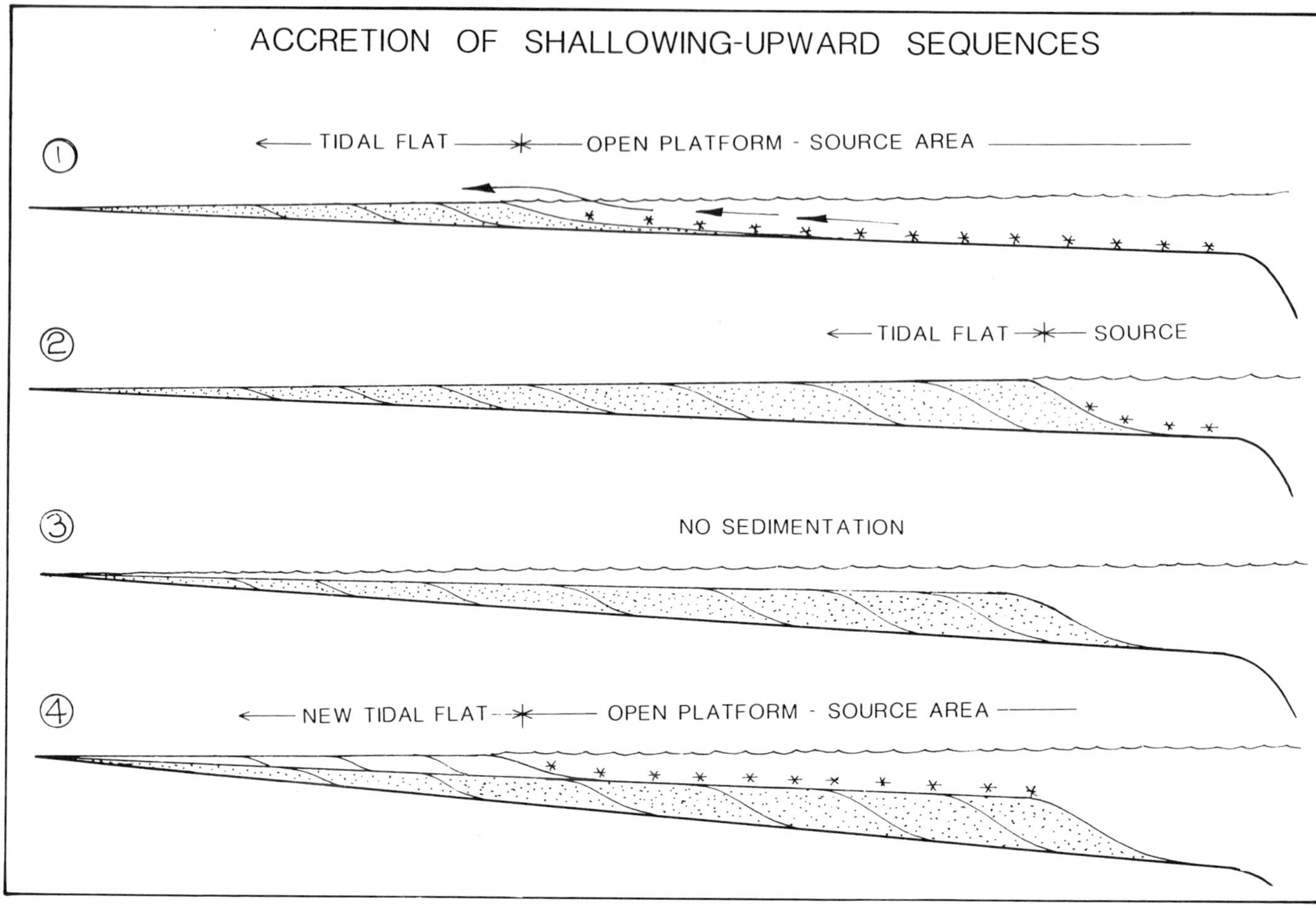

Figure 18
A sketch illustrating how two shallowing-upward sequences can be produced by progradation of a tidal flat wedge. These general conditions apply in the case of both eustatic and autocyclic models.

non-deposition. Sea level remains relatively stationary in this new position for a time (Fig. 18-4) and progradation begins again with a new shallowing-upward sequence forming over the old one. One variant of this model, calling for sudden platform-wide shifts in sea level has been formalized as "Punctuated Aggradational Cycles" by Anderson and Goodwin (1980).

While this model may be attractive as an explanation for large-scale shallowing-upward sequences it is less compelling for small-scale cycles, because each cycle must record either a sudden eustatic change in sea level or a tectonic event. If it is the cause of small-scale cycles then the problem becomes one of small-scale cyclicity on a global scale (Schwarzacher and Fischer, 1982) which is beyond the scope of this paper.

In the alternative *autocyclic model* the control is intrinsic and lies in the rate of carbonate sedimentation as controlled by source area (Fig. 18). The model was first proposed by Ginsburg (1971) and has been used for many years as a model in field seminars at the University of Miami. Similar schemes have subsequenly been suggested by Matti and McKee (1976), Mossop (1979) and Wilkinson (1982).

As in the eustatic model, sedimentation is envisaged as taking place on a gently inclined shelf under conditions of a gradually subsiding shelf or slowly rising sea level or some combination of both (Fig. 18-1). The source area of the sediments for the prograding wedge is the large subtidal area (Fig. 18-2), which is gradually reduced in size with seaward progradation. Eventually the situation is reached in which the source area is too small or too deep to provide sediments for the prograding wedge (Fig. 18-3), and so sedimentation stops. Relative sea level will continue to rise, however, and soon (Fig. 18-4) the whole platform will once again be subtidal and deep enough for sediment production and the cycle will begin again.

The second and related problem concerns precisely how tidal flats prograded over the vast areas of epeiric or epicontinental seas. The stratigraphic record illustrates that two situations are common. In one, individual component lithologies of remarkably uniform thickness occur over wide geographical areas. For these sequences Ginsburg (1982) has proposed either continuously prograding tidal-flat wedges which would create time transgressive units and leave large areas exposed to prolonged meteoric diagenesis, or repeated deposition of thin areally-extensive single-event units by wind-driven "tides".

In other examples individual lithologies cannot be correlated between adjacent wells only kilometres apart (e.g., Wong and Oldershaw, 1980) or transitions from supratidal to intertidal to subtidal facies can be walked out in a

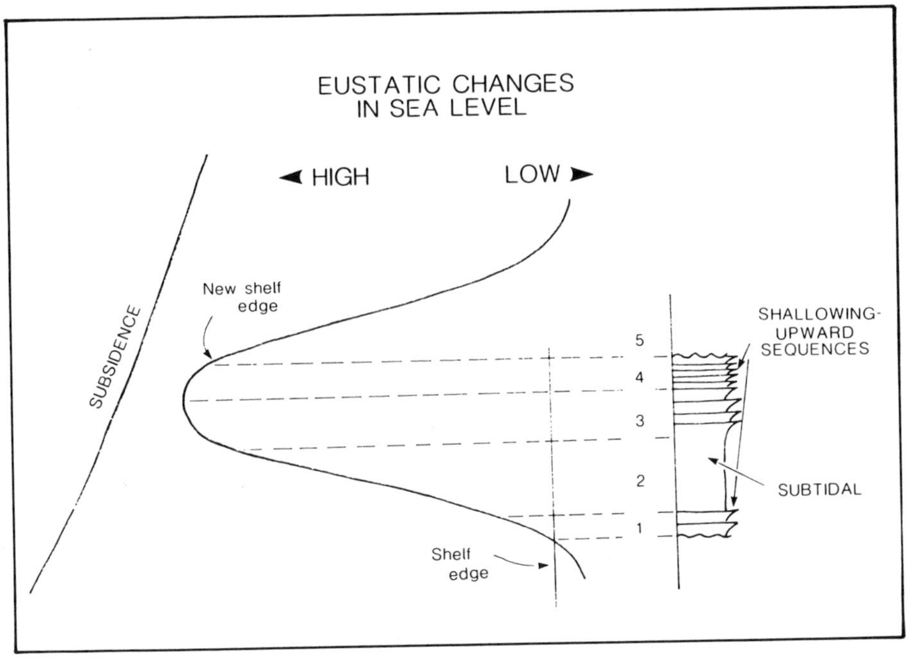

Figure 19
A diagram illustrating how a large-scale shallowing-upward sequence is produced under conditions of slow platform subsidence and a uniform rise and fall in eustatic sea level.

single bed over distances of only a few kilometres (e.g., Pratt and James, in press). The most appealing explanations for these carbonates is deposition on a platform dotted by a mosaic of exposed banks or islands separated by subtidal areas, with the whole complex shifting both laterally and vertically in response to hydrodynamic conditions through time.

This "island" model has an additional attraction in that a complex facies mosaic of numerous shallowing-upward sequences can be formed everywhere on a subsiding shelf at the same time, and is thus equally applicable to sequences formed under conditions of relatively uniform sea level or under conditions of sporadically changing sea level.

Large-Scale Sequences

These larger-scale packages, the upper parts of which are characterized by many small-scale sequences, are more complex, often involving an interplay of eustatic sea-level fluctuaitons, tectonics and platform geometry (see Aitken, 1978).

An example of how one such large-scale sequence might be developed is outlined in Figure 19. In this case, I have used the concepts formulated by the Exxon Seismic Stratigraphy Group (Vail pers. commun., 1984). Here subsidence is viewed as constant, with the rise and fall of sea level more or less symmetrical.

(1) As sea level begins to rise slowly over the shelf edge the platform is flooded. During this initial stage carbonate accretion can outpace relative sea level rise and shallowing-upward sequences develop.

(2) During the prolonged period of relatively rapid sea level rise platform subsidence is ongoing and the tendency is to maintain deep water over the shelf. Depending on the rates of each, either subtidal conditions may develop, or if both are slow a few thick shallowing-upward sequences may form.

(3) The rate of sea level rise slows, and shallowing-upward sequences develop easily.

(4) Sea level falls slowly, but subsidence is also continuing, so the net effect is a still stand or close to it. Numerous, thin, shallowing-upward sequences develop with long periods of subaerial diagenesis between.

(5) Sea level begins to fall rapidly, outpacing subsidence and quickly dropping below the shelf edge, which is now higher because of the intervening carbonate accretion; the whole platform is exposed.

The result of this sequence of events is a large-scale shallowing-upward sequence, possibly with a few cycles at the base but mainly a lower half of subtidal sediments and an upper half of numerous shallowing-upward sequences. Obviously this is one sequence formed in response to a given set of variables but other sequences can be easily generated using similar principles. What is important is that a large-scale asymmetrical cycle can be produced by a uniform rise and fall of sea level.

SUMMARY

In the past there has been a natural tendency to use obvious sedimentary structures (e.g., mud cracks, stromatolites) to infer that parts of a carbonate sedimentary sequence had been periodically exposed. Individual structures, however, often have counterparts in other sedimentary environments (e.g., syneresis cracks, subtidal stromatolites) resulting, in many cases, in questionable paleoenvironmental interpretations. With all the data now available on carbonate strandline deposition we can frequently use what have become natural associations of sedimentary features in a vertical succession and define, with precision, specific strandline facies and their interrelationships.

While this is true for low-energy shoreline sequences, it is much less so for high-energy shoreline sequences. To bring all aspects of this type of facies model to comparable levels of understanding much more data is needed on exposed or high-energy intertidal environments, not from the modern, but from the rock record. In addition, the time is ripe to test whether or not the diagenetic features which result from periodic subaerial exposure (cementation, microkarst, calcrete) can be commonly recognized in ancient sequences.

In conclusion, the shallowing-upward sequence is one of the commonest, and with the wealth of sedimentary features, one of the easiest models to apply to the carbonate facies spectrum. As a *norm* this model is constructed from a synthesis of over 40 well-documented fossil examples, but our understanding of the meaning of most diagnositc sedimentary features comes from half a dozen modern settings. Carbonate platform deposition, being what it is, a rapidly accreting system, once a shallowing-upward sequence has been recognized, one can *predict* that other similar sequences will be present vertically. If the sequences recur in an orderly fashion (are cyclic) then similar packages should be present laterally as well. Alternatively, if parts of the model are stacked in a less regular fashion, reflecting deposition as a complex of islands and banks on an open shelf, then specific laterally equivalent facies are less predictable. Recognition of the style of the model, which is in turn dependent upon the overall climatic (humid versus arid) and oceanographic (normal versus hypersaline) conditions, then allows the model to be both a *basis for physical and chemical interpretation* as well as a *guide* to the interpretation of other shallow water carbonates of the platform succession.

ACKNOWLEDGEMENTS

Various editions and parts of this manuscript have kindly been read by Roger Macqueen, Bob Stevens, David Kobluk, Brian Pratt, Bob Ginsburg, and Rick Sarg, and I thank them for their help.

REFERENCES

BASIC REFERENCES ON SHALLOWING-UPWARD CARBONATE SEQUENCES

Demicco, R.V., and Mitchell, R.W., 1982. Facies of the Great American Bank in the Central Appalachians. *In* Lyttle, P.T., ed., Central Appalachian Geology. NE-SE GSA'82 Field Trip Guidebook. American Geological Institute, p. 171-266.
An example of comparative sedimentology between Cambrian (Conococheague Limestone) - Ordovician (St. Paul Group) and modern tidal flat sequences with excellent photographs.

Fischer, A.G., 1964. The Lofer cyclothems of the Alpine Triassic. *In* Merriam, D.F., ed., Symposium on cyclic sedimentation. State Geological Survey of Kansas, Bulletin 169, v. 1, p. 107-149.
A regressive sequence of shallow water carbonates, superbly documented and illustrated.

Ginsburg, R.N., ed., 1975. Tidal deposits. New York, Springer-Verlag, 428 p.
The best all around, up-to-date reference on siliclastic and carbonate tidal deposits.

Hoffman, P., 1973. Recent and ancient algal stromatolites: seventy years of pedagogic cross-pollination. *In* Ginsburg, R.N., ed., Evolving concepts in sedimentology. Baltimore, MD., Johns Hopkins Press, p. 178-191.
A very readable essay on the evolution of our understanding of stromatolites and related sediments.

Irwin, M.L., 1965. General theory of epeiric clear water sedimentation. American Association of Petroleum Geologists, Bulletin, v. 49, p. 445-459.
The first integrated synthesis of shallow water carbonate sequences and their meaning.

Laporte, L., 1967. Carbonate deposition near mean sea-level and resultant facies mosaic: Manlius Formation (Lower Devonian) of New York State. American Association of Petroleum Geologists, Bulletin, v. 51, p. 73-101.
An extremely clear, well-written analysis of a shallowing-upward sequence.

Lucia, F.J., 1972. Recognition of evaporite-carbonate shoreline sedimentation. *In* Rigby, K.J., and Hamblin, K., eds., Recognition of ancient sedimentary environments. Society of Economic Paleontologists and Mineralogists, Special Publication 16, p. 160-192.
A well-written summary with many examples.

Merriam, D.F., ed., 1964. Symposium on cyclic sedimentation. State Geological Survey Kansas, Bulletin, 2 vols., 636 p.
40 papers, many of which are still basic references on cyclic sedimentary sequences.

Shaw, A.B., 1965. Time in stratigraphy. New York, McGraw-Hill, p. 1-71.
A pedantic, but thought-provoking analysis of epeiric sea carbonate sedimentation.

MODERN CARBONATE TIDAL FLATS

Bathurst, R.G.C., 1975. Carbonate sediments and their diagenesis: Developments in Sedimentology No. 12, New York, Elsevier, p. 178-209, p. 517-543.

Garrett, P., 1970. Phanerozoic stromatolites: non-competitive ecological restriction by grazing and burrowing animals. Science, v. 169, p. 171-173

Hardie, L.A., ed., 1977. Sedimentation on the modern carbonate tidal flats of Northwest Andros Island, Bahamas. Baltimore, MD., The Johns Hopkins Press, 202 p.

Hoffman, P., 1976. Stomatolite morphogenesis in Shark Bay, Western Australia. *In* Walter, M.R., ed., Stomatolites. New York, Elsevier, p. 261-273.

Kinsman, D.J.J., 1966. Gypsum and anhydrite of Recent Age, Trucial Coast, Persian Gulf. Second Symposium on Salt. Northern Ohio Geological Society, Cleveland, p. 1302-1326.

Logan, B.W., Davies, G.F., Read, J.F., and Cebulski, D., 1970. Carbonate sedimentation and environments, Shark Bay, Western Australia. American Association of Petroleum Geologists, Memoir 13, 223 p.

Logan, B.W., *et al.*, 1974. Evolution and diagenesis of Quaternary carbonate sequences, Shark Bay, Western Australia. American Association of Petroleum Geologists, Memoir 22, 358 p.

Kendall, C. St. G.C. and Skipwith, P.A. d'E., 1968. Recent algal mats of a Persian Gulf lagoon. Journal of Sedimentary Petrology, v. 38, p. 1040-1058.

Kendall, C. St. G.C. and Skipwith, P.A. d'E., 1969. Holocene shallow-water carbonate and evaporite sediments of Khor al Bazam, Abu Dhabi, Southwest Persian Gulf. American Association of Petroleum Geologists, Bulletin, v. 53, p. 841-869.

Purser, B.H., ed., 1973. The Persian Gulf. New York, Springer-Verlag, 471 p.

Shinn, E.A., 1983a. Tidal flat. *In* Scholle, P.A., Bebout, D.G., and Moore, C.H. eds., Carbonate depositional environments. American Association of Petroleum Geologists, Memoir 33, p. 171-211.

Shinn, E.A., Lloyd, R.M., and Ginsburg, R.N., 1969. Anatomy of a modern carbonate tidal flat, Andros Island, Bahamas. Journal of Sedimentary Petrology, v. 39, p. 1201-1228.

MODERN CARBONATE SAND BODIES

Ball, M.M., 1967. Carbonate sand bodies of Florida and the Bahamas. Journal of Sedimentary Petrology, v. 37, p. 556-591.

Hine, A.C., 1977. Lilly Bank, Bahamas: history of an active oolitic sand shoal. Journal of Sedimentary Petrology, v. 47, p. 1554-1583.

Halley, R.B., Harris, P.M., and Hine, A.C., 1983. Bank margin. *In* Scholle, P.A., Bebout, D.G., and Hine, A.C., eds., Carbonate depositional environments. American Association of Petroleum Geologists, Memoir No. 33, p. 463-507.

Imbrie, J. and Buchanan, H., 1965. Sedimentary structures in modern carbonate sands of the Bahamas. *In* Middleton, G.V., ed., Primary sedimentary structures and their hydrodynamic interpretation. Society of Economic Paleontologists and Mineralogists, Special Publication 12, p. 149-173.

LOW-ENERGY INTERTIDAL

The papers listed below are studies in which sequences or partial sequences have been well documented. I have omitted papers that simply mention that the rocks are deposited in very shallow environments.

1) Predominantly Muddy or Shaley Sequences

Aitken, J.D., 1966 Middle Cambrian to Middle Ordovician cyclic sedimentation, southern Rocky Mountains of Alberta. Bulletin of Canadian Petroleum Geology, v. 14, p. 405-411.

Aitken, J.D., 1978. Revised models for depositional grand cycles, Cambrian of the southern Rocky Mountains, Canada. Bulletin of Canadian Petroluem Geology, v. 26, p. 515-542.

Assereto, R. and Kendall, C.G. St. C., 1971. Megapolygons in Ladinian limestones of Triassic of Southern Alps: evidence of deformation by penecontemporaneous desiccation and cementation. Journal of Sedimentary Petrology, v. 43, 715-723.

Kahle, C.F. and Floyd, J.C., 1971. Stratigraphic and environmental significance of sedimentary structures in Cayugan (Silurian) tidal flat carbonates, Northwestern Ohio. Geological Society of American, Bulletin, v. 82, p. 2071-2098.

Macqueen, R.W. and Bamber, E.W., 1968. Stratigraphy and facies relationships of the Upper Mississippian Mount Head Formation, Rocky Mountains and Foothills, Southwestern Alberta. Bulletin of Canadian Petroleum Geology, v. 16, p. 225-287.

Mukherji, K.K., 1969. Supratidal carbonate rocks in the Black River (Middle Ordovician) Group of Southwestern Ontario, Canada. Journal of Sedimentary Petrology, v. 39, p. 1530-1545.

Reinhardt, J., and Hardie, L.A., 1976. Selected examples of carbonate sedimentation, Lower Paleozoic of Maryland. Baltimore, Maryland Geological Survey, Guidebook No. 5, 53 p.

Roehl, P.O., 1967. Stony Mountain Ordovician) and Interlake (Silurian), facies analogs of recent low-energy marine and subaerial carbonates, Bahamas. American Association of Petroleum Geologists, Bulletin, v. 51, p. 1979-2032.

Trettin, H.P., 1975. Investigations of Lower Paleozoic geology, Foxe Basin, Northeastern Melville Peninsula, and parts of northeastern and Central Baffin Island. Geological Survey of Canada, Bulletin 251, 177 p.

Walker, K.R., 1972. Community ecology of the Middle Ordovician Black River Group of New York State. Geological Society of America, Bulletin, v. 83, p. 2499-2524.

2) Grainy Sequences

Lohmann, K.C., 1976. Lower Dresbachian (Upper Cambrian) platform to deep-shelf transition in Eastern Nevada and Western Utah: an evaluation through lithologic cycle correlation. *In* Robison, R.A., and Rowell, A.J., eds., Paleontology and depositional environments: Cambrian of Western North America. Geological Studies, Brigham Young University, v. 23, p. 111-122.

Smith, D.L., 1972. Stratigraphy and Carbonate petrology of the Mississippian Lodgepole Formation in Central Montana. Summarized in J.L. Wilson, 1975, Carbonate facies in geologic history. New York, Springer-Verlag, p. 283-285.

3) Reef or Stromatolite-rich Sequences

Donaldson, J.A., 1966. Marion Lake map area, Quebec-Newfoundland. Geological Survey of Canada, Memoir 338, 85 p.

Havard, C. and Oldershaw, A., 1976. Early diagenesis in back-reef sedimentary cycles, Snipe Lake reef complex, Alberta. Bulletin of Canadian Petroleum Geology, v. 24, p. 27-70.

Hoffman, P., 1976. Environmental diversity of middle Precambrian stromatolites. *In* Walter, M.R., ed., Stromatolites. New York, Elsevier, p. 599-613.

Mountjoy, E.W., 1975. Intertidal and supratidal deposits within isolated Upper Devonian buildups, Alberta. *In* Ginsburg, R.N., ed., Tidal deposits. New York, Springer-Verlag, p. 387-397.

Read, J.F., 1973. Carbonate cycles, Pillara Formation (Devonian), Canning Basin, Western Australia. Bulletin of Canadian Petroleum Geology, v. 21, p. 38-51.

Wong, P.K. and Oldershaw, A.E., 1980. Causes of cyclicity in reef interior sediments, Kabob Reef, Alberta. Bulletin of Canadian Petroleum Geology, v. 28, p. 411-425.

4) Carbonate-Evaporite Sequences

Fuller, J.G.C.M., and Porter, J.W., 1969. Evaporite formations with petroleum reservoirs in Devonian and Mississippian of Alberta, Saskatchewan and North Dakota. American Association of Petroleum Geologists, Bulletin, v. 53, p. 909-927.

Meissner, F.F., 1972. Cyclic sedimentation in Middle Permian strata of the Permian Basin, West Texas and New Mexico. *In* Elam, J.C., and Chuber, S., eds., Cyclic sedimentation in the Permian Basin, Second Edition. West Texas Geological Society, p. 203-232.

Schenk, P.E., 1969. Carbonate-sulfate-redbed facies and cyclic sedimentation of the Windsorian Stage (Middle Carboniferous), Maritime Provinces. Canadian Journal of Earth Sciences, v. 6, p. 1037-1066.

Shearman, D.J., and Fuller, J.G., 1969. Anhydrite diagenesis, calcitization, and organic laminites, Winnipegosis Formation, Middle Devonian, Saskatchewan. Bulletin of Canadian Petroleum Geology, v. 17, p. 496-525.
For an alternate interpretation see Wardlaw, N.C., and Reinson, G.E., 1971, American Association of Petroleum Geologists, Bulletin, v. 55, p. 1759-1787.

Wilson, J.L., 1967. Carbonate-evaporite cycles in lower Duperow Formation of Williston Basin. Bulletin of Canadian Petroleum Geology, v. 15, p. 230-312.

Wood, G.V., and Wolfe, M.J., 1969. Sabkha cycles in the Arab-Darb Formation of the Trucial Coast of Arabia. Sedimentology, v. 12, p. 165-191.

HIGH-ENERGY INTERTIDAL

There appear to be few well described beach sequences in carbonate successions.

Inden, R.F., 1974. Lithofacies and depositional model for a Trinity Cretaceous sequence. *In* Perkins, B.F., ed., Geoscience and man, Vol. VIII. Aspects of Trinity Geology. Baton Rouge, LA., Louisiana State University, p. 37-53.

Inden, R.F. and Moore, C.H., 1983. Beach. *In* Scholle, P.A., Bebout, D.G., and Moore, C.H., eds., Carbonate depositional environments. American Association of Petroleum Geologists, Memoir No. 33, p. 211-267.

Purser, B.H., 1972. Subdivision et interpretation des sequences carbonates. Mémoir Bureau Rechearch Géologie et Mineralogie, v. 77, p. 679-698.

CYCLICITY

Anderson, E.J. and Goodwin, P.W., 1980. Application of the PAC hypothesis to limestones of the Helderberg Group. Society of Economic Paleontologists and Mineralogists, Guidebook, Eastern Section, 31 p.

Ginsburg, R.N., 1971. Landward movement of carbonate mud: new model for regressive cycles in carbonates (Abstract). American Association of Petroleum Geologists, Bulletin, v. 55, p. 340.

Ginsburg, R.N., 1982. Actualistic depositional models for the Great American Bank (Cambro-Ordovician). International Association of Sedimentologists, 11th International Congress on Sedimentology, Hamilton, Canada, Abstracts, p. 114.

Matti, J.C. and McKee, E.D., 1976. Stable eustacy, regional subsidence and a carbonate factory: self-generating model for onlap-offlap cycles in shallow water carbonate sequences (Abstract). Geological Society of America, Abstracts with Programs, v. 8, p. 1000-1001.

Mossop, G.D., 1979. The evaporites of the Ordovician Baumann Fiord Formation, Ellesmere Island, Arctic Canada. Geological Survey of Canada, Bulletin 298, 52 p.

Pratt, B.R. and James, N.P., in press. The St. George Group (Lower Ordovician) of Western Newfoundland: tidal flat island model for carbonate sediments in shallow epeiric seas. American Association of Petroleum Geologists, Bulletin.

Schwarzacher, W., and Fischer, A.G., 1982. Limestone-shale bedding and perturbations of the earth's orbit. *In* Einsele, G., and Seilacher, A., eds., Cyclic and event stratification. New York, Springer-Verlag, p. 72-96.

Wilkinson, B.R., 1982. Cyclic cratonic carbonates and phanerozoic calcite seas. Journal of Geological Education, v. 30, p. 189-203.

OTHER REFERENCES CITED IN TEXT

Aitken, J.D., 1967. Classification and environmental significance of cryptalgal limestones and dolomites with illustrations from the Cambrian and Ordovician of southwestern Alberta. Journal of Sedimentary Petrology, v. 37, p. 1163-1178.

Bauld, J., 1981. Geobiological role of cyanobacterial mats in sedimentary environments: production and preservation of organic matter. Bureau of Mineral Resources, Journal of Australian Geology and Geophysics, v. 6, p. 307-317.

Dunham, R.J., 1969. Early vadose silt in Townsend mound (reef), New Mexico. *In* Friedman, G.M., ed., Depositional environments in carbonate rocks: a symposium. Society of Economic Paleotologists and Mineralogists Special Publication 14, p. 139-181.

Ginsburg, R.N. and Hardie, L.A., 1975. Tidal and storm deposits, northwest Andros Island, Bahamas. *In* Ginsburg, R.N., ed., Tidal deposits. New York, Springer-Verlag, p. 201-208.

James, N.P., 1972. Holocene and Pleistocene calcareous crust (caliche) profiles; criteria for subaerial exposure. Journal of Sedimentary Petrology, v. 42, p. 817-836.

James, N.P., 1979. Facies models 10. Shallowing-upwards sequences in carbonates. *In* Walker, R.G. ed., Facies models. Geological Association of Canada, Geoscience Canada Reprint Series 1, p. 109-119.

Playford, P.E. and Cockbain, A.E., 1976. Modern algal stomatolites at Hamelin Pool, a hypersaline barred basin in Shark Bay, Western Australia. *In* Walter, M.R., ed., Stromatolites, developments in sedimentology. New York, Elsevier, p. 389-413.

Pratt, B.R., 1982. Stromatolite decline – a reconsideration. Geology, v. 10, p. 512-516.

Schlager, W., 1981. The paradox of drowned reefs and carbonate platforms. Geological Society of America, Bulletin, v. 92, p. 197-211.

Sepkowski, J.J., 1981. Flat-pebble conglomerates, storm deposits and the Cambrian bottom fauna. *In* Einsele, G., and Seilacher, A., eds., Cyclic and event stratification. New York, Springer-Verlag, p. 371-385. 371-385.

Shinn, E.A., 1983b. Birdseyes, fenestrae, shrinkage pores and loferites: a reevaluation. Journal of Sedimentary Petrology, v. 53, p. 619-628.

Read, J.F., 1976. Calcretes and their distinction from stromatolites. *In* Walter, M.R. ed., Stromatolites, developments in sedimentology. New York, Elsevier, p. 55-71.

Reproduced with the permission of the Geological Association of Canada. Published in *Facies Models,* R.G. Walker, editor, *Geoscience Canada,* Reprint Series No. 1, second edition, 1984, pp. 229-244.

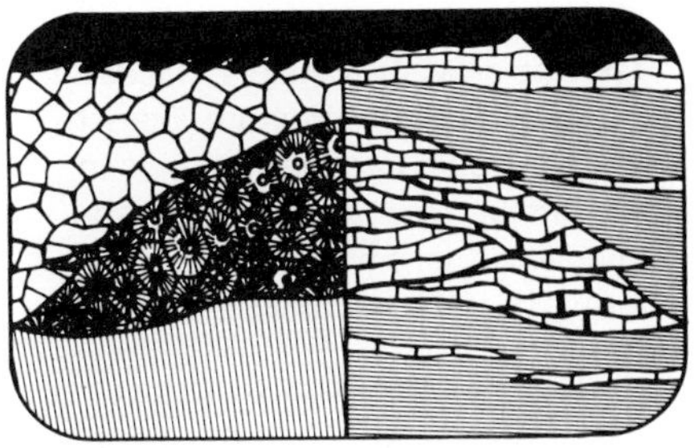

Reefs

NOEL P. JAMES
Department of Earth Sciences
Memorial University of Newfoundland
St. John's, Newfoundland A1B 3X5

Figure 1
A patch reef of Lower Cambrian age exposed in sea cliffs along the northern shore of the Strait of Belle Isle, Southern Labrador.

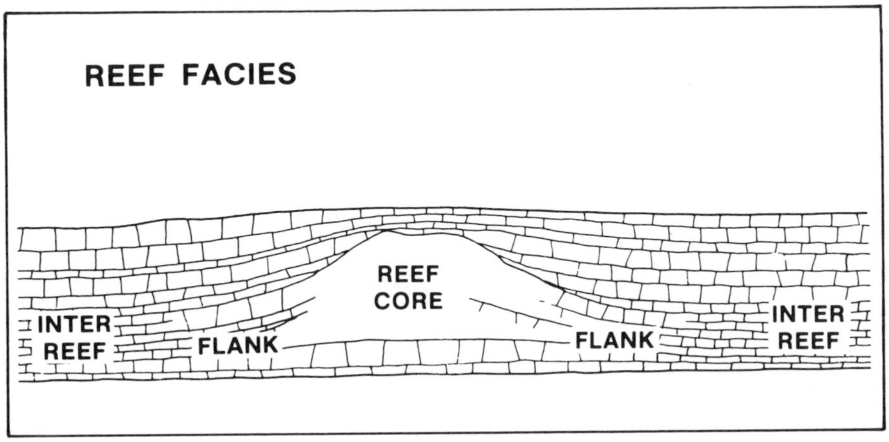

Figure 2
A sketch illustrating the three major reef facies in cross-section.

INTRODUCTION

A reef, rising above the sea floor, is an entity of its own making - a sedimentary system within itself. The numerous, large calcium carbonate secreting organisms stand upon the remains of their ancestors and are surrounded and often buried by the skeletal remains of the many small organisms that once lived on, beneath, and between them.

Because they are built by organisms, fossil reefs (Fig. 1) are storehouses of paleontological information and modern reefs are natural laboratories for the study of benthic marine ecology. Also, fossil reefs buried in the subsurface contain a disproportionately large amount of our oil and gas reserves compared to other types of sedimentary deposits. For these reasons, reefs have been studied in detail by paleontologists and sedimentologists, perhaps more intensely than any other single sedimentary deposit, yet from two very different viewpoints. This paper is an integration of these two viewpoints. I shall concentrate less on the familiar trinity of back-reef, reef, and fore-reef, but more on the complex facies of the reef proper.

Since the first edition of Facies Models, there has been much new information on both the sedimentology and paleontology of reefs. The model itself has been presented elsewhere (James, 1983) and amplified using numerous examples from the modern and fossil record. In this present version the model remains unchanged but many of the underlying concepts and implications that flow from it have been revised and/or enlarged.

THE ORGANISM-SEDIMENT MOSAIC

Reefs can generally be subdivided into three facies (Fig. 2).

1) *Reef-core facies* - massive, unbedded, frequently nodular and lenticular carbonate comprising skeletons of reef-building organisms and a matrix of lime mud.

2) *Reef-flank facies* - bedded lime conglomerates and lime sands of reef-derived material, dipping and thinning away from the core.

3) *Inter-reef facies* - normal shallow-water, subtidal limestone, unrelated to reef formation, or fine-grained siliciclastic sediments.

A useful, non-generic term for such a structure is "bioherm" - for discussion of this and other reef terminology the interested reader is referred to papers by Dunham (1970), Heckel (1974), Longman (1981), and James (1983).

Reef facies are best differentiated on the basis of several independent criteria including: 1) the relationship between, and relative abundance of large skeletons and sediments, i.e., the type of reef limestone, 2) the diversity of reef-building species, and 3) the growth form of the reef builders.

Types of Reef Limestone

The present state of any thriving reef is a delicate balance between the upward

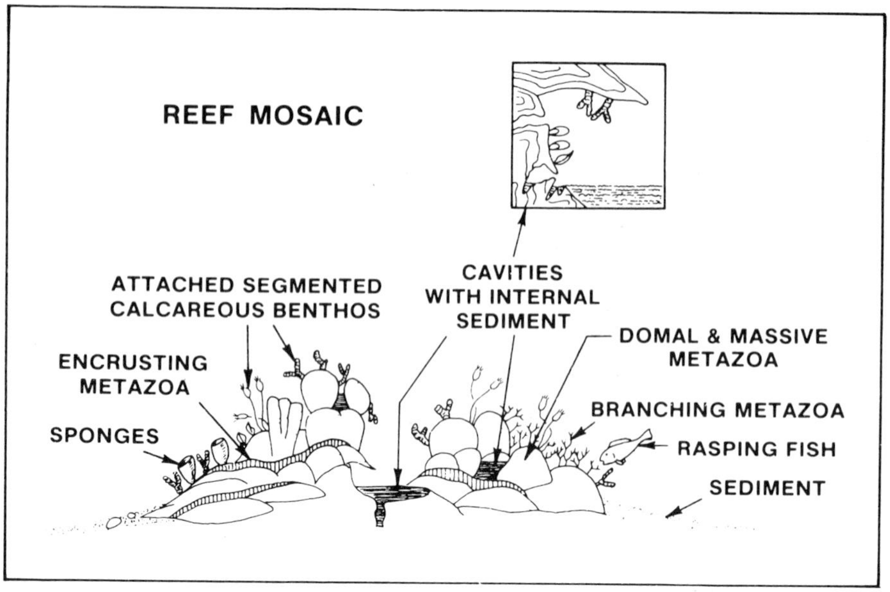

Figure 3
A sketch illustrating the different aspects of the organism/sediment mosaic that comprises a reef.

growth of large skeletal metazoans, the continuing destruction of these same organisms by a host of rasping, boring, and grazing organisms, and the prolific sediment production by rapidly growing, shortlived, attached calcareous benthos (Fig. 3).

The large skeletal metazoans (e.g., corals) generally remain in place after death, except when they are so weakened by bio-eroders that they are toppled by storms. The irregular shape and growth habit of these reef-builders result in the formation of roofed-over cavities inside the reef that may be inhabited by smaller, attached calcareous benthos. These cavities may also be partly or completely filled with fine-grained "internal" sediment. Encrusting organisms grow over dead surfaces and aid in stabilizing the structure. Branching reef-builders frequently remain in place, but just as commonly are fragmented into sticks and rods by storms to form skeletal conglomerates around the reef.

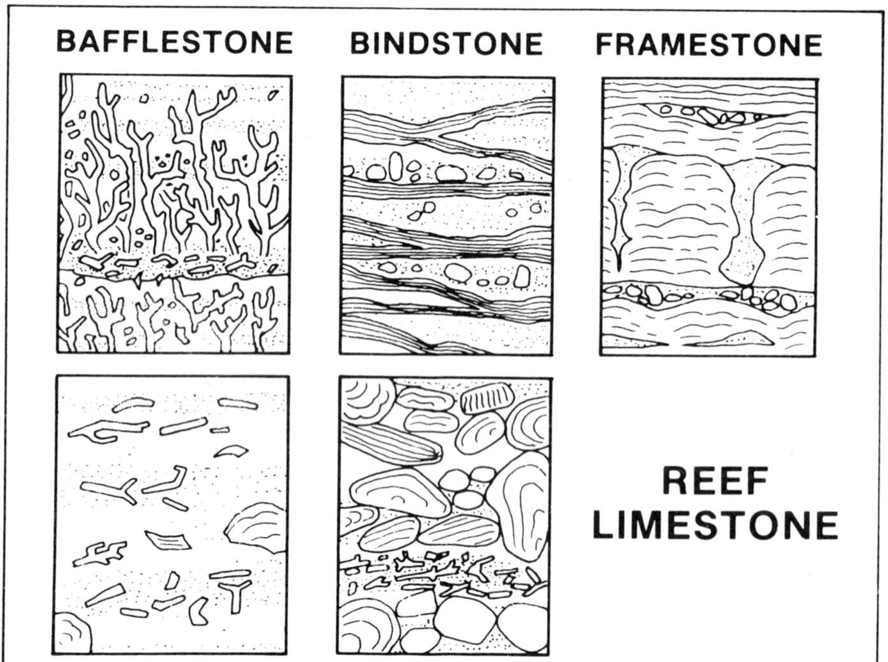

Figure 4
An interpretative sketch of the different types of reef limestone recognized by Embry and Klovan (1971). Autochthonous reef limestones are in the upper row, while allochthonous reef sediments are in the lower row.

Most reef sediment is produced by the post-mortem disintegration of organisms that are segmented (crinoids, calcareous green algae) or nonsegmented (bivalves, brachiopods, foraminifers). These organisms grow in the many nooks and crannies between larger skeletal metazoa. The remainder of the sediment is produced by various taxa that erode the reef: boring organisms (worms, sponges, bivalves) produce lime mud and rasping organisms that graze the surface of the reef (echinoids, fish) produce copious quantities of carbonate sand, and silt. This material is deposited around the reefs as an apron of sediment, and it also filters into the growth cavities to form internal sediment, which is characteristically geopetal.

Many different classifications have been proposed for the resulting reef carbonates, but the most descriptive and widely accepted is a modification of Dunham's (1970) classification of lime sand mud-rocks proposed by Embry and Klovan (1971) (Fig. 4). They recognize two kinds of reef limestone, allochthonous and autochthonous. The allochthonous limestones are the same as the finer grained sediments, but with two categories added to encompass large particles. If more than 10% of the particles in the rock are larger than 2 mm and they are matrix supported it is a *Floatstone*; if the rock is clast supported it is a *Rudstone*. The autochthonous limestones are more interpretative: *Framestones* contain in-place, massive fossils that formed the supporting framework; *Bindstones* contain in-place, tabular, or lamellar fossils that encrusted or bound the sediment together during deposition; *Bafflestones* contain in-place, stalked fossils that trapped sediment by baffling.

Many reefs also appear to be preferential sites for precipitation of synsedimentary cement (James and Choquette, 1983), and are hard limestone just below the growing surface. The abundance of early cement in many fossil reefs has led some workers to view these buildups as "cementation reefs" rather than biological - sedimentological structures.

Diversity Amongst Reef-Building Metazoans

Relative abundance of different organisms is one of the easiest observations that can be made on a fossil reef, and so potentially one of the most useful. Although intuitively there should be a simple relationship between diversity and environment, recent thinking suggests that this relationship is complex and not straightforward. It seems that diverse faunas (Fig: 5) probably develop when conditions for growth are optimum, i.e., nutrients are in adequate supply and daily chemical and physical stresses are low, *but* when the community is not able to reach competitive equilibrium because of frequent population reduction. If not disturbed, a community will reach competitive equilibrium where a few species dominate, i.e., low diversity. The relative diversity in such settings is probably a complex interplay between frequency of disturbance, population growth rate and nutrient supply. Many of the above concepts have been derived from the study of coral reefs (Connell, 1978) where high diversity is the result of growth in an environment which is relatively nutrient-poor and subject to periodic catastrophic disturbance by tropical storms. (Woodley *et al.*, 1982). An implication of this concept is that we should expect patchiness and evidence of extensive fragmentation and debris formation in the most diverse of fossil reef communities.

Low diversity reef communities are of at least 3 types: 1) those at competitive equilibrium; 2) new communities (those that have moved into a new environment; and 3) those communities subject to severe and continuing chemical and physical stress. Among the factors most likely to stress modern and fossil reef-building communities are: 1) temperature and salinity fluctuations - most modern and likely most ancient reef-builders grow or grew best in tropical sea water of normal salinity; 2) intense waves and swell - the skeletons of most reef-builders will be broken or toppled by strong wave surge: 3) low light penetration - in modern reef-building organisms rapid calcification takes place because light dependent symbionts take over some of the bodily functions of the host; and 4) heavy sedimentation - all reef-builders are sedentary filter-feeders or micro-predators and water filled with fine-grained sediments would clog the feeding apparatus.

Figure 5
A shallow-water (1 m deep) living reef, composed of branching (left), foliose (centre) and hemispherical (centre) corals, off Gouldin Cay, Bahamas.

The Growth Form of Reef-Building Metazoans

The relationship between organism shape and environment is one of the oldest and most controversial topics in biology and paleobiology. In terms of reef-building metazoans, observations from the rock record (Figs. 6 and 7) of the interrelationship between organisms and surrounding sediments, combined with studies of modern coral distribution on tropical reefs, allow us to make some generalizations about form and environment that are useful in reef facies analysis (Fig. 8).

The limitations of applying these concepts directly to fossil reefs has recently been emphasized by Stearn (1982), who pointed out that no general patterns are applicable to all reefs, and that variations in shape are the result of the interaction between environmental factors with the genetically dicatated growth pattern of the organism. Observations on growth form must be used in conjunction with other parameters, and are most useful in providing additional information when dealing with low diversity communities.

THE SPECTRUM OF REEF TYPES

Reefs can develop just about anywhere in the carbonate facies spectrum. As isolated structures they are dispersed across shallow carbonate platforms but they also grow, outpacing subsidence, in slope and basinal settings. As more contiguous elements they commonly form long, sinuous barriers along the margins of the same platforms, close to land along the edge of narrow shelves or as halos around positive structural elements.

Modern reefs are best developed and most successful on the windward sides of shelves, islands, platforms and atolls where wind and swell are consistent and onshore. The asymmetry of many ancient reefs and distribution of sediment facies suggests that this was so in the past as well. The reason for the preferential development of reefs on the windward side is by no means established but sedimentation is likely the most important. Shallow water reef-building species characteristically produce abundant fine sediment, yet the major reef-builders, because they are filter feeders and micropredators, are

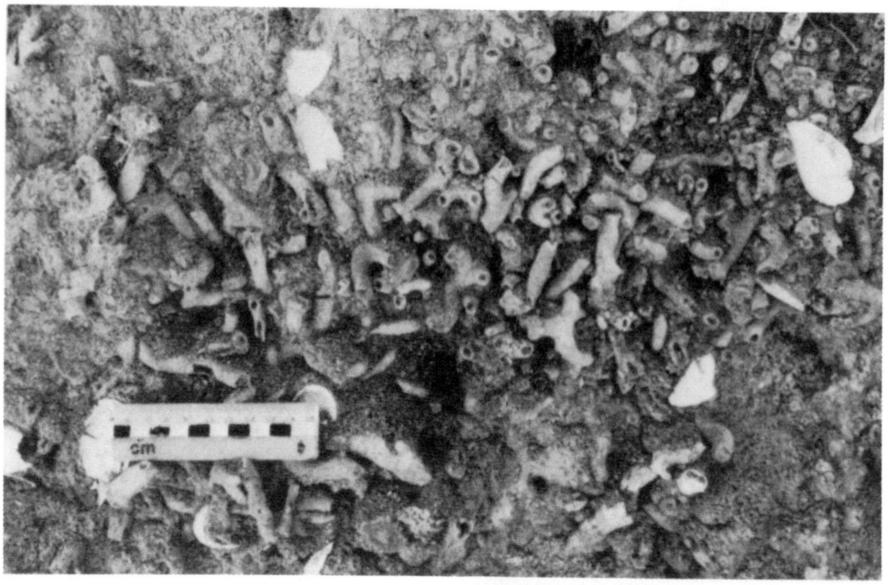

Figure 6
An accumulation of branching corals (Porites porites) *and bivalves in a late Pleistocene reef, Barbados, W.I.*

Figure 7
A small patch of domal shaped corals (Diploria *sp. in cross-section on a cliff exposure of Late Pleistocene reef limestone, Barbados, W.I.*

GROWTH FORM AND ENVIRONMENT OF REEF BUILDING SKELETAL METAZOA			
GROWTH FORM		ENVIRONMENT	
		Wave Energy	Sedimentation
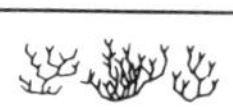	Delicate, branching	low	high
	Thin, delicate, plate-like	low	low
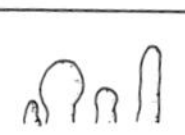	Globular, bulbous, columnar	moderate	high
	Robust, dendroid, branching	mod-high	moderate
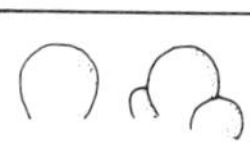	Hemispherical, domal irregular, massive	mod-high	low
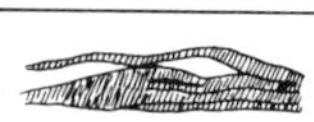	Encrusting	intense	low
	Tabular	moderate	low

Figure 8
The growth form of reef-building metazoans and the types of environments in which they most commonly occur. From James (1983).

intolerant of fine sediment. The open ocean and windward locations are the only places in which fine sediment is continuously swept away.

Reefs form a natural breakwater when they grow into the zone of onshore waves and swells and create a relatively quiet environment in the lee of the reef crest. Commonly, this restriction significantly changes water circulation on the shelf, platform, or lagoon behind the reef. In such a marginal location, the symmetrical reef facies model comprising a reef-core facies surrounded on all sides by reef-flank facies is no longer discernable. Instead facies are more asymmetrically distributed with the reef-core facies flanked on the windward side by the fore-reef facies and on the leeward side by the platform facies (often called the back-reef facies).

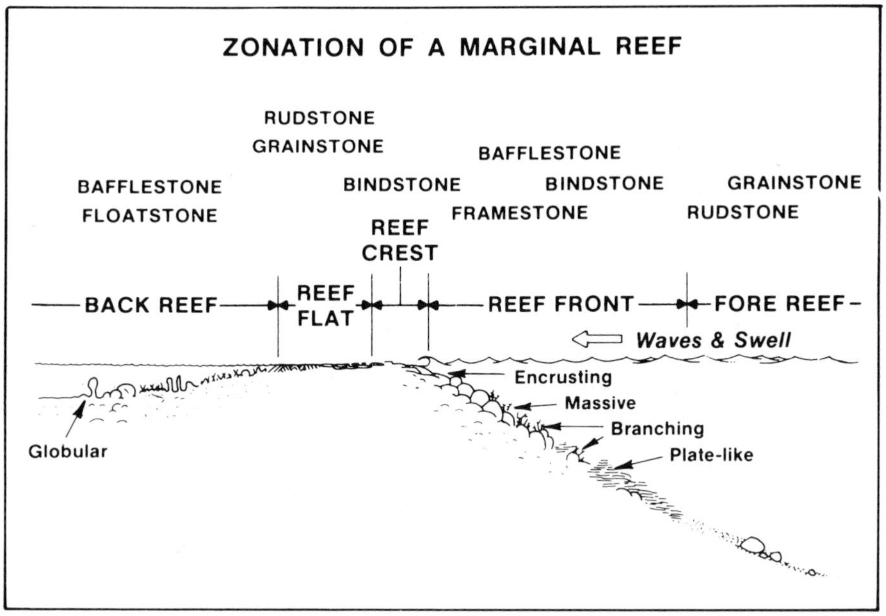

Figure 9
Cross-section through a hypothetical, zoned, marginal reef illustrating the different reef zones, spectrum of different limestones produced in each zone, and environment of different reef-building forms.

High-Energy Reefs

In these high energy settings the reef is distinctly zoned (Fig. 9)

Reef Crest Zone. This is the highest part of the reef at any stage in its growth, and if in shallow water, it is that part of the reef top that receives most of the wind and wave energy. Composition of the reef crest depends upon the degree of wind strength and swell. In areas where wind and swell are intense, only those organisms that can encrust, generally in sheet-like forms, are able to survive. When wave and swell intensity are only moderate to strong, encrusting forms still dominate but are commonly also bladed or possess short, stubby branches. In localities where wave energy is moderate, hemispherical to massive forms occur, with scattered clumps of branching reef-builders. The community is still of low diversity. The lithologies formed in these three cases would range from bindstones to framestones.

Reef Front Zone. This zone extends from the surf zone to an indeterminate depth, commonly less than 100 metres, where the zone of abundant skeletal growth grades into sediments of the fore-reef zone. Direct analogy between modern reefs, especially Caribbean reefs, and ancient reefs is difficult because today the sea floor from the surf zone to a depth of about 12 metres is commonly dominated by the robust branching form *Acropora palmata*, a species which developed only recently in the late Cenozoic. Such branching forms are rarely found in ancient reefs. Instead, the most abundant forms are massive, laminar to hemispherical skeletons, forming framestones and sometimes bindstones.

The main part of this zone supports a diverse fauna with reef-builders ranging in shape from hemispherical to branching to columnar to dendroid to sheet-like. Accessory organisms and various niche dwellers such as brachipods, bivalves, coralline algae, crinoids, and green segmented calcareous algae (Halimeda), are common. On modern reefs where the reef-builders are corals, this zone commonly extends to a depth of 30 metres or so. The most common rock type formed in this zone would still be framestone, but the variety of growth forms also leads to the formation of many bindstones and bafflestones.

Below about 30 metres wave intensity is almost non-existent and light is attenuated. The response of many reef-building metazoans is to increase their surface area, by having only a small basal attachment and a large but delicate plate-like shape. Rock types from this zone look like bindstones, but binding plays no role in the formation of these rocks and perhaps another term is needed.

The deepest zone of growth of coral and green calcareous algae on modern coral reefs is about 70 metres. The lower limit may depend upon many factors, perhaps one of the most important being sedimentation, especially in shale basins which border many reefs. This lower limit should therefore be used with caution in the interpretation of fossil reefs.

Sediments on the reef front are of two types: 1) internal sediments within the reef structure, generally lime mud giving the rocks a lime mudstone to wackestone matrix, and 2) coarse sands and gravels in channels running seaward between the reefs. These latter deposits have rarely been recognized in ancient reefs.

As a result of numerous observations on modern reefs it appears that most of the sediment generated on the upper part of the reef front and on the reef crest is transported episodically by storms up and over the top and accumulates in the lee of the reef crest. Sediments on the intermediate and lower regions of the reef front, however, are transported down to the fore-reef zone only when it is channelled by way of passes through the reef.

Reef Flat Zone. The reef flat varies from a pavement of cemented, large skeletal debris with scattered rubble and coralline algae nodules in areas of intense waves and swell, to shoals of well-washed lime sand in areas of moderate wave energy. Sand shoals may also be present in the lee of the reef pavement. Vagaries of wave refraction may sweep the sands into cays and islands. These obstructions in turn create small protected environments very near the reef crest. Water over this zone is shallow (only a few metres deep at most) and scattered clumps of reef-building metazoans are common. The resulting rock types range from clean skeletal lime grainstones to rudstones.

Back Reef Zone. In the lee of the reef flat, conditions are relatively tranquil and much of the mud formed on the reef front comes out of suspension. This, coupled with the prolific growth of mud and sand-producing bottom fauna such as crinoids, calcareous green algae, brachiopods, and ostracodes, commonly results in mud-rich lithologies. The two most common growth habits of reef-builders in these environments are stubby, dendroid forms, often bushy and knobby, and/or large globular forms that extend above the substrate to withstand both frequent agitation and quiet muddy periods.

The rock types characteristic of this environment are bafflestones or floatstones to occasional framestones with a skeletal wackestone to packstone matrix. In some reefs there are beds of nothing but dis-articulated branches in lime mud (e.g., *Amphipora* limestones of the Upper Devonian), but there is little evidence of much transport.

Fore-Reef Facies

This facies consists of thin to thick and massively bedded skeletal lime grainstones to lime packstones which are composed of whole or fragmented skeletal debris, blocks of reef limestones and skeletons of reef-builders. These grade basinward into shales or lime muds. In contrast to the reef facies, the beds are rarely dolomitized.

It should be remembered that this high-energy zonation, although most commonly observed on platform margin reefs, is also developed in on-platform isolated reefs and in reefs from slope or basinal settings that rise into the zone of breaking waves.

Low Energy Reefs

As wave energy decreases, so the style of reef growth changes; distinctive zonation becomes less noticable, shape of the reef building organisms is different and relative diversity decreases. The relationship between wave energy and reef type has been summarized for modern reefs by Geister (1980) and for fossil reefs by Wilson (1975) and is shown in Figure 10. Lower energy situations are also characterized by sluggish water exchange and so these changes may be amplified by variations in water salinity and nutrient content.

As a result most isolated reefs or patch reefs on carbonate platforms are poorly zoned and more like the ideal reef in Figure 1. They are generally circular to elliptical to irregular in plan and may be large enough to enclose a lagoon themsleves. Each reef is zoned with respect to depth, similar to the reef front in higher energy reefs.

Finally, in some settings, reefs as described above do not occur. These settings include the very inner parts of shallow platforms, the deeper (many 10s of metres) lagoons, and those platforms or parts of platforms covered by water of elevated or depressed salinity. There are, however, accumulations of carbonate sand and/or mud, built by organisms that are very close to reefs in composition. In modern seas these range from sea-grass banks to skeletal sand shoals, often bound by algae, to broad banks of corals and algae. Although not common today, these structures are an important part of the fossil carbonate spectrum and must be integrated into the reef model.

Reef Mounds

Many Phanerozoic carbonate sequences contain structures that some workers call reefs, some call mounds, and some call banks. They lack many of the characteristics we ascribe to reefs yet were clearly rich in skeletal organisms and had relief above the sea floor. The origin of these structures, which I have called *reef mounds* (Fig. 11), has probably caused more discussion than any other topic in the literature on reefs (Heckel, 1974).

Reef mounds are, as the name suggests, flat lenses to steep conical piles with slopes of up to 40° consisting of poorly sorted bioclastic lime mud with minor amounts of organic boundstone. With this composition they clearly formed in quiet water environments and from the rock record appear to occur in preferred locations: 1) arranged just downslope on gently-dipping platform margins (Fig. 12); 2) in deep basins; and 3) spread widely in tranquil reef lagoons or wide shelf areas. When viewed in section, reef mounds display a similar facies sequence in each case (Wilson, 1975) (Fig. 11).

Stage 1. Basal bioclastic lime mudstone to wackestone pile - muddy sediment with much bioclastic debris but no baf-

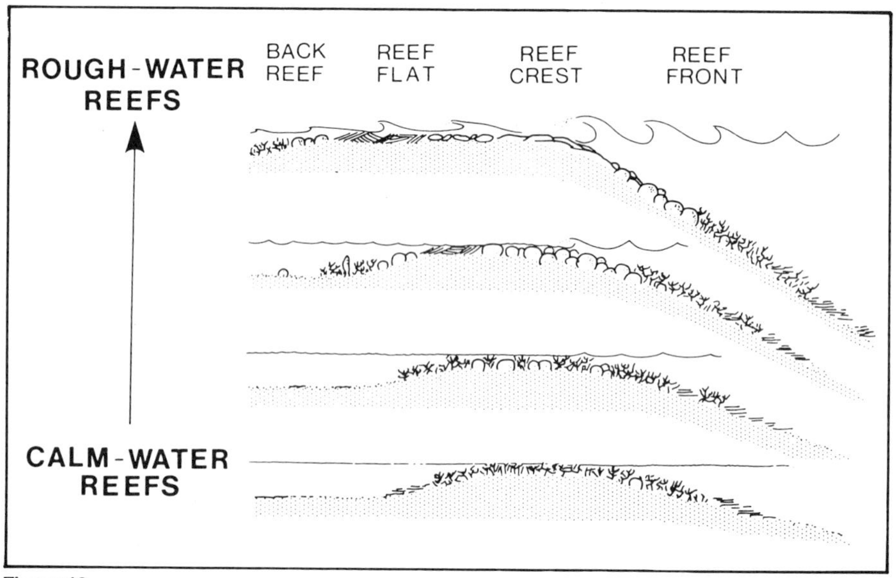

Figure 10
Generalized diagram of the different zonations expected from reefs growing under conditions ranging from calm water to rough water; see Figure 8 for growth forms.

fling or binding organisms.

Stage 2. Lime mudstone or bafflestone core - the thickest part of the mound, consisting of delicate to dendroid forms with upright growth habits in a lime mudstone matrix. The limestone is frequently brecciated, suggesting partial early lithification, dewatering and slumping, and contains stromatactis. Each geologic age has its own special fauna that forms this stage: Lower Cambrian - archaeocyathans; Middle to Lower Ordovician - sponges and algae; Middle Ordovician, Late Ordovician, Silurian, Early Carboniferous (Mississippian) - bryozoans; Late Carboniferous (Pennsylvanian) and Early Permian - platy algae; Late Permian to Middle Triassic - sponges and algae; Late Triassic - large fasciculate dendroid corals; Late Jurassic - lithistid sponges; Cretaceous - rudist bivalves.

Stage 3. Mound cap - a thin layer of encrusting or lamellar forms, occasional domal or hemispherical forms, or winnowed lime sands.

The massive, commonly well-bedded carbonates that flank the reef mounds comprise extensive accumulations of debris and chunks of archaeocyathans, pelmatozoans, fenestrate bryozoans, small rudists, dendroid corals, stromatoporoids, branching red algae or tabular foraminifers along with wholly to partly lithified lime mudstone. Volumetrically these flank beds may be greater than the core itself and almost bury it (Fig. 13).

Carbonate Mud Mounds

One end member of the reef mound category that deserves special attention is a group of structures called either Waulsortian mounds (from the name of a village in Belgium) or more commonly, carbonate mud mounds. These puzzling structures may be just as large and have sides just as steep as reef mounds, but they possess *no* large skeletons. They are made up only of crinoid fragments, sponge spicules or scattered bryozoans that together make up no more than 1/4 of the rock, the rest being lime mud (for a recent review see Pratt, 1982). These mud mounds seem to

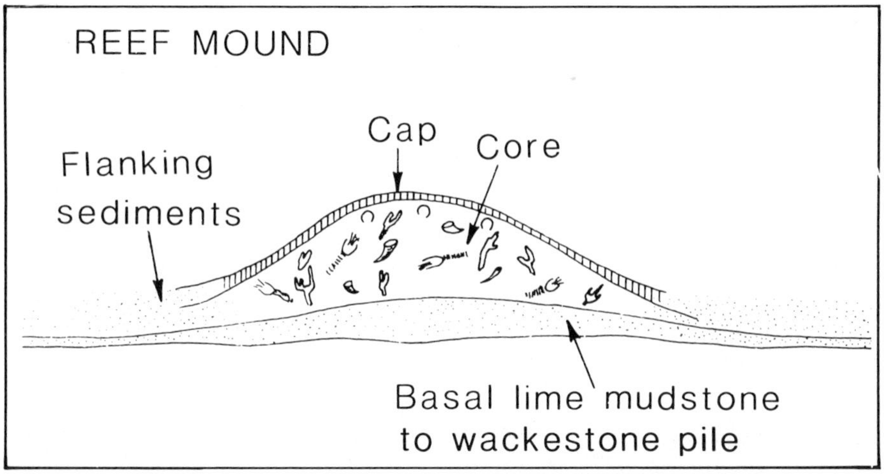

Figure 11
Cross-section through a hypothetical reef mound illustrating the geometry of the different facies.

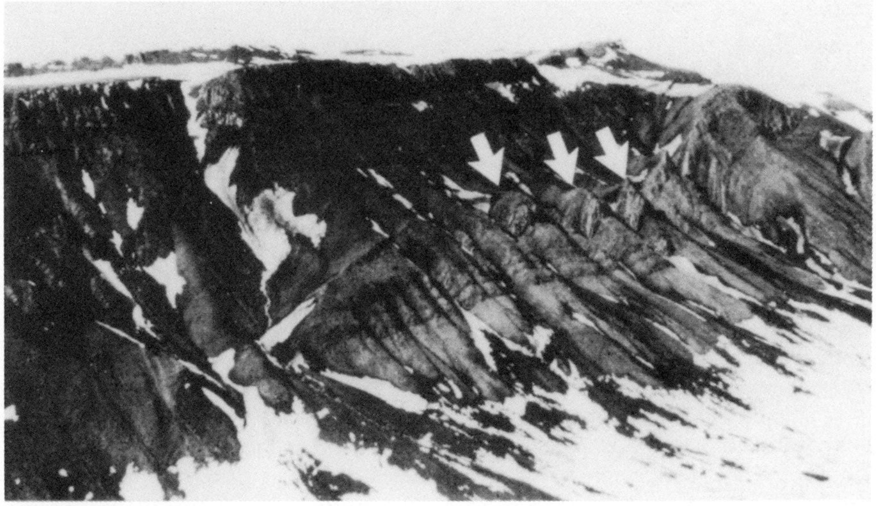

Figure 12
Massive reef limestone (right) of the Nansen Formation (Permo-Pennsylvanian) extending downward and basinward into dark, argillaceous limestones of the Hare Fiord Formation (left). Arrows point out small reef mounds developed on the seaward slopes of the reef front, western side of Blind Fiord, Ellesmere Island, N.W.T.

Figure 13 ▶
Muleshoe bioherm, a 60 m high reef mound of Late Missippian age exposed along the western escarpment of the Scaremento Mountins, New Mexico.

Figure 14
Stromatactis, *sub-horizontal layers of calcite spar illustrating irregular digitate tops and smooth bottoms, from Gros Morbe, reef-mound facies, Silurian West Point reef complex, Gaspé, Quebec.*

occur almost exclusively in deep water on carbonate slopes, and are not as common as on-shelf buildups.

Almost all workers have commented on the striking similarity of these structures regardless of their age, which suggets a common origin for all of them. There are, however, no obvious modern analogues. Structures that come closest are seagrass-stabilized, shallow-water mudbanks or deep-water lithoherms, but both also have considerable differences. The localization of the mud is probably some combination of *in situ* production, baffling, and binding by a variety of organisms, among which algae and cyanobacteria are the most important. The seafloor topography is clearly aided by rapid seafloor lithification.

Central to their genesis is the presence of *"Stromatictis"* (Fig. 14) which consists of masses of calcite-filled cavities which have a digitate roof and a flat floor commonly formed by geopetal sediment. This spar occurs in swarms and has a reticulate distribution (see Bathurst, 1982 for an extended discussion). It is clear that cavity development is penecontemporaneous with deposition, but its origin is still a mystery. Current hypotheses include: 1) dewatering, 2) compaction, slumping, and down slope creep of consolidated but unlithified sediment, 3) cementation and/or binding of sediment by algae to form crusts followed by erosion, and 4) decay and collapse of organic tissue (especially sponges) after partial lithification.

A final unsettling fact is that these mud mounds are almost entirely a Paleozoic phenomenon. They range in age from middle Cambrian to late Jurassic but confirmed stromatactis is found only in Paleozoic buildups. There is no clear answer as to why they do not occur in late Mesozoic and Cenozoic carbonates.

Stromatolite Reefs

During the Precambrian and earliest Paleozoic, prior to the appearance of herbivorous metazoans, stromatolites formed impressive build-ups (Fig. 15). These stromatolite complexes clearly had relief above the sea floor and in terms of morphology were surprizingly similar to later skeletal reefs. Mostly developed in shelf margin settings some exhibit excellent lateral zonation (Hoffman, 1974) while others clearly developed in a series of well-defined stages in which stromatolites had different growth forms (Cecile and Campbell, 1978). Although there has been much study of Precambrian stromatolites, much less attention has been paid to their role in the formation of reefs or reef mounds.

STAGES OF REEF GROWTH

While the composition of the reef core and the different facies can be determined from both modern and ancient examples, information as to the stages of reef growth can only come from the rock record.

It has long been recognized that there is an ecological succession in many Paleozoic reefs (Lowenstam, 1950), i.e., the replacement of one community of reef-building organisms by another as the reef grew. A synthesis by Walker and Alberstadt (1975) of reefs ranging in age from Early Ordocician to Late Cretaceous suggests that a similar community succession is present in reefs throughout the Paleozoic and Mesozoic. Application of this concept to Cenozoic reefs (Frost, 1977), which are dominated by scleractinian corals (the reef-builders in today's oceans), allows us to equate ancient reef community succession with observations on modern reef communities with some measure of confidence.

The reason for this ecologic succession is at present a topic of much debate. Some workers feel that the con-

Figure 15
A large stromatolite mound of Proterozoic age Kuuik Formation, Kilohigok Basin, Northwest Territories. Photograph by M. Cecile.

trol is extrinsic, reflecting a progressive replacement of deep-water communities by shallower water ones as the reef grows to sea level and into more turbulent water; however, there is often abundant evidence that the first two stages are developed in shallow water. Other workers feel that the control is intrinsic, reflecting a natural succession as the organisms gradually alter the substratum and change the energy flow pathways as the community develops; nevertheless, there is abundant evidence of increasing water turbulence as the structure grows.

THE MODEL

The reef model is an integration of information from the modern and the fossil record. The major stumbling block that makes the generation of a model difficult, the ever-changing character of the reef-building fauna and flora through time, is resolvable if the concept of community succession is used as a basis. In the model four separate stages of reef growth are recognized, and these stages, along with the types of limestone, relative diversity of organisms and growth-form of reef-builders in each, are summarized in Figure 16.

1) *Pioneer (Stabilization) Stage.* This first stage is most commonly represented by a series of shoals or other accumulations of skeletal lime sand composed of pelmatozoan or echinoderm debris in the Paleozoic and Mesozoic, and plates of calcareous green algae in the Cenozoic. The surfaces of these sediment piles are colonized by algae (calcareous green), plants (sea grasses) and/or animals (pelmatozoans) that send down roots or holdfasts to bind and stabilize the substrate. Once stabilized, scattered branching algae, bryozoans, corals, soft sponges and other metazoans begin to grow between the stabilizers.

2) *Colonization Stage.* This second stage is relatively thin when compared to the reef structure as a whole, and reflects the initial colonization by reef-building metazoans. The rock is generally characterized by few species, sometimes massive or lamellar forms but more commonly monospecific coppices or clumps of branching forms (Fig. 8). In Cenozoic reefs the one characteristic common to all corals in this stage of reef growth is that they are able to get

STAGES OF REEF GROWTH

STAGE	TYPE OF LIMESTONE	SPECIES DIVERSITY	SHAPE OF REEF BUILDERS
DOMINATION	bindstone to framestone	low to moderate	Laminate encrusting
DIVERSIFICATION	framestone (bindstone) mudstone to wackestone matrix	high	domal massive lamellar branching encrusting
COLONIZATION	bafflestone to floatstone (bindstone) with a mud stone to wackestone matrix	low	branching lamellar encrusting
STABILIZATION	grainstone to rudstone (packstone to wackestone)	low	skeletal debris

Figure 16
A sketch of the four divisions of the reef-core facies with a tabulation of the most common types of limestone, relative species diversity and shape of reef-builders found in each stage.

Figure 17
The diversification stage of an Upper Devonian reef, comprising domal stromatoporoids, and domal to branching tabulate corals, Blue Fiord Formation, south side of Eids Fiord, Ellesmere Island, N.W.T.

rid of sediment and clean their polyps, and so are able to grow in areas of high sedimentation. The branching growth form creates many smaller subenvironments or niches in which numerous other attached and encrusting organisms can live, forming the first stage of the reef ecosystem.

3) *Diversification Stage* This third stage, usually represented by the bulk of the reef mass, is the point at which most pronounced upward-building towards sea level occurs and easily definable, lateral facies develop. The number of major reef-building taxa is usually more than doubled, and the greatest variety in growth habit is encountered (Fig. 17). With this increase in form and diversity of framework and binding taxa comes increased nestling space, i.e., surfaces, cavities, nooks and crannies, leading to an increase in diversity of debris-producing organisms..

Once a reef reaches the diversification stage, and sometimes even earlier in the colonization stage, the structure is frequently high enough above the sur-

Figure 18
A small patch reef built by bryozoans, corals, and stromatoporoids, Long Point Formation (Middle Ordovician), Port-au-Port Peninsula, Newfoundland.

Figure 19
Reef and reef-flank deposits (R) ca. 100 m thick (Peechee Formation) of Upper Devonian age in the Flathead Range, southern Rocky Mountains, Alberta. Photograph courtesy B. Pratt.

rounding sea floor to affect water circulation and thus to alter sedimentation patterns. At this point not only are surrounding sedimentary environments altered but the reef itself develops a zonation, because its margins now reach from shallow to deep water. This zonation is, as outlined in an earlier section, dependent upon wave energy.

4) *Domination (Climax) Stage.* The change to this stage of reef growth is commonly abrupt. The most common lithology is a limestone dominated by only a few taxa with only one growth habit, generally encrusted to laminated. Most reefs show the effects of surf at this stage, in the form of beds of rudstone.

While these four stages are the norm, some reefs only display the upper two or three stages. Careful investigation, however, usually reveals that the reefs began growth on a hardground or lithified substrate.

The Complete Reef Structure

Reefs in the rock record vary widely in size. A complete Ordovician (Fig. 18) or Cretaceous reef, formed by a variety of organisms and displaying several stages of growth, may be only as large as a single coral head in a Devonian, Jurassic, or Pleistocene reef, kilometres long and hundreds of metres thick (Fig. 19).

While some large reefs display all 4 stages of development, most are internally stratified (Fig. 20) to form a series of superimposed or stacked reefs. Individual episodes of reef growth are commonly separated by periods of exposure, reflected in the rock by intensive diagenesis, calcrete horizons, karst, shales (paleosols) or by periods of non-deposition as indicated by hardgrounds, borings, or manganese and phosphate-impregnated bedding planes. When reef growth begins again after a hiatus, because the surfaces are both hard and elevated, it starts at the diversification stage, and so only two stages are present in the next package. Some workers tend to regard these unconformity-bounded layers as the different stages of reef growth which they are not. The different stages of reef growth are found as separate units between these layers.

THE MODEL AS A FRAMEWORK OR GUIDE FOR OBSERVATIONS

Reefs

The reef facies model is predicated on the assumption that a full spectrum of reef-building organisms are present. We see a full spectrum in the tropical oceans today, but this was not the case for much of the Phanerozoic. The critical element that is often missing, and without which the four stages of development in the reef core cannot occur, is the presence of skeletal metazoans that secrete large robust, branching, hemispherical or tabular skeletons. Without them the reef cannot exist in the zone of constant turbulence, usually wave induced, because smaller and more delicate forms would be broken and swept away (unless submarine cementation is very rapid, pervasive, and near-surface). This zone of turbulence is the optimum area for growth and diversity because sediment is constantly removed, water is clear, and nutrients are constantly swept past the sessile organisms. Such large skeletal metazoa were present only at certain times during the Phanerozoic (Fig. 21), and each period has its own specialized group of frame-builders: 1) Middle and Upper Ordovician - bryozoa, stromatoporoids, tabulate corals; 2) Silurian and Devonian - stromatoporoids, atabulate corals; 3) Late Triassic - corals, stromatoporoids; 4) Jurassic - corals, stromatoporoids; 5) Upper Cretaceous - rudist bivalves; and 6) Oligocene, Miocene Plio-Pleistocene - scleractinian corals. A more detailed review of these reefs during specific periods is given by James (1983).

Although reefs are found in platform margin, on-shelf, and basinal settings during these periods, reef mounds also develop, but only in environments which were inimical to the growth of larger metazoans. At many times during the Phanerozoic, however, the large frame-building metazoans were absent, and during these periods reef mounds were the only buildups in the facies spectrum (Fig. 21).

When viewed against the backdrop of the general reef facies model, I think of reef mounds as half-reefs or incomplete reefs because they represent only stages one, two, and occasionally four of the model. These structures did not develop the other upper stages either because the environment was not conducive to the growth of large skeletal metazoans or because these larger metazoans simply did not exist at the time when the structure formed.

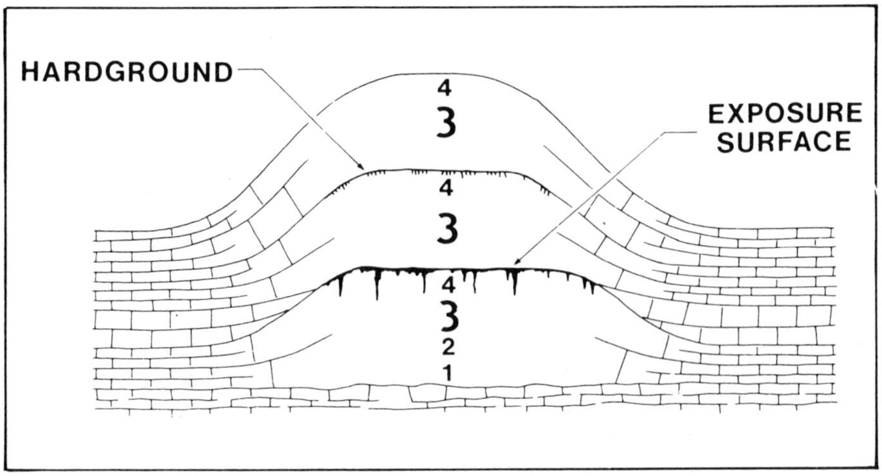

Figure 20
Sketch illustrating a large reef composed of a series of smaller stacked reefs, each with several growth stages (see Fig. 16) and each separated by exposure surfaces or hardgrounds. The size of the numbers represents relative thickness of the stages.

THE MODEL AS A BASIS FOR INTERPRETATION

The model dictates that reefs, especially large ones, have a series of growth stages. When interpreting an individual structure, which will differ from the model in a variety of ways, special attention must be paid to this internal stratigraphy. If some stages are missing, are thicker or thinner than should be expected, or if the sequence is inverted, then these differences should be a clear signal that extrinsic factors such as sea level fluctuations or changes in subsidence rates occurred during reef growth. In this way the reef is a much more sensitive indicator of such changes than surrounding subtidal facies.

On a larger scale a major question that often arises is whether the reef grew all at once and stood high above contemporaneous basinal strata only to be subsequently buried, or whether it grew in increments, with intervening periods of subaerial exposure and possible continuing off-reef deposition, and so never rose very much above the surrounding sea floor. These two hypotheses not only interpret the reef differently but dictate how surrounding basinal deposition took place and the nature of early diagenesis, and can only be separated by careful study of reef stratigraphy.

Finally reefs and associated platform facies are commonly important hydrocarbon reservoirs. In exploitation of these resources it is critical to determine whether the reservoir is a more or less uniform structure with good vertical as well as horizontal permeability or whether it is broken into a series of layers with poor vertical connection because of permeability barriers such as hardgrounds or paleosols (shales).

THE MODEL AS A PREDICTOR

If we know the age of a sequence of carbonate rocks and we have some idea of the paleotectonic setting then we can predict, from limited data, the types of reefs we might expect to be present in a shelf or platform setting.

Times When a Complete Spectrum of Reef Builders was Present

The edge of the shelf or platform is occupied by a marginal reef. The reef is well zoned if the front is steep and wave action intense because of the gradation from rough surface waters down the reef front into progressively more tranquil conditions. Zonation is weak, however, if the reef front slopes gradually seaward and the seas are relatively quiet. Patch reefs can occur anywhere on the platform and each is zoned with respect to depth. Reef mounds are found on the linear shallow parts of the shelf in areas of normal salinity but turbid water. Reef mounds also occur at

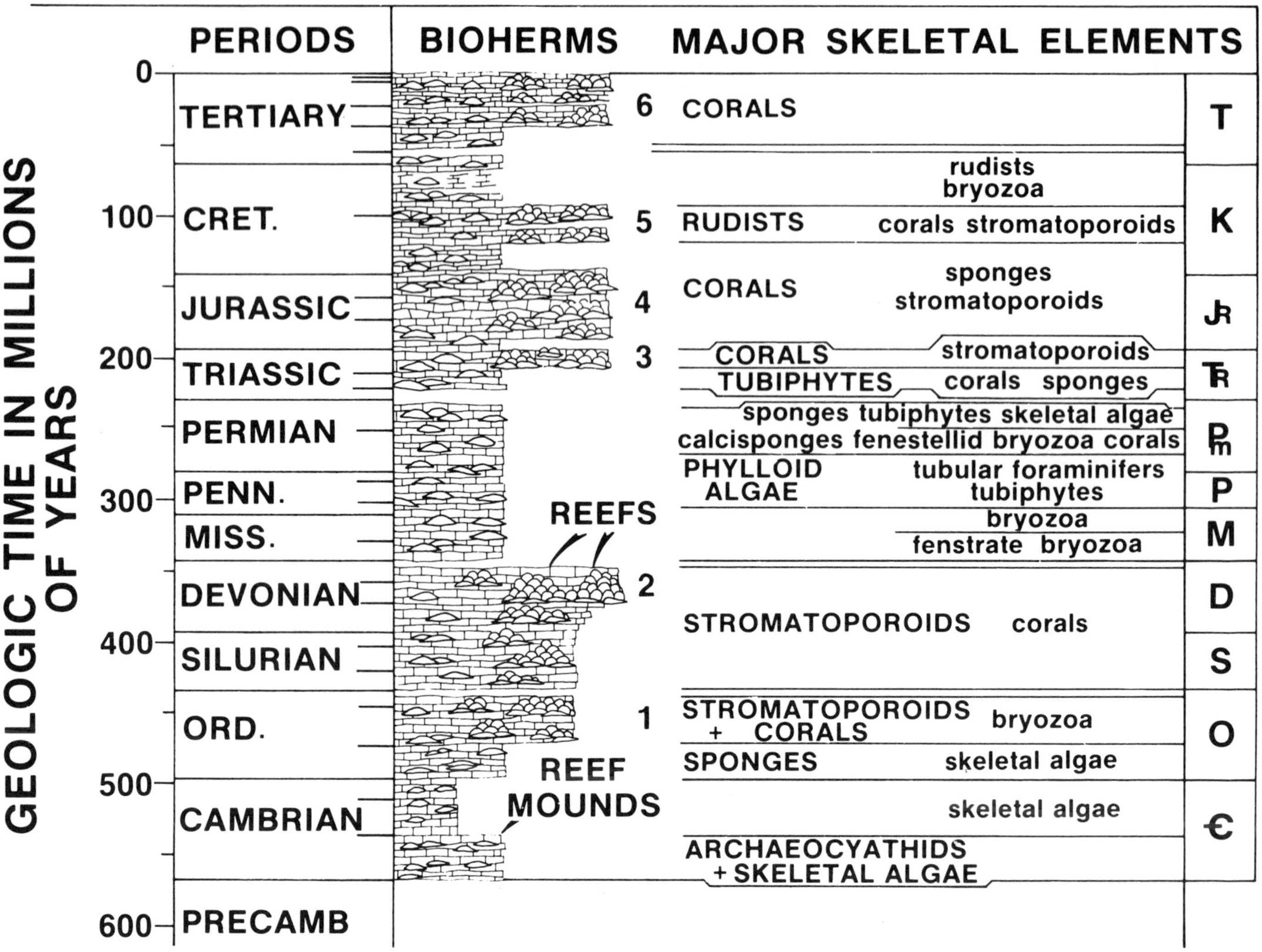

Figure 21
An idealized stratigraphic column representing the Phanerozoic and illustrating times when there appear to be no reefs or bioherms (gaps), times when there were only reef mounds and times when there were both reefs and reef mounds. The numbers indicate different associations of reef-building taxa discussed in the text. (From James, 1983).

depth, in front of the barrier reef down on the reef front or fore-reef.

Patch reefs or reef mounds commonly form a very widespread lithofacies compared to the barrier reef. The stratigraphic thickness of these reefs is dependent upon the rate of subsidence: if subsidence rate is low, reefs are thin; if subsidence rate is high, reefs may be spectacular in their thickness.

Times When Only Delicate, Branching and Encrusting Metazoa Prevail

The margin of the shelf or platform is normally a complex of oolitic or skeletal (generally crinoidal) sand shoals and islands. The only reef structures are reef mounds which occur below the zone of active waves down on the seaward slopes of the shelf or platform and if conditions are relatively tranquil behind the barrier, on the shelf itself. Mounds may display a zonation, with ocean-facing sides in shallow water armoured with accumulations of fragmented and winnowed skeletal debris.

Not only can we predict adjacent facies with some confidence but given only small outcrops or pieces of core we can, within limits, predict the style of the reef in question. For example, a core composed of lime mudstone to floatstone with scattered bryozoans and stromatactis from a Mississippian sequence, when viewed from the point of view of the model clearly points to a reef mound, which since reefs proper did not form at this time, may be 10's to 100's of metres thick. A similar rock in a Silurian sequence, while predicting a similar structure if surrounded by slope or inner shelf strata, would in most other cases predict that the sample is from the basal part of a large reef. Alternatively a small Devonian outcrop, composed of a bafflestone of fasciculate stromatoporoids grading up into a framestone of large tabulate corals and many different stromatoporoids would predict that a massive reef should lie above.

The clear limitations of the model are, however, when using small samples. Reefs are generally large and heterogeneous structures so that a drill core for example may pass through areas of sediment between large skeletons, a very likely possibility since 30% to 40% of modern reefs are sediment or pore space, and so give a false picture of the true deposit.

SUMMARY

The purpose of this article has been to marry the sedimentological and paleontological approaches to the study of reefs into a single facies model, useful to both disciplines. This model is an integration of data from two very different sources: from the modern sea floor, predominantly in the horizontal dimension; and from the rock record, predominantly in the vertical dimension as recorded in mountain exposures, quarries and drill core.

To alter and refine this model more information is needed from two areas. We must learn more about the succession of organisms and sediments that underlies the living surface of modern reefs, by drilling into these reefs. We must learn more about reefs from those parts of the stratigraphic record where reefs are known to occur, but have been little studied - the Precambrian, the Lower Paleozoic, and Cenozoic.

The trend in the past has been to compare specific fossil reefs with modern reefs. The comparative approach has just begun on fossil reefs, to compare and contrast the sedimentology and paleoecology of reefs formed by different groups of organisms at different times in geologic history.

ACKNOWLEDGEMENTS

Bob Stevens and David Kobluk kindly read earlier versions of this manuscript and helped clarify many of my ideas. Different parts of the systhesis have grown out of numerous field trips and discussions with Bob Ginsburg and Jim Wilson. By critically editing this version Roger Walker helped greatly in restructuring the basis for the model.

REFERENCES

FOSSIL REEFS - GENERAL

The following references are either syntheses or books with numerous papers on reefs of different ages.

Copper, P., 1974. Structure and development of early Paleozoic reefs. Proceedings of the Second International Coral Reef Symposium. Great Barrier Reef Committee, Brisbane, v. 6, p. 365-386.
An essay on the different types of Paleozoic reefs.

Dunham, R.J., 1970. Stratigraphic reefs versus ecologic reefs. American Association of Petroleum Geologists Bulletin, v. 54, p. 1931-1932.
A thoughtful essay on what is, and is not, a reef - from the point of view of a sedimentologist.,

Harris, P.M., ed., 1983. Carbonate buildups: a core workshop. Society of Economic Paleontologists and Mineralogists, Core Workshop No. 4, 593 p.
Twenty one, short, succinct, well-illustrated papers on reefs of all ages, with emphasis on depositional facies and diagenesis.

Hartman, W.D., Wendt, J.W., and Widenmayer, F., 1980. Living and fossil sponges. University of Miami, Sedimenta VIII, Comparative Sedimentology Laboratory, 274 p.
Notes for a short course with good descriptions of fossil reefs whose major components are sponges.

Heckel, P.H., 1974. Carbonate buildups in the geologic record: a review. *In* Laporte, L.F., ed., Reefs in time and space. Society of Economic Paleontologists and Mineralogists, Special Publication 18, p. 90-155.
A detailed account of reefs or buildups, their classification, zonation with geologic time and general models for reef formation - more than you ever wanted to know about fossil reefs and an excellent set of references.

Laporte, L.F., ed., 1974. Reefs in time and space. Society of Economic Paleontologists and Mineralogists, Special Publication 18, 256 p.
Seven papers dealing with various aspects of modern and fossil reefs.

Toomey, D.F., ed., 1981. European fossil reef models. Society of Economic Paleontologists and Mineralogists, Special Publication 30, 545 p.
Seventeen papers with emphasis on latest Paleozoic, Mesozoic, and Cenozoic structures; an excellent reference.

Walker, K.R., and Alberstadt, L.P., 1975. Ecological succession as an aspect of structure in fossil communities. Paleobiology, v. 1, p. 238-257.
The first half of this paper outlines succinctly the main theme of ecological succession in fossil reefs.

Wilson, J.L., 1975. Carbonate facies in geologic history. New York, Springer-Verlag, 471 p.
The best overall reference, especially chapters II, IV to VII, XI and XIII.

EXAMPLES OF FOSSIL REEFS

The following papers contain examples of different reef types from the fossil record and/or syntheses of reef types from specific periods; others can also be found in the general references above.

Precambrian

Cecile, M.P., and Campbell, F.H.A., 1978. Regressive stromatolite reefs and associated facies, middle Goulburn Group (Lower Proterozoic) in Kilohigok Basin, N.W.T.; an example of environmental control of stromatolite forms. Bulletin of Canadian Petroleum Geology, v. 26, p. 237-267

Hoffman, P., 1974. Shallow and deep-water stromatolites in lower Proterozoic platform-to-basin facies change, Great Slave Lake, Canada. American Association of Petroleum Geologists, Bulletin, v. 58, p. 856-867.

Cambrian

Ahr, W.M., 1971. Paleoenvironment, algal structures and fossil algae in the Upper Cambrian of central Texas. Journal of Sedimentary Petrology, v. 41, p. 205-216.

James, N.P., and Kobluk, D.R., 1978. Lower Cambrian patch reefs and associated sediments, southern Labrador, Canada. Sedimentology, v. 25, p. 1-32.

James, N.P., and Debrenne, F., 1980. Lower Cambrian bioherms; pioneer reefs of the Phanerozoic. Acta Palaeontologica Polonica, v. 25, p. 655-668.

James, N.P., 1981. Megablocks of calcified algae in the Cow Head Breccia, western Newfoundland: vestiges of a lower Paleozoic continental margin. Geological Society of America, Bulletin, v. 92, p. 799-811.

Ordovician

Pitcher, M., 1961. Evolution of Chazyan (Ordovician) reefs of eastern United States and Canada. Bulletin of Canadian Petroleum Geology, v. 12, p. 632-691.

Pratt, B.R. and James, N.P., 1982. Cryptalgal metazoan bioherms of early Ordovician age in the St. George Group, Western Newfoundland. Sedimentology, v. 29, p. 543-569.

Ross, R.J., Jaanuson, V., and Freidman, I., 1975. Lithology and origin of Middle Ordovician calcareous mudmound at Meiklejohn Peak, southern Nevada. United States Geological Survey, Professional Paper 871, 45 p.

Toomey, D.F., and Nitecki, M.H., 1979. Organic buildups in the Lower Ordovician (Canadian) of Texas and Oklahoma. Fieldiana, New Series, v. 2, 181 p.

Webby, B.D., 1984. Ordovician reefs and climate: a review. *In* Bruton, D.L., ed., Apsects of the Ordovician system. Paleontological Contributions from the University of Oslo, No. 8, p. 87-98.

Silurian

Bourque, P.A. and Gignac, H., 1983. Sponge-constructed stromatactic mud mounds, Silurian of Gaspé Québec. Journal of Sedimentary Petrology, v. 53, p. 521-532.

Fisher, J.H., ed., 1977. Reefs and evaporites – concepts and depositional models. American Association of Petroleum Geologists, Studies in Geology 5, 196 p.

Heckel, P., and O'Brien, D., eds., 1975. Silurian reefs of Great Lakes Region of North America. American Association of Petroleum Geologists, Reprint Series 14, 243 p.

Lowenstam, H.A., 1950. Niagaran reefs in the Great Lakes areas. Journal of Geology, v. 58, p. 430-487.

Narbonne, G.M., and Dixon, O.A., 1984. Upper Silurian lithistid sponge reefs on Somerset Island, Arctic Canada. Sedimentology, v. 31, p. 25-51.

Riding, R., 1981. Composition, structure, and environmental setting of Silurian biohems and biostromes in northern Europe. *In* Toomey, D.F., ed., European fossil reefs models. Society of Economic Paleontologists and Mineralogists, Special Publication 30, p. 41-85.

Scoffin, T.P., 1971. The conditions of growth of the Wenlock reefs of Shropshire, England. Sedimentology, v. 7, p. 173-219.

Shaver, R.H., *et al.*, 1978. The search for a Silurian reef model; Great Lakes Area. Indiana Geology Survey, Special Report No. 15, 36 p.

Devonian

Davies, G.R., ed., 1975. Devonian reef complexes of Canada, I and II. Canadian Society of Petroleum Geologists, Reprint Series 1, 229 p. and 246 p., respectively.

Klovan, J.E., 1974. Development of western Canadian Devonian reefs and comparison with Holocene analogues. American Association of Petroleum Geologists, Bulletin, v. 58, p. 787-799.

Krebs, W., 1971. Devonian reef limestones in the eastern Rhenish Shiefergebirge. *In* Muller, G., and Friedman, G.M., eds., Sedimentology of parts of central Europe. Frankfurt, Kramer, p. 45-81.

Krebs, W., and Mountjoy, E.W., 1972. Comparison of central European and western Canadian Devonian reef complexes. Montreal, Canada, 24th International Geological Congress, Section 6, p. 294-309.

Mountjoy, E.W., 1980. Some questions about the development of upper Devonian carbonate buildups (reefs), western Canada. Bulletin of Canadian Petroluem Geology, v. 28, p. 315-344.

Mountjoy, E.W., and Riding, R., 1981. Foreslope *Renalcis*-stromatoporoid bioherm with evidence of early cementation, Devonian Ancient Wall reef complex. Sedimentology, v. 28, p. 299-321.

Schmidt, V., McDonald, D.A., and McIlreath, I.A., 1980. Growth and diagenesis of Middle Devonian Keg River cementation reefs, Rainbow Field, Alberta. *In* Halley, R.B., and Loucks, R.G., eds. Carbonate reservoir rocks. Society of Economic Paleontologists and Mineralogists, Core Workshop 1, p. 43-64.

Playford, P.E., 1980. Devonian "Great Barrier Reef" of the Canning Basin, Western Australia. American Association of Petroleum Geologists, Bulletin, v. 64, p. 814-840.

Mississippian

Bolton, K., Lane, H.R., and LeMone, D.F., eds., 1982. Symposium on the paleoenvironmental setting and distribution of the Waulsortian facies. El Paso Geological Society and University of Texas at El Paso, 202 p.

Cotter, E., 1965. Waulsortian-type carbonate banks in the Mississippian Lodgepole Formation of central Montana. Journal of Geology, v. 73, p. 881-888

Lees, A., 1964. The structure and origin of the Waulsortian (Lower Carboniferous) "reefs" of west-central Eire. Philosophical Transactions of the Royal Society of London, Series B, No. 740, p. 485-531.

Pennsylvanian

Choquette, P.W., 1983. Platey-algal reef mounds, Parodox Basin. *In* Scholle, P.A., Bebout, D.E., and Moore, C.H., eds., Carbonate depositional environments. American Association of Petroleum Geologists, Memoir 33, p. 454-462.

Heckel, P.H., and Cocke, J.M., 1969. Phylloid algal mound complexes in outcropping Upper Pennsylvanian rocks of mid-continent. American Association of Petroleum Geologists, Bulletin, v. 53, p. 1084-1085.

Pray, L.C., Wilson, J.L., and Toomey, D.F., 1977. Geology of the Sacramento Mountains, Otero County, New Mexico. Midland, Texas, West Texas Geological Society, Guidebook, 216 p.

Toomey, D.F., and Winland, H.D., 1973. Rock and biotic facies associated with a Middle Pennsylvanian (Desmoinesian) algal buildup, Neca Lucia Field, Nolan County, Texas. American Association of Petroleum Geologists, Bulletin, v. 57, p. 1053-1074.

Permian

Davies, G.R., 1970. A Permian hydrozoan mound, Yukon Territory. Canadian Journal of Earth Sciences, v. 8, p. 973-988.

Dunham, R.J., 1972. Guide for study and discussion for individual reinterpretation of the sedimentation and diagenesis of the Permian Capital Geologic Reef and associated rocks, New Mexico and Texas. Midland, Texas, Permian Basin Section, Society of Economic Paleontologists and Mineralogists, Publication 72-14, 235 p.

Hileman, M.E., and Mazzulo, S.J., 1977. Upper Guadalupian Facies, Permian Reef Complex, Guadalupe Mountains, New Mexico and Texas. Midland, Texas, Permian Basin Section. Society of Economic Paleontologists and Mineralogists, Publication 77-16, 508 p.

Mazzullo, S.J., and Cys, J.M., 1979. Marine aragonite sea floor growths and cements in Permian phylloid algae mounds, Sacramento Mountains, New Mexico. Journal of Sedimentary Petrology, v. 49, p. 917-937.

Newell, N.E., *et al.*, 1953. The Permian Reef complex of the Guadalupe Mountains region, Texas and New Mexico. San Francisco, Freeman and Co., 236 p.

Newell, N.E., *et al.*, 1976. Permian Reef Complex, Tunisia. Brigham Young University, Geological Studies, v. 23, p. 75-112.

Triassic

Bosellini, A., and Rossi, D., 1974. Triassic carbonate buildups of the dolomites, northern Italy. *In* Laporte, L.F., ed., Reefs in time and space. Society of Economic Paleontologists and Mineralogists, Special Publication 18, p. 209-233.

Flugel, E., 1981. Paleoecology and facies of Upper Triassic reefs in the northern Calcareous Alps. *In* Toomey, D.F., ed., European fossil reef models. Society of Economic Paleontologists and Mineralogists, Special Publication 30, p. 291-361.

Stanley, G.D., 1979. Paleoecology, structure and distribution of Triassic coral buildups in western North America. University of Kansas Paleontology Contribution, Article 65, 58 p.

Zankl, H., 1971. Upper Triassic carbonate facies in the northern Limestone Alps. *In* Muller, G., and Friedman, G.M., eds., Sedimentology of parts of central Europe. Frankfurt, Kramer. p. 147-185.

Jurassic

Eliuk, L.S., 1979. Abenaki Formation, Nova Scotian Shelf, Canada; a depositional and diagenetic model for Mesozoic carbonate platforms. Bulletin of Canadian Petroluem Geology, v. 24, p. 424-514.

Gwinner, M.P., 1975. Origin of the Jurassic limestones of the Swabian Alb (southwest Germany). Contributions to Sedimentology 5. New York, Elsevier, 75 p.

Palmer, T.J., and Fursich, F.T., 1981. Ecology of sponge reefs from the Upper Bathonian (Middle Jurassic) of Normandy. Palaeontology v. 24, p. 1-25.

Rutten, M.D., 1956. The Jurassic reefs of the Yonne (southeastern Paris Basin). American Journal of Science, v. 254, p. 363-371.

Warme, J.E., Stanley, R.G., and Wilson, J.L., 1975. Middle Jurassic reef tract, central High Atlas, Morocco. Nice, France, International Association of Sedimentologists, Proceedings, Theme VIII, 11p.

Cretaceous

Bebout, D.G., and Loucks, R.G., 1974. Stuart City trend, Lower Cretaceous, south Texas. Bureau of Economic Geology, Report No. 78, 80 p.

Enos, P., 1974. Reefs, platforms, and basins of Middle Cretaceous in northeast Mexico. American Association of Petroleum Geologists, Bulletin, v. 58, p. 800-809.

Kauffman, E.G., and Sohl, N.F., 1974. Structure and evolution of Antillean Cretaceous rudist frameworks. Verhandl. Naturi. Ges. Basel., v. 84, p. 399-467.

Perkins, B.F., 1974. Paleoecology of a rudist reef complex in the Comanche Cretaceous Glen Rose Limestone of central Texas. *In* Perkins, B.F., ed., Aspects of Trinity Division Geology, Geoscience and Man VIII. Baton Rouge, LA., Louisiana State University, p. 131-173.

Philip, J., 1972. Paleoecologie des formations a rudistes du Cretace Superior-l'example du sud-est de la France. Paleogeography, Paleoclimatology, Paleoecology, v. 12, p. 205-222.

Scott, R.W., 1979. Depsitional model of Early Cretaceous coral-algal-rudist reefs, Arizona. American Association of Petroleum Geologists, Bulletin, v. 63, p. 1108-1128.

Cenozoic

Dabrio, C.J., Esteban, M., and Martin, J.M., 1981. The coral reef model of Nijar, Messinian (Uppermost Miocene) Almeria Province, southwest Spain. Journal of Sedimentary Petrology, v. 51, p. 521-541.

Esteban, M., 1979. Significance of the Upper Miocene coral reefs of the western Mediteranean. Paleogeography, Paleoclimatology, Paleoecology, v. 29, p. 169-189.

Frost, S.H., 1977. Cenozoic reef systems of Caribbean - prospects for paleoecological synthesis. American Association of Petroleum Geologists, Studies in Geology 4, p. 93-110.

Frost, S.H., 1981. Oligocene reef coral biofacies of the Vicentin, northeast Italy. *In* Toomey, D.F., ed., European fossil reef models. Society of Economic Paleontologists and Mineralogists, Special Publication 30, p. 483-541.

Pedley, H.M., 1979. Miocene bioherms and associated structures in the Upper Coralline Limestone of the Maltese Islands; their lithification and paleoenviroment. Sedimentology, v. 26, p. 577-593.

Pleistocene

Bathurst, R.G.C., 1975. Carbonate sediments and their diagenesis. Amsterdam, Elsevier, 658 p.
Chapters 3 and 4 nicely summarize specific modern reef environments.

Chappell, J., and Polach, H.A., 1976. Holocene sea-level change and coral-reef growth at Huon Peninsula, Papua New Guinea. Geological Society of America, Bulletin, v. 87, p. 235-239.

Crame, J.A., 1980. Succession and diversity in the Pleistocene coral reefs of the Kenya coast. Palaeontology, v. 23, p. 1-37

Geister, J., 1980. Calm-water reefs and rough water reefs of the Caribbean Pleistocene. Acta Paleontologica Polonica, v. 25, p. 541-556.

James, N.P., Stearn, C.S., and Harrison, R.S., 1977. Field guidebook to modern and Pleistocene reef carbonates, Barbados, West Indies. University of Miami, 3rd International Coral Reef Symposium, Fisher Island, 30 p.

Mesolella, K.J., Sealy, H.A., and Matthews, R.K., 1970. Facies geometries within Pleistocene reefs of Barbados, West Indies. American Association of Petroleum Geologists, Bulletin, v. 54, p. 1899-1917.

Stanley, S.M., 1966. Paleoecology and diagenesis of Key Largo limestone, Florida. American Association of Petroleum Geologists, Bulletin, v. 50, p. 1927-1947.

MODERN REEFS - GENERAL

Darwin, C., 1842. Structure and distribution of coral reefs. Reprinted by University of California Press from 1851 edition, 214 p.
The beginning of modern coral reef research by a young biologist.

Frost, S.H., Wiss, M.P., and Saunders, J.B., 1977. Reefs and related carbonates - ecology and sedimentology. American Association of Petroleum Geologists, Studies in Geology 4, 421 p.
Twenty-eight papers with copious references, mainly on modern reefs.

Ginsburg, R.N., and James, N.P., 1974. Spectrum of Holocene reef-building communities in the western Atlantic. *In* Ziegler, A.M., *et al.*, eds., Principles of benthic community analysis. University of Miami, Fisher Island Station, Short Course Notes, p. 7.1 - 7.22.
A succinct summary of the major reef types of the Atlantic-Caribbean area.

Hopley, D. 1982. The geomorphology of the Great Barrier Reef. New York, Wiley Interscience, 453 p.
Although concentrating on the Great Barrier Reef this book contains a wealth of information on modern reefs in general.

Stoddart, D.R., 1969. Ecology and morphology of recent coral reefs. Biology Review, v. 44, p. 433-498.
A superb summary of modern coral reefs - an excellent place to start.

Taylor, D.E., ed., 1977. Proceedings of Third International Coral Reef Symposium, Miami Florida 2. Geology: Fisher Island Station, Miami Beach, Florida, 628 p.
Many short papers and guidebooks with up to date research on reefs.

DIFFERENT TYPES OF MODERN REEFS

1) Pacific Atolls

Emery, K.O., Tracey, J.I., and Ladd, H.S., 1954. Geology of Bikini and nearby atolls. United States Geological Survey, Professional Paper 260-A, 265 p.

2) Marginal Reefs

Goreau, T.F., 1959. The ecology of Jamaican coral reefs: I. Species, composition and zonation. Ecology, v. 40, p. 67-90.

Goreau, T.F., and Goreau, N.I., 1973. The ecology of Jamaican coral reefs: II. Geomorphology, zonation and sedimentary phases. Bulletin of Marine Science, v. 23, p. 399-464.

James, N.P., Ginsburg, R.N., Marszalek, D., and Choquette, P.W., 1976. Facies and fabric specificity of early subsea cementation in shallow Belize (British Honduras) reefs. Journal of Sedimentary Petrology, v. 46, p. 523-544.

James, N.P., and Ginsburg, R.N., 1979. The seaward margin of belize barrier and atoll reefs. Blackwells, Oxford, International Association of Sedimentologists, Special Publication 3, 196 p.

Maxwell, W.G.H., 1968. Atlas of the Great Barrier Reef. Amsterdam, Elsevier, 258 p.

3) Pinnacle Reefs

Korniker, L.A., and Boyd, D.W., 1962. Shallow water geology and environment of Aacran reef complex, Campeche Bank, Mexico. American Association of Petroleum Geologists, Bulletin, v. 46, p. 640-673.

Logan, B.W., *et al.*, 1969. Carbonate sediments and reefs, Yucatan shelf, Mexico. American Association of Petroleum Geologists, Memoir 11, p. 1-196

4) Patch Reefs

Garrett, P., *et al.*, 1971. Physiography, ecology and sediments of two Bermuda patch reefs. Journal of Geology, v. 79, p. 647-668.

Maiklem, W.R., 1970. The Capricorn Reef complex, Great Barrier Reef, Australia. Journal of Sedimentary Petrology, v. 38, p. 785-798.

5) Reef Mounds

Turmel, R., and Swanson, R., 1976. The development of Rodriquez Bank, a Holocene mudbank in the Florida Reef Tract. Journal of Sedimentary Petrology, v. 46, p. 497-519.

6) Algal Reefs

Adey, W., and Burke, R., 1976. Holocene bioherms (algal ridges and bank-barrier reefs) of the eastern Caribbean. Geological Society of America, Bulletin, v. 87, p 497-519.

Ginsburg, R.N., and Schroeder, J.H., 1973. Growth and submarine fossilization of algal cup reefs, Bermuda. Sedimentology, v. 20, p. 575-614.

7) Stromatolites

Dravis, J.J., 1983. Hardened subtidal stromatolites, Bahamas. Science, v. 219, p. 385-387.

Hoffman, P., 1976. Stromatolite morphogenesis in Shark Bay, Western Australia. *In* Walter, M.R., ed., Stromatolites. Amsterdam, Elsevier, p. 261-273.

Playford, P.E., and Cockbain, A.E., 1976. Modern algal stromatolites at Hamelin Pool, a hypersaline barred basin in Shark Bay, Western Australia. *In* Walter, M.R., ed., Stromatolites. Amsterdam, Elsevier, p. 389-413.

8) Deepwater Mounds

Mullins, H.T., *et al.*, 1981. Modern deep-water coral mounds north of Little Bahama Bank; criteria for recognition of deep-water coral bioherms in the rock record. Journal of Sedimentary Petrology, v. 51, p. 999-1013.

Neumann, A.C., Kofoed, J.W., and Keller, G.H., 1977. Lithoherms in the Straits of Florida. Geology, v. 5, p. 4-11.

9) Linear Mud Banks

Enos, P., and Perkins, R., 1979. Evolution of Florida Bay from island stratigraphy. Geological Society of America, Bulletin, v. 90, p. 59-83.

Pusey, W.C., 1975. Holocene carbonate sedimentation on northern Belize Shelf. *In* Wantland, K.F., and Pusey, W.C., eds., Belize shelf-carbonate sediments, clastic sediments, and ecology. American Association of Petroleum Geologists, Studies in Geology 2, p. 131-234.

OTHER REFERENCES CITED

Bathurst, R.G.C., 1982. Genesis of stromatactis cavities between submarine crusts in Palaeozoic carbonate mud buildups. Journal of the Geological Society of London, v. 139, p. 165-181.

Connell, J.H., 1978. Diversity in tropical rain forests and coral reefs. Science, v. 199, p. 1302-1310.

Embry, A.F. and Klovan, J.E., 1971. A Late Devonian reef tract on northeastern Banks Island, N.W.T. Bulletin of Canadian Petroleum Geology, v. 19, p. 730-781.

Frost, S.H., 1977. Ecologic controls of Caribbean and Mediterranean Oligiocene reef coral communities. *In* Taylor, D.L., ed., Miami, Florida, Proceedings of Third International Coral Reef Symposium, p. 367-375.

James, N.P., 1983. Reefs. *In Scholle,* P.A., Bebout, D.G., and Moore, C.H., eds., Carbonate depositional environments. American Association of Petroleum Geologists, Memoir 33, p. 2346-440.

James, N.P., and Choquette, P.W., 1983. Limestone diagenesis: 6 – the sea floor diagenetic environment. Geoscience Canada, v. 10, p. 162-179.

Stearn, C.W., 1982. The shapes of Paleozoic and modern reef-builders: a critical review. Paleobiology, v. 8, p. 228-241.

Pratt, B.R., 1982. Stromatolitic framework of carbonate mud-mounds. Journal of Sedimentary Petrology, v. 52, p. 1203-1227.

Woodley, J.D. *et al.*, 1982. Hurricane Allen's impact on Jamaican coral reefs. Science, v. 214, p. 749-755.

Reproduced with the permission of the Geological Association of Canada. Published in *Facies Models*, R.G. Walker, editor, *Geoscience Canada*, Reprint Series No. 1, second edition, 1984, pp. 245-257.

Carbonate Slopes

IAN A. McILREATH
Petro-Canada
P.O. Box 2844
Calgary, Alberta T2P 3E3

NOEL P. JAMES
Department of Earth Sciences
Memorial University of Newfoundland
St. John's, Newfoundland A1B 3X5

INTRODUCTION

To any geologist who has seen them in the field, sediments that comprise the slope facies of carbonate complexes are often the most staggering and long remembered of all. The sheer size of the enormous limestone blocks chaotically intercalated with delicately laminated lime mudstones tests our understanding of sediment genesis and deposition more than for almost any other deposits.

While the sediments themselves are intriguing, they are also useful as the only remaining clues as to the nature and composition of a new dolomitized or tectonically obliterated platform margin. Furthermore, the very presence of huge blocks is an excellent indicator of a nearby carbonate platform or reef complex, and this principle has been successfully used to locate reefs in the subsurface. The lime sands and conglomerates of these deposits, where intercalated with organic-rich basinal sediments, can be reservoirs for oil and natural gas.

We cannot interpret these deposits with the same level of confidence as shallow water carbonate sediments because: 1) modern slope deposits are not easily accessible for field study (the limited use of small research submersibles and seismology is slowly changing this); 2) ancient slope deposits commonly occur in orogenic belts, where facies and tectonic relationships are so complicated that these deposits are often mistaken for tectonic breccias or mélanges; 3) slope sediments in the subsurface generally have not been serious exploration targets as long as adjacent platform margins remained the primary objective; 4) slope sediments are formed in a series of environments that transect major pressure, temperature and oxygen-level boundaries in the ocean and the precise effects of these physiochemical parameters on the sediments are poorly known.

As a result our present facies models are based on the rock record, with some additions from recent sediments. In addition, our understanding of downslope sediment emplacement is based in part upon processes and models determined for siliciclastic deposits (Dott, 1963; Middleton and Hampton, 1973). It has been easy to apply this comparative approach (Cook *et al.*, 1972; Cook and Mullins, 1983) because many similarities do exist, but there are also fundamental differences.

In the paper on "Turbidites and Associated Coarse Clastic Deposits" (this volume), Walker outlined the attributes of a turbidite model and then integrated all associated siliciclastic lithofacies that encompass the slope-to-basin transition into an overall, larger scale, submarine fan model. Similarly our approach in this article will be to outline the major aspects of the slope facies in carbonate sedimentary sequences, first by examining the major sediment types and their modes of emplacement, and second by relating these to general facies models, which are very much dependent upon the nature of the adjacent margin and the depositional setting.

Since the first edition of this summary appeared there has been an increased interest in synthesizing the various aspects of fore-reef and slope carbonates and excellent, well-illustrated summaries have been written by Read (1982), James and Mountjoy (1983), Enos and Moore (1983), and Cook and Mullins (1983).

CARBONATE SLOPE SEDIMENTATION

The slope facies is a transitional one between the rapid and active production of calcium carbonate in shallow water and the slow gentle rain of fine-grained pelagic sediments in the basin. The platform-to-basin transition may in places be abrupt, in the form of a steep cliff, but more commonly is a gently inclined slope decreasing in grade with depth and merging imperceptibly into basinal deposits at some distance, which may be 100s of km from the actual margin. Because the environment as a whole is an incline, short periods of gravity-induced catastrophic sedimentation alternate with long periods of relatively quiet pelagic sedimentation, or to paraphrase Ager (1973, p. 100), "long periods of boredom alternating with short periods of terror".

Pelagic Carbonates

Pelagic carbonates are those sediments deposited in the open sea and derived from the skeletons of planktonic microorganisms which inhabit the overlying water column. Such deposits include ooze and its lithified equivalent, chalk, and consist primarily of the skeletons of various nannofossil groups, especially coccoliths, the tests of planktonic and sometimes benthic foraminifers. Macrofossils such as pteropods, pelecypods, echinoderms and, in older units, ammonites, nautiloids, tentaculitids and styliolinids are present as accessory components. An excellent summary of such deposits can be found in Hsu and Jenkyns (1974), Scholle (1977), and Scholle *et al.* (1983).

True pelagic carbonates are apparently not known from the early Paleozoic and are first recognized from rocks of Upper Silurian age (Tucker, 1974). Planktonic foraminifers and coccoliths appear to have evolved in the Jurassic. During post-Jurassic time pelagic carbonate has increased to the point that in the last 100 Ma it comprises about 67% of worldwide carbonate deposition, and more than 50% of the present sea floor is covered with this type of carbonate sediment (Hay *et al.*, 1976).

Most chalks accumulate a a rate of between one and 30 cm per year. The sedimentary structures and colours depend upon the degree of circulation and oxygenation. Dark colours and preserved laminations reflect stagnation; lighter colours, more burrows and fewer preserved sedimentary structures reflect stronger bottom circulation.

Because of relatively narrow shelves and low sea level stands, Holocene pelagic sediments are restricted largely to deep ocean basins. However, during times of eustatic high stands of sea level, pelagic carbonates can and did accumulate on continental shelves, such as the North American mid-continent seaway in Cretaceous times.

The water depth of Recent pelagic carbonate deposition ranges from less than 100 m to greater than 4500 m. The limiting factors for such accumulations are the relative rates of sedimentation of carbonate versus non-carbonate components, physical erosion by submarine currents and chemical dissolution. Chemical dissolution is particularly important in carbonate slope facies because the environment passes, with depth, through several important increasing pressure and decreasing temperature boundaries (James and Choquette, 1983). Aragonite components, such as pteropods and benthic foraminifers, may be selectively removed by dissolution in water as shallow as 500 m (the aragonite compensation depth) while calcite components are completely dissolved at the carbonate compensation depth, between 4,000 and 5,000 m in today's oceans. Much less is known about the removal or recrystallization of Mg-calcite. This progressive removal by dissolution results in a residual sediment composed largely of siliceous skeletons, red hemipelagic clays and wind-blown silt.

In some ares of the modern ocean the production of siliceous plankton (silicioflagellates, diatoms and radiolaria) exceeds that of calcareous nanno- and microplankton. During the Paleozoic, when pelagic carbonate was reduced or absent, siliceous sediment was much more widespread in deep-water areas.

Hemipelagic Slope Sediments

Sediments that make up the fine-grained pelagic component of most slope deposits come not only from the water column but from the adjacent platform as well (Wilson, 1969). While the contribution at any one time from the water column is more or less constant, that portion derived from the platform is episodic. Most often storms stir up the wide, shallow, mud-floored areas of the shelf and the milk-white water streams out across the shelf margin to settle in deep water. A less voluminous but more regular transfer process exists at such near-vertical shelf-to-deep-oceanic-basin transitions as St. Croix, Virgin Islands, where warm sediment-rich shelf waters "float" over the cooler basinal waters by tidal exchange. These fine-grained, shallow-water derived slope sediments have been called "peri-platform ooze" by Schlager and James (1978) because they occur as an apron around the platform and because they are significantly different in their mineralogy and composition from the wholly pelagic sediments of the open sea.

In the Precambrian and Paleozoic most pelagic slope carbonates may well have been almost wholly peri-platform ooze.

The resultant hemipelagic slope deposits are monotonous, uniform dark grey, fine-grained lime mudstones, generally thin-bedded with flat planar contacts and internal micro-laminations (Fig. 1). Mudstone beds are often separated by partings into very thin beds of similar mudstone or beds of shale, forming characteristic "rhythmites" or "ribbon limestones". The original depositional textures and fabrics are often modified by downslope creep leading to sedimentary boudinage, while differential compaction and/or cementation frequently transforms the evenly-bedded sediments into a nodular limestone. The irregular nodules may, in some cases, be so packed together to form a jig-saw puzzle resembling an *in situ* breccia.

From the Mesozoic to Recent, pelagic components become progressively more important. The criteria for distinguishing the relative contribution of peri-platform vs pelagic ooze is speculative because most fine silt-mud size shallow water indicators are susceptible to early diagenesis (Enos and Moore, 1983).

Peri—Platform Talus

Directly seaward of the shallow water reefs or lime-sand shoals that form a platform margin, there is commonly a debris apron of limestone blocks (Fig. 2), skeletons of reef building metazoa, sediment bounded by submarine cement or encrusting organisms, lime sand and muds. These accumulations are the result of rock-fall and sand-streams from shallow water and, as illustrated in Figure 3, are very common along the seaward margins of modern reef complexes (James and Ginsburg, 1979; Mullins and Neumann, 1979; Schlager and Chermak, 1979; Land and Moore, 1977). The blocks themselves may be multi-generation in composition because the reefs, sand shoals and other deposits at the platform margin are characteristically susceptible to early lithification, either by submarine cementation, or if there are slight fluctuations in sea level, by complex subaerial diagenesis. In addition, parts of the talus wedge are commonly cemented on the sea floor (James and Ginsburg, 1979; Land and Moore, 1977). The lithified portions of these limestones become hard and brittle, and so

Figure 1
Peri-platform ooze; evenly-bedded, grey lime mudstone with thin interbeds of argillaceous lime mudstone, Cooks Brook Formation (Middle Cambrian), Humber Arm, Western Newfoundland.

are particularly susceptible to fracturing and fragmentation.

Large passes through a reef also act as conduits, funnelling back-reef sediments into this zone so that, along strike, areas of chaotic breccia may alternate with fans of lime sand. The later sediment is also commonly cemented, forming numerous hardgrounds.

Examination of sediment dispersal seaward of the platform in areas with low to intermediate slopes (up to 30°) indicates that this talus does not travel any significant distance away from the margin by day-to-day processes.

Carbonate Breccias and Conglomerates

These deposits have been called debris flows (deposits), submarine mass flows, mass breccia flows, breccia and mega-breccia beds, debris sheets, rudite sheets, debris avalanches, or olistostromes (in the non-tectonic sense). They are certainly the most impressive parts of the slope sequence. They originate in two very different areas, high up on the slope in shallow water or from lower down the slope profile.

Figure 2
Peri-platform talus; a block of shallow-water reef limestone (approximately 30m high) enclosed in thin-bedded, dark grey, peri-platform lime mudstones. Block occurs approximately 250m down slope from the toe of a near-vertical, 200m high platform margin. Note vertical orientation of bedding within the block, Cathedral Formation (Middle Cambrian), north face Mt. Stephen, British Columbia.

A) Breccias Derived from Shallow Water. These breccias are generally exposed in discontinuous to laterally extensive sheets, channels with lenticular cross sections or irregular masses. They stand out as resistant masses of light-coloured carbonate against a background of dark-coloured, well-bedded limestone and shale (Figs. 4 and 5). They are characterized by blocks of all sizes and shapes, but often equidimensional and somewhat rounded. Some of the blocks are so enormous that they have been mistaken for bioherms (see Mountjoy *et al.,* 1972). One exceptional clast in the Cow Head Group (Cambro-Ordovician) at Lower Head, Newfoundland is 0.2 km x 50 m in size, with surrounding blocks often 30 x 15 m in size, (Kindle and Whittington, 1958). The breccias commonly have a matrix of lime mud, lime sand or argillaceous lime mud.

The deposits are bedded, with a planar to undulating basal contact accentuated by differential compaction and a planar to irregular and often hummocky upper contact. The nature of the bedding contacts often cannot be determined accurately because the bedding planes are stylolitic, and so any original bedding-plane features are often destroyed. Bedding thicknesses range up to tens of metres. Davies (1977) made the interesting observation that the common occurrence of crinoids, bryozoa and ammonites at the upper surface of Permo-Pennsylvanian debris deposits on Ellesmere Island may represent an indigenous fauna inhabiting the "reef-like" upper surface of the deposit.

The polymict nature of the clasts reflects the complexity of the source area, namely the platform margin, which consists of partly lithified reefs and/or limesand shoals, downslope (yet still shallow) reef mounds, or peri-platform talus. Among the talus blocks, expected types include well-sorted and well-bedded lime sands which can be differentially submarine cemented (Fig. 6), individual colonies of reef builders, multigeneration reef rock, limestones with subaerial karst features, tidal flat lithologies, and even cemented talus that has been refractured to give breccia clasts within breccia.

The fabrics of "analagous" coarse siliciclastic deposits have been discussed by Walker (1976). In lime breccias they range from mainly chaotic to imbricated to horizontal to vertical, and rarely include graded or even reverse graded bedding. They range from clast-supported to most commonly matrix-supported with extremely poor sorting. The matrix ranges from shale to argil-

Figure 3
Peri-platform talus; looking across the steeply dipping fore-reef slope at a depth of 130m seaward of the Belize barrier reef complex. The slope is composed of blocks of limestone, plates of coral and lime sand composed of the plates of the green alga Halimeda; *the small block at the center (arrow) is about one metre high.*

Figure 4
Lime breccias; light grey, shallow-water reef-derived limestone breccias occurring in a 'channel' – a (approximately 8m thick); sheets – b (approximately 2m thick); and irregular masses – c (up to 12m thick), all enclosed in thin-bedded, dark grey, peri-platform lime mudstones, Cathedral Formation (Middle Cambrian), southface Mt. Field, British Columbia.

Figure 5
A sequence illustrating two different types of carbonate slope deposits; debris flows with large limestone clasts (right) and thin-bedded, graded calcarenites (the thin, grey limestone beds), interbedded with black fissile shale. This overturned sequence (top at lower left) of Middle Ordovician age occurs at Cape Cormorant, Port-au-Port Peninsula, Western Newfoundland.

laceous lime mud to lime mudstone with occasional lime sand. As Hopkins (1977) points out, however, what is often taken to be lime mud in outcrop turns out to be peloid lime sand in thin section, so that sand-sized matrix may be more common than supposed.

The exact mechanisms by which these sediments are transported are not yet clear. Submarine debris flows (Hampton, 1972) are sediment gravity flows in which granular solids such as boulders, pebbles and sand are more or less "floated" during transport by the yield strength of the matrix which is composed of interstitial fluid and fine sediment. Buoyancy of the fluid matrix also contributes to the support. Because not all such deposits have a clay mineral matrix, the transport mechanism is thought to be a combination of debris flow and grain flow (Middleton and Hampton, 1973). A major problem in this regard is that almost all experimental work to date has been done on clay-water mixtures; none of the experiments has been carried out on sediments with a clay-lime or lime mud matrix.

B) Breccias Derived from the Slope. The evenly-bedded calcilutites or lime muds of the slope facies are often prone to downslope creep. Individual beds can be seen to neck or wedge out, or whole intervals will move downslope within a series of slump folds (Fig. 7). Dislocation and movement of large masses of slope material downslope leads to the formation of breccias or submarine glide masses composed of numerous tabular clasts of slope limestone that have been bent or fractured, that are poorly-sorted and that exhibit random to subparallel orientations, often resembling shallow-water "flat pebble conglomerates" (Fig. 8). Enormous blocks of bedded slope sediments, often internally folded, are caught up in the breccias.

The source of these breccias is thought to be the large "intraformational truncation surfaces" (Fig. 9) or "cut-and-fill structures" (Wilson, 1969) which are sharp concave-up discontinuity surfaces that truncate underlying beds and are overlain by a downslope thickening wedge of sediment with an angular relationship on the truncated beds. In these deposits, reduction of shear stress occurs by displacement of

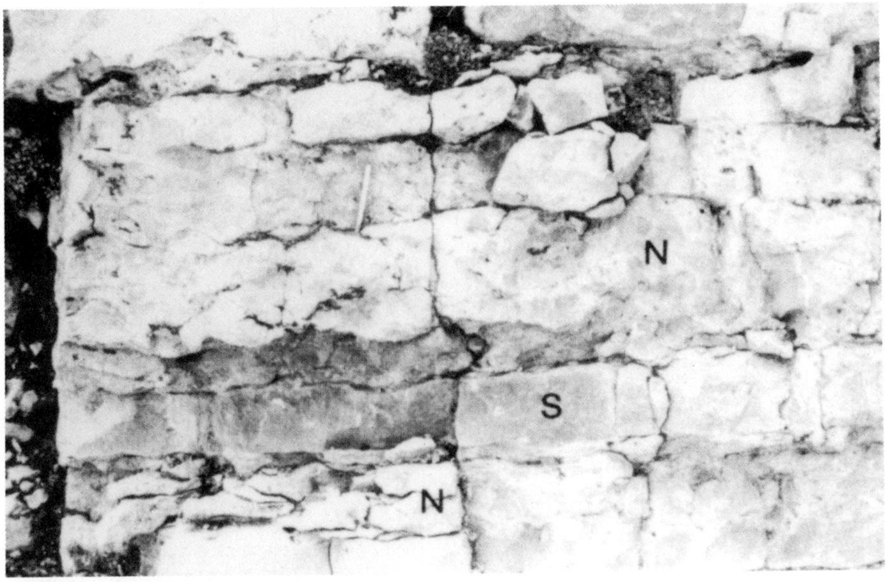

Figure 6
Thinly-bedded, nodular foreslope sequence comprising cemented nodules in compacted calcarenite (N) and a laterally continuous bed of cemented calcarenite (S). These calcarenites form the predominant foreslope facies below the Miette and Ancient Wall buildups (Upper Devonian), Alberta. Photo courtesy of J.C. Hopkins.

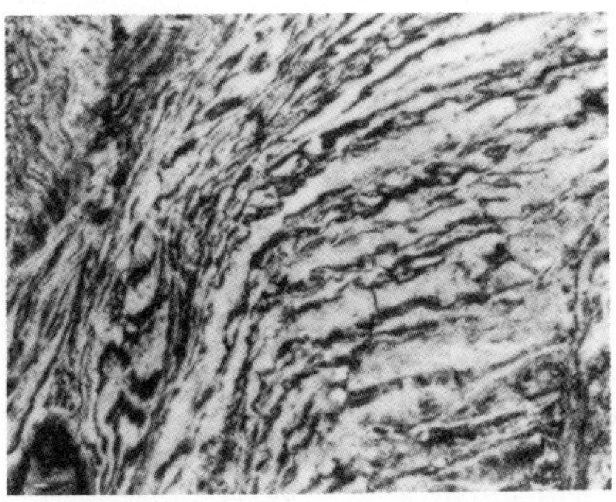

◀ **Figure 7**
Extensive syn-sedimentary distortion of bedding developed by creep in thin to very thin-bedded, upper basinal slope, peri-platform lime mudstones, Eldon Formation (Middle Cambrian), Wapta Mountain, British Columbia.

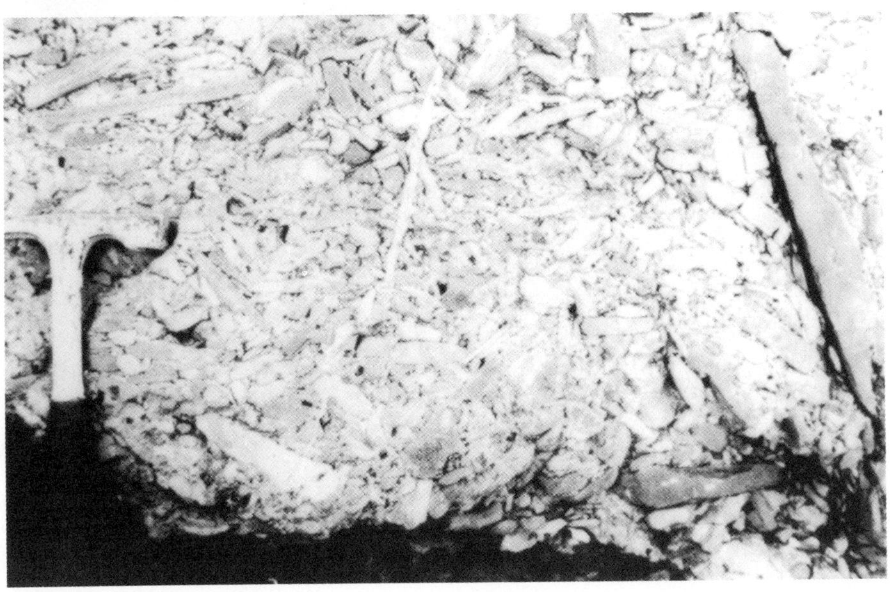

Figure 8
Slope-derived breccia; clasts of partly lithified peri-platform ooze (see Fig. 1) that have been eroded and transported as a clast-supported breccia, Cooks Brook Formation (Middle Cambrian), Humber Arm, Western Newfoundland.

coherent masses along discrete shear planes and not usually by deformation within the mass as occurs in slumps.

The tabular clasts of slope material, although derived by separation along bedding planes, clearly indicate that the slope sediments were partly consolidated very early, probably by submarine cementation. Cementation may have been similar to that in shallow-water with lithified and unlithified layers reflecting times of slow and rapid sedimentation respectively. If cementation and neomorphism took place below the thermocline, dissolution of aragonite and possible precipitation of calcite may have caused the same effect in layers of different original composition (Schlager and James, 1978). Alternatively, if the lime mudstone is interlaminated with shale, cementation of the carbonate may have taken place while the shale remained soft.

Deposits of the two end members, one originating high on the slope and the other down on the lower slope are sometimes found intermixed in extensive breccia masses (Fig. 10). Such deposits are similar to what Schlager and Schlager (1973) term marl-flaser breccia, characterized by a chaotic fabric of plastically deformed, dark grey, argillaceous lime mudstone lithoclasts separating irregular lenses of subangular limestone and other lithoclasts, with the deformed marls forming the flaser fabric. There are thought to be shallow-water derived breccia flows that incorporated lime mudstone clasts from the floor of the slope environment as they moved basinward and they may grade downslope into turbidites.

Graded Calcarenites

A large proportion of any slope sequence commonly consists of size-graded beds of clastic textured limestone, mainly of sand size, interpreted to be the carbonate equivalent of siliciclastic turbidites (Fig. 5). They are envisaged to be deposited from turbidity currents that formed by the sudden surge-type release of dense fluid rather than from a steady state flow such as described by Harms (1974). These deposits have also been called allodapic limestones (Meischner, 1964). Such sediments are well-bedded and characteristically have sharp planar bases that can be coplanar with, or locally scour and truncate underlying

Figure 9
Large intraformational truncation surface in agrillaceous and cherty limestones of the Hare Fiord Formation (Permo-Pennsylvanian), north side of Svartfjeld Peninsula, Ellesmere Island. Note smooth, curved concave-up (listric) geometry of the truncation surface and the lack of macro-scale deformation of beds below or above truncation surface. Shadow at lower left centre is of heliocopter: width of view 150m. Photo courtesy of G.R. Davies.

Figure 11
A bed of light grey calcarenite composed of a graded lower portion, planar laminated middle unit, and the upper portion having climbing ripples (A,B and C Bouma subdivisions respectively), capping a lime breccia, Sekwi Formation (Lower Cambrian), Cariboo Pass, Mackenzie Mountains. Photo courtesy of F.F. Krause.

Figure 10
A large deformed clast of well-bedded peri-platform ooze that was eroded, transported and redeposited as part of a debris flow, Cow Head Group (Middle Ordovician), Cow Head, Western Newfoundland.

slope beds. Sole marks and load structures are usually absent although in some cases they may be obliterated because of stylolitization and solution along bedding contacts. Calcareous turbidites can exhibit all five of the typical ABCDE division of the Bouma sequence but most commonly it is the A, and sometimes the B and C divisions that characterize the deposits (Fig. 11). The particles in the basal parts of division A are often cobble size and larger and the more common grain types are lithoclasts, skeletal debris and ooids, the petrology of which indicates a shallow water origin in contrast to the surrounding pelagic deposits.

The most obvious sources for these units are the unstable accumulations of lime sand and gravel that build up near the platform margin and are occasionally set into motion. It is also possible that they are the distal parts of carbonate debris flows representing the uppermost more dilute turbulent portions of the debris flow (Krause and Oldershaw, 1979). Davies (1977) has suggested a third origin, that skeletal material produced on the slope particularly by pelmatozoans may be easily remobilized.

Post-Paleozoic graded calcarenites derived from sediments further down the slope profile can be virtually indistinguishable compositionally from pelagic limestone. These calcarenites are generally rich in pelagic components such as coccoliths and foraminifers but may also contain lesser amounts of pteropods, sponge spicules, radiolarians, and coarser-grained skeletal debris (especially pelmatozoans). The sediments are size-sorted and may be mixed with clastic terrigenous or volcaniclastic sediment if they have travelled great distances. Although the sedimentary structures such as horizontal laminations, convolutions, occasional channels, flute and groove casts and trace fossils may be present, the A and B divisions of the Bouma sequence are commonly missing and they generally start with the C or D divisions.

Turbidity currents and debris flows appear to be the dominant transport mechanisms for the downslope movement of coarse detritus in modern carbonate slopes (Cook and Mullins, 1983; Enos and Moore, 1983).

Non-Graded Calcarenites

Massive to cross-bedded and ripple-marked calcarenites are an enigmatic type of deposit found in many slope sequences. These deposits are fine- to coarse-grained wackestones to grainstones with occasional large clasts or fossils. Individual beds have sharp bases and vary in geometry from lenticular to irregular masses. The fabric may be random or grains may be aligned parallel to the paleoslope.

The grains in these deposits are variable, ranging in composition from shallow-water derived particles to pelagic grains.

Non-graded calcarenites occurring below a carbonate margin of considerable slope and having an abundant supply of lime sand may have formed from grain flows or more likely, through a modified grain flow mechanism whereby the addition of lime mud matrix and turbulence may have aided dispersive pressures in supporting the grains during transport (Lowe, 1976). It is unlikely that liquified flow, or fluidized flow contribute significantly to the formation of these particular slope calcarenites. Another viable mechanism is reworking of previously deposited slope sediments by bottom currents. Perhaps the massive deposits having an apparent lack of sedimentary structures are nothing more than the product of downslope mass movement of well-sorted lime sands produced at a rapid rate near the platform margin such as occurs on modern leeward open margins during large storms (Hine *et al.*, 1981).

Sedimentary structures in the cross-bedded deposits indicate some sort of bottom currents, often running parallel to the slope (contour currents). Well-sorted, rippled ooid lime sands, sometimes with large scale bed forms, occur in the deeper parts of the slopes around the margins of the Tongue of the Ocean, Bahamas, and are also common on the slopes along the western parts of the Bahama Banks (Mullins and Neumann, 1979) where currents flow along and parallel to the slope at speeds of 60 cm/sec and more (although such velocities are high and not characteristic of today's oceans). These currents may rework bank-derived sands (Hine and Neumann, 1977) or rework pre-existing pelagic slope deposits, leaving only the larger foraminifers and pteropods together with lithoclasts of cemented pelagics to form a deep-water grainstone. They may also winnow the upper parts of turbidites, removing the finer layers and leaving a sequence composed only of shallow-water clasts, and divisions A and B of the Bouma sequence, capped by a cross-bedded lime sand.

Such clean, well-sorted sands are commonly sites of submarine cementation and hardground formation. In such areas precipitation of cement may lead to displacive expansion of grain-to-grain distance, resulting in fracturing and the formation of *in situ* breccias.

THE DEPOSITONAL SETTING

Even though there may be a continuous rain of pelagic sediment on the slope, most slope sediments come from the platform, as re-sedimented gravity flows or peri-platform ooze. Thus, the nature of slope deposition closely reflects events on the platform, or more specifically whether the platform is close to sea level, drowned or exposed (Fig. 12). If the platform surface is close to sea level, then there is a constant source of carbonate sediment being produced and so slope as well as adjacent basin deposits are correspondingly abundant. If sea level drops, the platform is exposed to meteoric diagenesis and the platform sediments are turned to limestone, thus shutting off the supply of sediment to the slope. Starved deep-water sedimentation can also reflect rapid sea level rise. If carbonate production is outpaced, a tranquil deep-water shelf consisting primarily of muddy, skeletal carbonates can develop.

The style of shelf-slope break in the fossil record also reflects the interaction between rates of carbonate production and relative sea-level movement (James and Mountjoy, 1983; Bosselini, 1984), the latter being due to the combined effect of subsidence (tectonic controls) and sea level fluctuations (Schlager and Ginsburg, 1981). If the rate of carbonate accretion is more or less equal to the rate of relative sea level rise, the shelf break will remain more or less in the same position and ultimately the relative relief will increase between the margin and the basin. When relative sea level rise is less than carbonate accretion, the slope facies prograde out over older slope deposits, and will consist of thick accumulations of many re-sedimented lime sands and conglomerates. Opposite circumstances create onlap margins which can be drowned or will move shelfwards in steps. In this situation, little sediment is transported seaward so slope and basin deposits are thin and mostly carbonate muds. Drowned and emergent margins due to large changes in relative sea level result ultimately in starved basinal sedimentation.

In the drowned situation, the style of carbonate shelf is close to what Ahr (1973) has termed a "carbonate ramp", where the high energy zone is the shoreline and the carbonate facies pass progressively into deeper water lithologies. Ginsburg and James (1974) have called these "open shelves" and good modern examples are present on the Yucatan Shelf, West Florida Shelf and in the Persian Gulf. If the ramp has a marked increase in slope at the seaward edge, slope facies will have abundant slumps, slope breccias and turbidites. This situation is termed by Read (1982) a "distally steepened ramp". The absence of a slope break results in a lack of significant slump and sediment gravity flow deposits in the deeper water facies and the formation of a "homoclinal ramp".

The nature of slope facies is also

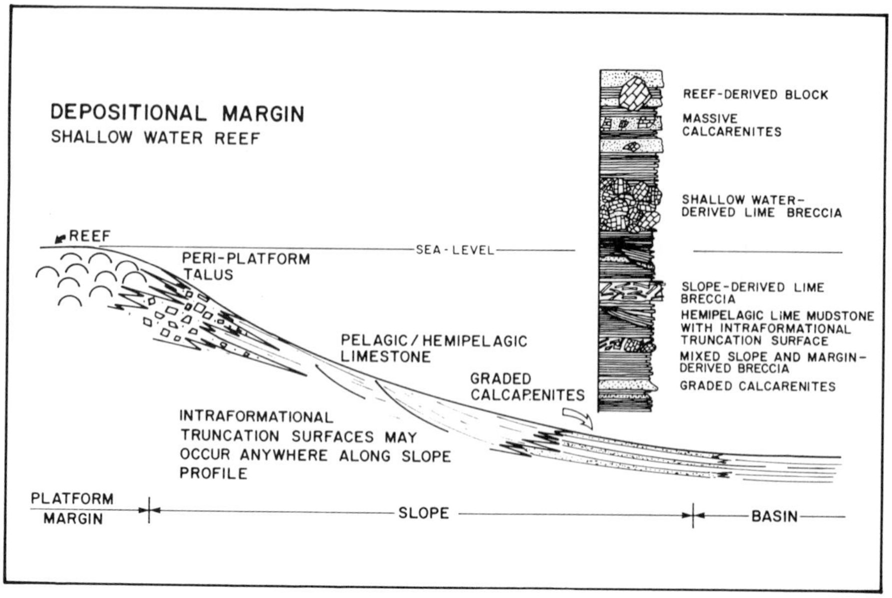

Figure 12

Variations in the style of slope sedimentation as a function of water depth on an adjacent carbonate platform.

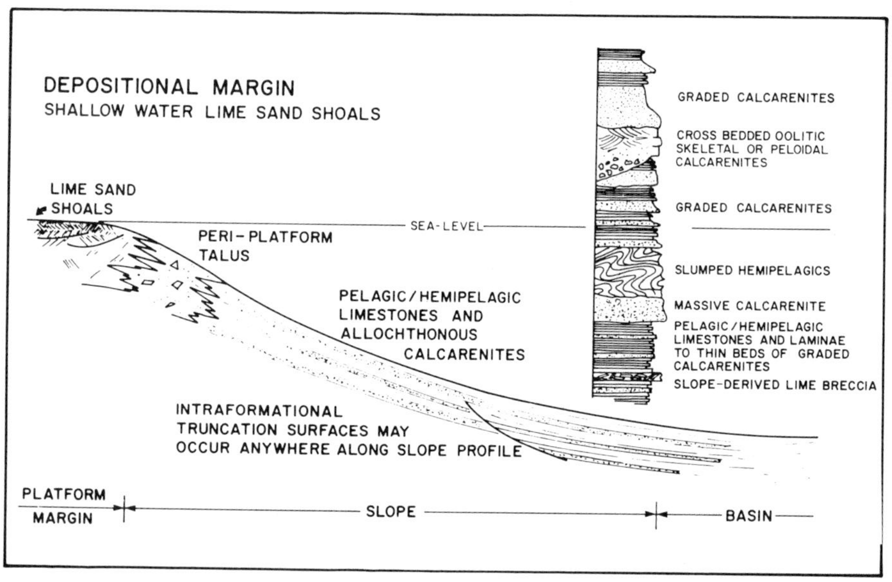

Figure 13

Schematic model for a shallow-water, reef dominated, depositional carbonate margin and illustration of a hypothetical sequence of deposits within the adjacent basin slope.

dependent on the direction and magnitude of off-shelf sediment transport resulting from the net effects of waves, storms and tides. Antecedent topography in the form of islands or subtidal rock ridges create energy barriers which can control the volume of sediment flux on or off the shelf. Oceanic circulation (bottom currents) can modify, through physical transport and erosion, pre-existing slope sediments as well as promote submarine cementation and hardground development on the slopes. For additional details on these parameters and how they affect slope sedimentation along the Northern Bahamas, the reader is referred to Hine *et al.* (1981) and Mullins and Neumann (1979).

In discussing the topic of sea level fluctuation and slope deposition, Schlager and Ginsburg (1981) make the important point that in siliciclastic systems, a lower sea level results in increased erosion and delivery of terrigenous material to the deep; the reverse is true in carbonate systems.

Carbonate slope sedimentation also differs from that adjacent to terrigenous shelves in another important way. Carbonate sand is delivered all along the platform margin. Deep sea fans are absent because no submarine dendritic drainage systems are developed that would funnel sediment into canyons. Instead, the reefs or sand shoals of the platform act as a line source, creating a continuous belt of overlapping turbidites and gravity-flow deposits at the toe-of-slope (Schlager and Chermak, 1979); the resulting deposit is more appropriately termed a debris apron (McIlreath, 1977b; Mullins *et al.*, 1984) which has a geometry that is distinct from a fan.

Carbonate sedimentation providing slope stability results in another distinction from terrigenous shelf-basins. Massive failure of a partly lithified carbonate slope will reduce the possibility of the simple evolution of a single flow (slump → debris flow → turbidity current) as postulated for siliciclastics (Middleton and Hampton, 1973).

FACIES MODELS

Because of the preceding variables we have chosen not to integrate the spectrum of carbonate slope deposits into one model. Instead, we have chosen to model carbonate slope sedimentation adjacent to a platform near sea level. In this situation, the style of slope and basinal sedimentation is dependent upon the relief between platform and basin, and the nature of the shallow portion of the margin. Where the margin itself is a facies transition with a gradual slope profile, then the sequence of slope deposits is very much different from the sequence where the margin is arupt. We have differentiated between these two types of margins and called them depositional and by-pass margins respectively (Fig. 12 to 16)

The nature of the slope sediments in each case also depends on whether the shallow-water margin is formed: 1) by metazoan, calcareous algal or stromatolitic reefs, and occurs either at the edge or slightly downslope below the zone of the most wave movement, or 2) by skeletal algal or oolitic lime-sand shoals. In the case of reefs, the nature of the reef-building biota changes with the

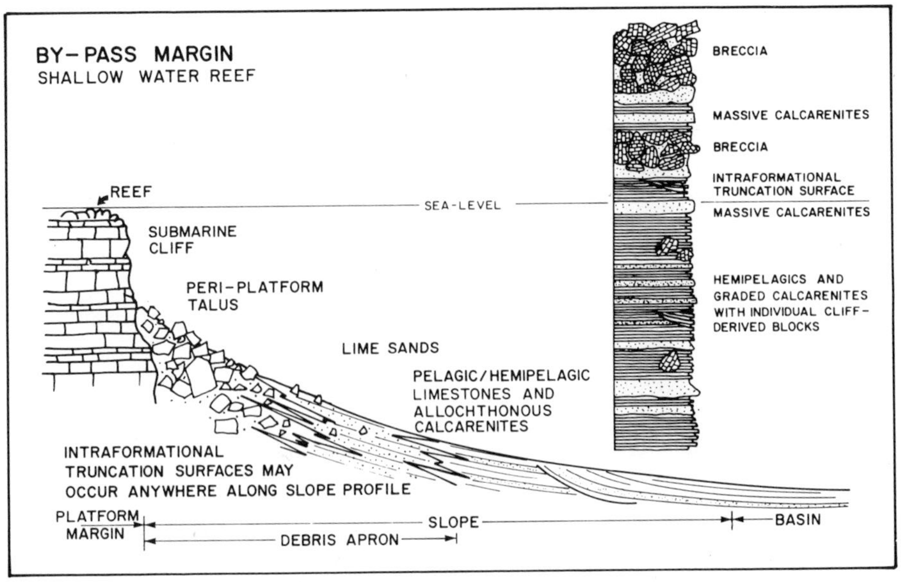

Figure 14
Schematic model for a depositional carbonate margin dominated by shallow-water lime sands and illustration of a hypothetical sequence of adjacent basinal slope deposits.

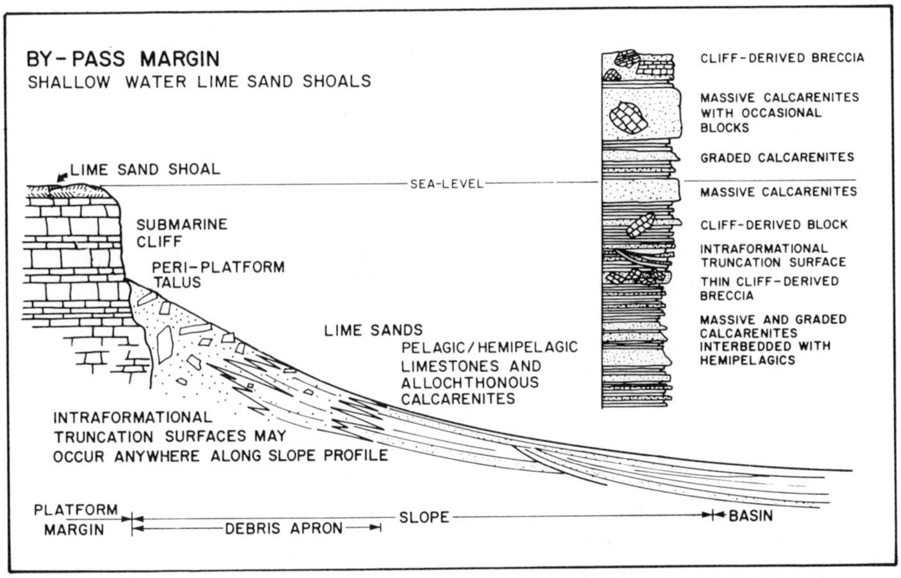

Figure 15
Schematic model for a shallow-water, reef-dominated, by-pass type of carbonate margin in a shallow-basin and illustration of a hypothetical sequence of deposits within the adjacent basin slope. In a deep basin there is an extensive by-pass slope below the peri-platform talus. Debris, including turbidites is funnelled through gullies onto the basin floor.

time and therefore the composition and nature of the resulting debris correspondingly changes.

It should be noted that none of the models is mutually exclusive and within a buildup or platform margin all four (depositional reef, depositional shoal, by-pass reef, by-pass shoal) may be present an any one time. They may be repeated in time and space; in the case of a buildup, it may even be possible to have all four occurring simultaneously in different places along the buildup margin. It is quite common in ancient platforms to progress from a ramp or depositional margin to a by-pass type and even to an erosional margin.

Depositional Margins

The slopes are generally gentle and decrease basinward to merge with the flat basin floor. Because ramps have even lower gradients, slope sediments are normally much finer and consist of pelagic carbonates and shales with minor slumping and slide development.

A) Shallow-Water Reef. The zone of peri-platform talus is relatively narrow but the full spectrum of allochthonous deposits is present downslope (Fig. 12). Because most of the allochthonous material comes from the reef or talus pile many of the allochthonous deposits generated high on the slope are deposited far down on the slope or in the basin. Consequently that zone seaward of the peri-platform talus is often composed of hemipelagic limestones and is often by-passed by the mass movements. This type of depositional slope occurs most frequently around reef complexes and basinward of platform-margin barrier reef systems along paleotopographic highs, structurally positive elements or hingelines in fairly stable cratonic or miogeosynclinal basins. Examples of this style of slope deposit occur in the Upper Devonian and the Cambrian of Western Canada (McIlreath, 1977a) and the Devonian of the Canning Basin, Australia (Conaghan *et al.*, 1976).

B) Shallow-Water Lime Sands. The slope flanking this style of margin is generally a calcarenite wedge of proximal-to-distal turbidite plain (Fig. 13). These slopes probably represent a depositional equilibrium in that sedimentation controls the slope angle and is active all along the profile. Turbidity currents and grain flows are the predominant transport mechanisms, and debris sheets and breccias rare. Some minor debris sheets composed of cemented lime-sand clasts or other slope-derived lithologies may be present. Hardgrounds and incipient brecciation are common.

Examples of this style of slope deposit includes the Pennsylvanian Dimple Limestone, Texas (Thomson and Thomasson, 1969); Silurian of California and Nevada (Ross, 1965); several of the Devonian Fairholme carbonate complexes of Western Canada, and Devonian encrinite banks in Arctic Canada.

By-Pass Margins

In these situations the magin is on top of a cliff or submarine escarpment so that sediments are transported directly from shallow to deep-water. They may bypass much of the slope along a wide front, or be funnelled through channels and canyons to accumulate at the toe-of-slope and adjacent basin. In shallower basins the submarine cliff is the actual by-pass slope, below which accumulates a debris apron of coarse peri-platform talus fining basinwards into peri-platform and pelagic oozes with occasional turbidites. Flows are triggered on the upper slope by oversteepening of peri-platform talus, collapse of the reef wall or by slumping of previously deposited sediments, and are seen to erode gullies on modern slopes (Schlager *et al.*, 1976). The cliffs may result from faulting, large fluctuations in sea-level or just rapid upbuilding of the platform as compared to the basinal deposits. This style of margin is particularly common along block-faulted oceanic margins or at the structural hingeline where a basin is subsiding faster than the adjacent platform.

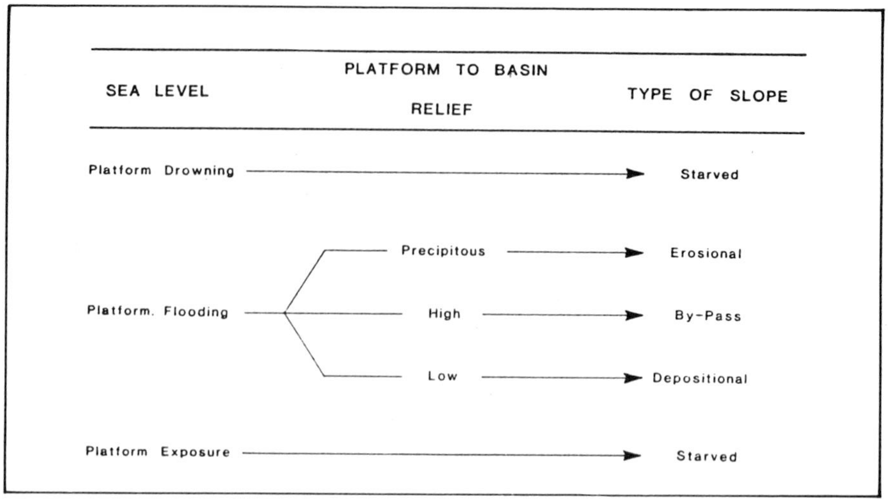

Figure 16
Schematic model for a by-pass type of carbonate margin dominated by shallow-water lime sands in a shallow-basin, and illustration of a hypothetical sequence of adjacent basinal slope deposits.

A) Shallow-Water Reef. Since the reef crowns the escarpment, the most characteristic and spectacular style of accumulation is the wedge of peri-platform talus (Fig. 14). This wedge of material may be enormous, especially if the area is subject to tectonics, with the main transport mechanisms being a combination of rock-fall, sandstreams and gravity-induced downslope mass movement. If the cliff is dissected by channels or canyons, the peri-platform talus may interdigitate along strike with carbonate submarine fans similar to those described for siliciclastic deposits (Evans and Kendall, 1977). Slumps creep and sliding are more active in the deposit than on the adjacent slope due to the variations in lithification. The talus wedge grades downslope into a relatively narrow zone of lime sands and then into pelagic calcilutites to form a debris apron.

This is the style of many modern slope deposits in Belize (Ginsburg and James, 1973), Puerto Rico (Conolly and Ewing, 1967), Jamaica (Goreau and Land, 1974), the Bahamas (Mullins and Newmann, 1979; Schlager and Chermak, 1979), and the Pacific Atolls (Emery *et al.*, 1954). Perhaps the most spectacular fossil example is in the Cretaceous of Mexico (Enos, 1977). Other fossil examples include the Cambrian Cathedral Escarpment, Western Canada, and Upper Devonian margins in the Front Ranges of Western Canada.

B) Lime Sand Shoals. If the shallow-water margin facies is lime sand, the peri-platform talus will also consist predominantly of lime sand intercalated with calcilutites (Fig. 15) with fewer limestone blocks than in the previous model (unless there have been substantial movements in sea level exposing the margin to subaerial diagenesis). Away from the escarpment the lime sands grade relatively quickly into slope or basinal pelagic lime muds. There are minor contributions from turbidites.

A fossil example of such a debris apron of calcarenite is the Cambrian Boundary Limestone (McIlreath, 1977b). Modern examples have been found in the northern Bahamian slopes (Mullins and Newmann, 1979).

The Models as a Norm

In these models we have not consciously placed the slope lithologies in any particular sequence because we feel that the sequence on such a broad scale represents more the complex interactions of sea level and tectonics at the shallow rim than any secondary sedimentary process on the slope. As a result unusual features are likely to record not so much the style of sedimentation on the slope, as the style of sedimentation and tectonics at the shallow margin.

The Models as a Framework and Guide for Description

The differences in carbonate deposition through time are, in large part, a function of the appearance and disappearance of different types of carbonate secreting organisms and thus affect the use of these facies models as a guide in two ways: 1) shallow-water benthic organisms build massive reefs, cause relief at the platform margin, and contribute major amounts of sediment to the slope; however, they are present only at specific times in geologic history; 2) the pelagic calcareous zooplankton and phytoplankton are insignificant in the early Paleozoic, minor in the middle and late Paleozoic, and prolific in the Mesozoic and Tertiary.

As a result, the hemipelagic slope deposition is almost entirely peri-platform ooze in the Precambrian and early Paleozoic, and perhaps one-half peri-platform ooze and one-half true pelagic carbonate in the Mesozoic and Tertiary. Interruptions in the fallout of peri-platform ooze in the Paleozoic sometimes resulted in shale interbeds, whereas in the Mesozoic and Tertiary interbeds are thinner but are wholly pelagic carbonate.

The Models as Predictors in New Situations

Based on a few observations, and bearing in mind the age of the deposits as well as their tectonic setting, one can extrapolate and formulate three critical conclusions: 1) examination of the overall sequence indicates the relative position on the slope and possible proximity of the platform; 2) the lithology of the lime-sand beds and relative calcarenite to hemipelagic ratio gives some idea as to the nature of the slope facies; depositional versus by-pass; and 3) the composition of the clasts indicates the nature of the margin, which has often been obliterated or is inaccessible.

The Models as Basis for Interpretation

The interpretation of carbonate sediment gravity flows has, to date, been based primarily on an analogy with siliciclastic deposits which have similar sedimentary characteristics. One of the important differences between carbonate and siliciclastic sediment gravity flows, however, is that a dispersal model for the hypothetical evolution of a single flow of carbonate platform-derived debris into deep water remains speculative. In contrast to the relatively unconsolidated sediments on continental shelves, carbonate sediments in similar environments tend to be stabilized by organisms and/or well-lithified. This results in distinctively different slope deposits being produced by a variety of gravity-driven transport processes rather than different types of deposits evolving from the same flow. Unequivocal examples of an ancient deposit evolving from the same flow are rare. It should be noted, however, that the concept of a singular flow spawning a series of deposits may apply where slide failure occurs in the lower portion of the slope, remobilizing and transporting these mixed deposits even further basinward.

SUMMARY

Carbonate slope sediments have, in the past, often been either ignored or interpreted as tectonic in origin. Their identification as deposits, separate from tectonically formed mélanges has come largely from a detailed analysis of not only the chaotic deposits but the fine-grained interbeds as well. Our understanding is increasing as more deposits are documented and the first timid steps are being taken beyond the reef into deeper water by submersibles. This latter aspect of carbonate sedimentology is still very much in its infancy.

Refinements of the models presented in this paper must come from two directions, experimentation and more observations from modern carbonate slope environments (in addition to those from current studies on Bahamian slopes). The hydrodynamic parameters and processes for gravity-induced mass movements involving only carbonate materials must be carefully documented and contrasted with the results from siliciclastic materials. A combination of detailed observations from submersibles, and high resolution seismic and bottom sampling is needed to make an inventory of the spectrum of sediments and structures that make up carbonate slope environments in the modern ocean.

ACKNOWLEDGEMENTS

Many of the original ideas presented here came about as a result of stimulating discussions with John Harper and Bob Stevens. John Hopkins, Dave Morrow and Federico Krause critically read the first manuscript and Roger Walker the present version. Gary Allwood and Dave Haden drafted the figures and Oralie Shearing typed the current manuscript.

REFERENCES

BASIC REFERENCES ON CARBONATE BASIN SLOPE FACIES

Cook, H.E., and Enos, P., eds., 1977. Deep-water carbonate environments: Society of Economic Paleontologists and Mineralogists, Special Publication 25, 336 p

Cook, H.E., and Mullin, H.T., 1983. Basin margin environment. *In* Scholle, P.A., Bebout, D.G., and Moore, C.H., eds., Carbonate depositional environments. American Association of Petroleum Geologists, Memoir 33, p. 540-617.

Enos, P., and Moore, C.H., 1983. Fore-reef slope environment. *In* Scholle, P.A., Bebout, D.G., and Moore, C.H., eds., Carbonate depositional environments. American Association of Petroleum Geologists, Memoir 33, p. 508-537.

Hsu, K.J., and Jenkyns, H.C., eds., 1974. Pelagic sediments on land and under the sea. International Association of Sedimentologists, Special Publication 1, 448 p.

James, N.P., and Mountjoy, E.W., 1983. Shelf-slope break in fossil carbonate platforms: an overview. *In* Stanley, D.J., and Moore, G.T., eds., The shelfbreak: critical interface on continental margins. Society of Economic Paleontologists and Mineralogists, Special Publication 33, p. 189-206.

Mountjoy, E.W., Cook, H.E., Pray, L.C., and McDaniel, P.N. 1972. Allochthonous carbonate debris flows - worldwide indicators of reef complexes, banks or shelf margins. Proceedings of 24th International Geological Congress, Section 6, p. 172-189.

Read, J.G., 1982. Carbonate platforms of passive (extensional) continental margins: types, characteristics and evolution. Tectonophysics, v. 81, p. 195-212.

Scholle, P.A.; 1977. Deposition, diagenesis and hydrocarbon potential of "deeper-water" limestones. American Association of Petroleum Geologists, Continuing Education Course Notes, Series No. 7, 25 p.

Scholle, P.A., Arthur, M.A., and Ekdale, A.A., 1983. Pelagic environment. *In* Scholle, P.A., Bebout, D.G., and Moore, C.H., eds., Carbonate depositional environments. American Association of Petroleum Geologists, Memoir 33, p. 620-692.

Wilson, J.L., 1969. Microfacies and sedimentary structures in "deeper-water" lime mudstones. *In* Friedman, G.M., ed., Depositional environments in carbonate rocks. Society of Economic Paleontologists and Mineralogists, Special Publication 14, p. 4-19.

EXAMPLES OF ANCIENT CARBONATE BASIN SLOPES

The following papers contain examples of ancient carbonate basin slopes, with a bias towards examples in Canada. However as with reefs, the location of the country in temperate to boreal latitudes during much of the Mesozoic and, unfortunately, the Cenozoic results in Canadian examples being mainly Paleozoic in age.

Bosellini, A., 1984. Progradation geometrics of carbonate platforms: examples from the Triassic of the Dolomites, northern Italy. Sedimentology, v. 31, p. 1-24

Conaghan, P.J., Mountjoy, E.W., Edgecombe, D.R., Talent, J.A., and Owen, D.E., 1976. Nubrigyn algal reefs (Devonian), eastern Australia: allochthonous blocks and megabreccias. Geological Society of America, Bulletin, v. 87, p. 515-530.

Cook, H.E., McDaniel, P.N., Mountjoy, E.W., and Pray, L.C., 1972. Allochthonous carbonate debris flows at Devonian bank ("reef") margins Alberta, Canada. Bulletin of Canadian Petroleum Geology, v.20, p. 439-497.

Davies, G.R., 1977. Turbidites, debris sheets and truncation structures in upper Paleozoic deep-water carbonates of the Sverdrup Basin, Arctic Archipelago. *In* Cook, H.E., and Enos, P., eds., Deep-water carbonate environments. Society of Economic Paleontologists and Mineralogists, Special Publication 25, p. 221-249.

Enos, P., 1977. Tamabra limestone of the Poza Rica trend, Cretaceous, Mexico. *In* Examples of ancient carbonate slope-to-basin systems. Cook, H.E., and Enos, P., eds., Deep-water carbonate environments. Society of Economic Paleontologists and Mineralogists, Special Publication 25, p. 273-314.

Hopkins, J.C., 1977. Production of fore-slope breccia by differential submarine cementation and downslope displacement of carbonate sands, Miette and Ancient Wall buildups, Devonian, Canada. *In* Cook, H.E., and Enos, P., eds., Deep-water carbonate environments. Society of Economic Paleontologists and Mineralogists, Special Publication 25, p. 155-170.

Hubert, J.F., Suchecki, R.K., and Callahan, R.K.M., 1977. Cow Head breccia: sedimentology of the Cambro-Ordovician continental margin, Newfoundland. *In* Cook, H.E., and Enos, P., eds., Deep-water carbonate environments. Society of Economic Paleontologists and Mineralogists, Special Publication 25, p. 125-154.

Hurst, J.M., and Surlyk, F., 1983. Depositional environments along a carbonate ramp to slope transition in the Silurian of Washington Land, North Greenland. Canadian Journal of Earth Sciences, v. 20, p. 473-449.

Kindle, C.H., and Whittington, H.B., 1958. Stratigraphy of the Cow Head Region, Western Newfoundland. Geological Society of America, Bulletin, v. 69, p. 315-342.

Krause, F.F., and Oldershaw, A.E., 1979. Submarine carbonate breccia beds - a depositional model for two-layer, sediment gravity flows from the Sekwi Formation (Lower Cambrian), Mackenzie Mountains, Northwest Territories. Canadian Journal of Earth Sciences, v. 16, p. 189-199.

Mackenzie, W.S., 1970. Allochthonous reef-debris limestone turbidites, Powell Creek, Northwest Territories. Bulletin of Canadian Petroleum Geology, v. 18, p. 474-492.

McIlreath, I.A., 1977a. Stratigraphic and sedimentary relationships at the western edge of the Middle Cambrian Carbonate Facies Belt, Field, British Columbia. Ph.D. Thesis, University of Calgary, Calgary, Alberta, 269 p.

McIlreath, I.A., 1977b. Accumulation of a Middle Cambrian, deep-water, basinal limestone adjacent to a vertical, submarine carbonate escarpment, southern Rocky Mountains, Canada. *In* Cook, H.E., and Enos, P., eds., Deep-water carbonate environments. Society of Economic Paleontologists and Mineralogists, Special Publication 25, p. 113-124.

Morrow, D.W., 1978. The Prairie Creek Embayment and associated slope, shelf and basin deposits. Geological Survey of Canada Paper 78-1A, p. 361-370.

Ross, D.C., 1965 Geology of the Independence Quadrangle, Inyo County, California. United States Geological Survey, Bulletin 1181-0, 64 p.

Srivastava, P.C., Stearn, C.W., and Mountjoy, E.W., 1972. A Devonian megabreccia at the margin of the Ancient Wall carbonate complex, Alberta. Bulletin of Canadian Petroleum Geology, v. 20, p. 412-438.

Thomson, A.F., and Thomasson, M.R., 1969. Shallow to deep water facies development in the Dimple Limestone (Lower Pennsylvanian), Marathon Region, Texas. *In* Friedman, G.M., ed., Depositional environments in carbonate rocks. Society of Economic Paleontologists and Mineralogists, Special Publication 14, p. 57-78.

Tyrrell, W.W., Jr., 1969. Criteria useful in interpreting environments of unlike but time-equivalent carbonate units (Tansill-Capitan-Lamar), Capitan Reef Complex, West Texas and New Mexico. *In* Friedman, G.M., ed., Depositional environments in carbonate rocks. Society of Economic Paleontologists and Mineralogists, Special Publication 14, p. 80-97.

Unfortunately a number of excellent papers on Mississippian and Permian basin slope carbonates occurring in New Mexico and Texas exist in difficult-to-obtain guidebooks of local societies. For example, the "Guidebook to the Mississippian Shelf-edge and Basin Facies Carbonates, Sacramento Mountains and Southern New Mexico Region", published by the Dallas Geological Society (1975).

EXAMPLES OF MODERN CARBONATE BASIN SLOPES

Andrews, J.E., 1970. Structure and sedimentary development of the outer channel of the Great Bahama Canyon. Geological Society of America, Bulletin, v. 81, p. 217-226.

Andrews, J.E., Shepard, F.P., and Hurley, R.J., 1970. Great Bahama Canyon. Geological Society of America, Bulletin, v. 81, p. 1061-1078.

Bushby, R.F., 1969. Ocean surveying from manned submersibles. Marine Technological Society Journal, v. 3, p. 11-24.

Conolly, J.R., and Ewing, M., 1967. Sedimentation in the Puerto Rico Trench. Journal of Sedimentary Petrology, v. 37, p. 44-59.

Emery, K.O., Tracey, J.I., Jr., and Ladd, H.S., 1954. Geology of Bikini and nearby atolls. United States Geological Survey Paper 260-A, p. 1-262.

Ginsburg, R.N., and James, N.P., 1973. British Hondoras by submarine. Geotimes, v. 18, p. 23-24.

Goreau, T.F., and Land, L.S., 1974. Fore-reef morphology and depositional processes, North Jamaica. *In* Laporte, L.F., ed., Reefs in time and space. Society of Economic Paleontologists and Mineralogists, Special Publication 18, p. 77-89.

Hine, A.C., Wilber, R.J., and Neumann, A.C., 1981. Carbonate sand bodies along contrasting shallow bank margins facing open seaways; northern Bahamas. American Association of Petroleum Geologists, Bulletin, v. 65, p. 261-290.

James, N.P., and Ginsburg, R.N., 1979. The deep seaward margin of Belize barrier and atoll reefs. International Association of Sedimentologists Sepcial Publication 3, 201 p.

Land, L.S., and Moore, C.H., 1977. Deep fore-reef and upper island slope, North Jamaica. *In* Frost, S.H., Weiss, M.P., and Saunders, J.B., eds., Reefs and related carbonates: ecology and sedimentology. American Association of Petroleum Geologists, Studies in Geology No. 4, p. 53-67.

Moore, C.H., Graham, E.A., and Land, L.S., 1976. Sediment transport and dispersal across the deep fore-reef and island slope (-55m to -305m), Discovery Bay, Jamaica. Journal of Sedimentary Petrology, v. 46, p. 174-187.

Mullins, H.T., and Neumann, A.C., 1979. Deep carbonate bank margin structure and sedimentation in the northern Bahamas. *In* Doyle, L.J., and Pilkey, O.H., eds., Geology of continental slopes. Society of Economic Paleontologists and Mineralogists, Special Publication 27, p. 165-192.

Mullins, H.T., Heath, K.C., VanBuren, H.M., and Newton, C.R., 1984. Anatomy of a modern open-ocean carbonate slope: northern Little Bahama Bank. Sedimentology, v. 31, p. 141-168.

Paulus, F.J., 1972. The geology of site 98 and the Bahama Platform. Initial reports of the deep sea drilling project. National Science Foundation, Washington, D.C., v. 11, p. 877.

Schlager, W., and Chermak, A., 1979. Sediment facies of platform-basin transition, Tongue of the Ocean, Bahamas. *In* Doyle, L.J., and Pilkey, O.H., eds., Geology of continental slopes. Society of Economic Paleontologists and Mineralogists, Special Publication 27, p. 193-208.

ADDITIONAL REFERENCES CITED IN TEXT

Ager, D.V., 1973. The nature of the stratigraphical record. New York, J. Wiley and Sons, Inc., 114 p.

Ahr, W.M., 1973. The carbonate ramp: an alternative to the shelf model. Transactions of the Gulf Coast Association of Geological Sciences , v. 23, p. 221-225.

Dott, R.H., Jr., 1963. Dynamics of subaqueous gravity depositional processes. American Association of Petroleum Geologists, Bulletin, v. 47, p. 104-128.

Evans, I., and Kendal, C.G. St. C., 1977. An interpretation of the depositional setting of some deep-water Jurassic carbonates of the central High Atlas Mountains, Morocco. *In* Cook, H.E., and Enos, P., eds., Deep-water carbonate environments. Society of Economic Paleontologists and Mineralogists, Special Publication 25, p. 249-261.

Ginsburg, R.N., and James, N.P., 1974. Holocene carbonates of continental shelves. *In* Burk, C.A., and Drake, C.L., eds., Geology of continental margins. New York, Springer-Verlag, p. 137-155.

Hampton, M.A., 1972. The role of subaqueous debris flow in generating turbidity currents. Journal of Sedimentary Petrology, v. 42, p. 775-793.

Harms, J.C., 1974. Brushy Canyon Formation, Texas: a deep-water density current deposit. Geological Society of America, Bulletin, v. 85, p. 1763-1784.

Hay, W.W., Southam, J.R., and Noel, M.R., 1976. Carbonate mass balance - cycling and deposition on shelves and in deep sea (Abst.). American Association of Petroleum Geologists, Bulletin, v. 60, p. 678.

Hine, A.C., and Neumann, A.C., 1977. Shallow carbonate-bank-margin growth and structure, Little Bahama Bank, Bahamas. American Association of Petroleum Geologists, Bulletin, v. 61, p. 376-406.

James, N.P., 1978. Facies models 10: reefs. Geoscience Canada, v. 5, p. 16-26.

James, N.P., and Choquette, P.W., 1983. Diagenesis 6, limestones - the sea floor diagenetic environment. Geoscience Canada, v. 10, p. 162-179.

Lowe, D.R., 1976. Grain flow and grain flow deposits. Journal of Sedimentary Petrology, v. 46, p. 188-199.

Meischner, K.D., 1964. Allodapische Kalke, Turbidite in riff-nahen Sedimentations - Becken. *In* Bouma, A.H., and Brouwer, A., eds., Turbidites. Amsterdam, Elsevier, p. 156-191.

Middleton, G.V., and Hampton, M.A., 1973. Sediment gravity flows; mechanics of flow and deposition. *In* Middleton, G.V., and Bouma, A.H., eds., Turbidites and deep-water sedimentation. Society of Economic Paleontologists and Mineralogists, Short Course, p. 1-38.

Schlager, W., and Schlager, M., 1973. Clastic sediments associated with radiolarites (Tauglboden-Schichten, Upper Jurassic, eastern Alps). Sedimentology, v. 20, p. 65-89.

Schlager, W., Hooke, R.L., and James, N.P., 1976. Episodic erosion and deposition in the Tongue of the Ocean, Bahamas. Geological Society of America Bulletin, v. 87, p. 1115-1118.

Schlager, W., and James, N.P., 1978. Low-magnesian calcite limestones forming at the deep-sea floor, Tongue of the Ocean, Bahamas. Sedimentology, v. 25, p. 675-702.

Schlager, W., and Ginsburg, R.N., 1981. Bahama carbonate platforms - the deep and the past. Marine Geology, v. 44, p. 160-181.

Tucker, M.E., 1974. Sedimentology of Palaeozoic pelagic limestones: the Devonian Griotte (Southern France) and Cephalopodenkalk (Germany). *In* Hsu, K.J., and Jenkyns, H.C., eds., Pelagic sediments: on land and under the sea. International Association of Sedimentologists Special Publication 1, p. 71-92.

Walker, R.G., 1976. Facies models 2: Turbidites and associated coarse clastic deposits: Geoscience Canada, v. 3, p. 25-36.

CARBONATE DIAGENESIS

Reprinted by permission of the Society of Economic Paleontologists and Mineralogists. Published in *Carbonate Cements*, N. Schneidermann and P.M. Harris, editors, SEPM Special Publication No. 36, pp. 79-95.

CARBONATE CEMENTATION—A BRIEF REVIEW

PAUL M. HARRIS, CHRISTOPHER G. ST. C. KENDALL[1] AND IAN LERCHE[1]
Gulf Research & Development Company, P.O. Box 37048, Houston, Texas 77236

ABSTRACT

Porosity prediction in subsurface carbonates remains difficult although our understanding of cementation in carbonate rocks has greatly increased during the past few years. The diagenetic setting is inherited from the depositional setting but continually modified during burial. In the marine setting, cementation occurs in the deep sea, in reefs, on shallow water platforms, and along shorelines. Carbonate and evaporite cementation typifies supratidal flats. During shallow burial, carbonate rocks contact fresh waters and additional cementation occurs along with solution, mineralogical stabilization and possibly dolomitization. Continued burial results in physical and chemical compaction, pressure solution and additional cementation.

Several factors play a role in cementation in carbonates, i.e., the rate and total volume of water movement and timing of cementation relative to hydrocarbon generation; fluid chemistry; and the effect of pH, CO_2 activity, temperature and pressure on mineral solubilities. Petrographic studies coupled with measurements of stable isotopic ratios, trace elements and fluid inclusions are made to assess various stages of cementation and their relation to burial history and fluid flow and composition. Such integrated studies are providing a more detailed understanding of diagenetic events, especially cementation, and ultimately will increase the ability to predict porosity distribution.

INTRODUCTION

With few exceptions, hydrocarbons are reservoired in sandstones or carbonates. In sandstones the porosity is generally interparticle while in carbonate rocks the porosity is associated with a wide variety of depositional and diagenetic fabrics (Choquette and Pray, 1970; Wardlaw, 1976). In efforts to find stratigraphic traps in clastics, despite the local importance of diagenesis, the greatest emphasis is normally placed on interpreting depositional facies and their regional trends. However, with carbonates, cementation and other diagenetic effects are equally important considerations. This paper briefly reviews the environments of cementation in carbonate rocks and many of the techniques commonly used in diagenetic studies.

Depositional porosity in carbonates is usually a function of sediment sorting and shape, which in turn are related to wave and current activity and the biologic community (salinity, turbidity, nutrient supply) at the depositional site. Where currents and waves are particularly active, lime mud is winnowed away leaving porous carbonate sands (Folk, 1962; Dunham, 1962; Enos and Sawatsky, 1981). In contrast, in low energy environments lime muds tend to collect and once the sediment has dewatered, the depositional porosity is much reduced (Shinn *et al.*, 1977; Enos and Sawatsky, 1981). This seemingly simple relation between mud percentage and bottom agitation is complicated since: (1) mud-sized carbonate in reef rock and hardgrounds may actually be micrite cement and not indicative at all of a low energy setting (MacIntyre, 1977); (2) mud-sized carbonate may be replaced during diagenesis by pseudospar and so interpreted as deposited in higher energy settings; and (3) during compaction and pressure solution (Shinn *et al.*, 1977; Wanless, 1979), carbonate grains floating in a mud matrix may be "squeezed" together biasing the texture toward a more grain-rich, mud-free lithology.

Many of the world's larger carbonate reservoirs have porosities which are partly primary in origin. A few examples are the Ghawar field in the Jurassic limestones of Saudi Arabia (Arabian American Oil Company Staff, 1959; Wilson, in press), the Jurassic Smackover limestones in southern Arkansas and northern Louisiana (Moore and Druckman, 1981; Baria *et al.*, 1982), the Devonian Leduc limestones at Red Water in Alberta (Klovan, 1964) and the Mississippian Madison limestones in North Dakota and Saskatchewan (Lindsay and Kendall, in press).

The relationship between porosity and diagenesis is complicated. Apart from mechanical and chemical compaction, three diagenetic processes controlling porosities are solution, cementation and dolomitization (Fig. 1). These processes require a permeable host rock and a mechanism to flush chemically active waters through the rock. Regionally this water movement is controlled by tidal pumping, hydrostatic head, or shale compaction, thermal expansion, smectite to illite conversion and rock fabric of the region.

Partial solution of massive reef limestones in the Miocene of Indonesia and the Cretaceous of the Middle East has led to the development of limestones with excellent reservoir characteristics (Vincellette and Soeparjadi, 1976; Epting, 1980; Frost *et al.*, 1983; Harris and Frost, 1984). However, where dissolution of carbonate sand grains is accompanied by calcite pre-

[1]Present Address: Department of Geology, University of South Carolina, Columbia, South Carolina 29208

cipitation in adjacent primary pores, the end product may be the same amount of porosity but a shift to moldic porosity and variable permeability. Examples of this can be seen in the Smackover Formation in Arkansas (Moore and Druckman, 1981) as well as the Jurassic Haynesville Formation of east Texas where porosities are 15% and higher, but the permeabilities generally are less than 0.5 millidarcies.

The relationship of dolomitization to porosity is complex: dolomitization may reduce, preserve or redistribute porosity. In some carbonate reservoirs, as in the Jurassic Arab limestone of the Ghawar field in Saudi Arabia, dolomite cements partially reduce the primary porosity. Such dolomite cements may be quite common (Kaldi and Gidman, 1982; Sibley, 1982). Contradicting the reputation dolomitization has for porosity enhancement, Halley and Schmoker (1983) have illustrated that at certain burial depths dolomites may have lower porosities than adjacent limestones. In many dolomitized reservoirs such as the Jurassic Smackover formation in Alabama and Florida (Baria *et al.*, 1982) and the Rainbow Lake reef carbonates in Alberta (Langton and Chin, 1968), moldic porosity is also important. Early dolomitization can preserve porosity, particularly in initially porous pelletal lime mudstones, by creating a rigid framework that inhibits compaction of the soft pellets (Roehl, 1967). In these cases the porosity exists either between slightly to moderately compacted soft pellets or between dolomite rhombohedrons. The association of porosity with dolomitized mud-rich carbonates is common in the Mississippian Mission Canyon (Lindsay and Kendall, in press) and Red River formations of the Williston Basin (Roehl, 1967; Longman *et al.*, 1983) along with the Permian San Andres and Grayburg reservoirs of the Permian Basin (Longacre, 1980; Bein and Land, 1982).

Fractures may increase porosity, but more commonly enhance permeability (Stearns and Friedman, 1972; Currie and Nwachukwu, 1974; McNaughton and Garb, 1975; Tillman, 1983). A classic example of this can be seen in the Ekofisk field of the North Sea where overpressured Danian Chalk has been fractured by underlying salt movement (Van den Bark and Thomas,

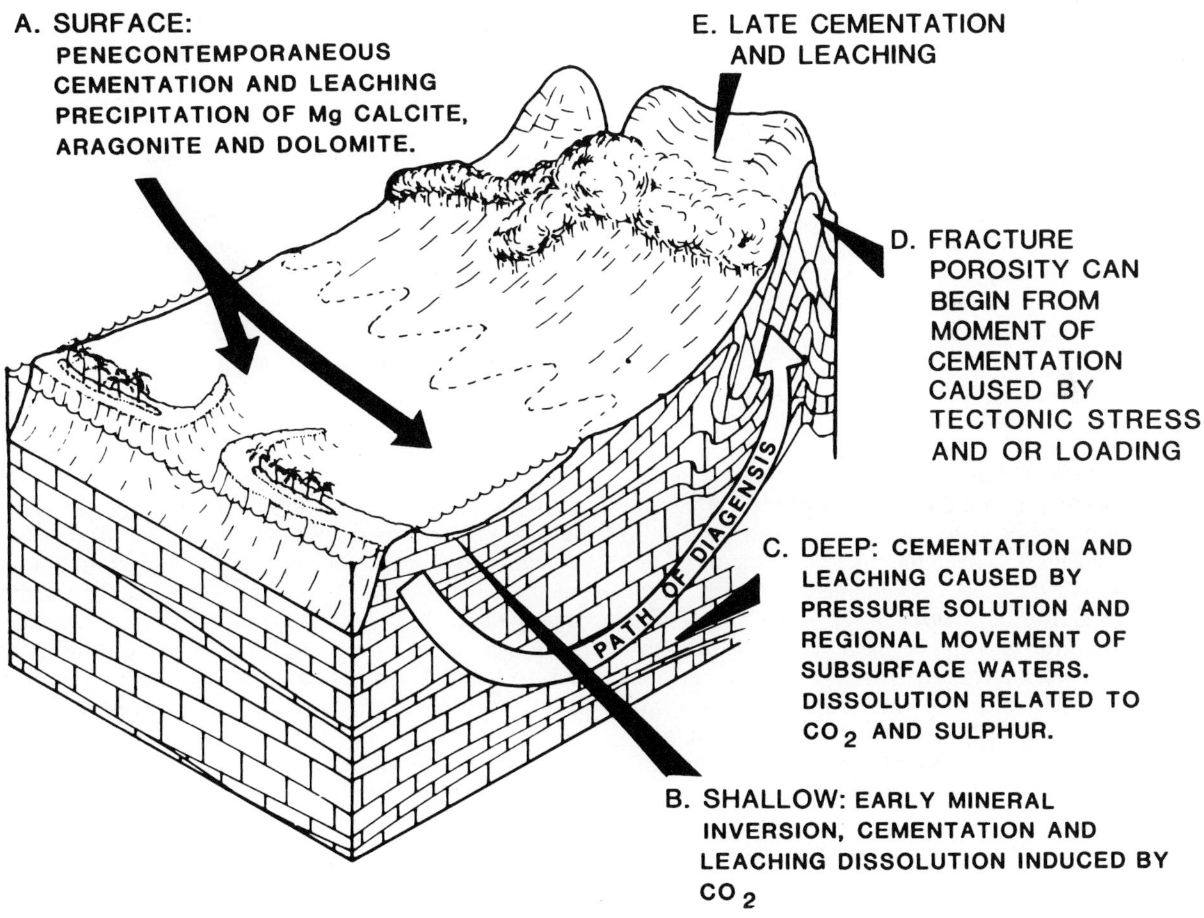

FIG. 1.—Generalized path of diagenesis for carbonate sediments. The diagenetic environments are discussed in more detail in the text.

1981). Here porosities in the reservoir are over 30% but, without fracturing, the permeability would be negligible. From our own observations of interbedded grain carbonates and lime muds, we note that lime muds are more likely to be fractured and the fracturing frequently increases reservoir continuity. Presumably the lime muds are denser and more competent than grain carbonates and so fracture more easily.

Porosity-filling cements in carbonates are characterized by measurements of $^{13}C/^{12}C$, $^{18}O/^{16}O$ and $^{87}Sr/^{86}Sr$ compositions as well as trace elements (see Veizer *et al.*, 1971; Veizer and Compston, 1974; Veizer and Hoefs, 1976; Hudson, 1977; Walls *et al.*, 1979; Campos and Hallam, 1979; Dickson and Coleman, 1980; Moore and Druckman, 1981; Arthur *et al.*, 1982). Microsampling techniques coupled with staining of thin sections (Friedman, 1959; Evamy, 1969) and cathodoluminescence (Meyers, 1974; Pierson, 1981; Frank *et al.*, 1982) are being increasingly used to document the isotopic and elemental signatures of individual cement zones within pores. Quantitative estimates based on mass balance calculations of the minimal amount of fluids that have passed through rocks as recorded by their stable isotopic composition have been presented by Magaritz and Taylor (1976) and Magaritz (1983). The δD composition of aqueous inclusions as well as in clays and chert, $\delta^{15}N$ composition of organic matter and oils, $\delta^{34}S$ composition of associated evaporites and organic matter, together with the uranium-thorium-potassium series of clays and organics can also be used as indicators of the paleoevolution of the rock.

Although distribution coefficients and fractionation factors may not be precisely known, the trends within co-genetic suites of diagenetic carbonates are used to obtain considerable insight into their diagenetic history (Land, 1980). In addition, examination of covariate trends of $\delta^{13}C$ and $\delta^{18}O$ compositions of components from a single stratigraphic unit is an extremely powerful approach.

The content of fluid inclusions in cements reflects the composition of the fluids during precipitation provided there have been no significant pressure and temperature changes between the time of formation of the inclusion and the time of sampling (Roedder, 1979; Hanor, 1980; Burruss, 1981). When such changes have occurred, corrections can be made to approximate the original composition (Burruss, 1981). The temperature at which the fluid inclusion was formed can be inferred by directly heating the inclusion and seeing at what point the fluid vaporizes; and the salinity of the precipitating fluid can also be estimated by freezing the inclusions (Klosterman, 1981; Moore and Druckman, 1981). Fluorescence studies with respect to different wavelengths of light determine whether the inclusions contain oil, condensate or brine (Burruss, 1981; Burruss *et al.*, 1983). Occasionally, direct composition measurements of the fluid in an inclusion are possible by gas chromatography.

To find efficiently hydrocarbon reservoirs in carbonates, porosity preservation needs to be understood relative to the generation, migration and accumulation of hydrocarbons. An important key to interpreting carbonate diagenesis and cementation is recognition of the respective timing and frequency of diagenetic events and their tie to sea level fluctuations, tectonic events, burial, development of trap and seal and oil migration. The extensive literature on cementation in carbonates is too vast for a complete review, but important studies include Folk (1959), Bricker (1971), Bathurst (1971), Milliman (1974), Zenger *et al.* (1980), Flügel (1982) and summary papers by Folk (1973), Moore (1979, 1980), and Longman (1980, 1981).

ENVIRONMENTS OF CEMENTATION

The cementation and diagenesis of carbonates can take place in many settings (Fig. 2): the marine environment during the deposition of the sediment, in a near-surface position where a freshwater lens penetrates the depositional environment, in a subaerial setting produced by a sea level fall, in a mixing zone between waters of varying chemistries, or in the brines of the deeper subsurface. We cannot apply absolute depths to these diagenetic realms, for the observed processes vary in importance and depth between basins.

The preservation of high porosities and permeabilities at depth within carbonate reservoirs is contrary to current thinking on the importance of cementation in surface environments. This would have the pores of these sediments plugged by cements early in their diagenetic history when unstable mineralogies in sediments are in contact with dynamic near-surface waters. Clearly the Florida and Yucatan aquifers, among many, argue against early porosity loss. Halley and Schmoker (1983) present evidence that in South Florida, despite near surface diagenesis, porosities remain high shallow in the section and decrease slowly with depth. In other settings, perhaps burial was so rapid that carbonates quickly passed through the near-surface phreatic zone and underwent little diagenesis. It is also possible all the carbonate is converted to low magnesian carbonate "upstream" in an aquifer and the supply of $CaCO_3$ originally contributed from metastable mineralogies was exhausted. Another possibility is that evaporites, shales, red beds or dense micrites formed protective impermeable seals updip, over and within the porous carbonates and thus prevented downward penetration of water which might cause cementation and porosity occlusion. A final possibility (suggested by Halley and Schmoker, 1983) is that diagenesis in shallow environments simply does not completely plug porosity, but that tight limestones develop during burial as the result of pressure solution, compaction and additional cementation.

Deep burial processes such as grain to grain interpenetration, stylolitization by physical compaction or pressure solution (Dunnington, 1967; Logan and Semeniuk, 1976; Wanless, 1979; Meyers, 1980; Pratt, 1982) and cementation (Moore and Druckman, 1981) are significant. Studies such as that of the Lake Valley formation (Meyers, 1978; 1980) show that about half of the cement volume is derived from burial processes. It is clear then that to predict porosity in subsurface carbonates one should consider: (1) the depositional environment where depositional porosity is generated, (2) the early diagenetic environments where the porosity can be enhanced, preserved or occluded, and (3) the burial history and associated diagenetic changes.

Marine Setting

Rapid marine cementation has been found to be common and some of the more spectacular examples include: the incorporation of war material and human skeletons into beach rock (Milliman, 1974); the presence of beer bottles in the subaqueous crusts from the Persian Gulf seafloor (Shinn, 1969); and the annual quarrying of beach rock for building purposes in some Pacific atolls, with the rapid formation of further beach rock following the removal of the quarried material (Moresby, 1835). Marine cementation is not restricted by water depth and can be traced from deep water to beaches (Fig. 3).

Deep Sea.—Deep sea cements are formed of magnesian calcite, calcite and aragonite (Gevirtz and Friedman, 1966; Bartlett and Greggs, 1969; Thompson, 1972; Schlager and James, 1978). Deep sea hardgrounds occur in areas where there is good bottom current movement. This is particularly true in crusts and lithoherms on the flanks of carbonate platforms (Mullins *et al.*, 1980) and the margins of submarine channels and canyons (Schlager and Chermak, 1979) where aragonite is precipitated. Deep water cementation may also occur in areas with low sedimentation rate or within enclosed basins with elevated salinities. An example of this latter case are the sediments found in the deeper parts of the Red Sea (Milliman *et al.*, 1969) and the Mediterranean (Fischer and Garrison, 1967) containing hardgrounds cemented by magnesian calcite.

Reefs.—(Fig. 4). In shallower water settings, particularly in reef margins, cementation is more widespread than in deeper water. Reef cements are bladed and micritic magnesian calcite or aragonite; the micritic cement is frequently pelleted (Ginsburg *et al.*, 1971; Friedman *et al.*, 1974; James *et al.*, 1976; MacIntyre, 1977; Mazzullo and Cys, 1979; Shinn *et al.*, 1983).

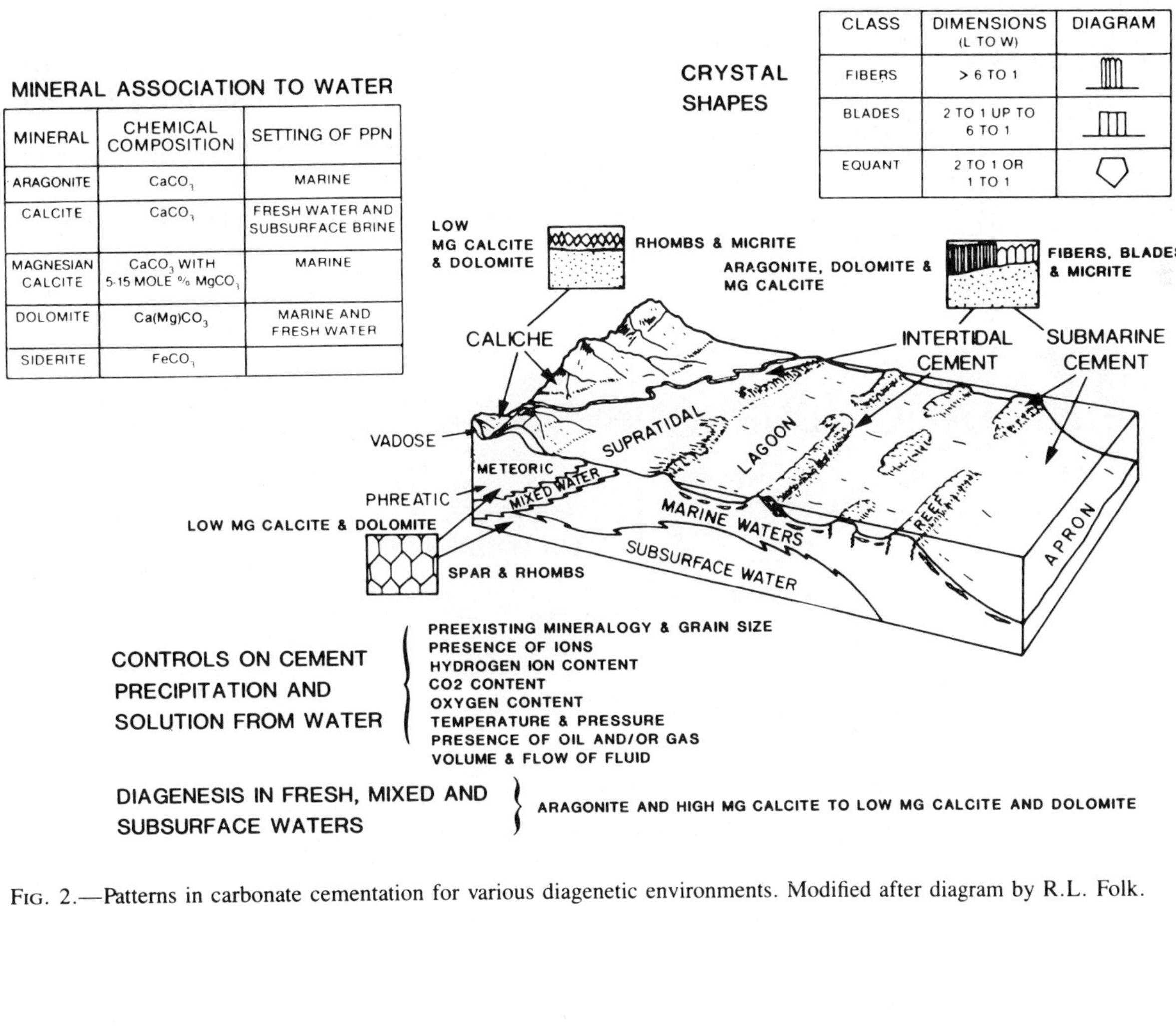

FIG. 2.—Patterns in carbonate cementation for various diagenetic environments. Modified after diagram by R.L. Folk.

The fabric of reef cement is controlled by its access to circulating water; the voids with greater permeability may be filled by isopachous or equant cements whereas the more confined voids commonly show much larger and irregular crystals. In some reef cavities, like those from the reefs of Belize, thick botryoidal crusts of aragonite occur in the larger cavities (Ginsburg and James, 1976; James *et al.*, 1976). The zone of active cementation in reefs appears to be confined in general to the outer few feet of the sediment forming the reef. Evidence of this localization of cementation on a larger scale has been recognized in the modern reefs from Belize (James and Ginsburg, 1979) and Devonian reefs from the Canning Basin (Playford, 1980), which are

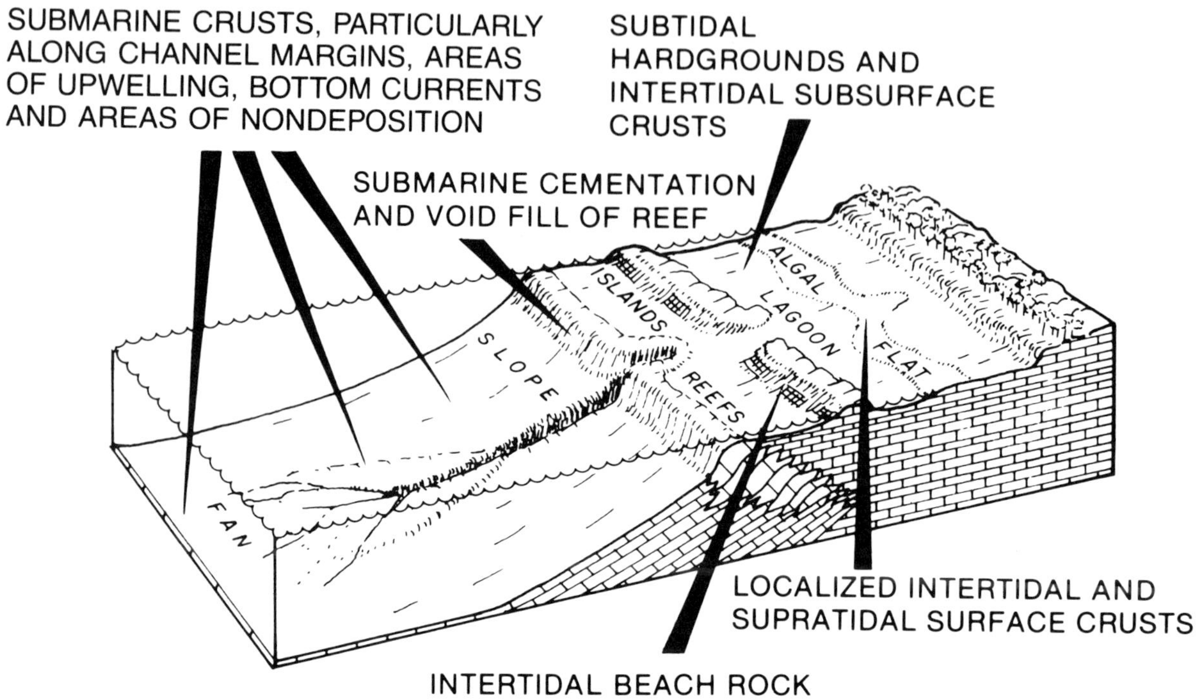

DIAGENETIC EFFECTS OF MARINE WATERS

MICRITIZATION

NEOMORPHIC GRAIN GROWTH

CEMENTATION DURING SEDIMENT DEPOSITION

FORMATION OF { OOLITES / GRAPESTONES / PELLET HARDENING

SUBMARINE / INTERTIDAL { CRUSTS

STABLE MARINE CEMENTS

ARAGONITE — FIBERS AND MICRITE
MG CALCITE — FIBERS, BLADES AND MICRITE

FIG. 3.—Environments of marine cementation.

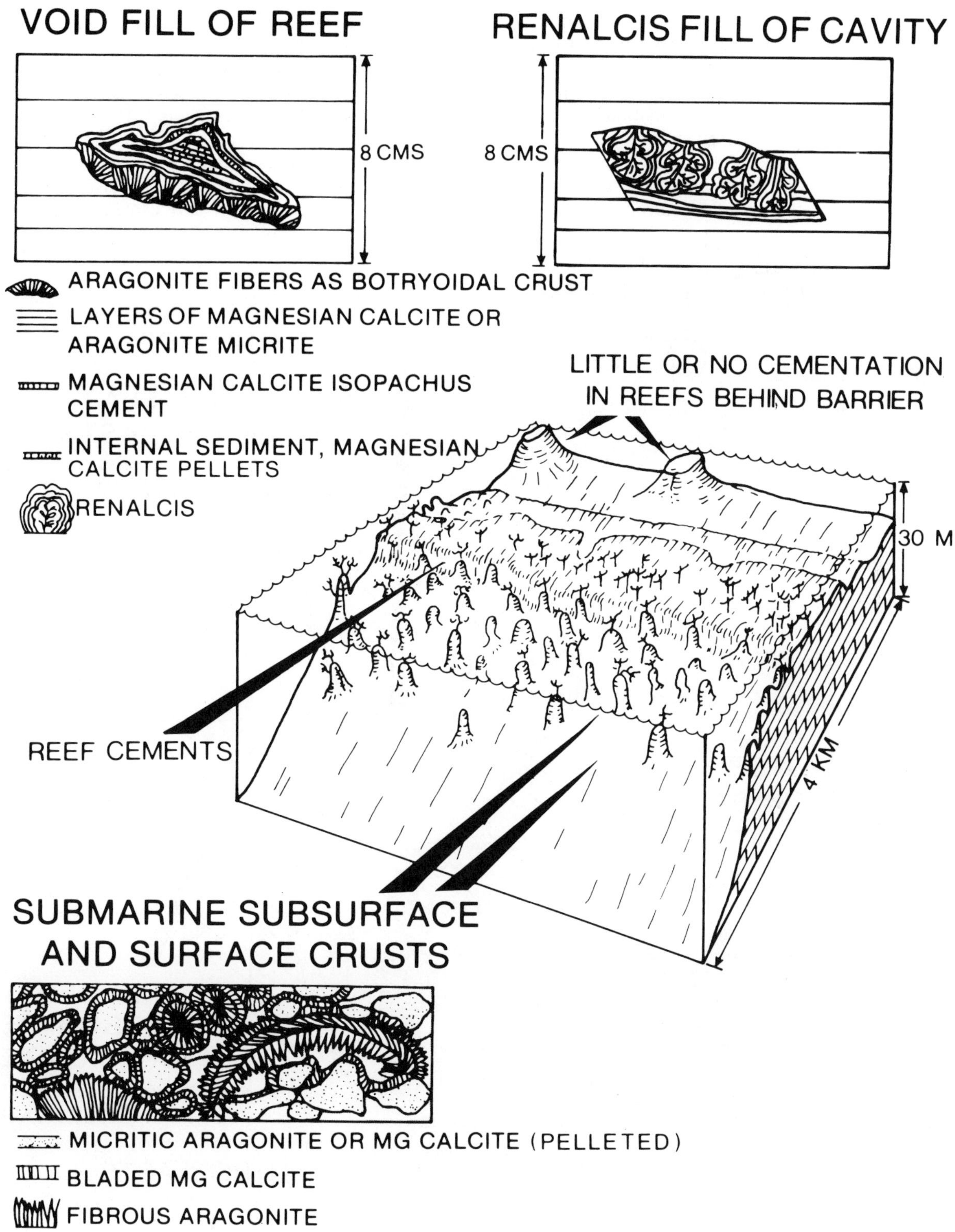

FIG. 4.—Reef cementation.

cemented on their seaward margin whereas the reefs to the lee of the major reef trend show little or no cementation.

Shallow water carbonate platforms.— Cementation penecontemporaneous with deposition of carbonate sediments also occurs across shallow water platforms (Fig. 5). For instance, oolites, which occur in zones of wave and tidal surge, are formed by the precipitation of calcium carbonate around a central nucleus (Illing, 1954; Loreau and Purser, 1973). Similarly grapestones, commonly found on a less agitated seafloor immediately to the lee of oolitic shoals, are formed in part by the precipitation of carbonate cement at points of contact between sandsized gains (Illing, 1954; Kendall and Skipwith, 1969; Winland and Matthews, 1974). Storms also break the cemented crusts that form on oolitic shoals, but do not completely disaggregate the cemented sediment, so partially cemented aggregates

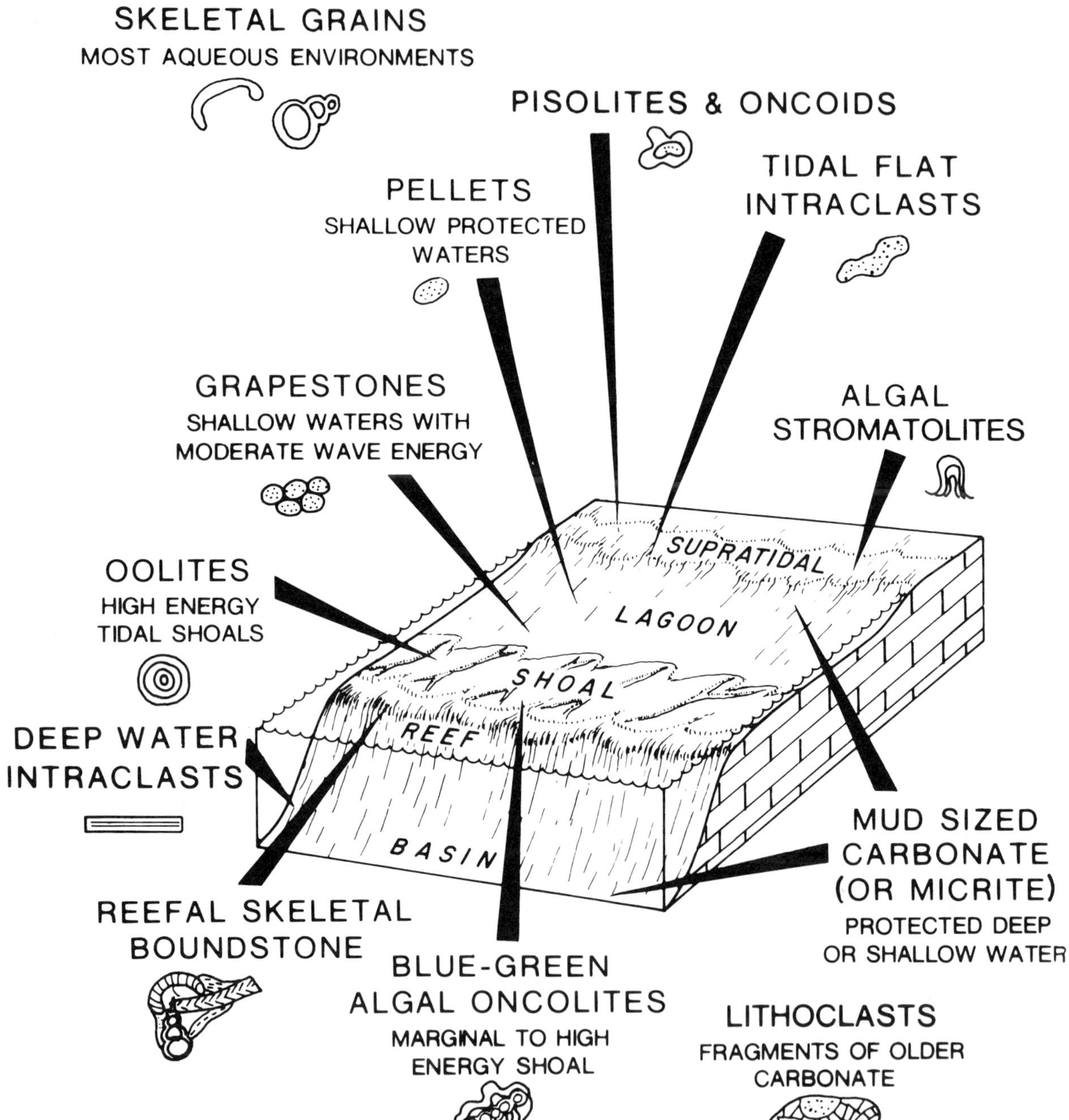

Fig. 5.—Carbonate grain types and their depositional setting.

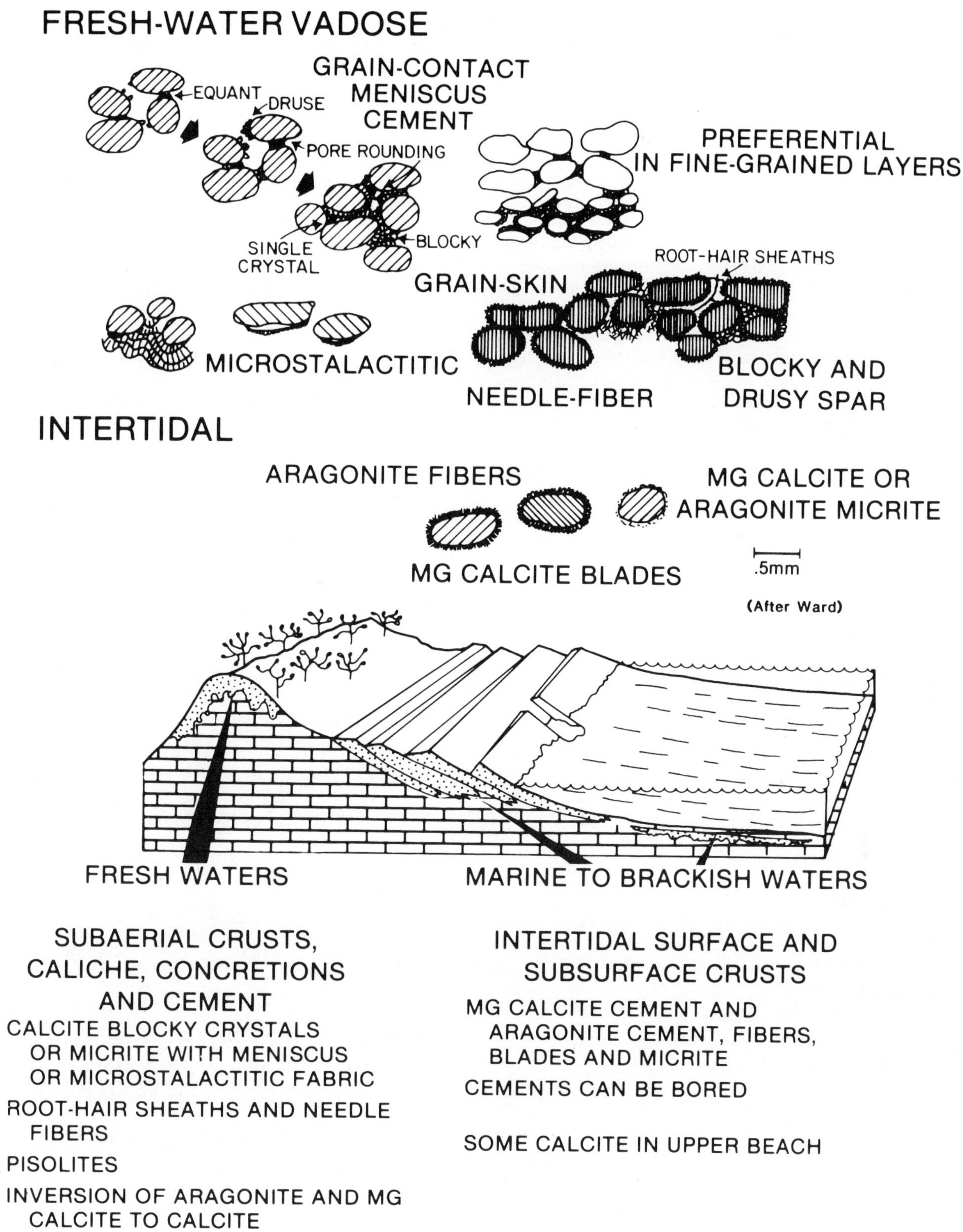

FIG. 6.—Near-shore and freshwater cementation of carbonates.

of the sediments (or intraclasts) remain. Further evidence of cementation can be seen in the way in which soft carbonate pellets excreted by marine organisms are progressively hardened by carbonate precipitation (Illing, 1954; Kendall and Skipwith, 1969). In shallower or more protected water than the hardened pellet zone, the lime mud pellets are not cemented.

Submarine crusts are best developed in grapestone sediments where the seafloor is cemented into thick layers tens of centimeters thick (Taft *et al.*, 1968; Purser, 1969, 1971; Harris, 1978; Dravis, 1979). In some cases, for instance in the Persian Gulf (Shinn, 1969; Assereto and Kendall, 1977), cementation causes the crust to expand and override itself forming compressional ridges or submarine tepees. These submarine crusts or hardgrounds can be recognized in the rock record by organisms encrusted on the hardened surface, borings cut through both sediment and cement and the morphology of the marine cements (Purser, 1969).

The cements involved in the cementation of oolites, grapestones, hardening of pellets and the development of hardgrounds are aragonite and magnesian calcite (Bathurst, 1971). The aragonite may be micritic or fibrous, and the magnesian calcite micritic or bladed. The micritic cements in submarine crusts are more common closer to the water sediment boundary whereas the fibrous and bladed cements are found toward the interior of the crusts and commonly occur as isopachous rims enveloping each grain (Shinn, 1969).

Cementation also occurs during the micritization of carbonate grains, associated with the boring activities of algae and bacteria (Bathurst, 1966; Alexandersson, 1972; Harris *et al.*, 1979). The mineralogy of the grain can sometimes be altered by the processes so that the identification of the altered grain can only be made by its shape. The importance of bacteria and algae in the micritization process may be quite important in that there is probably some organic intermediary involved in the development of marine cements.

Intertidal and Supratidal Marine

Cementation in the intertidal and supratidal zones of carbonate depositional environments is not uncommon (Bathurst, 1971; Milliman, 1974). Cementation in the intertidal zone occurs along beaches where a decrease in CO_2 by degassing due to tidal pumping may form beachrock (Moore, 1973, 1977; Hanor, 1978). The cements are micritic, bladed or fibrous with variable mineralogy (Fig. 6). Magnesian calcite and aragonite cements appear related to marine water tables; calcite cements in contrast seem related to an influx of freshwater from an adjacent landmass into the intertidal sediments.

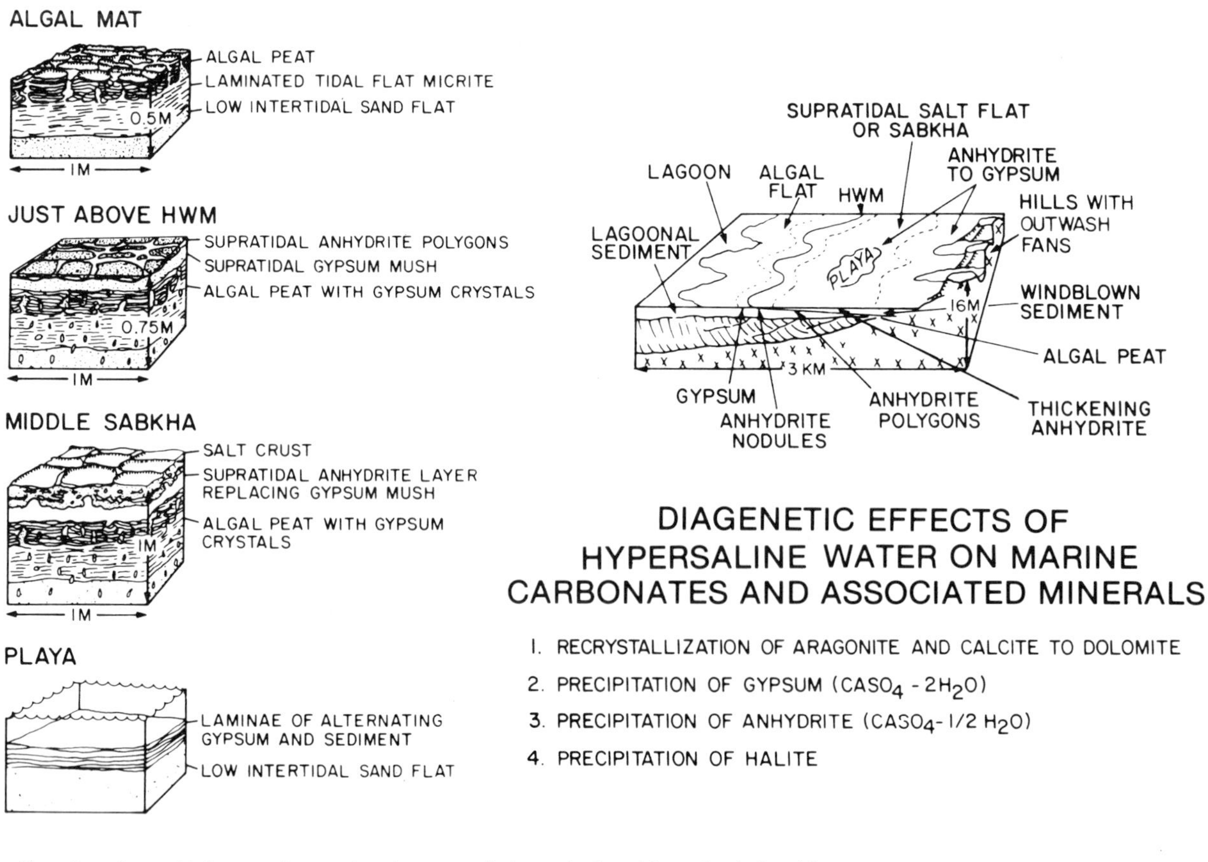

FIG. 7.—Supratidal evaporites and carbonates, their vertical and lateral relationships.

In regions of high evaporation and elevated salinities, the supratidal zone behind quiet water lagoons may be cemented as crusts of variable thickness (Fig. 3). These crusts show evidence of desiccation and expansion (Assereto and Kendall, 1977), and compressional ridges commonly outline polygons with saucer-like depressions. Areas where this type of cemented crust have been described include the southwestern Persian Gulf (Evamy, 1973), Shark Bay and Marion Lake, Australia (Davies, 1970; Warren, 1982) and the Bahamas (Shinn *et al.*, 1965).

Supratidal and intertidal crusts from arid and hypersaline areas commonly develop aragonite travertine coats. Examples of these cements can be seen in the southwest Persian Gulf (Purser and Loreau, 1973), the Permian of the Guadalupe Mountains (Kendall, 1969), Triassic of the Burgamasc Alps (Assereto and Kendall, 1977) and the Jurassic of the Moroccan High Atlas Mountains (Burri *et al.*, 1973). Marine travertines frequently fill the deformed compressional ridges found in these crusts. The deformed crusts are often composed of coated grains or pisolites which formed and/or collected in the splash zone and in the soil horizons of the high intertidal and supratidal environments (Esteban and Pray, 1977).

In addition to the cementation found in the supratidal zone, other diagenetic features are associated with flooding of marine waters. If salinities are elevated or magnesium to calcium ratios are increased, as during the precipitation of anhydrite minerals on supratidal flats of arid regions, the mineralogy of the sediments change (Fig. 7; Curtis *et al.*, 1963; Kendall and Skipwith, 1969; Butler *et al.*, 1982). Aragonite commonly gives way landward across supratidal flats as exemplified by the coastal flats of Abu Dhabi to calcium sulfate in the form of gypsum, anhydrite and finally, by replacement of previously precipitated anhydrite to gypsum. The source of the water may either be storm washover (Butler, 1970) or the evaporative pumping of water derived from the nearby marine environment (Hsü and Schneider, 1973). The precipitation of sulfate is accompanied sometimes by the aragonitic sediment being converted to dolomite (Adams and Rhodes, 1960; Hsü and Schneider, 1973). The dolomitization is initially confined to the finer grained sediments, also eventually replacing the coarser material.

The cements of the supratidal flats show a displacive fabric. The gypsum commonly appears as crystal laths; the anhydrite occurs as contorted ptygmatic layers or nodules. Some cementation and diagenesis may also be related to standing bodies of water in a supratidal setting where marine water is ponded to form supratidal playas. These playas show variable cement stratigraphy and diagenesis depending on their salinites (Logan, 1981; Warren, 1982). Where sulfates have precipitated directly from a standing body of seawater, they form horizontal layers parallel to the sediment water interface.

Shallow Burial

Shallow processes occur at near-surface temperature and pressure in meteoric or marine water. Cementation and dissolution of carbonates initiated in the marine realm continues during contact with freshwater (Fig. 8; Friedman, 1964; Matthews, 1968; Bricker, 1971; Longman, 1980), along with dewatering, grain reorientation and grain breakage during the earliest burial compaction. The source of the freshwater and the position of the freshwater table is controlled by the relief of the recharge area, climate and the position of the sea with respect to the environment during and immediately after deposition. Freshwater may enter the environment of deposition in the supratidal zone or the intertidal zone through the displacement of marine waters. Freshwater lenses within the sediment extend into the subsurface and may extend out into marine sediments beneath the sea, sometimes spilling into the sea as a freshwater spring. Examples of freshwater lenses and the resultant diagenesis are known from the Bahamas (Halley and Harris, 1979), Barbados (Matthews, 1968, 1971), Bermuda (Land, 1970), Florida (Matthews, 1974) and elsewhere. The thickness of a freshwater lens beneath the land mass with low relief can be substantial. For instance, the Florida peninsula is underlain by a freshwater table that extends 600 meters into the subsurface (Schmoker and Halley, 1982). The greatest cementation is expected at the water table and in the transition zone between freshwater and the marine lens or subsurface brine beneath it (Matthews, 1974; Meyers and Lohman, 1978; Longman, 1980). The mixing zone is important for it is here that the waters have their greatest variation in composition and position. Annual climatic changes cause variations in rainfall and so affect the position of the water table and mixing zone through the year.

The diagenetic overprint is surprisingly uniform throughout the vadose and phreatic zones of Pleistocene limestones from Barbados, the Bahamas and South Florida. Calcite and less commonly dolomite cements form as crusts of equant crystals in the phreatic zone, or may form random mosaics in both the vadose and phreatic zones where their orientation appears independent of the cemented grains. In random mosaics, the size of the crystals range from very fine to medium; the first generation of crystals tends to have more regular crystal facies and is finer than the later crystals. In the vadose zone, cementation takes place within the water meniscae which collect at the points of contact between grains (Fig. 6; Dunham, 1971; Halley and Harris, 1979). Similarily, cements form in the water draped on the lower sides of grains and generate microstalactites (Purser, 1971; Ward, 1975).

The trace element composition of freshwater cements formed at shallow depths is controlled by the temperature, Eh, pH and distribution coefficients of precipitating waters. In well oxygenated groundwaters, ferric ions are common and thus iron is not incorporated into the cements (Evamy, 1969). Manganese is precipitated preferentially to iron in near-surface oxygenated waters (Meyers and Lohmann, 1980), since Mn^{2+} is slightly more soluble in a wider range of Eh conditions than is Fe^{2+} (Garrels and Christ, 1965). Thus Mn^{2+} can occur in more oxidizing waters than Fe^{2+} leading to cements rich in Mn^{2+} but poor in Fe^{2+}. In waters from greater depths, oxygen is less common and in a reducing environment iron and manganese occur as divalent ions in solution which form ferroan and manganese-rich carbonate cements. In strongly reducing conditions both occur in solution. In high sulfide waters, iron is incorporated in pyrite giving preference to $MnCO_3$. $FeCO_3$ forms in low HS^- waters.

The diagenesis associated with freshwater includes solution, the inversion of aragonite to calcite, the exsolution of magnesium from magnesium calcite, cement precipitation and dolomitization. Aragonite and magnesian calcite are unstable in magnesium deficient waters regardless of whether these waters are fresh, brackish or saline. So in the vadose or shallow phreatic zones these minerals tend to change to calcite (Chave, 1962; Land, 1967). In some cases, the original mineral fabrics are completely preserved and in others they are completely destroyed. If the carbon dioxide content of a fluid is increased, the fluid will dissolve more carbonate and hold it in solution. This occurs in soil zones and leads to dissolution and landscape lowering and, where vadose and phreatic waters interact at the water table, to cave formation. The transformation of aragonite to calcite may cause partial plugging of the existing porosity by the formation of calcite cement. In carbonates where the pores are completely occluded by cement, it is generally thought a source of the cement other than the transformation of aragonite to calcite is required, perhaps being derived from the weathering and dissolution of adjacent carbonates (Dunham, 1969; Bathurst, 1971). In the inversion process, strontium is lost from aragonite. Similarly, magnesium may be lost from magnesian calcite to the groundwaters or concentrated to form euhedral crystals of dolomite (Lohmann and Meyers, 1977).

Once the sediment is buried below the freshwater table and comes into contact with shallow subsurface reducing fluids, cementation by calcite occurs at a decreased rate. The crystals tend to be very fine to medium grained and have irregular to equant shapes. Manganese and iron, if present, continue to substitute for calcium (Meyers and Lohmann, 1980), though the quantity of manganese and iron present in the waters is lower than those of the deeper subsurface because of the presence of oxygen. The $\delta^{18}O$ composition of these cements may be enriched in relation to shallower ce-

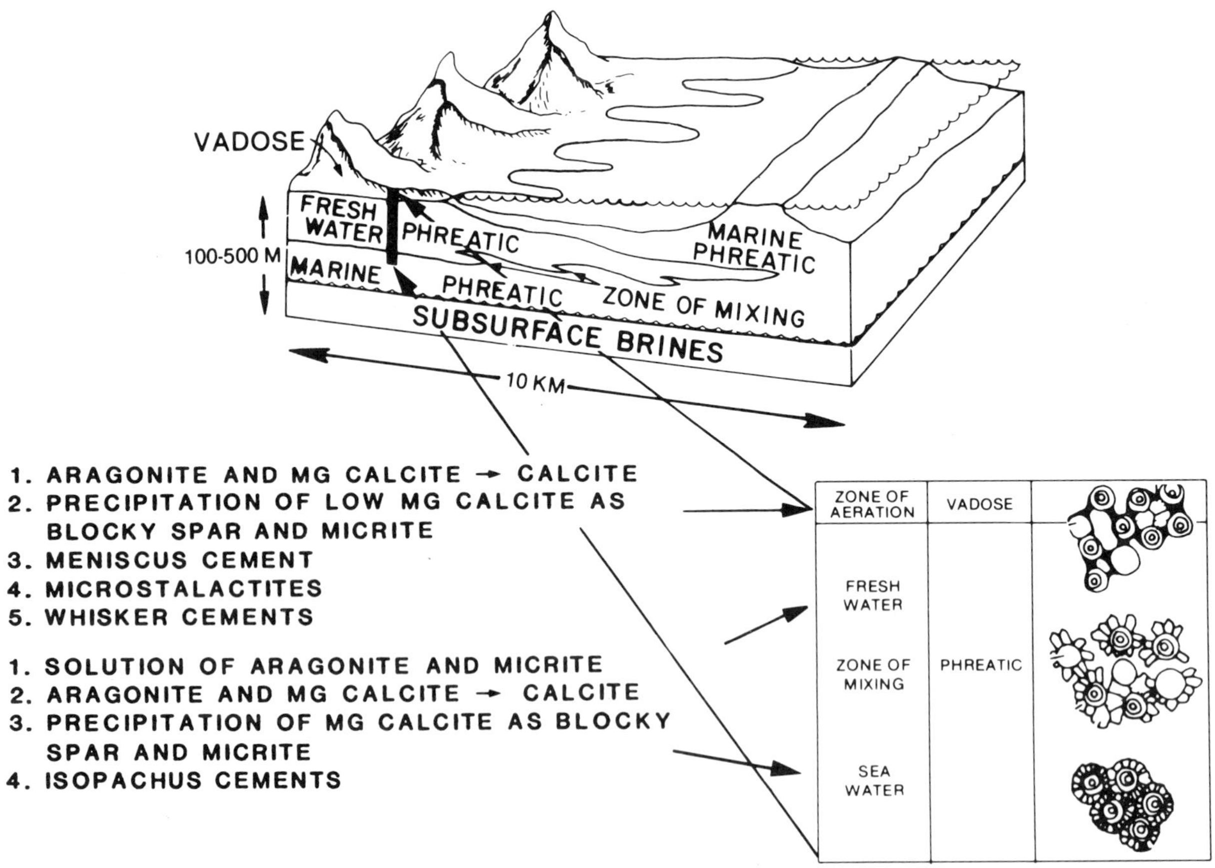

FIG. 8.—Freshwater diagenesis at the surface and during shallow burial.

ments due to contributions from rock oxygen. Conversely the $\delta^{18}O$ composition is depleted once the waters in contact with the sediment rise in temperature.

Deeper Burial

Whereas physical compaction of the kind Shinn *et al.* (1977) described from laboratory experiments obviously begins shortly after a carbonate is deposited and begins to dewater, we are unsure when true chemical compaction or pressure solution begins. Dunnington (1967) places the depth of stylolite formation between 600 to 900 m and Mossop (1972) at around 1000 m, but Schlanger (1964) gives evidence that in some cases burial need not be greater than 90 m, particularly in limestones with a high clay content. Meyers (1980) proves that considerable pressure solution in some carbonate rocks can occur under probably a few 10's of meters overburden, taking the stance that water chemistry may be as important as depth. At one extreme, Logan and Semeniuk (1976) and Wanless (1979, 1982) give evidence that pressure solution is pervasive in some limestone provinces and in some cases, as in the Devonian of the Canning Basin (Logan and Semeniuk, 1976), completely obscures many sedimentary relationships. In contrast, Pratt (1982) feels at least some of the fabrics identified by Wanless (1979) are the result of mechanical compaction. Similarly Playford (1980) persuasively downplays the significance of pressure solution in the Canning Basin. Nevertheless, there are many examples where pressure solution is important in carbonates and is destroying their porosity and permeability (Dunnington, 1967; Coogan, 1970; Mossop, 1972). Probably the best known example is that described by Dunnington (1967) where stylolites following bedding planes form permeability barriers in a producing oil field within Cretaceous limestones in Abu Dhabi.

Pressure solution fabrics take several forms (Park and Schott, 1968; Bathurst, 1971; Mossup, 1972; Wanless, 1979). Wanless (1979) recognized sutured-seam solution (common stylolites and grain-contact sutures), non-sutured seam solution (microstylolites, microstylolite swarms and clay seams) and non-seam solution (pervasive solution thinning of a unit). Solution surfaces are commonly marked by concentrations of dark clay, organic matter or iron oxides.

As Davies (1976), Bathurst (1980), Moore and Druckman *et al.* (1981) have noted, late stage cementation and dissolution of earlier-formed cements by warm subsurface fluids are volumetrically important diagenetic events. These late stage effects are dependent on fabric, texture, grain size and mineral stability, as well as on the pH, salinity and mineral content of the waters. The cement associated with subsurface brines is commonly an equant coarse-grained calcite spar, ferroan and with poikilotopic textures (Meyers, 1974; Grover and Read, 1978, 1983; Moore and Druckman, 1981; Crevello *et al.*, in press). These cements postdate earlier cements and compaction features, often filling cracks in broken grains or containing portions of earlier broken grains or cements. Baroque dolomite (Folk and Assereto, 1974; Radke and Mathis, 1980) is another common burial cement and is characterized by curved crystal faces with sweeping extinction.

Manganese can be an important trace element component of burial cements, but iron rich cements are more common. Subsurface waters are commonly reducing and any iron in the water tends to be incorporated in the calcite or dolomite. The changing manganese and iron compositions of subsurface fluids result in mineral zoning when cements are examined using cathodoluminescent petrography (Sommer, 1972; Meyers, 1974; Glover, 1977; Grover and Read, 1978, 1983; Pierson, 1981). Thermal and pressure effects and chemical solubilities play a dominant role in deeper burial. Oil maturation also proceeds as the temperature increases (Hunt, 1979; Waples, 1980), and hydrocarbons break down into CO_2 and water (Tissot and Espitalie, 1975; Tissot and Welte, 1978) which together may enhance carbonate porosity.

CONCLUSIONS

Carbonate cements form in a wide range of diagenetic environments. Depositional settings play an important role in determining depositional fabrics and the subsequent direction of diagenetic behavior of carbonate rocks. The challenge to the carbonate geologist is to unravel the diagenetic history and assess the importance of each stage of cementation in the alteration of porosity and permeability. Cement fabric and mineralogy reflect the chemistry of their precipitating and presently surrounding waters. The previous sections on environments of cementation point out that similarly appearing cements can form in different diagenetic environments and thus a more accurate interpretation of their origin must rely on geochemical as well as petrographic data. Such measurements of stable isotopic ratios, trace elements and fluid inclusions are being increasingly used in studies of carbonate cements in attempts to determine the composition of fluids and timing of precipitation in various diagenetic environments. Porosity in carbonates may be depositional, diagenetic or tectonically derived; it is lost through a combination of cementation along with chemical and physical compaction with depth.

ACKNOWLEDGMENTS

W.F. Buckingham, R.C. Burruss, W.J. Meyers, D.R. Prezbindowski, B.J. Russell, M. Magaritz and W.D. Wiggins read various drafts of the manuscript and provided numerous valuable suggestions. We thank Gulf Oil Corporation for typing, drafting and permission to publish.

REFERENCES

Adams, J.E., and Rhodes, M.L., 1960, Dolomitization by seepage refluxion: Amer. Assoc. Petroleum Geologists Bull., v. 44, p. 1912–1920.

Alexandersson, T., 1972, Micritization of carbonate particles: processes of precipitation and dissolution in modern shallow-marine sediments: Univ. Uppsala Geol. Inst. Bull., no. 5.3, v. 7, p. 201–236.

Arabian American Oil Company Staff, 1959, Ghawar Oil Field, Saudi Arabia: Amer. Assoc. Petroleum Geologists Bull., v. 43, p. 434–454.

Arthur, M.A., Scholle, P.A., and Halley, R.B., 1982, Report on the Second Soc. Econ. Paleontologists Mineralogists Research Conference: Stable isotopes in sedimentary geology: Jour. Sed. Petrology, v. 52, p. 1039–1044.

Assereto, R.L.A.M., and Kendall, C.G.St.C., 1977, Nature, origin and classification of peritidal tepee structures and related breccias: Sedimentology, v. 24, p. 153–210.

Baria, L.R., Stoudt, D.L., Harris, P.M., and Crevello, P.D., 1982, Upper Jurassic reefs of Smackover Formation, United States Gulf Coast: Amer. Assoc. Petroleum Geologists Bull., v. 66, p. 1449–1482.

Bartlett, G.A., and Greggs, R.G., 1969, Carbonate sediments, oriented lithified samples from the North Atlantic: Science, v. 166, p. 740–741.

Bathurst, R.G.C., 1966, Boring algae, micrite envelopes and lithification of molluscan biosparites: Geol. Jour., v. 5, p. 89–109.

______, 1971, Carbonate Sediments and Their Diagenesis: Developments in Sedimentology, v. 12, Elsevier Pub. Co., New York, 620 p.

______, 1980, Deep crustal diagenesis in limestones: Rev. Inst. Invest. Geol., Diputacion Provincial, Universidad de Barcelona, v. 34, p. 89–100.

Bein, A., and Land, L.S., 1982, San Andres carbonates in the Texas panhandle: Sedimentation and diagenesis associated with magnesium-calcium-chloride brines: The Univ. of Texas-Austin, Bureau Econ. Geology Rep. Inv. 121, 48 p.

Bricker, O.P. (ed.), 1971, Carbonate Cements: Johns Hopkins Univ., Studies in Geology No. 19, 376 p.

Burri, P., DuDresnay, R., and Wagner, C.W., 1973, Tepee structures and associated diagenetic features of intertidal carbonate sands (Lower Jurassic, Morocco): Sed. Geology, v. 9, p. 221–228.

Burruss, R.C., 1981, Hydrocarbon fluid inclusions in studies of sedimentary diagenesis, *in* L.S. Hollister, and M.L. Crawford (eds.), MAC Short Course in Fluid Inclusions: Applications to Petrology: Toronto, p. 138–156.

______, Cercone, K.R., and Harris, P.M., 1983, Fluid inclusion petrography and tectonic-burial history of the Al Ali 2 Well, evidence for the timing of diagenesis and oil migration, northern Oman Foredeep: Geology, v. 11, p. 567–570.

Butler, G.P., 1970, Holocene gypsum and anhydrite of the Abu Dhabi Sabkha, Trucial Coast: An alternative explanation of origin: Third Symposium on Salt, Northern Ohio Geol. Soc., v. 1, p. 120–152.

______, Harris, P.M., and Kendall, C.G.St.C., 1982, Recent evaporites from the Abu Dhabi Coastal Flats, *in* C.R. Handford, R.G. Loucks, and G.R. Davies (eds.), Depositional and Diagenetic Spectra of Evaporites—A Core Workshop: Soc. Econ. Peleontologists Mineralogists Core Workshop No. 3, p. 33–64.

Campos, H.S., and Hallam, A., 1979, Diagenesis of English Jurassic limestones as inferred from oxygen and carbon isotope analyses: Earth and Plan. Sci. Lett., v. 45, p. 23–31.

Chave, K.E., 1962, Factors influencing the mineralogy of carbonate sediments: Limnol. and Oceanog., v. 7, p. 218–223.

Choquette, P.W., and Pray, L.C., 1970, Geologic nomenclature and classification of porosity in sedimentary carbonates: Amer. Assoc. Petroleum Geologists Bull., v. 54, p. 207–250.

Coogan, A.H., 1970, Measurements of compaction in oolitic grainstone: Jour. Sed. Petrology, v. 40, p. 921–929.

Crevello, P.D., Harris, P.M., Stoudt, D.L., and Baria, L.R. (in press), Porosity evolution and burial diagenesis in a reef-debris reservoir, Smackover Formation (Jurassic), Hico Knowles Field, Arkansas, *in* P.O. Roehl and P.W. Choquette (eds.): Springer Verlag, Casebook of Petroleum Reservoirs.

Currie, J.B., and Nwachukwu, S.O., 1974, Evidence on incipient fracture porosity in reservoir rocks at depth: Bull. Can. Petroleum Geol., v. 22, p. 42–58.

Curtis, R., Evans, G.R., Kinsman, D.J.J., and Shearman, D.J., 1963, Association of dolomite and anhydrite in the recent sediments of the Persian Gulf: Nature, v. 197, p. 679–680.

Davies, G.R., 1970, Algal-laminated sediments, Gladstone Embayment, Shark Bay, Western Australia, *in* B. Logan *et al.* (ed.), Carbonate Sedimentation and Environments, Shark Bay, Western Australia: Amer. Assoc. Petroleum Geologists Mem. 13, p. 169–205.

______, 1976, "Bitumen" in post burial diagenetic calcite: Geol. Survey Canada, Paper 76-1C, p. 107–114.

Dickson, J.A.D., and Coleman, M.L., 1980, Changes in carbon and oxygen isotope composition during limestone diagenesis: Sedimentology, v. 27, p. 107–118.

Dravis, J.J., 1979, Rapid and widespread generation of recent oolitic hardgrounds on a high energy Bahamian platform, Eleuthera Bank, Bahamas: Jour. Sed. Petrology, v. 49, p. 195–208.

Dunham, R.J., 1962, Classification of Carbonate Rocks According to Depositional Texture: Amer. Assoc. Petroleum Geologists Mem. 1, Tulsa, p. 108–121.

______, 1969, Early vadose silt in Townsend Mound (Reef), New Mexico, *in* G.M. Friedman (ed.), Depositional Environments in Carbonate Rocks—A Symposium: Soc. Econ. Paleontologists and Mineralogists Spec. Pub. 14, p. 139–181.

______, 1971, Meniscus cement, *in* O.P. Bricker (ed.), Carbonate Cements: Johns Hopkins Univ. Studies in Geology, no. 19, p. 297–300.

Dunnington, H.V., 1967, Aspects of diagenesis and shape change in stylolitic limestone reservoirs: Proc. 7th World Petrol. Cong., Mexico, p. 339–352.

ENOS, P., AND SAWATSKY, L.H., 1981, Pore networks in Holocene carbonate sediments: Jour. Sed. Petrology, v. 51, p. 961–985.
EPTING, M., 1980, Sedimentology of Miocene carbonate buildups, central Luconia, offshore Sarawak: Geol. Soc. Malayasia Bull., v. 12, p. 17–30.
ESTEBAN, M., AND PRAY, L.C., 1977, Origin of the pisolite facies of the upper shelf area, *in* M.E. Hileman and S.J. Mazzullo (eds.), Upper Guadalupian Facies, Permian Reef Complex, Guadalupe Mountains, New Mexico and West Texas, 1977, Field Conference Guidebook, v. 1: Permian Basin Sec., Soc. Econ. Paleontologists and Mineralogists Pub. 77–16, p. 479–486.
EVAMY, B.D., 1969, The precipitational environment and correlation of some calcite cements deduced from artificial staining: Jour. Sed. Petrology, v. 39, p. 787–793.
______, 1973, The precipitation of aragonite and its alteration to calcite on the Trucial Coast of the Persian Gulf, *in* B.H. Purser (ed.), The Persian Gulf: Springer-Verlag, New York, p. 329–341.
FISCHER, A.G., AND GARRISON, R.D., 1967, Carbonate lithification on the seafloor: Jour. Geology, v. 75, p. 488–496.
FLÜGEL, E., 1982, Microfacies Analysis of Limestones: Springer-Verlag, Berlin, 633 p.
FOLK, R.L., 1959, Practical petrographic classification of limestones: Amer. Assoc. Petroleum Geologists Bull., v. 43, p. 1–38.
______, 1962, Spectral subdivision of limestone types, *in* W.E. Ham (ed.), Classification of Carbonate Rocks: Am. Assoc. Petroleum Geologists Memoir, 1, p. 62–84.
______, 1973, Carbonate petrography in the post-Sorbian age, *in* Evolving Concepts in Sedimentology: Johns Hopkins Univ. Studies in Geology, v. 21, p. 118–158.
______, AND ASSERETO, R.L.A.M., 1974, Giant aragonite rays and baroque white dolomite in tepee-fillings, Triassic of Lombardy, Italy [abs.]: Amer. Assoc. Petroleum Geologists, Abs. with Prog., Ann. Mtg., San Antonio, p. 34–35.
FRANK, J.R., CARPENTER, A.B., AND OGLESBY, T.W., 1982, Cathodoluminescence and composition of calcite cement in the Taum Sank Limestone (Upper Cambrian), southeast Missouri: Jour. Sed. Petrology, v. 52, p. 631–638.
FRIEDMAN, G.M., 1959, Identification of carbonate minerals by staining methods: Jour. Sed. Petrology, v. 29, p. 87–97.
______, 1964, Early diagenesis and lithification in carbonate sediments: Jour. Sed. Petrology, v. 34, p. 777–813.
______, AMIEL, A.J., AND SCHNEIDERMANN, N., 1974, Submarine cementation in reefs—Example from the Red Sea: Jour. Sed. Petrology, v. 44, p. 816–825.
FROST, S.H., BLIEFNICK, D.M., AND HARRIS, P.M., 1983, Deposition and porosity evolution of a Lower Cretaceous rudist buildup, Shuaiba Formation of eastern Arabian Peninsula, *in* Harris, P.M. (ed.), Carbonate Buildups—A Core Workshop: Soc. Econ. Paleontologists and Mineralogists Core Workshop No. 4, p. 381–410.
GARRELS, R.M., AND CHRIST, C.L., 1965, Solutions, Minerals and Equilibria: Freeman, Cooper and Co., San Francisco, p. 450.
GEVIRTZ, J.L., AND FRIEDMAN, G.M., 1966, Deep-Sea carbonate sediments of the Red Sea and their implications on marine lithification: Jour. Sed. Petrology, v. 36, p. 143–151.
GINSBURG, R.N., AND JAMES, N.P., 1976, Submarine botryoidal aragonite in Holocene reef limestones, Belize: Geology, v. 4, p. 431–436.
______, MARSZALEK, D.S., AND SCHEIDERMANN, N., 1971, Ultrastructure of carbonate cements in a Holocene algal reef of Bermuda: Jour. Sed. Petrology, v. 41, p. 472–482.
GLOVER, E.D., 1977, Cathodoluminescence, Iron and Manganese Content and the Early Diagenesis of Carbonates: Ph.D. Diss., Univ. Wisconsin, Madison, 444 p.
GROVER, G.A., JR., AND READ, J.F., 1978, Fenestral and associated vadose diagenetic fabrics of tidal flat carbonates, Middle Ordovician New Market Limestone, southwestern Virginia: Jour. Sed. Petrology, v. 48, p. 453–473.
______, AND ______, 1983, Paleoaquifer and deep burial related cements defined by regional cathodoluminescent patterns, Middle Ordovician carbonates, Virginia: Amer. Assoc. Petroleum Geologists Bull., v. 67, p. 1275–1303.
HALLEY, R.B., AND HARRIS, P.M., 1979, Freshwater cementation of a 1000-year old oolite: Jour. Sed. Petrology, v. 49, p. 969–988.
______, AND SCHMOKER, J.W., 1983, High porosity Cenozoic carbonate rocks of south Florida: Progressive loss of porosity with depth: Amer. Assoc. Petroleum Geologists Bull., v. 67, p. 191–200.
HANOR, J.S., 1978, Precipitation of beachrock cements: Mixing of marine and meteoric waters vs. CO_2-degassing: Jour. Sed. Petrology, v. 48, p. 489–501.
______, 1980, Dissolved methane in sedimentary brines: Potential effect on the PVT properties of fluid inclusions: Econ. Geology, v. 75, p. 603–609.
HARRIS, P.M., 1978, Holocene marine-cemented sands, Joulters Ooid Shoal, Bahamas: Trans. Gulf Coast Assoc. Geol. Societies, v. 28, p. 175–183.
______, HALLEY, R.B., AND LUKAS, K.J., 1979, Endolith microborings and their preservation in Holocene-Pleistocene (Bahama-Florida) ooids: Geology, v. 7, p. 216–220.
______, AND FROST, S.H. (1984), Middle Cretaceous carbonate reservoirs, Fahud Field and northwestern Oman: Amer. Assoc. Petroleum Geologists Bull., v. 68, p. 649–658.
HUDSON, J.D., 1977, Stable isotopes and limestone lithification: Jour. Geology Soc. London, v. 133, p. 637–660.
HUNT, J.M., 1979, Petroleum Geochemistry and Geology: W.H. Freeman and Co., San Francisco, 616 p.
HSÜ, K.J., AND SCHNEIDER, J., 1973, Progress report on dolomitization-hydrology of Abu Dhabi Sabkhas, Arabian Gulf, *in* B.H. Purser (ed.), The Persian Gulf: Springer-Verlag, New York, p. 409–422.
ILLING, L.V., 1954, Bahamian calcareous sands: Amer. Assoc. Petroleum Geologists Bull., v. 38, p. 1–95.
JAMES, N.P., AND GINSBURG, R.N., 1979, The seaward marginal Belize barrier and atoll reefs: Int. Assoc. Sedimentologists Spec. Pub. 3, 191 p.

______, ______, Marszalek, D.S., and Choquette, P.W., 1976, Facies and fabric specificity of early subsea cements in shallow Belize (British Honduras) reefs: Jour. Sed. Petrology, v. 46, p. 523–544.

Kaldi, J., and Gidman, J., ·1982, Early diagenetic dolomite cements: Examples from the Permian Lower Magnesian Limestone of England and the Pleistocene carbonates of the Bahamas: Jour. Sed. Petrology, v. 52, p. 1073–1085.

Kendall, C.G.St.C., 1969, An environmental reinterpretation of the Permian evaporite/carbonate shelf sediments of the Guadalupe Mountains: Geol. Soc. America Bull., v. 80, p. 2503–2526.

______, and Skipwith, P.A.d'E, 1969, Holocene shallow water carbonate and evaporite sediments of Khor al Bazam, Abu Dhabi, south west Persian Gulf: Amer. Assoc. Petroleum Geologists Bull., v. 53, p. 841–869.

Klosterman, M.J., 1981, Applications of Fluid Inclusion Techniques to Burial Diagenesis in Carbonate Rock Sequences: Louisiana State Univ., Applied Carbonate Research Program, Cont. 7, 101 p.

Klovan, J.E., 1964, Facies analysis of the Redwater Reef Complex, Alberta, Canada: Bull. Canadian Petroleum Geol., v. 12, p. 1–100.

Land, L.S., 1967, Diagenesis of skeletal carbonates: Jour. Sed. Petrology, v. 37, p. 914–930.

______, 1970, Phreatic versus vadose meteoric diagenesis of limestones, evidence from fossil water table: Sedimentology, v. 14, p. 175–185.

______, 1980, The isotopic and trace element geochemistry of dolomite: The state of the art, *in* D.H. Zenger, J.G. Dunham, and R.L. Ethington (eds.), Concepts and Models of Dolomitization: Soc. Econ. Paleontologists and Mineralogists, Spec. Pub. 28, p. 87–110.

Langton, J.R., and Chin, G.E., 1968, Rainbow Member facies and related reservoir properties, Rainbow Lake, Alberta: Bull. Canadian Petroleum Geol., v. 16, p. 104–143.

Lindsay, R.F., and Kendall, C.G.St.C. (in press), Depositional facies, diagenesis and reservoir character of the Mission Canyon Formation (Mississippian) of the Williston Basin at Little Knife Field, North Dakota, *in* P.O. Roehl and P.W. Choquette (eds.), Casebook on Carbonate Reservoirs: Springer-Verlag.

Logan, B.W., 1981, Modern Carbonates and Evaporite Sediments of Shark Bay and Lake McCloud, Western Australia: Geol. Soc. Australia, Australian Geological Convention, Field Excursion Guidebook, 63 p.

______, and Semeniuk, V., 1976, Dynamic Metamorphism, Processes and Products in Devonian Carbonate Rocks, Canning Basin, Western Australia: Geol. Soc. Australia Spec. Pub. 6, 138 p.

Lohmann, K.C., and Meyers, W.J., 1977, Microdolomites in cloudy prismatic calcites, a proposed criteria for former high-magnesium calcites: Jour. Sed. Petrology, v. 47, p. 1078–1088.

Longacre, S.A., 1980, Dolomite reservoirs from Permain biomicrites, *in* R.B. Halley and R.G. Loucks (eds.), Carbonate Reservoir Rocks: Soc. Econ. Paleontologists and Mineralogists Core Workshop No. 1, p. 105–117.

Longman, M.W., 1980, Carbonate diagenetic textures from near-surface diagenetic environments: Amer. Assoc. Petroleum Geologists Bull., v. 64, p. 461–487.

______, 1981, Carbonate Diagenesis and Diagenetic Environments: Amer. Assoc. Petroleum Geologists, Fall Education Conference, v. 2, 59 p.

______, Fertal, T.G., and Glennie, J.S., 1983, Origin and geometry of Red River Dolomite reservoirs, Williston Basin: Amer. Assoc. Petroleum Geologists Bull., v. 67, p. 744–771.

Loreau, J.P., and Purser, B.H., 1973, Distribution and ultrastructure of Holocene ooids in the Persian Gulf, *in* B.H. Purser (ed.), The Persian Gulf, Holocene Carbonate Sedimentation in a Shallow Epic Continental Sea: Springer-Verlag, New York, p. 279–328.

MacIntyre, I.G., 1977, Distribution of submarine cements in a modern Caribbean fringing reef, Galeta Point, Panama: Jour. Sed. Petrology, v. 47, p. 503–516.

Magaritz, M., 1983, Carbon and oxygen isotope composition of Recent and ancient coated grains, *in* T. Peryt (ed.), Coated Grains: Springer-Verlag, New York, p. 27–37.

______, and Taylor, H.P., Jr., 1976, Oxygen, hydrogen and carbon isotope studies of the Franciscan Formation, Coast Ranges, California: Geochim, Cosmochim. Acta, v. 40, p. 215–234.

Matthews, R.K., 1968, Carbonate diagenesis-equilibrium of sedimentary mineralogy to the subaerial environment, coral cap of Barbados, West Indies: Jour. Sed. Petrology, v. 38, p. 110–119.

______, 1971, Diagenetic environment of possible importance to the explanation of cementation fabric in subaerially exposed carbonate sediments, *in* O.P. Bricker (ed.), Carbonate Cements: Johns Hopkins Univ. Studies in Geology 19, p. 127–132.

______, 1974, A Process approach to diagenesis of reefs and reef associated limestones, *in* L.F. Laporte (ed.), Reefs in Time and Space: Soc. Econ. Paleontologists and Mineralogists Spec. Pub. 18, p. 234–256.

Mazzullo, S.J., and Cys, J.M., 1979, Marine aragonite seafloor growth and cements in Permian phylloid algal mounds, Sacramento Mountains, New Mexico: Jour. Sed. Petrology, v. 49, p. 917–936.

McNaughton, D.A., and Garb, F.A., 1975, Finding and evaluating petroleum accumulations in fractured reservoir rock, *in* V.S. Cameron (ed.), Exploration and Economics of the Petroleum Industry, New Ideas, New Methods, New Developments: Matthew Bender & Co., Inc., v. 13, p. 23–50.

Meyers, W.J., 1974, Carbonate cement stratigraphy of the Lake Valley Formation (Mississipian) Sacramento Mountains, New Mexico: Jour. Sed. Petrology, v. 44, p. 837–861.

______, 1978, Carbonate cements—Their regional distribution and interpretation in Mississippian limestones of southwestern New Mexico: Sedimentology, v. 25, p. 371–400.

______, 1980, Compaction in Mississippian skeletal limestones, southwestern New Mexico: Jour. Sed. Petrology, v. 50, p. 457–474.

______, and Lohmann, K.C., 1978, Microdolomite-rich syntaxial cements—Proposed meteoric-marine mixing zone phreatic cements from Mississippian limestones, New Mexico: Jour. Sed. Petrology, v. 48, p. 475–488.

______, AND ______, 1980, Geochemistry of regionally extensive calcite zones in Mississippian skeletal limestones, New Mexico [abs.]: AAPG-SEPM Abst. and Program, Annual Mtg., p. 91.
MILLIMAN, J.D., 1974, Marine Carbonates: Springer-Verlag, New York, 375 p.
______, ROSS, D.A., AND KU, T.L., 1969, Precipitation and lithification of deep-sea carbonates in the Red Sea: Jour. Sed. Petrology, v. 39, p. 724–736.
MOORE, C.H., 1973, Intertidal carbonate cementation, Grand Cayman, West Indies: Jour. Sed. Petrology, v. 43, p. 591–602.
______, 1977, Beach rock origin: Some geochemical, mineralogical, and petrographic considerations: Geoscience and Man, v. 18, p. 155–163.
______, 1979, Porosity in carbonate rock sequences, geology of carbonate porosity: Amer. Assoc. Petroleum Geologists, Continuing Education Course Notes 11, p. A1–A124.
______, 1980, Porosity in Carbonate Rock Sequences: Amer. Assoc. Petroleum Geologists, Fall Education Conference 1980, v. 1, 67 p.
______, AND DRUCKMAN, Y., 1981, Burial diagenesis and porosity evolution, Upper Jurassic Smackover, Arkansas and Louisiana: Amer. Assoc. Petroleum Geologists Bull., v. 65, p. 597–628.
MORESBY, R., 1835, Extracts from Commander Moresby's Report on the Northern Maldives: Jour. Royal Geol. Soc., v. 5, p. 398–404.
MOSSOP, G.D., 1972, Origin of the peripheral rim, Redwater Reef, Alberta: Bull. Canadian Petroleum Geol., v. 20, p. 238–280.
MULLINS, H.T., NEWMANN, A.C., WILBER, R.J., AND BOARDMAN, M.R., 1980, Nodular carbonate sediment on Bahamian slopes: Possible precursors to nodular limestones: Jour. Sed. Petrology, v. 50, p. 117–131.
PARK, W.C., AND SCHOTT, E.H., 1968, Stylolites, their nature and origin: Jour. Sed. Petrology, v. 38, p. 175–191.
PIERSON, B.J., 1981, The control of cathodoluminescence in dolomite by iron and manganese: Sedimentology, v. 28, p. 601–610.
PLAYFORD, P.E., 1980, Devonian "Great Barrier Reef" of Canning Basin, Western Australia: Amer. Assoc. Petroleum Geologists Bull., v. 64, p. 814–840.
PRATT, B.R., 1982, Limestone response to stress-pressure solution and dolomitization—Discussion and examples of compaction in carbonate sediments: Jour. Sed. Petrology, v. 52, p. 323–328.
PURSER, B.H., 1969, Syn-sedimentary marine lithification of Middle Jurassic limestones in the Paris Basin: Sedimentology, v. 12, p. 205–230.
______, 1971, Middle Jurassic syn-sedimentary marine cements from the Paris Basin, France, *in* O.P. Bricker (ed.), Carbonate Cements: Johns Hopkins Univ. Studies in Geol. 19, p. 178–184.
______, AND LOREAU, J.P., 1973, Aragonitic, supratidal encrustations on the Trucial Coast, Persian Gulf, *in* B.H. Purser (ed.), The Persian Gulf: Springer-Verlag, New York, p. 343–376.
RADKE, R.M., AND MATHIS, R.L., 1980, On the formation of saddle dolomite: Jour. Sed. Petrology, v. 50, p. 1149–1168.
ROEDDER, E., 1979, Fluid inclusion evidence on the environment of sedimentary diagenesis, a review, *in* P.A. Scholle and R.R. Schluger (eds.), Aspects of Diagenesis: Soc. Econ. Paleontologists and Mineralogists Spec. Pub. 25, p. 89–108.
ROEHL, P.O., 1967, Stony Mountain (Ordovician) and Interlake (Silurian) facies analogs of recent low-energy marine and subaerial carbonates, Bahamas: Amer. Assoc. Petroleum Geologists Bull., v. 51, p. 1979–2032.
SCHLAGER, W., AND JAMES, N.P., 1978, Low-magnesian calcite limestones forming at the deep-seafloor, Tongue of the Ocean, Bahamas: Sedimentology, v. 25, p. 675–702.
______, AND CHERMAK, A., 1979, Sediment facies of platform-basin transition, Tongue of the Ocean, Bahamas, *in* L. Doyle and O.H. Pilkey (eds.), Geology of Continental Slopes: Soc. Econ. Paleontologists and Mineralogists Spec. Pub. 27, p. 193–208.
SCHLANGER, S.O., 1964, Petrology of limestones of Guam: U.S. Geological Survey, Prof. Paper 403D, p. 1–52.
SCHMOKER, J.W., AND HALLEY, R.B., 1982, Carbonate porosity versus depth: A predictable relation for south Florida: Amer. Assoc. Petroleum Geologists Bull., v. 66, p. 2561–2570.
SHINN, E.A., 1969, Submarine lithification of Holocene carbonate sediments in the Persian Gulf: Sedimentology, v. 12, p. 109–144.
______, GINSBURG, R.N., AND LLOYD, R.M., 1965, Recent supratidal dolomite from Andros Island, Bahamas, *in* L.C. Pray and R.C. Murray (eds.), Dolomitization and Limestone Diagenesis: A Symposium: Soc. Econ. Paleontologists and Mineralogists Spec. Pub. 13, p. 112–123.
______, HALLEY, R.B., HUDSON, J.H., AND LIDZ, B.H., 1977, Limestone compaction—An enigma: Geology, v. 5, p. 21–24.
______, ROBBIN, D.M., LIDZ, B.H., AND HUDSON, J.H., 1983, Influence of deposition and early diagenesis on porosity and chemical compaction in two Paleozoic buildups: Mississippian and Permian age rocks in the Sacramento Mountains, New Mexico, *in* P.M. Harris (ed.), Carbonate Buildups—A Core Workshop: Soc. Econ. Paleontologists and Mineralogists Core Workshop No. 4, p. 182–221.
SIBLEY, D.F., 1982, The origin of common dolomite fabrics—Clues from the Pliocene: Jour. Sed. Petrology, v. 52, p. 1087–1100.
SOMMER, S.E., 1972, Cathodoluminescence of carbonates, I. Characterization of cathodoluminescence from carbonate solid solution: Chem. Geology, v. 9, p. 257–273.
STEARNS, D.W., AND FRIEDMAN, M., 1972, Reservoirs in fractured rock: Amer. Assoc. Petroleum Geologists Mem. 17, p. 82.
TAFT, W.H., ARRINGTON, F., HAIMOVITZ, A., MACDONALD, D., AND WOOLHEATER, C., 1968, Lithification of modern carbonate sediments at Yellow Bank, Bahamas: Bull. Marine Sci. Gulf and Carib., v. 18, p. 762–868.

Thompson, G., 1972, A geochemical study of some lithified carbonate sediments from the deep sea: Geochim. Cosmochim. Acta, v. 36, p. 1237–1253.
Till'man, J.E., 1983, Exploration for reservoirs with fracture-enhanced permeability: Oil and Gas Jour., February 21, p. 165–179.
Tissot, B.P., and Espitalié, J., 1975, The thermal evolution of organic matter in sediments: Rev. Inst. Fr. Petrology, v. 30, p. 743–777.
______, and Welte, D.H., 1978, Petroleum Formation and Occurrence, A New Approach to Oil and Gas Exploration: Springer-Verlag, Berlin, p. 527.
Van den Bark, E., and Thomas, O.D., 1981, Ekofisk: First of the giant oil fields in western Europe: Amer. Assoc. Petroleum Geologists Bull., v. 65, p. 2341–2363.
Veizer, J., and Compston, W., 1974, Sr^{87}/Sr^{86} Composition of seawater during the Phanerozoic: Geochim. Cosmochim. Acta, v. 38, p. 1461–1484.
______, and Hoefs, J., 1976, The Nature of O^{18}/O^{16} and C^{13}/C^{12} secular trends in sedimentary carbonate rocks: Geochim. Cosmochim. Acta, v. 40, p. 1387–1395.
______, Demovic, R., and Turan, J., 1971, Possible use of strontium in sedimentary carbonate rocks as a paleoenvironmental indicator: Sed. Geol., v. 5, p. 5–22.
Vincelette, R.R., and Soeparjadi, R.A., 1976, Oil-bearing reefs in Salawati Basin of Irian Jaya, Indonesia: Amer. Assoc. Petroleum Geologists Bull., v. 60, p. 1448–1462.
Walls, R.A., Mountjoy, E.W., and Fritz, P., 1979, Isotopic composition and diagenetic history of carbonate cements in Devonian Golden Spike Reef, Alberta, Canada: Geol. Soc. America Bull., v. 90, p. 963–982.
Wanless, H.R., 1979, Limestone response to stress: Pressure solution and dolomitization: Jour. Sed. Petrology, v. 49, p. 437–462.
______, 1982, Limestone response to stress: Pressure solution and dolomization—Reply: Jour. Sed. Petrology, v. 52, p. 328–332.
Waples, D.W., 1980, Time and temperature in petroleum formation, application of Lopatin's Method to petroleum exploration: Amer. Assoc. Petroleum Geologists Bull., v. 64, p. 916–926.
Ward, W.C., 1975, Petrology and diagenesis of carbonate eolianites of northeastern Yucatan Peninsula, Mexico, *in* Belize Shelf-Carbonate Sediments, Clastic Sediment and Ecology: Amer. Assoc. Petroleum Geologists Studies in Geology 2, p. 500–571.
Wardlaw, N.C., 1976, Pore geometry of carbonate rocks as revealed by pore casts and capillary pressure, Amer. Assoc. Petroleum Geologists Bull., v. 60, p. 245–257.
Warren, J.K., 1982, The hydraulical significance of holocene tepees, stromatolites and boxwork limestones in coastal salinas in South Australia: Jour. Sed. Petrology, v. 52, p. 1171–1202.
Wilson, A.O. (in press), Depositional and diagenetic facies in the Jurassic Arab-C and D Reservoirs, Qatif Field, Saudi Arabia, *in* P.L. Roehl and P.W. Choquette (eds.), Carbonate Petroleum Reservoirs: Springer-Verlag.
Winland, H.D., and Matthews, R.K., 1974, Origin and significance of grapestone, Bahama Islands: Jour. Sed. Petrology, v. 44, p. 921–927.
Zenger, D.H., Dunham, J.B., and Ethington, R.L., 1980 (eds.), Concepts and Models of Dolomitization: Soc. Econ. Paleontologists and Mineralogists, Spec. Publ. No. 28, 320 p.

Reprinted by permission of the Society of Economic Paleontologists and Mineralogists. Published in *Carbonate Cements,* N. Schneidermann and P.M. Harris, editors, SEPM Special Publication No. 36, pp. 97-106.

PREVENTION OF CARBONATE CEMENTATION IN PETROLEUM RESERVOIRS

CHARLES T. FEAZEL AND RICHARD A. SCHATZINGER
Phillips Petroleum Company, Bartlesville, Oklahoma 74004

ABSTRACT

By the time a carbonate unit has been buried to the depths of most petroleum reservoirs, the significant question is often not "how did the pores originate?", but rather, "why are they still there?" Preservation of porosity, regardless of its origin, is a consequence of one or more of the following mechanisms: (1) minimal burial, (2) reduced burial stress, generally due to overpressured pore fluids, (3) increased framework rigidity, which prevents compaction, (4) exclusion of pore waters by petroleum entry, (5) stable mineralogy, (6) permeability barriers, isolating porous intervals from external diagenetic fluids, and (7) pore resurrection, a consequence of the temporary filling of pores with cement that is subsequently removed. Examples from the stratigraphic record demonstrate that each of these pore-preserving mechanisms may control reservoir quality.

INTRODUCTION

This paper deals not with cements, but with their absence. Exploration for petroleum includes, among its many strategies, prediction of porous reservoirs at depth. Much effort has been expended on studies of the evolution and destruction of pore systems. Although many workers (notably Choquette and Pray, 1970) have catalogued the various types of pores found in carbonate rocks, a question seldom addressed is: "How are the pores preserved in petroleum reservoirs despite burial to depths exceeding five kilometers?" It is generally agreed, despite recent controversy (Moore and Druckman, 1981), that the majority of pore types described by Choquette and Pray (1970) resulted from early, shallow diagenesis. The following discussion concentrates on the processes that preserve such early pores as they pass through successive diagenetic environments.

MECHANISMS OF POROSITY PRESERVATION

We have identified seven mechanisms (Table 1) by which cementation in carbonate rocks may be retarded or prevented: (1) minimal burial, (2) reduced burial stress, (3) increased framework rigidity, (4) exclusion of pore water, (5) stable mineralogy, (6) permeability barriers, and (7) pore resurrection. Petroleum reservoirs from diverse geographic locations and geologic ages are cited in following paragraphs as examples of the control of reservoir quality by one or more of these pore-preserving mechanisms.

Minimal Burial

If carbonate sediments are never buried to significant depth (e.g. below the influence of surface-derived waters) they may retain much of their initial porosity, or have it considerably enhanced by meteoric processes. The shallow burial environment, the site of porosity generation and enhancement, is, however, a dynamic setting in which pore preservation is not favored. In this diagenetic environment porosity is destroyed by recrystallization, infiltration of clays, precipitation of vadose cements, or cements derived from biogenic gas.

The effect of shallow burial on reservoir potential thus depends on a delicate balance between porosity generating and porosity occluding processes. Despite the number of papers that emphasize the importance of porosity destruction during early, shallow diagenesis (Friedman, 1975, Longman, 1980, Bricker, 1971, and others), Halley and Schmoker (1983) and Scholle *et al.* (1983) have shown that significant porosity reduction in some sediments may not begin until carbonate strata are buried to depths of several hundred to nearly a thousand meters. Before these porous rocks can serve as petroleum reservoirs, they must be sealed. The likelihood of competent seals at shallow burial depth is minimal.

Although many examples of porous outcropping limestones that have never been deeply buried might be cited, from Cretaceous to Pleistocene in age (for example: Halley and Harris, 1979, Kerr, 1977, Land, 1973, and Land *et al.*, 1967), our goal is to understand porosity preservation at depth, not at the surface. In the following section we discuss rocks that may be deeply buried, but experience overburden pressures much like those of near-surface units.

Reduced Burial Stress

When pore pressures approach lithostatic pressures, the effective overburden stress resembles that of the shallow burial environment. Pore fluid overpressure reduces stress at grain contacts, and thereby prevents solution-transfer of calcium carbonate (pressure-solution at points of contact and reprecipitation of the dissolved material as pore-filling cement, Durney, 1972). In Maastrichtian and Danian chalk beneath the central North Sea, for example, porosity values of 40 percent are not uncommon at depths of over three kilometers (Van den Bark and Thomas, 1981; Feazel *et al.*, in

TABLE 1.—RELATIONSHIP BETWEEN THE PROCESSES, TIMING, AND RECOGNITION OF SEVEN MECHANISMS OF PORE PRESERVATION

MECHANISM	*PROCESS*	*TIMING*	*RECOGNITION*
Minimal Burial	Meteoric diagenesis Stable tectonics	Early	Stratigraphic setting Cement textures Fluid inclusions Chemical & isotopic signatures
Reduced Burial Stress	Increased pore pressure	Early to late	Pressure tests Log analysis
Increased Framework Rigidity	Lithification Grain support	Pre-compaction	Petrography
Exclusion of Pore Water	Hydrocarbon entry	Early to intermediate	Hydrocarbons in pores Log analysis (Sw)
Stable Mineralogy	Neomorphism Replacement	Early to intermediate	Petrography Trace elements
Permeability Barriers	Depositional encasement Diagenetic barrier	Early to late	Stratigraphic setting Petrography Geochemistry
Pore Resurrection	Occlusion followed by cement removal	Late	Petrography

press). Chalk normally loses porosity rapidly with increasing depth of burial (Fig. 1). The Niobrara chalk of the U.S. midcontinent, for example, has porosities of 40% where exposed, but where it was buried under 1500–2000 meters of Cretaceous and Tertiary deposits shed from the Rocky Mountains, porosity is reduced to 10%. Where the Niobrara is buried to a depth of 2400 meters at Fort Collins, Colorado, porosity is essentially zero (P. Scholle, 1977, and personal communication). The chalk at the Ekofisk Field in the North Sea is unusual because much of its original porosity has been retained. Scholle (1974) believed early dewatering of calcareous ooze reduces the 70–80% initial porosity to 50–60%. The presence in some Ekofisk intervals of porosity values near 50% indicates that little cementation followed mechanical compaction. Scholle (1975, 1977) observed that due to increased pore pressure in the chalk, the mechanism for both mechanical and chemical compaction is much less effective and porosity is preserved. The Central Graben of the North Sea is moderately overpressured, due to the low permeability of the sediments, to the effective seal provided by overlying Tertiary shales, and possibly to retardation of lateral fluid escape by boundary faults. The Ekofisk Field Chalk reservoir is pressurized to approximately 7100 psi at 3000 meters (10,000 feet) (Van den Bark and Thomas, 1981, Fig. 24). Normal pore pressure at that depth is approximately 4300 psi, with a lithostatic load of 9000 psi. Thus almost all the overburden above the Ekofisk structure is supported by the reservoir fluids. Net lithostatic pressure is approximately 1800 psi, which would normally correspond to

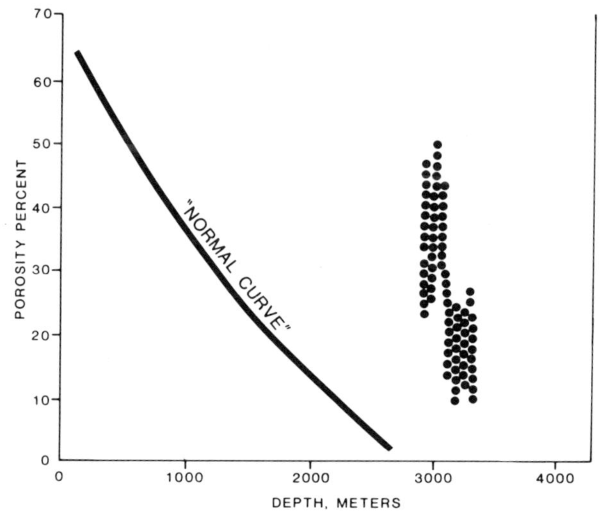

FIG. 1.—Porosity loss with depth for a "normal" pelagic sediment buried with marine or modified marine pore fluids (Scholle *el al.*, 1983). Data at right, from Ekofisk area wells, North Sea, show anomalously high porosity at 3 km burial depth.

a burial depth of only 610 meters (2000 ft.) (Scholle, 1977). The porosity of chalk buried to that depth (Hancock and Scholle, 1975) approximates that of the much deeper Ekofisk section. Without the confining structure and stratigraphy that led to overpressured conditions in the North Sea central graben, the pores would not have been available once the Jurassic source rock reached sufficient thermal maturity to expel petroleum. Oil entrapment as a pore-preserving mechanism will be discussed in a later section.

Increased Framework Rigidity

Any process that increases the strength of a potential reservoir unit, whether through fluid pressure buildup as described above, or by strengthening of the surrounding rock, decreases the probability of mechanical and chemical compaction. Increased rigidity of the rock framework generally results from two processes: (1) achievement of a grain-supported matrix by dewatering and early compaction (Shinn *et al.*, 1977), and (2) early lithification without pervasive pore-filling cementation (Sibley, 1982). The first process marks the transition from mechanical compaction to chemical compaction and may occur early in a rock's history. Figure 2 illustrates the results of the second process: a Holocene beachrock, cemented into a very rigid structure by a fine rim of aragonite crystals fringing the grains, with essentially no loss of primary porosity. Evidence for pre-compaction cement includes spalled fibrous calcite and undeformed delicate skeletal material (Fig. 3). Early dolomitization (pre-compaction: Kaldi and Gidman, 1982, Bebout *et al.*, 1981) of such rim cements (and pore-filling cements as well) leaves a very rigid framework that can withstand burial to depths exceeding five kilometers, even after leaching of the original grains (Fig. 4). The Smackover Formation in the eastern Gulf Coast contains numerous petroleum reservoirs (e.g. Chatom Field, Alabama) that owe their porosity to this mechanism.

Exclusion of Pore Water

Displacement of pore water by oil or gas prevents or significantly retards the precipitation of cement, as most minerals are insoluble in hydrocarbon fluids (Friedman and Sanders, 1978, Epstein and Friedman, 1983). The results of this process may include the preservation of Pennsylvanian aragonitic mollusc shells in

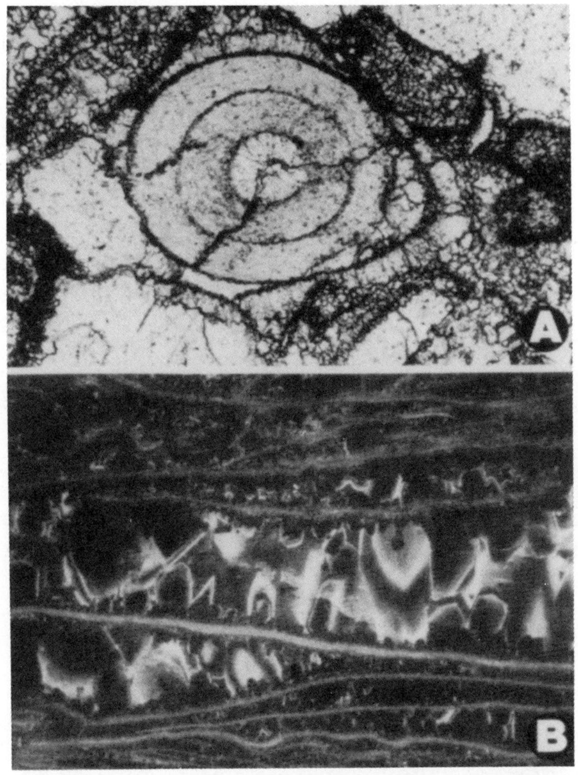

FIG. 3.—Pre-compaction cements. A. Marine calcite cement spalled from ooid by compaction. Pennsylvanian, Horseshoe Atoll, West Texas. Depth approximately 6500 feet. Photo width equals one millimeter. B. Unbroken delicate pelecypod valves preserved from compaction by early luminescent cement. Shublik Formation, Triassic, North Slope, Alaska. Depth approximately 9000 feet. Cathodoluminescence, photo width equals three millimeters.

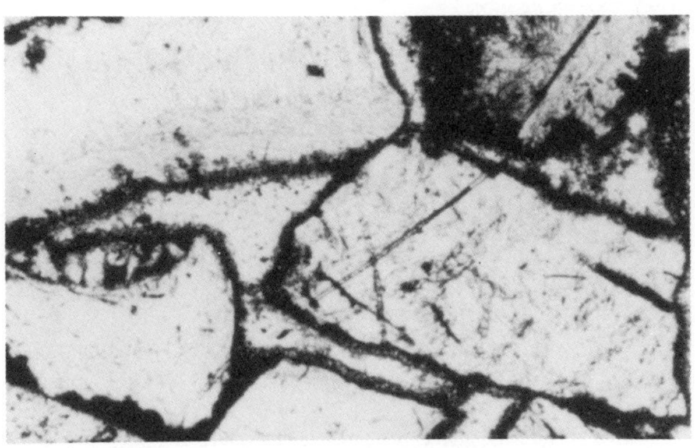

FIG. 2.—Increased framework rigidity caused by early aragonite cement (dark rims on bioclasts). Most primary porosity has been preserved. Only small volumes of cement were required for lithification. Holocene beachrock, U. S. Virgin Islands. Photo width equals one millimeter.

FIG. 4.—Framework rigidity caused by interstitial dolomitization between non-compacted peloids. Moldic pores created by leaching of peloids retain original shape despite deep burial. Smackover Formation, Chatom Field, Alabama, Depth exceeds 16,000 feet. From Feazel (in press).

tar (Squires, 1972). Similar shells of Lower Cretaceous age remain aragonitic where they are in contact with oil, but have recrystallized to calcite where they are in contact with water (Füchtbauer and Goldschmidt, 1964). Organic coatings of many different types on carbonate surfaces have been shown to inhibit diagenesis (Suess, 1970). In most reservoirs, following oil migration there remains a residual water saturation, which may be enough to allow diagenesis to continue after petroleum entry, but at greatly reduced rates.

In some cases (Davies, 1979, oral presentation) the last event before petroleum entry was corrosion of the pore walls. Likely agents for this corrosion were CO_2 and organic acids dissolved in waters moving ahead of the migrating oil. The moving organic fluids may thus have created their own enhanced reservoir porosity (Holland, 1982, Moore and Druckman, 1981). Once the petroleum was emplaced, however, most reservoir diagenesis ceased. An example is found in the Stuart City trend, South Texas (Fig. 5) in which pores contain, in paragenetic sequence: (1) marine cements, (2) meteoric cements, (3) subsurface overgrowths on the meteoric cements, containing hydrocarbon inclusions, (4) oil, forming a near-continuous film around the pore margins, and (5) late diagenetic calcite that apparently precipitated after oil migrated out of the pores (Prezbindowski, 1981).

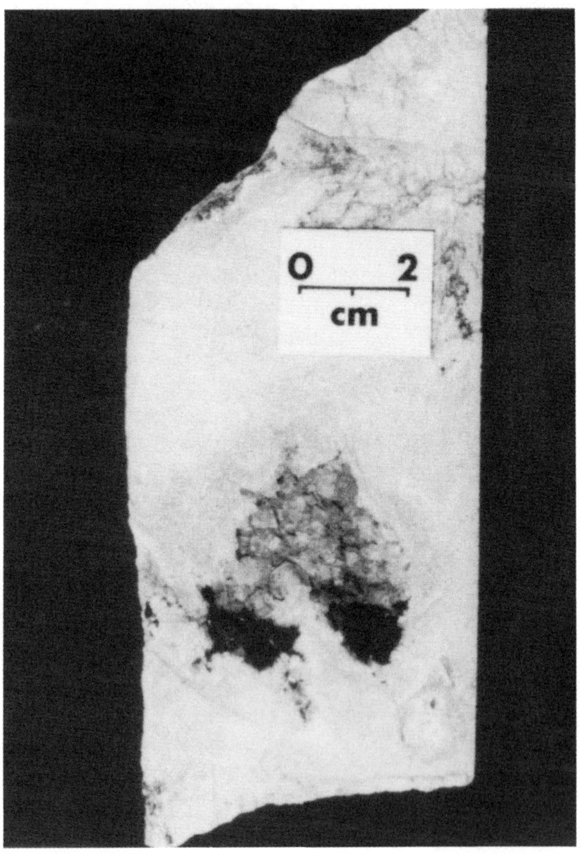

FIG. 5.—Fossil gas-oil contact in large vug. Migrating petroleum displaced pore water, effectively ending diagenesis. The last diagenetic event before petroleum entry was precipitation of coarse equant calcite on vug walls. Stuart City trend, south Texas, depth exceeds 13,000 feet.

Two estimates are possible of the conditions or timing of oil migration (Fig. 6). One involves the use of geochemical or isotopic paleothermometers (Milliken *et al.*, 1981) or fluid inclusion microscopy (Burruss, 1981) to estimate the temperature and salinity of fluids that precipitated cement on pore walls immediately before and during petroleum entry. The other technique involves calculation of the burial history and thermal maturity of petroleum source rocks (Harris *et al.*, 1980; Yukler and Kokesh, in press). Combining these two independent methods gives estimates of the time, depth, and temperature at which carbonate cementation stopped in a particular reservoir (McLimans, 1981).

Below the reservoir zone, however, cementation may continue unabated. Figure 7 illustrates the diagenetic variability found in a North Sea chalk reservoir, with samples from the oil-wet zone, the oil-water contact, and the water-wet zone. Clearly, oil in the pores preserved the chalk from further diagenesis. The abundance of skeletal calcite in the oil zone contrasts with the abundance of authigenic calcite in the water zone. The chalk within the diffuse oil-water contact appears transitional, and contains both visible skeletal elements and authigenic overgrowths. Pore pressure in all three zones is essentially identical. Thus overpressure (Scholle, 1977, cited previously under Reduced Burial Stress) alone is not responsible for porosity preservation in the chalk (Feazel *et al.*, in press). The small net lithostatic load may be sufficient to cause the generation of pore-filling cement by solution-transfer, unless aqueous fluids are prevented from reacting with the pore walls.

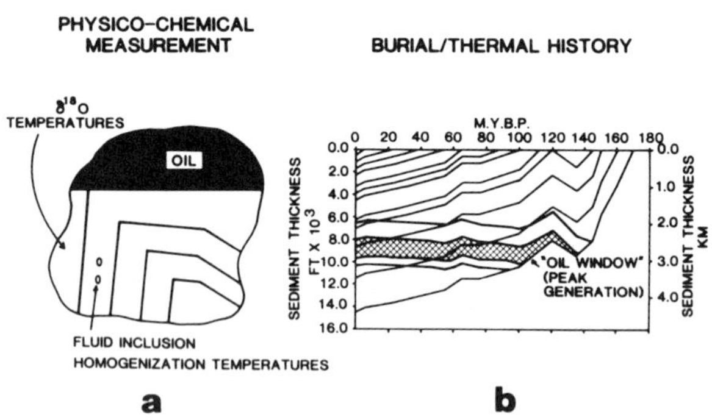

FIG. 6.—Independent techniques for determining the relative timing of hydrocarbon migration and diagenesis include: (a) estimation of paleotemperature and pore fluid composition from geochemical and isotopic signatures and fluid inclusions in cements precipitated before or during hydrocarbon migration; and (b) calculation of source rock maturity based on burial and thermal histories.

Stable Mineralogy

Sediment that starts its burial history with low diagenetic potential (Schlanger and Douglas, 1974) will be little affected by recrystallization. Syndepositional assemblages of stable minerals (low-Mg calcite or dolomite), or early neomorphic assemblages, will suffer less burial alteration than sediments with metastable components. The resistance of the rock framework to chemical alteration preserves the pores from compaction. Mineralogic stabilization is often associated with the increased framework rigidity mentioned earlier. The effects of these two mechanisms are the same; however, one is a chemical process, the other physical.

An example is the porous dolomite from the Smackover Formation shown in Figure 8. It matters little whether the dolomite formed by seepage-reflux, in the mixing zone between seawater and freshwater, or by some other mechanism: dolomitization was apparently early, and was accompanied by leaching of the intercrystalline sediment. Neither chemical alteration nor mechanical compaction altered this early intercrystalline porosity despite burial to a depth of 4.9 kilometers. Such dolomitic reservoirs might be cited throughout the Phanerozoic section (Ordovician, Williston Basin: Longman *et al.*, 1983; Silurian, Michigan Basin: Sears and Lucia, 1980; Devonian, Alberta: Mattes and Mountjoy, 1980; Permian, Permian Basin: Longacre, 1980; Triassic, Italy: Errico *et al.*, 1980; Jurassic, Florida: Ottman *et al.*, 1976; Miocene, East Malaysia: Scherer, 1980). The preservation of porosity in all these dolomite reservoirs is a consequence of simultaneous mineralogic stabilization and increased framework rigidity.

FIG. 7.—Preservation of porosity in chalk by hydrocarbon entry. Chalk from above (top), within (middle), and below base of hydrocarbons. Preservation of calcareous nannofossils suggest that primary porosity was greater than 15% in all three intervals ("predicted" values). Extensive secondary calcite in water-saturated chalk has all but obliterated any trace of skeletal particles. Pore pressures are approximately equal in all three intervals. Tor Field, North Sea, approximate depth 10,000 feet. From Feazel *et al.* (in press).

Permeability Barriers

Barriers to fluid permeability may enclose a lithologic entity, causing it to approximate a closed geochemical system. Truly closed systems, however, probably do not exist in sedimentary basins. The pores in a rock can be effectively sealed against fluid flow (physical barrier), but diffusion during geologic time will permit occlusion of even the most inaccessible pores unless some barrier to diffusion exists (geochemical barrier).

Diffusion may be self-limiting: as porosity decreases, the rate of diffusion (at constant concentration gradient) decreases by an amount equal to the porosity loss squared (Klinkenberg, 1951). Hence cemented margins often isolate porous carbonate reservoirs. (Cemented margins of Devonian reefs: McGillivray, 1977).

Most permeability barriers result from syndepositional processes. Porous carbonates may be surrounded

FIG. 8.—Intercrystalline porosity in dolomite. Replacement of original sediment by geochemically stable dolomite preserved pores from further alteration during burial. Smackover Formation, southeastern Mississippi, depth approximately 20,000 feet.

by impermeable carbonate facies (Permian algal reef with impermeable flanking beds: Mazzullo, 1982). Carbonate reservoirs may be sealed by shales (Cretaceous Stuart City shelf margin buildup: Bebout and Loucks, 1974; Pennsylvanian Horseshoe Atoll: Myers *et al.*, 1956; Miocene Philippine reefs: Longman, 1980; Miocene Indonesian reefs: Friedman, 1982). Evaporites may also encase porous carbonate bodies (Silurian reefs: Petta, 1980; Devonian reefs: Schmidt *et al.*, 1980). These permeability barriers presently seal hydrocarbon accumulations. However, prior to petroleum entry, they insured the preservation of porosity by isolation of the porous interval. The mechanism of porosity preservation is therefore distinct from that of oil entrapment: the former prevents inward migration of cement-precipitating fluids, the latter prevents the escape of hydrocarbons.

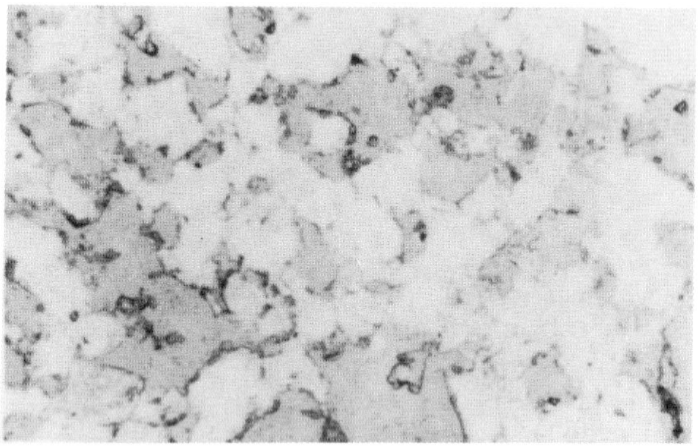

FIG. 9.—Resurrected pores in sandstone. Geometry of quartz grains (light) and epoxy-filled pores (dark) suggests the following sequence: (a) sand deposition, (b) calcite cementation and partial replacement of quartz and (c) resurrection of pore space by leaching of cement. Offshore west Africa, Cretaceous, approximately depth 7000 feet.

If such barriers effectively isolate a porous interval, little cement will precipitate. However, oil and gas may also be excluded if no migration conduit exists between the petroleum source and the potential reservoir. Such permeability barriers need to be breached at the time of petroleum entry, perhaps by fracturing (Longman, 1980). Once within the porous interval, petroleum can displace water by buoyancy. If fractures do not extend into overlying units, the petroleum will remain trapped. Skeletal buildups commonly grew from carbonate platforms. In such cases the preservation of porosity may require only that little cement be derived from the encasing shales or evaporites before oil can enter from below.

Pore Resurrection

A special category of moldic pores might be termed rejuvenated or resurrected. These are voids that formed relatively early in a rock's history, were later occluded by some cementing agent, and subsequently opened a second time by partial or complete removal of cement. Late resurrection increases the likelihood of such

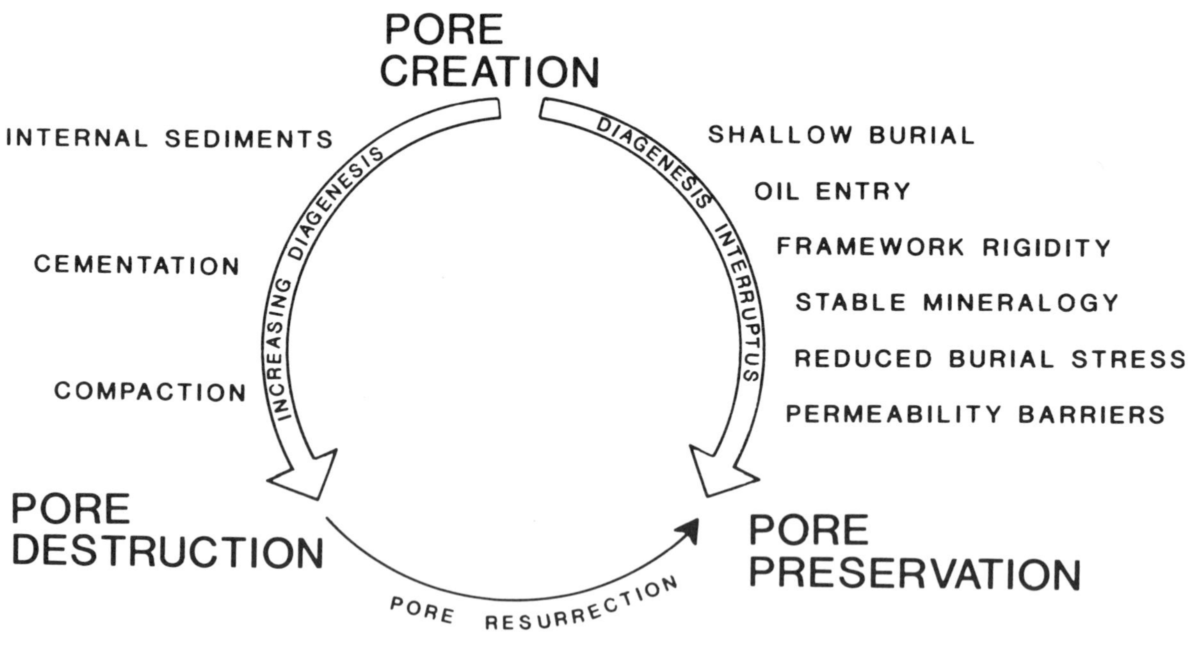

FIG. 10.—The presence of porosity in subsurface carbonates depends on a delicate balance between the mechanisms of porosity destruction and those of porosity preservation. Recognition of these mechanisms and their effects on reservoir properties may influence strategies for both exploration and development.

TABLE 2.—INFERENCES THAT CAN BE MADE ONCE PORE-PRESERVING MECHANISMS HAVE BEEN IDENTIFIED. PRESENCE OF SOME POROSITY ASSUMED.

Mechanism		Implications
MINIMAL BURIAL	Trend	Likely to follow original facies or paleotopography.
	Timing	Early pore preservation. Adjacent source rocks immature.
	Scale	Similar to scale of depositional facies.
	Effect on Reservoir	Bacterial degradation, water washing, or evaporation, probable low reservoir pressure.
REDUCED BURIAL STRESS	Trend	At major depocenters. Early preservation follows depositional patterns. May follow fracture pattern.
	Timing	Overpressure must pre-date compaction and significant cementation. When porosity was preserved compaction was minimal. Overpressure is a transient phenomenon; more common in Cretaceous and younger rocks.
	Scale	Regional.
	Effect on Reservoir	High porosity. High reservoir pressure; pressure maintenance necessary for sustained production. Flow may channel through fractures.
INCREASED FRAMEWORK RIGIDITY	Trend	a) Minor cementation-follows depositional trend. b) Dolomitization/recrystallization—may not follow depositional trend. Permeability variable.
	Timing	Determines trend. Ineffective after compaction or pervasive cementation.
	Scale	Stratigraphically restricted. Geographically restricted or widespread.
	Effect on Reservoir	Fractures and high permeability likely.
EXCLUSION OF PORE WATER	Trend	Structural, stratigraphic, or diagenetic.
	Timing	Pre-compaction; pre-cementation; a) Early: preserves primary pore network. b) Later: porosity enhancement. Secondary porosity. Fractures more important. Mature source rocks prabably deeper.
	Scale	Variable: Controlled by trend of structures or traps.
	Effect on Reservoir	High oil saturation implies good permeability (less post-oil cement). Pervasive cementation possible at base of oil column.
STABLE MINERALOGY	Trend	Depends on stabilization process: a) Initially stable mineralogy-may follow depositional trend. b) Replacement—relates to migration path of diagenetic fluid.
	Timing	a) Pre-burial; b) Early to late diagenetic.
	Scale	a) Initial—field to basin-wide. b) Replacement—cm scale to basin-wide.
	Effect on Reservoir	Depositional permeability may be preserved.
PERMEABILITY BARRIERS	Trend	a) Stratigraphic—follows depositional trend. b) Diagenetic—reflects past geochemical and hydrodynamic gradients.
	Timing	Sequence 1. Isolate porous interval. 2. Either: a. breach barrier so oil can enter, or b. generate oil within isolated interval.
	Scale	Small (diagenetic rims) to intermediate (pinnacle reefs).
	Effect on Reservoir	Fractures likely. Possible formation damage (plugging) as oil is produced and displaced by saturated brine.
PORE RESURRECTION	Trend	Diagenetic—in meteoric zone or at unconformities.
	Timing	a) Early—insignificant b) Late—more likely to be preserved.
	Scale	Dependent on original pore network.
	Effect on Reservoir	Permeability reflects original pore types.

pores' being preserved. This type of porosity is common in clastic rocks from which carbonate cements have been leached (Fig. 9). Its identification in carbonate rocks is less certain, as complete removal of a pore-filling cement is difficult to recognize.

An example was cited by Depowski and Peryt (in press), who described oil and gas reservoirs in the Zechstein dolomite (Upper Permian) of western Poland. They noted that highly productive intervals had been formed by (1) vadose leaching of the centers of ooids, oncoids, and bioclasts, (2) infilling of the pores with sulfate minerals, and (3) later dissolution of the sulfates by hydrothermal solutions.

Other examples (Moore and Druckman, 1981; Fischbuch and Havard, 1977) show that rejuvenated pores are generally solution-enhanced; they extend to some degree beyond their original borders. Such secondary porosity may be due to the migration of corrosive brines along preserved permeability-porosity networks. Therefore, it seems unlikely that resurrection of pores would occur in totally cemented, impermeable host strata.

SIGNIFICANCE TO EXPLORATION AND DEVELOPMENT

A successful exploration or exploitation program requires synthesis of stratigraphy, structure, paleogeography, sedimentation, and diagenesis. As Mazzullo (1982) observed, the best exploration approach asks why certain facies were deposited in particular locations at particular times. It may be equally important to understand the means by which pore systems are preserved following burial. Determination of pore-preserving mechanisms might explain why a pore system has a particular distribution, and represents another step in the hierarchy of models that predict the distribution of potential reservoir facies, whether depositional or diagenetic.

The following questions determine the effect that a given pore-preserving mechanism might have on exploration or development. Table 2 lists possible responses.

1. What trend is the pore system likely to follow?
   a. Is primary porosity preserved or has it been modified? Is preserved porosity facies-controlled?
   b. Is secondary porosity preserved? Is there any relationship to original facies, exposure surfaces, or geochemical gradients?
   c. Is preserved porosity related to structure, paleotopography, or unconformities?
2. What is the timing of pore preservation?
   a. When was cementation interrupted?
   b. When did compaction, leaching, permeability changes, or fracturing occur? What are their effects on the preserved pores?
   c. What is the relative timing of pore generation, hydrocarbon migration and entrapment, and pore preservation?
3. On what scale was the pore-preserving mechanism effective?
   a. Local (e.g. mineral stabilization by dolomitization of a supratidal belt)?
   b. Regional (e.g. mineral stabilization by dolomitization of a massive mid-cratonic platform)?
4. What effect might each pore-preserving mechanism have on reservoir properties?

SUMMARY

We have identified mechanisms by which cementation in carbonate rocks may be retarded or prevented. These mechanisms are responsible for the preservation of pores at reservoir depths, but are distinct from the mechanisms responsible for porosity creation and petroleum entrapment. Pores preserved by these mechanisms are recognized by petrographic, stratigraphic, geochemical, hydrodynamic, and electric log data. We have presented evidence from the rock record that such preserved pores exist, and that the mechanisms responsible for their preservation can be inferred with some certainty. An understanding of the preservation of reservoir porosity at depth has predictive utility to both exploration and development, particularly in terms of the trend, timing, and scale of the mechanisms of pore preservation, and the effect of each on reservoir properties.

REFERENCES

Bebout, D.G., and Loucks, R.G., 1974, Stuart City Trend Lower Cretaceous—A carbonate shelf margin model for hydrocarbon exploration: Univ. Texas-Austin, Bur. Econ. Geol. Rept. of Invest. 78, 80 p.

______, Budd, D.A., and Schatzinger, R.A., 1981, Depositional and diagenetic history of the Sligo and Hosston Formations (Lower Cretaceous) in South Texas: Univ. Texas-Austin Bur. Econ. Geol. Rep. of Invest. 109, 70 p.

Bricker, O.P. (ed.), 1971, Carbonate Cements: The Johns Hopkins Univ. Studies in Geology, v. 19, 376 p.

Burruss, R.C., 1981, Hydrocarbon fluid inclusions in studies of sedimentary diagenesis, *in* L.S. Hollister and M.L. Crawford (eds.), Short Course in Fluid Inclusions: Applications to Petrology: Mineral. Assoc. Canada Handbook 6, p. 138–156.

Choquette, P.W., and Pray, L.C., 1970, Geological nomenclature and classification of porosity in sedimentary carbonates: Amer. Assoc. Petroleum Geologists Bull., v. 54, p. 207–250.

Davies, G.R., 1979, Marine to late post-burial diagenesis of carbonate rocks, Arctic Archipelago: Oral presentation at AAPG Research Conf. on Deep Burial Diagenesis in Limestone, Tulsa.

Depowski, S., and Peryt, T.M., in press, Carbonate petroleum reservoirs in the Upper Permian Zechstein main dolomite (CA-1), Fore-Sudetic area, western Poland, *in* P.O. Roehl and P.W. Choquette (eds.), Carbonate Petroleum Reservoirs: a Casebook: Springer-Verlag.

DURNEY, D.W., 1972, Solution-transfer, an important geological deformation mechanism: Nature, v. 235, p. 315–317.
EPSTEIN, S.A., AND FRIEDMAN, G.M., 1983, Depositional and diagenetic relationships between Gulf of Elat (Aqaba) and Mesozoic of United States East Coast offshore; Amer. Assoc. Petroleum Geologists Bull., v. 67, p. 953–962.
ERRICO, G., GROPPI, G., SAVELLI, S., AND VAGHL, G.C., 1980, Malossa Field: A deep discovery in the Po Valley, Italy, *in* M.T. Halbouty (ed.), Giant Oil and Gas Fields of the Decade: 1968-1978: Amer. Assoc. Petroleum Geologists, Mem. 30, p. 525–538.
FEAZEL, C.T., KEANY, J., AND PETERSON, R.M., in press, Cretaceous and Tertiary chalk of the central North Sea: A major petroleum reservoir, in P.O. Roehl and P.W. Choquette (eds.), Carbonate Petroleum Reservoirs: A Casebook: Springer-Verlag.
FISCHBUCH, N.R., AND HAVARD, C.J., 1977, Facies, porosity relationships, and diagenetic processes which modify them, Swan Hills area, Alberta: Canadian Soc. Petrol. Geol. 5th Ann. Core Conf. Proc., Calgary, p. 41–65.
FRIEDMAN, G.M., 1975, The making and unmaking of limestones or the downs and ups of porosity: Jour. Sed. Petrology, v. 45, p. 379–398.
______, 1982, Reefs and porosity: Examples from the Indonesian Archipelago: Southeast Asia Petroleum Exploration Soc. Proc.
______, AND SANDERS, J., 1978, Principles of Sedimentology: New York, John Wiley and Sons, 147 p.
FÜCHTBAUER, H., AND GOLDSCHMIDT, H., 1964, Aragonitische Lumachelle im bisuminosen Wealden des Emslandes: Beitr. Mineral, Petrog., v. 10, p. 184–197.
HALLEY, R.B., AND HARRIS, P.M., 1979, Fresh-water cementation of a 1,000-year-old oolite: Jour. Sed. Petrology, v. 49, p. 969–988.
______, AND SCHMOKER, J.W., 1983, High-porosity Cenozoic carbonate rocks of South Florida: Progressive loss of porosity with depth: Amer. Assoc. Petroleum Geologists Bull., v. 67, p. 191–200.
HANCOCK, J.M., AND SCHOLLE, P.A., 1975, The petrology of chalk: Geol. Assoc. London Proc. 86, p. 499–535.
HARRIS, P.M., FROST, S.H., KENDALL, C.G., SHAFFER, B.C., AND SCHNEIDERMANN, N., 1980, Cretaceous sea level and stratigraphy, Eastern Arabian Peninsula [abs.]: Amer. Assoc. Petroleum Geologists Book of Abstracts, p. 64.
HOLLAND, M.T., 1982, Reservoir property implications of pore geometry modification accompanying sand diagenesis: Anahuac Formation, Louisiana: 57th Ann. SPE of AIME Fall Conf. (New Orleans), preprint no. SPE-10991, 12 p.
KALDI, J., AND GIDMAN, J., 1982, Early diagenetic dolomite cements: Examples from the Permian Lower Magnesian Limestone of England and the Pleistocene carbonates of the Bahamas: Jour. Sed. Petrology, v. 52, p. 1073–1085.
KERR, R.S., 1977, Facies, diagenesis, and porosity development in a Lower Cretaceous bank complex, Edwards Limestone, North-Central Texas, *in* D.G. Bebout and R.G. Loucks (eds.), Cretaceous Carbonates of Texas and Mexico, Applications to Subsurface Exploration: Univ. Texas Bur. Econ. Geol. Rept of Invest. 89, p. 216–233.
KLINKENBERG, L.J., 1951, Analogy between diffusion and electrical conductivity in porous rocks: Geol. Soc. America Bull., v. 62, p. 559–564.
LAND, L.S.,1973, Contemporaneous dolomitization of middle Pleistocene reefs by meteoric water, North Jamaica, in Coral Reef Project: Bull. Marine Science, v. 23, p. 64–92.
______, MACKENZIE, F.T., AND GOULD, S.J., 1967, Pleistocene history of Bermuda: Geol. Soc. America Bull., v. 78, p. 993–1006.
LONGACRE, S.A., 1980, Dolomite reservoirs from Permian biomicrites, *in* R.B. Halley and R.G. Loucks (eds.), Carbonate Reservoir Rocks: Notes for Soc. Econ. Paleontologists and Mineralogists Core Workshop No. 1, p. 105–117.
LONGMAN, M.W., 1980, Carbonate petrology of the Nido B-3A core, offshore Palawan, Philippines, in R.B. Halley and R.G. Loucks (eds.), Carbonate Reservoir Rocks: Notes for Soc. Econ. Paleontologists and Mineralogists Core Workshop No. 1, p. 161–183.
______, FERTAL, T.G., AND GLENNIE, J.S., 1983, Origin and geometry of Red River dolomite reservoirs, western Williston Basin: Amer. Assoc. Petroleum Geologists Bull., v. 67, p. 744–771.
MATTES, B.W., AND MOUNTJOY, E.W., 1980, Burial dolomitization of the Upper Devonian Miette buildup, Jasper National Park, Alberta, *in* D.H. Zenger, J.B. Dunham, and R.L. Ethington (eds.), Concepts and Models of Dolomitization: Soc. Econ. Paleontologists and Mineralogists Spec. Pub. 28, p. 259–297.
MAZZULLO, S.J., 1982, Types and controls of Permo-Pennsylvanian carbonate stratigraphic traps of shallow-marine origin in Permian Basin: exploration models: Oil and Gas Jour. v. 80, p. 124–141.
MCGILLIVRAY, J.G., 1977, Golden Spike D3A Pool, *in* I.A. McIlreath (ed.), The Geology of Selected Carbonate Oil, Gas, and Lead-Zinc Reservoirs in Western Canada: 5th Canadian Petrol. Geol. Core Conf. Proc., p. 67–88.
MCLIMANS, R.K., 1981, Application of fluid inclusion studies to reservoir diagenesis and petroleum migration: Smackover Formation, U. S. Gulf Coast, and Fateh Field, Dubai [abs.]: Amer. Assoc. Petroleum Geologists Bulletin, v. 65, p. 957.
MILLIKEN, K.L., LAND, L.S., AND LOUCKS, R.G., 1981, History of burial diagenesis determined from isotopic geochemistry, Frio Formation, Brazoria County, Texas: Amer. Assoc. Petroleum Geologists Bull., v. 65, p. 1397–1413.
MOORE, C.H., AND DRUCKMAN, Y., 1981, Burial diagenesis and porosity evolution, Upper Jurassic Smackover, Arkansas and Louisiana: Amer. Assoc. Petroleum Geologists Bull., v. 65, 597–628.
MYERS, D.A., STAFFORD, P.T., AND BURNSIDE, R.J., 1956, Geology of the late Paleozoic Horseshoe Atoll in West Texas: Univ. Texas-Austin, Bur. Econ. Geol. Publ. 5607, 113 p.
OTTMAN, R.D., KEYES, P.L., AND ZIEGLER, M.A., 1976, Jay Field, Florida-A Jurassic stratigraphic trap, in S.J. Mazzullo (ed.), Stratigraphic Traps in Carbonate Rocks: Amer. Assoc. Petroleum/Geologists Reprint Series, No. 23, p. 115–125.
PETTA, T.J., 1980, Silurian pinnacle reef diagenesis—Northern Michigan: effects of evaporites on pore space distribution, *in* R.B. Halley and R.G. Loucks (eds.), Carbonate Reservoir-Rocks: Notes for Soc. Econ. Paleontologists and Mineralogists Core Workshop No. 1, p. 32–42.

PREZBINDOWSKI, D.R., 1981, Carbonate rock-water diagenesis, Lower Cretaceous Stuart City Trend, South Texas: Ph.D. dissertation, Univ. of Texas, Austin, 236 p.

SCHERER, F.C., 1980, Exploration in East Malaysia over the past decade: in M.T. Halbouty (ed.), Giant Oil and Gas Fields of the Decade: 1968-1978: Amer. Assoc. Petroleum Geologists, Mem. 30, p. 423–440.

SCHLANGER, S.O., AND DOUGLAS, R.G., 1974, The pelagic ooze-chalk-limestone transition and its implication for marine stratigraphy: in K.J. Hsü and H.C. Jenkyns (eds.), Pelagic Sediments; On Land and Under the Sea: Internat. Assoc. Sedimentologists Spec. Pub. 1, p. 177–210.

SCHMIDT, V., McDONALD, D.A., AND McILREATH, I.A., 1980, Growth and diagenesis of Middle Devonian Keg River cementation reefs, Rainbow Field, Alberta, in R.B. Halley and R.G. Loucks (eds.), Carbonate Reservoir Rocks: Notes for Soc. Econ. Paleontologists and Mineralogists Core Workshop No. 1, p. 43–63.

SCHOLLE, P.A., 1974, Diagenesis of Upper Cretaceous chalks from England, Northern Ireland, and the North Sea: in K.J. Hsü and H.C. Jenkyns (eds.), Pelagic Sediments; On Land and Under the Sea: Internat. Assoc. Sedimentologists Spec. Pub. No. 1, p. 177–210.

______, 1975, Applications of chalk-diagenetic studies to petroleum—exploration problems (abs.): Amer. Assoc. Petroleum Geologists Bulletin, v. 59, p. 2197–2198.

______, 1977, Chalk diagenesis and its relation to petroleum exploration: oil from chalks, a modern miracle: Amer. Assoc. Petroleum Geologists Bull. v. 61, p. 982–1009.

______, ARTHUR, M.A., AND EKDALE, A.A., 1983, Pelagic environment, *in* P.A. Scholle, D.G. Bebout, and C.H. Moore (eds.), Carbonate Depositional Environments: Amer. Assoc. Petroleum Geologists, Memoir 33, p. 620–691.

SEARS, S.O., AND LUCIA, F.J., 1980, Dolomitization of Northern Michigan Niagara Reefs by brine refluxion and freshwater/seawater mixing, *in* D.H. Zenger, J.B. Dunham and R.L. Ethington (eds.), Concepts and Models of Dolomitlzation, Soc. Econ. Paleontologists and Mineralogists, Spec. Pub. 28, p. 215–236.

SHINN, E.A., HALLEY, R.B., HUDSON, J.H., AND LIDZ, B.H., 1977, Limestone compaction: An enigma: Geology, v. 5, p. 21–24.

SIBLEY, D.F., 1982, The origin of common dolomite fabrics: Clues from the Pliocene: Jour. Sed. Petrology, v. 52, p. 1087–1100.

SQUIRES, R.L., 1972, Paleoecology, biostratigraphy, and faunal chemistry of a Pennsylvanian age asphaltic limestone, Arbuckle Mountains, Oklahoma: Geol. Soc. America, Abstracts with Programs, v. 4, p. 675.

SUESS, E., 1970, Interaction of organic compounds with calcium carbonate—1. Association phenomena and geochemical implications: Geochim. et Cosmochim. Acta, v. 34, p. 157–168.

VAN DEN BARK, E. AND THOMAS, O.D., 1981, Ekofisk: first of the giant oil fields in western Europe: Amer. Assoc. Petroleum Geologists Bull. v. 65, p. 2341–2363.

YUKLER, M. A., AND KOKESH, F. C., in press, A review of models used in petroleum resource estimation and organic geochemistry: Advances in Petroleum Geochemistry, v. 1, Academic Press.

THE AMERICAN ASSOCIATION OF PETROLEUM GEOLOGISTS BULLETIN

APRIL 1980 VOLUME 64, NUMBER 4

Carbonate Diagenetic Textures from Nearsurface Diagenetic Environments[1]

MARK W. LONGMAN[2]

Abstract Understanding the processes and products of carbonate diagenesis is essential to exploration for, and optimum development of, hydrocarbon reservoirs in carbonate rocks. Much (and perhaps most) cementation and formation of secondary porosity (except fractures) in carbonates occurs at relatively shallow depths in one of four major diagenetic environments: the vadose zone, meteoric phreatic zone, mixing zone, and marine phreatic zone. Each of these zones may be divided into several parts on the basis of rate of water movement and saturation of the water with respect to calcium carbonate.

Most carbonates are deposited in and begin their diagenetic history in the marine phreatic environment. This zone may be divided into two end members of a continuous spectrum: a zone of relatively little water circulation in which micritization and minor *intra*granular cementation occur, and a zone of good water circulation near the sediment/water interface of shelf margins or the upper shoreface in which extensive *inter*granular and cavity-filling cementation occur. Fibrous aragonite and micritic Mg-calcite are the dominant cements.

With subaerial exposure, fresh water will replace sea water in the pores of shallow-water carbonates, and a zone of mixed fresh and marine waters may form. In long-lived mixing zones, dolomite may form if the water is of relatively low salinity, whereas bladed Mg-calcite may form if the water is relatively marine. Active water circulation in the mixing zone, which may be caused by seasonal rainfall, is necessary for dolomitization or cementation.

Diagenesis in the freshwater phreatic environment may involve leaching in the zone of solution, neomorphism of grains accompanied by extensive intergranular calcite cementation in the active saturated zone, or neomorphism of grains without cementation in the stagnant saturated zone. Syntaxial overgrowths on echinoderm fragments and interlocking crystals of equant calcite that coarsen toward pore centers are typical of cementation in the active freshwater phreatic zone.

The freshwater vadose environment is the zone with both air and meteoric water in the pores and may be divided into the zone of solution and the zone of precipitation. CO_2 from the atmosphere and soil contributes to solution which generally occurs near the soil zone and forms vugs, molds, and etched grains. When the water becomes saturated with respect to calcite, evaporation or CO_2 loss may cause precipitation of fine equant calcite in the form of pendant and meniscus cements. Grains may be altered to calcite, particularly in humid climates, and caliche crusts may be produced by evaporation and/or biologic (generally algal) factors.

Climate plays an important role in early diagenesis if subaerial exposure occurs. In arid climates, cementation in freshwater environments may be limited and primary intergranular porosity may be preserved. In humid climates, little primary porosity is likely to escape cementation, but significant amounts of secondary moldic and vuggy porosity may form.

Interpretation of diagenesis in carbonates is complicated by the fact that diagenetic environments may change many times in the history of a carbonate rock. By recognizing the processes leading to the formation or preservation of porosity, and the distribution of dia-

[1]Manuscript received, June 6, 1979; accepted, October 1, 1979.

[2]Cities Service Research Laboratory, Tulsa, Oklahoma 74150.

A summary paper such as this must rely heavily on the work of others and I wish to acknowledge the contributions of each of the authors cited in the references.

Special thanks to L. F. Baie, R. L. Folk, R. B. Halley, K. Helmold, R. G. Loucks, P. A. Mench, M. Meyer, J. Pol, D. R. Prezbindowski, and C. Scheiner for reviewing drafts of the manuscript and to the many people who contributed in discussions. Cities Service Co. provided time to prepare the paper and permission to publish.

Article Identification Number
0149-1423/80/B004-0001$03.00/0

genetic environments in which those processes acted, the distribution of porosity in subsurface carbonates can often be predicted.

INTRODUCTION

Because of the susceptibility of carbonate rocks to leaching and cementation, the best approach to exploration for nonstructural hydrocarbon reservoirs in carbonates is different from that for comparable sandstone reservoirs. In terrigenous reservoir rocks, porosity is usually intergranular between sand grains. Thus, highest priority in exploration has generally been given to interpreting depositional facies and trends of the sands with secondary emphasis on cementation and other diagenetic trends. In carbonate rocks, a remarkable variety of porosity types may be present, including intergranular, moldic, dolomitic, fracture, and so on as described by Choquette and Pray (1970). Not only is this porosity commonly better developed in the mud-rich facies than the grainy facies, but often the porosity forms completely independently of depositional facies (e.g., vuggy, fracture, late dolomitic, etc). Thus, emphasis in exploration for hydrocarbon reservoirs in carbonate rocks should be primarily on interpreting the diagenetic controls on porosity and on interpreting depositional facies when there is a relation between porosity and depositional fabric.

The goal of this paper is to summarize carbonate diagenesis by describing the types and textures of cements and porosity produced in the major diagenetic environments of the shallow subsurface to aid the exploration or exploitation geologist faced with the problem of interpreting diagenetic events and resultant porosity trends in carbonate rocks. Many petrologic and speleologic studies have shown that extensive (and perhaps most) leaching and cementation in carbonates occur at these shallow depths, although the extent of subsurface cementation and leaching is still uncertain.

Calcium carbonate dissolves hundreds of times faster than quartz in fresh water in normal earth-surface conditions. The solution and precipitation of calcium carbonate is influenced by partial pressure of CO_2, fluid chemistry, rate of fluid movement, crystal size, mineralogy, and a variety of other factors. In the shallow subsurface, these factors are present in recurring patterns over wide areas. The result is distinct, but variable, diagenetic patterns. Areas in which predictable and similar patterns of diagenesis occur may be thought of as diagenetic environments. Diagenetic environments may be independent of depositional environments and may change through time. Recognizing diagenetic products formed in the various diagenetic environments is a major part of carbonate petrology and can play a key role in predicting porosity trends in carbonate rocks.

Major diagenetic environments of the shallow subsurface are defined in this paper. Characteristic diagenetic textures produced in each environment are described and related to the processes responsible for their formation. The variability and complexity of diagenesis even within one diagenetic environment necessitate many generalizations. Thus, this paper is a summary and synthesis rather than a detailed account of all types of diagenesis and diagenetic environments.

Emphasis here is on diagenesis in carbonate sands because most diagenetic textures are more easily observed in these than in carbonate muds. However, many of the principles proposed for diagenesis in the carbonate sands should also apply to carbonate muds. Diagenetic characteristics which can be observed in standard petrographic thin sections are stressed because most exploration geologists have access to petrographic microscopes.

PREVIOUS WORK

Many concepts of diagenetic environments presented here are not new but represent a summary of the available literature. Early work by Sorby (1879) and Cullis (1904) set the stage for later studies of carbonate diagenesis. With the discovery of the major oil fields in carbonate rocks of the Middle East in the 1940s and 50s, carbonate sedimentology and petrology gained new importance. Papers on various aspects of carbonate diagenesis were published by Bathurst in the late 1950s (1958, 1959) and by Stehli and Hower (1961), but they were largely descriptive and contributed little to the knowledge of processes in carbonate diagenesis. Friedman (1964) made significant contributions to the knowledge of carbonate diagenesis, particularly with the concept of subaerial diagenesis, and a few years later (Friedman, 1968) made an early attempt to define diagenetic environments by contrasting marine and freshwater cementation. Characteristics of many different types of carbonate cements were described in Bricker (1971).

Matthews (1967, 1968) and Land (1970, 1971, 1973a, b) were the first to discuss in detail the concepts of diagenetic environments as defined in this paper and the diagenetic textures formed in each environment. Their work was followed by many detailed petrographic studies of diagenetic textures in Pleistocene and Holocene carbonates including those of Steinen (1974), Harrison (1974, 1975), Badiozamani et al (1977), and Harris (1978, 1979). Good summaries of diagenetic environments in carbonates have been published by

Folk (1973, 1974), Matthews (1974), Bathurst (1976), Friedman and Sanders (Ch. 6, 1978), and Moore (1979). And, although much work already has been done on specific aspects and problems in carbonate diagenesis, much remains to be done in defining and characterizing the products and processes of carbonate diagenesis in diagenetic environments of both the shallow and deep subsurface.

MAJOR DIAGENETIC ENVIRONMENTS

Four major diagenetic environments are typical of the shallow subsurface and may be defined on the basis of fluid chemistry and the distribution of that fluid in the pores. These are the freshwater[3] vadose zone (above the water table), freshwater phreatic zone (below the water table), mixed freshwater-marine phreatic zone, and marine phreatic zone (Fig. 1). Each of these zones may be divided into subzones as described in the following.

Until recently, most carbonate petrologists tended to lump the freshwater vadose and freshwater phreatic environments together in the so-called zone of subaerial diagenesis (Friedman, 1964; Bathurst, 1976). Although subaerial diagenesis successfully implies freshwater involvement, it is a vague term encompassing a wide range of diagenetic environments and reveals little about diagenetic textures. More information is contained in the terms freshwater vadose and freshwater phreatic because distinctive cement textures form in each of these diagenetic environments. These textures are described in this paper and other papers by Land (1970), Schroeder (1973), Harrison (1974), Matthews (1974), and Harris (1979).

The four zones mentioned previously generally occur in a consistent sequence governed by gravitational laws. Although any zone may be missing, and other zones may be present, from bottom to top the sequence consists of (1) marine phreatic, (2) mixing zone, (3) meteoric phreatic, and (4) meteoric vadose. Only under unusual circumstances of rapid rises in sea level or strong hydrostatic heads acting on layers of rock with marked permeability contrasts could meteoric water be forced under the more dense sea water. In equilibrium conditions the vadose zone may never be present below any other zone, although perched water may create pseudofreshwater phreatic zones in the vadose zone.

This characteristic sequence of diagenetic environments may produce predictable diagenetic sequences as changes in fluid chemistry occur in response to sea-level fluctuations or rainfall. For example, the normal sequence of diagenesis for a marine sediment during a relative lowering of sea level would be to move from the marine phreatic zone through the mixing zone into the meteoric phreatic and finally into the vadose environment. It would be rare for a sediment to move directly from the marine phreatic into the meteoric vadose environment, except in areas of very rapid relative uplift. During transgression, the sequence would be reversed and, if all nearby land areas were submerged, the sediments and rocks would end up in the marine phreatic environment. Knowledge of the interrelations of these diagenetic environments and relative changes in sea level aids in the interpretation of diagenesis in ancient limestones.

[3]The terms fresh water and meteoric water are used more or less interchangeably in this paper because most nonmarine waters causing diagenesis in the shallow subsurface are "fresh" and of meteoric origin.

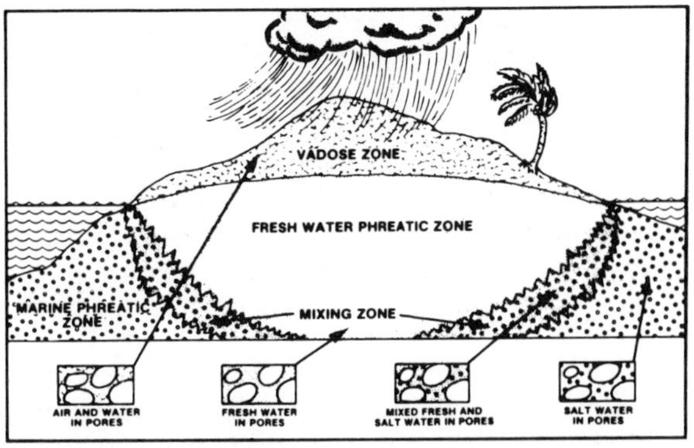

FIG. 1—Cross section showing distribution and relations of major diagenetic environments in shallow subsurface in ideal permeable carbonate sand island. Not to scale, but vertical distance would typically represent tens of meters and horizontal distance would be a few kilometers.

MARINE PHREATIC ZONE

When all pore space in a sediment or rock is filled with normal marine water, it is said to be in the marine-phreatic diagenetic environment. Most carbonates are deposited in marine environments, and thus begin their diagenetic history there.

Characteristics of marine waters are strongly influenced by water depth which, in turn, affects temperature and saturation with respect to $CaCO_3$. Patterns of carbonate diagenesis differ depending on whether the sediment is above or below the carbonate compensation depth (generally at a depth of several thousand meters in low-latitude regions). The following discussion refers only to diagenesis in the marine phreatic environment beneath warm shallow (<100 m) seas, and well above this compensation depth. Even within this shallow zone, marine phreatic diagenesis may be somewhat variable as the result of rates of water movement in the sediment. Matthews (1974) has pointed out that, although sea water is normally supersaturated with respect to calcium carbonate, only a very small amount of carbonate cement may be precipitated from one pore volume of water. Thus, for cementation to occur, many pore volumes of water must move through the sediment.

The marine phreatic environment may be divided into two zones: (1) the active marine phreatic zone in which water movement in combination with other processes results in cementation, and (2) the stagnant marine phreatic zone in which water movement through the sediment is relatively slow and cementation does not occur. Obviously these are end members of a continuous spectrum. Characteristics of these subzones are described in the following and are represented in Figure 2. Unfortunately other factors such as fluid chemistry and presence of organic material complicate this simple two-fold division; it is not uncommon to find marine sediments through which marine water circulates freely, but little cementation occurs (e.g., in a borehole in Enewetak Atoll which was cased to a depth of about 600 m, but in rocks with such good porosity and permeability that no tidal lag was observed; Ladd and Schlanger, 1960).

Active Marine Phreatic Zone

The zone in which sea water moves readily into the sediment resulting in marine cementation is herein called the active marine phreatic zone. Many studies of marine Holocene limestones have shown that aragonite and Mg-calcite are the only carbonates to precipitate inorganically from sea water at shallow depths (Shinn, 1969; Alexandersson, 1972; Bathurst, 1974; James et al, 1976), and that these cements do not occur everywhere. Most cementation apparently occurs where water is originally forced into the sediment (Shinn, 1969; Schroeder, 1972; James et al, 1976; Macintyre, 1977), where photosynthesis and respiration of the reef biomass increase the pH to more than 9 (Friedman et al, 1974), where CO_2 degassing occurs (Hanor, 1978), and/or where certain bacteria are present (Macintyre and Videtich, 1979). Tides, waves, and currents are the dominant forces capable of moving sea water into sediments, and they work best at shelf margins where a vertical obstruction such as a reef or carbonate buildup is present or in coarse-grained sediments of the upper shoreface (Fig. 3).

It is probable that too much emphasis is placed on the role of interstitial movement of marine wa-

MARINE PHREATIC ENVIRONMENT

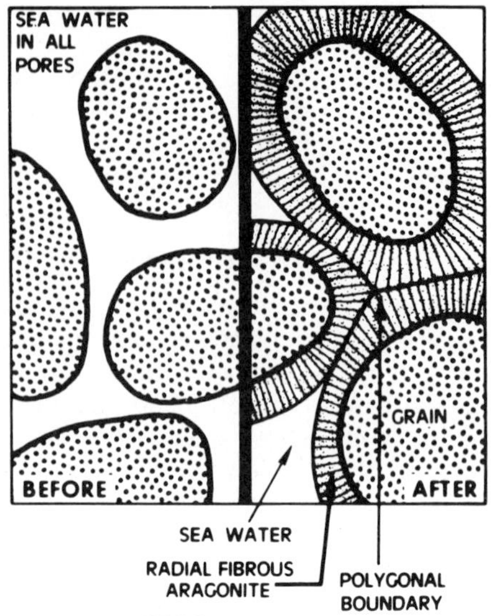

STAGNANT ZONE

PROCESSES

1. LITTLE OR NO WATER CIRCULATION THROUGH SEDIMENT
2. BACTERIAL (?) CONTROL ON CEMENTATION
3. WATER SATURATED WITH $CaCO_3$

PRODUCTS

1. LITTLE CEMENTATION EXCEPT IN SKELETAL MICROPORES
2. NO LEACHING
3. NO ALTERATION OF GRAINS
4. MICRITIZATION

ACTIVE ZONE

PROCESSES

1. WATER FORCED THROUGH SEDIMENTS BY WAVES, TIDES, OR CURRENTS
2. ALL PORES FILLED WITH SEA WATER
3. NO LEACHING IN SHALLOW MARINE ENVIRONMENTS

PRODUCTS

1. RANDOM ARAGONITE NEEDLES
2. ISOPACHOUS FIBROUS ARAGONITE
3. BOTRYOIDAL ARAGONITE
4. MICRITIC MG-CALCITE
5. ISOPACHOUS FIBROUS MG-CALCITE
6. MG-CALCITE PSEUDO-PELLETS
7. POLYGONAL BOUNDARIES BETWEEN ISOPACHOUS CEMENTS
8. INTERBEDDED CEMENTS AND SEDIMENTS
9. BORINGS IN CEMENTS
10. MOST CEMENTATION IN REEFS OR SURF ZONES

FIG. 2—Characteristics of marine phreatic diagenetic environment.

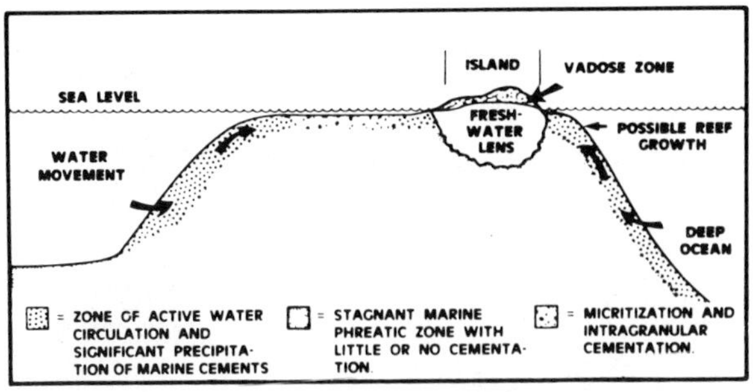

FIG. 3—Schematic cross section of large carbonate bank showing marine phreatic zone divided into areas with active water circulation (and thus cementation) and little water circulation (stagnant zones with little cementation). In active zone most cementation occurs within a few meters of sediment/water interface. Cross section modeled after Great Bahamas Bank with vertical scale in hundreds of meters and horizontal scale in kilometers.

ters in causing marine cementation because extensive areas such as that indicated by the fine stipple in Figure 3 (labeled stagnant marine phreatic zone) are commonly so highly porous and permeable that waters in the sediments move with every tide. Ladd and Schlanger (1960) found an example of this in the Marshall Islands. More emphasis needs to be placed on determining whether photosynthesis, respiration of organisms, CO_2 degassing, bacteria, or some other factor facilitates cementation near, but not more than several meters below, the sediment/water interface. In reality, probably any of the four factors listed, as well as others still unrecognized, can act in combination with the movement of water into the sediments to cause marine cementation, but the important point is that the vast majority of marine cementation occurs very near the sediment/water interface where water does actively move into the sediments.

Within the active marine phreatic zone, rigid substrates formed by organic binding are more likely to be cemented than unconsolidated substrates, partly because movement of the grains hinders the development of cements and partly because compaction of soft sediments squeezes water upward and prevents saturated sea water from circulating through the sediment. This means that reefs or grains stabilized by organic activity are the best sites for marine cementation. Most of the best known examples of marine cements are from reefs (Schroeder, 1972; Friedman et al, 1974; James et al, 1976; Macintyre, 1977). Some nearshore carbonate sand bars, specifically those characterized by slow sedimentation rates and relatively coarse grain size in arid climates, are also vulnerable to marine cementation (Shinn, 1969). However, the mobile ooid shoals of the Bahamas and lagoonal muds of Belize contain marine cements only locally, perhaps where some type of organic binding (algae?) was present.

Cement types—The dominant cements precipitated in the active marine phreatic environment are Mg-calcite and aragonite (Fig. 4; see Bricker, 1971, pt. 2, for many examples). Mg-calcite cement is commonly in the form of micritic or steep-sided rhombic crystals (Alexandersson, 1972; James et al, 1976; Macintyre, 1977) distributed as isopachous crusts on the grains. In some cases, continued growth of the upper faces of the rhombs produces thick rims of fibrous to bladed isopachous cement. In other cases the micritic Mg-calcite clusters to form pseudopellets. Such textures are common in recent marine-cemented reef rocks (Land and Goreau, 1970; Macintyre, 1977) and are present in marine phreatic beach rocks where Moore (1977) interprets them as being of organic origin.

Marine aragonite cement typically is present as fibrous crystals in a variety of textures (Fig. 5). Commonly the crystals nucleate on grains and grow outward to form an isopachous fibrous fringe. In some places aragonite forms a loose meshwork of needles. Ginsburg and James (1976) have described coarse botryoids of aragonite from the Belize barrier reef. The variety of shapes may result from the substrate on which the cement is growing, the nature of the cavity, and the rate of crystallization. Like the Mg-calcite, marine aragonite may form in a variety of rock types, including reefs (Ginsburg and James, 1976; James et al, 1976; Mazzullo and Cys, 1977), beach rocks (Moore, 1977; Hanor, 1978), and ooid shoals (Harris, 1978).

Other characteristics—The isopachous fibrous nature of most marine cements is a criterion that aids in their recognition in ancient rocks. The fibrous texture may be preserved in spite of neomorphism to calcite. Shinn (1975) described polygonal boundaries formed by the intersection of the isopachous fibrous cements (Fig. 4D) and suggested that such polygonal sutures could be used to distinguish former marine cements in ancient rocks. Used with caution, this is generally true, but isopachous bladed calcite formed in the freshwater phreatic environment can produce polygonal sutures similar to those in marine cement. Polygonal sutures should be considered a secondary characteristic of marine cements; their fibrous texture is the main criterion.

Distinguishing ancient marine fibrous Mg-calcite from aragonite cement is often possible on the basis of crystal morphology. Assereto and

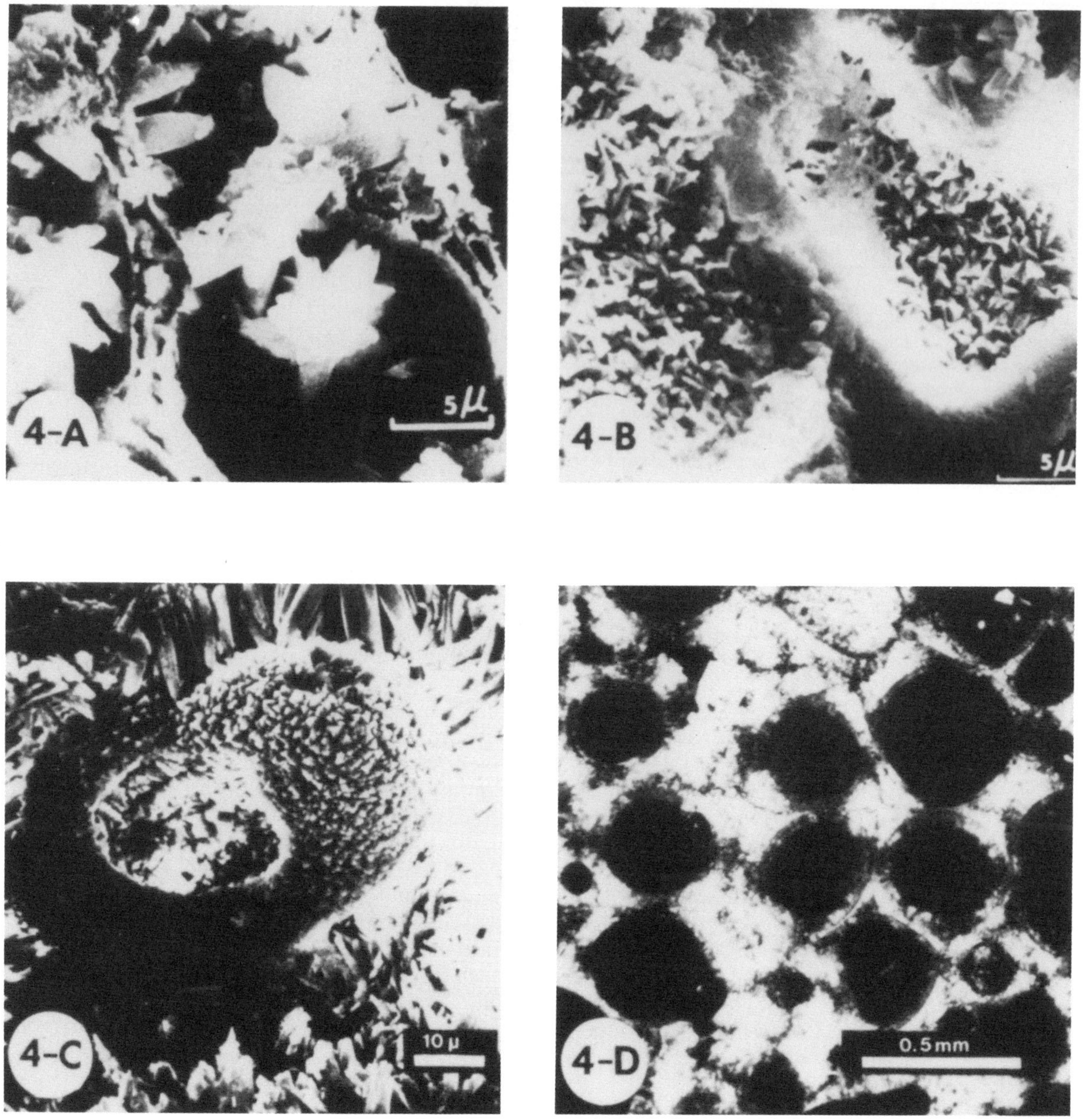

FIG. 4—Marine phreatic cements. **A**, Authigenic Mg-calcite crystals partly filling pores in red algae from reef wall off northern Jamaica in 25 m of water. Steep-sided rhombic shape and small crystal size are typical of marine Mg-calcite cement. **B**, Cement in pores of red algae recovered from stagnant marine phreatic environment of northern Belize Lagoon. Pore on left contains fibrous aragonite; other pore contains steepened rhombs of Mg-calcite. Such juxtaposition of different mineralogies indicates that very subtle factors control formation of Mg-calcite versus aragonite. **C**, This small foraminifer within cavity in coral was covered with micritic Mg-calcite while aragonite cement grew inward from coral skeleton in the active marine phreatic zone of Belize barrier reef. Picture shows that substrate sometimes can control cement mineralogy. **D**, Polygonal sutures formed by isopachous fibrous cements that have grown to cause compromise boundaries. Such straight sutures suggest fibrous cement formed in active marine phreatic environment although less regular polygonal sutures can form in freshwater phreatic environment. Sample from Jurassic Smackover Formation, northern Louisiana.

Folk (1976) and Loucks and Folk (1976) pointed out that orthorhombic aragonite commonly has square-ended fibers (Fig. 5C,D). If this shape is preserved after neomorphism to calcite, it provides evidence for a former aragonite mineralogy. Loucks and Folk applied this criterion to cements in Permian pisolites, and Mazzullo and Cys (1977) have used it to interpret cements in the Permian Capitan reef as former aragonite.

Fibrous Mg-calcite cements may also be recognized after neomorphism to calcite using criteria discussed by Lohmann and Meyers (1977). One of the strongest lines of evidence is the presence of microdolomite inclusions in the neomorphosed radial fibrous calcite which suggests both a former Mg-calcite composition and original precipitation in a marine diagenetic environment (presumably Mg-calcite cave carbonates could neomorphose in the same way and be confused with marine cements, but depositional textures in the rocks should allow marine cement to be distinguished from cave precipitates).

Other characteristics useful in identifying ancient marine phreatic cements include marine organisms encrusting the cements, and marine sediment in the cements. Extensive boring by sponges and other organisms in reefs combines with the influx of sediment carried by currents to make internal sediments a common part of many reefs (James et al, 1976; Macintyre, 1977). Microboring algae and fungi in early cements also seem to form most commonly in the marine phreatic envi-

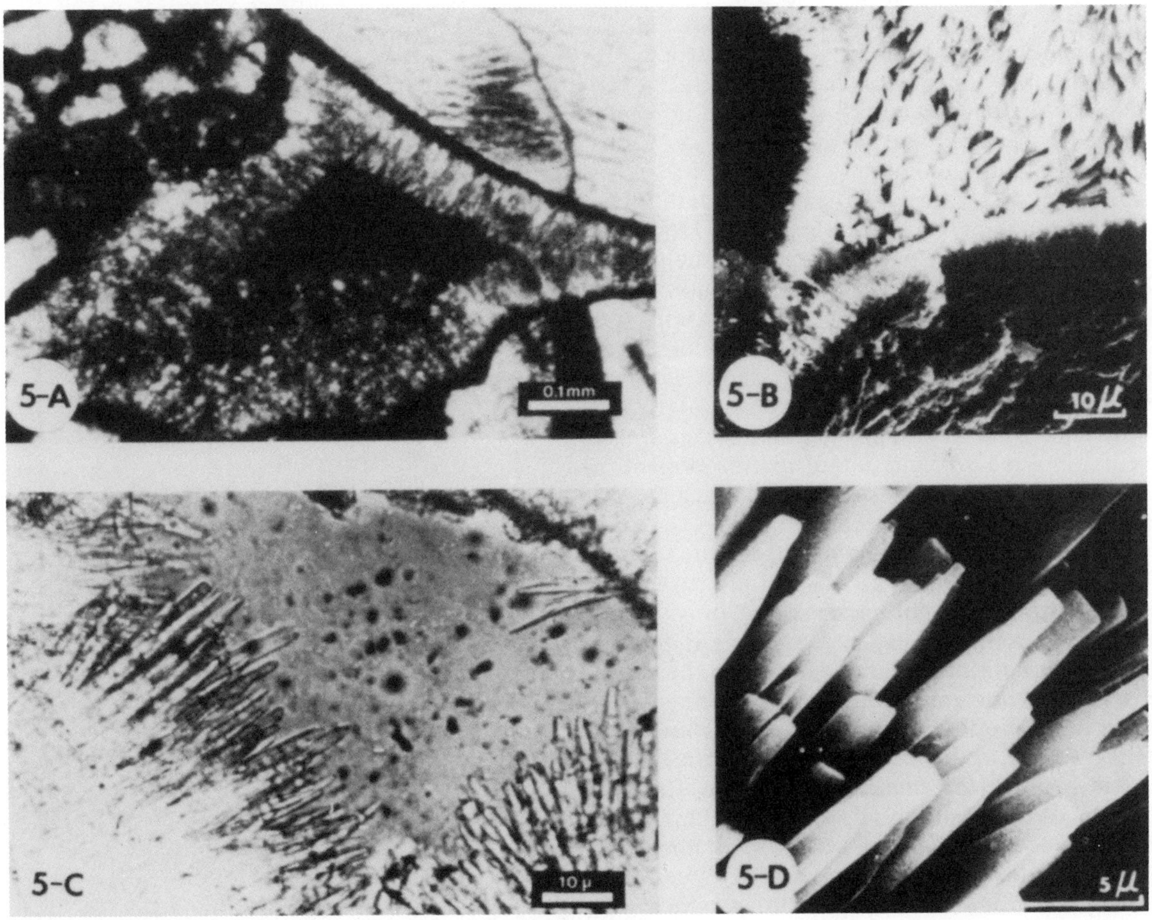

FIG. 5—Marine phreatic aragonite cements. **A**, Isopachous fringe of aragonite cement lining cavity in coralgal sand from active marine phreatic zone in Indo-Pacific. Geopetal sediment lies on top of cement. **B**, Scanning electron photomicrograph of isopachous aragonite cement coating oolites from active marine phreatic zone of Joulter's Cay, Bahamas. **C**, Thin-section photomicrograph of aragonite cement formed in marine phreatic environment. Chisel-shaped ends may aid in recognizing ancient aragonite cements after they alter to calcite. Sample from coral dredged from Gulf of Mexico off Mississippi. **D**, SEM photomicrograph of square-ended aragonite cement filling cavity in coral from active marine phreatic environment of Belize barrier reef.

ronment. Often these internal sediments and encrusting organisms will accumulate within, or on top of, the cements (Fig. 5A).

Stagnant Marine Phreatic Zone

As mentioned previously, the term "stagnant marine phreatic zone" is used here for the zone in which sediments are saturated with sea water, but in which little cementation occurs because marine water moves too slowly through the sediments or because some other factor such as fluid chemistry inhibits cementation. Sediments deposited in lagoons, deeper water parts of gently sloping carbonate ramps, or distal talus typically begin their diagenetic history in the stagnant marine phreatic environment. The combination of low depositional slopes and compaction of the fine-grained sediments (which expels water) allows little marine water to circulate into the sediment and partly explains the paucity of marine cements. However, some cementation may occur, particularly intragranular cementation in tiny skeletal pores. Study of lagoonal carbonate sands from northern Belize has revealed that many pores in the coralline algae and foraminifers contain marine cement (Fig. 4B). Clearly, within such small pores, water circulation must be very slow. However, ion diffusion may contribute to the cementation because of the very small pore sizes and volumes of cement involved by acting in combination with bacteria and microboring algae, which probably play an important role by locally removing organic films and/or increasing the pH. Macintyre and Videtich (1979) have suggested that bacteria may be causing the precipitation of marine cement in a cave off Belize, and it is probable that bacteria also contribute to cementation in skeletal cavities where decaying organic matter is present (Berner, 1971).

Micritization by boring algae and fungi is another important process in the stagnant marine phreatic zone, although it also occurs in the active marine phreatic zone. Most micritization occurs near the sediment/water interface (Kobluk and Risk, 1977), but it may also occur more than a meter beneath the interface (May and Perkins, 1979). Characteristics and processes of micritization have been summarized by Kobluk and Risk (1977) and need not be repeated here. Other than this micritization and the minor intragranular cementation, little diagenesis occurs in the stagnant marine phreatic environment.

VADOSE ZONE

The vadose diagenetic environment is the subaerial zone lying below the land surface and above the zone of saturation or water table. Both air and water may be present in the pores (Fig. 6). Because the zone is not water saturated, any water that is present tends to be held between grains by capillary forces or under grains as pendant drops. Characteristic cement textures may develop because of this inhomogeneous distribution of water. Another important process in the vadose zone is the frequent wetting and drying, depending on the amount of precipitation or evaporation.

Water in the vadose zone is generally meteoric

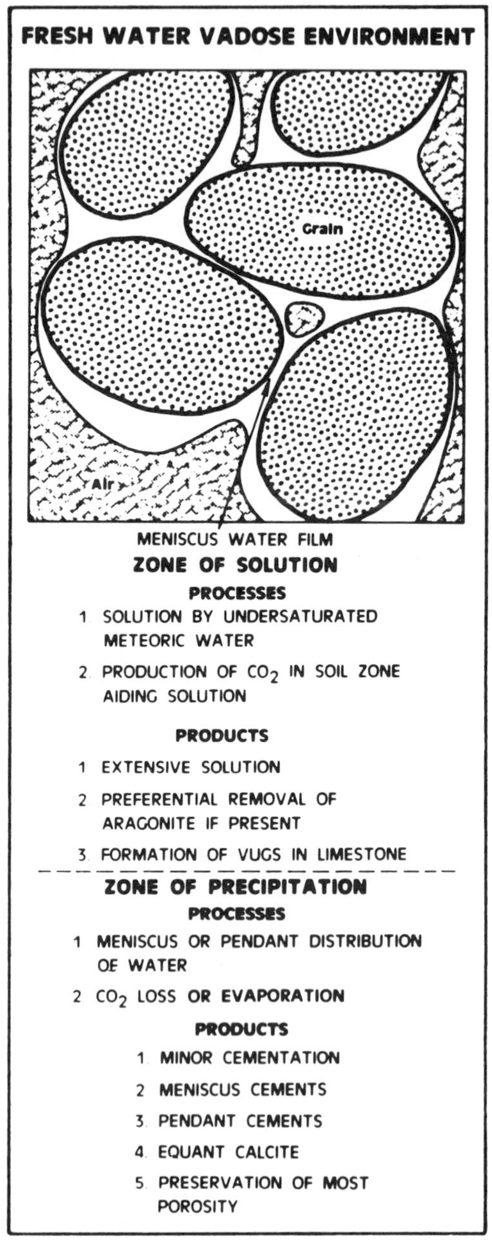

FIG. 6—Characteristics of vadose diagenetic environment.

(rain) water. Meteoric water is undersaturated with $CaCO_3$ as it falls, due in part to atmospheric CO_2, but may quickly reach equilibrium once it starts soaking through carbonate sediments. Thus, solution is an important process in the vadose zone. Contributing to the solution is the production of CO_2 by microbial breakdown of organic material in the soil zone, which is commonly present at the top of the vadose zone. However, solution is not the only diagenetic event occurring in the vadose zone. Once vadose waters reach saturation, they may begin precipitating calcite through evaporation or CO_2 degassing as occurs on Joulter's Cay in the Bahamas (Harris, 1979).

The vadose zone may be conveniently divided into two parts which represent the end members of a continuous spectrum. These are the soil zone, or zone of solution, and the zone of precipitation or capillary fringe zone. One or the other of these zones may be very small in any given vadose zone depending on local conditions such as climate, amount of vegetation, thickness of the soil zone, and thickness of the vadose zone as a whole. In some cases, the zone of precipitation may be missing completely. The boundary between the zones is gradational and may fluctuate considerably in a short time, particularly as the result of rainfall. Characteristics and processes of the vadose zone are shown in Figure 6 and relations of the zones are shown in Figure 7.

Climate plays a very important role in vadose diagenesis. Harrison (1975) has pointed out that the Pleistocene sediments of Barbados, now in the vadose zone, have altered only slightly where the climate is arid, but have completely altered to calcite in more humid regions. Solution and neomorphism in humid climates occur rapidly, partly because of the abundance of fresh water, and partly because a humid climate favors the development of a thick soil zone (up to several tens of meters in tropical areas). However, time is also an important control on the formation of the soil zone and carbonate shoals that become emergent for short periods may never have a well-developed soil zone.

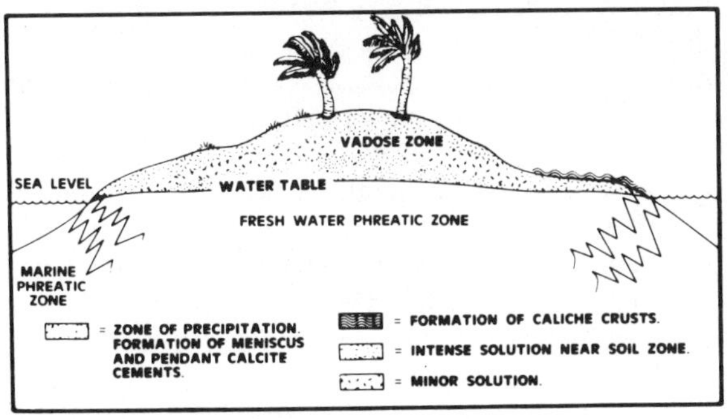

FIG. 7—Idealized cross section of freshwater vadose zone showing distribution of areas of solution and precipitation.

Recognizing diagenetic products of the vadose zone is complicated by the fact that perched (ground) water is common in inhomogeneous rocks and sediments. Robert Halley (personal commun.) has pointed out that such perched water may occur on any scale from that of individual pores to entire beds (especially those composed of finer grained sediments) to whole aquifers controlled by perched water tables. Within such areas of perched water, diagenetic textures are essentially identical to those in the freshwater phreatic zone as described later.

Zone of Solution

The zone of solution is at the top of the vadose zone, but may extend downward for tens or even hundreds of meters depending on the location of the water table and how quickly the meteoric water is able to become saturated with respect to calcium carbonate. The rate of saturation depends on composition of the host rock, rate of water movement which is controlled by permeability and amount of water, thickness of the soil zone, and the CO_2 partial pressure in the water. In many cases the zone of solution may extend to the water table, particularly where the host rock is stable calcite or where the vadose zone is thin.

The major diagenetic process active in the zone of solution is, not surprisingly, solution of the calcium carbonate. Solution is the result of two factors: (1) the initial undersaturation of rainwater with respect to calcium carbonate, due in part to atmospheric CO_2, and (2) the hydrolysis of CO_2 produced in the soil zone which increases the partial pressure of CO_2 in the water, thus increasing the amount of calcium carbonate the water can hold in solution. In areas without a soil zone, the former is more important, whereas in areas with a well-developed soil zone, the latter is much more important.

Any form of calcium carbonate may be dissolved in the zone of solution (Fig. 8A,B). Aragonitic grains are particularly susceptible to leaching and will tend to be removed preferentially if present with calcite grains or cement. Solution of limestone may also occur and reaches fantastic proportions in the remarkable karst topography present in many humid climates. Caves commonly form at the top of the freshwater phreatic zone where saturated vadose waters may become cooled or mixed with waters in equilibrium with a lower pCO_2 (Thrailkill, 1968). Degassing on the other hand, may cause waters in equi-

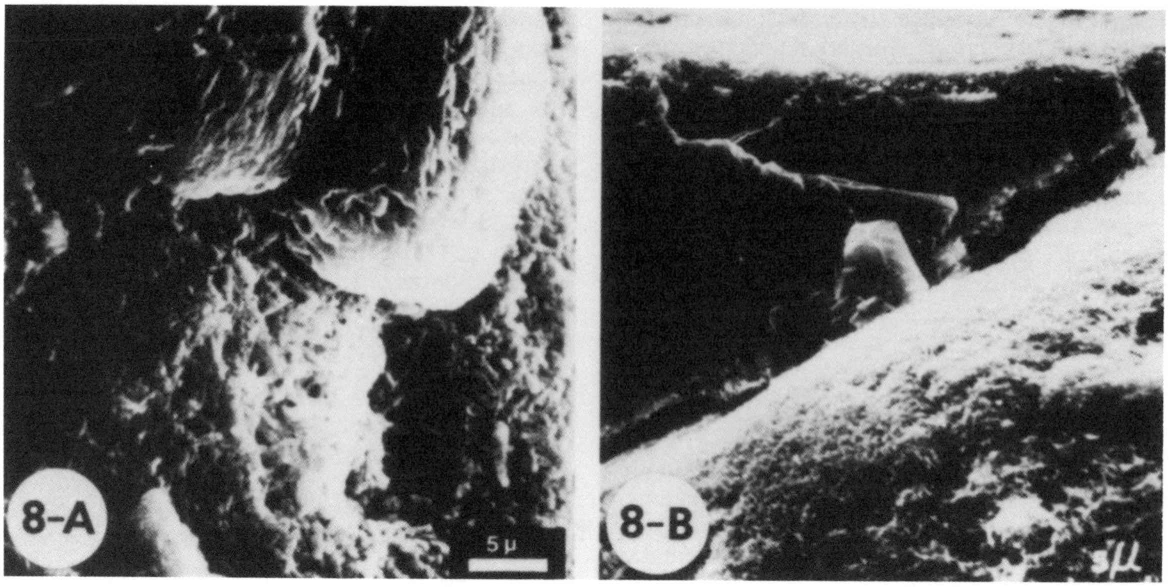

FIG. 8—Solution in vadose zone. **A**, Surfaces of aragonite ooid and calcite rhombs that have been partly dissolved in freshwater vadose zone on Joulter's Cay, Bahamas. Such solution is important process in upper parts of most vadose zones. **B**, Boundary between meniscus calcite cement and aragonite ooid. Solution of ooid followed cementation. From vadose environment on Joulter's Cay, Bahamas.

librium with the limestones to precipitate calcium carbonate near the water table.

In more moderate climates, solution of a homogeneous limestone may produce pitted outcrop surfaces and vugs. The only cementation that occurs in this zone is in the form of evaporitic and/or biologic caliche crusts at or near the air/rock interface. Such caliche crusts have been described in detail (James, 1972; Kahle, 1977; Knox, 1977; Robbin and Stipp, 1979) and need not be reviewed here. In the absence of such crusts and any other types of cementation, it is very difficult to distinguish leaching in the undersaturated vadose zone from leaching in the undersaturated phreatic zone.

Zone of Precipitation

As water moves through the zone of solution picking up calcium carbonate, it will eventually reach the point of saturation. Once saturated, evaporation, slight increases in temperature, or decreases in CO_2 partial pressure may cause calcium carbonate to precipitate. Cementation in the vadose zone is a relatively minor process but may occur below the zone of solution in the lower parts of the vadose zone. The amount of cementation is dependent on the degree of saturation, factors causing the precipitation (e.g., CO_2 loss from the system), flow rate which controls the amount of water from which material is being withdrawn for precipitation, and time.

Cement types—Cement morphology in the vadose zone tends to reflect the distribution of the pore waters. If pore water is distributed between the grains in a meniscus fashion, the cement will also assume a meniscus texture as first defined by Dunham (1971; Fig. 9). If water is present as droplets on the underside of grains, a pendant cement will form (Fig. 10). This texture has been described by Purser (1969), Müller (1971), Jacka and Brand (1977), and others. Another characteristic is the uneven distribution of cements in poorly sorted sediments; water, and thus cement, tends to be patchy and concentrated in the finer grained part of these sediments (Dunham, 1971).

Because meteoric water modified by the addition of calcium carbonate is the main precipitating fluid, cements tend to be calcite, commonly in the form of very fine equant crystals. Crystal terminations are rare; instead crystals tend to terminate at the air water interface and reflect the shape of the water on the grains (Dunham, 1971). Cements are not isopachous and pores tend to become more rounded, but smaller with increasing cementation. Continued precipitation of calcite may fill pores completely with crystals that coarsen toward pore centers (Schroeder, 1973). Under certain conditions, precipitation of calcite occurs around rootlets of terrestrial plants (Mathews, 1974).

In mineralogically unstable sediments rich in Mg-calcite, and in dolomitic terranes, meteoric

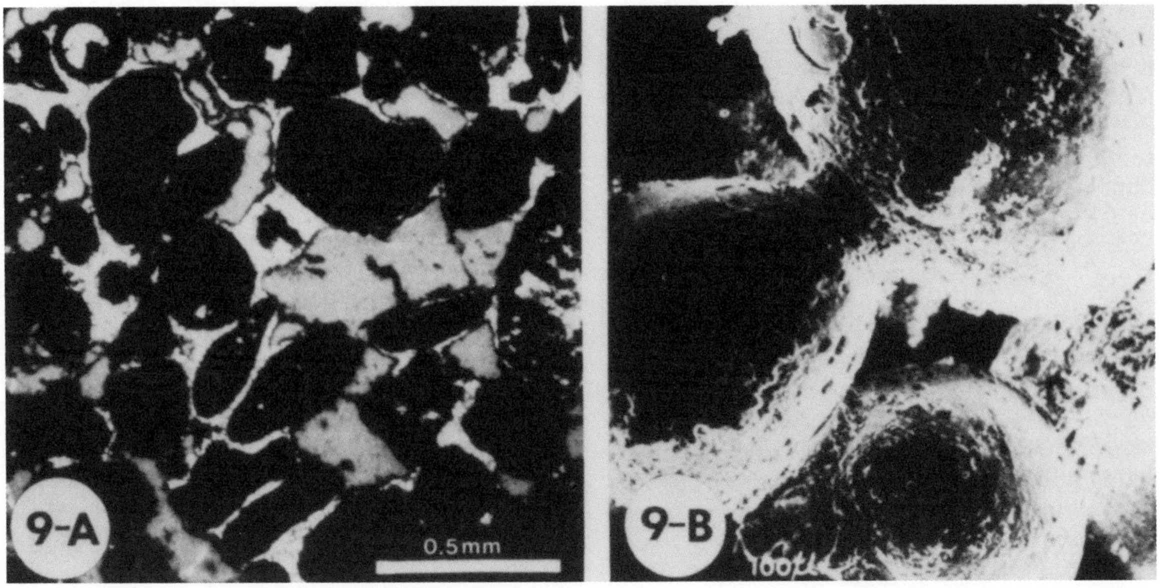

FIG. 9—Vadose meniscus cements. **A**, Meniscus calcite cements between grains in vadose calcarenite from Nassau, Bahamas. **B**, SEM photomicrograph of meniscus calcite cement between ooids from Joulter's Cay, Bahamas. Such meniscus cements form only in vadose environment where water hangs between grains.

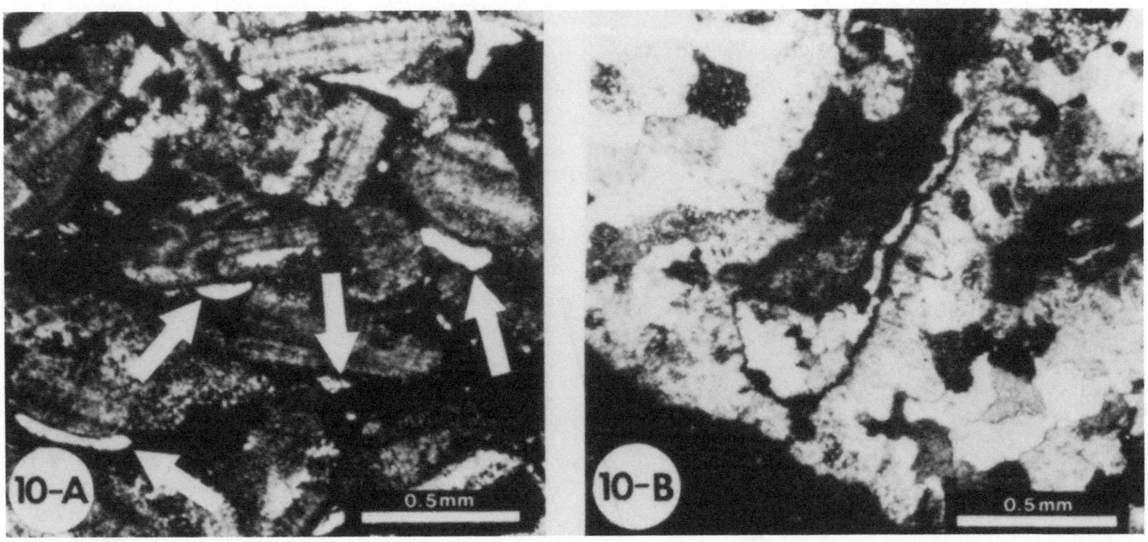

FIG.10—Vadose pendant cements. **A**, Pendant calcite cements (arrows) under mollusk fragments in Cretaceous calcarenite from central Texas. This texture is characteristic of meteoric vadose environment. **B**, Microstalactitic cement hanging on sliver of rock in cavity filled later by equant calcite in freshwater phreatic environment. Microstalactite clearly indicates period of vadose cementation.

water may pick up magnesium ions as it dissolves the carbonate. These magnesium ions may inhibit the precipitation of calcite and facilitate the precipitation of aragonite or even dolomite. This occurs most commonly in caves (Thrailkill, 1971). Strontium and other ions derived from immature sediments may also influence cementation, although the effects of such ions are not yet understood.

Cave cements—Two unusual types of vadose cement merit special mention. The first includes cave flowstone and other cave deposits. These have been described by Thrailkill (1971, 1976), Kendall and Broughton (1978), and many others. These deposits are typically bladed calcite with growth bands but may also be micritic or equant calcite. Study of cave carbonates should provide insight into many controls on cementation in more normal vadose diagenetic environments where the pores are intergranular. Of special interest in studying cave carbonates is the information that may be gained on the processes resulting in the formation of minerals other than calcite, such as dolomite and Mg-calcite.

Whisker calcite—The other unusual type of vadose cement is whisker calcite (Fig. 11) as described by Supko (1971), Steinen (1974), and others. Such cement typically, but not always, is present in subspherical secondary cavities as randomly distributed straight fibers 1 to 4 μ wide and hundreds of microns long. The cause of this crystal morphology is still unclear, but Harrison (1974) suggested that the carbonate may have accumulated as sheaths around root hairs. However, somewhat different fibrous aragonite crystals have been produced in laboratory-simulated vadose environments (Badiozamani et al, 1977) with no roots and the fabric may be completely inorganic. In any case, available evidence suggests that such whisker calcite forms only in the vadose environment.

Ancient Examples of Vadose Diagenesis

Secondary porosity formed by leaching is an important part of many carbonate hydrocarbon reservoirs. Much of this porosity may have formed in vadose environments, but leaching producing similar porosity may also occur in the freshwater phreatic environment and deeper subsurface. Thus, it is not usually possible to determine the diagenetic environment of leaching, except to say that water undersaturated with respect to calcium carbonate must have been present.

Cements deposited in vadose environments can sometimes be recognized in ancient rocks. Well-developed meniscus cements of equant calcite covering marine aragonite have been described from the Pleistocene of Bermuda (Schroeder, 1973). Other examples of vadose cements have been cited by Jacka and Brand (1977) and Grover and Read (1978). Pendant cements are the most easily recognized because of their distinct morphology and texture (Fig. 10). Fossil meniscus cements are harder to recognize because textures are commonly obscured by later cementation. Careful observation is generally necessary to distinguish them, but judging from their occurrence

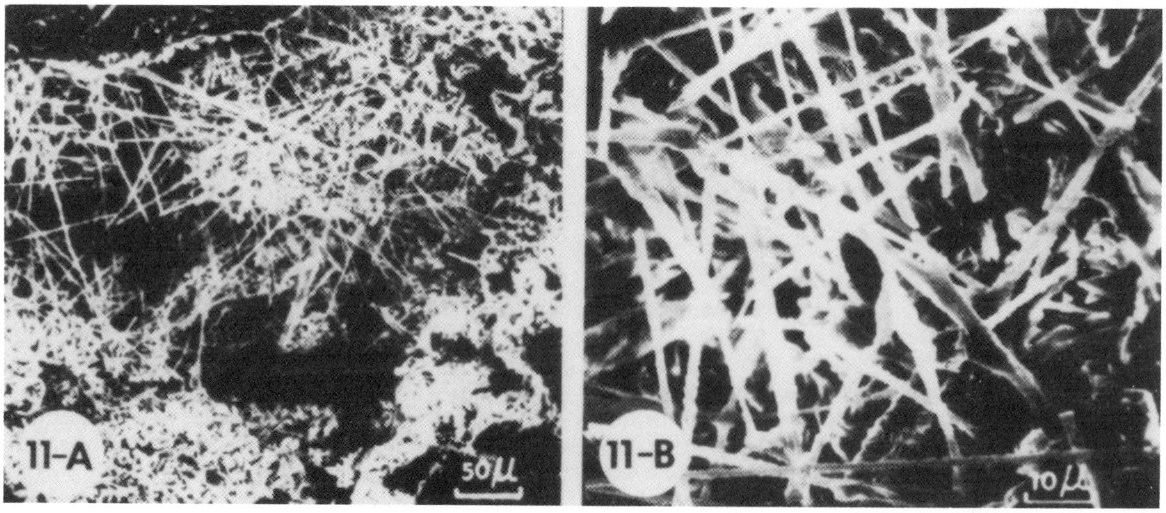

FIG. 11—Vadose whisker calcite. **A**, Whisker cement filling void in calcarenite from vadose environment. **B**, Higher magnification view of whisker cement showing straight fibers of probable calcite. Although origin of fibers is unclear, they seem to indicate vadose precipitation. Sample is from Pleistocene limestone on Barbados.

in Holocene and Pleistocene sediments, many ancient examples probably exist and have simply gone unrecognized.

On the other hand, Quaternary vadose diagenetic zones have been exceptionally common because of the widespread rapid fluctuations in eustatic sea level. During periods of prolonged stable sea level, carbonate sediments accumulate up to, but not much above, high tide level. Under such conditions the vadose zone will be almost nonexistent. This may partly explain the paucity of vadose cements in many ancient carbonate rocks. Only during periods of significant fluctuations (especially lowerings) of sea level, or uplift relative to sea level, is extensive vadose diagenesis likely to have occurred in rocks deposited in any marine environment (with the possible exception of some tidal-flat and supratidal deposits). Furthermore, rocks in which vadose cements form are very susceptible to erosion because of their position above sea level. Preservation of vadose cements is likely only if transgression or subsidence follows soon after their formation.

FRESHWATER PHREATIC ZONE

The freshwater phreatic zone lies between the vadose and mixed marine phreatic–freshwater zones. All pore space in this zone is filled with meteoric water containing variable amounts of dissolved carbonate. Much of the meteoric water may enter the phreatic zone directly through the vadose zone, but Harris (1971) has estimated that up to 99% of the water in the coastal phreatic zone of Barbados enters through streams, lakes, and other runoff from the interior of the island and does not pass directly through a soil zone or vadose zone. A major control on the amount of water that passes through the vadose zone is climate and the nature of the rainfall, that is, whether it is seasonal and torrential or year-round and gentle.

The top of the phreatic zone is marked by the water table whereas the lower boundary is gradational with marine waters in areas proximal to the seas. This gradational zone is described in a later section. In inland areas, the freshwater phreatic zone grades downward into the zone of subsurface fluids which may contain brines forced upward from greater depths by sediment compaction, clay dehydration, heating, or differences in density.

Whereas study of recent vadose diagenesis may be easily accomplished by gathering almost any sample from outcrops, study of recent meteoric phreatic diagenesis generally requires samples from wells. Direct observation is difficult because the rocks being affected lie below the water table. Information on water movement and chemistry is often limited. Thus, many ideas in this section are based on inferences and textures in older rocks.

The geometry of the freshwater phreatic zone is strongly controlled by topography, rainfall, and distribution of porosity and permeability in the rocks. In an ideal tropical island like that shown in Figure 1, the freshwater phreatic zone is typically lens shaped and lies directly under the island. Comparing the density of rainwater (0.9991 g/cc at 15°C) and sea water (1.025 g/cc at 15°C), and applying the "iceberg" principle (technically called the Ghyben-Herzberg theory) reveals that for every meter the water table rises above sea level, there may be up to 40 m of fresh water below the water table, if the system is in equilibrium and the rocks are permeable (Friedman and Sanders, 1978, p. 149). In nature, this ratio is rarely attained for two reasons: (1) because rainwater becomes more dense as it dissolves calcium carbonate, and (2) the system is too dynamic to reach equilibrium. Other complexities exist because meteoric water mixes with the sea water to produce brackish water. Typically, 1 m above the water table corresponds to about 32 m of fresh water below the water table. This means that the freshwater zone may extend to depths of hundreds of meters on large islands with high-standing water tables in rainy tropical climates. Under other circumstances, such as in arid climates, the freshwater phreatic lens may be completely missing.

In coastal areas, permeability pathways and the amount of hydrostatic head determine the depth and geometry of the freshwater zone. Very irregular freshwater wedges may be produced in areas of marked permeability variation in the rocks. With a sufficient hydrostatic head and the proper permeability pathways, fresh water may even be forced away from the shoreline under normal marine waters. Such may be the case off Florida today where a JOIDES core hole discovered interstitial fresh water 50 km offshore in 15 m of water (Matthews, 1974), but this could also be interpreted as relict meteoric water trapped in the rock during a time of lower sea level in the Pleistocene. Conditions such as that described by Matthews (1974) are fairly common today because of the wide fluctuations in eustatic sea level during the Quaternary and the resultant creation through diagenesis of strong permeability contrasts in different limestone units. However, under more "normal" conditions of stable sea level and relatively homogeneous permeabilities, freshwater lenses would rarely extend very far under the sea because of the unstable situation created by having lower density fresh water under sea water.

In inland areas, the geometry of the freshwater phreatic zone is controlled by the amount of rain-

INCREASING SATURATION WITH $CaCO_3$ / INCREASING DEPTH ↓	ZONE	CHARACTERISTICS
	1. ZONE OF SOLUTION OF CALCITE AND ARAGONITE.	FORMATION OF VUGGY AND MOLDIC POROSITY.
	2. ZONE OF SOLUTION OF ARAGONITE	FORMATION OF MOLDIC POROSITY
	3. ZONE OF ARAGONITE SOLUTION AND CALCITE PRECIPITATION	RAPID NEOMORPHISM OF ARAGONITE GRAINS TO EQUANT CALCITE. CEMENTATION BY EQUANT CALCITE.
	4. ZONE OF NO SOLUTION. PRECIPITATION OF CALCITE	RAPID CEMENTATION BY EQUANT CALCITE.
	5. STAGNANT ZONE. SATURATED WITH $CaCO_3$	LITTLE OR NO CEMENTATION. SLOW NEOMORPHISM OF ARAGONITE GRAINS WITH PRESERVATION OF SOME STRUCTURES.

FIG. 12—Idealized zonation in freshwater phreatic environment based on assumption that saturation of water with respect to $CaCO_3$ increases as water moves downward. Any zone may be very small or missing altogether.

fall, the migration of fluids in the subsurface, and the distribution of permeable beds. In areas where compaction of underlying sediments is important, the freshwater zone will be relatively shallow because subsurface waters tend to move upward. However, in areas that have already undergone maximum compaction and are being uplifted, the freshwater phreatic zones may extend to depths of hundreds of meters, depending on the hydrostatic head and permeability (Magara, 1978).

Diagenesis in the freshwater phreatic zone is very complex because of the variabilities of fluid migration rates, degree of saturation, composition of the host rock, and chemistry of the fluid. However, an ideal freshwater phreatic zone may be subdivided into five characteristic parts on the basis of the saturation state of the water with respect to $CaCO_3$ (Fig. 12). Water entering from the vadose zone may be strongly undersaturated with respect to $CaCO_3$ originally, but as it moves down through the freshwater phreatic zone it will become more and more saturated. Thus, at the top of the zone is an area where both calcite and aragonite are dissolved. Under this is a zone where only aragonite dissolves, followed by a zone of penecontemporaneous aragonite solution and calcite precipitation. Still lower are zones of active calcite cementation and no solution, and a stagnant zone of little water movement. Any zone may be small or missing, and permeability variations or mixing of waters may cause inversions in the sequence. The diagenetic processes affecting Mg-calcite in this sequence are still unclear, but the sequence provides a guide for interpretation of freshwater phreatic diagenesis in unstable marine sediments. Characteristics of the various subzones are summarized in Figure 13 and geometric relations are shown in Figure 14 and discussed in the following. For the sake of the dia-

FRESHWATER PHREATIC ENVIRONMENT

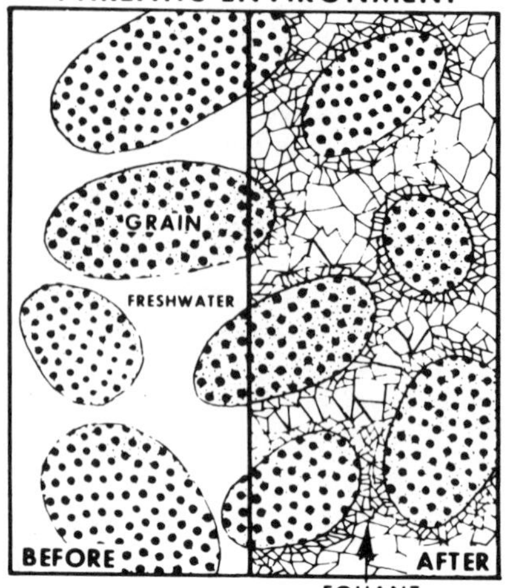

EQUANT CALCITE CEMENT
(CRYSTAL BOUNDARIES ARE GENERALLY MORE IRREGULAR THAN SHOWN

ZONE OF SOLUTION

PROCESS

SOLUTION BY UNDERSATURATED METEORIC WATER

PRODUCTS

1. DEVELOPMENT OF MOLDIC AND/OR VUGGY POROSITY
2. POSSIBLE NEOMORPHISM OF UNSTABLE GRAINS

STAGNANT ZONE

PROCESSES

1. LITTLE OR NO WATER MOVEMENT
2. WATER SATURATED WITH $CaCO_3$

PRODUCTS

1. LITTLE CEMENTATION
2. STABILIZATION OF MG-CALCITE AND ARAGONITE
3. LITTLE OR NO LEACHING
4. PRESERVATION OF POROSITY
5. NEOMORPHISM OF ARAGONITE GRAINS WITH SOME PRESERVATION OF TEXTURES

ACTIVE ZONE

PROCESSES

1. ACTIVE WATER CIRCULATION
2. SOME LEACHING OF ARAGONITE. LEACHING MAY BE ACCOMPANIED BY CALCITE REPLACEMENT.
3. RAPID CEMENTATION

PRODUCTS

1. ABUNDANT EQUANT CALCITE CEMENT
2. ISOPACHOUS BLADED CALCITE CEMENT
3. INTERLOCKING CRYSTALS
4. CRYSTALS COARSEN TOWARD CENTER OF PORES
5. COMPLETE REPLACEMENT OF ARAGONITE BY EQUANT CALCITE
6. SYNTAXIAL OVERGROWTHS ON ECHINODERMS
7. RELATIVELY LOW POROSITY

FIG. 13—Characteristics of freshwater phreatic environment.

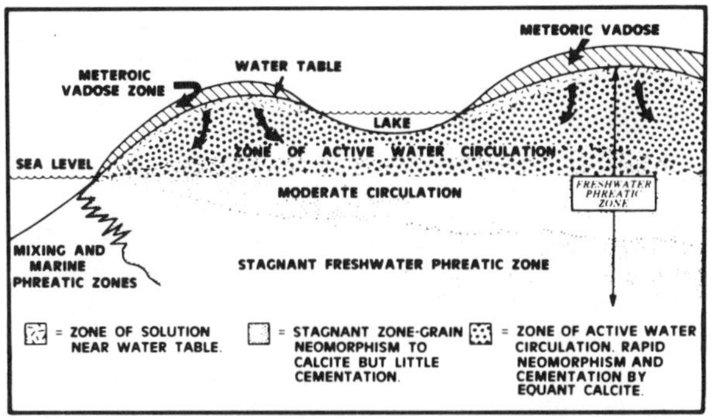

FIG. 14—Schematic cross section of idealized freshwater phreatic zone showing possible distribution of zone of solution, zone of active water circulation and cementation, and zone of stagnant water.

grams and discussion, the three zones of solution are combined.

Undersaturated Zone

Often meteoric water will pass through the vadose zone without becoming saturated with $CaCO_3$. This is particularly common where permeability pathways such as fractures, sinkholes, or caves are present. Once under the water table in the freshwater phreatic zone, the meteoric water will begin (or continue) dissolving until it reaches saturation. Either vuggy or moldic porosity may be produced depending on the mineralogy and crystal size of the components in the rock. Under certain conditions where vadose waters saturated with respect to calcite move across the water table and become cooler or are mixed with waters in equilibrium with a lower pCO_2, solution may also occur near the water table (Thrailkill, 1968; Ford and Ewers, 1978). Both aragonite and calcite may be affected by this solution (Fig. 15).

Solution in this zone produces textures similar to those produced in the vadose zone. Thus, the presence of moldic and vuggy porosity cannot be used as a criterion to distinguish meteoric vadose from meteoric phreatic diagenesis. However, cementation often follows the solution and the textures of cements produced in the two zones differ significantly as described in the following.

Active Saturated Zone

Water saturated with respect to $CaCO_3$ in the freshwater phreatic zone may be divided into two end members of a continuous spectrum—active and stagnant. All gradations between active and stagnant water exist, but by describing the characteristics of both,,the key characteristics of diagenesis in the freshwater phreatic zone can be summarized. Processes and products of diagenesis in the active zone are described in this section.

Water moving through the vadose zone picks up CO_2 and calcium carbonate. When it moves across the water table, its flow rate may decrease and often CO_2 is given off. If the water is saturated with respect to calcite, cementation may occur. Active water circulation occurs in the upper part of most freshwater phreatic zones where new meteoric water is added. Good circulation also

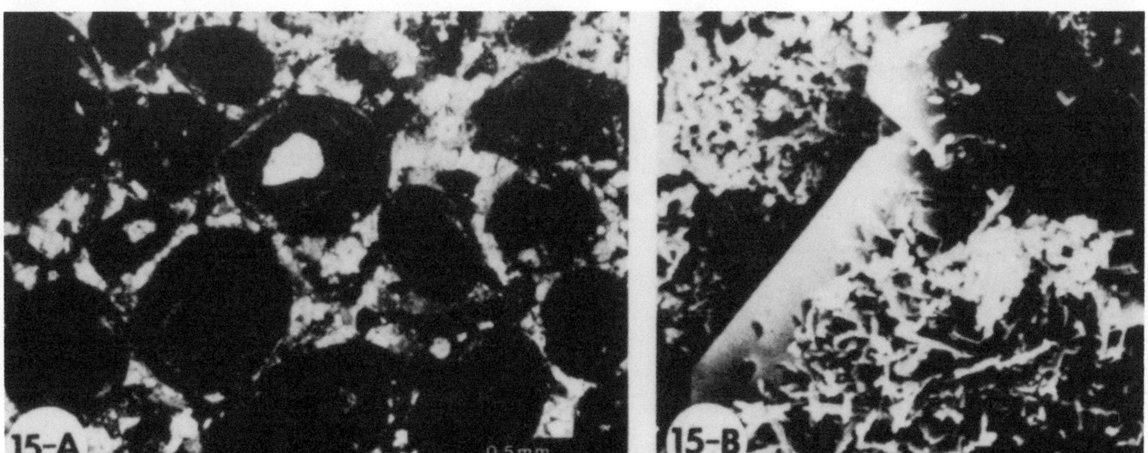

FIG. 15—Solution and neomorphism in freshwater phreatic zone. **A**, Solution of aragonitic oolites following cementation by equant calcite in active freshwater phreatic zone. Sample from Pleistocene Miami Oolite, south Florida. **B**, Aragonite needles in oolite being replaced by blocky equant calcite. Such neomorphism is common in freshwater phreatic environment, but can also occur in vadose zone in humid climates.

occurs along permeability pathways between recharge and discharge areas such as springs. Because additional calcium carbonate is always being added by the waters migrating in from the vadose zone, extensive and rapid cementation is characteristic of the actively circulating part of the freshwater phreatic zone.

Neomorphism of grains—Petrographic evidence and porosity-permeability studies, such as that of Pittman (1974), suggest that water in marine sediments rich in aragonite may be simultaneously undersaturated with respect to aragonite (thus, causing solution of aragonite), and saturated with respect to calcite (thus, allowing calcite to precipitate). Both solution of aragonite and precipitation of calcite are common in the freshwater phreatic environment. In theory, it could be argued that solution-reprecipitation of aragonitic grains in the active zone would produce the equant calcite mosaic so common in many replaced grains, whereas a similar process in the stagnant zone would result in the preservation of some relict structures of the aragonitic grains because slower water movement could permit solution-reprecipitation

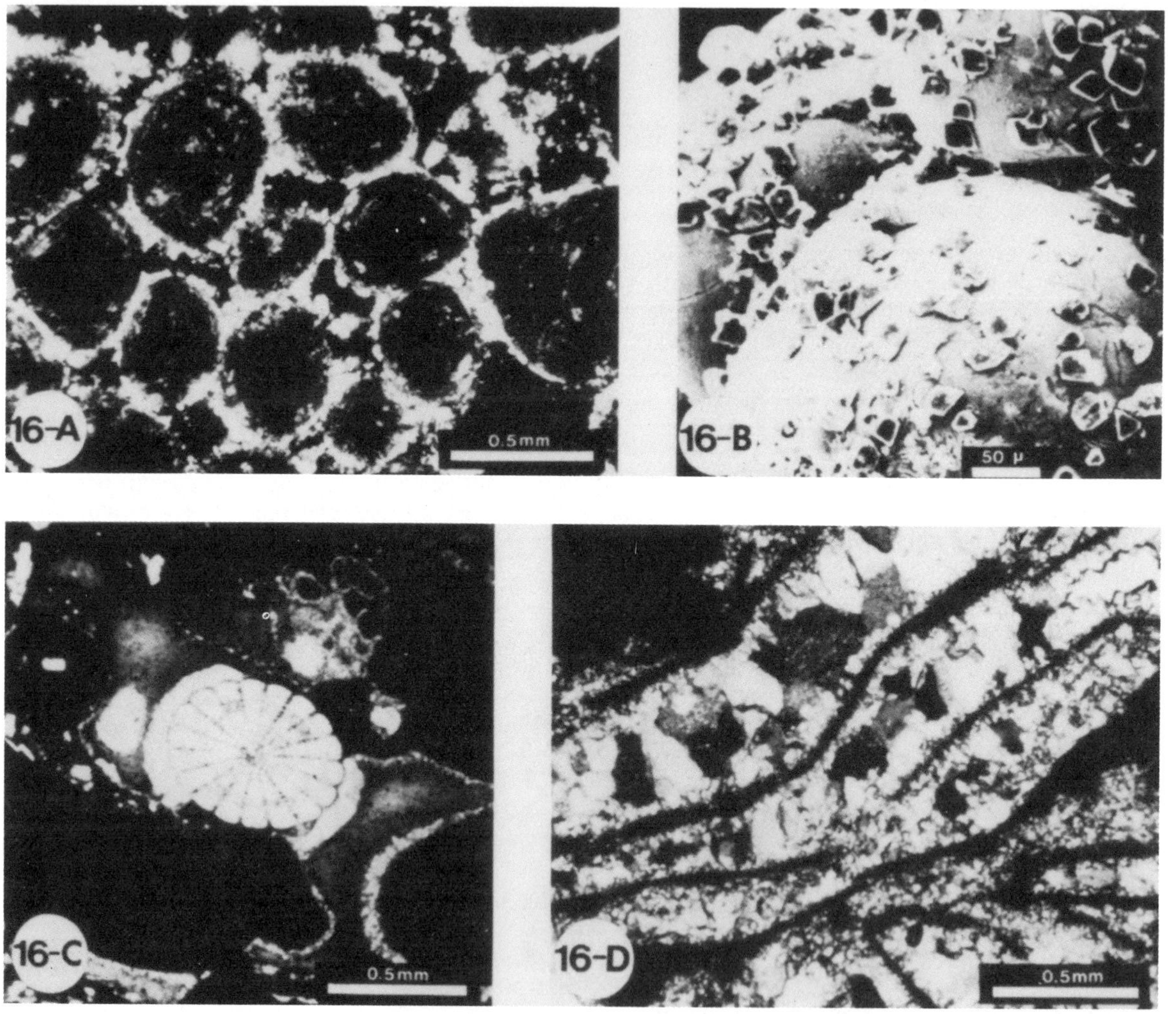

FIG. 16—Freshwater phreatic cements. **A**, Thin isopachous rhombic calcite deposited on ooids in freshwater phreatic zone. This is commonly first stage of meteoric phreatic cementation. There is also evidence of later meniscus cements in this photograph. **B**, SEM photomicrograph of scattered rhombic calcite on surfaces of ooids from freshwater phreatic environment. Sample from Joulter's Cay, Bahamas. **C**, Syntaxial overgrowths on echinoderm fragments are common in freshwater phreatic environment but apparently do not form in marine phreatic environment. Growth of syntaxial overgrowths is more rapid than that of nearby calcite fringe cements. **D**, Solution of mollusk fragments in this sample was followed by partial collapse of micrite envelopes and later cementation by equant calcite. Coarsening of cement toward pore centers is characteristic of early freshwater phreatic diagenesis.

across a narrow front without a significant void stage. However, this has yet to be documented. In any case, extensive rapid neomorphism of marine sediments to calcite occurs in the active freshwater phreatic environment.

Cement textures—Cement in the active freshwater phreatic environment is common and widespread (Fig. 16). Because meteoric water saturated with calcium carbonate and low in magnesium ions is the precipitating fluid, the cement tends to be calcite. Cement is distributed equally around all grains because all pores are saturated with water. The first cement may nucleate on crystals on the surfaces of the grains and grow outward to produce a bladed isopachous calcite rim. Competition for space between growing crystals often combines with decreasing nucleation rates due to low permeability to cause crystals to coarsen significantly toward pore centers.

Equant calcite is more common as a cement than bladed calcite. Often rhombs will nucleate on grain surfaces (Fig. 16A,B). Continued cementation results in crystals coarsening toward pore centers and produces an interlocking mosaic that fills all pore space. This texture has been observed in many ancient rocks (Cullis, 1904; Bathurst, 1976; Jacka and Brand, 1977). Loucks (1977) suggested that coarsening of calcite crystals toward pore centers is a distinctive characteristic of early cementation in a freshwater phreatic environment, and evidence from some Quaternary samples supports his interpretation. However, other factors such as substrate also need to be considered. Syntaxial overgrowths commonly form on echinoderm fragments. Rate of nucleation is also a critical factor, and it would be possible to have no significant coarsening of crystals in a relatively stagnant freshwater phreatic zone only slightly oversaturated with respect to calcite where the rate of nucleation was low.

Syntaxial overgrowths—Syntaxial overgrowths on echinoderm fragments have been described by Evamy and Shearman (1965) and many others. They can serve as an important clue for distinguishing the various diagenetic environments. Syntaxial overgrowths form most rapidly in the freshwater phreatic environment and generally surround the grains although they may be elongate along the C-axis. Because of the large crystal nucleus and absence of competitive crystal growth, they grow much faster than cements from adjacent nonechinoderm grains, as shown by the larger proportion of the pore they fill. Syntaxial overgrowths may also form from saturated vadose water but usually do not completely surround the grain.

For reasons not yet clearly understood, syntaxial overgrowths form very rarely, if ever, in marine phreatic environments. Examination of numerous thin sections from Holocene samples containing both echinoderm fragments and marine cements has revealed no obvious syntaxial overgrowths. Instead, fibrous aragonite or Mg-calcite surrounds the grains. Similar fibrous cements have never been observed forming on echinoderm fragments in freshwater diagenetic environments. Thus, the presence of fibrous cement coating non-micritized echinoderm fragments in ancient rocks suggests marine phreatic (or marine vadose) cementation, whereas syntaxial overgrowths suggest nonmarine cementation (unless marine cement has been replaced by the syntaxial overgrowth). Although syntaxial overgrowths form most commonly in freshwater phreatic environments, they may also form in vadose (rarely) and deeper subsurface diagenetic environments.

Ancient examples—Examples of ancient freshwater phreatic cements are widespread in carbonate rocks and have been described by Land (1970), Jacka and Brand (1977), Loucks (1977), Meyers (1978), and many others.

Stagnant Freshwater Phreatic Zone

In the deeper parts of many freshwater phreatic environments and in many arid climates, water movement in the freshwater phreatic environment is very slow. Meteoric waters have precipitated the excess calcium carbonate picked up in the vadose zone and reached equilibrium with the surrounding sediments by the time they get into the deeper parts of the freshwater phreatic zone. Thus, little additional cementation occurs. Recrystallization of aragonite to calcite may be an important process, but often occurs within the grains across a narrow front. Thus, structures of the aragonite grain are often preserved after neomorphism to calcite. Primary porosity is generally preserved in the absence of cementation. If cements are present, they generally form only a small rim of isopachous equant calcite.

MIXING ZONE ENVIRONMENT

The boundary between the marine phreatic and freshwater phreatic environments is marked by brackish water formed by mixing of waters of both. Examples of cements produced in the mixing zone are rare, probably because of a combination of three factors: (1) the narrowness and small volume of the zone between the much more extensive freshwater and marine phreatic zones, (2) the relatively stagnant state of the water, and (3) the impermanent nature of the zone in one place as it migrates back and forth in response to rainfall, sea level change, and so on. However, Runnells (1969) has shown that the mixing of natural waters can often result in either precipitation or solution of carbonates. This, combined with the spectrum of salinities that may exist in the mixing zone, means that several different processes may occur.

Diagenesis in the mixing zone is still poorly understood. Much of what is known was originally inferred by Land (1973a, b) from his work on the Pleistocene of Jamaica. The most interesting (although perhaps not the most important) diagenetic process in the mixing zone is dolomitization. Badiozamani (1973) has shown that mixing about 10% sea water with 90% fresh water can form a solution slightly undersaturated with calcite, and oversaturated with dolomite. Land (1973a, b) has shown on the basis of stable isotopes and trace elements that dolomite formed in such a mixing zone in the Pleistocene Hopegate and Falmouth Formations of Jamaica. Now many other people seem to be finding mixing zone dolomites in more ancient rocks (Badiozamani, 1973; Land et al, 1975).

Dolomitization does not occur in all mixing zone environments, but is probably restricted to those with good water circulation. Rough calculations show that at least several thousand pore volumes of mixed water must pass through the sediment to produce even a 30% dolomite. An alternative to actual water circulation is osmosis and diffusion of ions through stagnant water, although this is generally discounted by most geochemists because of the long distances involved, and the slow rates of such processes. Thus, for dolomitization to occur, some process contributing to good water circulation in the mixing zone is necessary. Tides produce some movement in exceptionally porous rocks (Ladd and Schlanger, 1960), but a better alternative for less porous rocks might be strongly seasonal rainfall. During the wet season, the freshwater lens would expand and displace the mixing zone downward (seaward). During the dry season, the freshwater lens would shrink and the mixing zone would move upward. This annual migration of the mixing zone might produce the circulation necessary for dolomitization over a prolonged period of time. Calculations show that between 2,000 and 10,000 pore volumes of mixed sea water–fresh water are needed to fill 30% primary porosity with dolomite, although the numbers vary greatly, depending on the amount of magnesium originally in the sediment and degree of removal of the magnesium from the water. Thus, it might be reasonable to assume that a mixing zone dolomite would take at least several thousand years to form if the mixing zone migrates through a given area of rock twice yearly. Unfortunately, no one has documented an example where dolomite formed in response to a seasonally fluctuating mixing zone, so this idea cannot yet be tested.

Other diagenetic processes occurring in mixing zones at the freshwater end of the spectrum include slight cementation, generally by micritic to bladed calcite, minor neomorphism of aragonite to calcite, leaching of aragonite, and neomorphism of Mg-calcite to calcite. In waters at the more marine end of the spectrum, Moore (1977) found isopachous rims of Mg-calcite cement forming by mixing of fresh and marine waters. Under these more saline conditions, neomorphism and leaching are rare because the waters tend to be saturated with $CaCO_3$.

OTHER DIAGENETIC ENVIRONMENTS

Having described the major diagenetic environments of the shallow subsurface, it is important to note that numerous other diagenetic environments also exist. The whole realm of the deep subsurface is an area of extensive diagenesis, including leaching, cementation, and dolomitization. A wide variety of subsurface diagenetic environments probably could be distinguished on the basis of fluid chemistry, rate of fluid migration, host and adjacent rock types, pressure, and temperature, but knowledge of this subsurface realm is just beginning to be accumulated.

In the shallow subsurface, other diagenetic environments are also present. Several of these are described briefly in the following, simply to bring their existence to attention. Detailed studies of most are available in the literature. In addition to these, a variety of other diagenetic environments may still be unrecognized. This list is not considered comprehensive.

Marine Vadose Zone

Within many marine beaches a marine vadose zone is present where salt water and air fill pore spaces. Because of rapid CO_2 degassing, precipitation of cements may be rapid (Hanor, 1978). Extensive cementation by aragonite and/or Mg-calcite is common (see Bricker, 1971, pt. 1, for examples). Because beaches at times may be saturated with sea water, and at others have both air and sea water in pores, cement textures often represent characteristics of both phreatic and vadose cements. However, emphasis in this section is on marine vadose cements which may be recognized by meniscus and pendant textures. Cements tend to be fibrous aragonite, micritic or fibrous Mg-calcite. Because of their abundance and easy accessibility, recent beach rocks have received extensive study (Moore, 1977; Hanor, 1978; see also reviews in Bathurst, 1974, 1976), and ancient examples of marine vadose cements have also been described (Purser, 1969).

Sabkha Brines

Under many arid tidal flats and some humid tidal flats, a phreatic zone saturated with dense hypersaline brines is present. Brines may form either by simple evaporation of sea water or by interaction of normal marine waters with preexisting evaporite and calcium carbonate minerals (Levy, 1977a). Detailed studies of these brines in the Persian Gulf (discussed in Purser, 1973) and the northern Sinai (Levy, 1977b) have revealed complex diagenetic patterns that include precipitation of evaporites and carbonates such as dolomite. Reviewing the literature on sabkha brines is beyond the scope of this paper, but this is clearly a shallow subsurface diagenetic environment that merits close examination because of the importance of tidal-flat dolomites as hydrocarbon reservoirs.

Diagenesis in Deep Marine Environments

Waters in the deep sea differ from those in the shallow marine environment in being undersaturated with $CaCO_3$. Solution of carbonate, especially aragonite, may occur. Deep-sea carbonates tend to be dominated by low magnesium calcite, either in the form of coccoliths (Scholle, 1977) or neomorphically altered skeletal debris (Schlager and James, 1978). Precipitation of calcite cement may occur (Hudson and Coleman, 1978) and, in general, diagenesis seems to resemble that present in the stagnant freshwater phreatic environment. Bathurst (1976, p. 561) summarized much of the literature on diagenesis under the deep seafloor, but many details of diagenesis in this relatively inaccessible environment still need to be clarified.

DIAGENETIC SEQUENCES

With the large number of diagenetic environments and the variable array of textures that may be produced in each zone, interpreting diagenetic sequences becomes a very complex puzzle. To complicate the picture further, a sediment may be exposed to a variety of environments over and over again as the water table or mixing zone migrates up and down in response to seasonal rains or any number of other processes. This should not discourage interpretation of diagenetic sequences, however, because it is the major events that result in either significant cementation or leaching to form secondary porosity so important to breaking or making a potential hydrocarbon reservoir. Sufficient evidence usually exists for interpreting these major events.

To point to a given diagenetic sequence and say that this is representative of diagenesis in most carbonate rocks would be foolish because of the wide variety of diagenetic environments to which a carbonate rock may be exposed. However, an ideal diagenetic sequence is presented in the diagrams of Figure 17. The intent of this diagram is to show how diagenetic textures produced in the various environments may be superimposed on older events and still be distinguished.

This sequence was not chosen at random but represents an ideal diagenetic sequence to which a typical shallow-marine carbonate sediment might be exposed in view of the consistent interrelations of diagenetic environments discussed earlier. Most carbonates begin their diagenetic history in the marine phreatic zone where intergranular marine cementation may (in the shelf margin or beach) or may not occur. With any slight lowering of sea level or stable sea level accompanied by the production of shallowing-upward depositional sequences as described by James (1977), the sediment may become saturated with fresh water and enter the meteoric phreatic environment. Because little diagenesis generally occurs in the mixing zone, it is not represented in the sequence. Extensive cementation and neomorphism in the freshwater phreatic zone may be followed by exposure to the vadose zone in which leaching occurs. Such a rock will then generally pass back through the meteoric phreatic environment on its way into the subsurface.

Stage 1 represents the initial deposition of our ideal shallow-marine carbonate sediment. The sediment is composed of red algae, foraminifers, mollusks, and an echinoderm fragment. Micrite is present between some of the grains and is assumed to be aragonite for simplicity. Geopetal sediment is present in the gastropod. Porosity makes up the rest of the field and is estimated to be about 45%. The sediment contains both Mg-calcite and aragonite grains as is typical of shallow-marine carbonates. Marine water saturates the pores.

Stage 2 represents the first stages of diagenesis likely to occur after deposition in the relatively stagnant marine phreatic diagenetic environment. Micritization of grains by boring algae and fungi results in the formation of micrite envelopes. Minor cementation occurs inside small pores of the red algae and forams, perhaps because of bacterial activity. No alteration of grains occurs nor is there any leaching. Porosity is reduced slightly by the intragranular cementation and by compaction.

Stage 3 reflects a change into the active marine phreatic environment in which intergranular cementation by fibrous aragonite and Mg-calcite occurs. The cement forms isopachous rims on all grains including the echinoderm fragment. A trend toward substrate control on cement mineralogy is reflected in the diagram, but there are

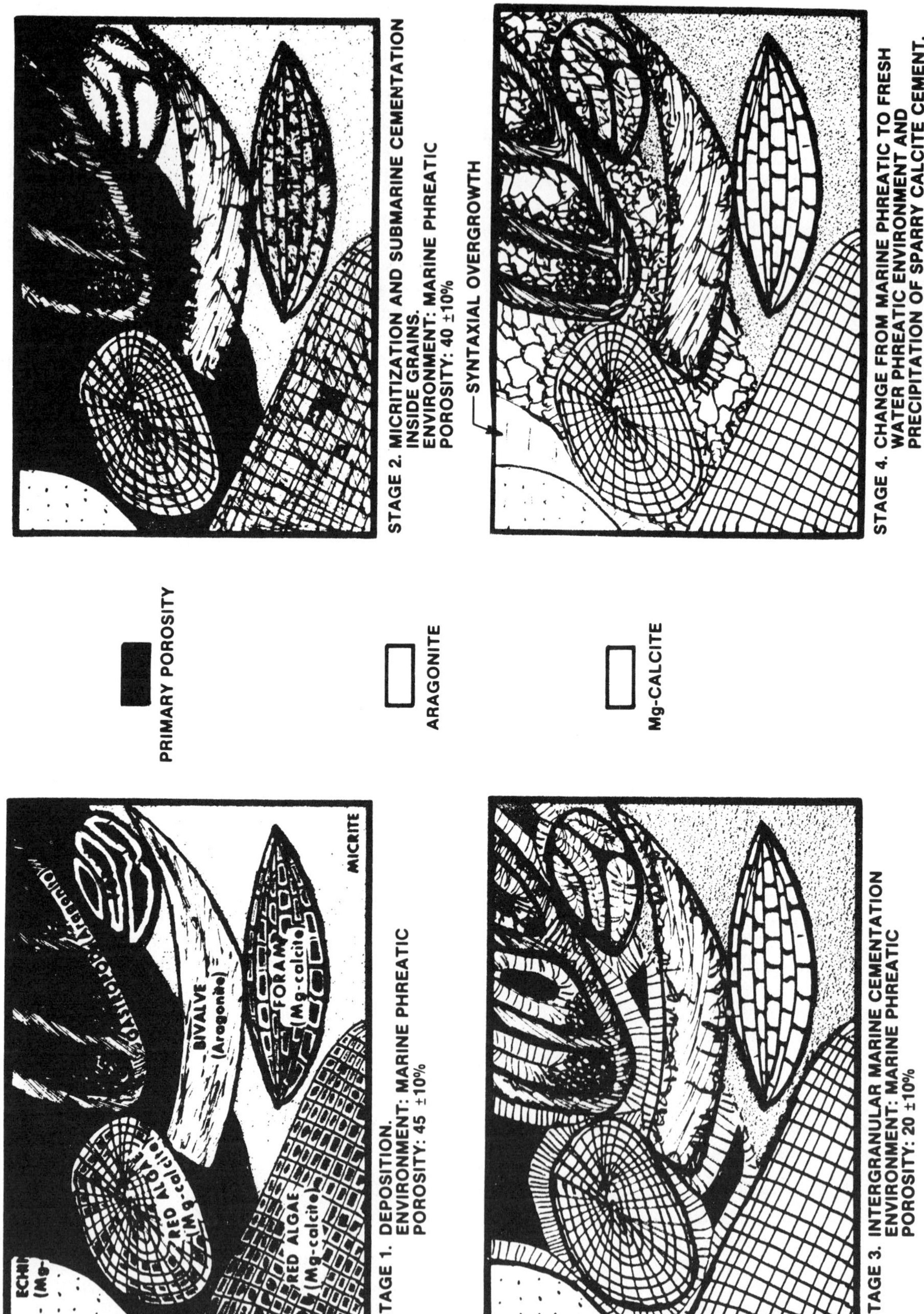

FIG. 17—Typical stages of diagenesis in carbonate rocks.

STAGE 5. LEACHING OF ARAGONITE AND SOME Mg-CALCITE. NEOMORPHISM OF SOME Mg-CALCITE TO CALCITE.
ENVIRONMENT: FRESH WATER PHREATIC
POROSITY: 20 ±10%

STAGE 6. FINAL RECRYSTALLIZATION OF UNSTABLE GRAINS AND MICRITE TO CALCITE. FILLING OF MOLDS WITH SPARRY CALCITE.
ENVIRONMENT: FRESH WATER PHREATIC
POROSITY: 5 ±5%

SECONDARY POROSITY

CALCITE

STAGE 7. FORMATION OF VUGGY POROSITY AND RECRYSTALLIZATION OF MICRITE.
ENVIRONMENT: FRESH WATER VADOSE
POROSITY: 10 ±5%

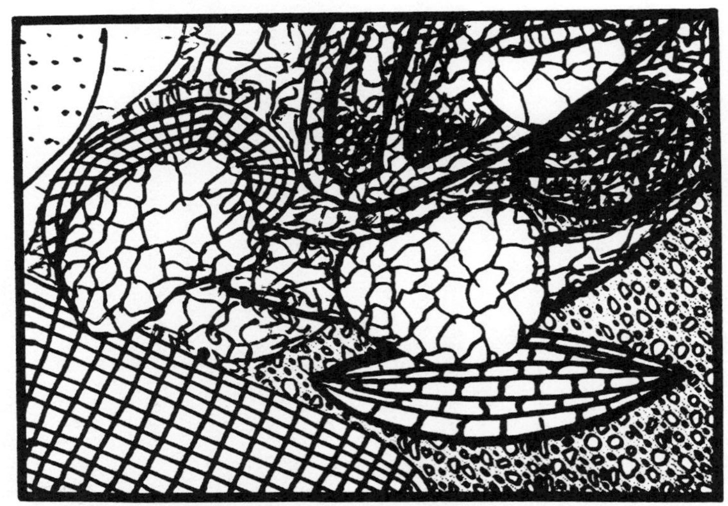

STAGE 8. FILLING OF VUGS WITH SPARRY CALCITE.
ENVIRONMENT: FRESH WATER PHREATIC
POROSITY: 5 ±5%

exceptions as shown by the aragonite cement on the smaller foraminifer. Porosity is reduced significantly by this marine cementation.

Stage 4 represents diagenesis in the freshwater phreatic zone and is intended to follow Stage 2 (micritization and intragranular marine cementation) without an intermediate stage of cementation in the active marine phreatic zone. Cementation in the active saturated freshwater phreatic zone occurs and all pore space is filled with equant calcite. The cement coarsens toward pore centers. Bladed calcite may form on some grains if cements nucleate on the grain and grow fastest along a preferred crystallographic axis. Syntaxial overgrowths form on echinoderm fragments and grow more rapidly than the cement on adjacent grains. Because the water is saturated with respect to calcium carbonate, no leaching of the grains occur prior to cementation. Prolonged exposure to these conditions would probably result in recrystallization of the unstable grains to calcite. Porosity is significantly reduced by the cementation.

Stage 5 represents continued diagenesis in the freshwater phreatic environment after a change into a zone undersaturated with aragonite. Aragonite and some Mg-calcite are leached from the original grains to produce moldic porosity. For reasons not yet clear, solution often attacks only certain Mg-calcite grains and leaves others untouched. Many Mg-calcite grains may neomorphose to normal calcite without undergoing a solution stage as shown for the smaller foram (upper right). Commonly cements such as the Mg-calcite in the pores of the red algae (bottom left) tend to be more stable than the red algae itself, and thus resist solution for a longer period. Reprecipitation of the dissolved material as equant calcite may occur almost simultaneously with solution; this occurs most commonly in fine-grained material like the micrite shown, but also in larger grains such as the mollusks (not shown). Porosity produced by leaching may increase to more than 20%.

Stage 6 reflects final stabilization of the initial sediment and cementation of all moldic porosity by calcite, thus creating a final product that bears a resemblance to many ancient limestones. Grain shapes and some textures are preserved in spite of neomorphism. Cement filling the mollusk molds may coarsen toward pore centers just as that in the intergranular pores. Depending on water chemistry and rates of water flow, micrite may recrystallize to coarser grains and form microspar. Porosity in this limestone is low because of the extensive cementation and neomorphism.

Diagenesis in our ideal limestone may end at this stage or, if our limestone is uplifted into the freshwater vadose zone, additional solution by undersaturated meteoric water may occur. This is shown in Stage 7. Solution in a relatively homogeneous limestone generally produces vuggy porosity that cuts across both grains and cements. Selective leaching of certain fossils may occur as shown in the red algae grain (lower left). Porosity generated by the solution may be small or large depending on proximity to the soil zone, rate of water flow, and duration of the solution phase.

Most rocks in which secondary vadose porosity is generated will move through the freshwater phreatic environment on their way to the subsurface. Much of the vuggy porosity may be filled with equant calcite at this time as shown in Stage 8. The result is once again a tight limestone.

In summary, the diagrams show an ideal and relatively simple sequence of events that could occur to change shallow-marine carbonate sediments into limestone. A tremendous number of variations on the theme is possible. The complexities of dolomitization and deep subsurface diagenesis are not represented in the diagrams but are important processes that need to be considered in interpreting ancient carbonates. Processes such as collapsing micrite envelopes, Mg/Ca variation in diagenetic fluids, rates of cementation, compaction of mud, and so on, are not considered in the sequence. Figure 18 shows just how variable diagenesis can be even within an area only a few millimeters square.

APPLICATIONS AND RELEVANCE

An exploration geologist might ask at this point how porosity ever gets preserved in carbonates because the last stage of the sequence and several intermediate stages lead to a tight rock. The answer is this sequence may be interrupted at any point, by stagnation of the diagenetic fluids, rapid burial, migration of nonreactive fluids such as hydrocarbons into the system, or replacement of fresh water with stagnant marine waters (e.g., during a transgression). Once a porous rock moves out of the shallow subsurface diagenetic environments into the deeper subsurface, evidence suggests that rates of diagenesis generally slow down, and porosity has a good chance of being preserved unless stylolitization occurs. Thus, porous carbonates are not uncommon in the subsurface. By recognizing the stage at which the diagenetic sequence was interrupted, and by interpreting the regional paleotopography and distribution of diagenetic environments, the distribution of such porous carbonates can often be predicted.

The most important factors affecting porosity

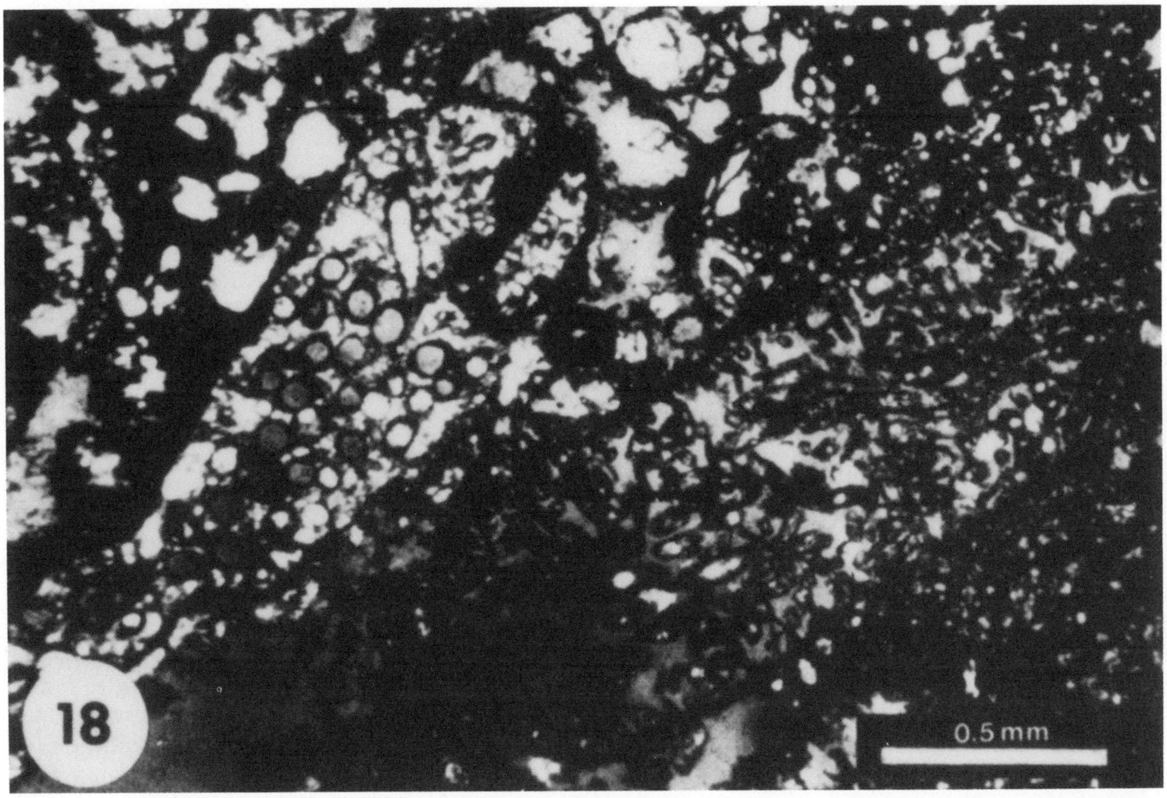

FIG. 18—Diagenesis is a complex and variable process as shown here. Two grains are *Halimeda* "leaves" but, in one, pores remained empty as leaf neomorphosed to equant calcite; in other, pores filled with equant calcite before all aragonite in leaf was removed by solution. Both leaves apparently were altered in freshwater phreatic environment. Sample from Windley Key Quarry, Florida.

preservation in carbonates are climate, changes in relative sea level, and rate of subsidence; secondary factors include the composition of the original sediment and rate of sedimentation. The importance of fresh water in carbonate cementation has already been emphasized and rainfall is the source of most of this fresh water. Thus, complete cementation of primary porosity almost always occurs during subaerial exposure of shallow-water carbonates in areas of high rainfall. Preservation of primary intergranular porosity generally occurs only under arid conditions or in carbonates unaffected by fresh water because of deposition (1) in deep water (e.g., chalks), or (2) during rapid subsidence. Most hydrocarbon production from primary intergranular pore space is from carbonate grainstones deposited in very arid climates as shown by the associated evaporites, which often act as seals. Examples include the production from the Jurassic carbonate rocks in Saudi Arabia, such as the giant Ghawar and Abqaiq fields, the Tertiary limestones in Kirkuk field in Iran, and parts of the Smackover Formation in the southeastern United States. Other rocks that produce from primary porosity include chalks (Scholle, 1977) such as those in the North Sea. These chalks escaped most early calcite cementation because the depth of water during deposition prevented meteoric-water influx, and the stable calcite mineralogy was relatively nonreactive with the interstitial fluids.

This does not mean that carbonates deposited in wet climates cannot be good reservoirs. Although primary porosity is generally completely filled in shallow-water carbonates exposed to much fresh water, meteoric water may also cause significant solution and the generation of secondary porosity. Most of this secondary porosity is generated during major lowering of relative sea level and the formation of unconformities. One of the best known examples is the famous Golden Lane field in Cretaceous limestones of Mexico in which Late Cretaceous and early Tertiary uplift resulted in caves that eventually became filled with oil (Coogan et al, 1972). This example reveals how climate, relative lowering of sea level, and subsequent subsidence can interact to form a significant reservoir.

Understanding of the nature of the porosity (e.g., primary intergranular, moldic, vuggy, etc) and the processes responsible for its formation can permit prediction of porosity trends in the subsurface. If primary porosity is the major target, emphasis should be on determining the distribution of the depositional facies in which that porosity is present, because distribution of primary porosity is always controlled by depositional facies. If secondary porosity is the target, its relation to depositional facies must be determined. Some types of secondary porosity, such as oomoldic and intercrystalline in tidal-flat dolomites, are facies controlled but many are not. If secondary porosity is not facies controlled as in the case of vugs, fractures, and much late dolomitization, the process responsible for forming the porosity must be determined.

The high susceptibility of carbonates to rapid leaching and cementation means that marked horizontal and vertical changes in porosity and permeability may occur over short distances. There is no reason to suggest that all porosity must be on top of structures; instead, off-structure stratigraphic and diagenetic traps are likely to be very common.

Diagenetic traps formed by cementation around structurally trapped oil have been described by Wilson (1977), and diagenetic traps may also form in other ways. Cementation by equant calcite in the active part of the upper freshwater phreatic zone could provide a seal above porous carbonates which were affected only by diagenesis in a stagnant freshwater phreatic or marine phreatic zone. Cementation of reefs in the active marine phreatic zone may provide a lateral seal either for porous back-reef sediments or fore-reef talus. In rare circumstances, a soil zone or caliche could provide a seal over carbonates which were leached in the vadose zone, but more often a fluvial flood-plain or marine shale may cover carbonates leached in the vadose zone. A more complex way of forming a diagenetic trap involves the formation of hydrologic "channels" which could produce either porous or cemented zones in otherwise homogeneous carbonates. Dolomitization often produces porosity locally and this porous dolomite may be capped by evaporites. Fracture porosity in carbonates may be thought of as a wild card that may form a potential reservoir in rocks deposited in any depositional environment and affected by any or all types of diagenesis. This list of ways to get diagenetic traps is not comprehensive; many others will undoubtedly be found. However, it shows that a good working knowledge of carbonate diagenesis may give the explorationist that new or different idea that will enable him to find hydrocarbons overlooked by others. And an understanding of diagenetic events and environments is even more important to the exploitation geologist trying to develop a field from which some cores are already available.

CONCLUSIONS

1. An understanding of diagenetic textures and processes is of key importance in exploring for hydrocarbons in carbonate rocks. Diagenetic processes can be related to diagenetic environments to provide models for interpreting the geometry of cemented and porous units.

2. Shallow-marine carbonates are generally susceptible to early diagenesis in any combination of four shallow subsurface diagenetic environments. These are the marine phreatic, meteoric vadose, meteoric phreatic, and mixed marine-meteoric phreatic environments.

3. The marine phreatic environment may be divided into a zone near the sediment/water interface where water circulation combines with some other factor such as CO_2 degassing, respiration by organisms, or bacteria to produce common aragonite and Mg-calcite cements, and a zone of little cementation in which waters are stagnant or the other factors contributing to marine cementation are absent.

4. The meteoric vadose zone may be divided into the near-surface zone where waters are undersaturated with $CaCO_3$, and (in some areas) a deeper zone where waters are saturated with $CaCO_3$. Solution is a dominant process in the undersaturated zone, particularly near soil zones, whereas meniscus and pendant calcite cements may precipitate in the saturated zone.

5. The meteoric phreatic zone may be divided into five subzones reflecting increasing saturation of water with $CaCO_3$. These are: (1) a zone of indiscriminate leaching of carbonates, (2) a zone in which only aragonite is leached, (3) a zone in which aragonite leaching is accompanied by calcite cementation, (4) an active saturated zone of calcite cementation and no leaching, and (5) a zone in which stagnant saturated water is present and produces grain neomorphism to calcite but little cementation. Where cementation occurs, bladed to equant calcite that coarsens toward pore centers is formed.

6. Mixing of fresh and marine waters in the mixing zone may produce a solution undersaturated with $CaCO_3$ but oversaturated with dolomite at the freshwater end of the spectrum. Solution may occur and dolomite may form in mixing

zones which have a good water flow to provide sufficient magnesium ions. Isopachous fibrous Mg-calcite may precipitate in waters at the marine end of the spectrum.

7. Changes from one diagenetic environment to another may occur many times in the diagenetic history of carbonates. If significant and recognizable textures are produced in each diagenetic environment without severely altering earlier textures, the diagenetic sequence may be determined. This may, in turn, provide clues to predicting porosity trends in carbonates.

8. Only diagenetic environments that commonly affect shallow-marine carbonates have been described here. Although most diagenesis of the shallow-marine carbonates apparently occurs in one or another of these diagenetic environments, the importance of cementation and leaching in deeper subsurface environments must not be overlooked.

9. Climate, changes in relative sea level, rates of subsidence and deposition, and composition of the sediments are the major controls on porosity in carbonate rocks. During subaerial exposure in arid climates, primary porosity in carbonate grainstones is likely to escape early cementation whereas, in wet climates, most primary porosity will be filled with cement, but secondary porosity related to leaching by meteoric water may form near unconformities.

10. Concepts of diagenetic environments presented here represent a summary of the available literature combined with a little logic and inductive reasoning. Many details of shallow diagenetic environments remain to be learned. As a better understanding of diagenesis in shallow-marine carbonates is developed, revisions and additions to the models presented here will undoubtedly be necessary.

REFERENCES CITED

Alexandersson, T., 1972, Intragranular growth of marine aragonite and Mg-calcite: evidence of precipitation from supersaturated sea water: Jour. Sed. Petrology, v. 42, p. 441-460.

Assereto, R., and R. L. Folk, 1976, Brick-like texture and radial rays in Triassic pisolites of Lombardy, Italy: a clue to distinguish ancient aragonitic pisolites: Sed. Geology, v. 16, p. 205-222.

Badiozamani, K., 1973, The Dorag dolomitization model—application to the Middle Ordovician of Wisconsin: Jour. Sed. Petrology, v. 43, p. 965-984.

——— F. T. MacKenzie, and D. C. Thorstensen, 1977, Experimental carbonate cementation: salinity, temperature and vadose-phreatic effects: Jour. Sed. Petrology, v. 47, p. 529-542.

Bathurst, R. G. C., 1958, Diagenetic fabrics in some British Dinantian limestones: Liverpool Manchester Geol. Jour., v. 2, p. 11-36.

——— 1959, Diagenesis in Mississippian calcilutites and pseudobreccias: Jour. Sed. Petrology, v. 29, p. 365-376.

——— 1974, Marine diagenesis of shallow water calcium carbonate sediments: Ann. Rev. Earth & Planetary Sci., v. 2, p. 257-274.

——— 1976, Carbonate sediments and their diagenesis (2d enlarged ed.): New York, Elsevier Pub. Co., 658 p.

Berner, R. A., 1971, Bacterial processes affecting the precipitation of calcium carbonate in sediments, *in* Carbonate cements: Johns Hopkins Univ. Studies Geology, no. 19, p. 247-251.

Bricker, O. P., ed., 1971, Carbonate cements: Johns Hopkins Univ. Studies Geology, no. 19, 376 p.

Choquette, P. W., and L. C. Pray, 1970, Geologic nomenclature and classification of porosity in sedimentary carbonates: AAPG Bull., v. 54, p. 207-250.

Coogan, A. H., D. G. Bebout, and C. Maggio, 1972, Depositional environments and geologic history of Golden Lane and Poza Rica trend, Mexico, an alternate view: AAPG Bull., v. 56, p. 1419-1447.

Cullis, C. G., 1904, The mineralogical changes observed in the cores of the Funafuti borings, *in* T. G. Bonney, ed., The atoll of Funafuti: Royal Soc. London, p. 392-420.

Cussey, R., and G. M. Friedman, 1977, Patterns of porosity and cement in ooid reservoirs in Dogger (Middle Jurassic) of France: AAPG Bull., v. 61, p. 511-518.

Dunham, R. J., 1971, Meniscus cement, *in* Carbonate cements: Johns Hopkins Univ. Studies Geology, no. 19, p. 297-300.

Evamy, B. D., and D. J. Shearman, 1965, The development of overgrowths from echinoderm fragments: Sedimentology, v. 5, p. 211-233.

Folk, R. L., 1973, Carbonate petrography in the post-Sorbian age, *in* Evolving concepts in sedimentology: Johns Hopkins Univ. Studies Geology, no. 21, p. 118-158.

——— 1974, The natural history of crystalline calcium carbonate: effect of magnesium content and salinity: Jour. Sed. Petrology, v. 44, p. 40-53.

——— and L. S. Land, 1975, Mg/Ca ratio and salinity: two controls over crystallization of dolomite: AAPG Bull., v. 59, p. 60-68.

Ford, D. C., and R. O. Ewers, 1978, The development of limestone cave systems in the dimensions of length and depth: Canadian Jour. Earth Sci., v. 15, p. 1783-1798.

Friedman, G. M., 1964, Early diagenesis and lithification in carbonate sediments: Jour. Sed. Petrology, v. 34, p. 777-813.

——— 1968, The fabric of carbonate cement and matrix and its dependence on the salinity of water, *in* G. Muller and G. M. Friedman, eds., Recent developments in carbonate sedimentology in central Europe: New York, Springer-Verlag, p. 11-20.

——— and J. E. Sanders, 1978, Principles of sedimentology: New York, John Wiley and Sons, 792 p.

——— A. J. Amiel, and N. Schneidermann, 1974, Submarine cementation in reefs: example from the Red Sea: Jour. Sed. Petrology, v. 44, p. 816-825.

Ginsburg, R. N., and N. P. James, 1976, Submarine bo-

tryoidal aragonite in Holocene reef limestones, Belize: Geology, v. 4, p. 431-436.
Grover, G., Jr., and J. F. Read, 1978, Fenestral and associated vadose diagenetic fabrics of tidal flat carbonates, Middle Ordovician New Market Limestone, southwestern Virginia: Jour. Sed. Petrology, v. 48, p. 453-474.
Hanor, J. S., 1978, Precipitation of beach rock cements: mixing of marine and meteoric waters vs. CO_2 degassing: Jour. Sed. Petrology, v. 48, p. 489-502.
Harris, P. M., 1978, Holocene marine-cemented sands, Joulters Ooid Shoal, Bahama: Gulf Coast Assoc. Geol. Socs. Trans., v. 28, p. 175-184.
——— 1979, Facies anatomy and diagenesis of a Bahamian ooid shoal, *in* Sedimenta VII: Univ. Miami Comp. Sed. Lab., 163 p.
Harris, W. H., 1971, Ground water–carbonate rock chemical interactions, Barbados, West Indies: PhD thesis, Brown Univ. 348 p.
Harrison, R. S., 1974, Near-surface subaerial diagenesis of Pleistocene carbonates, Barbados, West Indies: PhD thesis, Brown Univ., 332 p.
——— 1975, Porosity in Pleistocene grainstones from Barbados: some preliminary observations: Canadian Petroleum Geologists Bull., v. 23, p. 383-392.
Hudson, J. D., and M. L. Coleman, 1978, Submarine cementation of the Scheck limestone conglomerate (Jurassic, Austria): isotopic evidence: Neues Jahrb. Geologie u. Paläontologie Monatsh., v. 9, p. 534-544.
Jacka, A. D., and J. P. Brand, 1977, Biofacies and development and differential occlusion of porosity in a Lower Cretaceous (Edwards) reef: Jour. Sed. Petrology, v. 47, p. 366-381.
James, N. P., 1972, Holocene and Pleistocene calcareous crust (caliche) profiles: criteria for subaerial exposure: Jour. Sed. Petrology, v. 42, p. 817-836.
——— 1977, Shallowing-upward sequences in carbonates, facies models 8: Geosci. Can., v. 4, p. 126-136.
——— et al, 1976, Facies and fabric specificity of early subsea cements in shallow Belize (British Honduras) reefs: Jour. Sed. Petrology, v. 46, p. 523-544.
Kahle, C. F., 1977, Origin of subaerial Holocene crusts: role of algae, fungi, and sparmicritization: Sedimentology, v. 24, p. 413-436.
Kendall, A. C., and P. L. Broughton, 1978, Origin of fabrics in speleothems composed of columnar calcite crystals: Jour. Sed. Petrology, v. 48, p. 519-538.
Knox, G. J., 1977, Caliche profile formation, Saldanha Bay (South Africa): Sedimentology, v. 24, p. 657-674.
Kobluk, D. R., and M. J. Risk, 1977, Micritization and carbonate-grain binding by endolithic algae: AAPG Bull., v. 61, p. 1069-1082.
Ladd, H. S., and S. O. Schlanger, 1960, Drilling operations on Eniwetok Atoll: U.S. Geol. Survey Prof. Paper 260-Y, p. 863-903.
Land, L. S., 1970, Phreatic versus vadose meteoric diagenesis of limestones: evidence from fossil water table: Sedimentology, v. 14, p. 175-185.
——— 1971, Submarine lithification of Jamaican reefs, *in* Carbonate cements: Johns Hopkins Univ. Studies Geology, no. 19, p. 59-62.
——— 1973a, Contemporaneous dolomitization of middle Pleistocene reefs by meteoric water, north Jamaica: Bull. Marine Sci., v. 23, p. 64-92.
——— 1973b, Holocene meteoric dolomitization of Pleistocene limestones, north Jamaica: Sedimentology, v. 20, p. 411-422.
——— and T. R. Goreau, 1970, Submarine lithification of Jamaican reefs: Jour. Sed. Petrology, v. 40, p. 457-462.
——— M. R. I. Salem, and D. W. Morrow, 1975, Paleohydrology of ancient dolomites: geochemical evidence: AAPG Bull., v. 49, p. 1602-1625.
Levy, Y., 1977a, The origin of brine in coastal sabkhas, northern Sinai: Jour. Sed. Petrology, v. 47, p. 451-462.
——— 1977b, Description and mode of formation of the supratidal evaporite facies in northern Sinai coastal plain: Jour. Sed. Petrology, v. 47, p. 463-474.
Lohmann, K. C., and W. J. Meyers, 1977, Microdolomite inclusions in cloudy prismatic calcites: a proposed criterion for former high magnesium calcites: Jour. Sed. Petrology, v. 47, p. 1078-1088.
Loucks, R. G., 1977, Porosity development and distribution in shoal-water carbonate complexes—subsurface Pearsall Formation (Lower Cretaceous) south Texas, *in* Cretaceous carbonates of Texas and Mexico, applications to subsurface exploration: Texas Univ. Bur. Econ. Geology Rept. Inv. 89, p. 97-126.
——— and R. L. Folk, 1976, Fanlike rays of former aragonite in Permian Capitan reef pisolite: Jour. Sed. Petrology, v. 46, p. 483-485.
Macintyre, I. G., 1977, Distribution of submarine cements in a modern Caribbean fringing reef, Galeta Point, Panama: Jour. Sed. Petrology, v. 47, p. 503-516.
——— and P. A. Videtich, 1979, Pseudostalactites from submarine cave near Columbus Cay, Belize barrier-reef complex—evidence of extensive submarine lithification (abs.): AAPG Bull., v. 63, p. 489.
Magara, K., 1978, Compaction and fluid migration: practical petroleum geology, *in* Developments in petroleum science, v. 9: New York, Elsevier Pub. Co., 319 p.
Matthews, R. K., 1974, A process approach to diagenesis of reefs and reef associated limestones, *in* Reefs in time and space: SEPM Spec. Pub. 18, p. 234-256.
——— 1967, Diagenetic fabrics in biosparites from the Pleistocene of Barbados, West Indies: Jour. Sed. Petrology, v. 37, p. 1147-1153.
——— 1968, Carbonate diagenesis: equilibrium of sedimentary mineralogy to the subaerial environment; coral cap of Barbados, West Indies: Jour. Sed. Petrology, v. 38, p. 1110-1119.
May, J. A., and R. D. Perkins, 1979, Endolithic infestation of carbonate substrates below the sediment-water interface: Jour. Sed. Petrology, v. 49, p. 357-377.
Mazzullo, S. J., and J. M. Cys, 1977, Submarine cements in Permian boundstones and reef-associated rocks, Guadalupe Mountains, west Texas and southeastern New Mexico: SEPM Permian Basin Sec. Guidebook 77-16, v. 1, p. 151-200.
Meyers, W. J., 1978, Carbonate cements: their regional distribution and interpretation in Mississippian limestones of southwestern New Mexico: Sedimentology, v. 25, p. 371-400.
Moore, C. H., Jr., 1977, Beach rock origin: some geochemical, mineralogical, and petrographic considerations: Geosci. and Man, v. 18, p. 155-163.

——— 1979, Porosity in carbonate rock sequences, *in* Geology of carbonate porosity: AAPG Continuing Education Course Notes 11, p. A1-A124.

Müller, G., 1971, "Gravitational" cement: an indicator for the vadose zone of the subaerial diagenetic environments, *in* Carbonate cements: Johns Hopkins Univ. Studies Geology, no. 19, p. 301-302.

Pittman, E. D., 1974, Porosity and permeability changes during diagenesis of Pleistocene corals, Barbados, West Indies: Geol. Soc. America Bull., v. 85, p. 1811-1820.

Purser, B. H., 1969, Syn-sedimentary marine lithification of Middle Jurassic limestones in the Paris basin: Sedimentology, v. 12, p. 205-230.

——— ed., 1973, The Persian Gulf, Holocene carbonate sedimentation and diagenesis in a shallow epicontinental sea: New York, Springer-Verlag, 471 p.

Robbin, D. M., and J. J. Stipp, 1979, Depositional rate of laminated soilstone crusts, Florida Keys: Jour. Sed. Petrology, v. 49, p. 175-180.

Runnells, D. D., 1969, Diagenesis, chemical sediments, and the mixing of natural waters: Jour. Sed. Petrology, v. 39, p. 1188-1201.

Schlager, W., and N. P. James, 1978, Low-magnesium calcite limestones forming at the deep-sea floor, Tongue of the Ocean, Bahamas: Sedimentology, v. 25, p. 675-702.

Scholle, P. A., 1977, Chalk diagenesis and its relation to petroleum exploration: oil from chalks, a modern miracle?: AAPG Bull., v. 61, p. 982-1009.

Schroeder, J. H., 1972, Fabrics and sequences of submarine carbonate cements in Holocene Bermuda cup reefs: Geol. Rundschau, v. 61, p. 708-730.

——— 1973, Submarine and vadose cements in Pleistocene Bermuda reef rocks: Sed. Geology, v. 10, p. 179-204.

Shinn, E. A., 1969, Submarine lithification of Holocene carbonate sediments in the Persian Gulf: Sedimentology, v. 12, p. 109-144.

——— 1975, Polygonal cement sutures from the Holocene: a clue to recognition of submarine diagenesis (abs.): AAPG-SEPM Ann. Mtg. Abs., v. 2, p. 68.

Sorby, H. C., 1879, Anniversary address of the president: structure and origin of limestones: Geol. Soc. London Proc., v. 35, p. 56-95.

Stehli, F. G., and J. Hower, 1961, Mineralogy and early diagenesis of carbonate sediments: Jour. Sed. Petrology, v. 31, p. 358-371.

Steinen, R. P., 1974, Phreatic and vadose diagenetic modification of Pleistocene limestone: petrographic observations from subsurface of Barbados, West Indies: AAPG Bull., v. 58, p. 1008-1024.

Supko, P. R., 1971, "Whisker" crystal cement in Bahamian rock, *in* Carbonate cements: Johns Hopkins Univ. Studies Geology, no. 19, p. 143-148.

Thrailkill, J., 1968, Chemical and hydrologic factors in the excavation of limestone caves: Geol. Soc. America Bull., v. 79, p. 19-46.

——— 1971, Carbonate deposition in Carlsbad Caverns: Jour. Geology, v. 79, p. 683-695.

——— 1976, Speleothems, *in* M. R. Walther, ed., Stromatolites: Amsterdam, Elsevier, p. 73-86.

Wilson, H. H., 1977, "Frozen-in" hydrocarbon accumulations or diagenetic traps—exploration targets: AAPG Bull., v. 61, p. 483-491.

Reprinted by permission of the Society of Economic Paleontologists and Mineralogists. Published in *Carbonate Cements*, N. Schneidermann and P.M. Harris, editors, SEPM Special Publication No. 36, pp. 303-328.

BURIAL DIAGENESIS: OUT OF SIGHT, OUT OF MIND!

PETER A. SCHOLLE[1] AND ROBERT B. HALLEY
U.S. Geological Survey, P.O. Box 25046, Denver, Colorado 80225

ABSTRACT

Examination of porosity data from Holocene and Pleistocene carbonate strata indicates that there generally is little or no porosity loss in the zone of near-surface water circulation [that is, in the vadose, meteoric-phreatic, or mixing zone(s)]. Thus, the transition from very porous carbonate sediments to well-cemented, low-porosity carbonate rocks is a dominantly subsurface process. Indeed, both shallow- and deep-marine carbonate strata show a continuous loss of porosity with depth, indicating that porosity-reducing processes act continuously from the surface to depths in excess of 4 km.

Experimental, observational, and geochemical data show that porosity loss through burial diagenesis results from both physical and chemical compaction and from cementation. In near-surface sections, dewatering, grain reorientation, grain breakage, and other mechanical processes lead to sediment/rock porosities as low as 30 percent. Continued porosity loss requires mechanical compaction, chemical dissolution at grain contacts and along solution seams or stylolites, and/or re-precipitation of dissolved calcite as intergranular cement. Calcareous shales or marl seams (donor beds) can act as significant sources of dissolved carbonate which is precipitated as cement in adjacent limestones (recipient beds). Quantitantive studies of stylolites and solution seams commonly underestimate the total magnitude of pressure solution because they ignore contributions from associated thick calcareous shale sections and from thinner, regularly and irregularly distributed marl interbeds. Through these mechanisms, carbonate rock porosity may be reduced to values near zero in "semi-closed" systems without significant introduction of allochthonous cementing material.

In many young, subsiding basins, patterns of porosity loss with depth are crudely predictable. These patterns provide standards against which individual case studies of diagenesis may be compared and provide predictive tools for estimating porosity prior to drilling. In other areas, the standards allow identification of anomalously high porosity and focus attention on specific mechanisms which would act to preserve primary (or early diagenetic) porosity or to create secondary porosity at depth. Comparisons of oil field porosities with standard curves will allow further refinement of our understanding of diagenetic processes.

Predictive models are still in their infancy, however. There is a critical need to independently assess how rates of porosity loss with depth are affected by time, temperature, depositional setting, early diagenetic history, maturation history of organic matter, and other factors. In addition, overpressuring, early oil migration, dolomitization, and hydrothermal alteration are known to affect porosity-depth relationships. Refining our understanding of these factors may help geologists use their general knowledge of basin history to make valid predictions of carbonate reservoir quality in frontier areas.

INTRODUCTION

"Much (and perhaps most) cementation and formation of secondary porosity (except fractures) in carbonates occurs at relatively shallow depths in one of four major diagenetic environments: the vadose zone, meteoric phreatic zone, mixing zone, and marine phreatic zone" (Longman, 1980, p. 461). This statement from a recent summary of carbonate diagenesis represents a common approach to the understanding of limestone porosity. The reason for the widespread acceptance of this approach is that, for the past three decades, most studies of diagenesis have concentrated on early diagenetic processes (cementation and leaching) in Holocene and Pleistocene sediments. Because early diagenetic processes occur in near-surface settings (commonly on warm, tropical isles), they are accessible for study and in many cases are not obscured by later, complicating events. Many Holocene or Pleistocene limestones have undergone significant diagenetic alteration as a consequence of events acting in the four near-surface environments listed above, and it is natural (but not necessarily correct) to extrapolate this limited experience to the geologic record. A survey of the literature on carbonate diagenesis for the past decade will attest to the frequency with which that extrapolation is made.

There exists, however, a large body of data from rocks of all ages which can be used to argue forcefully against the concept that cementation and leaching of carbonate rocks are dominated by near-surface processes. We would not deny that early diagenetic events may leave an imprint which affects later diagenesis. However, we would argue here that, particularly for rapidly subsiding sedimentary basins, "most cementation" and much leaching take place in moderately deep to very deep subsurface settings (mesogenetic environments).

Observations by Pray (1960) and Zankl (1969) that compaction (largely in the form of breakage of fossils) is rare in limestones convinced many that cementation of carbonate rocks had to precede burial. Indeed,

[1]Present address: Gulf Research and Development Company, P.O. Box 37048, Houston, Texas 77236

Bathurst (1975, p. 439) wrote, "We must therefore infer: (1) that cementation begins early, before compaction, and (2) that about half the volume of $CaCO_3$ in many limestones has been carried into the sediment from outside."

More recently, experimental work by Shinn *et al.* (1977) and Shinn and Robbin (1983) has shown that significant compaction and sediment dewatering can take place with little or no grain breakage. Furthermore, several workers (e.g., Bathurst, 1975) have noted radical differences between the average porosities of near-surface limestones, which may have experienced some or all of the diagenetic environments listed by Longman, and the porosities of ancient limestones. These differences, which average about one order of magnitude, have led geologists in the past decade to search for evidence of burial cementation.

This search has been largely successful. In this paper we summarize data from regional studies, core and outcrop examination, and petrographic and geochemical analyses which document major deep-subsurface compaction fabrics and cements in carbonate rocks.

SOME GENERALITIES

Diagenesis in its broadest sense is any change in a sediment which occurs after its deposition. Burial diagenesis, then, is any change in a sediment which occurs below the zone of near-surface ground water circulation and above the zone of low-grade metamorphism. The depths of these transitions vary, of course, between basins and even within basins. Burial diagenetic processes largely are a consequence of the accumulation of overburden detritus in a subsiding sedimentary basin and the related changes in temperature, pressure, basin hydrology, and pore-fluid chemistry. This concept corresponds with the term "anadiagenesis" as used by Fairbridge (1967) and includes those changes which occur in the "mesogenetic zone" or "zone of deeper burial" as used by Choquette and Pray (1970). Burial diagenesis encompasses a range of processes including mechanical compaction, chemical compaction (pressure solution), cementation, replacement and other neomorphic processes, dissolution, and fracturing. We believe that burial diagenesis plays an important role in subsurface cementation and that understanding its principles will help us identify the sites of major porosity loss in carbonate strata.

The Mass Balance Problem

Dunham (1969) and Bathurst (1975) pointed out that cementation models based on a remote supply of calcium carbonate have difficulty in deriving the volume of cement required to reduce a carbonate sediment with 40 to 50 percent porosity to an ancient limestone with only a few percent pore space. Because calcium carbonate is only slightly soluble in water under typical subsurface conditions, several tens of thousands to several hundreds of thousands of pore volumes of water are required to fill pore space with cement. Pray (oral commun. 1966) estimated that between 10,000 and 50,000 pore volumes of solution were required to fill pore space completely with calcium carbonate. Dunham (1969, p. 161) calculated that between 10,000 and 100,000 pore volumes were needed, based on efficiencies of precipitation of 100 percent and 10 percent, respectively. Even these latter calculations are based on unrealistically high efficiencies of precipitation and do not take into account the constantly decreasing permeability of strata and the increasing difficulty of supplying water and cement to the pore space as cementation proceeds. If the scale of the problem is expanded from that of a single pore to an entire formation with an initial average porosity of about 50 percent, then it would take enormous volumes of water (50,000 times or more the volume of the formation) to cement this unit to near zero porosity. These problems led Friedman (1975, p. 395) to conclude that ". . . the problem of mass balance remains open. The amount of cement needed to convert carbonate sediment into limestone seems larger than can be presently accounted for. . .".

The above calculations were based on water compositions similar to those of near-surface waters, such as sea water or fresh water in carbonate aquifers. A calculation using typical subsurface brines yields similar results. The amount of dissolved calcium carbonate in subsurface waters is generally limited by the concentrations of carbonate and bicarbonate, which generally are on the order of a few hundred parts per million (Collins, 1975). Because carbonate makes up 60 percent of the weight of calcium carbonate, the total dissolved calcium carbonate in subsurface waters typically is less than 1 g per liter. This concentration is similar to concentrations in most near-surface waters. Assuming the ability to precipitate 10 percent of the dissolved calcium carbonate from solution, it would take more than 27,000 cm^3 of brine to precipitate 1 cm^3 of calcite (2.71 g). Thus, even in the subsurface, very large volumes of brine are required to reduce porosity in carbonate rocks if the calcium carbonate must be transported into the system.

The large volume of water required for cementation by importation of dissolved calcium carbonate places limits on the hydrologic settings in which this process can take place. Such cementation could only be effective in areas of active water flow. Such settings include the splash zone on beaches (Ginsburg, 1953), the intertidal zone (Moore, 1973), turbulent areas near reefs and carbonate sand shoals, and major aquifers, especially along their recharge zones and marginal mixing zones where ground water enters the sea (Back and Hanshaw, 1971; Runnells, 1969).

Flow models indicate that waters of compaction in subsiding basins provide insufficient volumes of fluid for major cementation (Bonham, 1980). In models of subsiding basins, formations may be envisioned as simply settling through water previously deposited with the underlying strata. Although the channeling of large volumes of compaction water through volumetrically small areas of permeable sediment may account for local well cemented zones, there is simply not enough compaction water available to cement formations on a regional scale. Therefore, the cementation of units during burial in subsiding basins presents a significant mass balance problem. This is true particularly for basins like South Florida which have a geologic history dominated by subsidence and which have not yet entered a phase of uplift. In basins which have undergone (or are undergoing) uplift, active aquifer systems can be formed which aid in cementation by moving large volumes of head-driven waters through the sedimentary section.

The mass balance problem for subsiding basins is largely circumvented by a model that uses a local supply of cementing material derived during chemical compaction. In this model, some or most of the cement is derived by pressure dissolution from nearby sources. This material is transferred to cementing sites by a combination of diffusion and fluid flow; in fact, diffusion alone may be sufficient where the distances between sites of dissolution and sites of reprecipitation

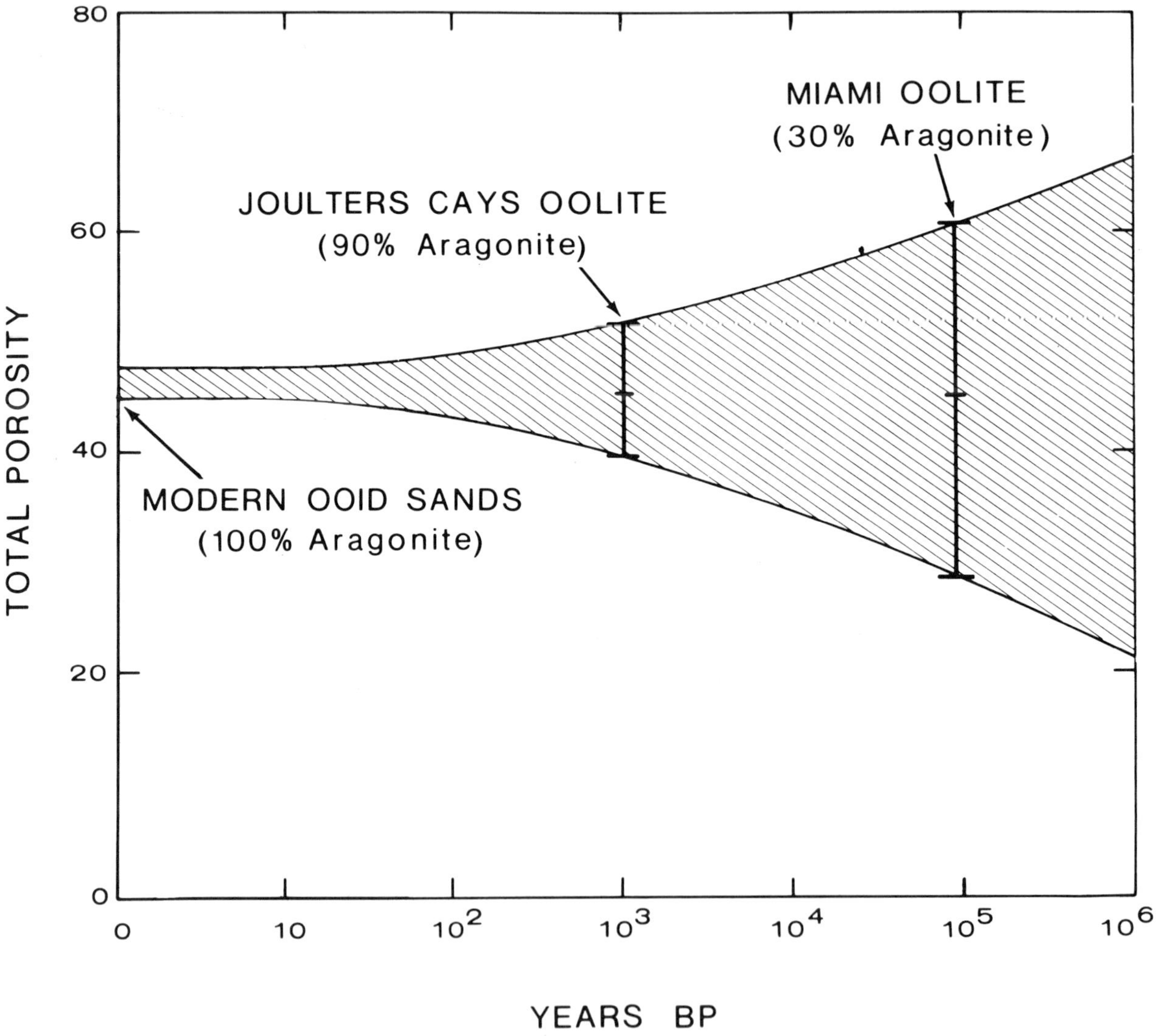

FIG. 1.—Diagram illustrating lack of porosity change during mineralogical stabilization of oolites in Florida and the Bahamas. A modern unconsolidated ooid sand is illustrated at the left of the diagram with 45 percent porosity and 100 percent aragonite. On Joulters Cays, about 10 percent of the aragonite has been converted to calcite cement, causing a wider range of porosity measurements, but the average porosity is still about 45 percent. The oolite facies of the Miami Limestone retains only about 30 percent aragonite, the other 70 percent having been converted to calcite. It has also developed vuggy porosity, causing individual porosity measurements to span a wide range, but the average porosity is still about 45 percent (from Halley and Evans, 1983).

are small. Diffusion is particularly important during late stages of cementation, when porosity is low and permeability is miniscule.

EVIDENCE FOR BURIAL DIAGENESIS

Porosity-Depth Curves

Perhaps the most direct evidence that near-surface diagenesis is not the major cause of porosity loss in carbonate strata comes from empirical examination of porosity values of carbonate rocks of all ages. Porosity data from modern carbonate strata are not abundant. Values for mudstones generally range from 60 to 70 percent (Ginsburg, 1956), and values for grainstones typically range from 40 to 50 percent (Enos and Sawatsky, 1979, 1981; Halley and Harris, 1979). These numbers compare well with those obtained in experimental studies by Vinopal and Coogan (1978).

Recent studies (Halley and Harris, 1979; Beach, 1982; and Halley and Evans, 1983) have shown that Holocene and Pleistocene carbonate sediments of the Florida-Bahamas area that have undergone progressive near-surface vadose and phreatic alteration retain high porosity (Fig. 1). Grainstone porosities in these diagenetically altered sediments still average about 45 percent with maximum porosity values of 59 percent. Data from other areas is sparse, but Harrison (1975) described mineralogically stabilized Pleistocene limestones from Barbados which averaged about 30 percent porosity and similar values were noted for exposed Pleistocene limestones of Bonaire (F. J. Lucia, oral commun., 1983).

Although the carbonate material in these sediments has been moved from one site to another through dissolution and reprecipitation, and although their permeability may have been greatly altered, their average porosity has remained high and, in some cases, essentially constant. The range of porosity values increases with the intensity of diagenesis (as measured by the conversion of aragonite to calcite), but the mean porosity values do not change. This indicates that there has been little or no net addition of carbonate material to the section despite as much as 120,000 years of near-surface alteration.

Sediments which are cemented in marine settings (at the seafloor or in marine phreatic environments) may undergo greater porosity loss due to the net addition of $CaCO_3$ from seawater. Individual thin units can show virtual obliteration of pore space by this process. However, these submarine cement fabrics are commonly recognizable in thin section, and they appear to affect only a small fraction of both modern and ancient strata.

Studies of thick carbonate sections have confirmed the idea that near-surface carbonate strata retain most of their pore space until they are more deeply buried. In chalk sections from both deep marine (Schlanger and Douglas, 1974) and shelf settings (Scholle, 1977), average porosities in excess of 40 percent were typically found in strata which had never been buried deeper than 500 m (Fig. 2). Only in rare, submarine-cemented hardground sections were lower porosities encountered in shallowly buried chalk sections. Burial, with increased lithostatic stress and increased temperature, is the only event which has lead to widespread porosity loss in these carbonate strata.

Some geologists have argued that chalks are atypical sediments because of their initial, stable, low-Mg calcite composition and their consequent low diagenetic potential. There is certainly some validity to this argument. However, it can also be reasoned that if typical shallow-water carbonate sediments are diagenetically stabilized in near-surface settings and still retain their original porosities, then they should also respond as stable strata with low diagenetic potential during burial. The relative scarcity of pre-Neogene aragonite and high-Mg calcite in platform strata supports the concept of near-complete diagenetic stabilization of shelf limestones before deep burial. Only in shaly sections, impermeable carbonate rocks, concretions, oil-saturated strata, or other unusual settings do unstable carbonate minerals persist in older strata.

Recently published data (Schmoker and Halley, 1982; Halley and Schmoker, 1983) have confirmed that chalks and shallow marine limestones behave similarly during burial. These data (Fig. 3) are based on borehole gravity measurements and interpretation of electric logs (as opposed to perm-plug measurements for the chalks) and thus represent the averaging of porosities of much larger volumes of rock, a significant advantage in vuggy carbonate sections. The data were collected from the more or less continuously subsiding Florida Platform and include Quaternary to Cretaceous strata. These limestones and dolomites have gone through the entire spectrum of near-surface diagenetic processes described by Longman (1980), because they were deposited during times of major eustatic sealevel fluctuations and are currently in an area of deep ground-water circulation. Nevertheless, near-surface (0 to 100 m depth) porosity values average about 40 percent, and values down to about 500 m generally exceed 30 to 35 percent. These values are not really so very different from those of chalks when one considers the differences in average grain size and the expected greater dewatering of chalks in near-surface settings. These data also align reasonably well with curves of porosity versus depth compiled by numerous authors for muddy terrigenous sediments (see review in Brown, 1969, p. 489).

Both chalks and shallow-marine carbonates show consistent and generally smooth patterns of porosity loss with burial (Figs. 2, 3). Rather than showing unusual stability, chalks apparently lose porosity at a faster rate than shallow-water carbonates. This presumably is a consequence of the finer grain size, lack of early cement framework, and virtual absence of dolomite in chalks. The fact that curves of porosity versus depth for individual chalk units are reasonably smooth (that is, there are no discrete steps) indicates that processes which destroy pore space act continuously from the surface to depths in excess of four kilometers. These processes are not tied in any way to near-surface circulation of water but rather relate to the inexorable increases of temperature and pressure with depth. Likewise, the data from platform carbonates of Florida indicate a generally continuous decrease in porosity from surface to depth with only minor discontinuities. These discontinuities may be related to inhomogeneities in primary facies, mineralogy, and early diagenesis, to variations in pore water chemistry as a function of depth, or to strong aquifer flow at intermediate depths.

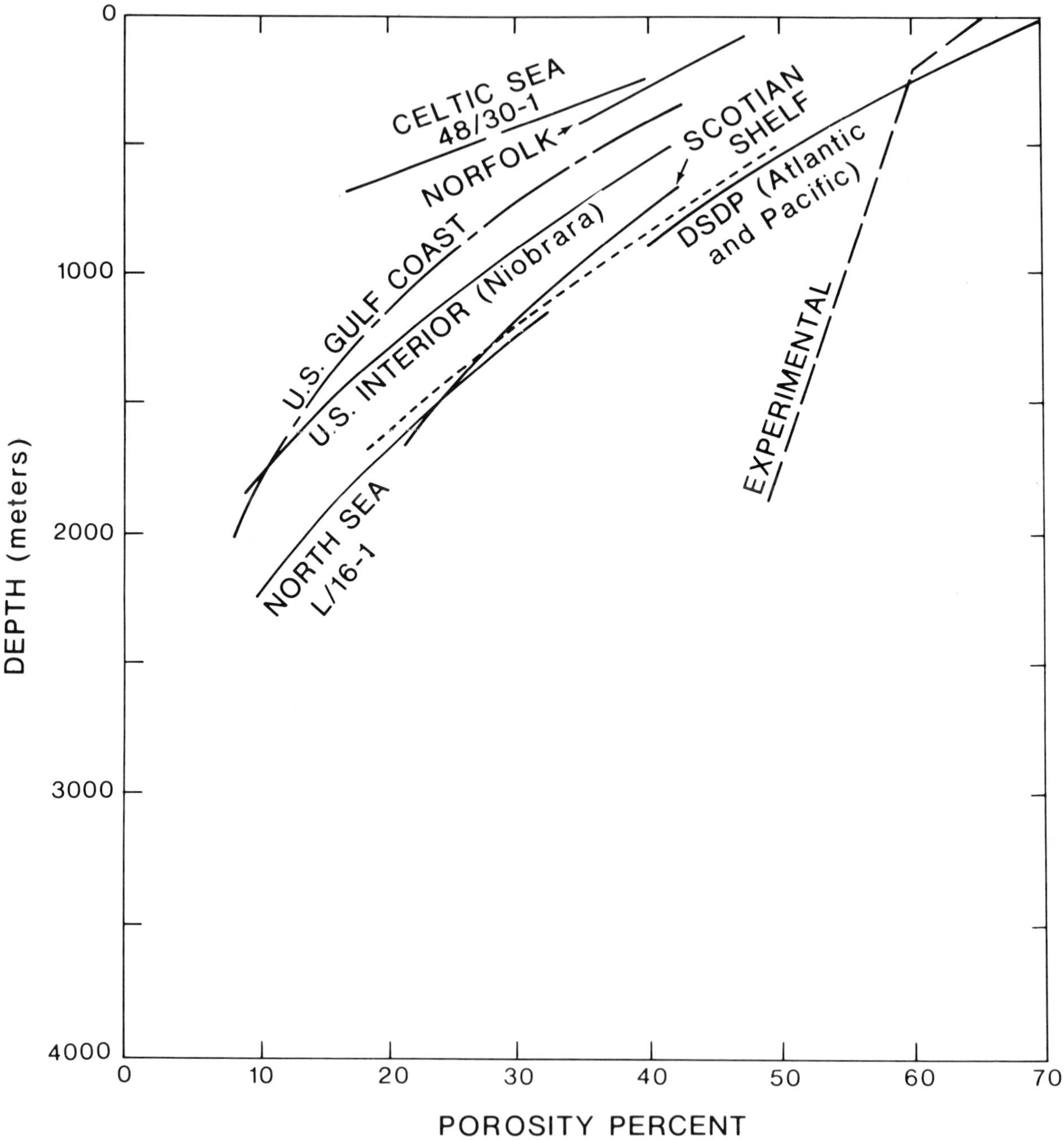

FIG. 2.—Plot of porosity changes as a function of maximum burial depth for onshore chalk sections from North America (Niobrara Limestone) and Europe (Norfolk Chalk, Scotian Shelf, North Sea, and Celtic Sea wells); Deep-Sea Drilling Project (DSDP) holes; and experimental studies (modified from Scholle, 1977). DSDP results from Schlanger and Douglas (1974); experimental work for Hamilton (1959).

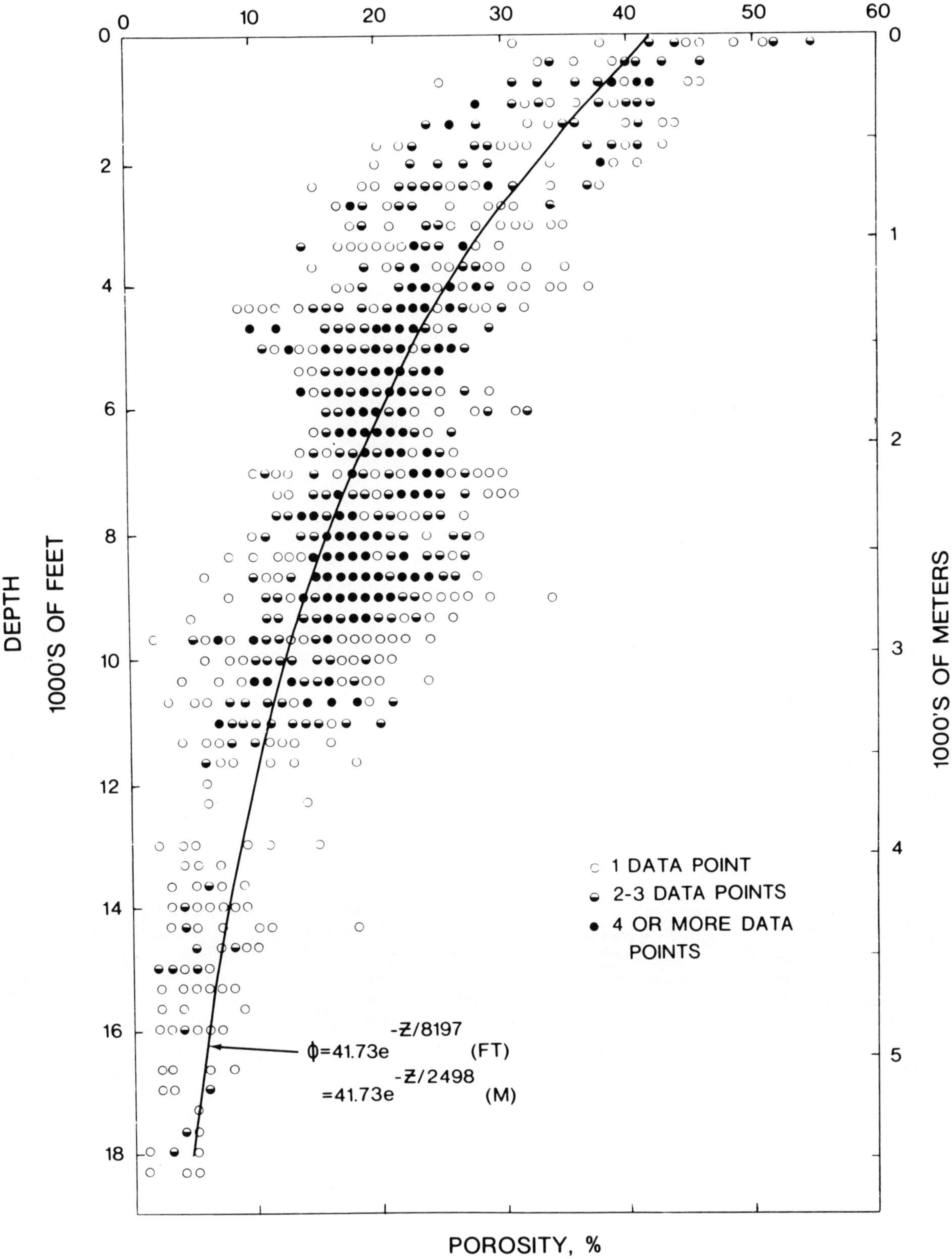

FIG. 3.—Porosity-depth values from 15 wells in south Florida and the least-square exponential fit to all data (from Schmoker and Halley, 1982).

Both chalk and platform carbonate data show rates of porosity loss which decrease with increasing depth but which eventually reach porosity of less than 5 percent (pore space), values typical of ancient carbonate rocks. In chalks, such values generally are reached at depths in excess of three kilometers in sections with marine pore fluids and at one to two kilometer depth in sections with Mg-poor ("fresh") pore fluids (Scholle, 1977; Lockridge and Scholle, 1978). In southern Florida, average porosity values of 5 percent or less were encountered in platform limestones at or just below four kilometers, whereas the average values for platform dolomites at that depth were still above 10 percent (Schmoker and Halley, 1982, Fig. 9).

By presenting this analysis of porosity data, we do not mean to imply that a fundamental or predictive model of porosity-depth relations is yet in hand. The data shown in Figures 2 and 3 were collected from a limited number of relatively young basins, most of which have had a continuous subsidence history with little or no tectonic uplift. Clearly, data exist for low-porosity strata from older basins which were never deeply buried, which were tectonically deformed, or which have had major artesian pore-fluid systems. Furthermore, porosity loss with depth is not irreversible. Several processes (discussed later in this paper) can lead to generation of secondary porosity and can act in deep, basinal settings.

Nevertheless, the porosity-depth data from Figures 2 and 3 indicate that answers to the problem of carbonate cementation in subsiding basins must be sought mainly by gaining an understanding of subsurface processes. Undoubtedly, as our knowledge in this area grows, we will find increasing complexities. Effects of variations in primary depositional facies, carbonate mineralogy, early diagenetic patterns, basin hydrodynamics, geothermal gradients, pressure gradients, organic carbon contents and types, geologic age (time) and many other factors will probably be shown to affect the rates and timing of subsurface porosity loss. Indeed, it may yet be found that variations in early diagenesis can account for significant differences in the response of strata to burial diagenetic effects, thereby confirming a correlation between near-surface diagenetic facies and hydrocarbon reservoirs. Yet even such conclusions will not obviate the fact that near-surface diagenetic phenomena are of significance only in the way they modify deeper-seated processes. Porosity-reducing diagenesis is primarily a subsurface process.

Compaction

Progressive changes of porosity with depth are accompanied by a series of changes in the physical appearance of the strata. These changes include the development of features attributable to both mechanical and chemical compaction. In most instances, mechanical compaction effects—that is, dewatering, grain reorientation, and even grain breakage—dominate the early burial history of carbonate strata. Chemical compaction (solution-reprecipitation) effects such as stylolitization and grain interpenetration generally are most significant later in the diagenetic history. However, in many cases, both processes can be shown to have been active throughout the burial of carbonate strata.

Mechanical compaction.—Dewatering of sediments and mechanical reorientation of grains to produce material with lower porosity is a particularly important process in fine-grained carbonate sediments. Such sediments commonly have initial porosities in excess of 60 percent, with some especially watery sediments ranging from 70 to 80 percent (Richards and Hamilton, 1967; Brown, 1969; Schlanger and Douglas, 1974; Enos and Sawatsky, 1981). Experimental data (Terzaghi, 1940; Hathaway and Robertson, 1961; Fruth *et al.*, 1966; Robertson, 1967; Ebhardt, 1968; Morelock and Bryant, 1971; Shinn *et al.*, 1977; Bhattacharyya and Friedman, 1979; Shinn and Robbin, 1983) and field data (Ginsburg, 1957; Richards and Hamilton, 1967; Schlanger and Douglas, 1974; Enos and Sawatsky, 1981) have shown that extensive loss of porosity can occur with the equivalent of only a few tens to hundreds of meters of burial. Indeed, Ginsburg (1957) and Brown (1969) described decreases in porosities of Florida Bay and Bahamian muds from 80 percent to 65 percent within the "first few inches" of burial; porosities then remain relatively constant to the surface of the Pleistocene. Much of this near-surface porosity loss appears to result from simple dewatering and grain reorientation and repacking as indicated by the lack of chemical changes observed in compaction experiments and the rapid rate at which these significant porosity changes are achieved (Shinn and Robbin, 1983). A common conclusion of these experimental and field studies on carbonate muds (or mud-supported sediments) is that, at porosity values of 30 to 40 percent, mechanical compaction either becomes ineffective or requires very great differential stresses and extensive grain breakage. This holds true at least on the relatively short-time scale of the existing laboratory studies. Presumably at porosity values of about 40 percent, a reasonably strong grain-supported fabric is established, even in micritic sediments. In coarser grained sediments which were grain supported from the time of deposition, physical compaction effects generally are less severe, especially where grains have significant mechanical strength.

Mechanical compaction effects, of course, are not necessarily terminated once a grain-supported fabric is achieved. Where grains are fragile, where little early-diagenetic cement has formed to weld grain contacts, where rates of overburden loading are rapid, or where significant tectonic stresses act on sediment which is only slightly lithified, extensive plastic or brittle grain deformation may take place. The rate of sediment loading appears to be especially important in this context. For example, chalks that show the most extensive flattening of foraminiferal tests or calcispheres generally are those which come from areas of extremely rapid overburden accumulation or significant tectonic deformation. The strongly compacted chalks of Northern Ireland, described by Wolfe (1968), come from an area in which as much as two kilometers of basalt was extruded over the chalk section in only a few million years. Other European chalks which were buried at slower rates, although commonly to greater maximum depths, show far less evidence of mechanical compaction (Scholle, 1977). Similar grain deformation effects and major porosity losses through mechanical compaction were reported by Mimran (1975; 1977) from deformed chalks of southern England. In those studies a direct correlation was established between intensity of deformation and degree of compaction.

In the Western Interior of North America, as well, chalk deposition was widespread during the Late Cretaceous. However, it is mainly in areas of very rapid burial that mechanical compaction effects are strikingly visible. Figure 4 shows a sample of the Niobrara Formation from near Ft. Collins, Colorado, an area where nearly three kilometers of Pierre Shale was laid down atop the Niobrara within a few million years after Niobrara deposition. This area lies along the major boundary faults of the Front Range of the Colorado Rockies, a zone where major normal fault development took place relatively soon after Pierre deposition. Compactional flattening of the microfossil constituents is pronounced, in our experience, mainly in areas which combine rapid sediment loading with later faulting. Such profound crushing of grains presumably is a consequence of application of tectonic or overburden stresses at rates which effectively exceed the ability of the sediment to respond through chemical dissolution and reprecipitation. This may be particularly prevalent where rapid sediment loading resulted in overpressuring, which may have delayed both mechanical and chemical compaction. Where overpressures were released rapidly (through faulting or other mechanisms) the effective rates of differential stress may have been even greater than could have been achieved simply by high initial rates of deposition or deformation.

The absence of grain breakage in most fine-grained carbonate units has led a number of workers to dismiss or downplay the general importance of mechanical compaction (Weller, 1959; Pray, 1960; Zankl, 1969; Bathurst, 1970). Yet experimental work reviewed by Brown (1969) and that of Shinn *et al.* (1977), and Shinn and Robbin (1983) has shown that modern carbonate muds and mud-supported carbonate sands can be compacted rapidly from original porosities of about 70 percent to values less than 40 percent (in some cases to 30 percent, as cited in Bathurst, 1980) without development of strong deformation fabrics. Compactional flattening of burrows and development of wispy, organic-rich seams were noted, but only minor breakage of fossils was observed in these experiments (Shinn *et al.*, 1977; Shinn and Robbin, 1983). Other experiments by Fruth *et al.* (1966) and Bhattacharyya and Friedman (1979) have shown that the higher the content of muddy matrix, the less noticeable are compaction effects. As long as allochems are not in contact with one another they show little or no effect of mechanical deformation, presumably because the micritic matrix responds to stress by minor flowage, providing an essentially nondirectional stress field around the allochems.

Mechanical compaction accounts for between one third and one half the porosity reduction needed to convert mud-supported sediments to ancient limestones. Dewatering, grain reorientation, and minor grain breakage combine to produce a stable framework of interlocking and self-supporting particles capable of resisting additional overburden pressure. Thus, by these mechanisms a significant part of Bathurst's (1975) volume problem in carbonate cementation is eliminated.

Mechanical compaction in coarser, grain-supported rocks is probably less extensive. Because grainstone and packstone sediments have an initially grain-supported fabric, there is little chance for porosity reduction through dewatering. Grain reorientation can sometimes achieve higher packing densities and thus reduce pore space, but in most cases grain deformation (plastic or brittle) is necessary for significant porosity reduction. Most field studies of grain-supported rocks do not (and cannot) distinguish between mechanical and chemical compaction, because these processes tend to work concurrently in grain-supported materials. Nevertheless, these studies commonly note plastic deformation of allochems resulting in planar rather than point contacts between grains and brittle deforma-

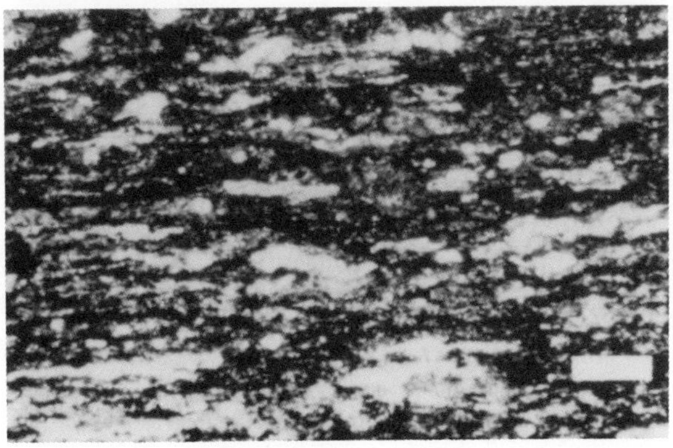

FIG. 4.—Thin-section photomicrograph of the Niobrara Limestone from the Ideal Cement Quarry near Ft. Collins, Colo. Note pervasive compactional flattening of foraminiferal tests. Scale bar is 0.11 mm.

tion of grains (shattering at grain contacts). These features are most readily noted in oolitic strata because the primary grain shapes are uniform and predictable. Observations by Kahle (1966), Coogan (1970), Coogan and Manus (1975), Conley (1977), and many others have demonstrated that plastic deformation and brittle deformation of oolitic grains are both common and produce significant loss of pore space.

Specific studies of mechanical compaction in non-oolitic, grain-supported rocks are scarcer, because mechanical and chemical compaction effects are so closely interrelated. Meyers (1980) and Meyers and Hill (1983) have presented excellent quantitative data on the extent and timing of compaction in Mississippian strata of the southwestern U.S. Interestingly, studies of this same area convinced Pray (1960) that compaction did not occur in Pennsylvanian calcilutites. Meyers and Hill, on the other hand, found that compaction was "a major process of porosity destruction in more than 90 percent of all coarse grainstones and packstones" (1983, p. 231). Compaction, including both mechanical effects and intra- and inter-granular dissolution but not stylolitization, accounted for as much as a 38 percent reduction of pore space, leaving as little as 24 percent pore space to be filled by intergranular cements. Furthermore, stratigraphic evidence indicates "that compaction began under a few tens to a few hundreds of feet of overburden and was completed under less than 6,500 feet (2,000 m) of overburden" (Meyers, 1980, p. 457). Plastic and brittle deformation of bryozoans is quantitatively most important, with breakage of brachiopods and other elongate shells being of lesser significance. Bioherms and strata which had been extensively cemented in marine phreatic environments or in meteoric waters escaped mechanical compaction (Meyers, 1978; 1980). Strata which had been exposed to "microetching" below subaerial exposure surfaces showed the most intense crushing. It should be noted, however, that mechanical compaction was considered to be less important than chemical compaction in bringing about pore space reduction in these Mississippian units (Meyers, 1980).

Evidence for physical compaction in shallow-water carbonates has been used as an exploration tool by Beach and Schumacher (1982). In this work, uncompacted island sediments forming part of Stanley field in the Williston basin were recognized to be surrounded by compacted marine carbonates. Beach and Schumacher (1982) concluded that compaction was important in the development of Stanley field and influenced the reservoir in three ways. First, only marine sediments were mechanically compacted, thus isopach maps of the upper Mission Canyon and overlying Charles Formations reflect distribution of depositional facies in the upper Mission Canyon. Second, compaction implies lack of early cementation, thus suggesting why primary porosity is preserved in the reservoir facies. Third, compaction reduced interparticle pore volume in marine sections, thus reducing porosity in marine reservoir facies and improving the effectiveness of the updip trap. To our knowledge, this is the first direct application of mechanical compaction in shallow-water carbonates to an exploration play.

In the fore-reef facies of the Permian (Guadalupian) Capitan Formation we have observed major mechanical compaction effects. At the top of the slope, all sediment was extensively bound by organic coatings and massive marine phreatic cements; in this zone there is virtually no evidence of mechanical compaction. In the middle slope environment, marine cements are largely lacking, and coarse grainstones display extensive shattering of grains at or near grain contacts (Fig. 5). These fractured grains are cemented by coarse, equant crystals of clear, "late" calcite. In lower slope settings, finer grained sediments were deposited and both mechanical and chemical compaction features are common. Thus, the nature and extent of compaction features depends, in part, on the amount and fabric of the early diagenetic cement framework, the grain size and packing of the sediment, and the amount and rate of overburden addition.

Chemical compaction.—Chemical compaction, solution transfer, or pressure solution is the second major factor in subsurface porosity reduction. Chemical solution phenomena in limestones have been studied for many years (see excellent historical discussion in Bathurst, 1975, p. 468–473), but the specific term "chemical compaction" was introduced by Lloyd (1977) to describe porosity loss in limestones which takes place without the addition of new or allochthonous carbonate material. In the extreme, chemical compaction can yield a completely cemented rock with no introduction of externally-derived cement. All that is required is the expulsion of existing pore fluids

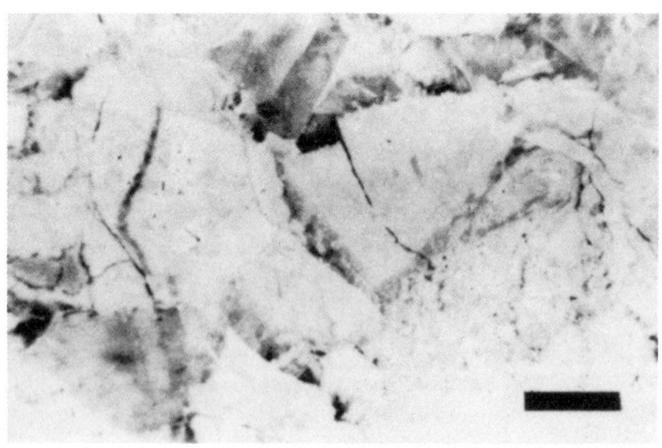

FIG. 5.—Polished slab of Upper Permian (Guadalupian) Capitan fore-reef limestone from McKittrick Canyon, Tex. Sample comes from about 600 ft below the crest of the reef, below the zone of intense penecontemporaneous submarine cementation, and thus shows extensive grain fracturing and compaction due to rapid burial. Scale bar is 1.5 cm.

in order to accomodate the volume of cement. The theoretical basis for such chemical compaction has been reviewed by Durney (1972), Bathurst (1975), de-Boer (1977), and Robin (1978).

Quantitative estimates of the importance of pressure solution really started with the detailed studies of stylolites by Stockdale (1922; 1926) and were greatly advanced by the studies of Dunnington (1967) and Mossop (1972). The importance of non-stylolitic pressure solution phenomena has been recognized in recent years, however, and has led to major improvements in our ability to evaluate the magnitude of chemical compaction.

Pressure solution features have been classified by numerous workers in recent years (Park and Schot, 1968a, 1968b; Trurnit, 1968; Mossop, 1972; Logan and Semeniuk, 1976; Garrison and Kennedy, 1977; Wanless, 1979; Buxton and Sibley, 1981). Three types of chemical compaction features are common, however, to all carbonate rocks: intra- or inter-granular solution features, solution seams, and stylolites. Intra- and inter-granular solution features are found at contacts between grains and at sites within grains where lithostatic stresses are concentrated. These features may be only weakly developed or may eventually lead to the "fitted fabrics" of Buxton and Sibley (1981), in which essentially all the grains in a rock have irregular to linear sutured contacts. Such fabrics apparently form during early stages of chemical dissolution in porous grainstones or in packstone and wackestone fabrics where there is partial matrix support. Solution seams (also termed horsetail groups, residual seams, wispy stylolites, clay seams, microstylolites, or non-sutured seams) represent discrete surfaces of dissolution which extend continuously across many grains and generally form in fine-grained limestones and clay-bearing carbonate rocks. Solution seams are commonly associated with lumpy or nodular fabrics (stylonodular fabric of Logan and Semeniuk, 1976; flaser structure of Garrison and Kennedy, 1977; and fitted nodular fabric of Wanless, 1979). Stylolites are similar to solution seams but have higher amplitudes and generally form in coarser-grained carbonate rocks and/or at lithologic discontinuities in carbonate sections (e.g., Elliott, 1982). Stylolites apparently can form throughout burial, but large-amplitude throughgoing stylolites are commonly formed during late stages of diagenesis but before complete loss of rock porosity (e.g., Dunnington, 1954; Gillett, 1983; Bathurst, 1980). All types of solution surfaces are marked by concentrations of dark, insoluble material (clay, organic matter, iron oxides, etc.).

Qualitative studies showing the wide distribution of chemical compaction features are too numerous to attempt any systematic listing. As even casual visitors to marble-walled lavatories can attest, stylolites are everywhere. Quantitative studies of precisely how much porosity loss is associated with chemical compaction are less common, however. Such studies are done by measuring and totalling the amplitudes of stylolites and other solution features, or by measuring the amount of insoluble residue accumulated along stylolite surfaces. Several factors make such quantitative studies difficult. First, measurement of stylolite amplitudes (or solution seam thicknesses) provides only a minimum estimate of the amount of material removed (Stockdale, 1926; Mossop, 1972). Some estimates use the amount of insoluble material at a solution contact and an average insoluble content for the whole rock to derive the thickness of removed carbonate, but these estimates must assume an initial homogeneity for the rock which is doubtful. Weyl (1959) and Oldershaw and Scoffin (1967) confirmed by observation that stylolites form preferentially in clay-rich layers. Thus, estimates based on measurements of insoluble residues at solution contacts must be considered to be unreliable or maximum values. These values are especially unreliable in cases where the "insoluble" material which accumulates along solution surfaces (the stylocumulate of Logan and Semeniuk, 1976) is not truly insoluble but merely less soluble than the associated limestone. "The irregular distribution of stylocumulate along pressure-solution surfaces and the textural evidence of corrosion of quartz by carbonate idens, suggests that relatively insoluble stylocumulate also is dissolved under appropriate chemical conditions" (Logan and Semeniuk, 1976, p. 137). Finally, the interrelationship of mechanical and chemical compaction makes distinguishing between them difficult in many cases. The distinction is an important one, however, because mechanical compaction simply moves grains closer together, increasing bulk density and reducing porosity. Chemical compaction, on the other hand, not only reduces porosity to some degree by changing packing density (see Manus and Coogan, 1974) but also may reduce porosity by liberating dissolved $CaCO_3$ for local cementation at sites of lesser stress, generally in inter- or intra-granular pore spaces (solution-reprecipitation or solution transfer).

Dunnington (1967) has provided some very interesting and careful observations on chemical compaction of carbonate rocks, primarily from the Middle East. He reported that solution features formed as a consequence of both overburden and tectonic stresses, and that stylolites parallel to bedding (overburden related) resulted in thinning of carbonate sections by as much as 40 percent, with 20 to 35 percent reductions being common. "Volume reductions of up to 35 to 42 percent having resulted from stylolitization" were reported by Park and Schot (1968a, p. 72). Coogan (1970) compiled extensive data from North American oolitic grainstones and concluded that as much as 90 percent (more commonly 50 to 70 percent) of initial porosity had been destroyed by compaction (both mechanical and chemical) in the Jurassic Smackover Formation at approximately three kilometers depth in the

Gulf Coast. Slightly lower values were obtained for deeply buried Mississippian limestones of the Appalachian Basin, whereas the relatively shallowly buried Ste. Genevieve Formation of Indiana showed only about 10 percent compaction. Mossop (1972) demonstrated as much as 24 percent volume reduction through pressure solution in lagoonal strata of the Upper Devonian Redwater atoll of Alberta. Atoll margin carbonates (which were coarser grained and may have had early marine phreatic cements) were compacted by an average of 13 percent. Finally, as noted earlier, Meyers and Hill (1983) have measured as much as 38 percent porosity reduction (from about 42 percent to 27 percent pore space) in Mississippian shelf carbonates of southern New Mexico.

The data clearly show that compaction is effective in reducing pore space in carbonate strata. Bulk volume reduction coupled with pressure solution-derived carbonate cementation can account for the entire change in porosity during the transition from sediment to indurated rock in many sections. The more pertinent and less answerable question, however, is "How universal is this process?" To answer this question, it is necessary to evaluate available evidence on factors which control the rate and extent of chemical compaction.

Controls on compaction.—The amount of over burden needed to produce significant pressure solution is poorly understood. Dunnington (1967) quoted a figure of 2,000 to 3,000 feet (600 to 900 m) as the depth at which pressure solution became "effective". Stratigraphic reconstructions by Buxton and Sibley (1981) allow the conclusion that extensive chemical compaction has taken place in Devonian limestones of Michigan at burial depths of no more than 1,500 m. Depths of "tens to hundreds of meters" were determined by Meyers and Hill (1983, p. 231) to have been adequate for the initiation of chemical compaction in areas with non-marine pore fluids. Similarly shallow depths of formation have been noted in a variety of sources quoted by Bathurst (1975, p. 473). Beall and Fischer (1969) observed that the transition from predominantly mechanical to chemical compaction occurred at about 250 m in North Atlantic DSDP cores. Finally, Neugebauer (1973, 1974) presented experimental data which showed that under the least favorable conditions solution transfer was ineffective at depths less than 300 m, was of only minor significance at depths less than 1000 m, but became extremely important at depths between 2,000 and 4,000 m.

Water chemistry, as noted earlier in this paper, affects both timing and rates of solution transfer processes. Theoretical and experimental work (Neugebauer, 1973; 1974) and field studies (Scholle, 1977; Meyers and Hill, 1983) all indicate that meteoric (Mg-poor) waters accelerate the effects of pressure solution when compared with marine or marine-derived pore fluids. In this context, it is worth noting that basins typically go through at least two phases. By definition, the early history is dominated by subsidence. During this phase, pore waters are dominantly marine-derived and sediments typically follow the porosity-depth curves shown in Figures 2 and 3. Many basins of the world, however, particularly pre-middle-Mesozoic ones, have undergone deformation and/or uplift of the basin margins. In this situation, a completely different hydrologic system typically will evolve. This system involves head-driven artesian fresh water being injected from the uplifted margins towards the basin center. Clearly, this has a major effect on the burial diagenesis of carbonate units for the burial depth vs. porosity curves of Figures 2 and 3. These curves are unrealistic for most, if not all, basins which have such a multi-stage history. These burial depth-porosity curves remain important, however, because they are applicable to the early development of all basins.

There are other effects of pore-fluid composition on burial diagenesis. The exclusion of all waters from the pore space or the prevention of contact between carbonate grains and pore fluids (as in oil-wet reservoirs or in strata in which hydrophobic organic compounds coat grains) greatly reduces stylolite formation, as noted by Dunnington (1967) in platform carbonates and by Scholle (1977) for basinal chalks. Early introduction of oil or gas into the pores can prevent solution transfer entirely, because chemical compaction requires a water phase for the removal of ions. Thus, stylolites may be common in the water zones but absent in the gas and oil zones of a reservoir (Dunnington, 1967; Feazel and Schatzinger, 1983). Carbonate reservoirs which were filled with hydrocarbons before deep burial may preserve porosity while the surrounding carbonate units are thinned and cemented by chemical compaction.

Pore fluid pressures can play an obvious role in affecting the rates of both mechanical and chemical compaction. Mechanical compaction clearly depends on differential stresses to bring about bulk volume reduction. Chemical compaction is just as dependent on differential stresses, however. Because stressed minerals are generally more soluble than unstressed ones, dissolution takes place at sites of maximum stress such as grain contacts (Riecke Principle). As hydrostatic pressures begin to approach lithostatic pressures, differential stresses are reduced, regardless of burial depth. Thus, in overpressured settings, chemical compaction can be greatly retarded. This has been demonstrated in the North Sea (Harper and Shaw, 1974; Scholle, 1977), where chalk porosities as high as 42 percent have been preserved at burial depths in excess of 3,000 m in areas of strong overpressuring. The differential stresses and compaction pressures on these chalks are like those at only about 1,000 m of burial, despite their much greater actual depth. The rest of the stresses are borne by the non-directional hydrostatic pressure.

The related effects of time and temperature variations on compaction are much more poorly understood. Studies on chalks have not shown a major effect of variations in regional geothermal gradients on rates of porosity loss with depth. In the North Sea, chalks are subject to geothermal gradients 1.5 to 2 times as high as those for equivalent age chalks in the U.S. Western Interior. Yet the porosity-depth curves for the two areas (where they are normally pressured) are virtually identical (Lockridge and Scholle, 1978).

On the other hand, continuing studies by J.W. Schmoker (oral comm., 1983) indicate that there is a consistent correlation between porosity-depth curves and time/temperature index in several basins he has examined. In these studies, older units and ones subjected to higher temperatures had lower porosities than otherwise similar strata. In experimental studies as well, elevated temperatures resulted in significantly increased rates of compaction (Ebhardt, 1968). However, data isolating the specific effects of either time or temperature are sparse. It is virtually impossible to find basins of early Mesozoic or older age which have remained under constant burial and temperature conditions and in which one can eliminate the possible effects of meteoric pore fluids. As with many chemical reactions, the effects of time and temperature can substitute for one another in a complex fashion. That is, application of low temperatures for a long time may result in the same degree of compaction as high temperatures applied for a short time, all other conditions being equal.

Primary depositional facies and early diagenetic alteration can also play a significant role in controlling rates of chemical compaction. Grain size and sorting, grain types, mineralogy, and bedding (or lithologic contrast) features inherited from depositional environments will affect the relative importance of mechanical versus chemical compaction as well as the rates of porosity loss with burial.

Enos and Sawatsky (1981) showed that porosity and permeability are negatively correlated in modern carbonate sediments. The highest porosities and the lowest permeabilities are found in carbonate muds, whereas the reverse is true for carbonate sands. In most ancient rock sequences this porosity-permeability relationship is reversed. That is, the most porous ancient limestones are usually the most permeable ones. Part of the explanation for this reversal is the tendency for fine-grained carbonate rocks to become cemented during burial diagenesis more quickly than their coarser-grained counterparts. Although the reasons for this selective cementation are not fully understood, they presumably involve effects related to grain size, such as surface area and reactivity. In addition, clay minerals and organic matter in finer-grained carbonate strata accelerate pressure solution and porosity loss. The result is that, as a crude approximation, those limestones which contained the largest pores and had the highest permeability originally tend to be the ones which retain porosity and permeability in the subsurface. Thus, reefs, carbonate sand bodies, talus deposits, and other originally coarse-grained facies tend to retain porosity preferentially in the subsurface. Finer-grained facies such as lagoonal or basinal muds, on the other hand, preferentially lose porosity during burial diagenesis. Furthermore, the nature or style of burial diagenesis can be influenced by the primary sediment composition and texture. Buxton and Sibley (1981) and Elliott (1982), for example, demonstrated that the style of chemical compaction (intergranular, solution seam, or stylolite dissolution) varies to a considerable degree with the starting lithology.

Effects of variations in early diagenesis on the style and magnitude of burial diagenesis are more difficult to evaluate. Small amounts of early cementation, as shown in Figure 6, are incapable of preventing intergranular dissolution (see also the discussion in Bathurst, 1975, p. 465–466). Nevertheless, variations in the nature and intensity of early dissolution and cementation must be expected to exert some influence on the rate and timing of burial diagenesis (e.g., Meyers and Hill, 1983; Heckel, 1983). Indeed, much of the variance in the porosity at any given depth (Fig. 3) is probably attributable to such early diagenetic variations and their effects on burial diagenesis.

Little is known about the detailed response of various early cement fabrics to burial stresses. However, any primary sedimentologic or early diagenetic property of a limestone which retards late mechanical or chemical compaction (or burial cementation) will help to preserve porosity in the subsurface. Purser (1978) noted the correlation of carbonate reservoir rocks and evidence for early diagenesis. This led him to conclude that early lithification tends to retard or prevent subsequent chemical compaction and thereby reduces the volume of calcium carbonate released for cementation.

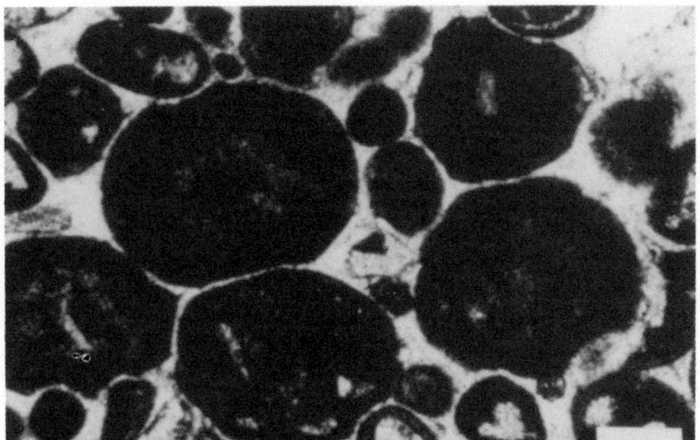

FIG. 6.—Thin-section photomicrograph of the Upper Mississippian Ste. Genevieve Limestone from Tennessee illustrating chemical compaction postdating cementation. Thin rims of submarine cement surround grains, yet ooids show extensive grain interpenetration due to pressure solution. Scale bar is 0.38 mm long.

A similar conclusion had been reached by Becher and Moore (1976) for a Jurassic oolitic reservoir in Arkansas.

Although the particular mechanisms which retard later porosity loss are not known, four consequences of early fresh-water alteration are well documented: (1) mineralogical stabilization; (2) lithification; (3) decreased permeability; and (4) preserved porosity. These factors presumably retard subsurface diagenesis in the following ways: (1) mineralogical stabilization decreases the average grain solubility and thus the potential for chemical compaction of these sediments; (2) lithification increases the resistance to mechanical compaction; and (3) low permeability decreases water flow through the rock, thus limiting the amount of allochthonous cement introduced.

The influence of early diagenesis on late diagenesis appears to be a fruitful topic for future experimental and field research. The literature is filled with papers claiming that porosity is preserved in (and hydrocarbon production obtained from) rocks cemented, leached, or diagenetically stabilized in one environment or another. However, if the data in Figure 3 are to be believed, what makes deposits from one early diagenetic environment untimately "favorable" for oil production as compared with those from another environment is how these early diagenetic fabrics affect the response of the strata during burial. In near-surface settings nearly all strata have adequate porosity and permeability to act as hydrocarbon reservoirs. By the time these units have been buried to depths at which abundant hydrocarbons are generated, however, only a small fraction still have characteristics of reservoir rocks. We do not yet know how near-surface variations in porosity (and perhaps even more importantly, permeability) are translated into the deep subsurface. It may be a "you can pay me now or you can pay me later" situation. That is, strata with significant early cementation and low near-surface porosities may be more resistant to mechanical and chemical compaction than strata which retained high porosities during near-surface alteration. Thus, porosities of these two types may converge or even cross at some depth.

An example of this is provided by the data on dolomites from Schmoker and Halley's (1982) study of Florida Platform carbonates (shown here in Fig. 7). Dolomites have lower porosities than age- and depth-equivalent limestones in near-surface sections. Porosities of the two rock types are similar at about 1,700 m, and at greater depths limestones become progressively less porous than dolomites. Thus, the prevalence of porous hydrocarbon reservoirs in dolomites apparently results less from the dolomitization process or the diagenetic environment of dolomite formation and more from the greater resistance of dolomites to burial diagenetic loss of pore space when compared with limestones. With further study, we will likely find that the limestone data set (Fig. 3) can be broken down into subsets by depositional and early diagenetic environments and that these subsets will show separate and distinct trends.

As noted earlier, the presence of clay minerals can significantly enhance formation of pressure solution features. Many carbonate strata, especially those deposited in outer shelf to abyssal environments, originally have a strong primary bedding rhythmicity consisting of couplets of limestone and marl (e.g., Fischer, 1980; Scholle *et al.*, 1983). These rhythmic beds range in thickness from a few millimeters to as much as a meter and represent between 20,000 and 100,000 years of deposition (Gilbert, 1900; Fischer, 1980). Superficially similar cycles are found in many platform carbonate sections (see papers in Merriam, 1964, and Einsele and Seilacher, 1982, for example).

These calcareous shale or marl interbeds in otherwise pure carbonate sequences are significant because they commonly are sites of intense pressure dissolution. In young pelagic sediments or in shallowly buried sections of older, rhythmically-bedded carbonate strata, carbonate contents of these shaly intervals commonly exceed 50 or 60 percent. Yet in more deeply buried sections, pressure solution may have removed most or all of this carbonate. The marly intervals of chalk sections generally show a progressive increase in the number of wispy solution seams and a progressive decrease in thickness and carbonate content with increasing depth of burial. Complicating geologic factors have never allowed a quantitative evaluation of pressure dissolution, but qualitative observations (e.g., Scholle, 1977) on sections throughout the world indicate that the process is ubiquitous and volumetrically important, at least in pelagic limestones. The stratigraphic effects of this process have been modeled by Hallam (1964) and Arthur (1979), and the results are shown graphically in Figure 8. Basically, there is a progressive "unmixing" of lithologies with a net export of carbonate material dissolved along solution seams from the marly layers (donor beds of Bathurst, 1975). The material derived from these zones is presumably added to adjacent carbonate layers as pore-filling cement (discussed further in the next section). Eventually the material becomes segregated into interbedded pure shales and carbonates. This process acts not only in thin shale interbeds in carbonate sections but also in thick sections of calcareous shales. Perry and Hower (1970), for example, demonstrated significant loss of $CaCO_3$ (shown as CaO) from Tertiary shales in the U.S. Gulf Coast. These shales have CaO contents in excess of 10 percent at depths of about 7,000 ft (2,100 m); by 12,000 ft (3,600 m) the average CaO value has decreased to about 2 percent, suggesting massive export of calcium carbonate.

There are several major consequences to the processes noted above. First, pressure solution can cause significant changes in lithology during burial. Interpretations of paleo-compensation-depths (e.g., Hesse

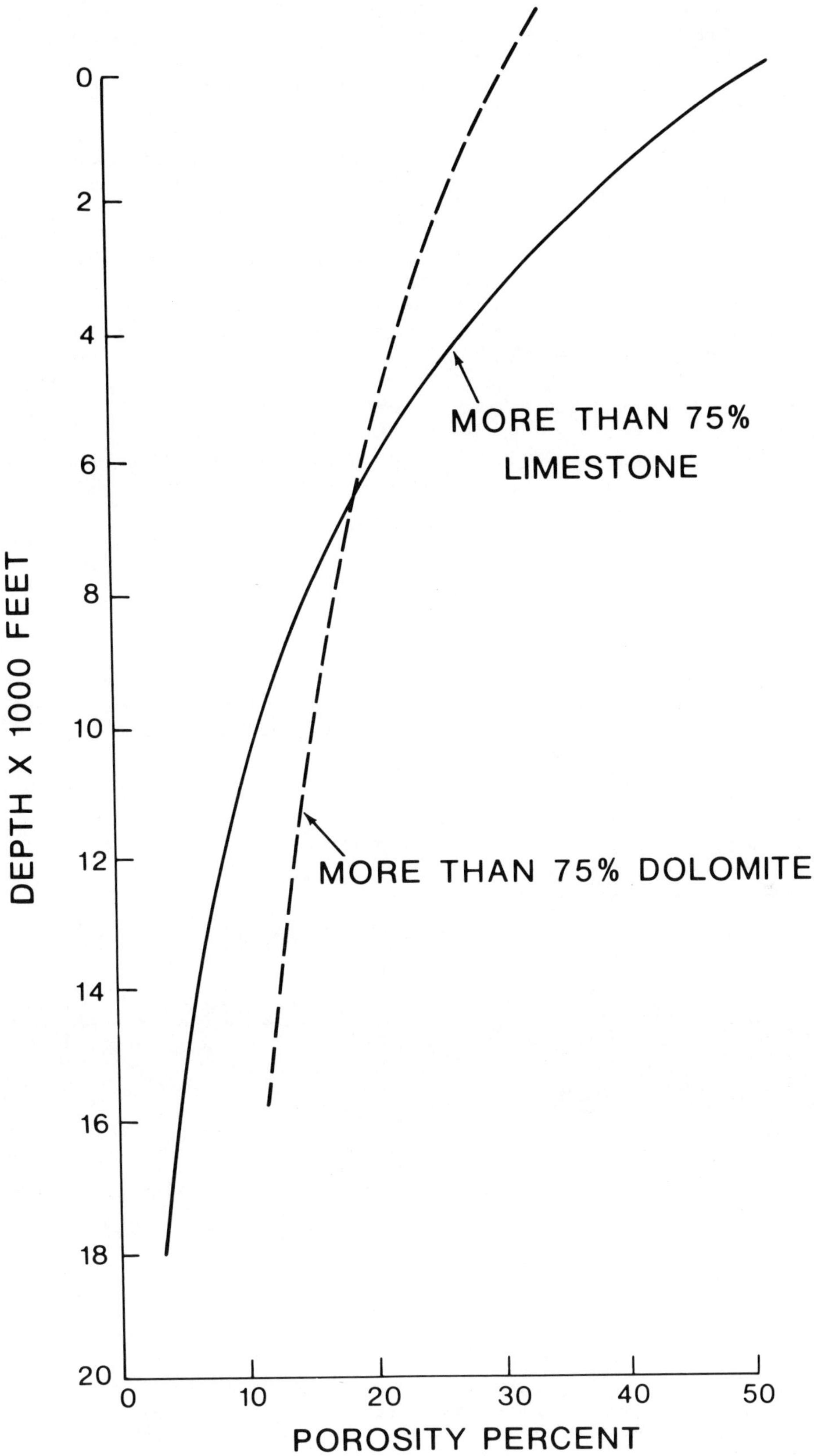

FIG. 7.—Average porosity versus depth for dolomite and limestone in south Florida (from Schmoker and Halley, 1982).

and Butt, 1976; Scholle, 1971) commonly are based on the presence or absence of carbonate material in hemipelagic interlayers between carbonate or clastic turbidites. These relationships clearly can be altered by complete diagenetic removal of carbonate minerals from the hemipelagic interlayers, making environmental interpretations difficult in deeply buried or tectonically deformed sections. Second, the amount of pressure solution in a carbonate-shale section may be seriously underestimated if the shales have been completely purged of their carbonate components. Because of coalescence, individual solution seams may no longer be recognizable, and yet significant amounts of carbonate may have been liberated to act as porosity-reducing cement in associated carbonate beds. Thus, although carbonate rocks in the section may show little or no sign of physical or chemical compaction, associated thin shale interbeds may have been sites of major dissolution, stratigraphic thinning and carbonate export.

It should be clear from the above discussion that we are not now (and indeed we may never be) in a position to prove the quantitative significance of pressure solution and solution transfer as mechanisms for subsurface reduction of porosity. The inability of mechanical compaction alone to account for widely observed and previously discussed porosity loss with increased burial depths, the abundance of pressure solution features, and the apparent impossibility of moving hundreds of thousands of pore volumes of water through most basins in order to introduce externally-derived cement indicate that pressure solution is the predominant porosity-reducing mechanism in burial diagenesis. But the obvious inability to obtain quantitative data in most sections, the fact that much of the dissolution can be taken up in shaly layers of unknown primary carbonate content, and the fundamental difficulty in assessing the world from limited and small-scale data all point toward a permanent inability to provide absolute proof of the efficacy of pressure solution and solution transfer.

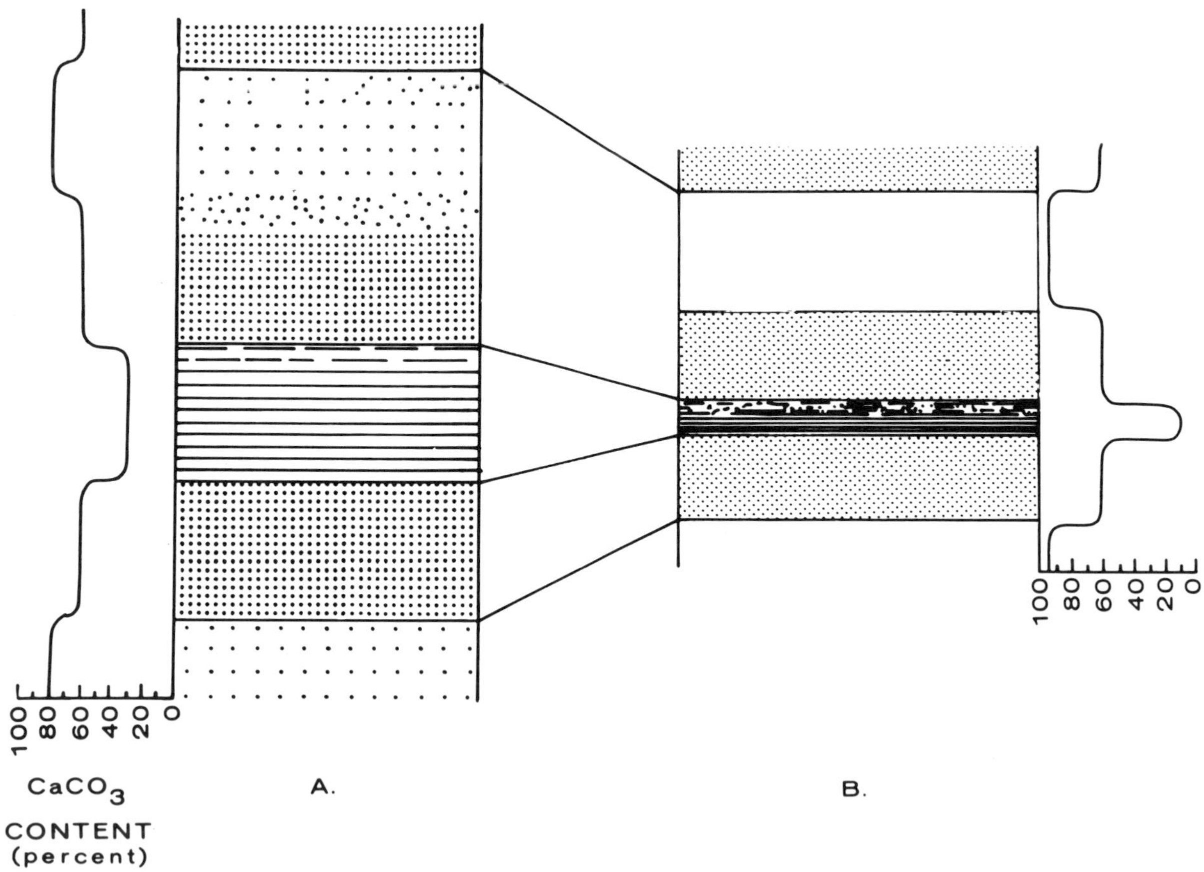

FIG. 8.—Diagrammatic section of typical rhythmically bedded carbonate-shale strata before burial (A) and after burial (B). Note thinning of section due to porosity loss, and note progressive loss of calcium carbonate from shalier units (lined) and increase of carbonate (added calcite cement) in originally more calcareous layers (stippled). Modified from Hallam (1964) and Arthur (1979).

Large-Scale Isotopic Data

As noted earlier, isotopic analyses of cements can provide specific evidence for the temperature and timing of burial cementation. Examination of the entire assemblage of isotopic analyses from carbonate rocks of diverse ages and locations can provide an additional perspective on diagenesis, however. Keith and Weber (1964) and Hudson (1975) noted the fact that most Phanerozoic marine limestones "have the same range of ^{13}C as modern marine shells and carbonate sediments, around 0 per mil on the PDB scale." (Hudson, 1975, p. 20).

Considerable data have now been gathered to show that near-surface meteoric diagenesis typically results in a significant enrichment of ^{12}C in limestones. In some cases, carbon isotopic shifts of as much as 6 per mil have been noted in association with soil zones and vadose zones (Gross, 1964; Hudson, 1975, Allan and Matthews, 1982), although this effect decreases rapidly with depth. Thus, Hudson (1975, p. 22) concluded that the vast majority of ancient limestones had not been significantly affected by near-surface vadose cementation but rather were "cemented mainly during the compactional phase of burial diagenesis. . . The source of this cement is in other limestones or in shales of the sedimentary basin; it is released by pressure solution along stylolites and microstylolites and by solution of aragonite shells."

Although this approach outlined above is useful in demonstrating the overall importance of burial cementation, it only shows that most cementation is accomplished without the introduction of large amounts of CO_2 from external sources. It does not distinguish between possible internal sources of carbonate cement. However, in recent years far more numerous carbon and oxygen isotopic analyses of bulk limestones have been completed on carefully measured stratigraphic sections (Scholle and Arthur, 1980). These data show that there have been major variations in the carbon isotopic composition of sea water through time. These variations are reflected in correlatable changes in the carbon isotopic signature of pelagic limestones. From a diagenetic viewpoint, what is significant, however, is that these isotopic variations occur over intervals of centimeters to meters and are essentially equally well represented in porous and non-porous sections. Figure 9, for example, shows a comparison of four chalk sections of equivalent (Late Cretaceous) age. Although the average porosity values for these sections range from less than 1 percent (Peregrina Canyon, Mexico) to more than 35 percent (Norfolk, England), the isotopic signatures are virtually identical in shape and intensity. Because these two- to three-per mil variations in isotopic composition take place over distances of meters to a few tens of meters and because the variations remain unaffected by diagenetic porosity loss, there is a strong implication that the sources of cement are not only internal but are probably nearby. That is, cements in the carbonate beds most likely are derived from adjacent or very nearby marl-shale layers. If significant volumes of cement had been introduced from outside the local system, this would have resulted in a damping of the primary fluctuations in the well-cemented sections.

Although these carbon isotopic variations are not identical for every section measured, they are sufficiently uniform that one can generalize that in most cases studied to date burial cements are of very local derivation. The fact that there are exceptions (such as the Gubbio section discussed in Scholle and Arthur, 1980) indicates that this isotopically-based approach may be used more fully in the future to refine our understanding of differences in styles of cementation in different basins.

Cementation

Significant loss of porosity occurs during grain-contact dissolution alone, or in most cases, during redeposition of some or most of the removed material as cement. Cement redeposition is particularly important where the porosity is below about 20 percent (Manus and Coogan, 1974; Mitra and Beard, 1980). A number of authors have noted pressure-solution features are often near zones of low porosity (Dunnington, 1967; Harms and Choquette, 1965; Nelson, 1981). A particularly well-documented example from the Devonian of Canada described by Wong and Oldershaw (1981) demonstrates not only the spatial link between pressure dissolution and cementation, but also documents the timing between stylolite growth and cement precipitation using zoned cements to illustrate temporal relationships. However, there are very few examples in which the logical links between pressure solution, cementation, and porosity loss can be demonstrated. Pressure solution features are widespread in many rocks, as is evidence of late cementation, but connecting the two processes in most instances requires a certain leap of faith.

Part of the reason for these difficulties is the rather poor understanding of the controls governing the dissolution and reprecipitation of carbonate minerals in the subsurface. The chemistry of these reactions is exceedingly complex. The reactions typically occur at elevated temperatures in water of widely varying composition and typically proceed very slowly over geologically significant periods of time. Likewise, the effects of grain size, fabric, mineralogy, pore fluid composition, and many other variables are known in only a general manner. For example, Neugebauer (1973) has suggested that magnesium in solution, because it interferes with the precipitation of calcite, retards the effects of pressure solution in chalks. In contrast, experimental studies by Baker *et al.* (1980), who studied the pressure solution and hydrothermal recrystallization of fine-grained carbonates, showed that clay, diatomite, and basaltic glass inhibited stabilizing reactions. Baker *et al.* (1980) found that variations in calcium, magnesium, alkalinity, and pH of the reacting solutions have only a small effect on the rates of alteration. Interestingly, Baker *et al.* (1980) also found that the presence of clay inhibited pressure dissolution, a result which appears opposite that noted by many earlier workers.

Land (1980) attempted to model changes in ^{18}O, ^{13}C, and strontium in deep-sea carbonates during burial diagenesis. He found particular difficulty modeling the strontium content of deep-sea oozes. In particular, he noted a lack of knowledge of the factors that control values of distribution coefficients. This lack of knowledge was partly remedied by Lorens (1981), who described a rather strong effect of precipitation rate on the distribution coefficient of strontium in calcite. However, Mucci and Morse (1983), monitoring experiments that lasted from two to twenty-four hours, found that the concentrations of magnesium and strontium in calcite were independent of precipitation rate. The question of factors affecting distribution coefficients remains unresolved. Nevertheless, Baker *et al.* (1982) successfully modeled the distribution of strontium in pore water and sediments; in their model, lithification is essentially isochemical, and the cement is derived from within the formation. Advective flow was unnecessary to explain its lithification. To explain the magnesium distribution of lithified calcite, however, required a distribution coefficient many times smaller than previously published values.

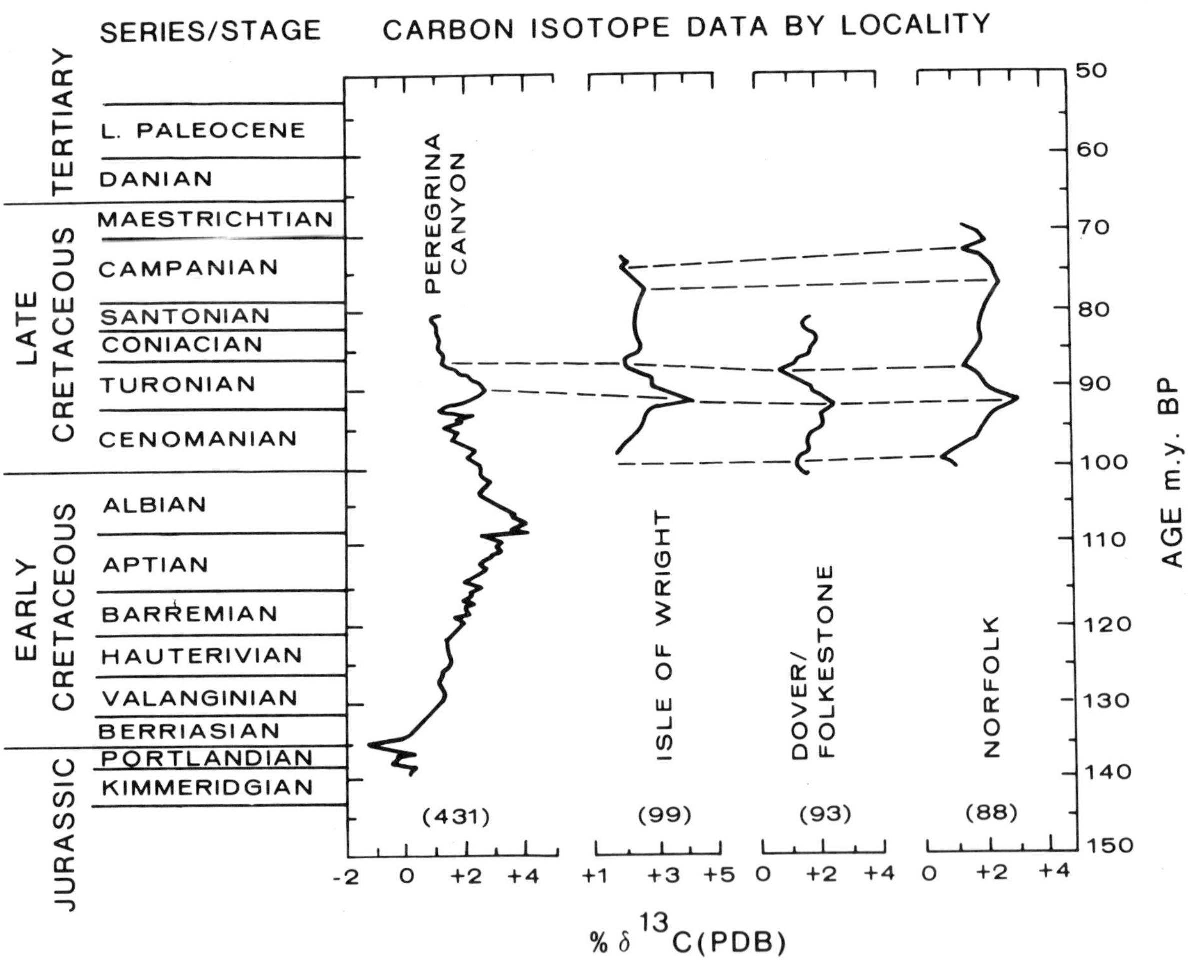

FIG. 9.—Carbon isotopic ratios of whole-rock samples of pelagic limestone from 4 localities (Peregrina Canyon, Mexico; and the Isle of Wight, Dover-Folkestone, and Norfolk, England). Porosities for these units (all cemented during burial) are less than 1 percent for Peregrina Canyon; 10–20 percent for the Isle of Wight; 25–40 percent for Dover-Folkestone; and 35–45 percent for Norfolk. Despite these great variations in porosity, the details of the isotopic cycles are virtually identical in shape and magnitude for all 4 sections, indicating that little or no cement was introduced from outside; rather, cements were apparently derived on a very local scale. Modified from Scholle and Arthur (1980).

Despite our relatively rudimentary understanding of subsurface cementation and the difficulties of cementation models, our ability to recognize burial cements has increased significantly during recent years. The increased use of special petrographic and geochemical methods is adding to a growing list of well documented examples of burial cementation.

CRITERIA FOR RECOGNITION OF BURIAL CEMENTS

The recognition of burial cements hinges on the identification of petrographic relationships which suggest precipitation after other recognizable events or processes in the burial history of a given rock. Burial cements post-date early cements and compaction features, and may post-date stylolites and fractures. Although these relationships do not necessarily establish a depth of burial for cements, or perhaps any burial in special cases, they commonly indicate precipitation some time after the onset of diagenesis and after some burial.

If pore spaces are sufficiently large, burial cements are commonly coarse grained and poikilotopic, because they grow slowly at relatively low levels of supersaturation. Occasionally, grains incorporated by growing burial cements may be seen to be more closely packed near the periphery of the cement crystal, indicating progressively tighter packing of the surrounding sediment during burial and cementation. Recent studies of burial cements have used a number of increasingly advanced techniques for the identification of burial cements and estimation of temperatures and depths at which they form. These include combinations of staining and cathodoluminescent petrography, analyses of fluid inclusions, and stable isotope analyses.

Ferroan Cements

Shortly after burial, often within a few meters or tens of meters below the sediment-water interface, dissolved oxygen is depleted from pore waters in marine sediments by bacterial respiration and organic matter oxidation (Claypool and Kaplan, 1974). Continued oxidation and sulfate reduction may cause pore water to become reducing, which promotes the co-precipitation of ferrous-iron carbonate within calcite and/or dolomite. Friedman (1959) describes staining techniques for the identification of ferroan calcites and dolomites.

Although burial cements are commonly observed to be ferroan, iron content alone is not sufficient for their identification. There is no theoretical reason why ferroan cements could not form early in near-surface anoxic waters (Evamy, 1969). However, such occurrences seem to be rare; most ferroan cements described from carbonate rocks have been attributed to late or burial cementation. Ferroan calcite cements have been described by Oldershaw (1971), Meyers (1974), Loucks (1977), Grover and Read (1978), and Wong and Oldershaw (1981), among others. Calcite cements formed during burial may be iron-free because there may be no source for iron. This can be the case in very thick carbonate sequences or carbonates partially isolated by evaporites. Examples of iron-poor burial calcite cements are illustrated by Moore and Druckman (1981).

Saddle dolomite (Radke and Mathis, 1980), also called baroque dolomite (Folk and Assereto, 1974), is another burial carbonate cement which is often ferroan. It is characterized by curved crystal faces and sweeping extinction and frequently contains iron-rich and iron-poor zones. Examples of late-stage, burial dolomite cements are illustrated by Choquette (1971), Radke and Mathis (1980), Moore and Druckman (1981), and Wong and Oldershaw (1981). Radke and Mathis suggest that saddle dolomite is indicative of temperatures between 60 and 150 degrees Centigrade.

Cathodoluminescence

Fluctuating oxidation states (redox potentials) and changing manganese and iron ratios in subsurface fluids cause some burial cements to exhibit marked zonation when examined using cathodoluminescent petrography (Meyers, 1978; Pierson, 1981; Frank and others, 1982; Fairchild, 1983). These zones can sometimes be mapped and correlated with regional stratigraphic discontinuities, as in the classic study of cement stratigraphy within the Lake Valley Formation of New Mexico by Meyers (1974). In this manner, the age of the cement zones may be bracketed and approximate depths of burial during formation may be estimated. Similarly, Grover and Read (1978, 1983) identified burial cements in Middle Ordovician carbonate ramp limestones from Virginia.

Luminescent cements in the Lake Valley Formation are shown by Meyers (1978) to have formed at relatively shallow depths, generally less than 500 meters. Grover and Read (1983) consider luminescent cements to form at intermediate depths, deep enough not to be influenced by oxidizing meteoric waters, but shallow enough to avoid strongly anoxic waters that allow incorporation of ferrous iron in calcite. In their opinion, luminescent cements represent burial cements separating early meteoric cements from late, deep burial cements. Grover and Read (1983) describe dull luminescence in iron-bearing cements interpreted to have formed at burial depths of 2,000–3,000 meters. Even deeper cementation (4,000 to 7,500 m) is linked to heavy metal mineralization during the late Paleozoic. These two studies (Meyers, 1974, 1978; Grover and Read, 1983) illustrate the growing role of cathodoluminescent petrography in studies of burial cementation.

Cathodoluminescent cements are often studied in conjunction with isotope and fluid inclusion data to best understand their origin. As an example, Figure 10 shows cements which are homogeneous in plane light but exhibit distinct luminescent zones. These zones

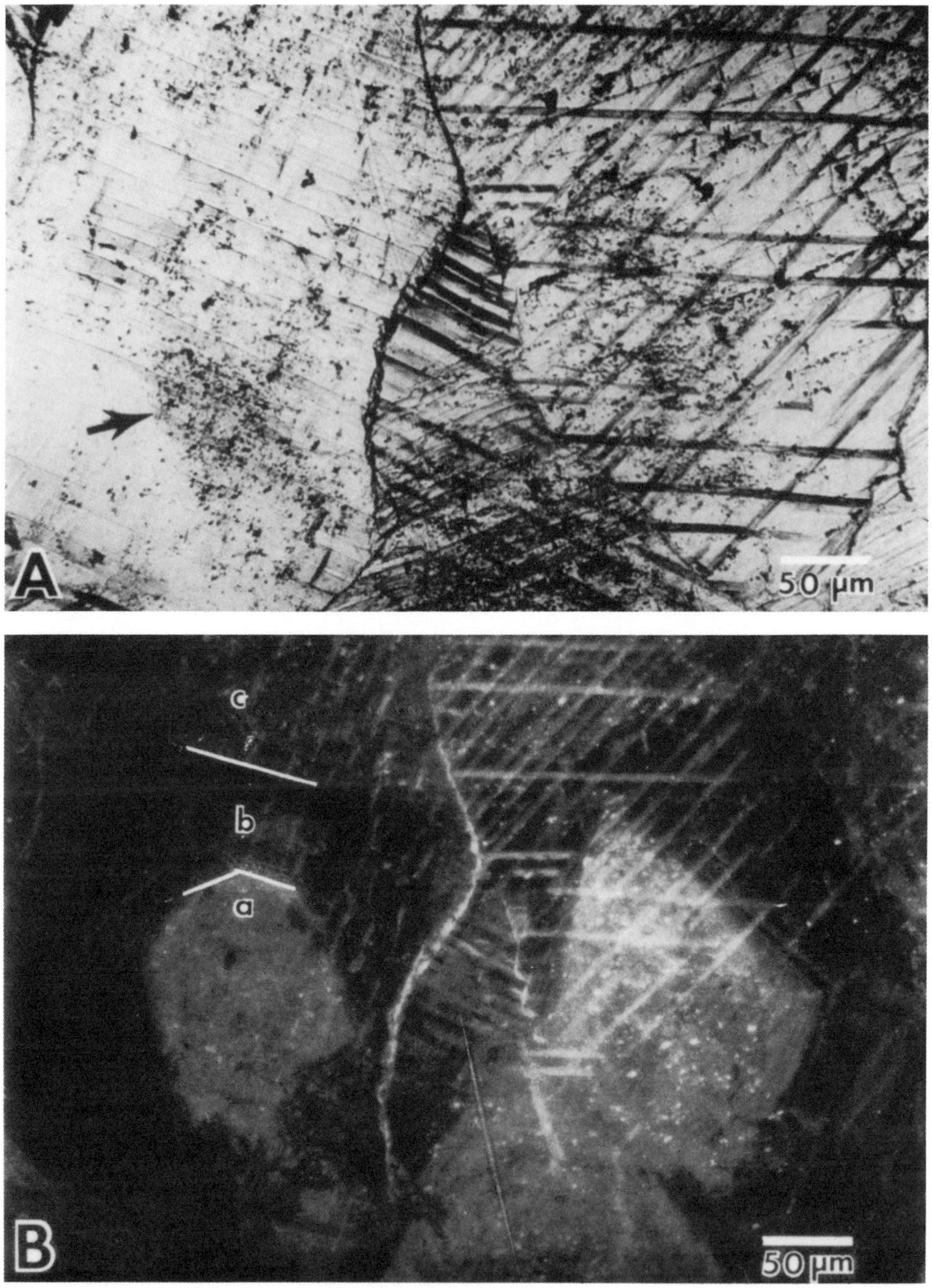

FIG. 10.—Photomicrographs illustrating pronounced zonation of calcite cement visible during cathodoluminescent petrography. Inclusion-rich zone visible in plane light (arrow in A) luminesces brightly under the cathode ray (zone a in B). Clear calcite overgrowths in A are divisible into 2 zones in B, a first zone (b) of alternately light and dark bands, and a second zone (c) of uniformly dull luminescence. ^{18}O analyses for the three cements are approximately −2, −4, and −6 percent (relative to PDB standard) from oldest to youngest, suggesting precipitation from successively lighter water or at successively higher temperatures. The latter interpretation is supported by fluid inclusions from the youngest cement, which homogenize near 175°C. Sample is from Upper Permian sponge-algal reef, near Ziyun, Guizhou Province, Peoples Republic of China.

have characteristic oxygen isotope ratios as well as luminescent characteristics, and most likely record calcite cementation at progressively higher temperatures. The brightly luminescing cores (zone a) of these cement crystals are cloudy with inclusions, have ^{18}O of about −2, and probably represent an early marine cement. Zone b contains alternately light and dark bands and has ^{18}O of about −4 per mil (relative to PDB standard). Zone c has rather uniform dull luminescence, ^{18}O of about -6, and contains fluid inclusions that homogenize at about 175°C. Combining these observations leads to the recognition of at least three distinct episodes of cementation, each at successively deeper burial depths and higher temperatures.

Fluid Inclusions

Primary fluid inclusions in carbonate cements have been analyzed in an attempt to determine the temperature of cement formation. By reference to a reasonable geothermal gradient, the temperature leads to a depth of burial at the time of cement precipitation. Roedder (1979), in a review of fluid inclusion studies in sedimentary rocks, cited only two studies from calcite or dolomite, both unpublished. Several fluid inclusion studies in carbonate rocks have been completed during the last ten years, most as unpublished Master's theses. Nelson (1974) performed over fifty homogenizations in a study of fluid inclusions of the Fernvale Limestone (Ordovician) of northern Arkansas. Homogenization temperatures ranged from 85 to 173 degrees Centigrade. Nelson concluded that two-phase fluid inclusions were trapped in the limestone during recrystallization of an earlier cement. Although homogenization determinations often took several hours to perform, technological advances in fluid inclusion stages have shortened the time required for a determination and have improved their accuracy.

Klosterman (1981) measured 222 homogenization temperatures from coarse calcite cements in the Smackover Formation (Jurassic) of Arkansas. She found a wide range of fluid inclusion homogenization temperatures (58°–158°C) and interpreted the range to represent the gradual precipitation of cement and entrapment of inclusions during changing subsurface temperatures. Depressed freezing temperatures and low initial melting temperatures of the inclusions indicated the cement precipitated from high-salinity brines. Klosterman (1981) cited several difficulties with the fluid inclusion analyses, including determination of inclusion origin, pressure corrections, small inclusion size, metastable phases, and lack of phase data for organic-inorganic systems. To these problems, we might add difficulties associated with stretching or leaking of inclusions, inadvertent operator selection of an unrepresentative sample population of inclusions, and alteration of inclusions during sample preparation, particularly when dealing with low-temperature inclusions (less than 100 degrees Centigrade). The possibility exists that inclusions may increase in volume by stretching when they are heated above their entrapment temperature (Poland, 1982) or during freezing analyses (Lawler and Crawford, 1983). Stretching results in artificially increased homogenization temperatures and has yet to be thoroughly investigated for calcite or dolomite. Stretching in flourite has been studied by Poland (1982), and her study indicates that small inclusions, such as those typically found in sedimentary calcites, may resist stretching at elevated temperatures and pressures in sedimentary basins. She also suggests that the ease at which stretching occurs may be related to the Mohs hardness scale, with soft minerals like calcite and anhydrite near the most stretchable end of the scale. Despite these shortcomings, fluid inclusion analyses are a developing key data source for the recognition of burial cements, as exemplified by the work of Moore and Druckman (1981).

Stable Isotopes

Stable isotope studies have played a key role in the recognition of burial cements. The fractionation of oxygen isotopes between water and precipitating calcite at a given temperature is well known (Epstein *et al.*, 1953). Stable isotopes of oxygen have been particularly useful in interpreting the temperature during the precipitation of calcite, given reasonable constraints on the oxygen isotopic composition of water. Conversely, with ^{18}O of calcite and temperatures of precipitation derived from fluid inclusions, the ^{18}O of precipitating waters may be estimated. Examples of the use of oxygen isotopic data have been previously cited and the reader is referred to Hudson (1975), Scholle (1977), Dickson and Coleman (1980), and Moore and Druckman (1981).

Stable carbon isotopes are not fractionated as extensively by temperature as oxygen isotopes and therefore are not as useful in determining temperature during cement precipitation. However, as suggested earlier, they may provide a key to the distance of migration of calcium carbonate during subsurface cementation.

Variations in the stable strontium isotopic composition of seawater have been documented by Veizer and Compston (1974) and Burke *et al.* (1982). These data provide an original strontium isotopic ratio for carbonate sediments of a given age, which may then be compared with strontium isotopic composition after diagenesis. From this comparison, inferences may be drawn about the strontium composition of waters involved with cementation and about their previous history. For example, Stueber and Pushkar (1983) have shown that some cements in the Smackover Formation yield a ratio of $^{87}Sr/^{86}Sr$ which is too large for cements precipitated from unaltered, buried Jurassic seawater. Rather, the precipitating waters have picked up large quantities of radiogenetic Sr (^{87}Sr), perhaps during feldspar dissolution within the underlying or laterally equivalent formations. Feldspar dissolution, a well documented late diagenetic process forming secondary porosity in sandstones (Schmidt and McDonald, 1979), preceded or coincided with the precipitation of burial cements in portions of the Smackover Formation. Stable strontium isotope analyses show promise for determining the timing of some of the subsurface processes that occur during burial diagenesis of carbonate rocks.

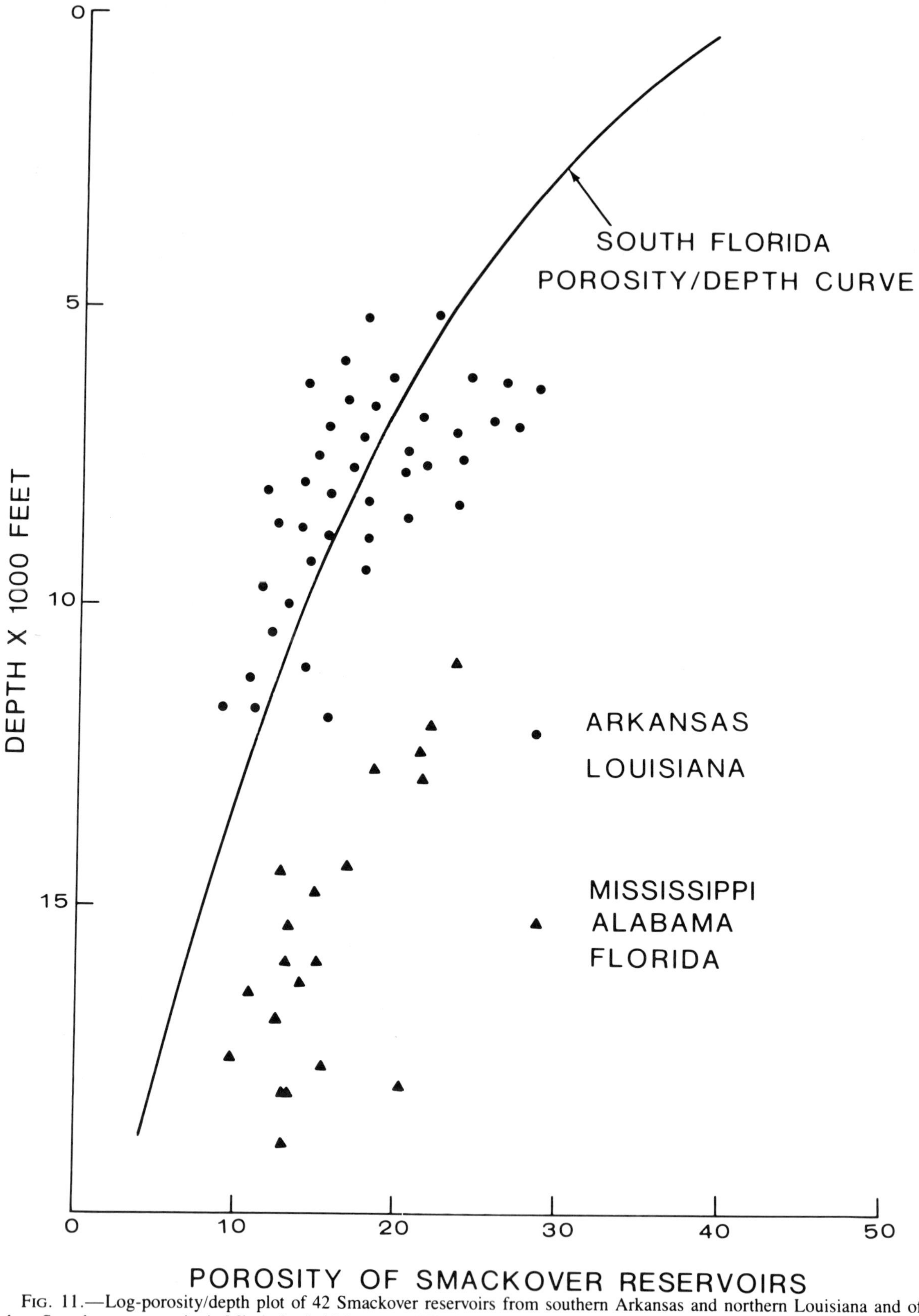

FIG. 11.—Log-porosity/depth plot of 42 Smackover reservoirs from southern Arkansas and northern Louisiana and of 19 deep Smackover reservoirs in Mississippi, Alabama, and Florida. Smooth curve is porosity/depth relation for south Florida from Figure 3. Note that porosity in these deep reservoirs is much higher than would be predicted by the Arkansas-Louisiana data or by the South Florida trend.

ESTABLISHING NEW PERSPECTIVES

The depth-porosity trends discussed in this work have value as predictive tools for estimating average porosity for large volumes of carbonate rock within some sedimentary basins. Undoubtedly, these predictive models will become more useful as controls on subsurface diagenesis are better defined. Beyond the goal of immediate porosity prediction, these models may offer insights, first, to the evolution of porosity within a basin and, second, to unusual and perhaps poorly understood but potentially important geologic processes controlling porosity distribution in carbonate rocks.

In the first case, the depth-porosity trends reveal isolated instants in the history of a subsiding basin. However, to the extent that similar porosity-reducing processes have been operative in the basin, the trend becomes a predictive line tracing the change in porosity, with respect to time, of any unit within the basin. A formation with little porosity may be traced back in time (upward) and an estimate made of the time when it was porous. This information can be combined with knowledge of source rock maturation and hydrocarbon migration (for example, Waples, 1980) to determine if there might be reservoirs in the unit which were charged with hydrocarbons at some time in the past. It is interesting to note, for instance, that the porosity-depth trend in Smackover reservoirs of northern Louisiana and southern Arkansas is very similar to that of non-reservoir rocks in south Florida (Fig. 11; also see Halley, 1981). This might be interpreted to suggest that hydrocarbons moved into Smackover reservoirs when the formation had a porosity-depth profile like that found in south Florida today and effectively blocked any further porosity loss. In this way, the depth-porosity trend of the reservoirs was preserved while the rest of the formation may have undergone continued alteration.

In the second case, porosity-depth trends provide tools for recognizing unusual porosity occurrences. As an example, compare the porosity-depth relationship for the Smackover Formation in Arkansas and Louisiana with similar data from the deeply buried Smackover of Florida, Mississippi, and Alabama (Fig. 11). Clearly, these deep Smackover reservoirs are much more porous than might be expected from the data for shallower reservoirs (Arkansas and Louisiana) or for non-reservoir rocks (Fig. 3). As yet, there is no clear consensus about what factors are responsible for the unusually porous reservoirs in the deep Smackover. Several studies have considered the origin of Smackover reservoirs in some detail. Chimene (1976), Moore and Druckman (1981), and Wagner and Matthews (1982) described the setting, facies, and diagenesis of several fields in Arkansas and Louisiana. Some of the deeper Smackover fields of Florida, Mississippi, and Alabama have been described by Ottman *et al.* (1976), Oglesby (1976), and Mancini and Benson (1981). The deep fields include many which are largely dolomite, some of which are limestone, and a few of which are largely sandstone with carbonate matrix. Some of the deep reservoirs are overpressured, many are not. What unusual happenstance has endowed these deep reservoirs with such porosity at these depths? Has something in their geologic history led to the preservation of early porosity at depth? Or has porosity been secondarily created at depth? These and other questions await thorough documentation and testing. Perhaps the reasons for porosity at depth will turn out to be as numerous as the reservoirs themselves, each existing for its own set of unique circumstances. However, it is our belief that this is not the case. Figure 11 presents a surprisingly consistent relationship and suggests that only a few basic causes underlie the existence of these deep reservoirs. The existence of deep reservoirs in carbonate rocks is promising, and such reservoirs might reasonably be postulated for other areas once the underlying causes are understood. It is our belief that the processes of burial diagenesis hold the keys to this knowledge and that the prediction of porosity at depth in carbonate rocks is controlled as much by burial diagenesis as by facies and early diagenesis.

ACKNOWLEDGMENTS

The authors wish to thank D.L. Gautier, E.A. Shinn, and W.D. Wiggins for reviewing earlier versions of this manuscript and for offering suggestions that substantially improved the final product. This paper was presented at the Abu Dhabi National Reservoir Research Foundation Seminar on Stylolites, November 7th and 8th, 1983, and a version is published in the proceedings for that symposium.

REFERENCES

Allan, J.R., and Matthews, R.K., 1977, Carbon and oxygen isotopes as diagenetic and stratigraphic tools: Surface and subsurface data, Barbados, West Indies: Geology, v. 5, p. 16–20.

Arthur, M.A., 1979, Sedimentologic and geochemical studies of Cretaceous and Paleogene pelagic sedimentary rocks: Pt. 1, the Gubbio Section: Ph.D. Thesis, Princeton University, Princeton, New Jersey.

Back W., and Hanshaw, B.B., 1971, Rates of physical and chemical processes in a carbonate aquifer: Advances in Chemistry, v. 106, p. 77–93.

BAKER, P.A., GIESKES, J.M., AND ELDERFIELD, H., 1982, Diagenesis of carbonates in deep-sea sediments—evidence from Sr/Ca ratios and interstitial dissolved Sr data: Jour. Sed. Petrology, v. 52, p. 71–82.

______, KASTNER, M., BYERLEE, J.D., AND LOCKNER, D.A., 1980, Pressure solution and hydrothermal recrystallization of carbonate sediments—An experimental study: Marine Geology, v. 38, p. 185–203.

BATHURST, R.G.C., 1970, Problems of lithification in carbonate muds: Proc. Geol. Assoc., v. 81, p. 429–440.

______, 1975, Carbonate sediments and their diagenesis: Amsterdam, Elsevier Scientific Publishing Co., 658 p.

______, 1980, Deep crustal diagenesis in limestones: Rev. Inst. Invest. Geol. Deputacion Provincial Univ. Barcelona, v. 34, p. 89–100.

BEACH, D.K., 1982, Depositional and diagenetic history of Pliocene-Pleistocene carbonates of northwestern Great Bahama Bank; evolution of a carbonate platform: Ph.D. dissertation, University of Miami, Coral Gables, Florida, v. 1, 447 p.

______, AND SCHUMACHER, ANN, 1982, Stanley field, North Dakota: A new model for a new exploration play, *in* J.E. Christopher, and John Kaldi (eds.), 4th International Williston Basin Symposium, p. 235–243.

BEALL, A.O., AND FISCHER, A.G., 1969, Sedimentology, *in* Initial Reports of the Deep-Sea Drilling Project, v. 1: U.S. Govt. Printing Office, Washington, p. 521–593.

BECHER, J.W., AND MOORE, C.H., 1976, The Walker Creek field: A Smackover diagenetic trap: Gulf Coast Assoc. Geol. Socs. Trans., v. 26, p. 34–56.

BHATTACHARYYA, A., AND FRIEDMAN, G.M., 1979, Experimental compaction of ooids and lime mud and its implication for lithification during burial: Jour. Sed. Petrology, v. 49, p. 1279–1286.

BONHAM, L.C., 1980, Migration of hydrocarbons in compacting basins: Amer. Assoc. Petroleum Geologists Bull., v. 64, p. 549–567.

BROWN, P.R., 1969, Compaction of fine-grained terrigenous and carbonate sediments—A review: Bull. Canadian Petroleum Geology, v. 17, p. 486–495.

BURKE, W.H., DENISON, R.E., HETHERINGTON, R.B., KOEPNICK, R.B., NELSON, H.F., AND OTTO, J.B., 1982, Variation of seawater $^{87}Sr/^{86}Sr$ throughout Phanerozoic time: Geology, v. 10, p. 516–519.

BUXTON, T.M., AND SIBLEY, D.F., 1981, Pressure solution features in a shallow buried limestone: Jour. Sed. Petrology, v. 51, p. 19–26.

CHIMENE, C.A., 1976, Upper Smackover reservoirs, Walker Creek field area, Lafayette and Columbia Counties, Arkansas, *in* J. Braunstein (ed.), North American Oil and Gas Fields: Amer. Assoc. Petroleum Geologists Memoir 244, p. 177–204.

CHOQUETTE, P.W., 1971, Late ferroan dolomite cement, Mississippian carbonates, Illinois Basin, U.S.A., *in* O.P. Bricker (ed.), Carbonate Cements: Baltimore, Johns Hopkins University Studies in Geology, no. 19, p. 339–346.

______, AND PRAY, L.C., 1970, Geologic nomenclature and classification of porosity in sedimentary carbonates: Amer. Assoc. Petroleum Geologists Bull., v. 54, p. 207–250.

CLAYPOOL, G.E., AND KAPLAN, J.R., 1974, The origin and distribution of methane in marine sediments in natural gases in marine sediments: New York, Plenum Press, p. 99–139.

COLLINS, A.G., 1975, Geochemistry of Oilfield Waters: New York, Elsevier Scientific Publishing Co., 496 p.

CONLEY, C.D., 1977, Origin of distorted ooliths and pisoliths: Jour. Sed. Petrology, v. 47, p. 554–564.

COOGAN, A.H., 1970, Measurements of compaction in oolitic grainstone: Jour. Sed. Petrology, v. 40, p. 921–929.

______, AND MANUS, R.W., 1975, Compaction and diagenesis of carbonate sands, *in* G.V. Chilingar and K.H. Wolf (eds.), Compaction of Coarse-grained Sediments, I. Developments in Sedimentology—18A: New York, Elsevier Scientific Publishing Co., p. 79–166.

DE BOER, R.B., 1977, Pressure solution: Theory and experiments: Tectonophysics, v. 39, p. 287–301.

DICKSON, J.A.D., AND COLEMAN, M.L., 1980, Changes in carbon and oxygen isotope composition during limestone diagenesis: Sedimentology, v. 27, p. 107–118.

DUNHAM, R.J., 1969, Early vadose silt in Townsend mound (reef), New Mexico, *in* G.M. Friedman (ed.), Depositional Environments in Carbonate Rocks: A Symposium: Soc. Econ. Paleontologists and Mineralogists Spec. Pub. 14, p. 139–181.

DUNNINGTON, H.V., 1954, Stylolite development post-dates rock induration: Jour. Sed. Petrology, v. 24, p. 27–49.

______, 1967, Aspects of diagenesis and shape change in stylolitic limestone reservoirs: 7th World Petroleum Congress, Proc., Mexico, v. 2, p. 339–352.

DURNEY, D.W., 1972, Solution transfer, an important geological deformation mechanism: Nature, v. 235, p. 315–317.

EBHARDT, G., 1968, Experimental compaction of carbonate sediments, *in* G. Muller and G.M. Friedman (eds.), Recent Developments in Carbonate Sedimentology in Central Europe: New York, Springer-Verlag Press, p. 58–65.

EINSELE, G., AND SEILACHER, A. (eds.), 1982, Cyclic and Event Stratification: New York, Springer-Verlag Press, 536 p.

ELLIOTT, T.L., 1982, Carbonate facies, depositional cycles, and the development of secondary porosity during burial diagenesis: Mission Canyon Formation, Haas Field, North Dakota, *in* J.E. Christopher and J. Kaldi (eds.), 4th International Williston Basin Symposium: Saskatchewan Geol. Soc. Spec. Pub. 6, p. 131–151.

ENOS, PAUL, AND SAWATSKY, L.H., 1979, Pore space in Holocene carbonate sediments [abs.]: Amer. Assoc. Petroleum Geologists Bull., v. 63, p. 445.

______, AND ______, 1981, Pore networks in Holocene carbonate sediments: Jour. Sed. Petrology, v. 51, p. 961–985.

EPSTEIN, S., BUCHSBAUM, R., LOWENSTAM, H.A., AND UREY, H.C., 1953, Revised carbonate-water temperature scale: Geol. Soc. America Bull. v. 62, p. 417–426.

EVAMY, B.D., 1969, The precipitational environment and correlation of some calcite cement deduced from artificial staining: Jour. Sed. Petrology, v. 39, p. 787–793.

FAIRBRIDGE, R.W., 1967, Phases of diagenesis and authigenesis, *in* G. Larsen and G.V. Chilingar (eds.), Diagenesis in Sediments: Amersterdam, Elsevier Scientific Publishing Co., p. 19–89.

FAIRCHILD, I.J., 1983, Chemical controls of cathodoluminescence of natural dolomites and calcites: New data and review: Sedimentology, v. 30, p. 579–583.

FEAZEL, C.T., AND SCHATZINGER, R.A., 1983, Prevention of carbonate cementation in petroleum reservoirs [abs.]: Am. Assoc. Petroleum Geologists Bull., v. 67, p. 459.

FISCHER, A.G., 1980, Gilbert-bedding rhythms and geochronology, *in* E.L. Yochelson (ed.), The Scientific Ideas of G.K. Gilbert: Jour. Soc. America Spec. Pub. 183, p. 93–104.

FOLK, R.L., AND ASSERETO, R., 1974, Giant aragonite rays and baroque white dolomite in tepee-fillings, Triassic of Lombardy, Italy [abs.]: Amer. Assoc. Petroleum Geologists, Abstracts with Programs, Annual Meeting, San Antonio, p. 34–35.

FRANK, J.R., CARPENTER, A.B., AND OGLESBY, T.W., 1982, Cathodoluminescence and composition of calcite cement in the Taum Sank Limestone (Upper Cambrian), Southeast Missouri: Jour. Sed. Petrology, v. 52, p. 631–638.

FRIEDMAN, G.M., 1959, Identification of carbonate minerals by staining methods: Jour. Sed. Petrology, v. 29, p. 87–97.

______, 1975, The making and unmaking of limestones or the downs and ups of porosity: Jour. Sed. Petrology, v. 45, p. 379–398.

FRUTH, L.S., JR., ORME, G.R., AND DONATH, F.A., 1966, Experimental compaction effects in carbonate sediments: Jour. Sed. Petrology, v. 36, p. 747–754.

GARRISON, R.E., AND KENNEDY, W.J., 1977, Origin of solution-seams and flaser structure in Upper Cretaceous chalks of southern England: Sed. Geology, v. 19, no. 2, p. 107–137.

GILBERT, G.K., 1900, Rhythms and geologic time: Amer. Assoc. Adv. Sci. Proc., v. 49, p. 1–19.

GILLETT, S.L., 1983, Major, through-going stylolites in the Lower Ordivician Goodwin Limestone, southern Nevada: Petrography with dating from paleomagnetism: Jour. Sed. Petrology, v. 53, p. 209–219.

GINSBURG, R.N., 1953, Beach rock in south Florida: Jour. Sed. Petrology, v. 23, p. 85–92.

______,1956, Environmental relationships of grain size and constituent particles in some south Florida carbonate sediments: Amer. Assoc. Petroleum Geologists Bull., v. 40, p. 2384–2427.

______,1957, Early diagenesis and lithification of shallow-water carbonate sediments in south Florida, *in* R.J. Leblanc and J.G. Breeding (eds.), Regional Aspects of Carbonate Deposition: Soc. Econ. Paleontologists and Mineralogists, Spec. Pub. 5, p. 80–100.

GROSS, M.G., 1964, Variations in the O^{18}/O^{16} and C^{13}/C^{12} ratios of diagenetically altered limestones in the Bermuda Islands, Jour. Geology, v. 72, p. 170–194.

GROVER, G., JR., AND READ, J.F., 1978, Fenestral and associated vadose diagenetic fabrics of tidal flat carbonates, Middle Ordovician New Market Limestone, Southwestern Virginia: Jour. Sed. Petrology, v. 48, p. 453–473.

______, 1983, Paleoaquifer and deep burial related cements defined by regional cathodoluminescent patterns, Middle Ordovician carbonates, Virginia: Amer. Assoc. Petroleum Geologists Bull., v. 67, p. 1275–1303.

HALLAM, A., 1964, Origin of the limestone-shale rhythm in the Blue Lias of England: A composite theory: Jour. Geology, v. 72, p. 157–169.

HALLEY, R.B., 1981, Evolution of carbonate porosity during burial—Bahamas, Florida and Gulf Coast: Holocene to Jurassic [abs.]: Amer. Assoc. Petroleum Geologists Bull., v. 65, p. 2466–2467.

______, AND HARRIS, P.M., 1979, Freshwater cementation of a 1,000-year-old oolite: Jour. Sed. Petrology, v. 49, p. 969–988.

______, AND EVANS, C.C., 1983, The Miami Limestone: A guide to selected outcrops and their interpretation: Miami Geol. Soc., Miami, Fla., 67 p.

______, AND SCHMOKER, J.W., 1983, High-porosity Cenozoic carbonate rocks of south Florida: Progressive loss of porosity with depth: Amer. Assoc. Petroleum Geologists Bull., v. 67, p. 191–200.

HAMILTON, E.L., 1959, Thickness and consolidation of deep-sea sediments: Geol. Soc. America Bull., v. 70, p. 1399–1424.

HARMS, J.C., AND CHOQUETTE, P.W., 1965, Geologic evaluation of a gamma-ray porosity device, *in* 6th Annual SPWLA Logging Symposium, Transactions: Soc. Prof. Well Log Analysts, v. 2, p. C1–C37.

HARPER, M.L., AND SHAW, B.B., 1974, Cretaceous-Tertiary carbonate reservoirs in the North Sea: Offshore North Sea Technology Conference, Stavanger, Norway, Paper G IV/4, 20 p.

HARRISON, R.S., 1975, Porosity in Pleistocene grainstones from Barbados: Some preliminary observations: Bull. Canadian Petroleum Geol., v. 23, p. 383–392.

HATHAWAY, J.C., AND ROBERTSON, E.C., 1961, Micro-texture of artificially consolidated aragonite mud: U.S. Geol. Survey Prof. Paper 424-G, p. 301–304.

HESSE, R., AND BUTT, A., 1976, Paleobathymetry of Cretaceous turbidite basins of the East Alps relative to the calcite compensation level: Jour. Geology, v. 84, no. 5, p. 505–533.

HECKEL, P.H., 1983, Diagenetic model for carbonate rocks in midcontinent Pennsylvanian eustatic cyclothems: Jour. Sed. Petrology, v. 53, p. 733–760.

HUDSON, J.D., 1975, Carbon isotopes and limestone cement: Geology, v. 3, p. 19–22.

KAHLE, C., 1966, Compaction in oolites: Compass, v. 44, p. 19–29.

KEITH, M.L., AND WEBER, J.N., 1964, Carbon and oxygen isotopic composition of selected limestones and fossils: Geochim. Cosmochim. Acta, v. 28, p. 1787–1816.

KLOSTERMAN, M.J., 1981, Applications of fluid inclusion techniques to burial diagenesis in carbonate rock sequences: Louisiana State University, Applied Carbonate Research Program, Technical Series, Contribution #7, 101 p.

LAND, L.S., 1980, The isotopic and trace element geochemistry of dolomite: The state of art, *in* D.H. Zenger, J.B. Dunham, and R.L. Ethington (eds.), Concepts and Models of Dolomitization: Soc. Econ. Paleontologists and Mineralogists, Spec. Pub. 28, p. 87–110.

LAWLER, J.P., AND CRAWFORD, M.L., 1983, Stretching of fluid inclusions resulting from a low-temperature microther-

mometric technique: Econ. Geology, v. 78, p. 527–529.
LLOYD, R.M., 1977, Porosity reduction by chemical compaction—stable isotope model [abs.]: Amer. Assoc. Petroleum Geologists Bull., v. 61, p. 809.
LOCKRIDGE, J.P., AND SCHOLLE, P.A., 1978, Niobrara gas in eastern Colorado and northwestern Kansas: Rocky Mtn. Assoc. Geologists 1978 Symposium, p. 35–49.
LOGAN, B.W., AND SEMENIUK, V., 1976, Dynamic metamorphism; processes and products in Devonian carbonate rocks, Canning Basin, Western Australia: Geol. Soc. Australia Spec. Pub. 16, 138 p.
LONGMAN, M.W., 1980, Carbonate diagenetic textures from near surface diagenetic environments: Amer. Assoc. Petroleum Geologists Bull., v. 64, p. 461–487.
LORENS, R.B., 1981, Sr, Cd, Mn, and Co distribution coefficients in calcite as a function of calcite precipitation rate: Geochim. Cosmochim. Acta, v. 45, p. 553–561.
LOUCKS, R.G., 1977, Porosity development and distribution in shoal-water carbonate complexes—subsurface Pearsall Formation (Lower Cretaceous) south Texas, *in* D.G. Bebout and R.G. Loucks (eds.), Cretaceous Carbonates of Texas and Mexico: Texas Bureau Econ. Geology, Report of Investigations 89, p. 97–126.
MANCINI, E.A., AND BENSON, D.J., 1981, Smackover carbonate petroleum geology in southwest Alabama: Oil and Gas Journal, Dec. 28, 1981, p. 266–275.
MANUS, R.W., AND COOGAN, A.H., 1974, Bulk volume reduction and pressure-solution derived cement: Jour. Sed. Petrology, v. 44, p. 466–471.
MERRIAM, C.W., 1964, Cambrian rocks of the Pioche mining district, Nevada: U.S. Geol. Survey Prof. Paper 469, 59 p.
MEYERS, W.J., 1974, Carbonate cement stratigraphy of the Lake Valley Formation (Mississippian) Sacramento Mountains, New Mexico: Jour. Sed. Petrology, v. 44, p. 873–861.
______, 1978, Carbonate cements: their regional distribution and interpretation in Mississippian limestones of southwestern New Mexico: Sedimentology, v. 25, p. 371–400.
______, 1980, Compaction in Mississippian skeletal limestones, southwestern New Mexico: Jour. Sed. Petrology, v. 50, p. 457–474.
______, AND HILL, B.E., 1983, Quantitative studies of compaction in Mississippian skeletal limestones, New Mexico: Jour. Sed. Petrology, v. 53, p. 231–242.
MIMRAN, Y., 1975, Fabric deformation induced in Cretaceous chalks by tectonic stresses: Tectonophysics, v. 26, p. 309–316.
______, 1977, Chalk deformation and large-scale migration of calcium carbonate: Sedimentology, v. 24, p. 333–360.
MITRA, S., AND BEARD, W.C., 1980, Theoretical models of porosity reduction by pressure solution for well-sorted sandstone: Jour. Sed. Petrology, v. 50, p. 1347–1360.
MOORE, C.H., JR., 1973, Intertidal carbonate cementation, Grand Cayman, West Indies, Jour. Sed. Petrology, v. 43, p. 591–602.
______, AND DRUCKMAN, Y., 1981, Burial diagenesis and porosity evolution, Upper Jurassic Smackover, Arkansas and Louisiana: Amer. Assoc. Petroleum Geologists Bull., v. 65, p. 597–28.
MORELOCK, J., AND BRYANT, W.R., 1971, Consolidation of marine sediments, *in* R. Rezak and H.J. Vernon (eds.), Contributions on the Geological and Geophysical Oceanography of the Gulf of Mexico: Texas A & M University Oceanog. Studies, v. 3, p. 181–202.
MOSSOP, G.D., 1972, Origin of the peripheral rim, Redwater Reef, Alberta, Bull. Canadian Petroleum Geol., v. 20, p. 238–280.
MUCCI, A., AND MORSE, J.W., 1983, The incorporation of Mg^{2+} and Sr^{2+} into calcite overgrowths: influences of growth rate and solution composition: Geochim. Cosmochim. Acta, v. 47, p. 217–233.
NELSON, R.A., 1981, Significance of fracture sets associated with stylolite zones: Amer. Assoc. Petroleum Geologists Bull., v. 65, p. 2417–2425.
NELSON, R.C., 1974, Fluid inclusion study of the Fernvale Limestone, northern Arkansas: Unpubl. Masters Thesis, University of Missouri-Columbia, Columbia, Missouri, 95 p.
NEUGEBAUER, J., 1973, The diagenetic problem of chalk—the role of pressure solution and pore fluid: Neues Jahrbuch für Geol. Paläontol. Abh., v. 143, p. 223–245.
______, 1974, Some aspects of cementation in chalk, *in* K.J. Hsü, and H.C. Jenkyns (eds.), Pelagic Sediments: On Land and Under the Sea: Internat. Assoc. Sedimentologists Spec. Pub. 1, p. 149–176.
OGLESBY, T.W., 1976, A model for the distribution of manganese, iron and magnesium in anthigenic calcite and dolomite cements in the upper Smackover Formation in eastern Mississippi: Unpubl. MS Thesis, University of Missouri-Columbia, Columbia, Missouri, 122 p.
OLDERSHAW, A.E., 1971, The significance of ferroan and non-ferroan calcite cements in the Halkin and Wenlock Limestones (Great Britain), *in* O. P. Bricker (ed.), Carbonate Cements: Johns Hopkins University Studies in Geology, no. 19, p. 225–232.
______, AND SCOFFIN, T.P., 1967, The source of ferroan and non-ferroan calcite cements in the Halkin and Wenlock Limestones: Geol. Jour., v. 5, p. 309–320.
OTTMANN, R.D., KEYES, P.L., AND ZIEGLER, M.A., 1976, Jay Field, Florida—A Jurassic stratigraphic trap, *in* Braunstein, Jules, ed., North American Oil and Gas Fields: Amer. Assoc. Petroleum Geologists Memoir 24, p. 276–286.
PARK, W.C., AND SCHOT, E.H., 1968a, Stylolitization in carbonate rocks, *in* G. Müller and G.M. Friedman (eds.), Recent Developments in Carbonate Sedimentology in Central Europe: New York, Springer–Verlag, p. 64–74.
______, AND ______, 1968b, Stylolites: their nature and origin: Jour. Sed. Petrology, v. 38, p. 175–191.
PERRY, E., AND HOWER, J., 1970, Burial diagenesis in Gulf Coast pelitic sediments: Clays and Clay Minerals, v. 18, p. 165–177.

Pierson, B.J., 1981, The control of cathodoluminescence in dolomite by iron and manganese: Sedimentology, v. 28, p. 601–610.
Poland, E.L., 1982, Stretching of fluid inclusions in flourite at confining pressures up to 1 kilobar: MS Thesis, University of California, Berkeley, California, 90 p.
Pray, L.C., 1960, Compaction in calcilutites [abs.]: Geol. Soc. America Bull., v. 71, p. 1946.
Purser, B.H., 1978, Early diagenesis and the preservation of porosity in Jurassic limestones: Jour. Petroleum Geol., v. 1, p. 83–94.
Radke, R.M., and Mathis, R.L., 1980, On the formation of saddle dolomite: Jour. Sed. Petrology, v. 50, p. 1149–1168.
Richards, A.F., and Hamilton, E., 1967, Investigations of deep-sea sediment cores, III. Consolidation, *in* A.F. Richards (ed.), Marine Geotechnique: Proc. Internat. Res. Conf. on Marine Geotechnique, Urbana, Illinois, Univ. of Illinois Press, p. 93–117.
Robertson, E.C., 1967, Laboratory consolidation of carbonate sediment, *in* A.F. Richards (ed.), Marine Geotechnique: Internat. Res. Conf. on Marine Geotechnique, Urbana, Illinois, Univ. Illinois Press, p. 118–127.
Robin, P.F., 1978, Pressure solution at grain-to-grain contacts: Geochim. Cosmochim. Acta, v. 42, p. 1383–1389.
Roedder, E., 1979, Fluid inclusion evidence on the environment of sedimentary diagenesis, a review, *in* P.A. Scholle, and P.R. Schluger (eds.), Aspects of Diagenesis: Soc. Econ. Paleontologists and Mineralogists Special Publication 25, p. 89–108.
Runnells, D.D., 1969, Diagenesis, chemical sediments and the mixing of natural waters: Jour. Sed. Petrology, v. 39, p. 1188–1201.
Schlanger, S.O., and Douglas, R.G., 1974, The pelagic ooze-chalk-limestone transition and its implication for marine stratigraphy, *in* K.J. Hsü and H.C. Jenkyns (eds.), Pelagic Sediments: On Land and Under the Sea; Internat. Assoc. Sedimentologists Spec. Pub. 1, p. 117–148.
Schmidt, V., and McDonald, D.A., 1979, The role of secondary porosity in the course of sandstone diagenesis, *in* P.A. Scholle, and P.R. Schluger (eds.), Aspects of Diagenesis: Soc. Econ. Paleontologists and Mineralogists Spec. Pub. 26, p. 175–207.
Schmoker, J.W., and Halley, R.B., 1982, Carbonate porosity versus depth: A predictable relation for south Florida: Amer. Assoc. Petroleum Geologists Bull., v. 66, p. 2561–2570.
Scholle, P.A., 1971, Diagenesis of deep-water carbonate turbidites, Upper Cretaceous Monte Antola Flysch, northern Appennines, Italy: Jour. Sed. Petrology, v. 41, p. 233–250.
______, 1977, Chalk diagenesis and its relation to petroleum exploration: Oil from chalk, a modern miracle?: Amer. Assoc. Petroleum Geologists Bull., v. 61, p. 982–1009.
______, and Arthur, M.A., 1980, Carbon isotope fluctuations in Cretaceous pelagic limestones: Potential stratigraphic and petroleum exploration tool: Amer. Assoc. Petroleum Geologists Bull., v. 64, p. 68–87.
______, ______, and Ekdale, A.A., 1983, Pelagic environment, *in* P.A. Scholle, D.G. Bebout, and C.H. Moore (eds.), Carbonate Depositional Environments: Amer. Assoc. Petroleum Geologists Memoir 33, p. 619–690.
Shinn, E.A., Halley, R.B., Hudson, J.H., and Lidz, B.H., 1977, Limestone compaction: An enigma: Geology, v. 5, p. 21–24.
______, and Robbin, D.M., 1983, Mechanical and chemical compaction in fine-grained shallow-water limestones: Jour. Sed. Petrology, v. 53, p. 595–618.
Stockdale, P.B., 1922, Stylolites: their nature and origin: Indiana Univ. Studies, v. 9, p. 1–97.
______, 1926, The stratigraphic significance of solution in rocks: Jour. Geology, v. 34, p. 399–414.
Stueber, A.M., and Pushkar, P., 1983, Application of strontium isotopes to origin of Smackover brines and diagenetic phases, southern Arkansas [abs.]: Amer. Assoc. Petroleum Geologists Bull., v. 67, p. 553–554.
Terzaghi, R.D., 1940, Compaction of lime mud as a cause of secondary structure: Jour. Sed. Petrology, v. 49, p. 437–462.
Trurnit, P., 1968, Analysis of pressure solution contacts and classification of pressure-solution phenomena, *in* G. Müller, and G.M. Friedman (eds.), Recent Developments in Carbonate Sedimentology in Central Europe: Berlin, Springer, p. 75–84.
Veizer, J., and Compston, W., 1974, $^{87}Sr/^{86}Sr$ composition of seawater during the Phanerozoic: Geochim. Cosmochim. Acta, v. 38, p. 1461–1484.
Vinopal, R.J., and Coogan, A.H., 1978, Effect of particle shape on the packing of carbonate sands and gravels: Jour. Sed. Petrology, v. 48, p. 7–24.
Wagner, P.D., and Matthews, R.K., 1982, Porosity preservation in the Upper Smackover (Jurassic) carbonate grainstone, Walker Creek field, Arkansas: Response of paleophreatic lenses to burial processes: Jour. Sed. Petrology, v. 52, no. 1, p. 3–18.
Wanless, H.R., 1979, Limestone response to stress: Pressure solution and dolomitization: Jour. Sed. Petrology, v. 49, p. 437–462.
Waples, D.W., 1980, Time and temperature in petroleum formation: Application of Lopatin's method to petroleum exploration: Amer. Assoc. Petroleum Geologists Bull., v. 64, p. 916–926.
Weller, J.M., 1959, Compaction of sediments: Amer. Assoc. Petroleum Geologists Bull., v. 43, p. 273–319.
Weyl, P.K., 1959, Pressure solution and the force of crystallization—a phenomenological theory: Jour. Geophys. Res., v. 64, p. 2001–2025.
Wong, P.K., and Oldershaw, A., 1981, Burial cementation in the Devonian, Kaybob Reef complex, Alberta, Canada: Jour. Sed. Petrology, v. 51, p. 507–520.
Wolfe, M.J., 1968, Lithification of a carbonate mud: Senonian chalk in northern Ireland: Sed. Geology, v. 2, p. 263–290.
Zankl, H., 1969, Structural and textural evidence of early lithification in fine-grained carbonate rocks: Sedimentology, v. 12, p. 241–256.

Reprinted by permission of the Journal of Petroleum Geology. Published in *Journal of Petroleum Geology.* V. 1, No. 2 (1978), pp. 83-94.

EARLY DIAGENESIS AND THE PRESERVATION OF POROSITY IN JURASSIC LIMESTONES

Bruce H. Purser*

The study of Quaternary sediments has shown that carbonates are susceptible to early lithification, or to dissolution, in many different environments. Most of these surface effects, although reducing primary porosity, rarely destroy it completely. Although similar diagenetic phenomena also occur in ancient rocks, the great majority of limestones do not exhibit early diagenetic fabrics; much sparitic cement is possibly related to carbonate liberated by pressure-solution within the limestone mass.

In spite of the fact that carbonate sediments frequently have high initial porosities, most of it is destroyed by sparitic cement. The exceptional circumstances favouring the preservation of effective porosity may be related to the presence of an early diagenetic phase. A number of reservoirs exhibit early diagenetic cements, dissolution and dedolomitisation fabrics. It is concluded that early lithification tends to retard or prevent subsequent pressure-compaction and thus reduces the volume of carbonate available for porosity destruction. The recognition of early diagenetic phenomena in ancient limestones therefore has important economic implications.

Introduction

Carbonate rocks are notorious for their irregular porosity distribution; not only are the great majority of limestones and dolomites devoid of effective porosity but the occasional reservoirs frequently exhibit rapid lateral variations, even within a given field. In many (possibly most) instances this absence of porosity is due to diagenetic effects.

In understanding the distribution of that relatively minor part of limestone sequences which is porous, the basic problem (as Choquette & Pray, 1970 have observed) concerns not the creation of porosity but rather its *preservation;* carbonate sediments prior to diagenesis are just as porous as non-carbonates, the vast bulk of this primary porosity in carbonates being destroyed subsequently by sparitic cement. Preservation of porosity therefore is closely dependent on the absence of pore-filling cement.

The absence of cement presumably may be due to one or other of several factors of which the following are the more obvious:

—early emplacement of hydrocarbons from which carbonate cannot precipitate;
—isolation from a source of carbonate cement by impermeable sediments which block hydrodynamic flow;
—absence of a source of carbonate cement.

While one or other of these three basic controls may explain the preservation of a particular reservoir, the author believes that the question of *source* is probably the most important factor determining the preservation of porosity. The objective of this contribution therefore is to outline the reasons for this conclusion.

There are at least three major potential sources of carbonate which may precipitate from solution in the form of pore-destroying cement:

Surface cementation is a frequent phenomenon in Quaternary carbonates whose petrographic aspects are outlined in the following section. Present in virtually all vadose and shallow phreatic environments, these cements are responsible for the lithification of important quantities of carbonate sediment. However, studies of Quaternary diagenesis in the Caribbean (Land, 1967; Mathews, 1968; Harris, 1977), and along the shores of the Arabian

**Dept. des Sciences de la Terre, Bâtiment 505, Faculté des Sciences, Université de Paris XI, 91405 ORSAY, France.*

Gulf (pers. obs.) all indicate that this lithification, while reducing primary porosity, rarely destroys it completely; most Pleistocene limestones would be effective reservoirs. This lack of complete destruction would seem to be related to an equilibrium which follows the early transformation of the sedimentary aragonite into a stable calcite.

Subsurface, deep phreatic cementation from interstitial fluids requires an effective hydrodynamic flow plus an important source of carbonate, possibly by "canibalisation" (dissolution) of limestones elsewhere in the basin, or from fluids squeezed from marls. While this process exists in certain basins, there are several major problems which introduce a doubt concerning its efficiency. Solutions flowing through porous limestone media rapidly attain equilibrium and are capable of precipitating carbonate mainly near their source (Weyl, 1958). Furthermore, the complete cementation of a porous limestone mass requires the passage of great volumes of fluid whose transit would become increasingly difficult as permeability was reduced during cementation. In sum, an "external" carbonate source, although possible, has serious difficulties in being introduced into the limestone mass.

Local source carbonate liberated during pressure-dissolution of adjacent carbonates must constitute a potential source for pore-destroying cements. There exist numerous examples where pressure-solution both by stylolites and, especially, by grain compaction, has reduced the thickness of rock sequences by at least 20% (Dunnington, 1967; Park & Schot, 1968; Purser, 1975 etc.). As in quartz sandstones, it is highly probable that a major part of the dissolved rock has reprecipitated locally, destroying pre-existing porosity. If this hypothesis is substantially correct, it is clear that any factor which tends to *inhibit pressure-solution* will favour the preservation of porosity.

Examination of certain carbonate reservoirs, both in Europe and in the Middle East, together with study of published information relating to the petrography of porous limestones, has convinced the author that many reservoirs are characterized by a minimum degree of compaction. Although stylolites frequently occur in reservoir formations, the porous zones generally have fewer stylolites and, more importantly, grain contacts are generally "normal"; overpacked fabrics in reservoirs are the exception. One of the principal factors responsible for this lack of compaction would seem to be related to the diagenetic history of the porous horizon, notably the frequent presence of an early (pre-burial) diagenetic phase. The identification of this first-phase diagenesis has been facilitated greatly by studies of Quaternary diagenesis, essential aspects of which are summarised in the following section.

EARLY (PREBURIAL) DIAGENESIS: NATURE & ORIGINS

One of the major contributions of modern research on recent carbonate sedimentary basins concerns the widespread synsedimentary lithification of these sediments under a variety of contrasting environments. These studies have revealed that carbonate sediments may lithify on a regional scale within submarine, intertidal and continental areas. Furthermore, each of these three major environments imparts a particular diagenetic fabric to the sediment, facilitating the interpretation of certain ancient analogues. The essential petrographic aspects of these three major diagenetic environments, summarised in Fig. 1, are as follows.

Submarine diagenesis in modern seas results in the lithification of sands and reefs; it rarely affects muds. Common in both the Arabian Gulf (Shinn, 1969) and in the Bahamas (Taft, *et al.*, 1968), it generally results in the formation of thin, superficial crusts, rarely exceeding one metre in thickness, whose upper surfaces are bored and encrusted with organisms. These surfaces have much in common with ancient hard-grounds. The aragonite or magnesian calcite cements responsible for lithification most often have an acicular or micritic habit and generally are distributed in an isopachous manner around the intergranular void. This disposition reflects the complete water saturation of these submerged substrates.

Inter and supratidal diagenesis has much in common with that of the submarine domain, aragonite and magnesian calcite cements generally being precipitated from sea water. However, because beaches are only partly saturated with water, cementation tends to be irregular. Fabrics include grain-contact cement (meniscus cement of Dunham, 1971), microstalactitic cements (Purser, 1969) and various other features relating to the partial saturation in this marine vadose zone. Beach-rocks, like submarine hard-grounds, rarely exceed one metre in thickness, and are underlain by loose sediment. As one proceeds up the beach into the supratidal zone the increasing effects of meteoric waters are expressed as dissolution of aragonite grains and by the "solid state" recrystallisation of these elements into calcite.

Continental (non-marine) diagenesis contrasts with that of the marine environment both by the presence of massive dissolution phenomena and by the contemporaneous precipitation of

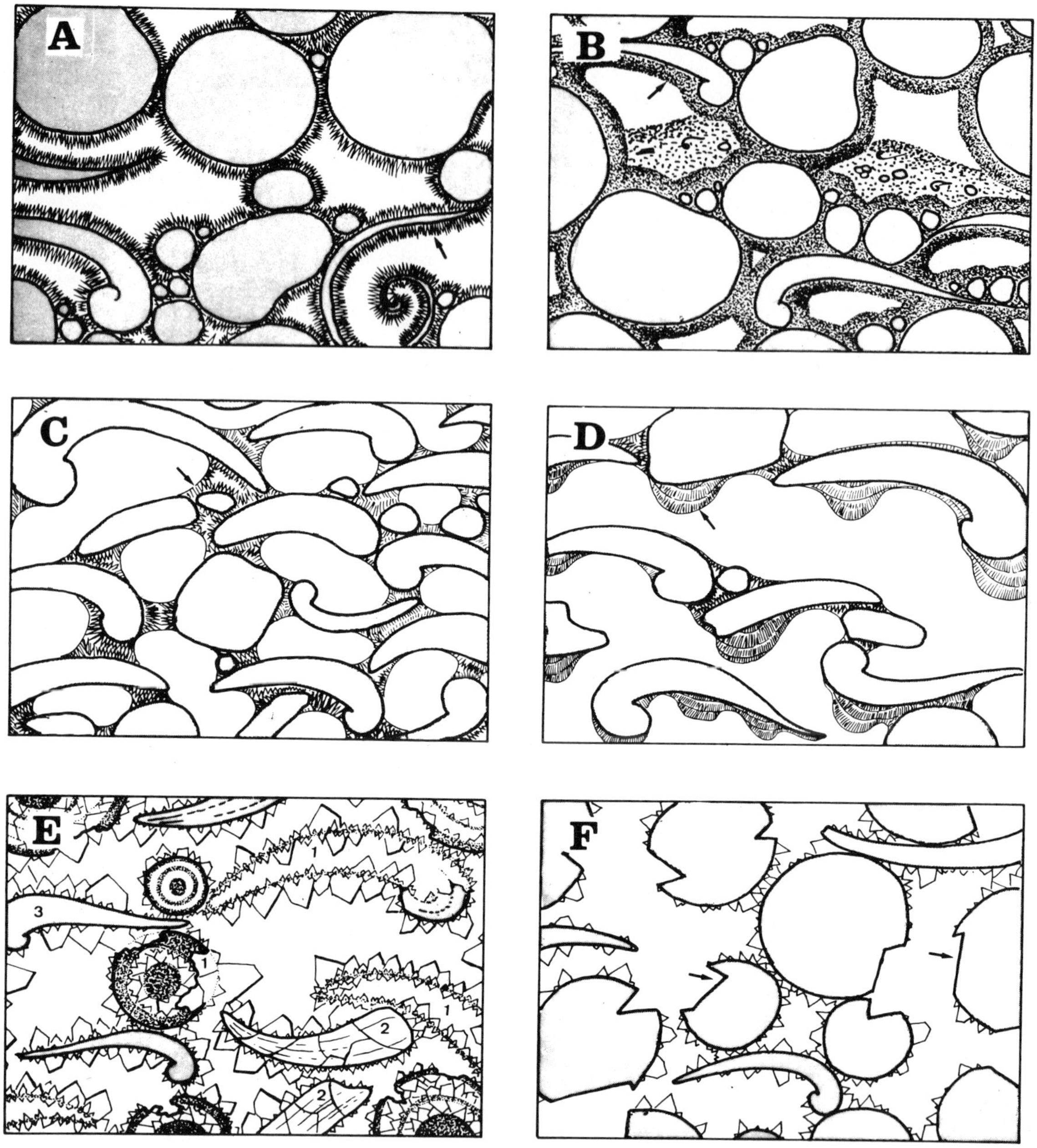

Fig. 1 Essential Petrographic Aspects of Recent Carbonate Diagenesis. A & B — Submarine Cements: Isopachous fibrous (A) or micritic (B) aragonite or magnesian calcite cements reflect sediments saturated with marine waters; C & D — Inter-Supratidal Cements: grain-contact (C) and microstalactitic (D) fibrous cements reflect partial saturation by marine waters; E & F — Continental Diagenesis: characterised by dissolution (1) or recrystallisation (2) of aragonite grains and unaltered calcite (3) particles; cement is blocky calcite. In F, dissolution of dedolomite is indicated by angular or flat-sided secondary pore-spaces. (All drawings from thin-slides).

blocky, sparitic calcite cement. Emergence of marine, carbonate sediments almost invariably is accompanied by lithification. The thickness of sediments thus affected is variable, depending partly on relief; generally speaking, virtually all sediments situated above the water table are lithified while those below it exist in various states ranging from uncemented (Abu Dhabi, pers. obs.), slightly cemented (Joulters Cay, Bahamas, Harris, 1977), to strongly cemented (Bermuda, Land, 1970). Depending on the degree of emergence, near-surface continental diagenesis may affect a greater thickness of sediment than the equivalent marine diagenesis.

An important feature of continental diagenesis concerns the calcitisation of dolomite (dedolomite) and the subsequent dissolution of this latter to produce a secondary, rhombic porosity (Evamy, 1967). This process of "dedolomitisation" seems to be a common soil phenomenon whose activity requires both low Mg/Ca ratios and low temperatures and partial pressures (De Groot, 1967). As such, it is restricted to surface diagenetic environments. The presence of dedolomite or rhombic porosity in the subsurface therefore implies an early, pre-burial diagenesis.

Ancient sequences may be affected both by pre-burial (early) and by post-burial (late) diagenetic processes, the latter nearly always being predominant. The distinction between these two (or more) phases is not always possible, mainly because sparitic cements may precipitate from non-marine waters at any time before and during burial. Dissolution of aragonite grains, on the other hand, generally seems to reflect surface (pre-burial) diagenesis, while the presence of thin, first phase, fibrous cements probably are reliable indications of early, essentially marine lithification. Early cements generally occur below discontinuity surfaces.

In conclusion it is stressed that surface diagenesis of Quaternary sediments, both in the marine and in the continental domains, although reducing initial porosity, rarely destroys it completely. Furthermore, the lithified but porous limestones frequently overlie, or include, intercalations of unlithified, friable sediments. During subsequent burial one may logically expect that these two phases (lithified and friable) will react differently to the effects of overburden or tectonic pressures.

EARLY DIAGENETIC FABRICS IN MIDDLE JURASSIC LIMESTONES OF THE PARIS BASIN

Most of the Quaternary diagenetic fabrics summarised in the previous section have also been observed in ancient sequences, notably within the Middle Jurassic limestones of the Paris Basin. That their presence has tended to prevent compaction can be demonstrated by the following observations.

Non-compacted lithoclasts (Fig. 2A). The sand or gravel-sized elements, by definition, are sedimentary rock fragments which are dispersed through many limestones, including the Middle Jurassic formations of the Paris Basin and the "Arab Zones" of Qatar and Saudi Arabia (Powers, 1962). When individual clasts are composed of grainstone (which often is the case) the early cement responsible for their lithification can readily be distinguished (see Fig. 2B). Compacted, lithoclast-bearing limestones, on the contrary, rarely have early cements, the granular interpenetration (compaction) of this matricial sediment contrasting with the lack of compaction within the clasts. In these early lithified rock fragments, the primary, sedimentary packing is preserved.

Non-compacted Middle Jurassic hard-grounds. The Dogger calcarenites of the Paris Basin are composed of numerous cyclic sequences most of which terminate in a hard-ground. Mollusc borings which penetrate these surfaces truncate both grains and the first phase of fibrous cement, permitting identification of the synsedimentary, diagenetic phase. The habits of these cements indicate that certain hard-grounds have formed in submarine environments while others have formed as beach-rocks.

Submarine hard-grounds (Fig. 2C) are characterised by the presence of a thin (30-50 microns), isopachous fringe of fibrous cement which may be cut by borings. This early cement affects only 10-30cm of sediment, thus indicating the formation of crusts on the Jurassic sea floor. The oolitic or bioclastic grains within these paleo-crusts are *uncompacted.* Similar sediments lacking early cement, situated directly above or below the zone of synsedimentary lithification, on the contrary, are generally highly overpacked and stylolites are frequent.

Intertidal or supratidal hard-grounds (Fig. 2D) are very common within an Upper Bathonian, pelletoidal limestone (the Calcaire de Comblanchien). They are interpreted as paleobeach-rocks, not only because of the frequent presence of bird's-eye structure and algal laminations, but also because of the presence of vadose cements; early, fibrous cement frequently exhibits stalactitic morphology, being thickest on the lower surfaces of larger grains, in this respect closely resembling the cements of many modern beach-rocks. This pre-burial cement affects numerous, 30-70cm thick horizons of pelletoidal

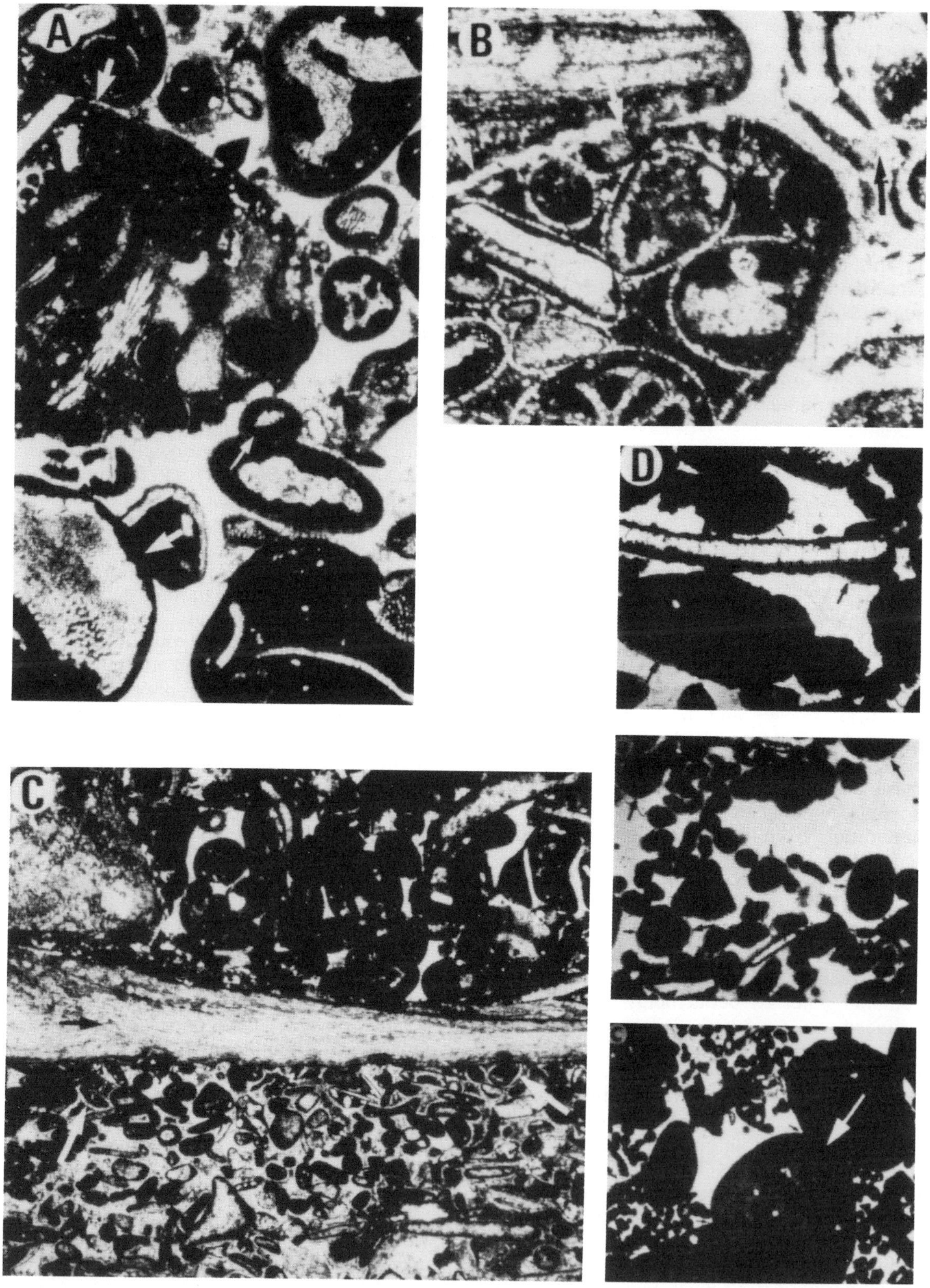

Fig. 2 Early Marine Diagenetic Fabrics in Middle Jurassic Limestones of the Paris Basin. A & B—Lithoclasts; note the lack of granular compaction within the clasts and overpacked fabrics (arrows) in surrounding sediment, this latter lacking early cement; C—Submarine Hard-Ground: an oyster (arrow) cemented to hard substrate separates (below), uncompacted oolite with early fibrous cement (arrows), and highly compacted oolite (above) lacking early cement; D—Intertidal Beach-Rock: three photos showing (in upper photo) early microstalactitic cement (arrows), and thin isopachous cement around birds'eye void (middle photo): NB lack of compaction contrasting with lower photo, lacking early cement, having long grain contacts (arrow).

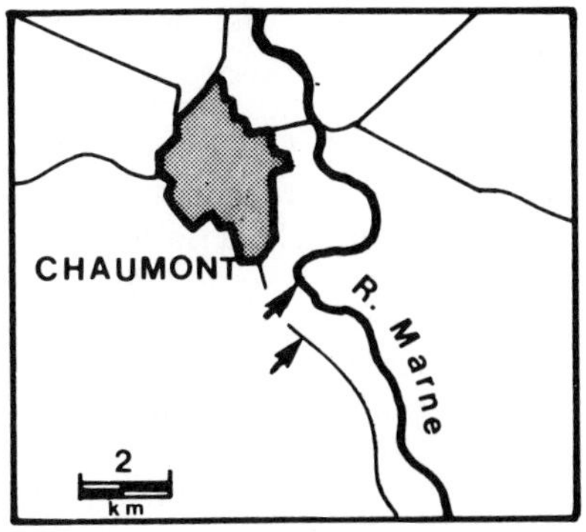

Fig. 3 Early Cementation of Bathonian Oolite, Marne Valley, Paris Basin. The upper 15m of oolite is affected by soil phenomena at the top (A) and early fibrous cements (B). This uncompacted, porous zone, lacking stylolites, contrasts with the underlying oolite (C) lacking early cements.

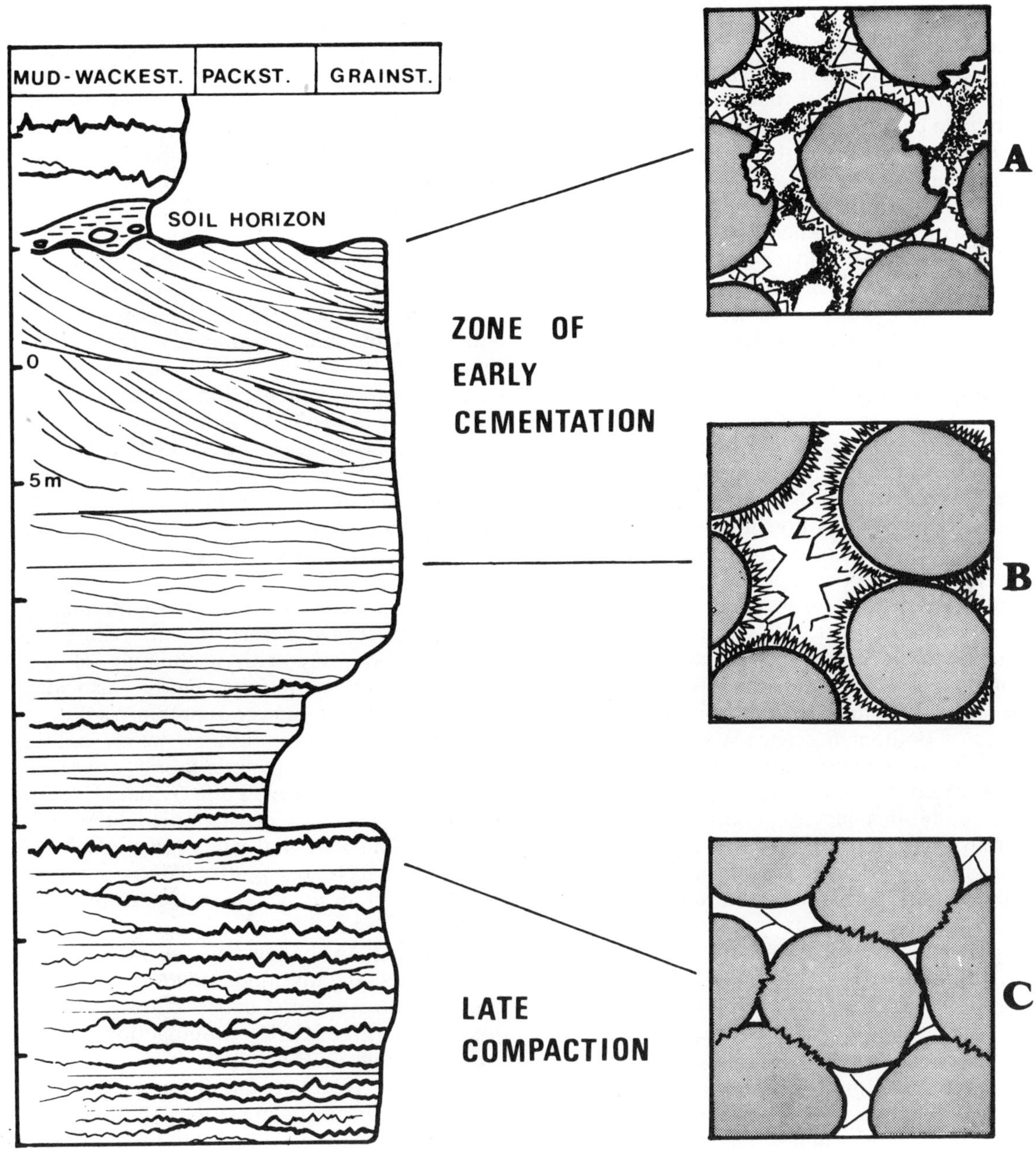

grainstone, these paleocrusts being underlain by sediments of similar composition and texture lacking early cement. As within the submarine hard-grounds, these paleobeach-rocks, although having occasional large stylolites, lack granular compaction.

Thin slides of the scattered stylolites located within the zone of beach-rock cementation reveal that some (if not all) stylolites coincide with thin horizons lacking early cement. This would suggest that these Bathonian beach-rocks, as in many modern analogues, have consisted initially of limestone crusts and interbedded friable sediments. Pelletoidal grainstones situated below the zone of beach-rock cementation are strongly overpacked and stylolites are abundant. Careful measurement of stylolite amplitudes and degree of grain interpenetration suggest that those sediments which have been buried in a friable state have been reduced in thickness (by pressure-solution) in the order of 40%.

The Middle Jurassic limestones of the Paris Basin (average thickness 200m) contain abundant stylolites and overpacked fabrics and it is clear that much carbonate has been liberated during pressure-solution. It therefore is not surprising that most sequences, including the numerous, thin, uncompacted zones of early lithification, are strongly cemented with sparitic calcite.

Porous Bathonian oolite of the Marne Valley (Fig. 3). In addition to the numerous, but thin layers of calcarenite having submarine or intertidal diagenesis, oolitic limestone measuring up to 15m in thickness is affected locally by early cement. A spectacular example outcrops on both sides of the Marne Valley 2km SSE of the town of Chaumont.

The top of the Bathonian "Oolithe blanche" is limited by an erosion surface overlain by some 50cm of lignitic clay containing reworked pebbles of Bathonian oolite. Synsedimentary lithification of the upper 15m of the formation is confirmed by the presence of several diagenetic fabrics, including both non marine and fibrous cements. The presence of root-traces and other soil phenomena (caliche) localised on the top of the oolite suggests that this early diagenesis was the result of local emergence, possibly comparable to that of many Caribbean cays. The relief may explain the relatively thick (15m) development of early cement and other diagenetic features.

The zone affected by pre-burial cementation has no stylolites, grain packing is essentially normal, and the rock is *porous.* This contrasts with similar oolitic limestone lower in the sequence lacking early cement. These latter have abundant stylolites, grains are overpacked, and the rock is tightly cemented with sparitic calcite. Similarly, the lagoonal carbonate muds discordantly overlying the soil horizon at the top of the oolite have many stylolites. That localised, early lithification has prevented compaction of the top 15m of the "Oolithe blanche" is confirmed by a Bathonian sequence in the Meuse Valley, some 50km to the N. Here, the contact between the oolitic barrier-facies and the overlying lagoonal muds is conformable and there is no trace of emergence and early lithification of the oolite. This latter exhibits numerous stylolites and overpacked fabrics from top to bottom. The formation is tightly cemented throughout.

Middle Jurassic reservoirs of Coulommes field (Fig. 4). The Middle Jurassic limestones of the Paris Basin produce from two distinct formation: an upper (Callovian), oolitic and bioclastic formation, the "Dalle nacrée", which underlies Callovo-Oxfordian cap-rock marls, and a lower (Bathonian) pelletoidal formation, the Calcaire de Comblanchien, already mentioned in connection with paleobeach-rocks.

In the Coulommes field, situated near Meaux (some 50km E of Paris) the Callovian oolitic formation includes a lenticular porous zone. Numerous thin slides (examined by the author) reveal that the reservoir has well developed, early diagenetic cements (Fig. 5 A, B) and that the porosity is partly secondary. Geometry of the pore spaces suggests that they have been formed by slight enlargement of a rhombohedral crystal mould, almost certainly resulting from the dissolution of dedolomite. Other pore spaces are primary. This reservoir has occasional stylolites but grain packing is essentially normal. However, the lenticularity of this uppermost reservoir unit (Zone A) coincides not only with a thinning of the formation, but—more importantly—also with the appearance of numerous pressure-solution phenomena. In these thinner, *dense* oolitic grainstones, early diagenetic fabrics are rare.

The two lower reservoir units (Zones C & D), within the Calcaire de Comblanchien, are developed in pelletoidal sands and muds. That these units have been affected by synsedimentary diagenesis is demonstrated not only by the presence of vadose, microstalactitic cements (Fig. 6A) but also by the nature of the pore spaces; slightly modified rhombic voids clearly indicate early dedolomitisation and dissolution (Fig. 5C, D). Compaction within these reservoir units is minor.

In sum, the relatively minor parts (5-10%) of the Middle Jurassic limestones of the Paris Basin which have been affected by pre-burial lithification are only slightly compacted. Certain of these units are effective reservoirs. The bulk of the Jurassic grainstones, buried in

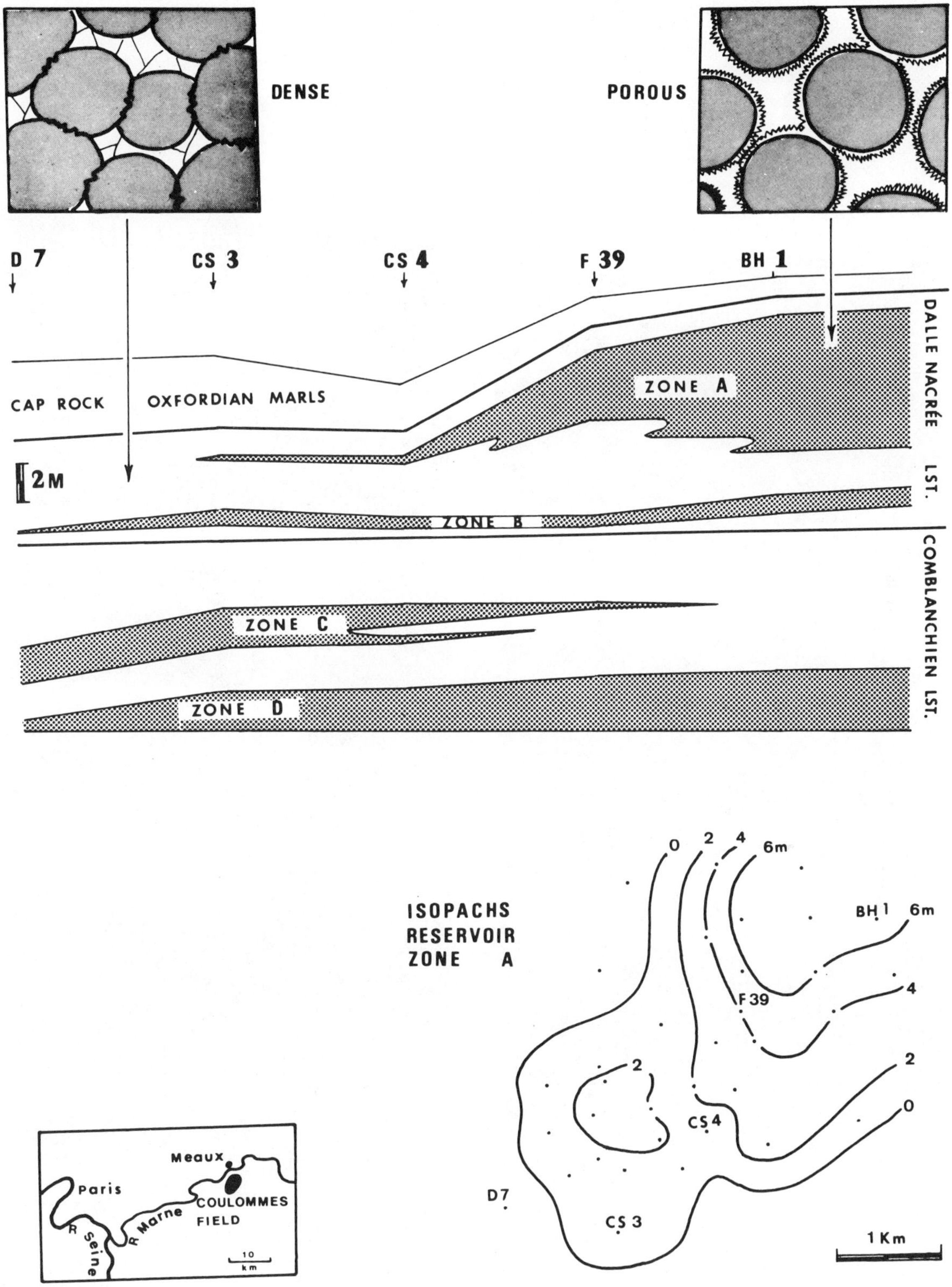

Fig. 4 Early Diagenesis in Middle Jurassic Reservoirs at Coulommes Field, Paris Basin. The lenticularity of the uppermost reservoir (Zone A) is due mainly to localised early cementation which seems to have prevented late compaction of the oolitic grainstone. Lower zones C & D are developed in pellet sands and muds (see Fig. 5).

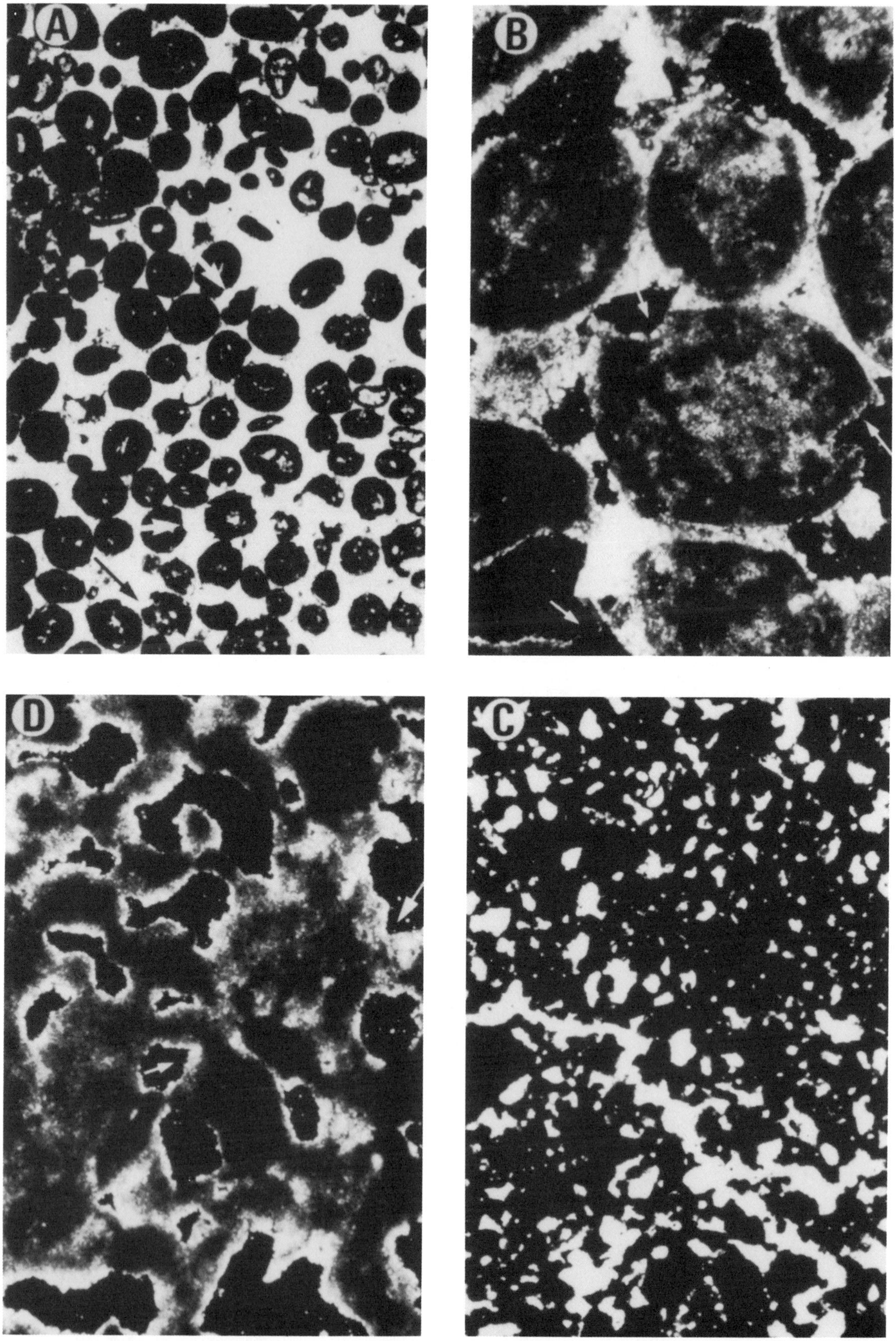

Fig. 5 Middle Jurassic Reservoir Fabrics, Coulommes Field, Paris Basin; A & B Zone A: oolitic reservoir showing dedolomite porosity and lack of compaction; C & D Zone C: pellet muds showing well developed rhombohedral porosity (arrows); all photos from well BHI, A × 10; B × 40; C × 10; D × 30.

a friable state, on the contrary, have abundant stylolites and overpacked pressure-solution fabrics indicating that important volumes of carbonate have been liberated. These limestones are tightly cemented with sparitic calcite.

EARLY DIAGENESIS OF CERTAIN UPPER JURASSIC RESERVOIRS: THE SMACKOVER OF LOUISIANA AND THE ARAB ZONES OF ARABIA

The Smackover Formation of N. Louisiana, studied by Bishop (1968), seems to have much in common with the Callovian oolitic limestones of the Paris Basin. In North Haynesville Field (according to Bishop, *op. cit.*), this formation grades upwards from argillaceous wackestones and pelletoidal packstones into oolitic and pelletoidal grainstones, the sequence being limited by a well-defined discontinuity surface. Most of the formation has numerous stylolites and over-packed fabrics, including much of the terminal carbonate sand unit. Only the uppermost parts of this latter are porous. This relatively thin (max. 18m), lenticular reservoir unit, according to Bishop, lacks stylolites and grain packing tends to be normal. Although discussed essentially in terms of sedimentary processes by Bishop, it would appear that the top of the Smackover Formation has suffered an early diagenetic lithification. Isopach and facies maps presented by Bishop indicate that North Haynesville Field is located near the culmination of a carbonate sand shoal susceptible to local emergence and thus to early lithification of its crestal parts. Again, preservation of porosity would appear to have been conditioned by pre-burial diagenesis which has prevented pressure-solution, this latter strongly affecting the remaining parts of the formation.

The "Arab D" reservoirs of Qatar and Saudi Arabia. Personal research and published literature (Powers, 1962 & Wilson, 1975) suggests that the relationships between early diagenesis and the preservation of porosity, deduced from studies of Middle Jurassic carbonates of the Paris Basin, may also exist within the Upper Jurassic calcarenitic reservoirs of the Arabian Gulf region.

Arab D calcarenites, both in Qatar and in Saudi Arabia, exhibit abundant traces both of early lithification and dissolution (Fig. 6B, C). The presence of numerous lithoclasts and the seemingly widespread distribution of an early fibrous cement, visible in many of the photographs published by Powers, strongly suggest early lithification of parts of this calcarenite. This early lithification is not entirely surprising; the regressive nature of the Arab zones, certain of which terminate in stromatolites and "chicken wire" anhydrites, suggests emergence at the end of each regressive phase. In Quaternary equivalents, emergence almost invariably leads rapidly to lithification.

The geometry of individual pore spaces within certain Arab D calcarenites reveals secondary enlargement. The pores occasionally have flat sides or acute angles strongly suggesting that they have been produced by dissolution of a pre-existing crystal, possibly of rhombic habit (Fig. 6D). In other words, parts of the Arab D reservoirs (examined by the author) are probably the products of dissolution of dedolomite.

An essential feature of these Arab D reservoirs is the very low degree of granular compaction. Although stylolites occur throughout much of the sequence (Dunnington, 1967) creating thin, horizontal, impermeable barriers, the reservoirs as a whole are not greatly affected by compaction. In sum, the presence of an early diagenetic phase within these limestones may well be responsible for the preservation of effective porosity.

Conclusions

The presence of early diagenetic phenomena within the porous Jurassic limestones of the Paris Basin, Louisiana, and in Arabia, suggests that early lithification has tended to retard or prevent porosity-destroying pressure-solution. This hypothesis is confirmed by the lateral passage from porous to dense limestone, the latter generally lacking early diagenetic fabrics but exhibiting a relatively high degree of pressure-solution. This has liberated important quantities of carbonate from which pore-filling, sparitic cement probably has precipitated.

If this conclusion is substantially correct, it is clear that the recognition of early diagenetic fabrics, facilitated by studies of Quaternary sediments, may be important to the understanding of limestone reservoirs.

Speculating, one might also suspect that the frequent occurrence of effective porosity in dolomites and in reefoid limestones may be due partly to this same phenomenon. Many (possibly most) dolomites seem to have formed under near-surface conditions. Furthermore, the mineral dolomite is somewhat more resistant to lithostatic or tectonic pressures than is calcite; stylolites are less frequent in dolomites than in limestones (pers. obs. & discussions with numerous colleagues). Local-source carbonate cement, liberated by pressure-solution, is therefore less readily available in dolomites than in limestones.

Studies of modern reefs reveal that they are particularly susceptible to early lithification (Schroeder, 1972; Ginsburg *et al.*, 1973). Early

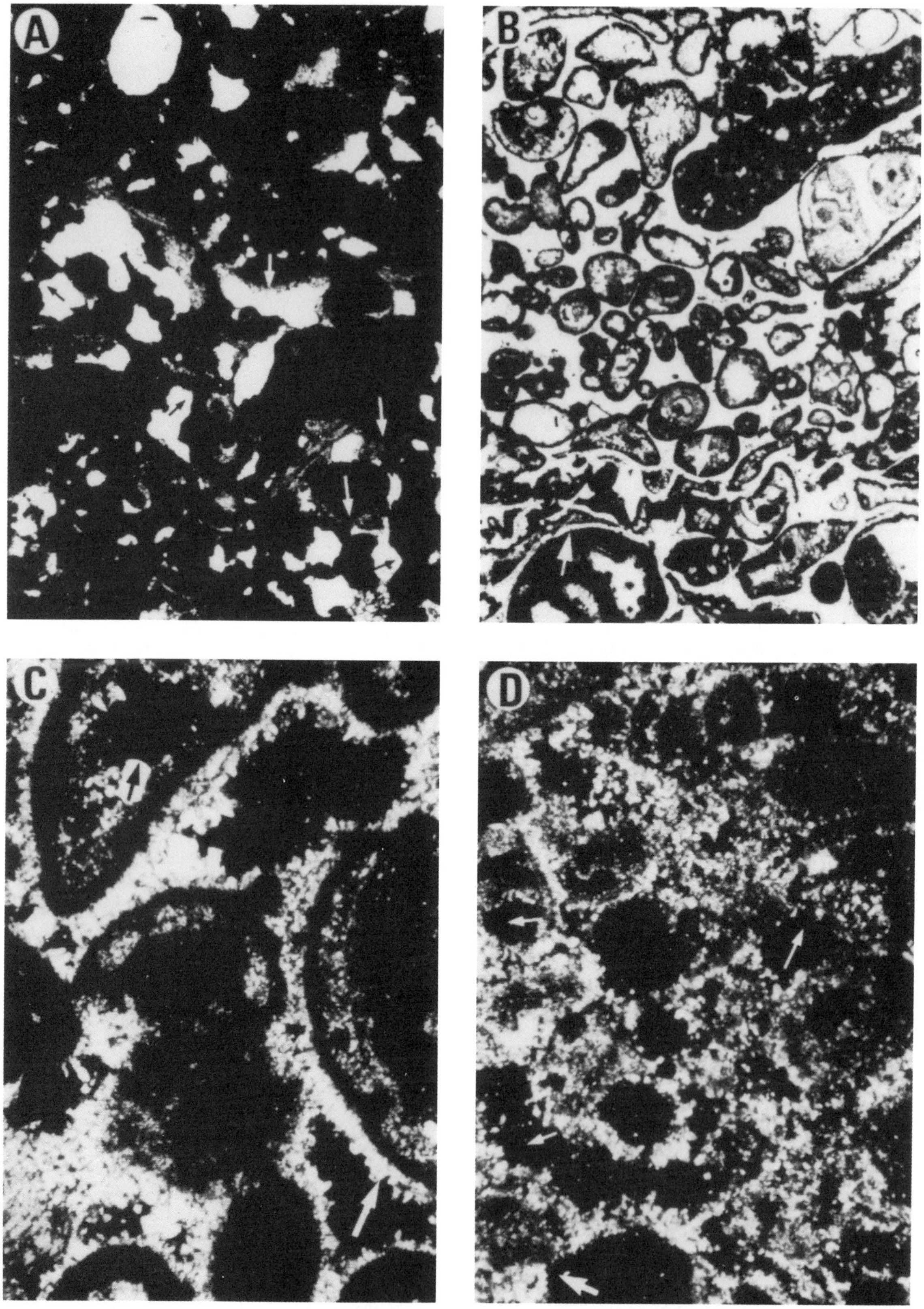

Fig. 6 Jurassic Reservoir Fabrics, Paris Basin & Arabia. A (× 20) Coulommes Field Zone D: pellet grainstone with early microstalactitic cements (white arrows) and rhombohedral porosity (black arrows). NB: lack of compaction. B (× 15) Arab D Reservoir: grainstone from Qatar showing early cement (arrows) and absence of compaction; C (× 50) Arab D Reservoir: showing early cement (white arrows), dissolution of bioclast (black arrows) and lack of compaction; pore-spaces are black (polarised light); D (× 30) Arab D Reservoir: secondary porosity (black) with acute angles suggesting crystal moulds (white arrows).

diagenetic cement may be essential to the construction and survival of these wave-resistant masses. More importantly, this cement, together with frequent dolomitisation, may be a significant factor in the *preservation* of porosity.

BIBLIOGRAPHY

BISHOP, W. F., 1968. Petrology of the Upper Smackover limestone in North Haynesville Field, Claiborne Parish, Louisiana. *Am. Assoc. Petrol. Geol. Bull.*. **52,** 1, 92-128.

CHOQUETTE, P. W. & PRAY, L. C., 1970. Geological nomenclature and classification of porosity in sedimentary carbonates. *Am. Assoc. Petrol. Geol. Bull.*. **54,** 2, 207-250.

DUNHAM, R. J., 1971. Meniscus cement. *In* Carbonate cements. Bricker, O. P., Ed., Johns Hopkins Press, Baltimore and London 297-300.

DUNNINGTON, H. V., 1967. Aspects of diagenesis and shape-change in stylolitic limestone reservoirs. *7th World Petrol. Congr. (Mexico) Proc.*. **2,** 339-352.

DE GROOT, K., 1967. Experimental de-dolomitisation. *Journ. Sed. Pet.*. **37,** 1216-1220.

EVAMY, B. D., 1967. Dedolomitisation and the development of rhombohedral pores in limestones.*Journ. Sed. Pet.*. **37,** 1204-1215.

GINSBURG, R. N., 1973. Sedimentation and diagenesis in the deep fore-reef, British Honduras barrier and atoll reefs. *Am. Assoc. Petrol. Geol. Bull.*. **57,** Abstr., 781.

HARRIS, P. M., 1977. Sedimentology of the Joulters Cays ooid sand shoal, Great Bahama Bank. *Thesis.* University of Miami, 452p.

LAND, L. S., 1967. Diagenesis of skeletal carbonates. *Journ. Sed. Pet.*. **37,** 914-930.

LAND, L. S., 1970. Phreatic versus vadose meteoric diagenesis of limestones: evidence from a fossil water-table. *Sedimentology,* **14,** 175-185.

MATHEWS, R. K., 1968. Carbonate diagenesis: equilibrium of sedimentary mineralogy to the subaerial environment, coral cap of Barbados, W.I. *Journ. Sed. Pet.,* **38,** 1110-1119.

PARK, W. & SCHOT, E. H., 1968. Stylolitisation in carbonate rocks. *In* Carbonate sedimentology in Central Europe. Müller, G. & Friedman, G. M., Eds., Springer-Verlag, 66-74.

POWERS, R. W., 1962. Arabian Upper Jurassic carbonate reservoir rocks. *Am. Assoc. Petrol. Geol. Memoir,* **1,** 122-192.

PURSER, B. H., 1969. Synsedimentary lithification of Middle Jurassic limestones in the Paris Basin. *Sedimentology,* **12,** 205-230.

PURSER, B. H., 1975. Sédimentation et diagenèse précoce des séries carbonatées du Jurassique moyen de Bourgogne. *Thesis,* Univ. de Paris-Orsay, 383p.

SCHROEDER, J. H., 1972. Fabrics and sequences of submarine carbonate cements in Holocene Bermuda cup reefs. *Geol. Rundsch.* **61,** 708-730.

SHINN, E. A., 1969. Submarine lithification of Holocene carbonate sediments in the Persian Gulf. *Sedimentology,* **12,** 109-144.

TAFT, W. H. *et al.*, 1968. Lithification of modern carbonate sediments at Yellow Bank, Bahamas. *Bull. Marine Sci. Gulf Caribbean,* **18,** 762-828.

TAYLOR, J. M. & ILLING, L. V., 1969. Holocene intertidal calcium carbonate cementation, Qatar, Persian Gulf. *Sedimentology,* **12,** 69-107.

WEYL, P. K., 1958. The solution kinetics of calcite. *Journ. Geol.,* **66,** 163-176.

WILSON, J. L., 1975. Carbonate facies in geological history. Springer-Verlag, 471p.

The American Association of Petroleum Geologists Bulletin
V. 66, No. 12 (December 1982), P. 2561-2570, 9 Figs., 2 Tables

Carbonate Porosity Versus Depth: A Predictable Relation for South Florida[1]

JAMES W. SCHMOKER and ROBERT B. HALLEY[2]

ABSTRACT

This study examines the porosity of limestones and dolomites in the South Florida basin. Porosity data are derived from borehole-gravity measurements and from suites of acoustic, neutron, and density logs. Both types of wire-line measurements sample large volumes of rock relative to petrographic methods and can be examined at vertical scales approaching those of aquifers and hydrocarbon reservoirs. Investigation depths range from the surface to about 18,000 ft (5,500 m) and span the transition from high-porosity near-surface carbonate sediments of Pleistocene age to much denser Mesozoic carbonate rocks with porosities of only a few percent.

Carbonate porosity in the South Florida basin was affected by a variety of diagenetic processes. However, a number of factors that could complicate porosity-depth relations are of limited importance in southern Florida. The basin contains little clastic material; present depths of burial are about equal to maximum depths of burial; the influences of tectonism, geopressures, and hydrocarbon accumulations are minimal.

Curves of porosity versus depth, reflecting large-scale porosity-loss processes in the subsurface, are derived for a composite carbonate section and for carbonate strata of different ages and compositions. The decrease of porosity with depth for a composite carbonate section representing a wide range of depositional environments and subsequent diagenetic histories can be characterized by the exponential function $\phi = 41.73e^{-z/8197}$ (ft) [$\phi = 41.73e^{-z/2498}$ (m)], where ϕ is the porosity (%) and z is the depth below ground level (feet or meters). Average porosity is reduced by a factor of two in a depth interval of about 5,700 ft (1,740 m).

Carbonate strata of different ages that are buried to equal depths show no systematic porosity differences. This implies that the effect of time on porosity in these rocks is probably subordinate to that of burial depth. The data also show a faster than expected rate of porosity decrease with depth for rocks of Eocene age and younger. If it is assumed that the decrease in the volume of evaporites in these rocks indicates less saline pore fluids, porosity loss in shallow-water carbonates may be inversely related to the magnesium content of pore waters.

Dolomite porosity is lower than limestone porosity in the near surface, but does not decrease as rapidly with depth. Below about 5,600 ft (1,700 m), dolomite is more porous than limestone. It is hypothesized that most dolomitization occurred relatively early and either reduced original porosity or selectively favored lower-porosity limestones. With continued burial, dolomite was more resistant than limestone to associated porosity-reducing effects.

INTRODUCTION

Newly deposited carbonate sediments typically have porosities of 40 to 70%, whereas ancient carbonate strata with porosities greater than a few percent are uncommon (Choquette and Pray, 1970). Thus, the dominant trend of porosity evolution in the subsurface is that of porosity destruction (Choquette and Pray, 1970), but there is little quantitative documentation of these porosity-reducing processes for shallow-water carbonates (Choquette and Steinen, 1980). The problem of theoretically predicting the porosity of carbonate rocks, and in particular of shallow-water carbonate rocks, is a difficult one (Scholle, 1981). The majority of carbonate studies are concerned with either recent or ancient end members, or with individual hydrocarbon reservoirs representing single (and possibly atypical) points along the continuum of porosity diagenesis.

This article describes porosity data from limestones and dolomites of peninsular Florida that span the transition from high-porosity near-surface sediments of Pleistocene age to much denser Mesozoic rocks with porosities of only a few percent. Investigation depths range from the surface to about 18,000 ft (5,500 m). Porosity data are derived from borehole-gravity measurements and from suites of acoustic, neutron, and density logs. Both types of wire-line measurements sample large volumes relative to petrographic methods, and are averaged here over vertical intervals of 5 to 70 ft (1.5 to 21.3 m) that approximate the scale of aquifers and hydrocarbon reservoirs.

Empirically derived curves of porosity versus depth for the South Florida basin and for carbonate strata of different ages and composition are given for a composite carbonate section. These curves reflect large-scale porosity-loss processes in the subsurface.

GEOLOGIC SETTING

The wire-line data of this report are from wells within, and along the northern shelf edge, of the South Florida basin (Fig. 1). The South Florida basin comprises an area of about 77,000 sq mi (200,000 sq km) in southern peninsular Florida

[1]Manuscript received, March 1, 1982; accepted, July 7, 1982.

[2]U.S. Geological Survey, Box 25046, MS 921, Federal Center, Denver, Colorado 80225.

We thank F. W. Meyer, U.S. Geological Survey, Miami, and S. L. Robbins, D. L. Smith, and R. J. Martinez, U.S. Geological Survey, Denver, for their help in obtaining the borehole-gravity data used in this study. T. C. Hester and J. A. Medeski, U.S. Geological Survey, Denver, assisted in the processing of the conventional wire-line data.

and adjacent waters to the west (Oglesby, 1965). The basin contains up to 20,000 ft (6,100 m) of Cenozoic and Mesozoic rocks, most of which are limestones, dolomites, and anhydrites (Pressler, 1947).

General depositional conditions in the basin continued more or less unchanged from the Early Cretaceous through much of the Tertiary, with carbonates and evaporites deposited by shallow seas upon a slowly subsiding landmass (Palacas, 1978). Within this framework, however, there were repeated cycles of transgression and regression (Banks, 1960; Oglesby, 1965; Winston, 1972; Puri and Winston, 1974, p. 9; Randazzo and Hickey, 1978).

The cyclic transgressions and regressions typical of the basin throughout its history produced lateral shifts of shallow-shelf, subtidal, intertidal, and supratidal environments. It follows that the wire-line data of this report, which sample the complex carbonate assemblage of the South Florida basin both laterally and vertically, represent most types of shallow-water carbonate depositional environments.

The present temperature gradient in the South Florida basin, based on corrected onshore bottom-hole well temperatures, averages about 1.0° F/100 ft (1.8° C/100 m) (Geothermal Survey of North America Subcommittee, 1976). In the upper part of the section, the geothermal gradient is locally affected by the circulation of fresh and salt water (Kohout, 1967). Zones of geopressure, to our knowledge, were not encountered in wells drilled to date.

Commercial hydrocarbon production in the South Florida basin occurs only along a northwest-southeast-trending fairway known as the Sunniland producing trend (Fig. 1). Here, 10 fields with total recoverable reserves of about 100 million barrels of oil produce at an average depth of 11,500 ft (3,500 m) from the Sunniland Limestone of Early Cretaceous (Trinity) age (Pontigo et al, 1979). Production is from low-relief depositional and compactional structures which, in Sunniland time, bordered the northeastern rim of a Cretaceous basin (Applin, 1960; Feitz, 1976). The likely source of the oil is organic matter within the Sunniland Limestone (Palacas, 1978).

The South Florida basin is particularly well-suited for the study of regional carbonate porosity because (1) it contains a nearly continuous sequence of carbonate rocks ranging in depth from the surface to about 20,000 ft (6,100 m) and in age from Pleistocene through Cretaceous; (2) it contains little clastic material, either disseminated in carbonate rocks or as separate facies; anhydrite is the only non-carbonate rock present in significant amounts; (3) present depths of burial are about equal to maximum depths of burial, so complicating effects of subsidence and subsequent uplift upon porosity are limited; and (4) the influence of tectonism, geopressures, and hydrocarbon-containing pore fluids has been minimal.

The South Florida basin is thus a setting where carbonate porosity can be determined on a regional basis over a wide range of ages and depths, with porosity evolution affected by a minimum of complicating factors relative to many other basins.

STRATIGRAPHIC SETTING

Lateral lithology changes and a lack of consistent marker horizons make it difficult to subdivide and correlate rocks of the South Florida basin. Stages are established in only a few wells and stage boundaries are often approximate. Detailed stratigraphic nomenclature is developed only for those relatively few units of direct economic importance.

For these reasons, the stratigraphic positions of the wire-line data are defined here in terms of stage, or higher-rank time-stratigraphic boundaries. The divisions adopted for this report are shown in Figure 2. Boundary depths were derived at key well locations from Applin and Applin (1965), Puri and Winston (1974), and Palacas (1978), and extrapolated to other well locations on the basis of wire-line-log character. Figure 3 locates the stratigraphic positions of wire-line measurements from the wells located in Figure 1, and identified in Table 1.

POROSITY DETERMINATION

The upper parts of wells 1 and 2, and wells 3 through 5 (Fig. 1; Table 1), were logged by the authors with the U.S. Geological Survey LaCoste and Romberg[3] borehole gravity meter. Porosities were computed from the borehole-gravity data using standard methods (McCulloh, 1966; Robbins, 1981), with grain densities determined from lithologic logs

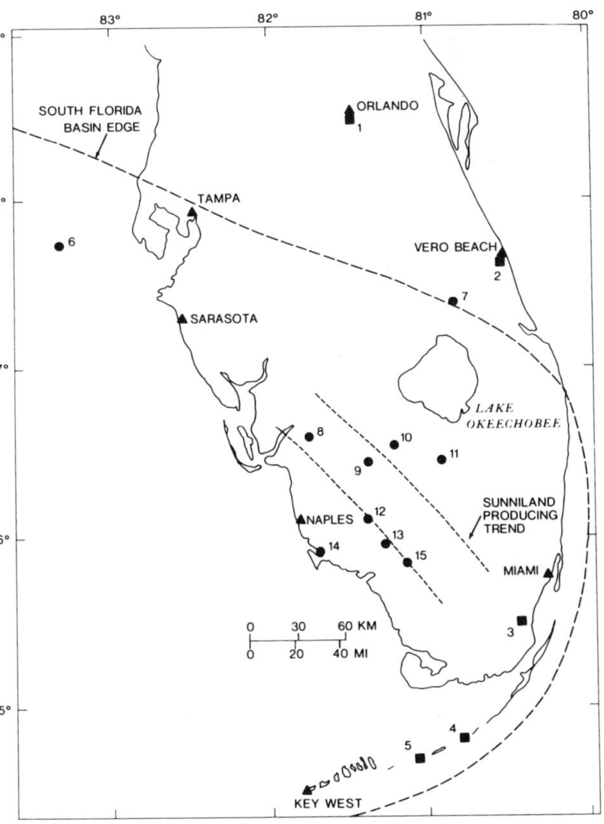

FIG. 1 — Map of peninsular Florida showing locations of wells where borehole-gravity (squares) and conventional-porosity-log (circles) data were obtained. Also shown are the approximate boundaries of the South Florida basin and the Sunniland producing trend. Wells are identified in Table 1.

[3]Use of brand names in this report is for descriptive purposes only and does not imply endorsement by the U.S. Geological Survey.

prepared by the well-site geologists. Details of these surveys and porosity computations are given by Schmoker et al (1979). The borehole-gravity data of this study are limited to the upper 2,740 ft (835 m) of section.

Borehole-gravity data are not significantly influenced by the wellbore rugosity or formation alteration that often occurs when drilling high-porosity formations. Also, the borehole gravity meter has a large radius of investigation and can sample heterogeneously distributed pore volumes typical of high-porosity carbonate rocks. For these reasons, it was thought that porosities of the shallow, very high-porosity carbonate strata of the study area could be more representatively determined from borehole-gravity measurements than from conventional wire-line logs.

Considering the fundamental accuracy of borehole-gravity measurements (Schmoker, 1978) and possible errors in determining grain density in limestone-dolomite sequences, individual porosity values calculated from the borehole-gravity data are estimated to have an experimental error of about ± 3 porosity units.

The porosity and lithology of strata in the lower parts of wells 1 and 2 and in wells 6 through 15 (Fig. 1; Table 1) were determined from suites of conventional acoustic, neutron, and density logs. The conventional-log data range downward in depth from 2,260 ft (689 m); there is little depth overlap with the borehole-gravity measurements. Although data are limited, the transition from borehole-gravity to conventional-log porosity determinations appears smooth when expected differences between limestone and dolomite porosity are taken into account. (These differences are discussed in a following section.)

Porosity was computed from log-determined values of formation density and grain density, with formation density obtained from the density log and grain density obtained from lithology estimates derived from the Schlumberger M-N plot (Burke et al, 1969). The M-N plot utilizes acoustic, neutron, and density logs to help identify the components of ternary mineral mixtures. Pores were assumed to be water saturated.

Because the intent was to study limestones and dolomites unaffected by non-carbonate minerals, all intervals that were indicated to contain shale or anhydrite by the M-N plot were rejected. The limestone-dolomite percentages of the carbonate intervals comprising the final data set are estimated to have an experimental error of about ±20%. Considering the resulting uncertainty in grain density, and the basic accuracy of the density log, individual interval-porosity values calculated from the acoustic, neutron, and density-log data are estimated to have an experimental error of about ±3 porosity units.

ERA	PERIOD	EPOCH	NAME USED IN THIS REPORT
CENOZOIC	QUATERNARY	RECENT PLEISTOCENE	EOCENE & YOUNGER
	TERTIARY	PLIOCENE MIOCENE OLIGOCENE EOCENE PALEOCENE	
			PALEOCENE
MESOZOIC	CRETACEOUS	LATE	GULF
		EARLY	WASHITA
			FREDERICKSBURG
			TRINITY
			OLDER THAN TRINITY

FIG. 2 — Age terminology used in this report.

DATA SET

Table 2 summarizes the wire-line data set used in this study. Initial wire-line coverage totaled 82,638 ft (25,188 m)

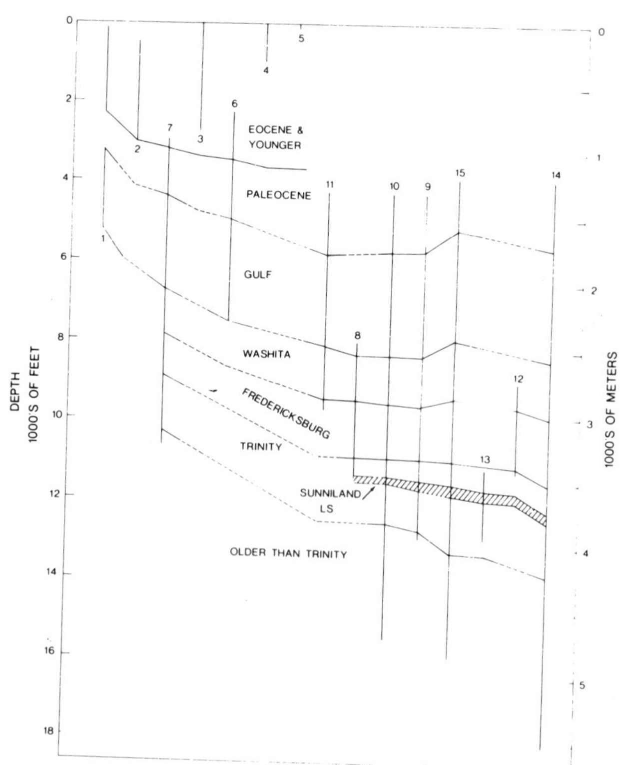

FIG. 3 — Ages and depths sampled by wire-line measurements of this study. Numbered wells are located in Figure 1 and identified in Table 1.

of section in the South Florida basin. Obvious non-carbonate facies were eliminated by visual inspection of the wire-line logs, leaving 50,002 ft (15,241 m) of wire-line data that were subdivided into 1,695 intervals of variable thickness (5 to 70 ft; 1.5 to 21.3 m) but roughly homogeneous composition and log character. The lithology of each subdivision was quantitatively determined from lithology logs (borehole-gravity data) or the M-N plot (conventional-log data), and intervals with evidence of non-carbonate minerals were rejected. This left as the final composite data set 1,302 intervals from 15 wells representing 41,459 ft (12,637 m) of clean carbonate rocks. Each time-stratigraphic unit used in this report is represented in the composite data set (Table 2).

The rather irregular distribution of interval thicknesses in the composite data set is shown in Figure 4a. The large peak is due to the arbitrary subdividing of uniform lithologies (primarily in the Gulfian) into 50-ft (15.2-m) intervals. Interval thicknesses are roughly comparable to the vertical scale of hydrocarbon reservoirs. A full range of limestone-dolomite percentages, including end members, is represented in the composite data set (Fig. 4b). Figure 4c shows the distribution of interval-porosity values, which range from near zero to 55%. Taken as a whole, the intervals studied represent a very porous assemblage of carbonate rocks.

The 1,302 intervals form a data set from which general porosity relations, representative of the basin as a whole, can be calculated. Such statistical results represent averages for thick carbonate sequences and are not site specific. Individual carbonate strata may show a considerable variation above or below the regionally representative regression curve.

Wells in which wireline data were obtained probably were not drilled at random, but according to criteria that may have biased the sampling of basin facies. Also, it is possible that the methods of wireline analysis used here resulted in a systematic rejection or acceptance of intervals that should or should not have remained in the data set. Having considered these possibilities, we nevertheless believe that the data of this study represent a reasonable sampling of carbonate porosity in the South Florida basin, from which we can draw conclusions with basin-wide applicability.

POROSITY VERSUS DEPTH

Exponential Representation

The least-squares exponential fit to the data set described in the previous section is shown in Figure 5. The locations of porosity-depth values also are plotted in Figure 5 to show the envelope of data represented by the exponential curve. In the regression-curve calculations, intervals are considered to be of equal importance in describing carbonate porosity and are not weighted according to thickness.

An exponential fit is a reasonably good visual representation of the data and serves to quantify comparisons and discussions. The correlation coefficient is -0.81 and the standard error of estimate—the root-mean-square of the porosity deviations about the exponential curve—is 5.1 porosity units.

Based on statistical parameters (correlation coefficient, variance, and standard error of estimate), there is little differ-

Table 1. Wells Where Wire-Line Data Were Obtained

Borehole Gravity Data

1. Orange County Public Utilities Sand Lake Road Wastewater Treatment Facility Test Injection Well W13287, Orange County, Florida. Sec. 32, T23S, R29E.
   Data interval = 155-2,317 ft (47-706 m).
   = 3,150-5,150 ft (960-1,570 m) (conventional logs).
2. Hercules Injection Test Well W14167, Indian River County, Florida. Sec. 25, T33S, R38E.
   Data interval = 470-2,350 ft (143-716 m).
   = 2,350-3,000 ft (716-914 m) (conventional logs).
3. Miami-Dade Water and Sewer Authority Test Injection Well W13768, Dade County, Florida. Sec. 21, T56S, R40E.
   Data interval =20-2,740 ft (6-835 m).
4. Florida Keys Aqueduct Authority W13752, Monroe County, Florida. Sec. 26, T64S, R35E.
   Data interval =40-970 ft (12-296 m).
5. Florida Keys Aqueduct Authority W12799, Monroe County, Florida. Sec. 11, T66S, R32 E.
   Data interval = 8-140 ft (2-43 m).

Conventional Porosity Logs

6. Texaco OCS G2523 1, Offshore Gulf of Mexico, Florida. Tampa Area Block N639, E54.
   Data interval = 2,258-7,498 ft (688-2,285 m).
7. Shell Oil Co. 34-1 Sloan, Okeechobee County, Florida. Sec. 34, T35S, R36E.
   Data interval =2,926-10,644 ft (892-3,244 m).
8. Exxon Co. 16-2 City of Ft. Myers, Lee County, Florida. Sec. 16, T44S, R26E.
   Data interval =7,996-11,424 ft (2,437-3,482 m).
9. Shell Oil Co. 1 Oleum Corporation, Collier County, Florida. Sec. 23, T46S, R30E.
   Data interval =4,230-12,982 ft (1,289-3,957 m).
10. Shell Oil Co. 19-2 Alico Land & Development, Hendry County, Florida. Sec. 19, T45S, R31E.
    Data interval =4,226-15,484 ft (1,288-4,720 m).
11. Shell Oil Co. 1 Gulf and Western, Palm Beach County, Florida. Sec. 7, T46S, R35E.
    Data interval = 4,236-9,718 ft (1,291-2,962 m).
12. Wainoco Inc. 35-2 Collier Co., Collier County, Florida. Sec. 15, T50S, R30E.
    Data interval = 9,010-11,250 ft (2,746-3,429 m).
13. Bass Enterprises Production Co. 5-2 Oleum Corporation, Collier County, Florida. Sec. 5, T52S, R31E.
    Data interval = 11,178-12,892 ft (3,407-3,929 m).
14. Bass Enterprises Production Co. 12-2 Pumpkin Bay Collier Company, Collier County, Florida. Sec. 12, T52S, R27E.
    Data interval = 3,926-18,150 ft (1,197-5,532 m).
15. Mobil Oil Corp. 1 R. M. Trien, Collier County, Florida. Sec. 14, T53S, R33E.
    Data interval = 3,872-15,980 ft (1,180-4,871 m).

ence in the fit between linear and exponential regression lines. However, high porosities of the near surface and residual porosities below about 16,000 ft (4,900 m) are not well represented by a linear regression line. For these reasons, the exponential function is used here to quantify the decrease of carbonate porosity with depth. Similar exponential functions have been used since the 1930s to quantify shale compaction (e.g., Athy, 1930; Hedberg, 1936; Maxant, 1980).

The equation for the exponential regression curve shown in Figure 5 is:

$$\phi = 41.73\, e^{-z/8197}\ (\text{ft}) \qquad (1a)$$
$$\phi = 41.73\, e^{-z/2498}\ (\text{m}) \qquad (1b)$$

where ϕ is the porosity (%) and z is the depth below ground surface (feet or meters). The constant 41.73 represents initial porosity at the surface, and is at the lower end of the porosity range of modern carbonate sediments (Enos and Sawatsky, 1981), indicating that near-surface processes did not greatly affect original pore space (Halley and Schmoker, in prep.). The constant 8,197 (ft) or 2,498 (m) represents the depth interval over which porosity is reduced by the factor 1/e.

Equation 1 is analogous to the equation describing radioactive decay. It is useful to describe radioactive decay in terms of a half life, and porosity "decay" can similarly be described in terms of a "half depth." In the South Florida basin, the half depth of carbonate porosity—the depth interval over which porosity is reduced by a factor of two—is 5,682 ft (1,732 m).

The broad significance of these results is that there is a close relation between porosity and depth in carbonate rocks representing a wide range of ages, lithologies, depositional environments, and diagenetic processes. The scatter in the porosity trend is not negligible, and is due to the secondary dependence of porosity on non-depth-related factors as well as to experimental errors. Nevertheless, a predictable decrease of porosity with depth exists in the data.

Sunniland Trend

The average porosity of producing intervals in the Sunniland Limestone is about 18% and the average producing depth is 11,500 ft (3,500 m) (Feitz, 1976; Tyler and Erwin, 1976). The expected porosity of south Florida carbonate rocks at this depth is about 10% (equation 1). Thus, the porosity of productive Sunniland Limestone is higher than the expected regional value.

This anomalously high porosity could have been a primary reason for the initial localization of hydrocarbons. High porosity may have been preserved by oil, emplaced when the rocks were shallower and more porous. Processes associated with the generation and migration of hydrocarbons may have created late secondary porosity. Unfortunately, such interesting and important questions can not be resolved using the data of this report.

Comparison to Published Porosity Data

Quantitative comparison of the south Florida porosity-depth curve (equation 1), representing lithified rocks, to laboratory sediment-compaction data (e.g., Terzaghi, 1940; Fruth et al, 1966; Robertson, 1967; Morelock and Bryant, 1971) is of limited significance. Laboratory experiments do not duplicate very well the stress fields, changes in pore-fluid composition, and time spans involved in the lithification of carbonate rocks.

The only published porosity-depth data analogous to this study that we are aware of relate to chalks and pelagic calcareous sediments. Neugebauer (1973, 1974) examined (theoretically) the loss of porosity in chalks, but did not quantify his results to the point of calculating porosity-depth

Table 2. Summary of Wire-Line Data Set

	Ft	*M*	*No. of Intervals*
Total wire-line coverage	82,638	25,188	--
Coverage after visual screening	50,002	15,241	1,695
Final data set	41,459	12,637	1,302
Eocene and younger			
Borehole gravity	7,240	2,207	167
Conventional logs	946	288	39
Paleocene	4,766	1,453	209
Gulf	17,790	5,422	384
Washita	4,206	1,282	210
Fredericksburg	3,568	1,088	158
Trinity	1,095	334	51
Older than Trinity	1,848	563	84
	41,459	12,637	1,302

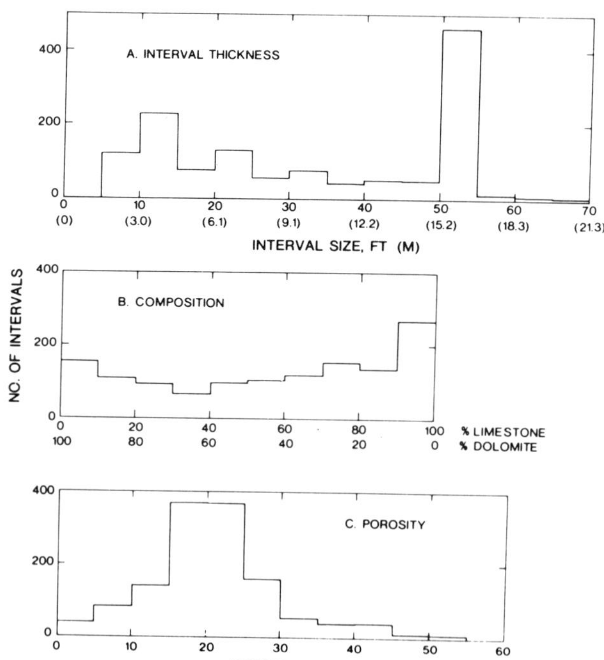

FIG. 4 — Distributions of interval thickness (A), carbonate composition (B), and porosity (C) for the wire-line data of this study.

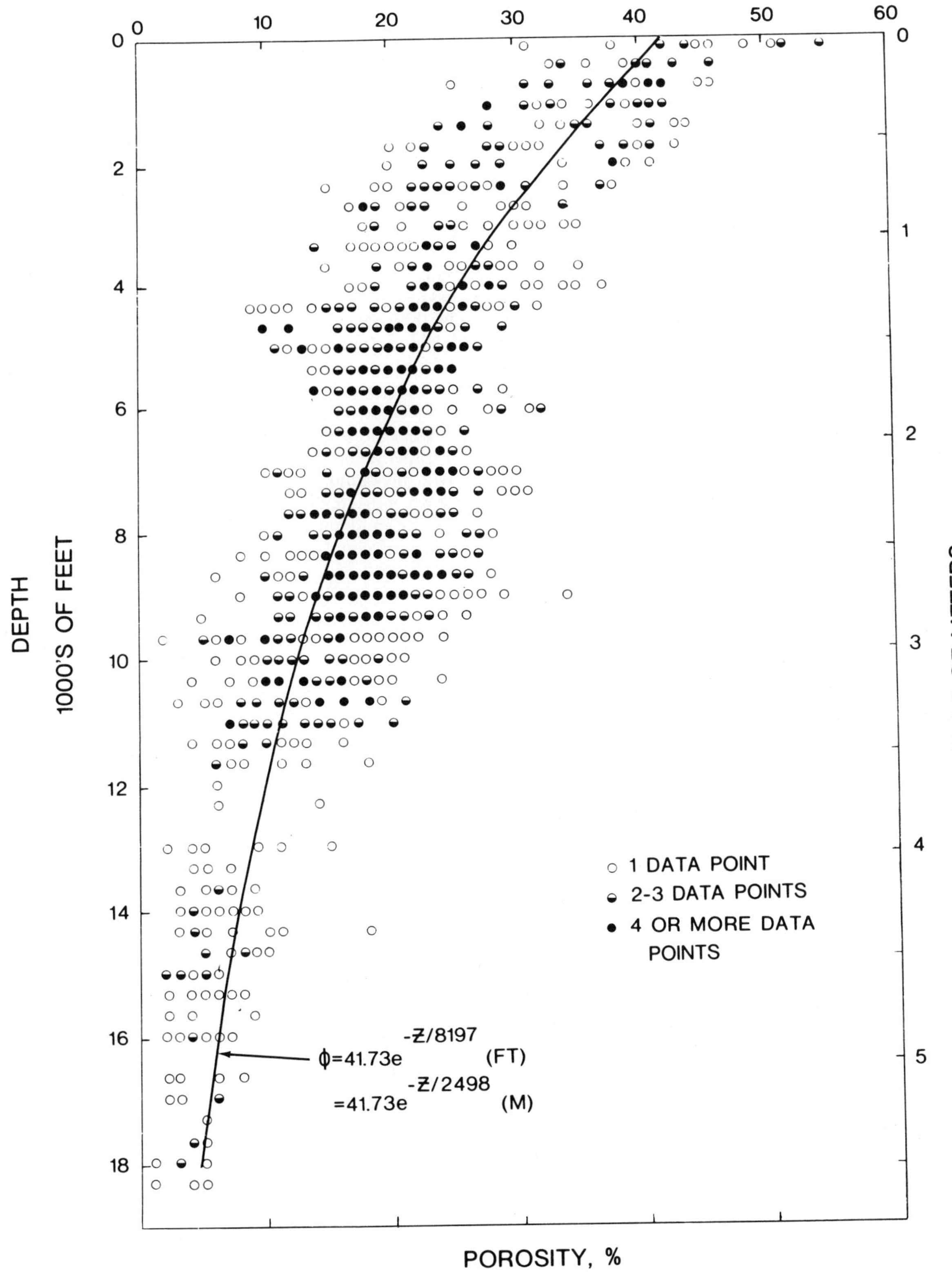

FIG. 5 — Porosity-depth values and the least-square exponential fit to all data (equation 1). The exponential regression curve has a correlation coefficient of -0.81 and a standard error of estimate of 5.1%.

relations. Representative empirical curves of porosity versus depth for chalks (Scholle, 1977; Lockridge and Scholle, 1978) and for deep-sea calcareous sediments (Schlanger and Douglas, 1974; Hamilton, 1976) are shown in Figure 6. Porosity-depth curves for the composite south Florida data set (equation 1) and for south Florida Gulf-age rocks only, are also shown for comparison in Figure 6. In the wells studied here, the Gulfian section consists primarily of chalks, but its porosity-depth curve does not differ substantially from that of the composite data set.

The chalk and pelagic-ooze data shown in Figure 6 form a porosity envelope which is greater in the near surface than the south Florida curve, but which decreases more rapidly with depth. The results of the present study do not agree with published porosity-depth data representing chalks and oozes.

Because of a more porous initial depositional texture and the greater chemical stability of low-magnesium calcite, chalks may be less subject to early diagenetic alteration than most shallow-water carbonates, and may be buried with a generally higher initial porosity (Scholle, 1977), as indicated by Figure 6.

Precise reasons for the more rapid porosity decrease with depth of the deep-water carbonates (Fig. 6) are not obvious to us. In a general sense, differences in trace and minor elements, grain size and shape, and rock texture, as well as external factors such as pore-fluid composition, geothermal gradient, and rate of subsidence, could cause variations in the response of shallow-water and deep-water carbonates to increasing burial depth. However, the relative importance of these and other factors on the depth dependence of carbonate porosity is poorly understood.

DEPTH VERSUS AGE

Although Figure 5 shows a strong relation between porosity and depth, it is possible that there is also a correlation between porosity and time, because older rocks tend to be buried more deeply. This hypothesis can be tested by comparing porosity measurements from rocks of equal depths but different ages. If porosity is time dependent, older rocks should be consistently less porous than younger rocks at equal depths.

In the data set of this study, there are depth overlaps between carbonate strata of different ages. As Stephenson (1977) pointed out, however, one must be careful not to ascribe impossibly large effects to a single variable. There is some question whether the age differences at given depths in the south Florida data are great enough to conclusively prove or disprove a porosity dependence on time.

With this caution in mind, porosity-depth curves representing each of the time-stratigraphic units used in this report are compared in Figure 7. In zones of depth overlap, the older unit is more porous than the younger unit in three cases, the

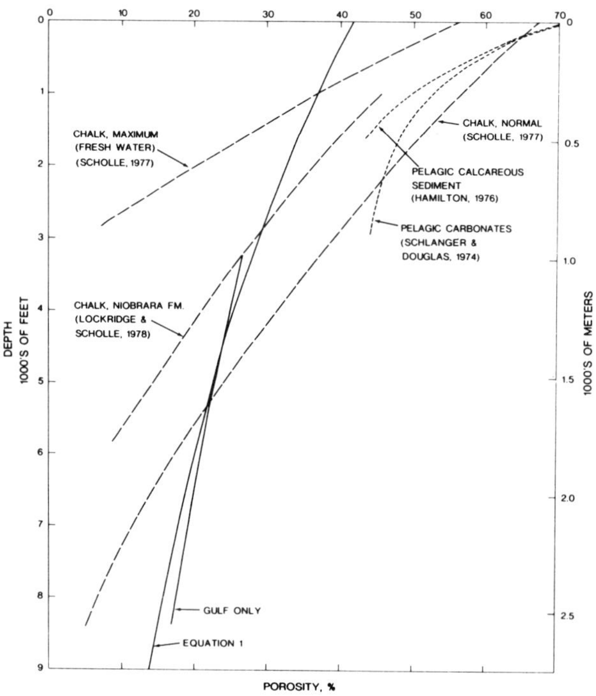

FIG. 6 — Representative curves of porosity versus depth for chalks and pelagic calcareous sediments compared to equation 1 and to exponential curve representing chalk-rich Gulf-age rocks of south Florida.

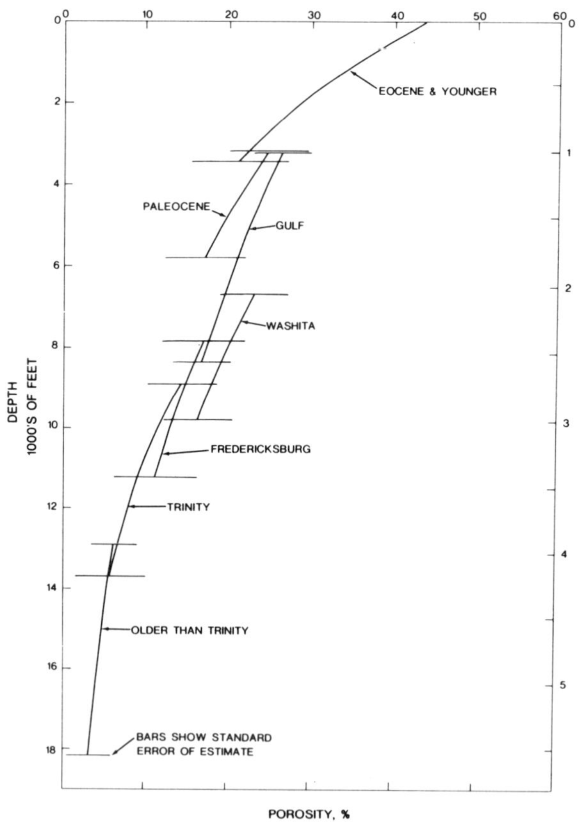

FIG. 7 — Exponential porosity-depth curves for the time divisions of Figure 2. There is no systematic porosity difference between older and younger rocks where they overlap in depth.

older unit is less porous in two cases, and in one case there is essentially no porosity difference. Furthermore, porosity differences in zones of depth overlap are within the expected range of scatter indicated by the standard error of estimate. Thus, there is no systematic porosity difference between older and younger carbonate rocks at equal depths, which suggests that the porosity decrease shown in Figure 5 is primarily a function of depth rather than of time.

EFFECT OF PORE-WATER COMPOSITION

A widespread change in depositional conditions in south Florida occurred between Eocene and Paleocene time. Evaporites, common in pre-Eocene strata (except chalks of Gulfian age), are very limited in Eocene and younger rocks (Puri and Winston, 1974). To the extent that evaporites reflect saline conditions, depositional environments became less saline in the Eocene and remained less saline. Today, Eocene and younger carbonate rocks contain fresh- and brackish-water aquifers in much of south Florida, and may have contained similar waters during periods of emergence since the Eocene.

Theoretical and experimental evidence indicates that resistance to pressure solution and, thus, to loss of porosity with depth increases as the magnesium content of the pore water increases (Neugebauer, 1973, 1974; Folk, 1974; Scholle, 1977).

Thus, rocks of Eocene age and younger in south Florida might be expected to show an accelerated loss of porosity with depth if they contained, on the average, fresher pore waters with less magnesium than older carbonate rocks. It appears from Figure 8 that the rate of porosity decrease with depth of Eocene and younger rocks is indeed faster than expected when compared to Paleocene and older rocks. Thus, the general line of evidence is internally consistent and supports the idea that pressure solution in shallow-water carbonates is inversely related to the magnesium content of pore waters.

An alternate hypothesis is that Eocene-age and younger rocks of south Florida form a closed system, sealed at the base by Paleocene anhydrites, in which carbonate is leached near the surface by fresh waters and precipitated at depth. The strength of this argument depends in part on the assumption that a near-surface porosity of 45% (Fig. 8) is exceptionally high because of leaching. However, we favor the idea that the near-surface porosity of 45% is representative of the normal transition from unconsolidated to consolidated sediments (Halley and Beach, 1979; Halley and Harris, 1979) and does not support the concept of mass transport from the surface to depth by freshwater circulation.

COMPARISON OF LIMESTONE AND DOLOMITE POROSITY

As pointed out by workers such as Landes (1946), Murray (1960), and Weyl (1960), the change in porosity associated with the transition from limestone to dolomite does not follow simple rules and is difficult to predict. Further complicating the picture, the porosity of dolomite relative to limestone at reservoir depths is also a function of burial diagenesis. Differences in mechanical properties (Hugman and Friedman, 1979) and susceptibility to pressure solution (Glover, 1968) may lead to different rates of porosity decrease with depth for the two rock types.

The porosity-depth data of this study provide an opportunity to empirically examine factors affecting the porosity of dolomite relative to limestone. Two data subsets with carbonate compositions of 75 to 100 % limestone (489 intervals) and 75 to 100% dolomite (336 intervals) were derived from the original data set. The cutoff value of 75% is a compromise that maintains both reasonable end-member purity and adequate data-set size. To quantify comparisons of the limestone and dolomite porosity-depth data, exponential regression curves were fit to the two data subsets (Fig. 9).

There is a marked visual difference between the limestone and dolomite porosity-depth curves of Figure 9. Statistical tests (Snedecor and Cochran, 1980, p. 385) confirm that both the slopes and intercepts of these curves differ at a confidence level greater than 99%. The correlation coefficient of the curve representing limestone intervals is -0.90 and the standard error of estimate is 4.8%. The curve representing dolomite intervals has a poorer correlation coefficient (-0.62), a larger standard error of estimate (5.5%), and visually does not fit the data as well. The characterization of dolomite porosity in south Florida is more uncertain than the characterization of limestone porosity.

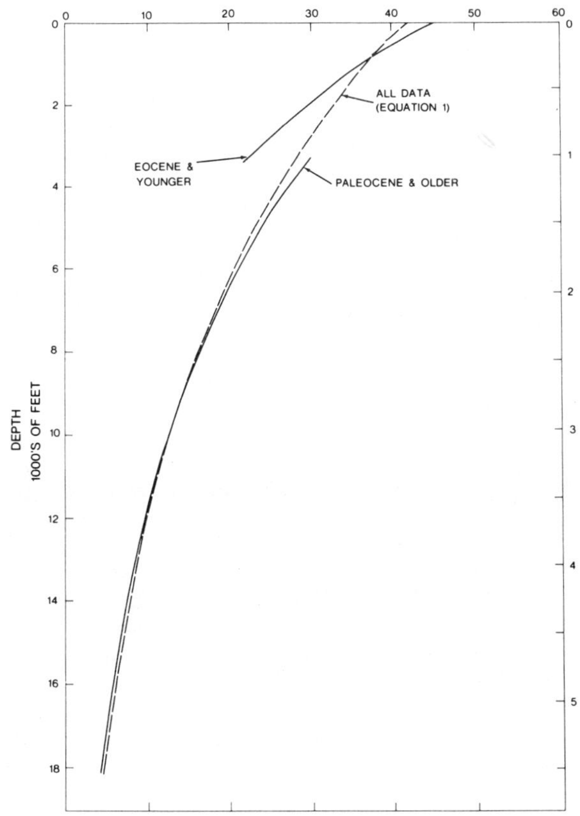

FIG. 8 — Exponential porosity-depth curves for carbonate strata of Eocene age and younger and Paleocene age and older, showing a faster than expected porosity decrease with depth for Eocene age and younger rocks.

Dolomite porosity is lower than limestone porosity in the near surface, but does not decrease as rapidly with depth (Fig. 9). The two porosity curves intersect at about 5,600 ft (1,700 m), below this depth dolomite is more porous than limestone.

Because the porosities of both curves are greater than 20% above their intersection point, the porosity difference between limestone and dolomite at relatively shallow depths may be of secondary importance compared to other reservoir properties. At greater depths, however, the better porosity of dolomite relative to limestone could be of considerable economic significance. Figure 9 reinforces the widely held supposition, based on practical experience and supported by the fact that dolomites account for about 80% of North American carbonate reservoirs (Zenger et al, 1980, p. iii), that dolomite is commonly a better reservoir rock than limestone.

The relation between limestone and dolomite porosity shown in Figure 9 could be a function of depositional and diagenetic conditions peculiar to the South Florida basin. It is thought more likely, however, that the observed relation results, in large part, from basic differences in the response of limestone and dolomite to normal burial-related parameters. More specifically: (1) the dolomitization process, occurring relatively early near the surface, either reduced original porosity or selectively favored lower-porosity limestone; and (2) dolomite was more resistant than limestone to the porosity-reducing effects of burial. Because this sequence of events is related more to carbonate properties than to basin history, the general form of the two porosity curves could be representative of areas other than south Florida. This model places particular emphasis on the importance of diagenesis in the deep subsurface in the evolution of dolomite porosity relative to that of limestone.

SUMMARY AND DISCUSSION

This study examines the porosity of limestones and dolomites in the South Florida basin using data derived from borehole-gravity measurements and from suites of acoustic, neutron, and density logs. These wire-line measurements are averaged into intervals of relatively uniform log character with vertical scales approaching those of aquifers and hydrocarbon reservoirs, and are combined into regional data sets that are further smoothed by regression-line representations. Thus, the results presented here are based on a large sampling scale that is uncommon in carbonate studies.

Results are summarized as follows:

1. The decrease of carbonate porosity with depth for a composite section representing a wide range of depositional environments and diagenetic histories can be characterized by an exponential curve (equation 1).
2. Average porosity of the composite carbonate section is reduced by a factor of two in a depth interval of about 5,700 ft (1,740 m); published curves for deep-water carbonates show a greater rate of porosity decrease with depth.
3. Although the evidence is not conclusive because of limited age differences, the decrease of carbonate porosity appears to be primarily a function of depth rather than of time.
4. Porosity loss in the shallow-water carbonate rocks of south Florida appears inversely related to the magnesium content of pore waters.
5. Dolomite porosity is lower than limestone porosity in the near surface, but does not decrease as rapidly with depth. Below about 5,600 ft (1,700 m), dolomite is, on the average, more porous than limestone.

The data developed here indicate that regional-porosity relations in the South Florida basin are not primarily dependent upon the imprint of the highly variable surface and near-surface environment, but relate instead to more predictable parameters of burial. Although the particular numerical results presented here should be applied with considerable caution to areas other than south Florida, the form of the porosity-depth relations may result from basic properties of carbonate rocks, and thus represent a more general case.

A direct implication of the data presented here is that the porosity of Cenozoic and Mesozic carbonate rocks in a given exploration province may be empirically predictable on a regional scale as a function of depth. A less direct implication of the same data is that most shallow-water carbonate sediments enter the subsurface with high porosities relative to the requirements of hydrocarbon reservoirs (Figs. 5, 9), regardless of early depositional and diagenetic history. A change in geologic emphasis, from the establishment of correlations between early history and porosity *evolution* in the near

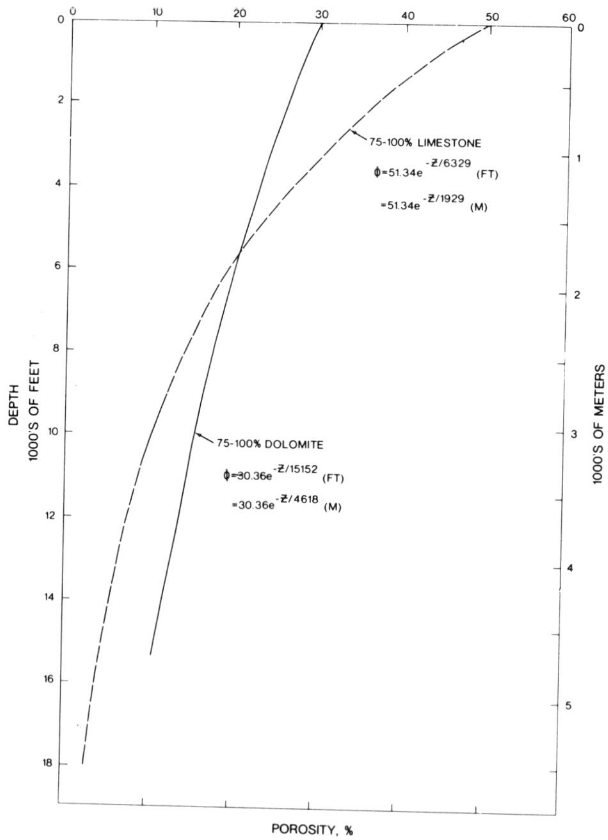

FIG. 9 — Exponential porosity-depth curves for limestones and dolomites. Dolomite porosity is lower in near surface but does not decrease as rapidly with depth.

surface to the establishment of correlations between early history and porosity *preservation* in the subsurface, could prove to be of significant economic value.

A next step in this study (and one being pursued) is to examine regional carbonate-porosity trends in other basins, using similar large-scale measurement methods. Such studies, whether confirming or refuting the hypotheses presented here, should contribute to the understanding of porosity in carbonate rocks.

REFERENCES CITED

Applin, P. L., 1960, Significance of changes in thickness and lithofacies of the Sunniland Limestone, Collier County, Florida: U.S. Geol. Survey Prof. Paper 400-B, p. B209-B211.

——— and E. R. Applin, 1965, The Comanche Series and associated rocks in the subsurface in central and south Florida: U.S. Geol. Survey Prof. Paper 447, 84 p.

Athy, L. F., 1930, Density, porosity, and compaction of sedimentary rocks: AAPG Bull., v. 14, p. 1-24.

Banks, J. E., 1960, Petroleum in Comanche (Cretaceous) section, Bend area, Florida: AAPG Bull., v. 44, p. 1737-1748.

Burke, J.A., A. W. Schmidt, and R. L. Campbell, Jr., 1969, The lithoporosity cross plot: The Log Analyst, v. 10, no. 6, p. 25-43.

Choquette, P. W., and L. C. Pray, 1970, Geologic nomenclature and classification of porosity in sedimentary carbonates: AAPG Bull., v. 54, p. 207-250.

——— and R. P. Steinen, 1980, Mississippian non-supratidal dolomite, Ste. Genevieve Limestone, Illinois basin: Evidence for mixed-water dolomitization, *in* D. H. Zenger, J. B. Dunham, and R. L. Ethington, eds., Concepts and models of dolomitization: SEPM Spec. Pub. 28, p. 163-196.

Enos, P., and L. H. Sawatsky, 1981, Pore networks in Holocene carbonate sediments: Jour. Sed. Petrology, v. 51, p. 961-985.

Feitz, R. P., 1976, Recent developments in Sunniland exploration of south Florida: Gulf Coast Assoc. Geol. Socs. Trans., v. 26, p. 74-78.

Folk, R. L., 1974, The natural history of crystalline calcium carbonate: Effect of magnesium content and salinity: Jour. Sed. Petrology, v. 44, p. 40-53.

Fruth, L. S., Jr., G. R. Orme, and F. A. Donath, 1966, Experimental compaction effects in carbonate sediments: Jour. Sed. Petrology, v. 36, p. 747-754.

Geothermal Survey of North America Subcommittee of the AAPG Research Committee, 1976, Geothermal gradient map of North America: AAPG and the U.S. Geol. Survey, scale 1:5,000,000.

Glover, J. E., 1968, Significance of stylolites in dolomitic limestones: Nature, v. 217 (March 2), p. 835-836.

Halley, R. B., and D. K. Beach, 1979, Porosity preservation and early freshwater diagenesis of marine carbonate sands (abs.): AAPG Bull., v. 63, p. 460.

——— and P. M. Harris, 1979, Fresh-water cementation of a 1,000-year-old oolite: Jour. Sed. Petrology, v. 49, p. 969-988.

——— and J. W. Schmoker, 1983, High-Porosity Cenozoic carbonate rocks of south Florida: progressive loss of porosity with depth: AAPG Bull., v. 67, no. 2.

Hamilton, E. L., 1976, Variations of density and porosity with depth in deep-sea sediments: Jour. Sed. Petrology, v. 46, p. 280-300.

Hedberg, H.D., 1936, Gravitational compaction of clays and shales: Am. Jour. Science, v. 31, no. 184, p. 241-287.

Hugman, R. H. H., III, and Melvin Friedman, 1979, Effects of texture and composition on mechanical behavior of experimentally deformed carbonate rocks: AAPG Bull., v. 63, p. 1478-1489.

Kohout, F. A., 1967, Ground-water flow and the geothermal regime of the Florida Plateau: Gulf Coast Assoc. Geol. Socs. Trans., v. 17, p. 339-354.

Landes, K. K., 1946, Porosity through dolomitization: AAPG Bull., v. 30, p. 305-318.

Lockridge, J. P., and P. A. Scholle, 1978, Niobrara gas in eastern Colorado and northwestern Kansas: Rocky Mountain Assoc. Geol. 1978 Symp., p. 35-49.

Maxant, J., 1980, Variation of density with rock type, depth, and formation in the Western Canada basin from density logs: Geophysics, v. 45, p. 1061-1076.

McCulloh, T. H., 1966, The promise of precise borehole gravimetry in petroleum exploration and exploitation: U.S. Geol. Survey Circ. 531, 12 p.

Morelock, Jack, and W. R. Bryant, 1971, Consolidation of marine sediments, *in* Richard Rezak and H. J. Vernon, eds., Contributions on the geological and geophysical oceanography of the Gulf of Mexico: Texas A & M Univ. Oceanographic Studies, v. 3, p. 181-202.

Murray, R. C., 1960, Origin of porosity in carbonate rocks: Jour. Sed. Petrology, v. 30, p. 59-84.

Neugebauer, J., 1973, The diagenetic problem of chalk—the role of pressure solution and pore fluid: Neues Jahrb. Geologie u. Palaontology Abh., v. 143, no. 2, p. 223-245.

——— 1974, Some aspects of cementation in chalk, *in* K. J. Hsu and H. C. Jenkyns, eds., Pelagic sediments: on land and under the sea: Int. Assoc. Sedimentologists Spec. Pub. No. 1, p. 149-176.

Oglesby, W. R., 1965, Folio of South Florida basin: Florida Geol. Survey Map Series 19, 3 p., 10 maps.

Palacas, J. G., 1978, Preliminary assessment of organic carbon content and petroleum source rock potential of Cretaceous and Lower Tertiary carbonates, South Florida basin: Gulf Coast Assoc. Geol. Socs. Trans., v. 28, p. 357-381.

Pontigo, F. A., Jr., A. V. Applegate, and J. H. Rooke, 1979, South Florida's Sunniland oil potential: Oil and Gas Jour. v. 77, no. 31, p. 226-232.

Pressler, E. D., 1947, Geology and occurrence of oil in Florida: AAPG Bull., v. 31, p. 1851-1862.

Puri, H. S., and G. O. Winston, 1974, Geologic framework of the high transmissivity zones in south Florida: Florida Bur. Geol. Spec. Pub. 20, 101 p.

Randazzo, A. F., and E. W. Hickey, 1978, Dolomitization in the Floridan Aquifer: Am. Jour. Science, v. 278, no. 8, p. 1177-1184.

Robbins, S. L., 1981, Reexamination of the values used as constants in calculating rock density from borehole gravity data: Geophysics, v. 46, p. 208-210.

Robertson, E. C., 1967, Laboratory consolidation of carbonate sediment, *in* A. F. Richards, ed., Marine geotechnique: International Research Conference on Marine Geotechnique, Urbana, Illinois, Univ. Illinois Press, p. 118-127.

Schlanger, S. O., and R. G. Douglas, 1974, The pelagic ooze-chalk-limestone transition and its implications for marine stratigraphy, *in* K. J. Hsu and H. C. Jenkyns, eds., Pelagic sediments: on land and under the sea: Int. Assoc. Sedimentologists Spec. Pub. No. 1, p. 117-148.

Schmoker, J. W., 1978, Accuracy of borehole gravity data: Geophysics, v. 43, no. 3, p. 538-542.

——— et al, 1979, Preliminary porosity estimates of south Florida Cenozoic carbonate rocks based on borehole gravity measurements: U.S. Geol. Survey Open-File Report 79-1652, 17 p.

Scholle, P. A., 1977, Chalk diagenesis and its relation to petroleum exploration: oil from chalks, a modern miracle?: AAPG Bull., v. 61, p. 982-1009.

——— 1981, Porosity prediction in shallow vs. deepwater limestones: Jour. Petroleum Technology, v. 33, p. 2236-2242.

Snedecor, G. W., and W. G. Cochran, 1980, Statistical methods (7th ed.): Ames, Iowa, Iowa State Univ. Press, 507 p.

Stephenson, L. P., 1977, Porosity dependence on temperature: limits on maximum possible effect: AAPG Bull., v. 61, p. 407-415.

Terzaghi, R. D., 1940, Compaction of lime mud as a cause of secondary structure: Jour. Sed. Petrology, v. 10, p. 78-90.

Till, R., 1974, Statistical methods for the earth scientist: New York, John Wiley & Sons, 154 p.

Tyler, A. N., and W. L. Erwin, 1976, Sunoco-Felda field, Hendry and Collier Counties, Florida, *in* J. Braunstein, ed., North American oil and gas fields: AAPG Mem. 24, p. 287-299.

Weyl, P. K., 1960, Porosity through dolomitization—conservation-of-mass requirements: Jour. Sed. Petrology, v. 30, p. 85-90.

Winston, G. O., 1972, Oil occurrence and Lower Cretaceous carbonate-evaporite cyclothems in south Florida: AAPG Bull., v. 56, p. 158-160.

Zenger, D. H., J. B. Dunham, and R. L. Ethington, eds., 1980, Concepts and models of dolomitization: SEPM Spec. Publ. 28, 320 p.

Reproduced with the permission of the Geological Association of Canada. Published in *Diagenesis 1; Dolomite, Part 1, Geoscience Canada,* V. 9, No. 1, pp. 5-13.

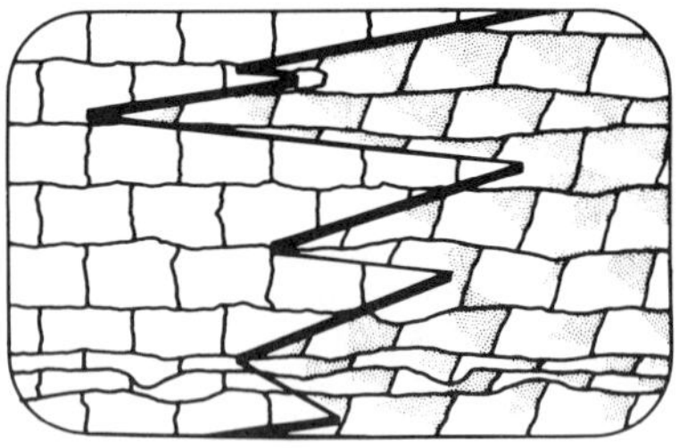

Diagenesis 1. Dolomite - Part 1: The Chemistry of Dolomitization and Dolomite Precipitation

D.W. Morrow
Institute of Sedimentary and Petroleum Geology
3303 - 33 Street N.W.
Calgary, Alberta T2L 2A7

Introduction

Dolostones form a large part of the Proterozoic and Phanerozoic successions in many areas of Canada but are particularly abundant in lower Paleozoic strata (Fig. 1). These strata are predominantly carbonate unlike the overlying Mesozoic and underlying Proterozoic strata that are dominated by terrigenous sediments. This reflects the fact that during early Paleozoic time Canada was part of the Laurentian landmass that was situated in the tropical equatorial belt (Ziegler *et al.*, 1979). The economic importance of dolostones parallels their abundance in the stratigraphic record (Fig. 1). Dolostones form important reservoirs for oil (e.g., the Leduc Formation or D3) and gas (e.g., Manetoe Facies, Fig. 2) and are the main host rock in Canada for lead-zinc mineral deposits (e.g., the Presqu'ile Facies of the

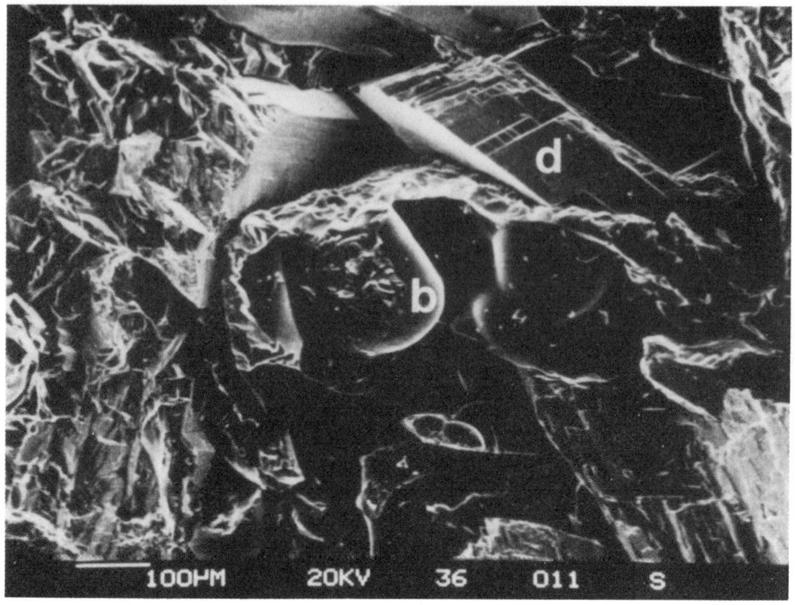

Figure 2 *An SEM photomicrograph of the gas-bearing dolomite of the Manetoe Facies in the Yukon and Northwest Territories. Well-formed (i.e. idiotopic) crystals of white dolomite (D) project into a vug that is lined in part by reniform bitumen (B). The surrounding dolomite has very low porosity and permeability.*

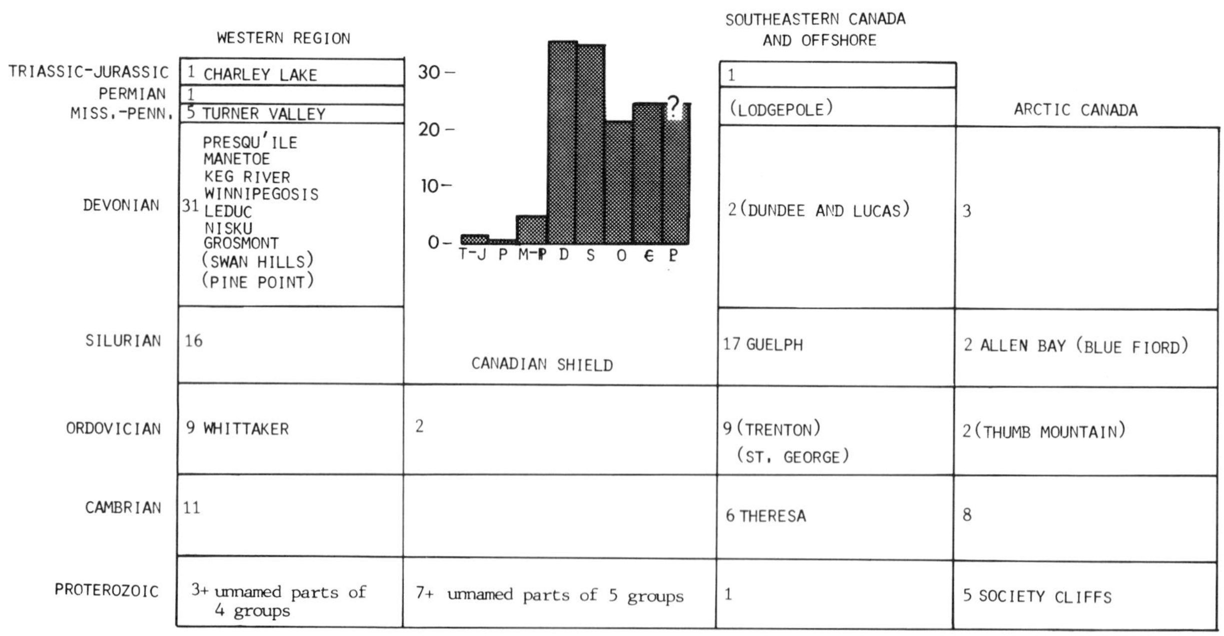

Figure 1 *A chart showing the numerical distribution of Canadian dolostones that have been formally named. Units that are named in the chart have economic importance both as reservoirs for oil and gas or as hosts for Pb-Zn deposits. Names in brackets are partly dolomitized limestone units in which the dolomite phase has economic importance.*

Pine Point deposit). The importance of dolostones in Canada is a true reflection of the world wide importance of dolostones as reservoirs for hydrocarbons and as hosts for lead-zinc deposits (e.g., see Davies, 1979 and Macqueen, 1979).

Given the major, albeit indirect, role that the mineral dolomite has played in contributing to the well being and growth of the Canadian and world economies one might assume that the salient facts concerning the process of dolomitization and the precipitation of dolomite are well established. Unfortunately, although there is a considerable amount of experimental and theoretical data concerning the chemistry of the dolomitization process and the precipitation of the mineral dolomite from solution, no concensus exists regarding the general chemical conditions required for the dolomitization process. The absence of a widely accepted theory concerning the chemistry of dolomitization is due primarily to the difficulty in precipitating dolomite from appropriate solutions at temperatures less than 100° C. This has prevented direct interpretation of the physico-chemical conditions that prevailed during the precipitation of most dolostones in nature with the exception of those that were formed at high temperature. The inability to precipitate dolomite under controlled laboratory conditions constitutes the essence of the problem of the origin of dolomite.

The interpretation of the origin of a particular dolostone is therefore dependant upon inferences that are derived mainly from comparisons with a variety of dolomitization models, such as the Evaporative Pumping Model, the Reflux Model, the Coorong Model, the Solution-Cannibalization Model, the Mixed-Water or Dorag Model, the Compaction Model and other, less commonly encountered ones. These models will be discussed in a following paper on dolomite entitled *Dolomitization Models and Ancient Dolostones*. Taken together, they present a bewildering complexity of choice and it is commonly not apparent or universally agreed upon as to what are the essential features of these models that renders them agents of dolomitization (e.g., high Mg/Ca solution ratios, low salinities, high salinities, high [CO_3^{2-}], large organic content) or even why such features promote dolomitization. For example, does a lowering of salinity favour dolomitization from a kinetic standpoint (Folk and Land, 1975) or largely from a thermodynamic standpoint (Badiozamani, 1973)? It is not sufficient merely to state that a drop in salinity promotes dolomitization, we must know why and under what circumstances this is true if we are to make full use of this insight in our interpretations. In other words the formidable array of current models for dolomitization is not clothed by any certain or complete understanding of the precipitational conditions of the mineral dolomite and, like the emperor who had no clothes, it stands before us embarrassingly naked. In this review, I will assemble and stitch together like some demented tailor, the experimental, field and theoretical evidence concerning the chemistry of dolomitization into what I hope will pass as an example of sartorial elegance.

Dolomite Crystal Structure and Compositional Variability

The mineral dolomite has an ideal chemical composition of $CaMg(CO_3)_2$ and a symmetry structure $\overline{3}$ in the trigonal subsystem of the hexagonal crystal stystem. The unit cell is a rhombohedron elongate parallel to the "c" crystallographic axis (Fig. 3) similar to that of calcite but more highly ordered, and has a lower degree of symmetry because of the substitution of magnesium atoms for half of the calcium atoms. Like calcite, the dolomite crystal is built up of layers of cations (Ca^{2+}, Mg^{2+}) alternating with anions (CO_3^{2-}) perpendicular to the c-axis direction with the distinction that half of the cation layers are magnesium layers. It is this regular alternation of calcium and magnesium layers with intervening carbonate layers that uniquely characterizes the mineral dolomite.

Dolomite crystals in nature and those synthesized experimentally are better represented by the formula $Ca_{(1+x)}Mg_{(1-x)}(CO_3)_2$ and commonly depart considerably from an ideal composition and structure. Three distinct factors contribute to this departure.

The degree of order of a compositionally ideal (i.e., stoichiometric) dolomite crystal is affected by the degree to which calcium and magnesium are segregated into their respective layers. If calcium atoms from calcium layers are inter-

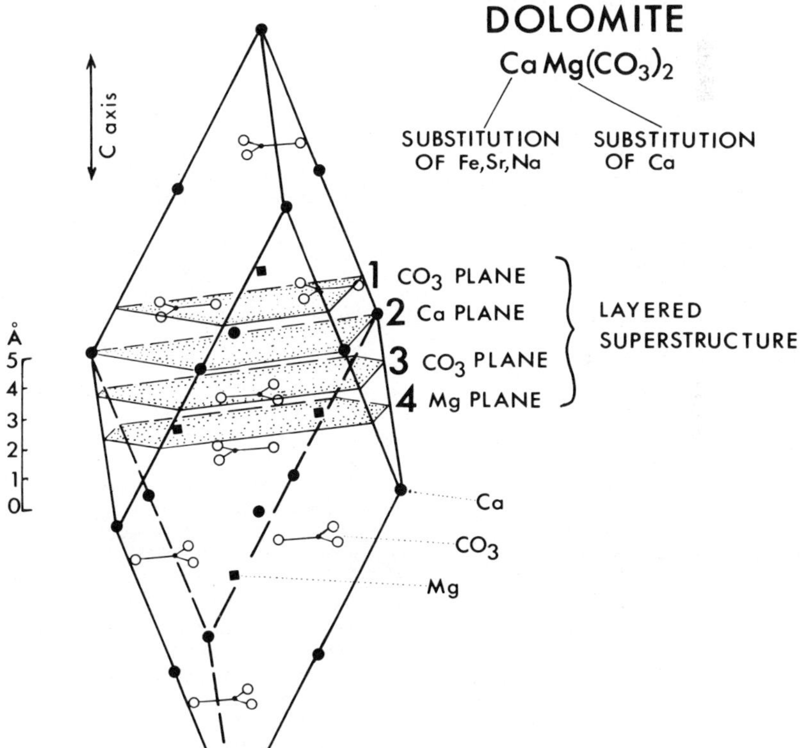

Figure 3 *Diagram of four rhombohedral unit cells of dolomite showing its layered superstructure perpendicular to the "c" axis direction composed of regularly alternating sheets of calcium, magnesium, and carbonate. Commonly Ca and Mg are incompletely segregated into their respective layers and there is an excess of calcium. Also many foreign ions substitute for calcium. These defects affect the thermodynamic properties of dolomite.*

changed with an equal number of magnesium atoms from magnesium layers the degree of order of the crystal is lowered although its composition is unaffected. Most natural dolomites display some degree of mixing of calcium and magnesium between cation layers (Goldsmith and Graf, 1958; Carpenter, 1980). Secondly, dolomites commonly depart from a stoichiometric composition because of an excess of calcium which is accomodated in the magnesium layers (Goldsmith and Graf, 1958; Lumsden and Chimahusky, 1980). Finally, a multitude of cations, primarily Fe, Sr, Na and Mn, substitute for calcium in many dolomites (e.g., Veizer *et al.*, 1978). Iron in particular may be present in concentrations of up to 4 or 5 mole per cent in ferroan dolomites.

These departures from the ideal composition also diminish the ordering of the crystals (Carpenter, 1980). Intracrystal variations in composition and degree of order are also common, particularly in larger crystals (e.g., Fritz and Katz, 1972). There is even some evidence that compositional variations have influenced the shape of the coarse crystals with curved faces that are common in some late diagenetic dolostones (Radke and Mathis, 1980). The degree of disorder and of departure from stoichiometry also affects the unit cell dimensions and the solubility of dolomite (Land, 1980).

Many authors have applied the name protodolomite or even pseudodolomite (e.g., Gaines, 1977, 1978; Deelman, 1978; Gidman, 1978) to dolomites that are not ideal in structure or composition. Such names tend to obscure the similarity of these non-ideal dolomites to ideal dolomite and Land (1980) suggested retaining the name dolomite for all such phases that have a composition near CaMg $(CO_3)_2$ and display evidence, such as superstructure reflections in x-ray patterns, of the degree of order characteristic of dolomite.

The Stability of Dolomite in Natural Environments

Minerals that have survived unchanged for long periods of time in natural environments do so for one of two reasons. The mineral may be stable thermodynamically, which is to say it will not be affected by small changes in the environment, or the mineral is unstable thermodynamically but is inert. Inert minerals are in the process of change to other species that have a lower free energy of formation, but the process may be extemely slow, as, for example in the conversion of diamond to graphite on the earth's surface. Such compounds are commonly called metastable to indicate the fact that they exist for variably long periods of time within the stability field of other compounds of similar composition.

The appropriateness of the terms 'metastable' and 'inert' are obviously dependant on the period of time over which a reaction is being considered, which in this case, is geologic time.

The factors (e.g., catalysts, temperature and concentration) that influence the speed of a chemical reaction are known collectively as reaction kinetic effects. The dolomite-calcite relationship is one of the more spectacular natural examples of the influence of kinetic factors. Under a wide range of surface and near-surface conditions dolomite is favoured thermodynamically over coexisting calcite so that at equilibrium in solution the dolomitization reaction, as it is commonly expressed (e.g., Bathurst, 1975)

$$2\,CaCO_3 + Mg^{2+} \rightleftharpoons Ca\,Mg\,(CO_3) + Ca^{2+} \quad (1)$$

is characterized by a free energy change of $\Delta G°_r$ = -1.83 kilocal/mole for the transformation of aragonite to dolomite of ideal composition and structure (Table I p. 11), which indicates that this reaction has a thermodynamic drive towards dolomitization.

We may use the relationship $\Delta G_r = \Delta G°_r + RT \ln Q$ (Q = activity ratio of Ca^{2+}/Mg^{2+}, R = the gas constant, and T = absolute temperature), to derive a value of 21.96 for the activity ratio Ca^{2+}/Mg^{2+} for reaction (1) at equilibrium. Solutions with smaller values of Q will favour the replacement of aragonite by ordered dolomite. Seawater is one such solution with a Q value of only 0.16 (Garrels and Thompson, 1962; Wigley and Plummer, 1976) indicating a strong thermodynamic drive towards the replacement of aragonite and calcite by dolomite in the worlds oceans. What perversity of nature has left the vast present day submarine blankets of aragonite undolomitized?

The Mg^{2+} in equation (1) is necessarily supplied in solution to the reaction site from an external source. Lippman (1973) has emphasized that dolomite precipitates from ions in solution and that the $CaCO_3$ must dissolve before dolomite can form if the Ca^{2+} and CO_3^{2-} ions are derived from the $CaCO_3$ alone. It is also possible that part of the CO_3^{2-} in dolomite may have been supplied by the solution from an external source and Lippman (1973) has proposed the reaction

$$CaCO_3 + Mg^{2+} + CO_3^{2-} \rightleftharpoons CaMg\,(CO_3)_2 \quad (2)$$

in which the CO_3^{2-} anions as well as Mg^{2+} are supplied by solution to the reaction site as an alternative to the traditional equation of reaction (1). For the dolomitization of aragonite, the free energy for reaction (2) at equilibrium is $\Delta G°_r$ = -13.24, favouring the formation of ordered dolomite even more strongly than reaction (1). Reaction (2) also has the advantage that the calcium liberated by dissolution of $CaCO_3$ is removed entirely from solution by dolomite whereas dolomitization by reaction (1) poses the problem of Ca^{2+} removal from the reaction site. If this evolved calcium is not removed its presence will raise the Ca/Mg ratio of solution and lower the thermodynamic drive towards dolomitization.

It may be assumed that, like other diagenetic reactions involving carbonates, dolomitization is entirely a wet chemical process (see Lippman, 1973; Bathurst, 1975; Land, 1980). Solid state reactions between carbonate minerals in the absence of an aqueous phase proceed prohibitively slowly at temperatures less than several hundred degrees centigrade (e.g. Fyfe and Bischoff, 1965). The existence of calcium-rich, non-stoichoimetric dolomite with compositions that fall in the region of the immiscibility gap between calcite and dolomite (Fig. 4) is evidence of slow reaction rates. However, recent work has shown that there may be some stable phases of intermediate composition between the low

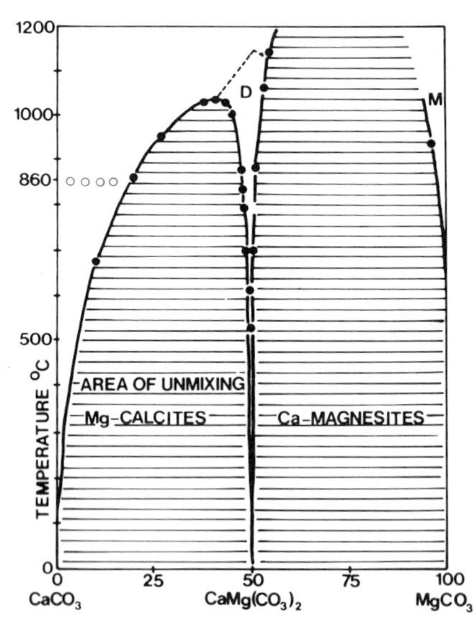

Figure 4 *Phase diagram of (water-free) $CaCO_3$, $CaMg(CO_3)_2$ (dolomite), and $MgCO_3$ (Magnesite) at temperatures between 0° and 1200° C. Only pure dolomite and calcite are thermodynamically stable at earth surface temperatures (after Deelman 1978; Goldsmith and Graf, 1958; and Goldsmith and Heard, 1961.). Recent work by Reeder and Wenk (1979) may cause some modification of this simple scenario.*

temperature calcite and dolomite stable end members (Reeder and Wenk, 1979).

Water of various compositions occupies even submicron-sized lamellar spaces between adjacent crystals in carbonate rocks and participates in their diagenesis (Pingitore, 1976; and Wardlaw, 1976). Most of these natural solutions are mixtures of waters of meteoric and marine origin modified by burial processes and commonly are supersaturated with respect to dolomite even when undersaturated with calcite, except for dilute solutions undersaturated in both phases (Fig. 5).

In summary we can state that dolomite is stable in most natural solutions and that the existence of aragonite and calcite in these solutions is a metastable condition.

The Role of Kinetic Effects on the Crystallization of Dolomite

Several kinetic factors that impede the precipitation of dolomite from supersaturated solution have been identified. These factors involve the molecular mechanics of precipitation and difficulties associated with the formation of the regularly alternating sheets of calcium and magnesium atoms in growing dolomite crystals. Many of these kinetic effects interlock and it is difficult to separate cause and effect. However, there are three major underlying kinetic barriers to the precipitation of dolomite:

1) Rapid crystallization from supersaturated, concentrated (i.e., saline) solutions impedes the segregation of calcium and magnesium ions into their respective layers so that magnesium layers are contaminated by calcium atoms and vice versa and disordered Ca-Mg carbonates are formed (Folk, 1979; Folk and Land, 1975; Lippman, 1973, p. 179);

2) Pronounced differences in the hydration behavior in aqueous solution of the constituent ions of dolomite promote precipitation of calcium-rich phases (Lippman, 1973); and

3) The low activity or concentration of CO_3^{2-} relative to Ca^{2+} or Mg^{2+} in most natural solutions inhibits the precipitation of magnesium-bearing carbonate minerals in general (Lippman, 1973).

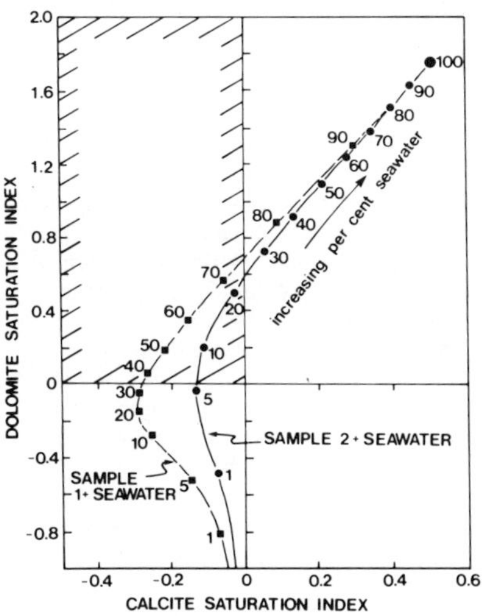

Figure 5 *Calcite and dolomite saturation index changes for mixing of sample 1 (PCO_2 = 10^{-2} atm; broken line) and sample 2 (PCO_2 = 10^{-3} atm; solid line) with sea water. Shaded region is where Badiozamani (1973) and others have argued that dolomitization is favoured (diagram after Wigley and Plummer, 1976).*

Some have argued that the growth of dolomite crystals is predominantly by the sequential addition of successive cation-anion layers (e.g., Deelman, 1975, but see also Towe, 1975). However, in modern evironments of dolomite formation, dolomite is invariably present in the sediment as rhombs that range in size from less than 1μm to about 10μm. No crystals have been caught in the act of adding layers as planar sheets perpendicular to the c-axis. Instead dolomite crystals appear to grow as rhombs at more or less equal rates in all directions in spite of its layered construction (Fig. 3).

The speed of crystallization influences the degree of order because ions that adhere to incorrect lattice positions on the surfaces of rapidly growing crystals are more likely to be entombed in these positions than if the crystals grow slowly. Slower growth permits a longer period of exposure to the solution of an incorrectly situated ion. This increases the likelihood that it will be displaced by an ion of proper type for this lattice site. Calcium and magnesium ions are easily interchanged in dolomite's crystal lattice because of the small differences of lattice energy that are involved (Folk and Land, 1975).

The strength of the electrostatic ion of magnesium to water molecular bond is about 20 per cent greater than that of its calcium counterpart and much greater than that of the carbonate ion (Lippman, 1973, Table 20). The relatively strong attachment of the magnesium ion to its water of hydration shell (Fig. 6), is even more evident in the data of Samoilov (1965, see Horne, 1969, p. 106) in which magnesium is cited as being about 40 times as likely to carry its water of hydration sheath with it during migration through the liquid. The multitude of mestastable, naturally occurring hydrous magnesium carbonates, such as hydromagnesite or nesquehonite, that crystallize in the thermodynamic stability field of anhydrous magnesite, is also evidence of the relative strength of the hydration sheath of the magnesium ion. Magnesite, like dolomite, is highly super-saturated in seawater and many environments of modern dolomite formation, such as the Persian Gulf sabkhas or ephemeral lakes associated with the Coorong Lagoon in Australia (Alderman, 1965; Patterson, 1972), are accompanied by magnesite precipitation (plus assorted hydrous magnesium carbonates).

Calcium ions, which are less strongly hydrated than magnesium ions, are more readily incorporated into a growing Ca-Mg carbonate and will tend to occupy sites within magnesium layers as well (Fig. 3).Again, rapid rates of crystal growth reduces the time available for magnesium ions in solution to displace errant calcium ions before these defects are buried by continued crystal growth. Thus, more rapid crystal growth or crystallization rates favour the development of more calcium rich phases.

Another important kinetic effect related to the strength of the magnesium ion hydration sheath, is the inability of the relatively unhydrated carbonate ion to batter its way through this hydration barrier and come in contact with the crystal surface (Fig. 6). Only a part of the total CO_3^{2-} in solution has sufficient energy of motion (translational or vibrational) to break through the hydration barrier surrounding magnesium ions on the surface of a growing dolomite crystal. This low activity of the CO_3^{2-} ion compared to the activities of the Ca^{2+} and Mg^{2+} ions in many neutral and acid

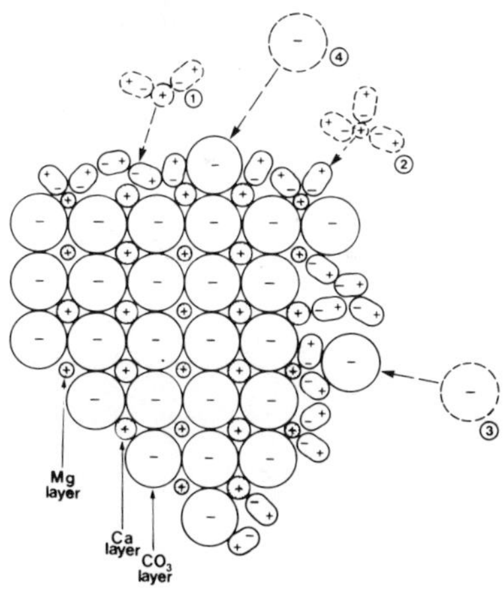

Figure 6 *A dolomite structure, showing water molecules preferentially bonded to the cations of the crystal surface. Calcium ions (1) are less strongly hydrated than magnesium ions (2) so calcium ions tend to be incorporated into magnesium layer positions causing the growth of a disordered Ca-rich carbonate. Carbonate ions are unhydrated but must have sufficient energy to displace the water molecules adjacent to the cation layer (e.g. 3 and 4). For simplicity an NaCl type of crystal structure is shown (after Lippman, 1973, Fig. 32).*

natural solutions, such as in sea water, is therefore a kinetic hindrance to the formation of dolomite although the thermodynamic solubility product of dolomite is exceeded by a wide margin (Lippman, 1973, p. 158-161).

The three main parameters that will affect the kinetics of the growth of dolomite are the Mg^{2+}/Ca^{2+} ratio, the salinity, and the CO_3^{2-}/Ca^{2+} ratio. The block diagram shown in Figure 7 is a plot of these solution parameters and the dolomite-calcite field boundary is shown as a plane separating a dolomite field on the near side and a calcite field on the far side. This plane represents a generalized kinetic boundary that lies entirely within the region of thermodynamic stability of dolomite. The lower surface is the bivariate plot of Folk and land (1975). A solution in the calcite field can move into the dolomite field by increasing its Mg/Ca ratio, decreasing its salinity or by increasing its CO_3^{2-} content.

An increase in the solution's Mg/Ca ratio favours the uptake of magnesium into a precipitating Mg-Ca carbonate. This is because an increase in the amount of magnesium in solution probably increases the number of magnesium ions relative to calcium ions that are energetic enough to shed their water of hydration and be incorporated into a growing Ca-Mg carbonate (Fig. 8). The laboratory syntheses by Glover and Sippel (1967) and Erenburg (1961) of magnesium calcites containing progressively greater mole per cent $MgCo_3$ contents with increasing Mg/Ca ratios in solution, constitute a good illustration of the effect of independently varying the Mg/Ca ratio of the solution. The sabkha environment of the Persian Gulf is a natural setting where the influence of the solution Mg/Ca ratio on dolomitization of aragonitic sediments is particularly well documented (e.g., Kinsman, 1965 and Patterson, 1972). At high temperature ($> 100°$ C) the rate of dolomite precipitation is increased by higher Mg/Ca ratios up to an optimum value (Gaines, 1980). This may apply also at lower temperatures.

The disordering effect caused by rapid precipitation may be overcome by a reduction in the rate of precipitation by dilution of these solutions. This is difficult to verify by laboratory experiments, because of the slowness of crystallization from solutions more dilute than sea water, but has been inferred from natural examples such as modern dolomitization from brackish waters in the Floridan Aquifer (Hanshaw *et al.*, 1971; Fanning *et al.*, 1981), in the sediments of Lake Balaton, Hungary (Muller *et al.*, 1972), and from Pleistocene examples where the geologic history is known with a high degree of certainty (e.g., Land, 1973a, 1973b).

Calculations of the thermodynamic stability of various carbonate minerals in mixtures of sea water and fresh groundwater show that there is a large region in the mixing continuum in which calcite is undersaturated but dolomite remains supersaturated (Badiozamani, 1973; Wigley and Plummer, 1976; and Fig. 5). Badiozamani (1973) speculated that the replacement of calcite by dolomite would be considerably easier within this mixing region because of the thermodynamic drive causing calcite to go into solution.

It is certainly true that mixing a wide variety of solution compositions can induce limestone dissolution. However, it is less clear why this fact in itself should induce dolomitization. The dissolution of aragonite and precipitation of dolomite takes place in solutions that are saturated or supersaturated with both phases in settings such as the sabkha environment (Kinsman, 1964) or evaporite lagoons or ponds (Morrow, 1978; Friedman, 1980) because of an appropriate rise in the solution Mg/Ca ratio through the precipitation of gypsum. It seems likely therefore, that the undersaturation of calcite that results from the mixing of saline with dilute solutions is not an important factor for dolomitization. The dilution of saline solutions favours dolomite precipitation simply because of kinetic effect of the attendant decrease in the rate of precipitation.

An increase in the CO_3^{2-} ion concentration of the solution, particularly in excess of the Ca^{2+} concentration, may also be an important variable for promoting dolomitization in that a greater number of CO_3^{2-} ions with sufficient energy to shoulder aside the hydration shield of magnesium ions would be present. A variety of laboratory experiments by Lippman (1968a) and Davies *et al.* (1977) have indicated the importance of the CO_3^{2-} concentration. Liebermann (1967) also synthesized dolomite from artificial solutions more saline than sea water and with a Mg/Ca ration approximately that of sea water but with a relatively high pH and CO_3^{2-} concentration.

Sources for naturally occurring alkaline solutions containing high CO_3^{-2} concentrations include continental groundwaters that have been involved in the weathering of siliceous rocks or that have dissolved continental alkalic carbonate minerals, or marine and con-

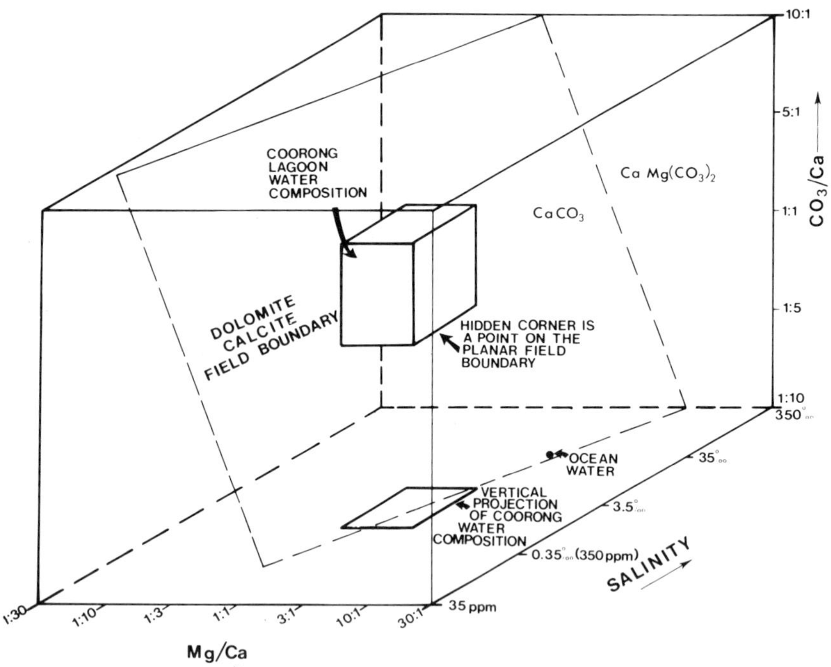

Figure 7 *Block diagram showing the effect of variation in the three parameters, the Mg/Ca solution ratio, the salinity, and the CO_3/Ca solution ratio. The plane represents the kinetic boundary between dolomite and calcite or aragonite and it includes the hidden corner of the Coorong Lagoon waters as a point on the plane. The basal plane is after Folk and Land (1975). Note that the vertical projection of Coorong Lagoon waters falls largely on the calcite-aragonite side of the stability boundary on the basal plane. A decrease in salinity, an increase in the Mg/Ca ratio or an increase in the CO_3/Ca ratio favours the precipitation of dolomite.*

tinental waters that have been involved in anaerobic bacterial sulphate reduction (Lippman, 1973). The Coorong Lagoon is the best documented modern occurrence of dolomitization under the influence of continental groundwater (Von der Borch, Lock and Schwebel, 1975; Von Der Borch, 1976; Von der Borch and Lock, 1979).

The composition of groundwater in the Coorong Lagoon Area is shown in Figure 7. Note that the projection of this solution composition onto the Folk and Land (1975) bivariate diagram falls largely outside the region of dolomite kinetic stability. This may be an indication that the high CO_3^{2-} concentration in the Coorong ground waters also promotes the precipitation of dolomite and Figure 7 has been constructed on this basis.

Generation of CO_3^{2-} by anaerobic sulphate reduction may have occurred in many burial environments from depths less than a metre to several hundred metres below the sediment-water interface. This process has been suggested as a factor in the dolomitization of some recent burrow fillings (Brown and Farrow, 1978). A supply of SO_4^{2-} ions in excess of the Ca^{2+} concentration is probably necessary because the initial rise in alkalinity will precipitate $CaCO_3$ before proceeding to dolomitization (Lippman, 1973).

Other Factors Controlling the Rate and Extent of Dolomitization

Most of the major models for dolomitization such as the Sabkha Model, the Reflux Model, the Mixed-Water Model and the Coorong Model may be explained in terms of the parameters shown in Figure 7, but other parameters under specific circumstances may be important. An increase in temperature such as occurs during burial will reduce the kinetic inhibitions to dolomitization (see Gaines, 1974; Mattes and Mountjoy, 1980). The length of time and the supply of magnesium available for dolomitization will govern the amount of dolomite formed (Hanshaw *et al.*, 1971). These problems will be discussed in relation to specific models of dolomitization in following paper.

The relative ease of dissolution of the various polymorphs of $CaCO_3$ is also an important rate-determining step in the dolomitization process (Gaines, 1974). The dolomitization of aragonitic sediments proceeds considerably more rapidly than for calcitic sediments (e.g., Schmidt, 1966). The rapid conversion of aragonite to low-magnesian calcite in humid climates has been cited as a factor in the resistance to dolomitization of Pleistocene shelf limestones exposed during low stands of sea level (Sibley, 1980).

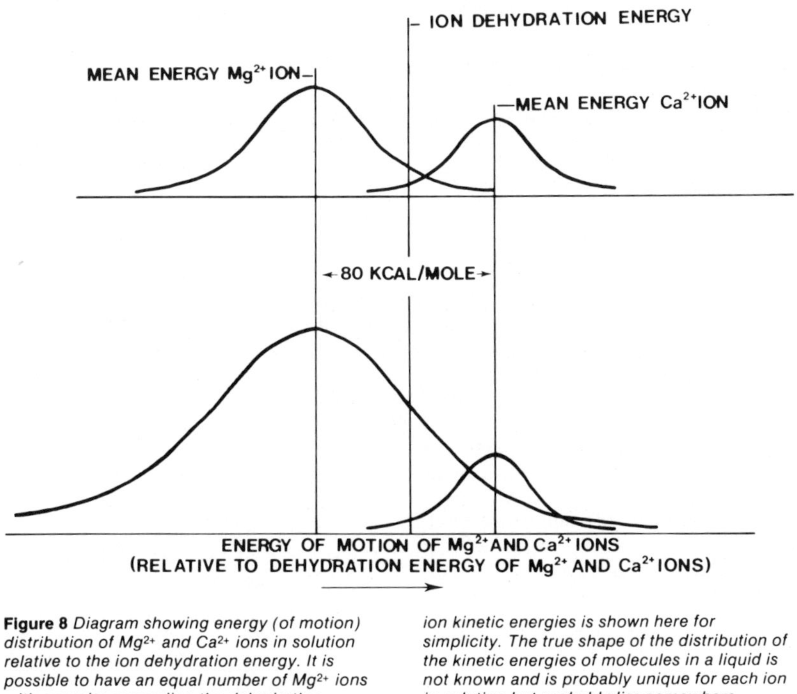

Figure 8 *Diagram showing energy (of motion) distribution of Mg^{2+} and Ca^{2+} ions in solution relative to the ion dehydration energy. It is possible to have an equal number of Mg^{2+} ions with energies exceeding the dehydration barrier by increasing the amount of Mg^{2+} in solution relative to the amount of Ca^{2+} in solution. A similar argument also applies to the CO_3^{2-} ion in solution. A normal distribution of ion kinetic energies is shown here for simplicity. The true shape of the distribution of the kinetic energies of molecules in a liquid is not known and is probably unique for each ion in solution but probably lies somewhere between a normal distribution and the assymetric Boltzman distribution for the energies of molecules in a gas.*

The presence of certain cations, such as lithium and possibly iron, have a catalytic effect on the precipitation of dolomite (Carpenter, 1980; Gaines, 1980). Carpenter (1976) has also suggested that because the proportion of strongly hydrated magnesium ions decreases markedly with increasing salinity, dolomitization should be favoured by high salinities and it does appear that dolomites that have formed under conditions of higher salinity and higher solution Mg/Ca ratios tend to be less calcium-rich and hence more stoichiometric (Fuechtbauer and Goldschmidt, 1966; Glover and Sippel, 1967). But the degree of disorder introduced by the faster rate of precipitation from more saline solutions overshadows the effect of the diminution of the hydration barrier of the magnesium ion with increasing salinity so that, although the Ca/Mg carbonates that are formed approach a dolomite composition, they tend to be disordered (Glover and Sippel, 1967; Folk and Land, 1975; Morrow, 1978). More experimental data is needed to separate the effects of precipitation rate versus the Mg/Ca ratio of the solution.

Dolomitization and the Volume Problem

The chemical reaction for dolomitization is most commonly given in the form shown in reaction (1) in which a mole of dolomite replaces two moles of calcite. The alternative dolomitization reaction given in reaction (2) was proposed by Lippman (1973, p. 180) largely because of the advantages of a high CO_3^{2-} concentration in solution. In reaction (1) the dolomitizating solution supplies the Mg^{2+} required and removes the Ca^{2+} released to solution, whereas in reaction (2) the solution supplies both Mg^{2+} and CO_3^{2-} but is not required to remove any reaction products.

Both of these equations involve considerable volume changes. A loss of volume of 6% to 13% (aragonite or calcite replacement) accompanies dolomitization according to reaction (1), a circumstance commonly cited by workers in the oil industry. On the other hand, dolomitization according to reaction (2), if applied literally, involves a tremendous volume gain of about 75% to 88% (Lippman, 1973, Table 28).

However, most dolomitized rocks do not appear to have undergone volume changes during dolomitization indicating a volume for volume replacement rather than a mole for mole replacement (Friedman and Sanders, 1967). Depositional fabrics such as those displayed by algal laminites (Gebelein and Hoffman, 1973) or by oolitic sediments (Powers, 1962) are not disrupted by dolomitization. The increase in permeability that accompanies total dolomitization may be related to the idiomorphic shape and uniformity of size of the dolomite crystals (Wardlaw, 1979).

The conservation of volume during the dolomitization reaction may be an important constraint. For a volume-for-volume reaction the net dolomitization reaction may be written

$$(2\text{-}x)\ CaCO_3 + Mg^{2+} + x\ CO_3^{2-} \rightleftharpoons CaMg\ (CO_3)_2 + (1\text{-}x)\ Ca^{2+} \qquad (3)$$

where x = 0.11 and 0.25 for the dolomitization of aragonite and calcite respectively. Any departure from a volume for volume replacement will result in changes in these numbers which represent the net exchange of ions between the solution and the solid phases. Reaction (3) is characterized by a free energy change at equilibrium of $\Delta G°_r$ = -3.085 for dolomitization of aragonite (Table I) which is intermediate between the free energies of reaction of (1) and (2). Reaction (3) is a composite of reactions (1) and (2) which may be regarded as end members of a series of reactions dependant on the degree to which there is a gain or loss of rock volume during dolomitization.

Patterson (1972, p. 325) has proposed the reaction

$$(2\text{-}x)\ CaCO_3 + Mg^{2+} + x\ HCO_3^- \rightleftharpoons CaMg\ (CO_3)_2 + (1\text{-}x)\ Ca^{2+} + x\ H^+ \qquad (4)$$

based on chemical mass balance considerations for dolomitization of sabkha sediments which is similar to reaction (3) except that carbonate is supplied in the form of bicarbonate because of the slightly acid diagenetic environment of the sabkha. Patterson suggested that the average value of x in reaction (4) is close to 0.1 based on the organic content of the sabkha sediments. This value of x also satisifies the conservation of volume criterion.

Reaction (3) implies that the equilibrium constant for dolomitization depends on the CO_3^{2-} concentration as well as the Mg^{2-} and Ca^{2+} concentrations. But, because of kinetic factors, the concentrations of Mg^{2+} and CO_3^{2-} in the solution must be higher than that necessary merely to satisfy the thermodynamic requirements of dolomitization and the dolomitizing solution must play an active role as a transporter of ions to and from the reaction site. This has important implications for carbon and oxygen stable isotope studies of dolostones.

The release of Ca^{2+} to solution during dolomitization can lead to the local precipitation of gypsum as has been documented in Persian Gulf sabkhas (Patterson, 1972). This may be an alternative explanation also for the minute gypsum or anhydrite crystal inclusions in the coarsely crystalline white dolostones of the Mississippi Valley and Pine Point Districts which Beales and Hardy (1977, 1980) suggested were remnants of former evaporites.

The Dolomitization Process - Conclusions

1) Dolomite is thermodynamically stable in most natural solutions at earth surface conditions and a thermodynamic drive exists for the conversion of calcite and aragonite to dolomite.

2) The precipitation of dolomite is hindered by kinetic factors that are a consequence of the high degree of order of the dolomite crystal lattice and of the differences in hydration behavior of the Ca^{2+}, Mg^{2+} and CO_3^{2-} ions.

3) The extent to which these factors inhibit dolomitization appear to be governed largely by the interplay of three critical parameters: a) The rate of crystallization governs the degree of order of precipitating Ca-Mg carbonates. Dilution of supersaturated natural solutions is one method of reducing the crystallization rate; b) the Mg/Ca ratio of the solution determines the degree to which Mg^{2+} is incorporated into precipitating Ca-Mg carbonates. This ratio commonly rises during the precipitation of gypsum and/or anhydrite to a level favourable for dolomitization; and c) A CO_3/Ca ratio greater than one will aid in the precipitation of dolomite. This ratio may rise in ground water involved in weathering of igneous rocks or where anaerobic bacterial reduction of sulphate occurs.

4) A conservation of volume constraint implies a reaction of the form:

$$(2\text{-}x)\ CaCO_3 + Mg^{2+} + x\ CO_3^{2-} \rightleftharpoons CaMg\ (CO_3)_2 + (1\text{-}x)\ Ca^{2+}$$

where x equals 0.11 or 0.25 for dolomitization of aragonite or calcite respectively. All three of the constituent ions of dolomite are exchanged between the solution and the solid phases emphasizing the role of fluid movements and compositions during dolomitization.

It is hoped that this review has succeeded in providing a coherent synthesis of evidence and theories specifically concerning the chemistry of the dolomitization process. The main kinetic factors controlling dolomite precipitation may have been identified but a more determined effort should be made to precipitate dolomite in the laboratory in quantities large enough to verify the influence of these kinetic effects and perhaps even to gain a degree of quantitative understanding of them. Only then will we be in a position to fully understand the significance of the isotopic composition and of the trace element contents of dolostones because of their dependence on the chemistry of the dolomitization process. Knowledge of the chemical environment of dolomitization also provides a clearer insight into the process of dolomitization in specific models for dolomitization.

Acknowledgements

Douglas Kent of the University of Saskatchewan and Alan Kendall of Amoco Canada Ltd. critically read the manuscript. I particularly wish to thank Kendall for his assistance in clarifying some of the ideas presented here and Ian McIlreath, the editor of this series, for his comments and encouragement.

Table 1 *Free Energies of Dolomitization Reactions (1), (2) and (3) at Equilibrium*

	$\Delta G°_r$ calcite (kilocal/mole)	$\Delta G°_r$ aragonite (kilocal/mole)
Reaction (1)	-1.38*	-1.83
Reaction (2)	-13.01	-13.24
Reaction (3)	-4.29	-3.08

* Calculation of free energies from data of Helgeson *et al.* (1978) using fully ordered dolomite. These numbers vary depending on the degree of order of the dolomite, but ordered dolomite is the most favoured phase thermodynamically.

References

Dolomite Crystal Structure and Compositional Variability

Carpenter, A.B., 1980, The chemistry of dolomite formation I: the stability of dolomite: in D.H. Zenger, J.B. Dunham and R.G. Ethington, eds., Concepts and Models of Dolomitization: Soc. Econ. Paleontol., Mineral. Spec. Publ. 28, p. 111-121.

Gaines, A.M., 1977, Protodolomite redefined: Jour. Sed. Petrology, v. 47, p. 543-546.

Goldsmith, J.R. and D.G. Graf, 1958, Structural and Compositional variations in some natural dolomites: Jour. Geol., v 66, p. 678-693.

Land, L.S., 1980, The isotopic and trace element geochemistry of dolomite: the state of the art: in D.H. Zenger, J.B. Dunham and R.A. Ethington, eds., Concepts and Models of Dolomitization: Soc. Econ. Paleontol., Mineral. Spec. Publ. 28, p. 87-110.

Lippman, F., 1973, Sedimentary Carbonate Minerals: New York, Springer Verlag, 228p.

The Stability of Dolomite in Natural Environments

Carpenter, A.B., 1980, The chemistry of dolomite formation I: the stability of dolomite: in D.H. Zenger, J.B. Dunham and R.G. Ethington, eds., Concepts and Models of Dolomitization: Soc. Econ. Palentol., Mineral. Spec. Publ. 28, p. 111-121.

Garrels, G.M. and M.E. Thompson, 1962, A chemical model for sea water at 25° C and one atmosphere total pressure: Amer. Jour. Sci., v. 258, p. 402-418.

Gaines, A.M., 1977, Protodolomite redefined: Jour. Sed. Petrology, v. 47, p. 543-546.

Goldsmith, J.R. and H.C. Heard, 1961, Subsolidus phase relations in the system $CaCO_3$: Jour. Geol., v. 69, p. 45-74.

Helgeson, H.C., J.M. Delany, H.W. Nesbitt and D.K. Bird, 1978, Summary and critique of the thermodynamic properties of rock-forming minerals: American Jour. Sci., v. 278-A, 229 p.

Land, G.S., 1980, The isotopic and trace element geochemistry of dolomite: the state of the art: in D.H. Zenger, J.B. Dunham and R.A. Ethington, eds., Concepts and Models of Dolomitization: Soc. Econ. Plaentol., Mineral. Spec. Publ. 28, p. 87-110.

Lippman, F., 1973, Sedimentary carbonate minerals: New York, Springer Verlag, 228p.

The Role of Kinetic Effects on the Crystallization of Dolomite

Badiozamani, K. 1973, The Dorag dolomitization model-application to the Middle Ordovician of Wisconsin: Jour. Sed. Petrology, v. 43, p. 965-984.

Davies, P.J., B. Bubela, and J. Ferguson, 1977, Simulation of carbonate diagenetic processes: formation of dolomite, huntite and monohydro-calcite by the reactions between nesquehonite and brine: Chem. Geol., v. 19, p. 187-214.

Folk, R.G., 1974, The natural history of crystalline calcium carbonate: effect of magnesium content and salinity: Jour. Sed. Petrology, v. 44, p. 40-53.

Folk, R.L. and L.S. Land, 1975, Mg/Ca ratio and salinity: two controls over crystallization of dolomite: American Assoc. Petroleum Geol. Bull., v. 59, p. 60-68.

Friedman, G.C., 1980, Dolomite is an evaporite mineral: evidence from the rock record and from sea-marginal ponds of the Red Sea: in K.H. Zenger, J.B. Dunham and R.L. Ethington, eds., Concepts and Models of Dolomitization; Soc. Econ. Paleontol., Mineral. Spec. Publ. 28, p. 69-80.

Gaines, A.M., 1980, Dolomitization kinetics; recent experimental studies: in D.H. Zenger, J.B. Dunham and R.L. Ethington, eds., Concepts and Models of Dolomitization: Soc. Econ. Paleontol., Mineral. Spec. Publ. 28, p. 81-86.

Glover, E.D. and R.F. Sippel, 1967, Synthesis of magnesium calcite: Geochim. Cosmochim. Acta, v. 31, p. 603-613.

Hanshaw, B.C., W. Back, and R.G. Deike, 1971, A geochemical hypothesis for dolomitization by groundwater: Econ. Geology, v. 66, p. 710-724.

Land, L.S., 1973a, Contemporaneous dolomitization of Middle Pleistocene reefs by meteoric water, North Jamaica: Bull. Marine Sci., v. 23, p. 64-92.

Land, L.S., 1973b, Holocene meteoric dolomitization of Pleistocene limestones, North Jamaica: Sedimentology, v. 20, p. 411-424.

Liebermann, O., 1967, Synthesis of dolomite: Nature, v. 213, p. 241-245.

Lippman, F., 1968, Synthesis of $BaMg(CO_3)_2$ (Norsethite) at 20° C and the formation of dolomite in sediments: in G. Muller and G.M. Friedman, eds., Recent developments in carbonate sedimentology in Central Europe: New York, Springer Verlag, p. 33-37.

Lippman, F., 1973, Sedimentary carbonate minerals: New York, Springer Verlag, 228 p.

Morrow, D.W., 1978, The influence of the Mg/Ca ratio and salinity on dolomitization in evaporite basins: Can. Petrol. Geol. Bull., v. 26, p. 389-392.

Muller, G., G. Irion and V. Forstner, 1972, Formation and diagenesis of inorganic Ca-Mg carbonates in the lacustrine environment: Naturwissenschaften, v. 59, p. 158-164.

Patterson, R.J., 1972, Hydrology and carbonate diagenesis of a coastal sabkha in the Persian Gulf: Ph.D. Diss., Princeton Univ., Princeton, New Jersey, 498 p.

Von der Borch, C.C., 1976, Stratigraphy and formation of Holocene dolomitic carbonate deposits of the Coorong area, South Australia: Jour. Sed. Petrology, v. 46, p. 952-966.

Other Factors Controlling the Rate and Extent of Dolmitization

Carpenter, A.B., 1976, Dorag dolomitization model: by K. Badiozamani: Jour. Sed. Petrology, v. 43, p. 965-984: Jour. Sed. Petrology, v. 46, p. 258-261.

Fuchtbauer, H. and H. Goldschmidt, 1965, Beziehugen zwishen calcium-gehalt und bildungsbedingungen der dolomite: Geol. Rundschau, v. 55, p. 29-40.

Gaines, A.M., 1974, Dolomitization kinetics: recent experimental studies: in D.H. Zenger, J.B. Dunham and R.L. Ethington, eds., Concepts and Models of Dolomitization: Soc. Econ. Paleontol., Mineral. Spec. Publ. 28, p. 111-121.

Sibley, D.F., 1980, Climatic control of dolomitization, Seroe Domi Formation (Pliocene), Bonaire, N.A.: in D.H. Zenger, J.B. Dunham, and R.L. Ethington, eds. Concepts and Models of Dolomitization: Soc. Econ. Paleontol., Mineral. Spec. Publ. 28, p. 247-258.

Dolomitization and the Volume Problem

Friedman, G.M. and J.E. Sanders, 1967, Origin and occurrence of dolostones: in G.V. Chilingar, H.J. Bissell, and R.W. Fairbridge, eds., Carbonate Rocks: Developments in Sedimentology, 9A: Amsterdam, Elsevier, p. 267-348.

Lippman, F., 1973, Sedimentary carbonate minerals: New York, Springer Verlag, 228 p.

Powers, R.W., 1962, Arabian Upper Jurassic carbonate reservoir rocks: in W.E. Ham, ed., Classification of Carbonate Rocks - A Symposium - American Assoc. Petroleum Geol. Memoir 1, p. 122-192.

Wardlaw, N.C., 1979, Pore systems in carbonate rocks and their influence on hydrocarbon recovery efficiency: in Geology of carbonate porosity: Amer. Assoc. Petroleum Geologists Continuing Education Course Note Series II., p. E1-E24.

Other References Cited in the Text

Alderman, A.R., 1965, Dolomitic sediments and their environment in the south-east of South Australia: Geochim. Cosmochim. Acta, v. 29, p. 1353-1365.

Bathurst, R.G.C., 1975, Carbonate Sediments and Their Diagenesis: Developments in Sedimentology No. 12: Amsterdam, Elsevier, 658 p.

Beales, F.W., and J.W. Hardy, 1977, The problem of recognition of occult evaporites with special reference to southeast Missouri: Econ. Geol., v. 72, p. 487-490.

Beales, F.W. and J.L. Hardy, 1980, Criteria for the recognition of diverse dolomite types with an emphasis on studies on host rocks for Mississippi Valley type ore deposits: in D.H. Zenger, J.B. Dunham and R.L. Ethington, eds., Concepts and Models of Dolomitization: Soc. Econ. Paleontol., Mineral Spec. Publ. 28, p. 197-213.

Brown, R.J. and G.E. Farrow, 1978, Recent dolomitic concretions of crustacean burrow origin from Goch Sunart, West Coast of Scotland: Jour. Sed. Petrology, v. 48, p. 825-834.

Davies, G.R., 1979, Dolomite reservoir rocks: processes, controls, porosity development: in Geology of Carbonate Porosity: American Assoc. Petroleum Geologists Continuing Education Course Note Series II, p. C1-C17.

Deelman, J.C., 1975, Dolomite synthesis and crystal growth: Geology, v. 3, p. 471-472.

Deelman, J.C., 1978, Discussion: protodolomite redefined: Jour. Sed. Petrology, v. 48, p. 1004-1067.

Erenburg, B.G., 1961. Artificial mixed carbonates in the $CaCO_3$-$MgCO_3$ series: Zhurnal Skrukt. Khim. v. 2, p. 178-182.

Fanning, K.A., R.H. Byrne, J.A. Breland, R.R. Betzer, W.S. Moore, R.J. Elsinger and T.E. Pyle, 1981, Geothermal springs of the west of Florida Continental Shelf: evidence for dolomitization and radionuclide enrichment: Earth and Planet. Science Letters, v. 52, p. 345-354.

Fritz, P. and A. Katz, 1972, The sodium distribution of dolomite crystals: Chem. Geol., v. 10, p. 237-244.

Fyfe, W.S. and J.L. Bischoff, 1965, The calcite-aragonite problem: in L.C. Pray and R.C. Murray, eds., Dolomitization and Limestone Diagenesis: A Symposium - Soc. Econ. Paleont., Mineral., Spec. Publ. 13, p. 3-13.

Gaines, A.M., 1978, Reply-protodolomite redefined, Jour. Sed. Petrology, v. 48, p. 1009-1011.

Gebelein, C.D. and Hoffman, P., 1973, Algal origin of dolomite laminations in stromatolitic limestone: Jour. Sed. Petrology, v. 43, p. 603-613.

Gidman, J., 1978, Discussion - protodolomite redefined: Jour. Sed. Petrology, v. 48, p. 1007-1008.

Horne, R.A., 1969, Marine Chemistry: New York, Wiley-Interscience. 568 p.

Kinsman, D.J.J., 1965, Gypsum and anhydrite of recent age, Trucial Coast, Persian Gulf: in J.L. Rau, ed., Second symposium on salt, v. 1. Cleveland, Ohio, Northern Ohio Geol. Soc., p. 302-326.

Lumdsden, D.N. and J.S. Chimahusky, 1980, Relationship between dolomite nonstoichoimetry and carbonate facies parameters: in D.H. Zenger, J.B. Dunham and R.A. Etherington, eds., Concepts and Models of Dolomitization: Soc. Econ. Paleont., Mineral. Spec. Publ. 28, p. 123-137.

Macqueen, R.W., 1979, Base metal deposits in sedimentary rocks: Some approaches. Geoscience Canada, v. 6, p. 3-9.

Mattes, B.W. and E.W. Mountjoy, 1980, Burial dolomitization of the Upper Devonian Miette Buldup, Jasper National Park, Albert: in D.H. Zenger, D.B. Dunham and R.L. Ethington, eds., Concepts and Models of Dolomitization: Soc. Econ. Paleontol., Mineral. Spec. Publ. 28, p. 259-297.

Pingitore, N.E., Jr., 1976, Vadose and phreatic diagenesis: Processes, products and their recognition in corals: Jour. Sed. Petrology, 46, p. 60-67.

Radke, B.M. and M.L. Mathis, 1980, On the formation and occurrence of saddle dolomite: Jour. Sed. Petrology, v. 50, p. 1149-1168.

Reeder, R.J. and H.R.Wenk, 1979, Microstructures in low temperature dolomites: Geophy. Res. Letters, v. 6, p. 77-80.

Samoilov, O.Y., 1965, Structure of aqueous electrolyte solutions and hydration of ions: Consultants Bureau, New York.

Schmidt, V., 1965, Facies, diagenesis, and related reservoir properties in the Gigas Beds (Upper Jurassic), northwestern Germany: in L.C. Pray and R.C. Murray, eds., Dolomitization and Limestone Diagenesis: Soc. Econ. Paleontol., Mineral. Spec. Publ. 13, p. 124-168.

Towe, K.M., 1975, Dolomite synthesis and crystal growth: Comment: Geology, v. 3, p. 612.

Veizer J., J. Lemieux, B., Jones, J.R. Gibling, and J. Savelle, 1978, Paleosalinity and dolomitization of a Lower Paleozoic carbonate sequence, somerset and Prince of Wales Island, Arctic Canada: Canadian Jour. Earth Sci., v. 15, p. 1448-1461.

Von der Borch, C.C., D.E. Lock and D. Schwebel, 1975, Groundwater formation of dolomite in the Coorong region of South Australia: Geology, May, p. 283-285.

Wardlaw, N.C., 1976, Pore geometry of carbonate rocks as revealed by pore casts and capillary pressur: American Assoc. Petroleum Geologists Bull., v. 60, p. 245-257.

Wigley, T.M.L. and L.N. Plummer, 1976, Mixing of carbonate waters: Geochim. Cosmochim, Acta, v. 40, p. 989-995.

Ziegler, A.M., C.R. Scotese, W.S. McKerrow, M.E. Johnson and R.K. Bambach, 1979, Paleozoic paleogeography: Ann. Rev. Earth Planet Sci. v. 7, p. 473-502.

MS received November 27, 1981

Reproduced with the permission of the Geological Association of Canada. Published in *Diagenesis 2; Dolomite, Part 2, Geoscience Canada,* V. 9, No. 2, pp. 95-107.

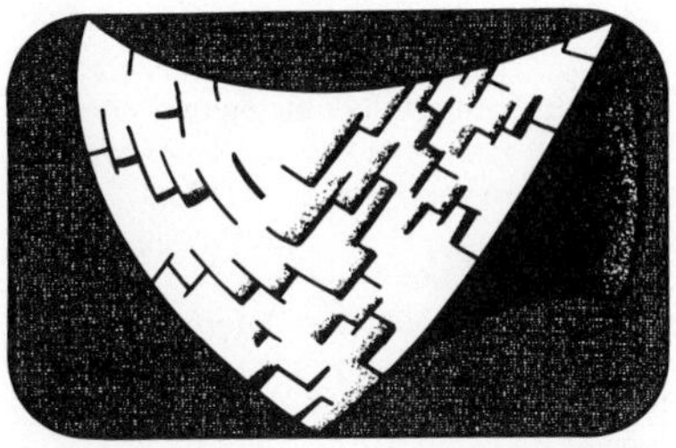

Diagenesis 2. Dolomite - Part 2 Dolomitization Models and Ancient Dolostones

D.W. Morrow
Institute of Sedimentary and Petroleum Geology
Geological Survey of Canada
3303 - 33rd St. N.W.
Calgary, Alberta T2L 2A7

Introduction

The first paper of this review series (Morrow, 1982) was a synthesis of what is presently known concerning the chemistry of dolomitization. In that paper three parameters that are probably of major importance to the dolomitization process were identified: the Mg/Ca solution ratio; the ionic strength or salinity of the solution; and the CO_3/Ca solution ratio. Other factors, such as temperature, that influence the dolomitization process also were discussed.

In this paper I wish to focus the reader's attention on a different aspect of dolomitization - the great variety of models for dolomitization that have been proposed to explain the origin of the wide spectrum of dolostone types. In approximate historical order of appearance, these are the Hypersaline Lagoon and Reflux Model, the Burial Compaction Model, the Solution Cannibalization Model, the Coorong Model, the Sabkha Model, and the Mixed Water (Dorag) or Dilution Model (Fig. 1). The Sabkha and Coorong Models are specific to well defined environmental settings. The other models are more generalized and emphasize particular aspects of the dolomitizing process that may be operative in a variety of environmental settings. For example, in the recently proposed Dilution Model, the emphasis is on the dilution of seawater by fresh groundwater which may occur in a multitude of situations both penecontemporaneous with, or long after sedimentation.

A number of conditions must be satisfied for these models to act as agents for dolomitization (Fig. 1). These conditions, in their simplest form are as follows:

1) The amount of Mg^{2+} available for dolomitization must be adequate to form a given mass of dolostone (i.e., source of Mg^{2+}).

2) A mechanism is required to deliver the available Mg^{2+} to the dolomitization site. This mechanism may also be required to deliver a small amount of CO_3^{2-} ions to the dolomitization site and carry away Ca^{2+} ions.

3) The composition of a proposed dolomitizing solution must be conducive to dolomitization. The first paper of this review series provided some criteria to test this proposition.

Each model will be examined in the light of these conditions to place constraints on their application to the stratigraphic record and on their viability in general. A variety of specific factors may contribute to these conditions in particular instances. For example, the delivery mechanism (condition 2) may depend on porosity and permeability relationships, the hydraulic head of the solution and the time over which the mechanism is operative.

The application of these models to individual dolostone masses depends on inferences concerning the attributes of these dolostones. These include considerations of scale and spatial relationships, major stratigraphic and paleogeographic relationships, internal facies relationships, petrographic relationships, and of geochemical data. It is at this stage, where statements are made concerning the application of specific dolomitization models based on the physical and chemical attributes of dolostone bodies, that the confusion concerning the dolomitization process becomes

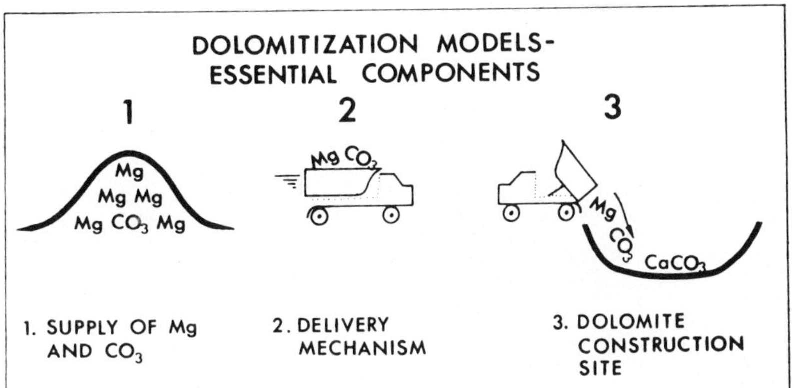

Figure 1 *A construction site analogue of the process of dolomitization illustrating the three essential elements of any dolomitization model. The external supply of magnesium, and, to a lesser extent, the supply of carbonate must be sufficient to form the observed amounts of dolomite. There must be a delivery mechanism, commonly in the form of fluid flow, adequate to deliver these ions to the dolomitization site and lastly there must exist chemical conditions favourable for the precipitation of dolomite at the 'dolomite construction site'. Failure to meet any of these conditions results in an inadequate dolomitization model.*

readily apparent to the geological public. An attempt is made here to determine the relative degree of certainty that can be placed on the diverse criteria used to substantiate the choices of dolomitization models that have been applied to many ancient dolostones in Canada (Fig. 2) and elsewhere.

Dolomitization Models

The Hypersaline Lagoon and Reflux Model. One of the first models proposed for dolomitization on a large scale was the Seepage Refluxion or more simply, the Reflux Model (King 1947; Adams and Rhodes, 1960). In this model (Fig. 3) open marine water evaporates as it passes landward across a hypersaline shelf lagoon. The resulting increase in density of these evaporated waters causes them to infiltrate the underlying sediment and to move seaward by seepage (i.e., reflux) through the seaward-dipping beds. The cycle of fresh marine replenishment of lagoon waters and seaward seepage of brines is the mechanism by which Mg^{2+} is continually transported to the sediment. The source of Mg^{2+} is the evaporated sea water itself. The dolomitization reaction probably occurs after a great deal of the gypsum (or anhydrite) has precipitated and the Mg/Ca ratio approaches 9.0 whereupon the effect of the Mg/Ca ratio overshadows the inhibiting effect of salinity (Morrow, 1978). Possibly the increase of the CO_3/Ca solution ratio that occurs during gypsum or anhydrite precipitation would also aid dolomitization.

Modern examples of the seepage refluxion hypothesis are all small in scale and operate at or near the shoreline. Müller and Teitz (1971) and Kocurko (1979) have documented shoreline dolomitization in the Canary Islands and in San Andres Island by evaporated sea water that collects in ephemeral pools in the "spray zone," a few metres above high tide. The shallow but permanent evaporating pool at Bonaire (Deffeyes *et al.*, 1965; Murray, 1969) is perhaps the best modern analogue of the large scale reflux model envisaged by Adams and Rhodes (1960).

The Mg^{2+} content of refluxing solutions in the reflux model is extremely high but it is not clear what the rate of subsurface flow of these dense solutions is likely to be. Deffeyes *et al.* (1965) calculated a rate of seaward reflux flow of 2×10^{-6} cm^3/sec. cm^2 at Bonaire but subsequent work (Murray, 1969) showed that the refluxing brines flow seaward for only a few weeks each year and most of this flow is confined to more permeable conduits bypassing most of the limestone. Model studies or computer simulations might shed light on the feasibility of large scale reflux scenarios such as that envisaged by Adams and Rhodes (1960). At present the dynamics of water movement beneath and adjacent to evaporitic shelf lagoons is unknown and a potentially fertile field of investigation. There are many examples of Canadian dolostones explained by variants of the reflux hypothesis (Fig. 2).

DOLOMITIZATION MODELS - CANADIAN EXAMPLES

Model	Example (Formation and author)
1. Hypersaline Basin and Seepage Reflux	Keg River - Pine Point - Presqu'ile (Skall, 1975) Muskeg - Keg River (Davies and Ludlum, 1973; Bebout and Maiklem, 1973) Prairie Evaporite - Winnipegosis (Wardlaw and Reinson, 1971) Lodgepole (Young and Greggs, 1975) Leduc (Wilson, 1975)
2. Sabkha	Charlie Lake (Roy, 1972) St. George (Collins and Smith, 1975) Turner Valley (Murray and Lucia, 1967) Muskeg-Winnipegosis (Shearman and Fuller, 1969)
3. Mixed-Water	Guelph (Sears and Lucia, 1980) Nisku (Exploration Staff, Chevron Standard Limited, 1979) Abenaki (Eluik, 1978) Allen Bay (Land *et al.*, 1975; Veizer *et al.*, 1978) Thumb Mountain (Morrow and Kerr, 1978) St. George? (Collins and Smith, 1975) Stone (Morrow, 1975; Macqueen and Thompson, 1978) Manetoe (Morrow, 1975)
4. Coorong	Belcher Group (Ricketts, in press)
5. Burial-Compaction	Presqu'ile (Griffin, 1965; Beales and Jackson, 1969) Leduc (Illing, 1959) Miette (Mattes and Mountjoy, 1980) Guelph (Jodry, 1969)
6. Solution Cannibalization	Yeoman (Kendall, 1977)
7. Tectonic or Hydrothermal	Black River - Trenton (Sanford, 1962) Keg River - Rainbow (Langton and Chin, 1968) Presqu'ile (Jones, 1980)

Figure 2 *Canadian examples of the dolomitization models discussed in this paper. Note that some units, such as the Presqu'ile Formation, have been assigned to more than one model.*

The Burial Compaction Model. Compaction of fine-grained sediments during burial involves the progressive explusion of pore water. Commonly large shale masses contain carbonate reef bodies and are underlain by or are bordered by thick shelf carbonate sequences. A part of this Mg^{2+}-bearing compaction water may pass through adjacent limestones and cause dolomitization (Fig. 3). The Mg^{2+} required for dolomitization is derived primarily from the pore water. Illing (1959), Jodry (1969) and Griffin (1965) were early proponents of this model (Fig. 2).

The popularity of the Burial Compaction Model waned for a long period because of the upsurge in studies of Holocene penecontemporaneous dolomites. But recent developments in the understanding of diagenetic changes in clay minerals and associated amorphous material, and the parallel maturation of organic material to hydrocarbons during burial has led to a renewed interest in this model (Davies, 1979; Mattes and Mountjoy, 1980). Clay minerals undergo a sequence of depth-related mineralogical changes that provides an additional source of pore water and of a variety of ions, including Mg^{2+}, released from clay minerals to the pore fluids. These transformations have been cited as additional causes for burial-compaction dolomitization.

There is little doubt that the kinetic inhibitions that bedevil the formation of dolomite at near-surface conditions are greatly reduced in deep burial environments largely because of the increase of temperature with depth. Also, the time available for dolomitization is much greater in deep subsurface environments than in the near-surface environments of early dolomitization (Mattes and Mountjoy, 1980).

However, the problem of the supply of magnesium for dolomitization is an important constraint in deep burial environments. About 5.5 x 10^{-3} moles of Mg^{2+} are required to dolomitize 1 cm^3 of calcite at 10 per cent porosity. If this Mg^{2+} is supplied by compaction of adjacent shales and if the pore solution has the composition of seawater, then about 32 cm^3 of shale, compacted from 70 per cent to two per cent porosity, are required to provide this amount of Mg^{2+}. (These mass balance considerations were first discussed by the present author at the C.S.P.G. Andy Baillie Symposium, 1979). This assumes that all the water of compaction passes through the carbonates. In reality more than half of shale compaction water is lost vertically upward into the body of water from which the shale was deposited and most of the remainder passes downward (Magara, 1976, p. 92-95). On a basinwide scale, therefore, a more reasonable estimate would be that on the order of hundreds of cm^3 of compacted shale are required for the dolomitization of 1 cm^3 of laterally adjacent limestone. Therefore shale compaction water is insufficient to form large masses of dolomite such as the Paleozoic platform dolomite sequences of Western Canada and elsewhere.

In environments of burial greater than 2000m the transformation of montmorillonitic clays to illite is accompanied by the release of Si^{4+}, Fe^{2+}, Ca^{2+}, and Mg^{2+} ions to the pore solution so that authigenic ankerite, ferroan dolomite and chlorite are precipitated as cements in some sand-shale sequences (Boles, 1981). It seems very unlikely, however, that there is a sufficient volume of fluid at these depths to move these ions laterally and to participate in the solution of large amounts of limestone and contemporaneous precipitation of dolomite. Shale porosity is less than 20 per cent at burial depths exceeding 2000 m and processes such as aquathermal pressuring or the release of structured water from clays do not add significantly to this volume of interstitial pore water (Magara, 1976). The existing evidence strongly indicates that the late diagenetic release of ions from clay minerals is accompanied by local cementation rather than the long distance transportation of ions to shelf carbonates facing shale basins.

Another factor limiting the dolomitizing potential of these deep burial diagenetic solutions is the fact that Ca^{2+} ions are released to solution in amounts nearly

DOLOMITIZATION MODELS

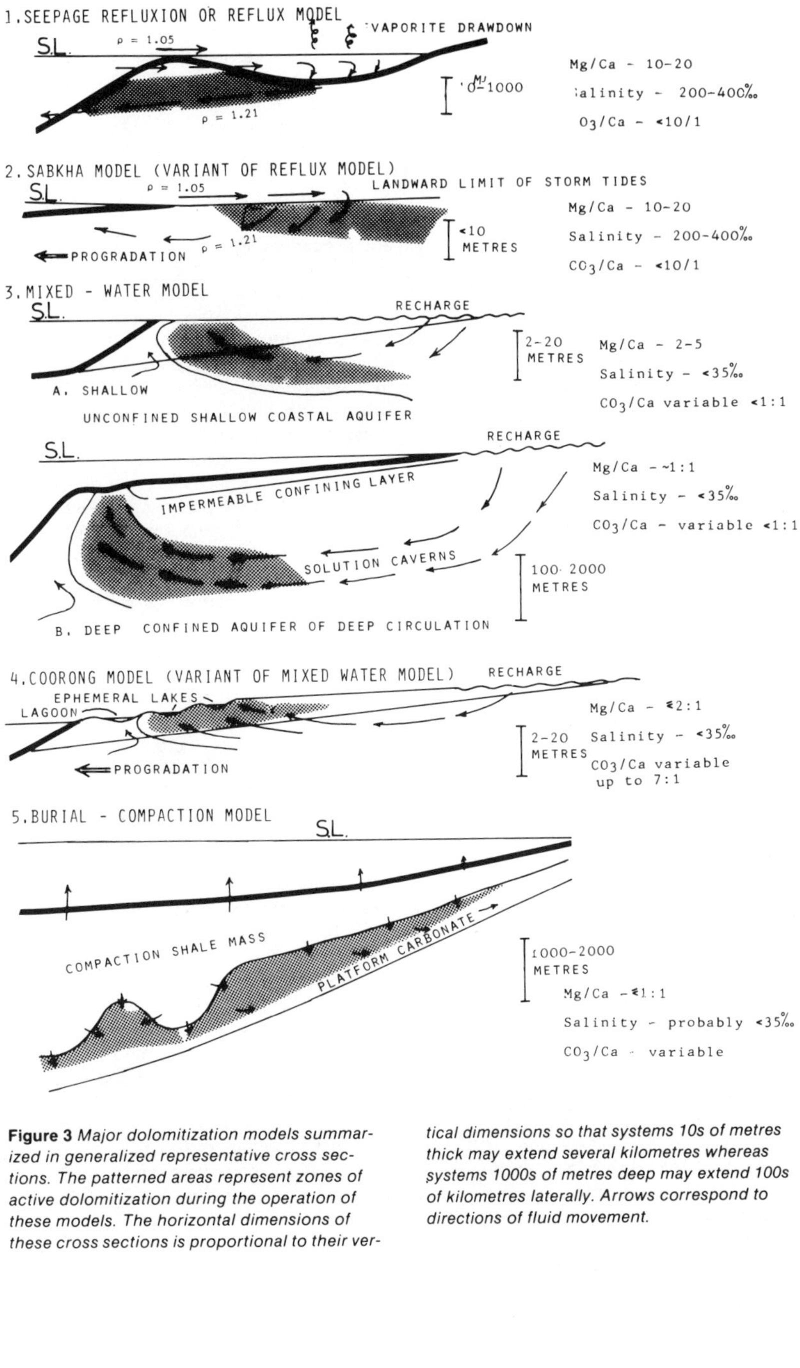

Figure 3 *Major dolomitization models summarized in generalized representative cross sections. The patterned areas represent zones of active dolomitization during the operation of these models. The horizontal dimensions of these cross sections is proportional to their vertical dimensions so that systems 10s of metres thick may extend several kilometres whereas systems 1000s of metres deep may extend 100s of kilometres laterally. Arrows correspond to directions of fluid movement.*

equal to, or greater than, Mg^{2+} (Boles, 1981; Foscolos and Kodama, 1974). There is little, if any, excess Mg^{2+} available for dolomitization. In the most favourable situation one might expect the void spaces in porous carbonates facing shales (e.g., Wong and Oldershaw, 1981) to be partly occluded with ferroan dolomite, but little actual dolomitization of preexisting limestone.

Although more work needs to be done to quantify the many variables involved in the Burial-Compaction Model, as a first approximation it does not appear that regional bodies of dolomite can form by this mechanism. The observation that many deeper water slope and basinal carbonates commonly are not dolomitized even where they are overlain by thick shale masses tends to confirm this impression (Wilson, 1975). Small amounts of dolomite may be associated with this model but in general the greater ease of dolomitization at the higher temperatures of deeper burial is countered by the progressive loss of the raw materials for dolomitization (i.e., the pore fluids and their ions) as burial-compaction proceeds. Little is known concerning the chemistry of solutions in environments of deep burial.

The Coorong and Sabkha Models. The Coorong Lagoon was known to be an area of dolomite formation since the later 1920s (Mawson, 1929) but was ignored until Alderman and Skinner (1957) inaugurated a series of investigations. Recent work (Von der Borch, 1976; Von der Borch and Lock, 1979; Von der Borch, *et al.*, 1975 and Muir *et al.*, 1980) has established the Coorong Lagoon as a model for early dolomitization in many ancient sequences of aphanitic dolomite that are not associated with evaporites (Fig. 3). The lagoon and a series of ephemeral lakes extend 200 km parallel to the south coast of Australia and are confined between Pleistocene fossil dunes that extend parallel to the coast. Surface drainage follows the existing topography, but the subsurface groundwater flows seaward perpendicular to the surface drainage (Fig. 3). The ephemeral lakes are filled during the humid winter months by groundwater seepage and are evaporated to partial or complete dryness during the summer months.

Extremely fine-grained calcian dolomite (yoghurt) mud and magnesite forms in the landward ephemeral lakes under the influence of continental groundwater or in the zone of mixing between sea water and continental water. This dolomite is in the form of 0.5 to 1 μm, amorphous, spherular aggregates and may have precipitated directly as a gel (Von der Borch and Jones, 1976). Aragonite and calcite predominates in the Coorong Lagoon itself which is filled with sea water. Dolomite (and magnesite) precipitation in the ephemeral lakes is aided by groundwater dilution and by the high CO_3^{2-} concentration of the alkaline groundwater (Morrow, 1982). The source of Mg^{2+} ions has not been identified with certainty; dolomites forming close to the coast may have derived their Mg^{2+} directly from sea water but dolomites forming further inland probably utilize Mg^{2+} in the groundwater, some of which may have been derived from the weathering of basic volcanic rocks (Von der Borch, *et al.*, 1975). In either case the delivery mechanism is the continual subsurface groundwater flow.

Successive regressions of a Coorong-type system of lagoons and ephemeral lakes on a subsiding continental margin can build thick sequences of evaporite-free peritidal dolomite (Muir *et al.*, 1980). Because of the predominant role of fresh groundwater, the Coorong Lagoon Model may be regarded as a specific type of the more generalized mixed-water or dilution model. Only one documented Canadian example of this model exists (Fig. 2) but more undoubtedly will be found.

The other well-documented setting for penecontemporaneous dolomite, the sabkha environment, is restricted to those dolomites with accompanying evaporite minerals. Illing *et al.* (1965) in collaboration with geologists of the Shell Oil Company provided the first comprehensive report on the Holocene intertidal and supratidal dolomite that had replaced aragonitic sediments exposed on sabkha surfaces bordering the Qatar Peninsula on the Persian Gulf. This was followed by the studies of Kinsman (1966), Butler (1969), de Groot (1973), Bush (1973), Hsü and Schneider (1973), Patterson (1972) and Patterson and Kinsman (1981, 1982) of the Abu Dhabi Sabkhas east of the Qatar Peninsula.

Dolomite in the sabkhas of the Persian Gulf is most abundant as 1 to 5 μm euhedral crystals in the high intertidal zone close to the strandline and is strongly controlled by the network of flood channels across the sabkhas. Storm driven flood tides reach furthest inland along these channel courses (Patterson, 1972). The frequency of flooding decreases landward across the sabkha but the Mg/Ca ratio of the floodwaters rises uniformly landward by means of gypsum precipitation so that the zone of optimum dolomite formation is less than a kilometre wide straddling the boundary between high intertidal and supratidal areas (Patterson, 1972; Bush, 1973; and Fig. 3). The dense Mg^{2+}-bearing hypersaline floodwater brines sink downward and flow seaward through the sediment by seepage refluxion. Dolomitization of the underlying intertidal and subtidal sediments occurs to a depth of 2 to 3 m beneath the sabkha surface in regions landward of the continuous algal mat (Patterson, 1972; Bush, 1973; de Groot, 1973). Hsü and Siegenthalar (1969), Hsü and Schneider (1973) and McKenzie *et al.* (1980) have developed an alternative model for sabkha dolomitization based on a process which they term *evaporative pumping*. In this process, a continual flow of seawater moves landward through the sabkha sediments to replace groundwater lost by evaporation at, or near the sabkha surface. This flow pattern is the reverse of seepage refluxion (Fig. 3) but the field evidence for evaporative pumping in the sabkhas of the Persian Gulf is contradictory (see Hsü and Schneider, 1973; McKenzie *et al.*, 1980; Patterson, 1972; Patterson and Kinsman, 1981). It is unlikely that evaporative pumping has contributed greatly to sabkha dolomitization because evaporation of sabkha groundwater is slowed considerably by a number of factors such as absorption of water at night and high relative humidity of the overlying air mass. Consequently this mechanism does not provide the volume of flow of ion-bearing groundwater necessary for dolomitization. The Sabkha Model may be regarded as a specific example of the Reflux Model modified slightly by the process of evaporative pumping. Many Canadian examples of dolomitization by the Sabkha Model have been identified (Fig. 2).

The Mixed-Water or Dilution Model. This model, in its general form, arose in the period following 1970. Before that time there were a few reports of dolomite that formed in low salinity environments such as in lakes (e.g., Klähn, 1928) or in caves (e.g., Holland *et al.*, 1964) but Hanshaw *et al.* (1971) documented the first large-scale system of deep burial dolomitization under the influence of fresh groundwater in the deep confined Tertiary carbonate aquifer of Florida (Fig. 3). They emphasized the role of active groundwater circulation in supplying marine-derived Mg^{2+} for dolomitization.

Land (1973) argued convincingly from field relationships and petrographic and chemical evidence, that the Middle Pleistocene reef rocks of Jamaica were dolomitized penecontemporaneously by the interaction of marine and fresh groundwater in a shallow coastal aquifer and suggested that many ancient platform dolomites which lack evaporites and are depleted in both trace elements (e.g.,

Na^+, Sr^{2+}) and heavy isotopes (^{18}O, ^{13}C) may have formed in a similar manner. This suggestion has been followed recently by many authors (e.g., Badiozamani, 1973; Land *et al.*, 1975; Veizer *et al.*, 1978; Randozzo and Hickey, 1977; Dunham and Olson, 1980; Choquette and Steinen, 1980). In most of these examples thick platform dolomite sequences consist of many individual regressive carbonate sediment wedges that have been penecontemporaneously dolomitized by the shallow coastal freshwater aquifers that accompanied their seaward progradation (Fig. 3). The vast differences in the chemical and isotopic composition between dolomites that have formed under the influence of freshwater and dolomites that have formed from hypersaline brines has led to an emphasis on these types of data as criteria in choosing between the Mixed-Water Model and other models. However, uncertainties in the interpretation of chemical and isotopic data preclude their use in model discrimination in the absence of other data (Land, 1980). Modern settings where the Mixed-Water Model may be operative include the Coorong area (Morrow, 1982) and Andros Island (Gebelein *et al.*, 1980).

In the Mixed-Water Model, dolomitization occurs within the zone of mixing of fresh groundwater with phreatic seawater. The Mg^{2+} ions for dolomitization are derived primarily from seawater and the delivery mechanism is the continual circulation of seawater induced by the flow of fresh groundwater (Land, 1973). Dilution of saline solutions causes slow precipitation which favours dolomite precipitation (Folk and Land, 1975; Morrow, 1982) and the high CO_3^{2-} concentration in many dilute continental groundwaters may also promote the precipitation of dolomite (Lippman, 1973; Morrow, 1982). This model has been used to explain the origin of many ancient dolostones in Canada (Fig. 2).

Solution Cannibalization and Pressure Solution Dolomite
Solution Cannibalization (Goodell and Garman, 1969) refers to the derivation of Mg^{2+} for dolomitization from the dissolution of magnesian calcite and reprecipitation of low-magnesian calcite. Some examples of partial dolomitization, such as dolomitized burrow fillings (Kendall, 1977) have been explained by solution cannibalization but for more complete dolomitization other sources of Mg^{2+} are required (Hsü, 1966).

Recently the phenomenon of pressure solution along solution seams and stylolites has been cited as a major cause of dolomitization (Logan and Semeniuk, 1976; Wanless, 1979). The source of Mg^{2+} ions in pressure solution dolomitization is by solution cannibalization of Mg^{2+} from pre-existing magnesian calcite during pressure solution (Wanless, 1979). Complete dolomitization is thought to result from the concentration of dolomite along solution seams as a relatively insoluble residue. Intriguing as this dolomitization model is, it suffers from the difficulty in reconciling the deep burial setting of most stylolites (e.g., Mossop, 1972) with the common pre-burial conversion of magnesian calcite to low-magnesian calcite. For large amounts of dolomite other sources of Mg^{2+} are required.

Tectonic or Hydrothermal Dolomite
Many coarsely crystalline, white sparry dolostones that have cross-cutting relationships with the enclosing strata have been designated as 'tectonic' or 'hydrothermal' dolomites in older studies, particularly in instances where these dolostones occur near known faults (e.g., Sanford, 1962). However, the common absence of other hydrothermal minerals implies that many of these dolostones have other origins. The relationship of some dolostone masses to faults may merely reflect the fact that groundwater movement in aquifers tends to be much more rapid along subsurface fault zones (i.e., groundwater cascade zones), enhancing diagenesis.

Ancient Dolostones - Criteria for Recongition of Dolomitization Models
Criteria that may guide the researcher in his choice of dolomitization models to apply to ancient dolostones can be divided into three broad categories - stratigraphic and facies criteria, petrographic criteria and geochemical criteria. Few, if any, of these criteria are unequivocal in themselves and it is only after the accumulation of much complementary data that a truly definitive choice can sometimes be made.

Stratigraphic and Facies Criteria
Size and Shape. Small intraformational bodies of dolomite may have local sources of Mg^{2+}, derived from the sediment itself as in the Solution-Cannibalization Model (e.g., Kendall, 1977; and Gebelein and Hoffman, 1973) but large masses, such as entire formations or major parts of a formation, invariably require external sources of Mg^{2+}. Dolostone masses that conform to rock stratification may indicate that dolomitization occurred before burial, but non-conformable dolostone bodies that intersect primary stratification and facies at high angles strongly indicate that dolomitization is post-lithification and post-burial, particularly where these dolostones are associated with solution-collapse breccias (e.g., Collins and Smith, 1977; Morrow, 1975). The form and location of some dolostone masses conform to known faults indicating a late, post-burial mode of origin (e.g., Skall, 1975; Jones, 1980).

Associated Evaporites and Shales. Laterally adjacent or overlying evaporites provide sources of Mg^{2+}-bearing hypersaline solutions capable of dolomitizing by seepage-refluxion particularly where shelf carbonates separate evaporite bodies from more open marine sediments (e.g., Skall, 1975; Wilson, 1975, p. 317). Dolomitized reefal masses encased in evaporites is a common situation where early diagenetic brine reflux may be suspected (e.g., Sears and Lucia, 1980).

Late diagenetic movements of Mg^{2+}-bearing connate brines derived by expulsion during burial compaction from evaporite masses into laterally adjacent carbonates (Jodry, 1969), or overlying carbonates (Jones, 1980) may also cause dolomitization. Similarly shales adjacent to dolomitized carbonates are also suspect as sources of dolomitizing solutions during their compaction.

Unconformities. Dolostone bodies are commonly developed beneath regional unconformities (Sonnenfeld, 1964). Many of these dolostones are best explained by the Mixed-Water Model where the unconformity surface represents the area of subaerial recharge for a fresh water aquifer (e.g., Collins and Smith, 1975 and Fig. 3). In other instances, such unconformities are considered merely to represent a period of karst development that enhances the porosity of the underlying carbonates so that dolomitizing solutions, whatever their origin, subsequently were localized in the strata beneath the unconformity (e.g., Skall, 1975).

Facies Development. Models for penecontemporaneous dolomitization such as the Sabkha and Coorong Models are verified most readily by the identification of associated facies characteristic of these environments. The sabkha depositional model with its peritidal and supratidal vertical sequence of dolomite that culminates in an interval of displacive anhydrite, forms a very distinctive cyclic sequence of lithologies and sedimentary structures (e.g., Wood and Wolfe, 1969). Similarly, environmental conditions of the Coorong Model generates an evaporite-free, cyclic sequence of very finely crystalline, penecontemporaneous dolomite with different sedimentary structures (Muir *et al.*, 1980). In many instances reef buildups or carbonate mud mounds are

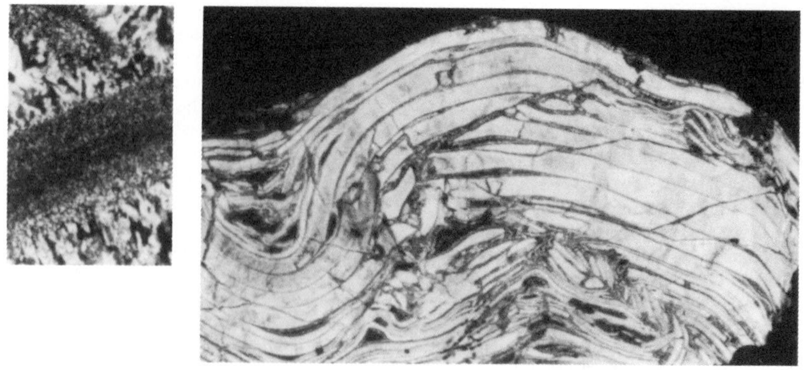

Figure 4 *A hand specimen of a bright yellow, supratidal flat dolomicrite that has undergone synsedimentary folding after partial lithification, from the Rowatt Formation, a Precambrian unit in the Belcher Islands, Hudson Bay. Thicker laminae are about one centimetre thick. The photomicrograph on the left shows the aphanocrystalline character of this type of early replacement dolomite with preservation of microscopic detail (photographs supplied by B.D. Ricketts).*

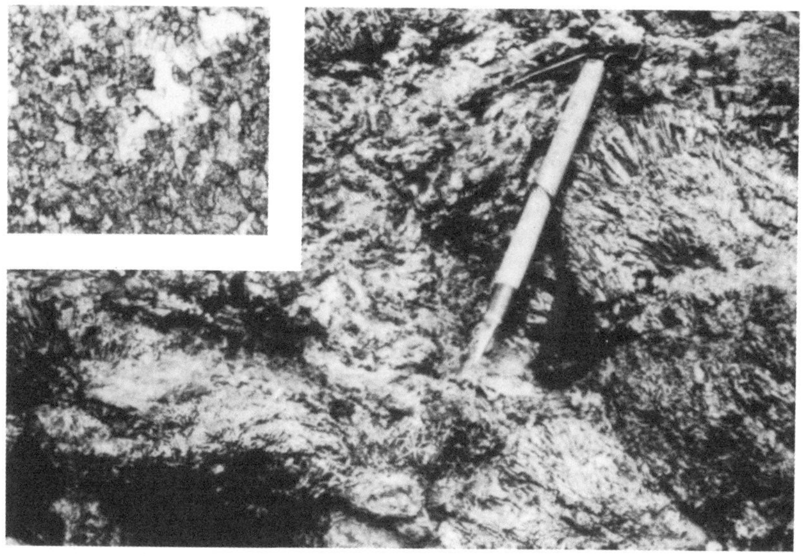

Figure 5 *Medium to coarsely crystalline 'late' diagenetic sucrosic dolomite in the Ordovician Allen Bay Formation of the Arctic Islands. Macroscopically identifiable Halysitid-type colonial corals populate this outcrop from the south coast of Devon Island. Microscopic details of these corals have been destroyed by dolomitization that formed the hypidiotopic fabric of subhedral crystals shown in the photomicrograph in the upper left with a field of view of 1.2 mm (photographs supplied by U. Mayr).*

facies that coincided with paleotopographic 'highs' and thus were susceptible to dolmitization by the Reflux Model or by the Mixed-Water Model after exposure by relative changes in sea level (e.g., Eliuk, 1978; Sears and Lucia, 1980; Exploration Staff, Chevron Exploration Staff Ltd., 1979).

Paleogeography. There is a strong correlation between paleogeography, as expressed by rock facies, and dolomitization reflecting a common tectonic control (Wilson, 1975; Dunham and Olson, 1980). Shelf areas tend to be exposed during regression and become the sites for fresh-water recharge and mixed-water dolomitization. Deeper water carbonates which are never exposed, are not subject to this type of dolomitization. Thick sequences of shelf dolomite may be built up as a consequence of multiple regressions. The seaward progradation of each regressive cycle is accompanied by seaward expansion of the coastal aquifer so that a dolomitization front also moves seaward as progradation continues (Fig. 3).

Petrographic and Textural Criteria

Crystal Size. Previous reviews (e.g., Folk, 1973) have emphasized that petrographically, dolostones are of two basic types separated by a crystal size boundary of 10 to 20 μm. Dolostones more finely crystalline than this were characterized as primary or very "early", pre-burial replacements in which sedimentary and organic fabrics are well preserved (Fig. 4), whereas more coarsely crystalline dolostones that occur mainly as strataform replacements of limestone with poorly preserved fossil remains were considered to be 'secondary', 'sucrosic' or 'late replacement' dolomites (Fig. 5). A third major petrographic variety of dolomite, which has received considerable attention in recent years, is the white, very coarsely crystalline dolomite or megacrystalline (i.e. ranging from millimetres to centimetres) that, in many places, is associated with lead-zinc mineralization (Fig. 6). Much of this type of dolomite is void-filling and displays curved crystal faces (e.g., Radke and Mathis, 1980).

Dolomicrite. The very finely crystalline fabric preservation type of dolomite in many ancient dolostones is directly analogous to the fine grained dolomitic sediments that form as early replacements of aragonite in the intertidal and supratidal areas of modern carbonate tidal flats such as in the Caribbean (dolomitic crusts), in the sabkhas of the Persian Gulf and in the Coorong region of Australia. This petrographic type of dolomite usually is inferred to have replaced aragonite mud before burial (e.g., Schmidt, 1966 and Fig. 4).

Medium Crystalline Sucrosic Dolomite. The origin of the finely to medium crystalline, sucrosic dolomites (generally hypidiotopic - an aggregation of subhedra) is much more problematic because of the apparent absence of modern analogues. However, this type of dolomite does occur as replacements of Pliocene-Pleistocene limestones on some Caribbean Islands (Land, 1973; Supko, 1977; Sibley, in press). Unlike similar dolostones in more ancient rocks, the geological history of these Caribbean dolostones is known with a high degree of certainty and this history in conjunction with other data, suggests that most, if not all, of the Caribbean dolomite originated by the Mixed-Water Model (e.g., Land, 1973).

Crystals of this type of dolomite commonly contain cloudy centres surrounded by clear rims and the Caribbean dolostones are no exception. Sibley (1980) suggested that this phenomenon, in the Caribbean examples at least, is due to the evolution of the diagenetic pore fluids from a state of near calcite saturation to a state of calcite under-saturation, as grain growth continues under the influence of mixed-water diagenesis. It is likely that the interlocking "hypidiotopic" texture of subhedra and anhedra that characterizes ancient sucrosic dolostones is the result of grain growth from an initial state where very finely crystalline dolomite euhedra are scattered throughout. These crystals grow and eventually join along irregular compromise boundaries (e.g., Morrow, 1978). The euhedral, micrometre sized dolomite rhombs that are scattered in the recent sediments forming Andros Island (Gebelein *et al.*, 1980) may record the beginning of a progression towards a totally dolomitized fabric of sucrosic dolostone (Fig. 7).

Although many ancient, stratiform, platform dolostone sequences of sucrosic, dolostone have been interpreted to be the consequence of mixed-water diagenesis (e.g., Randazzo and Hickey, 1978; Dunham and Olson, 1980), the Reflux Model has been invoked in situations where sucrosic dolostones are adjacent to evaporites (e.g., Sears and Lucia, 1980) or adjacent to shales (Mattes and Mountjoy, 1980).

Megacrystalline White Dolomite. This type of dolomite is less abundant than the more finely crystalline varieties but is economically important as a host rock for lead-zinc deposits (e.g., Presquile Dolomite) and hydrocarbon accumulations (e.g., Manetoe Dolomite). The origin of megacrystalline, white dolomite is the least certain of the three major petrographic types because it has not been observed in recent sediments or in Quaternary rocks. However, the universal association of white dolomite, both as a void-filling and as a replacement mineral, with solution-collapse breccias and solution cavities indicates that it formed after lithification and probably after at least some burial. Several workers have emphasized the common association of lead-zinc mineralization and white dolomite (e.g., Beales and Hardy, 1980; Davis, 1979; Radke and Mathis, 1980) suggesting that both the sulphides and the dolomite precipitated together by means of a single, overall-sulphate reduction reaction at temperatures between 60° and 150° C (Barton, 1967; Radke and Mathis, 1980). But any theory linking the origin of lead-zinc mineralization with the associated white sparry dolomite in spe-

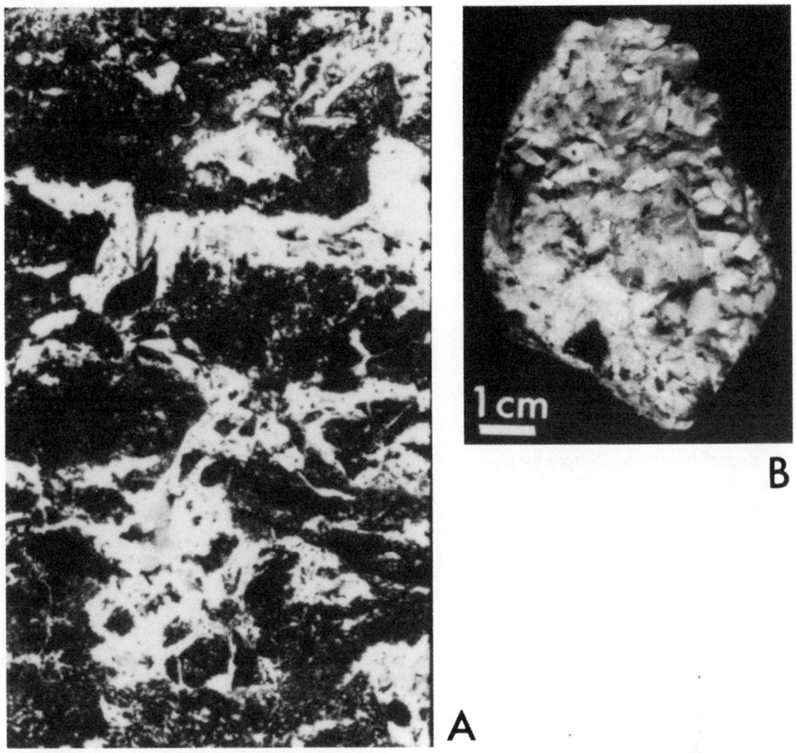

Figure 6 *Megacrystalline white sparry dolomite of the Manetoe Facies (Formation) in the Nahanni Formation of the Yukon Territory. Near centimetre-sized crystals of white sparry dolomite have partly filled spaces created by solution-collapse in the core slab (A). Photograph B, at the same scale, is the exposed inner surface of one of these vugs displaying crystals with markedly curved crystal faces. This type of dolomite is totally fabric destructive. This sample is a core specimen taken 11,707.0 feet below sea level in the gas-producing Columbia* et al. *Kootaneelee H-38 well.*

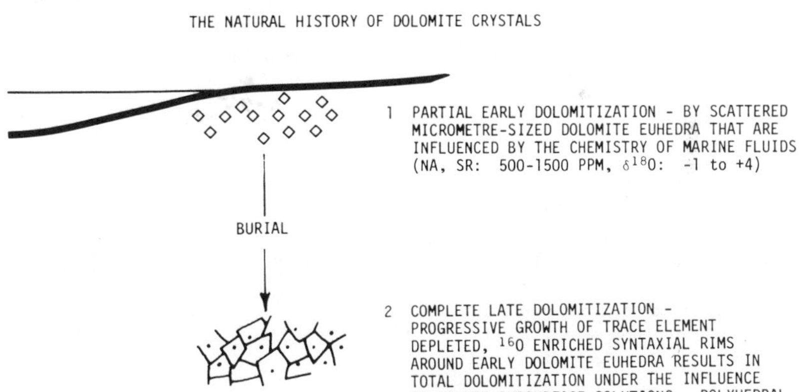

Figure 7 *Many dolomites conform to this type of diagenetic history. The dilute subsurface solutions are probably mixed-water or burial-compaction in origin. It is not known whether trace element depletion in crystal rims could be the result of slow precipitation from more saline solutions of special compositions.*

cific areas must also account for the widespread distribution of dolomite beyond the mineralized area (e.g., Skall, 1975; Gorody, 1980). It must also account for the fact that mineralization in most examples post-dated the emplacement of much of the white dolomite, as, for example in the Pine Point deposit (Skall, 1975) and in the Viburnum Trend of the Mississippi Valley district (Gorody, 1980; Hagni and Trancynger, 1977). The lead-zinc deposits in the St. George Formation in Newfoundland are a notable exception to this generalization (Collins and Smith, 1975). Clearly, the processes responsible for the emplacement of widespread white dolomite masses beyond the mineralized areas are not dependent on the processes of mineralization, although in some places the converse may be true (e.g., Collins and Smith, 1975; Davis, 1977). Probably only the relatively small amount of white sparry dolomite that is actually intergrown with the ore minerals was coprecipitated with these minerals.

Variations of the Mixed-Water Model have been cited as the mode of origin of white sparry dolomites in many recent studies (e.g., Lyle, 1977; Gorody, 1980; Morrow, 1975) but the Reflux Model (Davis, 1977) and the Burial Compaction Model (Griffin, 1965; Jackson and Beales, 1967) have also been cited, in addition to the model based on the late diagenetic sulphate reduction reaction (Radke and Mathis, 1980). Unfortunately, it is evident that little overall agreement exists between workers concerning the origin of megacrystalline white dolomite either in general, or with regard to specific occurrences.

Paragenesis and Texture. Detailed petrographic relationships and textural data in some instances can be used to infer the timing of dolomitization with respect to early events such as the recrystallization of aragonite and high-magnesian calcite (e.g., Schmidt, 1965; Kendall, 1977; Sibley, 1980) or to later events such as the development of stylolites (e.g., Macqueen and Thompson, 1978; Wong and Oldershaw, 1981). Early dolomitization is often controlled by original permeability and the commonly noted preferential dolomitization of finer grained sediments (Murray and Lucia, 1967) is probably due in large part to their great abundance of nucleation sites. Increasingly, multiple episodes of dolomitization are being recognized based mainly on the recognition of a wide variety of petrographic types of dolomite within the threefold petrographic classification outlined above. Many of these varieties have been related to the operation of specific dolomitization models (e.g., Sears and Lucia, 1980; Mattes and Mountjoy, 1980), so that a given dolostone may have been influenced by the operation of a succession of such models. However, it is not possible at present to relate any of these petrographic types to any particular model in the absence of other data, but the petrographic character of a given dolomite may help to limit the choice of models.

Geochemical Criteria

Geochemical Parameters. Geochemical parameters include compositional variations of the major elements Ca and Mg and the degree of concomitant crystal disorder, variations in the content of minor and trace elements such as Fe, Sr, Na, Mn and other less commonly studied elements (Ba, Si, S, Zn, Pb), and the isotopic composition of carbon and oxygen. A considerable body of literature has arisen in the past two decades dealing with these parameters. The most effective of these studies have combined stratigraphic and petrographic data with geochemical data. Stratigraphic and petrographic data provide a foundation of more readily accessible facts upon which a framework of interpretations involving geochemical data may be built.

Compositional Variations of Major Elements and Crystal Disorder. Compositional variation of major elements in dolomites is commonly expressed as a mole percent of $CaCO_3$ (Fig. 8). Calcium is almost invariably present in excess of Mg because of the relative ease with which Ca is incorporated into growing dolomite crystals (see Morrow, 1982). There is a growing recognition that the degree of enrichment of Ca in dolostones reflects the Mg/Ca ratio of the precipitational solution (e.g., Füchtbauer and Goldsmith, 1965; Morrow, 1978; Lumsden and Chimahusky, 1980). Typically, dolomicrites or finely crystalline dolostones that are associated with evaporites have Ca contents that are lower than for finely crystalline dolostones not associated with evaporites (i.e., Groups 2 and 3 in Fig. 8). Dolomites in both of these groups are early diagenetic in that dolomitization occurred within the influence of the depositional environment (e.g., Patterson, 1972; Morrow, 1978; and Lumsden and Chimahusky, 1980). The relatively high Mg/Ca ratios of solutions in evaporitic settings (e.g., Reflux and Sabkha Models) induces the formation of nearly stoichiometric dolomites, because the abundance of Mg^{2+} ions compensates for their relative difficulty in precipitating from solution to form magnesium-bearing compounds (see Morrow, 1982). On the other hand, the lower Mg/Ca ratios of solutions in more humid, non-evaporitic settings (e.g., Mixed Water and Coorong Models) results in the precipitation of more calcium-rich dolomites.

Medium to coarsely crystalline ancient dolomites tend to be more stoichiometric (Group 1 in Fig. 8). One explanation is

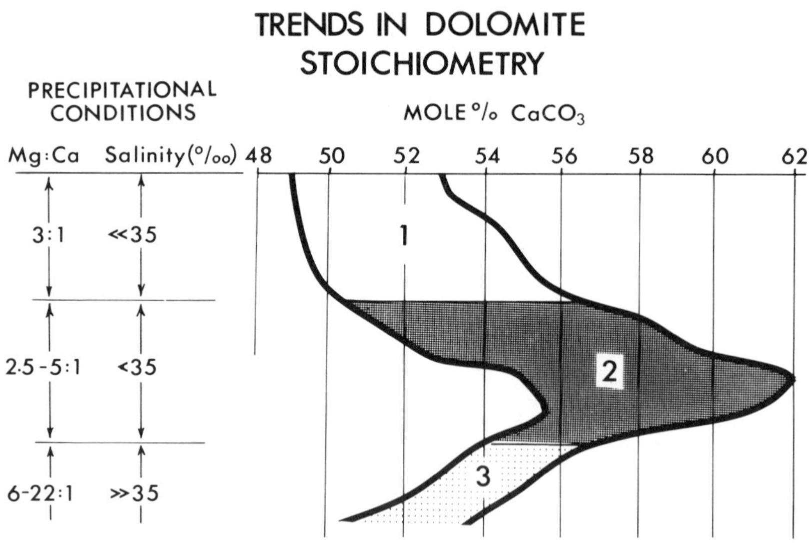

Figure 8 *A summary diagram incorporating data compiled by Morrow (1978), Mattes and Mountjoy (1980), and Lumsden and Chimahusky (1980) concerning the stoichiometry of dolomite both modern and ancient. Group 1 is composed of ancient sucrosic and sparry dolomites, Group 2 is composed of finely crystalline modern and ancient dolomites not associated with evaporites and Group 3 are finely crystalline modern and ancient dolomites associated with evaporites.*

that scattered, early diagenetic dolomite crystals in incompletely dolomitized limestones continue to grow very slowly in late diagenetic subsurface environments, under the influence of very dilute solutions and possibly elevated temperatures. Slow precipitation may overshadow the influence of low solution Mg/Ca ratios and cause the growth of the trace element depleted, more stoichiometric, coarsely crystalline rims that enclose Ca and trace element-enriched cores of dolomite crystals (e.g., Land *et al.*, 1975; Morrow, 1978). The net effect of continued crystal growth during later diagenesis is the formation of dolostones whose bulk composition approaches that of stoichiometric dolomite (e.g., Fritz and Jackson, 1971 and Fig. 7). Certainly this interpretation explains the depletion in most trace elements that invariably occurs towards the periphery of dolomite crystals and implies that they are the products of variants of the Mixed-Water or the Burial-Compaction Models.

Some data concerning the departure from ideal cation ordering within dolomite crystals has been gathered, particularly for dolomites forming in modern environments (e.g., Patterson, 1972). Unfortunately few ancient dolostones have been analyzed for this parameter. Comparisons of the degree of cation ordering as determined by the analysis of superstructure reflections on x-ray patterns with the degree of Ca enrichment might impart some insight into how closely these parametres are correlated. For example, crystal disorder not associated with Ca enrichment would reinforce the conclusion that finely crystalline, stoichiometric dolomite was rapidly precipitated from saline solutions with a high Mg/Ca ratio.

Trace Element Variations. The concentration of trace elements in dolomite ideally provides a means for estimating the composition of the precipitational medium because the fractionation of an element, such as Sr, from a liquid phase into a solid phase is proportional to its concentration in the liquid phase (amongst a variety of other factors). This implies that dolostones with greater concentrations of trace elements, such as strontium or sodium, precipitated from more saline solutions; but quantitative interpretation of solution compositions based on dolomite trace element data are not feasible at present. Estimates of low temperature partition coefficients based on high temperature experiments (Katz and Matthews, 1977; Jacobson and Usdowski, 1976) or derived from element concentrations in Holocene dolomites (Patterson, 1972; Land and Hoops, 1973; Behrens and Land, 1972) have not been satisfactory (Land, 1980; Veizer *et al.*, 1978) and direct determinations from precipitation experiments are of course not possible at reasonable temperatures.

The extreme difference in the contents of trace elements in ancient versus modern dolomites is the salient fact to emerge from trace element studies. The two most commonly studied elements, Sr and Na, are present in concentrations approaching, or greater than, 1000 ppm in modern dolomite but in ancient dolomites, they commonly are less than a few hundred ppm (Land *et al.*, 1975; Mattes and Mountjoy, 1980). Other elements such as S, K, Al, Ba, Zn, Mn and Fe have been analyzed only sporadically in ancient dolostones and, little, if any data exists for modern dolomites.

The phenomenon of intracrystal variations or zonations with respect to trace elements within ancient dolomite crystals is commonly observed, even in crystals smaller than 10 μm (e.g., Choquette and Steinen, 1980). Studies of element zonations in dolomite crystals have also concentrated on the elements Na and Sr, and in addition, there is some data for Fe and Mn. A general pattern exists of high Na and Sr contents in the cores surrounded by rims with much lower contents. (Fritz and Katz, 1972; and Fig. 7). On the other hand, Fe, and possibly also Mn, tend to exhibit an opposite zonation with higher concentrations toward the periphery of crystals. Bulk sample analyses, of course, are weighted averages of these zoned crystals. Some studies have shown slight statistical differences in the bulk trace element contents of a variety of dolostone types (Veizer *et al.*, 1978).

Any explanation of dolomitization must account for both the element zonation of ancient dolomite crystals, and their reduced trace element contents, particularly for the aphanocrystalline and fine to medium crystalline, sucrosic replacement dolomites that commonly have cloudy centres and clear rims. It is difficult to avoid the conclusion that as dolomite crystal growth continues, it does so under the influence of progressively more dilute solutions that dissolve pre-existing calcite and precipitate dolomite. It seems likely that the cores of ancient dolomite crystals, in largely or totally dolomitized rocks, represent individual crystals that formed in shallow early diagenetic environments under the influence of seawater or modified seawater solutions. Subsequent burial of these rocks into deeper subsurface environments may have been accompanied by the continued precipitation from more dilute solutions of dolomite overgrowths that are relatively inclusion-free, more stoichiometric and lower in Sr and Na (Fig. 7). Such dilute and subsurface solutions are consistent with the operation of either the Mixed-Water Model or the Burial-Compaction Model. Ion filtration during burial compaction of shales can yield dilute solutions that may participate in the dolomitization of nearby carbonates, although there are limitations on the potential amount of dolomite that can be formed.

The commonly observed high Fe concentration of up to several mole per cent in the rims of larger crystals, particularly in those that have precipitated as open space fillings with complex zonation (as revealed by staining and cathodoluminescence) have been related qualitatively to diagenesis in the reducing subsurface phreatic zone, with fluctuations in the supply of Fe^{2+} (Wong and Oldershaw, 1981; Gorody, 1980). These observations are also consistent with either a Mixed-Water or Burial-Compaction Model.

Coarsely crystalline, ferroan and non-ferroan, white, sparry dolomites directly associated with Pb-Zn mineralization or with hydrocarbons have also been interpreted to have been precipitated from saline host fluids, based primarily on the occurrence of very saline fluid inclusions in associated minerals (Radke and Mathis, 1980). Certainly subsurface precipitation of dolomite at the inferred temperature of greater 100° C is less subject to kinetic inhibitions. However, it is puzzling that the few occurrences of white, sparry dolomite associated with Pb-Zn mineralization which have been analyzed systematically for their trace element contents (e.g., Fritz and Jackson, 1971) are characterized by low trace element contents in conformance with the general rule regarding trace elements in dolostones. Saline solutions should precipitate dolomites markedly enriched in trace elements. This major contradiction concerning the origin of white megacrystalline, sparry dolomite cannot fail to stimulate further research, and trace element studies will undoubtedly play a key role in resolving this paradox. Distribution coefficients for some elements, such as strontium (Jacobson and Usdowski, 1976) have been determined for the high temperature (>100° C) precipitation of dolomite which might enable quantitative modelling of trace element distributions.

Trends of trace element concentrations within dolostone bodies have also been used as indicators of the direction of fluid migration during diagenesis (Land, 1980). This type of data may allow discrimination between models for dolomitization that require different directions of fluid migration during dolomitization, such as the Burial-Compaction Model versus the Mixed-Water Model.

Oxygen and Carbon Isotopes. Land (1980) has recently summarized the status of stable isotope studies ($\delta^{18}O$, $\delta^{13}C$) of dolomite and has pointed out the considerable difficulties in their interpretation. Qualitative interpretation of oxygen and carbon isotope data involves an appraisal of the influence of the variables of temperature and solution compositions. Higher temperatures and dilution by meteoric water favour the incorporation of the light ^{16}O isotope into the solid phase so that dolomites that are greatly enriched in ^{16}O may have been influenced by high subsurface temperatures (Mattes and Mountjoy, 1980) or by dilute ^{16}O rich groundwater (Land *et al.*, 1975). The interplay of these factors is complicated by the probable existence of secular changes in the isotopic composition of ancient seawater (Veizer and Hoefs, 1976).

One certainty with regard to oxygen isotopes in ancient dolostones is that their bulk $\delta^{18}O$ is much lower than that of modern dolomites (Land, 1980). Probably dolomite crystals are zoned with respect to their ^{16}O content in a manner analogous to that described for trace elements. Although intracrystal variations in ^{16}O have not been described, there does appear to be a correlation between increasing crystal size and decreasing $\delta^{18}O$ values for dolomite crystals (Fritz and Jackson, 1971; Land *et al.*, 1975; Mattes and Mountjoy, 1980) which is consistent with the peripheral enrichment of ^{16}O. This may corroborate the interpretation that the growth of dolomite crystals continues under the influence of progressively more dilute solutions. Regional trends of $\delta^{18}O$ within dolostones may also corroborate trends in their trace element distributions (Land *et al.*, 1975).

In contrast to oxygen isotopes, the ratio of stable carbon isotopes ($\delta^{13}C$) in most ancient dolostones is virtually the same as in Holocene dolomites with perhaps only a slight tendency for ancient dolostones to be enriched in ^{12}C (Land, 1980). This may simply reflect the fact that the amount of carbon in dissolved species in natural solutions is very small compared to the amount of carbon in the carbonate rocks that are being replaced. There are a few dolostones enriched in ^{12}C that may have been involved in bacterial reduction reactions.

Summary and Conclusions

The description and classification of dolostones previous to about 1960 was a comparatively simple matter as dolostones were commonly classified as 'primary' or 'secondary' depending primarily on whether the sedimentary fabric of the rock was preserved or destroyed by dolomitization. The additional category of 'hydrothermal' dolomite was used and is still used (Jones, 1980) occasionally to characterize masses of coarsely crystalline, space-filling white dolomite. Little was understood concerning processes of dolomitization and consequently most dolostones were subjected to comparatively casual investigations in spite of their economic importance.

The decades following 1960 were marked by the discovery of many examples of modern dolomitization in a variety of environmental settings (e.g., sabkha, ephemeral lake, deep carbonate aquifer) and the characterization of the chemistry of these environments has generated a greatly increased understanding of the chemistry of dolomitization. At the same time, new analytical techniques (e.g., x-ray fluorescence, electron microprobe, etc.) have permitted a better overall characterization of the mineral dolomite itself in both ancient rocks and modern occurrences. The rapid increase in knowledge of other aspects of geoscience such as fluid dynamics and chemical changes during the burial compaction of sediments and the hydrology and chemistry of coastal aquifers have had a profound influence on the development of theories concerning dolomitization. This multidirectional increase in knowledge has resulted in the development of a variety of dolomitization models and an increase in the number of criteria available to discriminate between models.

The objectives of this review have been twofold. First, to outline clearly the models for dolomitization that have arisen and to review the criteria commonly employed in choosing between these models. The multitude of models for dolomitization that are now recognized are moderately to well defined. It seems evident to this reviewer that variants of the Mixed-Water Model are the most widely applicable — the first among equals. This is not to say that the Mixed-Water Model in itself constitutes the answer to the 'dolomite problem' only that many ancient dolostones may have formed by the operation of this model. The other models are certainly viable but perhaps not to the same degree and in less regional settings. The importance of a model such as the Burial-Compaction Model lies not in its potential for forming vast amounts of dolomite but rather in its capacity for modifying pre-existing porosity.

Unfortunately the criteria for choosing between models are not as definitive as could be desired and many of these criteria have been interpreted in more than one way. Indeed it could be argued that the same criteria in some dolostones such as the Presqu'ile Formation (or facies) have been used to support a succession of different models as these models successively rose in popular esteem. Hopefully, the time of fads is past and the choice of a model for dolomitization by present day workers is based on an equitable consideration of all available criteria and by practicing the spirit embodied in the phase "multiple working hypothesis." Obviously much work remains to be done.

Acknowledgements

Douglas Kent of the University of Saskatchewan and Alan Kendall of Amoco Canada Ltd. critically read the manuscript and offered many useful suggestions. I would also extend my appreciation to Ian McIlreath, the editor of this series, for his suggestions and encouragement. I thank A.E. Foscolos for reading the manuscript.

References

Reflux Model

Adams, J.E., and M.G. Rhodes, 1960, Dolomitization by seepage refluxion: American Assoc. Petroleum Geol. Bull., v. 44, no. 12, p. 1912-1920.

Deffeyes, K.S., F.J. Lucia and P.K. Weyl, 1965, Dolomitization of recent and Plio-Pleistocene sediments by marine evaporite waters on Bonaire, Netherlands Antilles: in A.C. Pray and R.C. Murray, eds., Dolomitization and Limestone Diagenesis: Soc. Econ. Paleontol., Mineral Spec. Publ. 13, p. 71-88.

King, R.H., 1947, Sedimentation in Permian Castile Sea: American Assoc. Petroleum Geol. Bull., v. 31, p., 470-477.

Morrow, D.W., 1978, The influence of the Mg/Ca Ratio and salinity on dolomitization in evaporite basins: Canadian Soc. Petroleum Geol., Bull., v. 26, p. 389-392.

Murray, R.C., 1969, Hydrology of south Bonaire, N.A.-a. rock selective dolomitization model: Jour. Sed. Petrology, v. 39, p. 1007-1013.

Burial-Compaction Model

Griffin, D.G., 1965, The Devonian Slave Point, Beaverhill Lake, and Muskwa Formations of northeastern British Columbia and adjacent areas: British Columbia Dept. Mines and Petroleum Resources Bull., v. 50, p. 90.

Illing, L.V., 1959, Deposition and diagenesis of some upper Paleozoic carbonate sediments in western Canada: 5th World Petroleum Cong., New York, Proc., sec. 1, p. 23-52.

Jodry, R.L., 1969, Growth and dolomitization of Silurian reefs, St. Clair County, Michigan: American Assoc. Petroleum Geologists, Bull., v. 52, p. 957-981.

Mattes, B.W. and E.W. Mountjoy, 1980, Burial dolomitization of the Upper Devonian Miette Buildup, Jasper National Park, Alberta: in D.H. Zenger, J.B. Dunhjam and R.L. Ethington, eds., Concepts and Models of Dolomitization: Soc. Econ. Paleontol. Mineral. Spec. Publ. 28, p. 259-297.

Wilson, J.G., 1975, Carbonate facies in geologic history: Springer Verlag, New York, N.Y., p. 471.

Wong, P.K. and A. Oldershaw, 1981, Burial cementation in the Devonian, Kaybob Reef Complex, Alberta, Canada: Jour. Sed. Petrology, v. 51, p. 507-520.

Coorong and Sabkha Models

Bush, P., 1973, Some aspects of the diagenetic history of the sabkha in Abu Dhabi, Persian Gulf: in B.H. Purser, ed., The Persian Gulf-Holocene Carbonate Sedimentation and Diagenesis in a Shallow Epicontinental Sea: New York, Springer Verlag, p. 471.

Hsü, K.J. and C. Siegenthaler, 1969, Preliminary experiments on hydrodynamic movement induced by evaporation and their bearing on the dolomite problem: Sedimentology, v. 12, p. 11-25.

Illing, G.U., A.J. Wells, and C.M. Taylor, 1965, Penecontemporary dolomite in the Persian Gulf: in A.C. Pray and R.C. Murray, eds., Dolomitization and Limestone Diagenesis: Soc. Econ. Paleontol. Mineral Spec. Publ. 13, p. 89-111.

Kinsman, D.J.J., 1966, Gypsum and anhydrite of recent age, Trucial Coast, Persian Gulf: in J.L. Rau, ed., Second Symposium on Salt, v. 1: Northern Ohio Geol. Soc., p. 302-326.

Muir, M., D. Lock, and C. Von der Borch, 1980, The Coorong Model for penecontemporaneous dolomite formation in the Middle Proterozoic McArthur Group, Northern Territory, Australia: in D.H. Zenger, J.B. Dunham and R.L. Ethington, eds., Concepts and Models of Dolomitization: Soc. Econ. Paleontol. Mineral Spec. Publ. 28, p. 51-68.

Patterson, R.J., 1972, Hydrology and carbonate diagenesis of a coastal sabkha in the Persian Gulf: Ph.D. Thesis, Princeton Univ., Princeton, New Jersey, p. 498.

Patterson, R.J. and D.J.J. Kinsman, 1981, Hydrologic framework of a sabkha along Arabian Gulf: Amer. Assoc. Petroleum Geol. Bull., v. 65, p. 1457-1475.

Patterson, R.J. and D.J.J. Kinsman, 1982, Formation of diagenetic dolomite in coastal sabkha along Arabian (Persian) Gulf.: American Assoc. Petroleum Geol. Bull., v. 66, p. 28-43.

Von der Borch, C.C., 1976, Stratigraphy and formation of Holocene dolomitic carbonate deposits of the Coorong area, South Australia: Jour. Sed. Petrology, v. 46, p. 952-966.

Von der Borch, C.C. and J.B. Jones, 1976, Spherular Modern dolomite from the Coorong area, South Australia: Sedimentology.

Von der Borch, C.C. and D.E. Lock, 1979, Geological significance of Coorong dolomites: Sedimentology, v. 26, p. 813-824.

Mixed Water Model

Badiozamani, K., 1973, The Dorag dolomitization model-application to the Middle Ordovician of Wisconsin: Jour. Sed. Petrology, v. 43, p. 965-984.

Choquette, P.W. and R.P. Steinen, 1980, Mississippian non-supratidal dolomite, Ste. Genevieve Limestone, Illinois Basin: evidence for mixed-water dolomitization: in D.H. Zenger, J.B. Dunham and R.L. Ethington, eds., Concepts and Models of Dolomitization: Soc. Econ. Paleontol. Mineral. Spec. Publ. 28, p. 163-196.

Folk, R.L. and L.S. Land, 1975, Mg/Ca ratio and salinity: two controls over crystallization of dolomite: American Assoc. Petroleum Geol. Bull., v. 59, p. 60-68.

Hanshaw, B.C., W. Back and R.G. Deike, 1971, A geochemical hypothesis for dolomitization by groundwater: Econ. Geology, v. 66, p. 710-724.

Land, L.S., 1973, Contemporaneous dolomitization of Middle Pleistocene reefs by meteoric water, North Jamaica: Bull. Marine Sci., v. 23, p. 64-92.

Land, L.S., M.R.I. Salem, and D.W. Morrow, 1975, Paleohydrology of ancient dolomites-Geochemical evidence: American Assoc. Petroleum Geol., Bull., v. 59, p. 1602-1625.

Lippman, F., 1973, Sedimentary Carbonate Minerals: New York, Springer Verlag, p. 228.

Solution-Cannibalization and Pressure-Solution

Goodell, H.G. and R.K. Garman, 1969, Carbonate geochemistry of Superior deep test well, Andros Island, Bahamas: American Assoc. Petroleum Geol. Bull., v. 53, p. 513-536.

Hsü, K.J., 1966, Origin of dolomite in sedimentary sequences: a critical analysis: Mineralium Deposita 2, p. 133-138.

Logan, B.W. and V. Semenuik, 1976, Dynamic metamorphism; processes and products in Devonian carbonate rocks Canning Basin, Western Australia: Geol. Soc. Australia, Spec. Pat. 6, p. 138.

Wanless, H.R., 1979, Limestone response to stress: Pressure solution and dolomitization: Jour. Sed. Petrology, v. 49, p. 437-462.

Ancient Dolostones-Stratigraphic and Petrographic Criteria

Beales, F.W. and J.L. Hardy, 1980, Criteria for the recognition of diverse dolomite types with an emphasis on studies on host rocks for Mississippi Valley - type ore deposits: in D.H. Zenger, J.B. Dunham and R.L. Ethington, eds., Concepts and Models of Dolomitization. Soc. Econ. Paleontol. Mineral. Spec. Publ. 28, p. 197-213.

Collins, J.A. and L. Smith, 1977, Zinc deposits related to diagenesis and intrakaratic sedimentation in the Lower Ordovician St. George Formation, western Newfoundland: Canadian Soc. Petroleum Geol. Bull., v. 23, p. 393-427.

Davis, J.H., 1977, Genesis of the southeast Missouri lead deposits: Econ. Geology, v. 72, p. 443-450.

Eliuk, L.S. 1978, The Abenaki Formation, Nova Scotia Shelf, Canada - A depositional and diagenetic model for a Mesozoic carbonate platform: Canadian Soc. Petroleum Geol. Bull., v. 26, p. 424-514.

Exploration Staff, Chevron Standard Limited, 1979, The geology geophysics and significance of the Nisku Reef discoveries, West Pembina area, Alberta, Canada: Canadian Soc. Petroleum Geol. Bull., v. 27, p. 326-359.

Folk, R.L., 1973, Carbonate petrography in the Post-Sorbian age: in R.N. Ginsburg, ed., Evolving Concepts in Sedimentology, The John Hopkins University Studies in Geology, no. 21, p. 118-158.

Gebelein, C.D., R.P. Steinen, P. Garrett, E.J. Hoffman, J.M. Queen and L.N. Plummer, 1980, Subsurface dolomitization beneath the tidal flats of central west Andros Island, Bahamas: in D.H. Zenger, J.B. Dunham and R.L. Ethington, eds., Concepts and Models of Dolomitization: Soc. Econ. Paleontol. Mineral. Spec. Publ. 28, p. 31-49.

Jodry, R.L., 1969, Growth and dolomitization of Silurian reefs, St. Clair County, Michigan: American Assoc. Petroleum Geol. Bull., v. 52, p. 957-981.

Jones, R.M.P., 1980, Basinal isostatic adjustment faults and their petroleum significance: Canadian Soc. Petrol. Geol. Bull., v. 28, p. 211-251.

Lyle, J.R., 1977, Petrography and carbonate diagenesis of the Bonneterre Formation in the Viburnum Trend Area, Southeast Missouri: Econ. Geology, v. 72, p. 420-434.

Macqueen, R.W. and R.I. Thompson, 1978, Carbonate-hosted lead-zinc occurrences in northeastern British Columbia with emphasis on the Robb Lake deposit: Canadian Jour. Earth Sci., v. 15, p. 1737-1762.

Radke, B.M. and R.L. Mathis, 1980, On the formation and occurrence of saddle dolomite: Jour. Sed. Petrology, v. 50, p. 1149-1168.

Ricketts, B.D., in prep., The evolution of a Middle Pre-Cambrian dolostone sequence - a spectrum of dolomitization regimes.

Schmidt, V., 1965, Facies, diagenesis and related reservoir properties in the Gigas Beds (Upper Jurassic), northwestern Germany: in L.C. Pray and R.C. Murray, eds., Dolomitization and Limestone Diagenesis: Soc. Econ. Paleontol. Mineral Spec. Publ. 13, p. 124-168.

Sears, S.O. and F.J. Lucia, 1980, Dolomitization of northern Michigan Niagara reefs by brine refluxion and freshwater/seawater mixing: in D.H. Zenger, J.B. Dunham and R.L. Ethington, eds., Concepts and Models of Dolomitization: Soc. Econ. Paleontol. Mineral. Spec. Publ. 28, p. 215-235.

Sibley, D.F., 1980, Climatic control of dolomitization, Seroe Domi Formation (Pliocene), Bonaire, N.A.: in D.H. Zenger, J.B. Dunham, and R.L. Ethington, eds., Concepts and Models of Dolomitization: Soc. Econ. Paleontol. Mineral. Spec. Publ. 28, p. 247-258.

Sibley, D.F., in press, The origin of common dolomite textures: Clues from the Pliocene: Jour. Sed. Petrology.

Skall, H., 1975, The paleoenvironment of the Pine Point lead-zinc district: Econ. Geology, v. 70, p. 22-47.

Supko, P.R., 1977, Subsurface dolomites, San Salvador, Bahamas: Jour. Sed. Petrology, v. 47, p. 1063-1077.

Wong, P.K. and A. Oldershaw, 1981, Burial cementation in the Devonian, Kaybob Reef Complex, Alberta, Canada: Jour. Sed. Petrology, v. 51, p. 507-520.

Wood, G.V. and M.J. Wolfe, 1969, Sabkha cycles in the Arab-Darb Formation off the Trucial Coast of Arabia: Sedimentology, v. 12, p. 165-191.

Ancient Dolostone - Geochemical Criteria

Behrens, E.W. and L.S. Land, 1972, Subtidal Holocene dolomite, Baffin Bay, Texas: Jour. Sed. Petrology, v. 42, p. 155-161.

Fritz, P. and S.A. Jackson, 1971, Geochemical and isotopic characteristics of Middle Devonian dolomites from Pine Point, northern Canada: 24th Internatl. Geol. Congress, Montreal, Sec. 6, p. 230-243.

Fritz, P. and A. Katz, 1972, The sodium distribution of dolomite crystals: Chem. Geol., v. 10, p. 237-244.

Füchtbauer, H. and H. Goldschmidt, 1965, Beziehungen Zwishchen calciumgehalt und bildungsbeding-ungen der dolomite: Geol. Rundschau, v. 55, p. 29-40.

Land, L.S., 1980, The isotopic and trace element geochemistry of dolomite: the state of the art: in D.H. Zenger, J.B. Dunham and R.L. Ethington, eds., Concepts and Models of Dolomitization: Soc. Econ. Paleontol. Mineral. Spec. Publ. 28, p. 87-110.

Land, L.S. and G.K. Hoops, 1973, Sodium in carbonate sediments and rocks: a possible index to the salinity of diagenetic solutions: Jour. Sed. Petrology, v. 43, p. 614-617.

Land, L.S., M.R.I. Salem and D.W. Morrow, 1975, Paleohydrology of ancient dolomites: geochemical evidence: American Assoc. Petroleum Geol. Bull., v. 59, p. 1602-1625.

Lumsden, D.N. and J.S. Chimahusky, 1980, Relationship between dolomite nonstoichiometry and carbonate facies parametres: in D.H. Zenger, J.B. Dunham and R.L. Ethington, eds., Concepts and Models of Dolomitization: Soc. Econ. Paleontol. Mineral. Spec. Publ. 28, p. 123-137.

Morrow, D.W., 1978, The influence of the Mg/Ca ratio and salinity on dolomitization in evaporite basins: Petroleum Geol. Bull., v. 26, p. 389-392.

Veizer, J. and J. Hoefs, 1976, The nature of O^{18}/O^{16} and C^{13}/C^{12} secular trends in sedimentary carbonate rocks: Geochem. et Cosmochim. Acta, v. 40, p. 1387-1395.

Veizer, J., J. Lemieux, B. Jones, M. Gibling and J. Savelle, 1978, Paleosalinity and dolomitization of a Lower Paleozoic carbonate sequence, Somerset and Prince of Wales Islands, Arctic Canada: Canadian Jour. Earth Sci., v. 15, p. 1448-1461.

Other References Cited In Text

Alderman, A.R. and H.C.W. Skinner, 1957, Dolomite sedimentation in the southeast of South Australia: American Jour. Science, v. 255, p. 561-567.

Barton, P.B., Jr., 1967, Possible role of organic matter in the precipitation of the Mississippi Valley areas: in J.S. Brown, ed., Genesis of Stratiform Lead-Zinc-Barite-Fluorite Deposits: Econ. Geology Monograph 3, p. 371-378.

Beales, F.W. and S.A. Jackson, 1969, Pine Point-a stratigraphic approach: Canadian Inst. Mining and Metallurgy Trans., v. 61, p. 1-12.

Bebout, D.G. and W.R. Maiklem, 1973, Ancient anhydrite facies and environments, Middle Devonian Elk Point Basin, Alberta: Canadian Society Petroleum Geol. Bull., v. 21, p. 287-343.

Boles, J.R., 1981, Clay diagenesis and effects on sandstone cementation (case histories from the Gulf Coast Tertiary): in F.J. Longstaffe, ed., Clays and the Resource Geologist: Mineral. Assoc. Canada, Short Course Handbook, v. 7, p. 148-168.

Butler, G.P., 1969, Modern evaporite deposition and geochemistry of co-existing brines, the sabkha, Trucial Coast, Arabian Gulf: Jour. Sed. Petrology, v. 39, p. 70-89.

Davies, G.R., 1979, Dolomite reservoir rocks: processes, controls, porosity development: in Geology of Carbonate Porosity: American Assoc. Petroleum Geol. Short Course, note series no. 11, Houston, p. C1-C17.

Davies, G.R. and S.D. Ludlam, 1973, Origin of laminated and graded sediments, Middle Devonian of Western Canada: Geol. Soc. America Bull., v. 84, p. 3527-3546.

de Groot, K., 1973, Geochemistry of tidal flat brines at Umm Said, S.E. Qatar, Persian Gulf: in B.H. Purser, ed., The Persian Gulf - Holocene Carbonate Sedimentation in a Shallow Epicontinental Sea: New York, Springer Verlag, p. 377-394.

Dunham, J.B. and E.R. Olson, 1980, Shallow subsurface dolomitization of subtidally deposited carbonate sediments in the Hanson Creek Formation (Ordovician-Silurian) of Central Nevada: In D.H. Zenger, J.B. Dunham and R.L. Ethington, eds., Concepts and Models of Dolomitization: Soc. Econ. Paleontol. Mineral. Spec. Publ. 28, p. 139-161.

Foscolos, A.E. and J. Kodama, 1974, Diagenesis of clay minerals from Lower Cretaceous shales of northeastern British Columbia: Clays Clay Minerals, v. 22, p. 319-335.

Gebelein, C.D. and P. Hoffman, 1973, Algal origin of dolomite laminations in stromatolitic limestone: Jour. Sed. Petrology, v. 43, p. 603-613.

Gorody, A.W., 1980, Dolomitization and paleohydraulic history of the Lower Ordovician Mascot Formation, Upper Knox Group, in north central Tennessee: in Abstracts with Program, Geological Society America, v. 12, p. 435.

Hagni, R.D. and T.C. Trancynger, 1977, Sequence of deposition of the ore minerals at the Magmont Mine, Viburnum Trend, southeast Missouri: Econ. Geology, v. 72, p. 451-464.

Holland, H.D., T.V. Kirsipu, J.S. Heubner and U.M. Oxburgh, 1964, On some aspects of the chemical evolution of cave waters: Jour. Geology, v. 72, p. 36-67.

Hsü, K.J. and J. Schneider, 1973, Progress report on dolomitization hydrology of Abu Dhabi Sabkhas, Arabian Gulf: in B.H. Purser, ed., The Persian Gulf-Holocene Carbonate Sedimentation in a Shallow Epicontinental Sea, New York, Springer Verlag, p. 409-422.

Jackson, S.A. and F.W. Beales, 1967, An aspect of sedimentary basin evolution: The concentration of Mississippi-Valley-type ores during late stages of diagenesis: Canadian Soc. Petroleum Geol. Bull., v. 15, p. 383-433.

Jacobson, R.G. and H.E. Usdowski, 1976, Partitioning of strontium between calcite dolomite and liquids: Contr. Mineralogy and Petrology, v. 59, p. 171-185.

Katz, A. and A. Mathews, 1977, The dolomitization of $CaCo_3$: an experimental study at 252° C-295° C: Geochem. et Cosmochim. Acta, v.41, p. 297-308.

Kendall, A.C., 1977, Origin of dolomite mottling in Ordovician limestones from Saskatchewan and Manitoba: Canadian Soc. Petroleum Geol. Bull., v. 25, p. 480-504.

Klähn, H., 1928, Süsswasserkalkmagnesia Gesteine und Kalkmagnesiasüsswasser: Chemie der Erde, v. 3, p. 453-587.

Kocurko, J.M., 1979, Dolomitization by spray - zone brine seepage San Andres, Colombia: Jour. Sed. Petrology, v. 49, p. 209-214.

Langton, J.R. and G.E. Chin, 1968, Rainbow Member facies and related reservoir properties, Rainbow Lake, Alberta: Canadian Soc. Petroleum Geol. Bull., v. 16, p. 104-143.

Magara, K., 1976, Compaction and fluid migration: practical petroleum geology: Developments in Earth Science, v. 9: Amsterdam, Elsevier, 319 p.

Mawson, O., 1929, South Australian algal limestones in the process of formation: Quart. Jour. Geol. Soc. London, v. 85, p. 613-621.

McKenzie, J.A., K.J. Hsü and J.F. Schneider, 1980, Movement of subsurface waters under the sabkha, Abu Dhabi; UAE, and its relation to evaporative dolomite genesis: in D.H. Zenger, J.B. Dunham and R.L. Ethington, eds., Concepts and Models of Dolomitization: Soc. Econ. Paleontol. Mineral. Spec. Publ. 28, p. 11-30.

Morrow, D., 1975, The Florida Aquifer: a possible model for a Devonian paleoaquifer in northeastern British Columbia: in Report of Activities, Part B, Geological Survey of Canada Paper 75-1B, p. 261-266.

Morrow, D.W. and J.W. Kerr, 1978, Stratigraphy and sedimentology of Lower Paleozoic Formations near Prince Alfred Bay, Devon Island: Geol. Survey Canada Bull. 254, p. 122.

Mossop, G.D., 1972, Origin of the peripheral rim, Redwater Reef, Alberta: Canadian Soc. Petroleum Geol. Bull., v. 20, p. 238-280.

Müller, S. and G. Teitz, 1971, Dolomite replacing "cement A" in biocalcarenites from Fuerteventura, Canary Islands, Spain: in D.P. Bricker, ed., Carbonate Cements: Baltimore, John Hopkins Press, p. 376.

Murray, R.C. and F.J. Lucia, 1967, Cause and control of dolomite distribution by rock selectivity: Geol. Soc. America Bull., v. 78, p. 21-36.

Randazzo, A.F. and E.W. Hickey, 1978, Dolomitization in the Floridan Aquifer: Amer. Jour. Science, v. 278, p. 1177-1184.

Roy, K.J., 1972, The Boundary Member; a buried erosional remnant of Triassic age in northeastern British Columbia: Canadian Soc. Petroleum Geol. Bull., v. 20, p. 27-56.

Sanford, B.V., 1962, Sources and occurrences of oil and gas in the sedimentary basins of Ontario: Proc. Geol. Assoc. Canada, v. 14, p. 59-89.

The American Association of Petroleum Geologists Bulletin
V. 61, No. 7 (July 1977), P. 982-1009, 13 Figs., 1 Table

Chalk Diagenesis and Its Relation to Petroleum Exploration: Oil from Chalks, a Modern Miracle?[1]

PETER A. SCHOLLE[2]

Abstract Chalks consist largely of stable low-magnesium calcite. Thus, they undergo diagenetic alteration different from that of more widely studied aragonite and high-magnesium calcite-bearing, shallow-marine carbonate deposits. Examination of outcrop and subsurface samples of chalks from the North Sea, onshore Europe, the Scotian Shelf, Gulf Coast, and the U.S. Western Interior indicates that chalks undergo significant diagenetic changes during their postdepositional history. Scanning-electron microscopy, light microscopy, oxygen-isotopic analysis, and trace-element analysis outline the major factors that control the patterns of chalk alteration.

The major mechanism of chalk cementation is pressure solution and local reprecipitation. Although small variations in initial grain size, faunal composition, or clay content can lead to significant bed-to-bed variations in cementation, overall patterns of chalk diagenesis appear to be related to two main factors: (1) maximum depth of burial, and (2) pore-water chemistry. With a few notable exceptions, the porosity (and permeability) of chalks decreases as a direct function of burial depth. The exceptions include cases where: (1) oil entered the rock, reducing or terminating carbonate reactions; (2) chalks are overpressured and therefore are not subject to the normal grain-to-grain stresses expected at those depths; and (3) tectonic stresses increase solution and cementation. In areas where fresh water entered the pores before major burial, chalks show a much steeper gradient of porosity loss versus burial depth as compared with regions where marine pore fluids were retained.

Under normal circumstances, a typical nannofossil chalk ooze will have 70% porosity at the sediment-water interface. At a depth of 1 km, porosity will be reduced to about 35%; at 2 km, to about 15%; and at 3 km, to essentially 0. Thus, one observes progressive lithification of chalks (and their isotopic alteration) as one moves downhole or toward areas of greater burial. Petrophysical and isotopic studies can predict maximum depths of burial, paleogeothermal gradients, and proximity to zones of deformation.

In areas such as the Ekofisk field in the North Sea, however, major quantities of oil are produced from chalks having as much as 40% porosity (largely primary) at depths greater than 3 km. This appears to be related largely to the widespread overpressuring of the Central graben in that area. Other such areas of anomalous porosity in thick chalk sections should be detectable by seismic methods. Significant hydrocarbon production from chalks can occur in three major settings: (1) overpressured or oil-saturated zones where these phenomena were initiated early in the subsidence history (e.g., the North Sea); (2) areas where chalks never have been buried deeply (e.g., the Scotian Shelf); and (3) cemented and fractured chalks in several possible settings (e.g., the Gulf Coast).

INTRODUCTION

Interest in chalks has increased during the last decade as a consequence of several factors. First, research interests in general have swung in the direction of deeper water sediments, mainly stimulated by the efforts of the Deep Sea Drilling Project (DSDP). Second, petroleum exploration has been progressing into ever deeper waters and greater drilling depths, to the point where deeper water facies are becoming common exploration targets. Finally, the past few years have seen major oil and gas discoveries in chalks. The Ekofisk and associated fields in the North Sea have total recoverable reserves of the order of 2.5 billion bbl of oil and 8 Tcf of gas, all in chalks (Tiratsoo, 1976, p. 99-100). Other, although far less productive, discoveries of hydrocarbons in chalks have been made in the North Sea, the Scotian Shelf of Canada (Wyandot Formation), Texas (Austin Chalk), and Colorado (Niobrara Limestone).

This paper will describe the basic physical properties of chalks and show how these properties are altered during subsequent diagenesis. Because postdepositional alteration plays the dominant role in determining whether a particular chalk will retain reservoir properties, a firm understanding of the diagenetic factors affecting chalks allows one to predict, with reasonable confidence, the regional patterns of chalk porosity and hydrocarbon potential. The section on "Hydrocarbon Production from Chalks" discusses

[1]Manuscript received, November 5, 1976; accepted, January 20, 1977.

[2]U.S. Geological Survey, Reston, Virginia 22092.

Work on this project has been supported at various stages by Cities Service Oil Co., Shell Development Co., National Science Foundation Grant GA-36696 (at The University of Texas at Dallas), and the U.S. Geological Survey.

Discussions with J. M. Hancock, W. J. Kennedy, R. Bromley, K. Cloud, R. E. Garrison, A. G. Fischer, M. A. Arthur, D. E. Hattin, W. A. Cobban, and R. M. Lloyd were very helpful in clarifying many of the ideas contained in this paper. E. A. Shinn, B. B. Hanshaw, and J. L. Wilson kindly reviewed preliminary copies of the manuscript. The technical assistance of A. White, R. Larson, D. Goodyear, and L. Tomlinson was indispensable to the project and is acknowledged gratefully.

Finally, I would like to thank the following companies for their kind permission to publish information on subsurface samples they supplied: Phillips Petroleum Co. (London), Cities Service International, Nederlandse Aardolie Maatschappij, Continental Oil Co. (London), Esso Exploration and Production UK, Shell UK Exploration and Production, Shell Canada, Mobil Oil Canada, Shell Development Co., Exxon Co. (USA), and Mountain Petroleum Co.

some examples of chalk reservoirs. A larger than usual number of references is included to facilitate further investigations into the literature.

METHODS OF STUDY

The data presented in this paper combine observations on a wide variety of European and North American Cretaceous chalks, including samples from 25 North Sea wells, 53 Gulf Coast wells from Texas to Florida, 5 wells in Kansas and Colorado, and 8 wells on the Scotian Shelf of Canada. Outcrop samples from England, Northern Ireland, Denmark, the Netherlands, Belgium, France, Germany, the United States, and Mexico were collected and processed.

Analytic work included examination of samples by transmitted light and scanning-electron microscopy. More than 400 porosity measurements were run on 1-in. (2.5 cm) core plugs or subsurface cuttings and 225 permeability measurements were made on outcrop samples. Porosity was determined by bulk-density versus volume measurement, and permeability was based on rate of air flow. Both types of analyses were obtained through the courtesy of Shell Development Co. in Houston, Texas. Other analytical work included 900 trace-element analyses done by atomic absorption at the U.S. Geological Survey, Shell Development Co., and The University of Texas at Dallas. In addition, 1,300 oxygen and carbon-isotopic analyses were obtained through Teledyne Isotopes, Geochron Laboratories, and Shell Development Co. All samples for isotope analysis were prepared using standard techniques of acid decomposition in phosphoric acid at 25°C. Both oxygen- and carbon-isotopic data are reported in per mil notation relative to the Chicago PDB-1 standard and sample correction was made according to Craig (1957). Standard reproducibility of such measurements is $0.2/10^3$ or better; this was confirmed by having duplicate standards run by the different laboratories involved.

CHALK DEPOSITION

Chalks are fine-grained limestones (micrites) composed largely of the minute skeletons of coccoliths (a nannofossil group of golden-brown algae), with only subordinate contributions of planktonic Foraminifera, calcispheres, and other, coarser skeletal material. Figure 1 shows a scanning-electron microscope (SEM) photograph of a typical porous chalk. Generally, chalks are considered to be soft, friable deposits. The DSDP reports refer to "nannofossil oozes" as the extremely soft, unconsolidated equivalents of chalks, and to "nannofossil limestones" as the lithified equivalents. The recent work of Matter (1974), Schlanger and Douglas (1974), Hancock and Scholle (1975), and others has shown, however, that all nannofossil deposits undergo progressive lithification as a function of depth of burial and other factors. Thus, any distinctions between soft, intermediate, and hard deposits are completely arbitrary and of no major significance; it simply gives three different names to the same deposit as it passes through a continuous spectrum of diagenetic stages. In this paper, the term "chalk" will be used broadly to cover the whole range of nannofossil deposits, with modifiers to describe the degree of lithification. The usage is compatible with the "type" Chalk of England, which locally ranges from nearly 50% to less than 10% porosity. However, the term "chalk" specifically is not extended to include soft or friable fine-grained limestones which may have formed by weathering or alteration. It also does not include rather pure, fine-grained limestones with evenly distributed, micron-scale porosity but lacking nannofossils (as with some Middle Eastern "chalks"). Such deposits may have radically different properties from true nannofossil chalks.

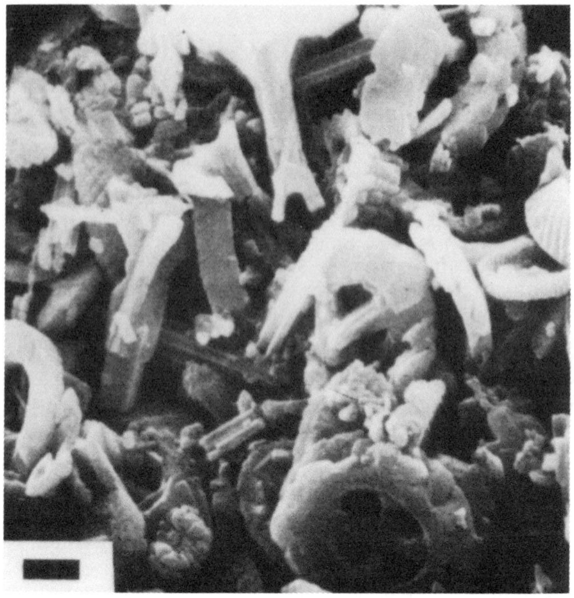

FIG. 1—Typical chalk containing abundant coccoliths and rhabdoliths. Porosity about 30 to 35%. Smoky Hill Chalk Member of Niobrara Limestone, Mountain Petroleum 1-16 State well, 1,680 to 1,690 ft (512 to 515 m) depth, Yuma Co., Colorado. Scale bar is 1μ long.

Coccoliths are pelagic organisms which have an extremely wide environmental and latitudinal distribution (e.g., McIntyre and Be, 1967). Yet their rate of production, at least in modern sediments, is slow enough that coccolith-rich sediments or chalks generally are present only in settings that have low rates of input of other materials. Examples can be found in shallow-water areas where other clastic terrigenous and other carbonate inputs are low, and coccolith-rich sediments are forming today (Scholle and Kling, 1972). In general, though, modern chalks are found in water depths between 100 and 4,000 m. For the past 100 m.y., such deeper water carbonate deposits have been estimated to account for as much as 67% of the world's total marine-carbonate deposition (Hay et al, 1976).

Sedimentation rates in Upper Cretaceous chalks have been estimated by Black (1953) to be 15 m/m.y. (1.5 cm/1,000 years). Hancock (1975b) calculated rates ranging from 3.6 to 54.3 m/m.y. (0.36 to 5.4 cm/1,000 years), the usual range for pure chalks being 20 to 40 m/m.y. (2.0 to 4.0 cm/1,000 years). Thus, some evidence indicates that rates of chalk deposition may have been considerably higher in the Cretaceous as compared with modern sediments (Black, 1953; Håkansson et al, 1974).

As modern sea-level stands are relatively low, most chalks today are confined to the ocean basins proper. During higher sea-level stands in the past, with broader epicontinental and shelf seas, and reduced terrigenous input from the continents, chalks were deposited in areas which today are continental or under only shallow water.

The high eustatic sea-level stands of the Late Cretaceous led to major flooding of continental areas (e.g., Wise, 1972; Bond, 1976) and chalks achieved their widest distribution then. Major regions of the North American Atlantic and Gulf of Mexico shelves received chalk deposition, as did the Western Interior seaway. Roughly 30% of the European continent was covered with chalks during the same interval. Chalks continued to be deposited in the deeper ocean basins during the Late Cretaceous.

The drop in eustatic sea levels from the Cretaceous high to the present relative low is marked by gradually less chalk in onshore-marine sections. The scarcity of pre-Cretaceous chalks, however, is due to other causes. The earliest well-documented coccoliths are in Lower Jurassic sedimentary rocks (Black, 1971) and not until Late Jurassic to Early Cretaceous time did they become sufficiently well established to be a rock-forming element. This implies that chalks, as properly defined, are restricted to Jurassic and younger sedimentary deposits. Chalks are likely to be important in onshore sections only in Late Cretaceous and early Tertiary intervals.

Specifics of the water depths at which various chalks have been deposited are difficult to estimate. In light of the postulated Late Cretaceous eustatic sea-level rise of about 300 to 500 m (Hays and Pitman, 1973; Bond, 1976); the general absence of bioherms, algal grains, and algal borings; and paleoecologic determinations of water depths based on siliceous sponges (Bignot and Defretin, 1967; Reid, 1968), Foraminifera (Burnaby, 1961), and other groups, 150 to 500 m can be proposed as a general depth for most European chalk deposition. Clearly, some localities, such as parts of the Central graben of the North Sea, had greater water depths, whereas marginal areas in continental Europe may have had shallower water conditions. More detailed discussion of these types of depth estimates were given by Håkansson et al (1974), Scholle (1974), and Hancock 1975b).

Gulf Coast chalks, such as the Austin, Selma, and Annona, probably were deposited in even shallower water as evidenced by the local presence of bioherms and erosional channels. Water depths in the Western Interior seaway are more difficult to determine, particularly because the restricted conditions make paleoecologic water-depth estimation difficult. Eicher (1969) proposed depths of at least 500 m and perhaps as much as 600 to 900 m for the Greenhorn sea. This must be considered an absolute maximum; most workers favor depths in the range of 100 to 300 m.

In general, most onshore chalk sections apparently were deposited at shelf or upper slope water depths, although deeper water chalks locally are exposed in subduction zones or geosynclinal areas (e.g., Fischer et al, 1967).

Chalks are present in a wide variety of environments largely because coccoliths, as a group, have a wide salinity tolerance (Bukry, 1974). Thin chalks are being deposited today in the Black Sea despite the 17 to 18 parts per thousand salinity (36 ppt is approximately normal marine). One species of coccolith is even able to survive salinities as low as 11 ppt in the Sea of Azov (Bukry, 1974). So chalks in the geologic record do not necessarily imply fully marine conditions. Indeed, there is some indication that the chalks of the U.S. Western Interior were deposited in waters of significantly lower than normal salinity because echinoderms, sponges, and brachiopods are absent (e.g., Kauffman, 1967).

More detailed discussion of chalk-depositional processes is beyond the scope of this paper, but have been covered amply in several papers including Kennedy (1970), Håkansson et al (1974), Scholle (1974), and Hancock (1975b).

PHYSICAL AND CHEMICAL PROPERTIES OF CHALKS

Bulk Composition

The abundance of coccoliths and coccolith fragments in chalks (more than 60% of the less than 20-μ fraction according to Håkansson et al, 1974) gives these rocks some unique properties. Coccoliths originally are composed of low-Mg calcite, generally the stablest polymorph of $CaCO_3$ at near-surface temperature, pressure, and water-chemistry conditions. This gives chalks an unusually great chemical stability for a limestone, especially when compared with modern, shallow-water reef or bank deposits. The scarcity of chemically unstable components (such as aragonite or high-magnesium calcite) in chalks plays a major role in its porosity retention in some settings.

Many chalks are virtually pure carbonate deposits. Analyses of samples from the upper Chalk of England, for example, show total HCl-insoluble residues often less than 1% (by weight) of the sediment, and generally less than 5% (e.g., Hancock, 1975b). The isotopic and trace-element composition of these ancient chalks is greatly affected by the diagenetic history of the rock. In recent sediments, however, chalks have the same or similar trace-element and isotopic compositions as individual coccoliths and planktonic Foraminifera which are their dominant constituents. Milliman (1974) gave average compositions for groups of modern organisms and from these values it can be estimated reasonably that most chalks had initial strontium trace-element concentrations of about 1,200 ppm. Modern carbon-isotopic values average +2 per mil (PDB), and modern oxygen-isotopic values for chalks range from +0.5 to +3.0 per mil (PDB). It should be remembered that with higher paleotemperatures during the Late Cretaceous, original oxygen-isotope values for Cretaceous chalks were more probably in the range of 0 to −1.5 per mil (PDB).

Some chalks contained a significant contribution of coarser carbonate skeletal debris including mollusks (especially inoceramids and oysters), echinoderms, bryozoans, sponges, and brachiopods. Yet even most of this fauna appears to have had an original calcitic composition. Only the mollusks appear to have added any significant amount of aragonite to chalks in a few areas. No aragonite or high-magnesium calcite was preserved in any of the samples analyzed during this study.

The acid-insoluble constituents most common in chalks include authigenic silica (cristobalite, quartz-chert, and chalcedony), clay minerals, detrital silt and very fine sand (mostly quartz with minor other minerals), phosphate, glauconite, pyrite, and hematite. The authigenic silica commonly forms discrete nodules or sheets although selective silicification of skeletal fragments is not unusual (Brown et al, 1969; Bromley et al, 1975; Shepherd, 1972). Nodular and tabular flints commonly follow or replace original sediment textures such as burrows or bedding surfaces. In chalks that have a high clay content, the silica commonly has not segregated into discrete nodules but rather remains disseminated through the sediments as cristobalitic spherules (lepispheres) as shown by Leclaire et al (1973) and Scholle (1974). The source of silica for these flints and other silica minerals is from the dissolution of the opaline skeletons of siliceous sponges, and to a lesser extent, diatoms and radiolarians. In those areas, such as the U.S. Gulf Coast or Western Interior, where siliceous sponges are absent, chert nodules are completely absent as well.

Even more important insoluble components in many chalks are clay minerals. Most North American chalks have insoluble residues in excess of 10 to 20% (e.g., Hattin and Darko, 1971; Frey, 1972; Cloud, 1975) and the Lower Chalk of England has as much as 60% insolubles, although it averages less than 20% (Hancock and Kennedy, 1967). The percentages of clay minerals generally vary inversely with distance to source area, and the purest chalks normally are at transgressive maxima when source areas were least productive and farthest away. Jeans (1968) proposed that a significant proportion of the clays of the English Chalk, which are predominantly montmorillonite with subordinate illite, are neoformed. Others (e.g., Valeton, 1960) have suggested a volcanic source for at least some of the clayey interlayers in chalks, and Borst (1972) has shown evidence of authigenic kaolinite formation during burial.

The other contributors to the insoluble fraction of chalks generally are of only minor importance. Yet there are areas where all occasionally can reach major proportions. Glauconite, quartz sand and silt, heavy minerals, and pelletal and fragmental phosphate can be significant in nearshore areas of chalk deposition. In some rare cases dolomite and siderite have been described (Cayeux, 1935; Wolfe, 1970; Kennedy and Juignet, 1974; Scholle, 1974; Cloud, 1975). Again, only in nearer shore sections do these minerals ever become very important. Pyrite commonly is an apparent early diagenetic replacement and/or void-filling mineral (e.g., Müller, 1951).

Sedimentary Structures

Bedding is a very pronounced feature of chalks wherever they have a significant clay content, for the clay provides a color contrast to accentuate bedding surfaces. Virtually all chalks examined in this project showed bedding, generally having individual units ranging from 0.2 to 1.0 m. In marly sections this bedding consists of rhythmic alternations (Fig. 2) of grayish, clay-rich marl and white, clay-poor chalk. In areas with very low overall clay content, bedding is developed much less

FIG. 2—Lower Chalk section at Culver Cliffs, Isle of Wight, England, with well-developed rhythmic bedding, consisting of alternation of chalks and marls. Bedding thickness ranges from 0.2 to 1.0 m.

strikingly but still is recognizable as fracture surfaces or zones of chert replacement. The bedding rhythms are primary (Kennedy, 1967) although they may be enhanced by diagenetic processes. The origin of these regular bedding features is not entirely clear although it has been proposed that the cycles reflect fluctuations in the "productivity of planktonic organisms and/or rhythmic fluctuations in the intensity of carbonate dissolution coupled with periodic variations in terrigenous influx" (Arthur and Roggenthen, 1976).

Within individual beds, primary sedimentary structures normally are few. This is largely because of the extensive reworking of the sediment by bottom-dwelling organisms. Burrows and borings are very common in most chalks (Bromley, 1967, 1968, 1975; Kennedy, 1967, 1970; Frey, 1972). On the other hand, in those areas, such as the U.S. Western Interior, where bottom conditions were unfavorable for the survival of bottom dwellers, there are units in which burrows are few and primary structures, especially horizontal lamination (Fig. 3), are common. Small-scale ripple cross-lamination and microchannel structures are present in these settings, and fecal pellets also may be preserved (Hattin, 1975).

Larger scale sedimentary structures have been described from some chalk outcrops, although these features are not particularly common. Cloud (1975) noted broad but extremely shallow channels in the Austin Chalk of Texas. Kennedy and Juignet (1974) described submarine banks, erosional channels, and slump features in the Cretaceous chalks of the French coast. In general, however, the rather great depositional depths of most chalks (100 m or greater) have excluded most wave- and current-generated structures. The only structures normally present are those produced by fluctuations in the supply of either the clastic or carbonate component or by burrowing organisms. The exceptions are in areas where soft, even soupy, chalk oozes were deposited in unstable slope settings, or in unusual, shallower water depositional sites subject to significant bottom currents.

Facies

Chalks, unlike shallow-water carbonate rocks, show only gradual lateral facies variations. Deeper water settings have broader areas of relatively uniform conditions and less abrupt transitions from one environment to the next. This translates into widespread, uniform facies with gradual transitions into adjacent lithologies.

Nevertheless, chalks obviously do pass eventually into other facies and Hancock (1975a) has described some of the most common patterns of transition. Chalk-facies transitions can be generalized into three types. The first reflects passage into shallower water facies such as calcarenites, terrigenous sandstones, or greensands. The second represents transitions into areas of increased clastic terrigenous input (but not shallower waters). This is shown by the passage of chalks, through marls, into shales and siltstones. The final type is common only in abyssal sediments and represents the transition from chalks well above the calcite compensation depth (CCD) in the oceans through progressive dissolution facies to red or black shales or cherts well below the CCD (e.g., Garrison and Fischer, 1969; Bosellini and Winterer, 1975). In predicting the presence of chalks in unexplored areas, it is necessary to examine all three of these factors in addition to the possibility of subsequent erosion.

Chalks and marly-chalk facies commonly can achieve considerable section thickness. Hancock and Scholle (1975) presented isopach maps of the Upper Cretaceous and Danian chalk of the North Sea and found areas having more than 1,200 m of section. Indeed, in some areas of the North Sea there is as much as 700 m of Maestrichtian chalk alone (Stenestad, 1972). Onshore sections in Europe and North America generally are considerably thinner and range from 250 to 450 m in many areas of England (e.g., Peake and Hancock, 1961; Gallois, 1965); 85 to 185 m for the Austin Chalk of central and southern Texas (Smith, 1973); and 0 to 300 m for the Niobrara Limestone of the U.S. Western Interior.

FIG. 3—Well-preserved horizontal lamination in shaly chalks of Smoky Hill Chalk Member of Niobrara Limestone near Russell Springs, Logan County, Kansas.

Grain Size

Grain-size analyses are difficult to perform on chalks because of the problems of disaggregation of samples and of analysis of very small particles. However, most coccoliths (and related groups of nannofossils) have skeletons in the size range of 1 to 20μ (Bukry, 1969). The coccoliths themselves are composed of individual crystal elements which have crystal sizes from 0.2 to more than 1μ (Neugebauer, 1975). Thus, grain-size analyses of chalks commonly yield bimodal or polymodal distributions with a silt-sized fraction composed of foraminiferal and other coarser fragments; a fraction in the 1 to 10-μ range representing whole coccolith shields; and, finally, visible through electron microscopy, a third fraction with coccolith crystal elements in the 0.2 to 1-μ range (Håkansson et al, 1974; Neugebauer, 1975).

Petrophysical Characteristics

Chalks are deposited with as much as 70 to 80% porosity. Early dewatering reduces this initially high porosity to about 50 to 60% within the first few tens to hundreds of meters of burial (Schlanger and Douglas, 1974). Beyond this point, considerable burial is required to reduce porosity significantly. Thus, many chalks, even those of Cretaceous age, have porosities of 35 to 50% (Scholle, 1974). Yet other chalks have much lower porosities, in some cases reduced to less than 1%. The causes of this porosity variation will be discussed later but, clearly, the range of values must be taken into account when considering the physical properties of chalks. Table 1 shows some porosity values for North American and European chalks. Because of the effect that burial diagenesis has on chalk porosities, these data should not be applied to chalks of the same or stratigraphically equivalent units at other localities.

Permeability is very low for all chalks, regardless of porosity, as might be expected for rocks of such small grain size. There is, however, a clear relation between matrix porosity and permeabil-

Table 1. North American and European Chalks

Area	Unit	Present Burial Depth (m)	No. of Samples*	Sr (ppm)	Mg (ppm)	δ^{18} (PDB)	C^{13} (PDB)	Porosity (%)
Western Alabama	Selma	0	11	710±55	2475±330	-2.7±0.4	1.1±0.4	37.6
Dallas-Waco area, Texas	Austin	0	12	1130±280	1200±430	-4.1±0.6	1.4±0.4	26.6
Austin-San Antonio, Texas	Austin	0	17	420±160	1425±330	-3.5±0.4	0.9±1.0	18.8
Uvalde area, Texas	Austin	0	5	420±270	1800±430	-3.9±0.6	0.3±0.6	15.7
Del Rio-Langtry area, Texas	Austin	0	11	680±125	1675±320	-5.5±0.5	1.1±0.5	13.3
Coahuila, Mexico	Austin-San Felipe	0	4	1190±290	1235±230	-6.3±0.7	0.9±1.0	9.4
#1 Woodrow Free well Washington Co., Texas	Austin	4554	5	700±310	1590± 60	-8.0±0.7	0.7±1.2	
#1 E.W. Jones well Bastrop Co., Texas	Austin	1992	5	845±175	2060±160	-4.6±0.5	0.9±0.9	
#1 Tuerpe well Atascosa Co., Texas	Austin	756	3	333± 40	1935±115	-3.1±0.7	0.7±0.6	
Ft. Collins area, Colorado	Niobrara	0	1			-7.4	1.8	
Pueblo area, Colorado	Niobrara	0	15	1000±210	2410±990	-6.8±0.9	1.1±0.5	
Las Animas area, Colorado	Niobrara	0	8	1010± 55	1445± 75	-6.2±0.5	1.1±0.7	
Central western Kansas	Niobrara	0	26	945±240	3090±2330	-5.3±0.6	1.5±0.9	39.9
Cimarron Co., Oklahoma	Hartland	0	12	600±180	1590±505	-7.2±0.4	0.4±0.7	
Cimarron Co., Oklahoma	Bridge Creek	0	8	330	1740	-4.2±0.7	0.2±0.7	
Oneida O-25 well Scotian Shelf, Canada	Wyandot	1245	3			-2.8±1.0		
Dauntless D-35 well Scotian Shelf, Canada	Wyandot	1700	6			-3.8±0.8	2.3±0.3	
Northern Ireland	White Limestone	0	53	560±155	1465±505	-5.4±1.6	1.4±0.4	7.5
Yorkshire, England	Mid. & Upper Chalk	0	16	855±180	1440±170	-4.2±0.6	2.3±0.4	21.7
Lincolnshire, England	Mid. & Upper Chalk	0	11			-3.8±0.5	2.4±0.6	26.7
Norfolk-Suffolk, England	Mid. & Upper Chalk	0	12	690±130	1800±320	-2.3±0.7	1.9±0.5	39.6
Kent, England	Mid. & Upper Chalk	0	15	650± 70	1250±200	-2.6±0.4	1.8±0.4	43.3
Isle of Wight, England	Mid. & Upper Chalk	0	68	540±110	2275±660	-3.0±0.6	2.6±0.6	26.0
Isle of Wight, England	Lower Chalk	0	24	640± 95	2025±370	-3.3±0.8	2.3±1.1	15.3
Sussex, England	Chalk	0	14	450±135	1680±150	-2.6±0.6	1.8±0.7	39.5
Wiltshire, England	Mid. & Upper Chalk	0	6			-3.0±0.7	1.9±0.7	34.9
Buckinghamshire, England	Mid. & Upper Chalk	0	9	515±155	1555±450	-2.6±0.4	2.1±1.3	
Jutland, Denmark	Maestrichtian Chalk	0	8	870±230	2395±505	-1.4±0.4	1.7±0.3	46.0
Breitenburg, W. Germany	Chalk	0	5			-2.7±0.1	2.3±0.1	42.8
Etretat area, France (Normandie)	Chalk	0	10	275± 90	2900±470	-2.8±0.5	1.4±0.6	33.3
Limburg area, Netherlands	Maestrichtian Chalk	0	23			-1.3±0.7	1.5±0.4	42.4
48/30-1 well, Celtic Sea	Chalk	399	23	560±160	1670±605	-4.5±1.0	2.0±0.7	
54/1-1 well, British North Sea	Chalk	1082	22	560±230	1790±455	-3.4±1.0	1.9±0.6	
53/4-1 well, British North Sea	Chalk	1196	13	450± 90	1545±440	-3.4±0.6	2.0±0.7	
49/17-7 well, Brit. North Sea	Chalk	841	6	355±170	1505±685	-2.6±0.7	1.5±0.8	
47/13-1 well, Brit. North Sea	Chalk	595	11	650±120	2923±1739	-3.0±0.9	2.4±0.3	
30/19-1 well, Brit. North Sea	Chalk	3338	20	1070±145	3550±580	-4.6±1.0	1.7±0.6	11.0
29/3-1 well, Brit. North Sea	Chalk	2893	6			-4.6±0.9	1.5±0.4	
22/21-1 well, Brit. North Sea	Chalk	3239	6			-4.7±0.7	1.7±0.4	
22/11-2 well, Brit. North Sea	Chalk	2707	6	695± 65	2025±265	-4.2±0.5	2.0±0.3	
2/4-1AX well, Norwegian N.Sea	Chalk	3192	5	1245±150	2938±1591	-4.7±0.8	1.5±0.2	15.0
2/4-2X well, Norwegian N.Sea	Chalk	3237	61	1005±155	1695±605	-4.7±0.8	1.4±0.4	27.0
I-1 well, Danish North Sea	Chalk	2804	6	985± 95	4065±240	-4.9±0.2	0.7±0.9	21.6
K/8-1 well, Dutch North Sea	Chalk	1582	12			-3.3±0.5	1.8±0.4	
L/13-1 well, Dutch North Sea	Chalk	1874	11			-3.8±1.0	2.2±0.3	
L/16-1 well, Dutch North Sea	Chalk	1731	16	1010±435	1615±395	-3.4±0.6	2.0±0.3	20.8

* Number of samples refers only to the number of isotopic analyses; either greater or lesser numbers of samples may be involved in other analyses. The different types of analyses were not always performed on the same samples so there is not always complete stratigraphic equivalency between all groups of data.

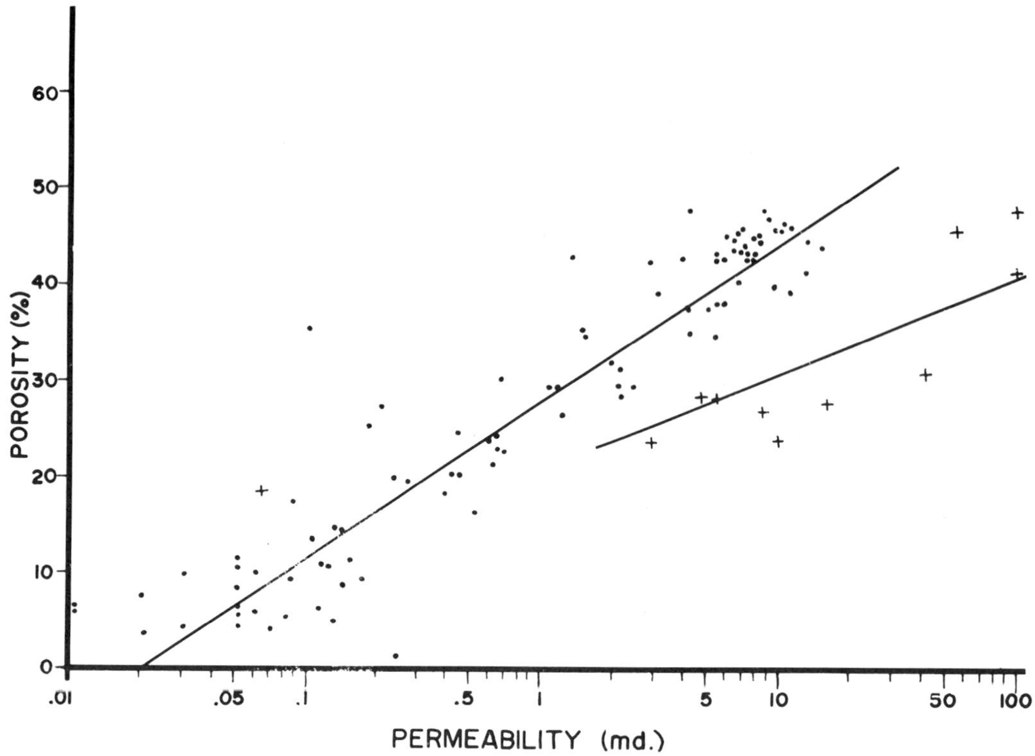

FIG. 4—Plot of porosity versus permeability (air flow) for chalks and calcarenitic chalks. Dots are values for pure chalks; crosses are values for coarser, calcarenitic chalks. Lines are least-squares fits to data.

ity in chalks as shown in Figure 4. It must be emphasized that this relation holds only for rather pure chalks. Calcarenitic chalks which have a significant contribution of coarser material may have much higher permeabilities for any given porosity, whereas clayey chalks (marls) have variable porosity-permeability relations.

The porosity and permeability measurements were conducted in the laboratory on 1-in. (2.5 cm) core plugs. In nature, considerably higher effective permeabilities are achieved because of large-scale fracturing. The Chalk of southern England is a major aquifer, for example, in spite of matrix permeabilities of 1 to 8 md. Flow rates as high as 2 km/day are present in natural systems in the English Chalk (Edmunds et al, 1973; Atkinson and Smith, 1974; Smith et al, 1976), virtually all of it moving through fractures. Foster (1975) suggested that effective permeabilities in fractured chalks are as much as 50 to 500 times greater than in equivalent unfractured units.

Dry-bulk densities in chalks range from 1.47 g/cc (corresponding to a porosity of 45%) to 2.71 (corresponding to a porosity of 0). Other physical properties of chalks have been summarized by Dessenne and Duffaut (1970) and Carter and Mallard (1974). They presented values of preconsolidation pressure, tensile strength, deformation modulus, and other engineering properties which show that chalks are rather weak sedimentary rocks.

EARLY DIAGENESIS

The initial physical and chemical properties of chalks begin to alter immediately after deposition. To understand the final patterns of porosity distribution in chalks, and to predict their reservoir potential, one must be able to decipher the entire history of alteration.

For most chalks, early diagenesis consists mainly of compactional dewatering. With the addition of overburden through continued sedimentation, as well as through the activities of burrowing organisms, water is expelled until eventually a grain-supporting framework is formed. However, for other chalks early diagenesis is a more complex process involving, in some cases, significant dissolution, cementation, and replacement.

In many deep-water settings dissolution takes place before or during sedimentation. In areas where bottom-water depths are below the lysocline particularly extensive dissolution takes place both during settling and while grains are exposed at the seafloor (e.g., Berger, 1973; Honjo, 1975;

Milliman, 1975). Sediments from shallower water depths have been described in which all the aragonite had been removed prior to burial (Hudson, 1967) and, of course, below the CCD essentially all carbonate minerals are eliminated. Modern studies have shown that not only are certain mineralogies (such as aragonite) removed preferentially, but also that there is selective dissolution of individual species, even where the organisms have low-Mg calcite skeletons (McIntyre and McIntyre, 1971; Adelseck et al, 1973; Wise, 1973). Thus, what is buried in the sediments is material which has been, to varying degrees, selected for chemical stability. The extent of this selective solution depends on the rate of sediment supply versus the rate of sediment removal. Therefore, deep-water areas and regions which have low rates of carbonate supply or accumulation commonly have the most extensive dissolution of the lower stability faunal elements.

In some places, this seafloor dissolution is accompanied by other diagenetic processes, including carbonate cementation and replacement of carbonate by glauconite or phosphate. Such zones of synsedimentary, seafloor cementation commonly are termed "hardgrounds" and have been described extensively by Bromley (1965; 1968), Bathurst (1971), Kennedy and Garrison (1975), and many others. In European chalk sections, these hardground zones have porosities reduced to 10 to 20% by cementation, even where the surrounding, unaffected sediments have porosities of as much as 50% (Scholle and Kennedy, 1974). The cementation can be shown to have taken place at, or very near, the seafloor because hardened pebbles are reworked into the overlying sediments, because the hardground surfaces are cut by borings which transect grain boundaries, and because the surfaces are encrusted by organisms which required a hard substrate.

Multiple hardground surfaces are present in packages which can be correlated over wide areas. Time-equivalent surfaces are mainly over paleotopographic highs and apparently reflect drops in sea level (Hancock, 1975b). However, the hardgrounds were cemented under fully marine conditions where submarine erosion either halted net sediment accumulation or produced slight net sediment removal. Long exposure at the sediment-water interface then eventually resulted in widespread cementation.

In European sections hardgrounds are fairly common. Although such seafloor cementation affects less than 5% of the total section there, it may be important in producing impermeable zones which can affect fluid-flow patterns (including oil migration). Similar hardground surfaces are present in North American Cretaceous sedimentary rocks (e.g., Rose, 1970) but generally are rare or absent in chalk sections.

BURIAL DIAGENESIS

Most chalks are little affected by seafloor cementation. Thus, any porosity losses in these rocks must be related to burial or postburial alteration, and it is these processes which were understood the least until the last 5 years. Now, however, a consistent picture has been established to relate chalk porosity losses to burial history.

It must be emphasized again that chalks have unusual chemical properties for a limestone. Because they consist largely of original low-magnesium calcite, they have great chemical stability. In shallow-water carbonate deposits, exposure to fresh water, for example, produces wholesale dissolution and cementation because of their large content of unstable aragonite and high-magnesium calcite. Burial effects are generally far less important in shallow-water limestones because of the dominance of these freshwater, early diagenetic effects.

Chalks, by contrast, start out initially with low permeabilities which hinder movement of large volumes of fresh water through the matrix. Even when fresh water is introduced over long time periods by fracture flow, probably only small numbers of pore volumes enter by diffusion into the matrix, and little alteration occurs. Some dissolution does take place at the surface and in near-surface fractures, producing an undulating topography with solution pits and pipes filled with loamy, insoluble residue (e.g., Havard and Steward, 1969; Thorez et al, 1971; Smith et al, 1976), at the same time deriving $CaCO_3$ which may be precipitated elsewhere as fracture fillings. In other areas, caliche or calcrete crusts can be formed on chalk surfaces (Yaalon and Singer, 1974).

Yet, in spite of this surficial alteration, even chalks of Cretaceous age, exposed for as much as 70 m.y. to fresh water, show little or no signs of matrix-porosity enlargement or reduction where there has been only shallow burial. The Upper Chalk of southern England, for example, has a primary matrix porosity of as much as 45% (Scholle, 1974) in areas of shallow burial. Likewise, the Niobrara Limestone of western Kansas has retained as much as 40% porosity (Table 1) where the overlying Pierre Shale is thinnest. In both, samples show little chemical (isotopic) or physical (as seen in SEM) alteration (e.g., see Black and Barnes, 1959), indicating that neither age nor freshwater exposure alone strongly affect chalk porosity.

Several other hypotheses have been put forward to explain regional differences in the degree of chalk lithification. Hancock (1963) proposed that original variations in the amount of fine-grained aragonite admixed with the chalks resulted in areas of greater and lesser cementation. Although this is certainly a possible mechanism, no positive evidence has been presented to support the hypothesis. The chalks involved have no original aragonite left in them, although there is evidence (molds and impressions) of some former aragonitic fauna. There is no indication of, and no apparent source for, any significant fine-grained aragonitic matrix.

Another theory has been that regional differences in the amount of interstitial silicification account for the variations in hardness (Peach et al, 1901). Through insoluble residues and scanning-electron microscopy it can be shown that this hypothesis is clearly neither true for the specific case in which it was proposed, nor is it valid as a general rule.

The data in the following sections will show that there is only one major factor which consistently correlates with regional porosity loss in chalks—burial depth. Secondary factors such as pore-fluid pressure, tectonic stresses, pore-water chemistry, and possibly thermal gradient influence rates and patterns of porosity loss, but all of these are subordinate, on a regional scale, to burial depth.

Principles

There are two main mechanisms which may operate during burial diagenesis: mechanical compaction and chemical compaction (also called solution reprecipitation or solution transfer). External introduction of cement is another possible process but, because it requires the movement of very large volumes of pore fluids, it is considered to be of minor importance in low-permeability sediments such as chalks. Mechanical and chemical compaction processes can take place without any external contribution, although pore fluids must be expelled to achieve porosity reduction.

Mechanical compaction moves grains closer together by breakage, reorientation, or repacking as overburden stresses are increased. Possible evidence for mechanical compaction in sediments includes visible grain breakage or deformation, flattening of sedimentary structures such as burrows, and compactional drape of originally soft sediments around shells, nodules, or other rigid objects. A strongly developed grain-orientation fabric, in some cases, also can be evidence of mechanical rotation of grains. The rate and extent of physical deformation are largely a function of the applied differential stresses in relation to the size, shape, and strength of the constituent grains and the amount of prior cementation. For deposits with the uniformity of physical properties shown by chalks, reasonable theoretical and experimental estimates can be made to relate overburden stresses (burial depth) to "mechanical" porosity reduction (Hamilton, 1959; Carter and Mallard, 1974).

Chemical compaction has been recognized in recent work as being of very considerable importance in the subsurface cementation of limestones (Bathurst, 1971). Good recent introductions to the mechanisms of chemical compaction (solution transfer), first described in the 1860s, have been provided by Bathurst (1971) and Durney (1972). The active mechanism in chemical compaction is pressure solution—the process of dissolution of minerals under nonhydrostatic stress, with dissolution occurring preferentially at grain and crystal contacts where external pressure is greatest. The dissolved calcium carbonate migrates mainly by grain-boundary diffusion (Durney, 1972), rather than by bulk flow, and generally is reprecipitated in nearby sites (probably within millimeters to meters from the source) with lower external pressure. Such precipitation can take place in intergranular or intragranular pore spaces, fissure cavities, or even (by displacement) in areas without open pores. Solution transfer can be a response either to overburden or tectonic stresses, and Weyl (1959) has shown that the process is favored by the presence of thin clay seams which facilitate diffusional transport.

The features that indicate that chemical compaction has taken place include stylolites, solution seams, embayed and partly dissolved grains and crystals, and associated draping of sediments. Bulk-volume reduction is a very significant consequence of solution transfer (as with mechanical compaction), especially where initial porosities were high (Manus and Coogan, 1974). The overall effect of this process is to produce a cemented sediment but a thinner section. In essence, one part of the section is cannibalized to provide the calcium carbonate to cement the remainder. This either can take place on an intergranular basis, or most of the material can be derived from discrete surfaces such as stylolites. In either case, large volumes of material are involved because chalks have such high initial porosities.

However, it is difficult to demonstrate conclusively that the cement in any given section is derived locally by solution transfer, rather than introduced from outside sources. Harms and Choquette (1965) showed that there was selective cementation adjacent to stylolite surfaces. Calculations based on the number and amplitude of stylolites or on the amount of insoluble residue along solution surfaces have given estimates of the amount of carbonate lost along such surfaces (but have not shown where the dissolved carbonate went). Dunnington (1967) noted common volume reductions of 20 to 35% (after induration) in strongly stylolitic rocks, with some examples even reaching a reduction of 40%. Mossop (1972) calculated a 24% volume reduction in fine-grained lagoonal carbonate deposits. So, the process is clearly capable of bringing about large bulk-volume reductions and generating large amounts of carbonate cement.

The burial depth at which solution transfer first becomes an important process is not well defined. Dunnington (1967) suggested 600 to 900 m as the minimum depth at which stylolites formed. Neugebauer (1973, 1974) presented interesting theoretical calculations to show that the effective depth is dependent in large measure on the composition of interstitial fluids. Under the least favorable conditions, production of cement by solution transfer is insignificant below 300 m, produces no more than 5% cement by 1,000 m, but results in virtually complete lithification in chalks by 2,000 to 4,000 m burial. Beall and Fischer (1969) indicated that the transition from dominantly mechanical to chemical compaction occurs below 250 m in DSDP cores from an area east of the Bahamas. Again, it must be kept in mind that water chemistry can play a major role in determining the effectiveness of pressure solution. When magnesium concentrations are above 0.01 M (240 mg/l), dissolution is retarded (Neugebauer, 1974). Thus, unless chalks are flushed extensively with magnesium-poor fluids, such as typical ground waters, they may resist pressure solution to considerable depths.

Physical Evidence

Evidence for burial diagenesis is widespread in chalks. The most striking confirmation comes from the general relation between porosity and present burial depth in a wide variety of settings. Beall and Fischer (1969, p. 583) noted even in the first leg of the DSDP program that "alteration [of nannofossil oozes] with increasing burial appears to be mainly one of compaction to harder and harder chalks . . . Under sufficient pressure, solution welding of coccoliths occurs, and a hard, fine-grained limestone results . . ." Other such observations were made by Tracey et al (1971), Davies and Supko (1973), Schlanger et al (1973), Matter (1974), and others through the DSDP reports.

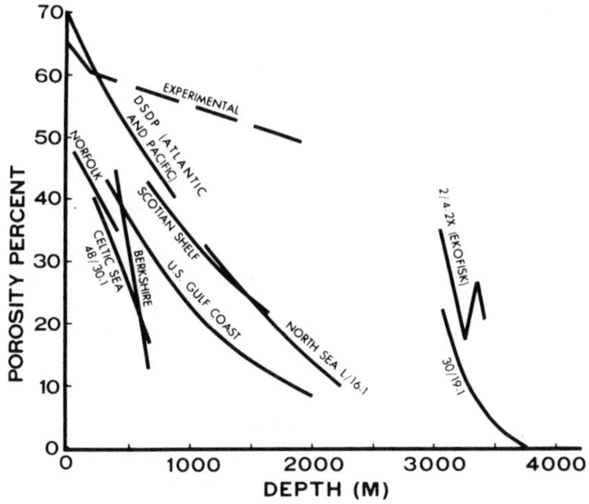

FIG. 5—Plot of porosity changes as function of burial depth in chalks.

These relations also are seen in onshore sections. For example, Edmunds et al (1973) and Carter and Mallard (1974) showed a consistent loss of porosity as depth of burial increases for onshore English sections. In the Gulf Coast, the Austin Chalk progressively loses porosity when traced downdip into the subsurface or along outcrop into areas of formerly greater burial (Cloud, 1975). In the North Sea, the Maestrichtian chalk, which has as much as 45 to 50% porosity on outcrop in Denmark, generally has between 0 and 25% porosity by the time it has been buried to 2,000 to 3,000 m.

Figure 5 summarizes the burial-porosity relations of chalks. The DSDP data are taken largely from Schlanger and Douglas (1974), the experimental data are from Hamilton (1959), and the Berkshire (England) data are from Edmunds et al (1973). All other data have been newly obtained during this research project.

Even a brief glance at Figure 5 shows a relation between depth of burial and porosity, but not a simple one. The experimental data of Hamilton (1959) have an unusually low slope relative to natural samples largely because the experimental setup produced mainly mechanical compaction. In nature, where chemical compaction (solution transfer) is probably the dominant process (especially at depth), far greater porosity loss can be achieved relative to overburden stress, but this takes place on a time scale too slow to be modeled easily in the laboratory.

The rather smooth curves which can be drawn through the data for the DSDP samples, the Scotian Shelf, and the North Sea L/16-1 well probably reflect the most typical relations for offshore chalks. In these areas, burial occurred, in all like-

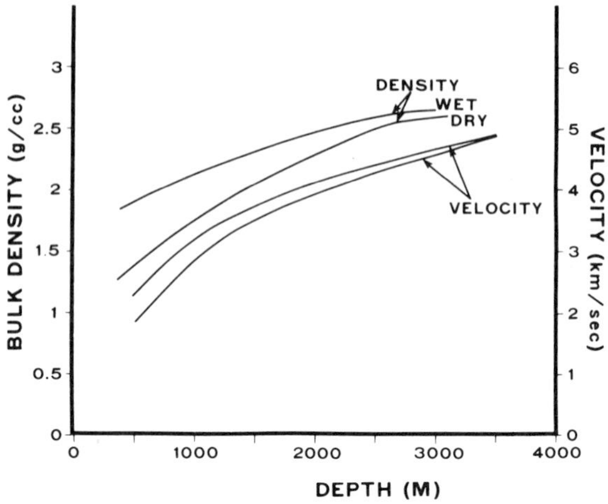

FIG. 6—Variation of wet and dry bulk density and sonic velocity for chalks as function of burial depth. Diagram is based on predicted "normal" curve of porosity loss with burial.

lihood, with marine pore fluids still present in the chalks. There is no evidence from these areas to show that uplift and flushing by nonmarine pore fluids occurred prior to burial. According to Neugebauer (1973, 1974), the presence of marine pore fluids (or other fluids having a significant magnesium content) retards pressure solution in chalks and thus helps to preserve porosity. The curves for DSDP, Scotian Shelf, and L/16-1 North Sea chalks, in fact, fall very close to the theoretical predictions made by Neugebauer (1974) for chalks with magnesium-bearing fluids.

The Norfolk, Celtic Sea, and Berkshire curves, on the other hand, represent onshore or shallow-marine sections where independent geologic evidence indicates that chalk sedimentation was followed by a period of subaerial exposure, erosion, and freshwater flushing prior to burial by Tertiary sediments. In these areas, again as predicted by Neugebauer (1974), rates of cementation are much higher, with virtually complete loss of porosity by about 1 km of burial.

The U.S. Gulf Coast data probably reflect an intermediate situation, with some areas receiving early freshwater input, but most of the region not being flushed with fresh water during its early burial history. The data from the 2/4-2X Ekofisk and 30/19-1 wells of the North Sea represent major anomalies in the pattern of rather rapid porosity loss as a function of burial depth. These anomalies, a key to exploration for oil in chalks, will be discussed in the section titled "Hydrocarbon Production from Chalks."

The pattern of burial-related porosity loss also apparently can be accelerated significantly where tectonic stresses are superimposed on overburden stresses (especially in the presence of fresh water). This is apparent from the data of Mimran (1975) and from observations during this project on samples from Europe, northern Mexico, and central Colorado.

Figure 6 shows the change in wet and dry bulk densities and sonic velocity as a function of burial depth assuming the "normal" porosity-loss curve shown by the DSDP–Scotian Shelf–North Sea L/16-1 data. The velocity data are derived from the compilation of Woollard (1962). Using these curves, one can make realistic predictions of the subsurface properties of buried chalks in coastal and offshore provinces. However, it must be remembered that all these curves are strongly smoothed. Although the general relations between porosity and burial depth are consistent, the variations in porosity between even adjacent beds in any given section can be very large indeed. Schlanger and Douglas (1974) pointed out some of these complexities and related them to primary sedimentologic variations. Possible factors of significance in determining the initial "diagenetic potential" of any given bed may include overall grain size, faunal composition, chemical stability of the fauna, and clay content. Schlanger and Douglas (1974) also pointed out that these factors may be related to major changes in oceanic conditions so that bed-to-bed porosity variations can be widespread and tied closely to oceanic acoustic stratigraphy.

In addition to the association of porosity loss with burial depth just shown, there is abundant physical evidence in chalks of both mechanical and chemical compaction, as well as cementation. Mechanical compaction appears to be, under most circumstances, the least important of the three phenomena after chalks reach about 50% porosity. However, prior to that stage, mechanical compaction is the most important factor. Håkansson et al (1974) noted only about 10% compaction (post consolidation) in the Maestrichtian chalk of northern Europe which never has been buried deeply. Likewise, shallowly buried chalks of southern England have few broken fossils (Scholle, 1974). Some increase in the number of broken and crushed fossils has been noted in chalks which have a deeper burial history (e.g., Matter, 1974; Schlanger and Douglas, 1974) but the effect rarely appears to be important enough to account for a very significant porosity loss during deeper burial diagenesis.

An exception was described by Wolfe (1968) in chalks from Northern Ireland. Here compaction values of as much as 30% were present, although there was significant bed-to-bed variation with considerable breakage of microfossils (see also photomicrographs in Scholle, 1974, p. 185-186) as well as crystal reorientation. Another example of extensive mechanical compaction in chalks was described by Mimran (1975; in press) from strongly folded chalks of southern England. Both cases seem to show that a rapid rate of deformation may play a role in increasing the importance of mechanical compaction relative to chemical compaction. In Northern Ireland, overburden stresses were applied by a thick capping of Tertiary basalts (Wilson, 1972) which were extruded in a relatively short period of time. In the deformed chalks of southern England, maximum stresses were related to tectonic events rather than to deposition of overburden, and again, may have been applied rapidly.

Solution seams, stylolites, grain embayments, and other forms of evidence of chemical compaction are common in most chalk sections. Solution seams (Fig. 7) are low-amplitude, irregular, wispy, and anastomosing zones of marly and clayey sediment which have been concentrated by pressure solution. Garrison and Kennedy (in press) described these features from numerous European sections and showed that they are produced by intrastratal dissolution and compaction in relatively soft sediments. This results in a nodular, lumpy texture with relict pieces of clay-poor chalk surrounded by darker, marlier zones similar to the horsetail seams described by Mossop (1972). Garrison and Kennedy (in press) also pointed out that dissolution is most intense in layers that have high primary clay content and that dissolution affects only the fine-grained, coccolith-rich matrix, not the coarser skeletal fragments. Thus, the end product shows accentuation of primary sedimentary variations and bedding rhythms.

As chalks begin to lithify through cementation, and grains are more rigidly held in place, dissolution becomes concentrated more sharply along individual thin planes. Thus, once porosities fall to less than about 25% there is a transition from solution seams to stylolites. This type of transition is uncommon in shallow-water carbonate deposits because they generally reach a stage of at least moderate lithification before significant burial occurs. In chalks, where burial diagenesis dominates, early diagenetic-solution seams and later diagenetic stylolites are very common, both in outcrop and subsurface areas.

On a smaller (intergranular or intercrystalline) scale, dissolution features are a bit more difficult to find. Wise (1973) illustrated partly dissolved coccoliths but believed that much of this dissolution occurred close to the sediment-water interface. Schlanger et al (1973), Matter (1974), Schlanger and Douglas (1974), and Matter et al (1975) presented evidence either of progressive loss of unstable coccoliths and planktonic Foraminifera or of direct dissolution effects visible in SEM. Because dissolution of unstable forms is often complete, and because associated cementation often obscures dissolution effects, the indirect evidence of progressive loss of the least stable faunal elements with increased burial often must be accepted as the principle line of evidence for intrastratal dissolution. This process has been simulated experimentally by Adelseck et al (1973).

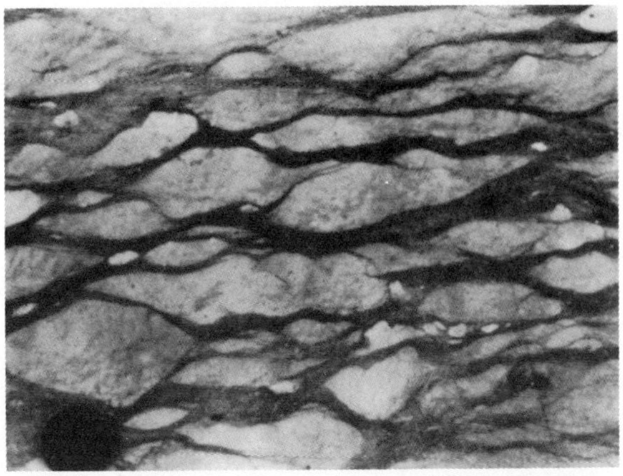

FIG. 7—Solution seams in lower Turonian (*Labiatus* Zone) Middle Chalk from Beachy Head, west of Eastbourne, Sussex, England. Early lithification of chalk and subsequent dissolution of carbonate material by pressure solution yields nodular, wispy texture. Photo courtesy of J. M. Hancock. Coin is 2.5 cm in diameter.

Evidence of progressive addition of cement during burial is much more widespread. In DSDP cores, cementation has been shown to be quite selective, with individual species or groups of organisms being especially prone to overgrowth cementation. In particular, discoasters receive the earliest cement, followed by individual groups of coccoliths, and finally the interstitial areas between coccolith (placolith) shields (Wise, 1973; Matter, 1974). Precipitation of calcite within foraminiferal chambers also can be important in some cases (e.g., Schlanger et al, 1973).

In onshore chalks, Neugebauer (1975) illustrated the fact that very small differences in the size and form of coccolith crystal elements can make a great difference in their susceptibility to overgrowth cementation. Abundant overgrowths are present on crystal elements equal to or larger than 1μ, whereas no overgrowths were formed on 0.2 to 0.4-μ crystals. Likewise, where rhombohedral cleavage surfaces bounded the coccolith crystals, overgrowths were inhibited. Scholle (1974, p. 188) illustrated a graphic example of such coccolith overgrowth in chalk from Northern Ireland, and Mapstone (1975) showed "spot welding" (cementation at grain contacts), selective overgrowth, and sparry calcite pore filling in North Sea chalks.

Such observations also have been made during this study in all chalks examined. Whether one looks in the Gulf Coast at the Austin Chalk from outcrop to deep burial, in the Western Interior at the Niobrara Limestone from Kansas to the Denver basin, or in the North Sea area at the Upper Chalk from England to the Central graben area, the results are always the same. Where the chalks now are (or once were) buried more deeply, cementation is far more advanced. Figures 8A-D and 9A-D illustrate two such cementation sequences. Unlike DSDP studies (where progressively older units are examined downhole) these photographs compare the same stratigraphic unit now (or formerly) buried to different depths at various localities. Figure 8A-D illustrates a series of progressively more deeply buried Upper Cretaceous chalks from the North Sea in which one can see gradual increase in crystal size and angularity, and progressive decrease in porosity as a result of increasing overgrowth cementation. This is the normal sequence of alteration in North Sea chalks outside the Central graben area. The comments of Mapstone (1975) about cementation stopping at "spot welding" or thoracospherids being common enough to prevent cementation, appear to be inapplicable to most, if not all, North Sea chalks.

Figure 9A-D shows a similar progressive cementation sequence in the Fort Hays Limestone Member of the Niobrara Limestone of the U.S. Western Interior. Although all of these samples are from outcrops, if one estimates the former thickness of overburden (mainly the Pierre Shale; Reeside, 1944), it becomes clear that samples from central Colorado once were overlain by at least 1,500 to 2,000 m of sediment, whereas those from Kansas probably had less than 200 to 300 m of overburden. Thus, the progressive diagenesis from Kansas to central Colorado is a function of a paleoburial gradient. Indeed, this progressive loss of porosity, as measured directly or by examination of SEM photographs, can be used as a rough index of former burial depths, especially when combined with oxygen isotopic data (Cloud, 1975; Scholle, 1975, 1976). However, one must remember that the analysis of only a small number of samples can lead to serious errors because of the large bed-to-bed variability.

The process of progressive cementation by solution transfer discussed previously is one which can, and does, proceed to the point of virtually complete elimination of porosity because it does not require the external input of water or calcium carbonate. Thus, porosities of less than 10%, and even less than 1%, are common, particularly in geosynclinal areas where significant overburden and tectonic stresses may have acted. Fischer et al (1967), Scholle (1971), and Wachs and Hein (1974) illustrated some typical geosynclinal chalks which have had complete cementation and even neomorphism (strain recrystallization). These limestones have reached a state of "near zero diagenetic potential" (Schlanger and Douglas, 1974, p. 133) as well as near-zero petroleum potential.

The abundant evidence of cement addition to chalks, coupled with the lack of evidence of extensive compaction, implies that cementation is a continuous process. It starts with early (sometimes even seafloor) precipitation of a framework of cement at grain contacts (e.g., Bathurst, 1970), forming a rigid, load-resistant scaffolding into which is inserted later diagenetic cement at a rate sufficient to prevent compaction under increasing overburden. Other nearby units which have less stable constituents or higher clay contents do not form this framework and are eliminated during burial, yielding their calcium carbonate (through solution transfer) for the cementation of the stable beds. Thus, one reaches a final picture of "donor" and "recipient" units. The "recipient" units are preserved in the geologic record with many signs of cementation and little compaction. Associated "donor" beds largely are eliminated, eventually appearing only as clay-rich seams.

Geochemical Evidence

Geochemical data have been gathered on both oceanic (DSDP) and onshore chalks to support the concept that cementation is a gradual process which takes place throughout the burial history. Most of the data are oxygen- and carbon-isotopic analyses combined with strontium and magnesium trace-element values. A detailed introduction to isotopic analysis is beyond the scope of this paper, but Bowen (1966), Garlick (1969), Schwarcz (1969), Milliman (1974, p. 30-33), and others have provided good basic explanations.

Briefly, oxygen- and carbon-isotopic values are reported as per mil deviation from a standard (generally, the PDB-1 belemnite is used). Initial oxygen-isotopic values of modern marine carbonate deposits range from +2.0 to −2.0 per mil (PDB); initial carbon-isotopic values extend from 0 to +4.5 per mil (PDB; Milliman, 1974). These values are a function of the isotopic composition of the water and the temperature at which the carbonate deposits were precipitated. Thus, oxygen- and carbon-isotopic ratios have been used in different studies to determine water composition

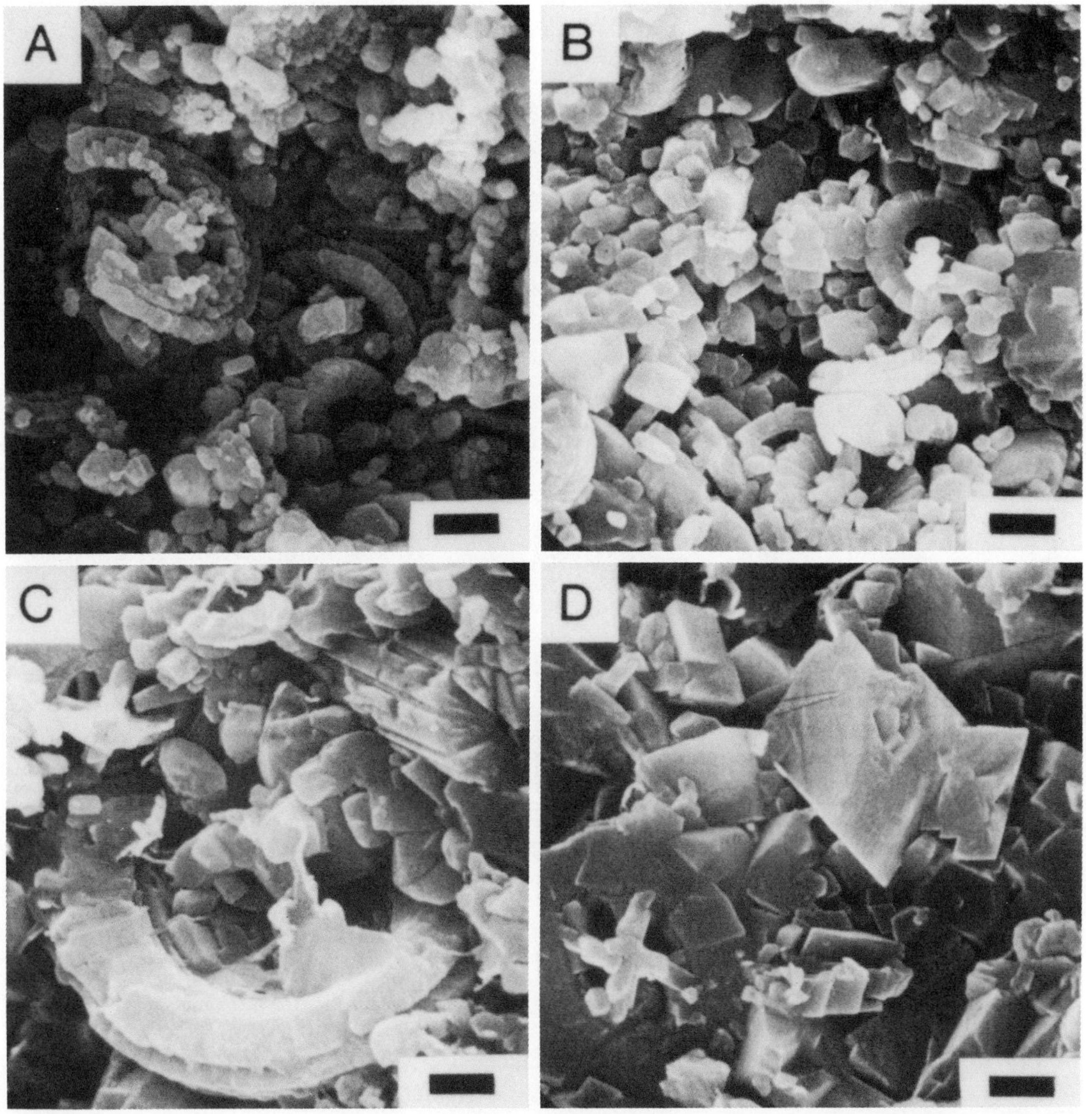

FIG. 8—Sequence of scanning-electron photomicrographs showing progressive cementation during burial. All samples are Upper Cretaceous chalk.

A, Sample from Conoco/NCB 49/21-3 well in British North Sea. Depth 433 m; porosity about 40%. Little interlocking of crystals; most grains are not euhedral and many have rounded outlines or multiple crystal faces. Scale bar is 2μ.

B, Sample from NAM/CLMS K/8-1 well in Dutch North Sea. Depth 1,760 m; porosity is about 25%. Some overgrowth of coccolith fragments and start of crystalline, interlocking texture. Scale bar is 2μ.

C, Sample from Conoco/Gulf/NCB 22/11-2 well in British North Sea. Depth 2,697 m; porosity about 10%. Very considerable recrystallization with dissolution and cementation of coccolith plates. Scale bar is 1μ.

D, Sample from Shell-UK 30/19-1 well in British North Sea. Depth 3,361 m; porosity about 5%. Completely interlocking, coarse, euhedral calcite fabric with coccoliths occasionally still visible. Scale bar is 1μ.

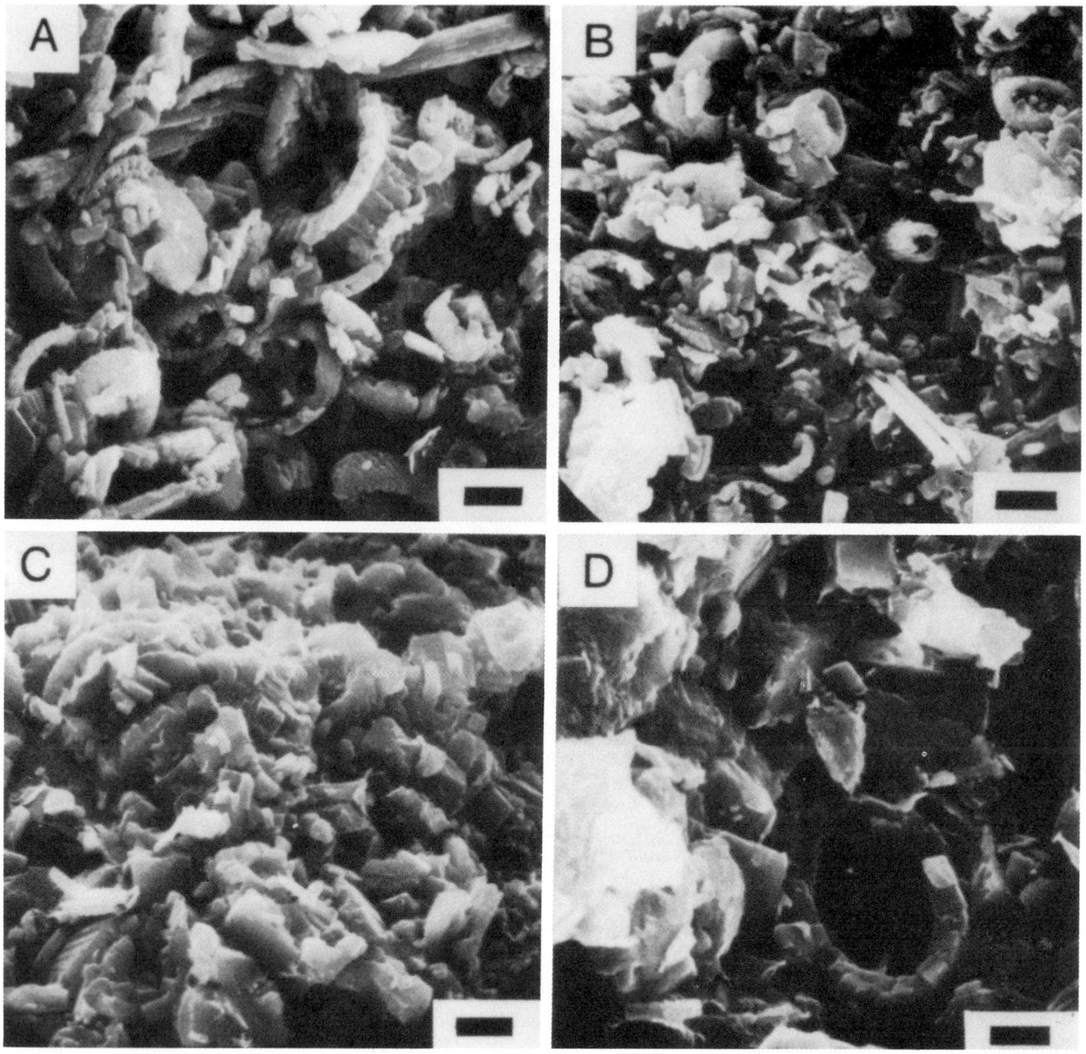

FIG. 9—Sequence of scanning-electron photomicrographs showing progressive cementation with increasing amounts of former overburden. All samples are from Upper Cretaceous Fort Hays Limestone Member of Niobrara Limestone.

A, Sample from along Smoky Hill River, southwest of WaKeeney, Trego County, Kansas; porosity about 40%. Abundant coccoliths and rhabdoliths with only slight traces of dissolution or cementation. Minimum former overburden about 245 m. Scale bar is 2μ.

B, Sample from same locality as 9A; porosity about 37%. Coccoliths well preserved, but some coarse, euhedral calcite overgrowths are within matrix. Scale bar is 2μ.

C, Sample from quarry west of Las Animas, Bent County, Colorado; porosity about 15 to 20%. Extensive interlocking of euhedral calcite crystals. Coccoliths still recognizable but often have overgrowths. Minimum former overburden about 760 m. Scale bar is 2μ.

D, Sample from quarry along Arkansas River, west of Pueblo, Pueblo County, Colorado; porosity about 10%; minimum former overburden about 1,525 m. Virtually complete obliteration of matrix pore space by interlocking calcite crystals. Coccoliths recognizable but strongly corroded and overgrown. Scale bar is 1.3μ.

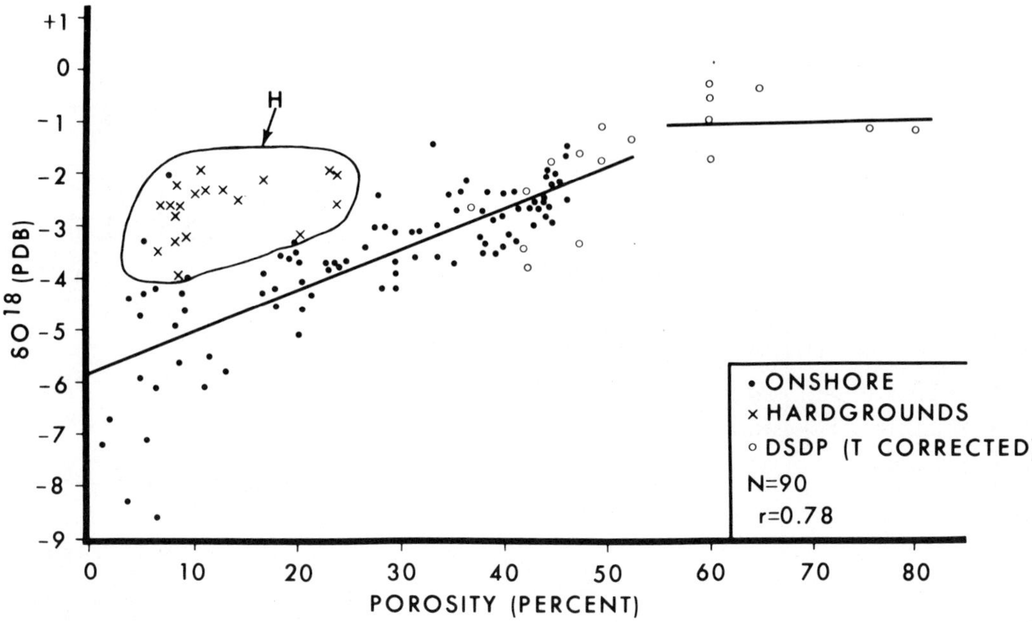

FIG. 10—Plot of porosity versus oxygen-isotopic ratios (relative to PDB standard) for chalks. Figure 11 gives schematic explanation.

when paleotemperatures are known (e.g., Rothe et al, 1974); or, more commonly, to determine paleotemperatures when water compositions are known (e.g., Savin and Douglas, 1973). Deposition from fresh water or at high temperature will yield limestones that have significantly negative oxygen-isotopic values, whereas freshwater limestones also may have strongly negative carbon-isotopic values (Gross, 1964; Milliman, 1974).

This same problem in differentiating between water composition and water temperature is present in diagenetic studies as well. Materials which do not recrystallize or reequilibrate still may show original depositional isotopic values, as shown in the studies of belemnite isotopic paleotemperatures (e.g., Bowen, 1966). However, because most limestones undergo extensive diagenetic transformation (by alteration of unstable minerals, solution transfer, or cement introduction), there can be wholesale alteration of original isotopic values during diagenesis. The newly precipitated carbonate material will be in equilibrium with the pore-fluid compositions and subsurface temperatures at the time of cement formation. Thus, in many limestones, isotopic values can give more of a clue to the diagenetic history than to original depositional environments (e.g, Hudson, 1975).

A second problem in the interpretation of isotopic values in carbonate rocks may be the slow, solid-state isotopic reequilibration of carbonate minerals with carbon dioxide in the pore fluids. Anderson (1969) has calculated the likely rates of reaction, which are a function of crystal size and temperature. Because chalks are extremely fine grained, reequilibration indeed may be important especially in Mesozoic units which have been buried deeply.

The results of the isotopic and trace-element analyses of chalks during this project are shown in Table 1. It gives average values for different localities and stratigraphic units; the individual analyses will be published in later papers. The averaged data eliminate some of the bed-to-bed fluctuations introduced by variations in primary paleotemperatures and focuses attention on large, diagenetically induced variations from section to section.

Figure 10 gives the relations between porosity and oxygen-isotopic values in some typical chalks, and Figure 11 shows a schematic explanation of these relations. The DSDP data are from Tertiary chalks recovered in hole 116 (Leg 12) from the Hatton-Rockall basin of the North Atlantic. All other data are from bulk analyses of Upper Cretaceous chalks of England and Northern Ireland. The DSDP data have been plotted as being 2.0 per mil more negative than the actual analyses indicated in order to compensate for the roughly 10°C colder depositional paleotemperatures of these Tertiary chalks relative to the Upper Cretaceous onshore samples. The exactness of this correction is not important here, for it is only

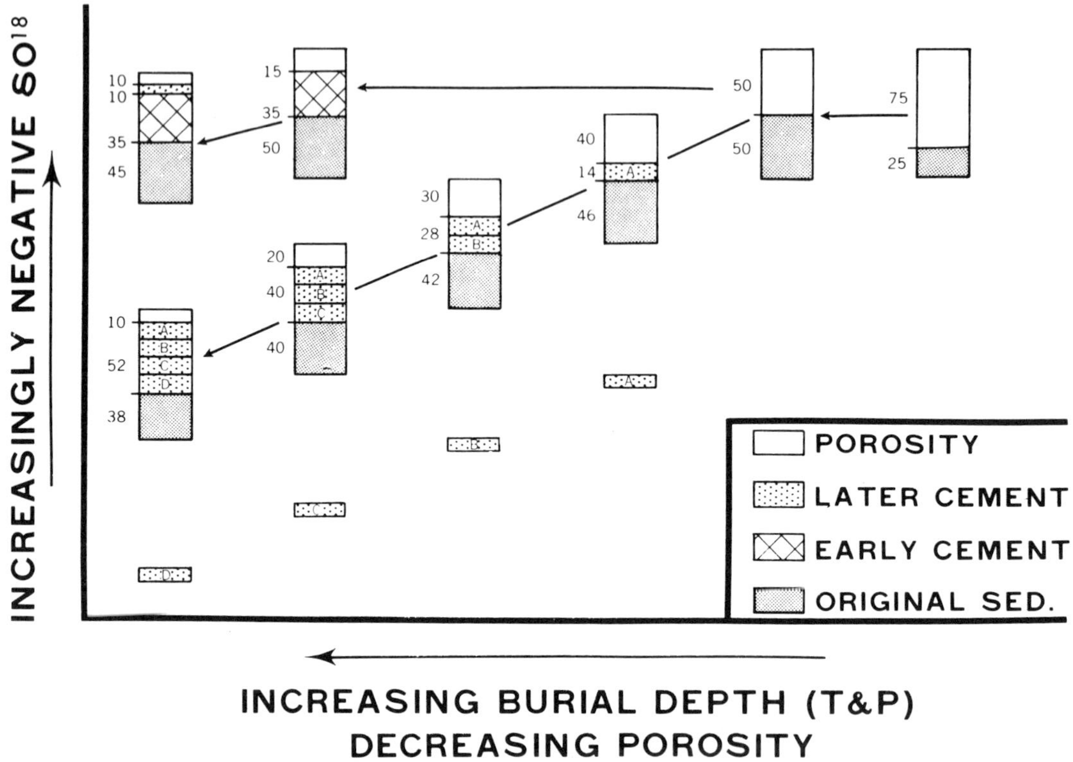

FIG. 11—Schematic diagram of relations between porosity loss in chalks and shifting oxygen isotopic values. Numbers on diagram are in percent. Lettered boxes represent generations of cement.

done to show that there is a roughly continuous spectrum of isotopic values during the diagenetic sequence.

Even a cursory examination of Figure 10 shows that, as porosity decreases, oxygen-isotopic values become more negative. Carbon-isotopic values, not plotted, show no variation during this porosity loss. In detail, chalks having as much as 80% initial porosity dewater to about 50% porosity with no significant shift in isotopic values. Below the 50% porosity level, porosity loss is accomplished mainly by dissolution and reprecipitation of carbonate material. The observed shift in isotopic values is a result of the dissolution of coccoliths and Foraminifera and reprecipitation of this carbonate material from pore fluids at elevated subsurface temperatures. Progressive reequilibration of matrix carbonate material also may be important.

These relations are generalized in Figure 11 which shows a series of hypothetical stages in the burial of a chalk. At each progressive step there is less original sediment and more cement. The cements precipitated at greater depths and temperatures are progressively more negative, yielding a moving-average isotopic analysis of the bulk sediment which becomes more negative with increased overburden and cementation. In reality, of course, this process is continuous and not broken into such discrete steps.

Because the time- and depth-related isotopic compositional fluctuations of pore waters in chalks cannot be determined independently, these relations are impossible to quantify. However, as the carbonate rocks are getting isotopically lighter, the waters must be getting isotopically heavier with respect to oxygen; or other minerals, such as clays, must be involved in the reactions (Lawrence, 1973). If the process is to continue, there must be isotopic exchange either with silicate minerals or with pore fluids outside the system; perhaps with fluids expelled from underlying units. In this system the carbon-isotopic values do not shift measurably because, unlike oxygen, there is very little noncarbonate carbon in the water-sediment system with which exchange can take place (e.g., Hudson, 1975).

The only exception to the rule that progressive lithification involves a shift in oxygen-isotopic values is in the case of hardgrounds. Figures 10

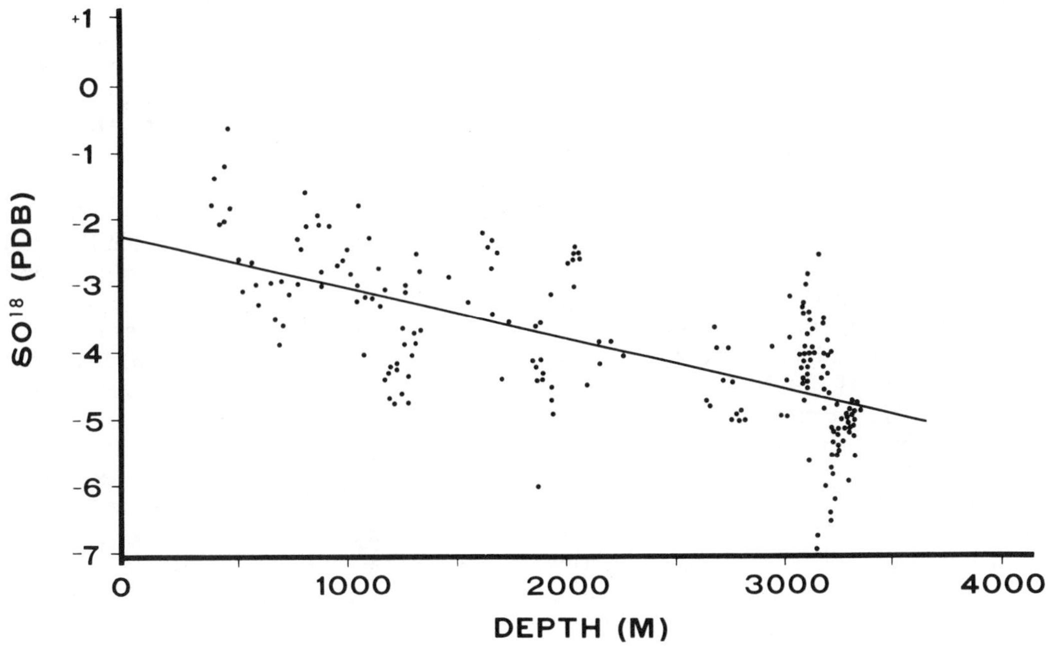

FIG. 12—Plot of oxygen-isotopic ratios of North Sea chalk samples versus burial depth.

and 11 show that hardgrounds, which are cemented at the seafloor, have most of their porosity occluded by low-temperature cement which has relatively heavy (normal marine) oxygen-isotopic values. Because most of the porosity is destroyed early in the diagenetic history, subsequent diagenetic effects are small and only minor late diagenetic alteration is present in hardgrounds. Indeed, in sections that have been buried deeply it is quite simple to tell those beds that were cemented at the seafloor from those cemented later, on the basis of isotopic values (Scholle and Kennedy, 1974). This argues against widespread solid-state isotopic reequilibration of matrix carbonate minerals at elevated subsurface temperatures.

Figure 12 shows the relations between present burial depth and oxygen-isotopic values from North Sea Upper Cretaceous and lower Paleocene chalks. Because there were good correlations between burial depth and porosity (Fig. 5) and between porosity and oxygen-isotopic values (Fig. 10), it is not surprising that there is also a consistent relation between depth of burial and oxygen-isotopic values, in spite of the fact that samples from different areas were used. Similar relations are present in chalks from the Gulf Coast and Western Interior but are not presented here.

The considerable scatter of data points in Figure 12 probably is attributable to several factors. First, samples were taken from chalks of Cenomanian to Danian age. During that time interval considerable paleotemperature fluctuations are thought to have taken place (e.g., Bowen, 1966; Saito and Van Donk, 1974; Douglas and Savin, 1975), and these would leave an imprint of variation in the oxygen-isotopic values. Second, different localities may have had different amounts of water input, different thermal gradients, or different subsurface-pressure regimes, all of which could produce variations in isotopic values. Finally, there are considerable bed-to-bed variations in degree of lithification related to primary sedimentary factors—these too could yield variations in final oxygen-isotopic values. Yet, in spite of the data scatter, there is a good enough relation so that isotopic information, combined with petrographic and petrophysical analyses, can be used to reconstruct maximum burial depths for chalks on a regional or local basis (Hancock and Scholle, 1975; Scholle, 1976).

Strontium and magnesium trace-element data, shown in Table 1, are much more difficult to interpret. In some areas, such as England and Northern Ireland, there appears to be a loss of strontium and relatively constant magnesium with increasing burial diagenesis. Yet, in the North Sea basin, the most deeply buried chalks have the highest values of both strontium and magnesium. In all probability, this reflects variations in subsurface patterns of water flow, with relative stagnation in the Central graben of the

North Sea leading to retention of high initial amounts of strontium and magnesium. In the Austin Chalk of Texas, Cloud (1975) showed the lowest strontium values along the San Marcos arch, an area in which one might expect the strongest freshwater flushing. Yet, the oxygen-isotopic values of these chalks show no signs of isotopic reequilibration with fresh waters, leaving a major problem of interpretation which has not been solved.

Many studies have been made of the variations of stable isotopes and trace elements in DSDP samples. Anderson and Schneidermann (1973), Coplen and Schlanger (1973), Cook (1974), and Matter et al (1975) gave extensive data and varied interpretations. In general, the same trends of increasingly negative oxygen-isotopic values as a function of increased burial depth and decreased porosity are present in DSDP cores as were found in the onshore and North Sea samples described earlier. In trace-element work, Thompson (1972), Matter et al (1975), and Sayles and Manheim (1975) showed a progressive decrease in the strontium content of chalks with deeper burial. Cook (1974), on the other hand, found higher strontium values in the lower part of the section than in the upper part.

An interesting piece of additional information from DSDP work comes from studies of the isotopic values of individual Foraminifera and coccoliths done by Douglas and Savin (1975). They showed that the isotopic values of nannofossils, especially discoasters, are significantly altered by 200 m of burial, apparently through the addition of overgrowth cement. Foraminifera, which generally act as donors of carbonate material, are affected less and retain original isotopic values.

In summary, the geologic and geochemical data presented confirm the following ideas: (1) chalk cements are derived internally by selective dissolution; (2) cement is reprecipitated at selected donor sites; (3) the solution-transfer process, rather than mechanical compaction, is responsible for most porosity loss in chalks after initial dewatering; (4) solution-transfer acts continuously from just after deposition to deepest burial depths; and (5) the rates of cementation in chalks are a function of burial depth and water chemistry, and those rates are predictable.

HYDROCARBON PRODUCTION FROM CHALKS

As mentioned earlier, hydrocarbon production now has been established in chalks from a wide variety of areas. In the North Sea, the Ekofisk, Eldfisk, Torfelt, SE Torfelt, Edda, Albuskjell, Dan, and other fields have major production from Maestrichtian and Danian chalks (e.g., Harper and Shaw, 1974; Rickards, 1974; Byrd, 1975; Childs and Reed, 1975). In the Middle East (Abu Dhabi), the coarse chalks of the Neocomian Thamama Group of the Zakum field, and the Aptian Shu'aiba Formation of the Bu Hasa field have yielded hydrocarbons (Illing et al, 1967). However, it has not yet been shown that these are true nannofossil chalks. Minato et al (1967) did illustrate that some of the units in the Khafji oil field of the Neutral Zone were nannofossil bearing. In the Gulf Coast, oil and gas production from chalks is found from South Texas to Alabama. The Austin, Annona, Selma, Saratoga, and "Monroe" Formations all produce from chalky facies (e.g., Doyle, 1955; Shreveport Geol. Soc., 1968; Long, 1976). On the Scotian Shelf of Canada significant shows of oil have been noted in the Wyandot Chalk, especially in the N-50 and E-48 wells. In Kansas and Colorado, recent gas production has been established in the Smoky Hill Chalk Member of the Niobrara Limestone in several wells (J. A. Lockridge, written commun., 1975).

Production from chalks ranges from prolific (as in the North Sea where rates of 10,000 BOPD/well are common) to sparse (as in the Austin Chalk where production is generally 30 to 370 BOPD/well initially, and the maximum drops to 60 or 70 BOPD/well within 6 months; Long, 1976). Production depths range from less than 200 m to more than 4,500 m, with most of the extensive North Sea production being from about 3,050 m. Reservoir rock-matrix porosities range from as high as 45% in the North Sea to 5% or less in the Gulf Coast.

The previous discussion of chalk diagenesis pointed out that virtually all primary porosity should be eliminated before about 3,000 m of burial. Similar observations have been made on Middle Eastern chalks by J. L. Wilson (written commun., 1976). How, then, is it possible to have matrix porosities averaging about 30% in thick chalk sections in the North Sea at 3,000 m or greater depth? This anomaly is shown graphically in Figure 5. There clearly must be mechanisms of porosity retention or enhancement which can affect the reservoir properties of chalks. These will be discussed from a regional point of view for the North Sea and the Gulf Coast areas.

North Sea

Overall reviews of the geologic framework of the North Sea have been provided by Kent (1975) and Ziegler (1975). The geologic history is controlled largely by a major structural feature, the Central graben (shown on Fig. 13), which is probably a failed rift arm of the North Atlantic. The

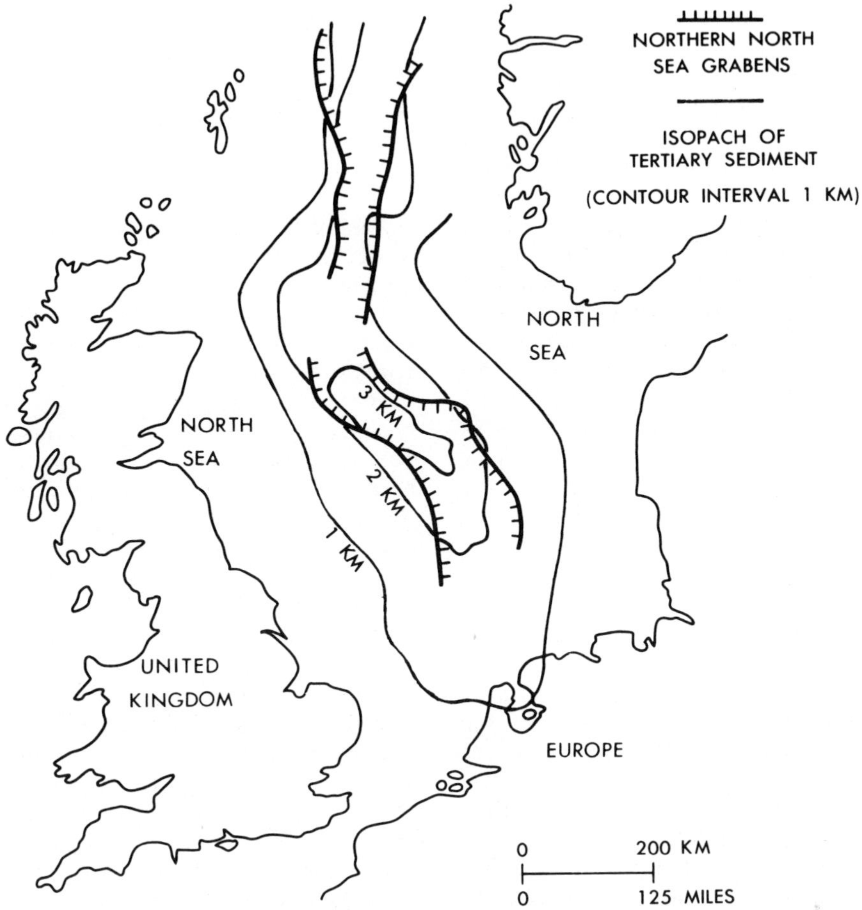

FIG. 13—Index map of North Sea area showing isopachs of Tertiary sediment and location of central rift.

entire North Sea basin, and especially the Central graben, has undergone major subsidence since at least the Permian. Chalks were deposited across the entire North Sea basin during the Late Cretaceous; especially great thicknesses accumulated in the Central graben, the Anglo-Dutch basin, and various smaller inversion basins. An isopach map of the distribution of the Upper Cretaceous in the North Sea was given in Hancock and Scholle (1975). Figure 13 of this paper shows an isopach of Tertiary sediments in the North Sea. Again, subsidence and deposition were concentrated in areas within the central rift.

Essentially all oil production from chalk facies in the North Sea is within the Central graben. The chalk section is between 400 and 1,400 m thick in this area and is buried beneath 2,000 to 3,000 m of overburden. Reservoir characteristics of the chalks in the Central graben oil fields were given in World Oil (1971), Owen (1972), Harper and Shaw (1974), Rickards (1974), Scholle (1974), Byrd (1975), Childs and Reed (1975), Hancock and Scholle (1975), and Mapstone (1975).

Anomalously high porosity values (relative to most chalks) are present in chalks throughout the Central graben. In Figure 5, data from the Shell/Esso 30/19-1 well (dry hole) and the Phillips Ekofisk 2/4-2X well (oil producer) are shown. In both wells, and in other wells not plotted, porosity is highest near the top of the chalk section. There is considerable bed-to-bed variation but in the Danian chalks porosities of as much as 40% are quite common and values of 25 to 35% are the average. A significant decrease in porosity (to less than 5%) is present in some wells (e.g., Owen, 1972) in association with a major regional disconformity at the Cretaceous Tertiary boundary (Rickards, 1974). In the upper Maestrichtian, porosities again are in the range of 20 to 30%, decreasing down section. In even the most porous Danian chalks, matrix permeabilities do not exceed 3 md and generally are less than 1 md.

Scanning-electron microscopy (Harper and Shaw, 1974; Rickards, 1974; Scholle, 1974; Mapstone, 1975) has shown that most of the pore space in the North Sea chalks is preserved primary porosity. There is some small-scale dissolution (as in most chalks) and cementation at grain contacts ("spot welding") is common (Rickards, 1974; Mapstone, 1975). Yet the major part of the pore space remains uncemented, and these chalks, despite more than 3,000 m of burial look very similar to onshore chalks which have been buried to less than 1,000 m.

Isotopic and trace-element analyses also indicate extensive preservation of primary chemical composition. Strontium content of Ekofisk chalks averages 1,230 ppm, about that expected for unaltered chalks. The oxygen-isotopic values in the most porous, oil-saturated Danian chalks average −3.6 per mil (PDB), again showing far less alteration than one would expect for samples so deeply buried (e.g., Fig. 12).

The unusually great primary-porosity preservation has been explained by Mapstone (1975) as being due to the presence of abundant thoracosphaerids in the Danian section as well as by the development of a framework of "spot welds" at grain contacts which resisted further compaction. Although the concept of selective overgrowth (or lack of overgrowth) on different types or species of nannofossils is valid, thoracosphaerids do not seem to be sufficiently common in most beds to explain the remarkably high porosities. In the Maestrichtian section, particularly, thoracosphaerids are absent but porosity is still high. Likewise, the "spot-welding" concept is valid but does not explain particularly the anomalously high porosity in the North Sea chalks. "Spot welds" have been recognized in chalks from many areas in which anomalous porosity preservation did not take place. Clearly, a more general explanation is needed.

The high porosities in chalks of the Central graben correlate most directly with an area of overpressuring. Rickards (1974) and Byrd (1975) published pressure-depth plots for Ekofisk wells based on sonic-log travel time. These figures showed that the thick Tertiary shales overlying the chalk are strongly overpressured. The chalk section also shows overpressuring, especially in its upper parts. The anomalous pressures presumably are due to low permeabilities in the thick shales and chalks of the Central graben. Lateral pore-fluid escape may be retarded further by the fault boundaries of the graben. T. D. Eames (oral commun., 1974) has pointed out that such overpressures are common in wells throughout the Central graben area, and are restricted to the graben itself. Thermal gradients, on the other hand, are high over much of the northern part of the North Sea and show a considerably weaker relation with the Central graben (Evans and Coleman, 1974; Cooper et al, 1975).

The direct relation between overpressuring and anomalously high porosity in chalks (also noted by Harper and Shaw, 1974) can be explained most readily by considering, as discussed earlier, that porosity loss in chalks is produced by compaction, both mechanical and chemical. In overpressured areas, as hydrostatic pressure begins to approach lithostatic pressure, the differential or effective stresses at grain contacts are reduced greatly. Thus, with elevated hydrostatic pressures, the driving mechanism for both mechanical and chemical compaction is reduced sharply and primary porosity can be preserved to very considerable depths of burial. Mimran (in press) even suggested the process of "confining pressure solution" which produces rounded crystal surfaces such as those shown in Ekofisk chalks by Scholle (1974, p. 197).

In the Ekofisk-Torfelt area, Harper and Shaw (1974) reported pore pressures of 7,100 psi at 3,050 m. Normal pore pressure at that depth would be about 4,300 psi, and lithostatic load would be about 9,000 psi. Thus, a major part of this lithostatic load is supported by the fluid itself. Net-lithostatic pressure is reduced to about 1,900 psi which would correspond to a depth of burial of about 3,500 ft (1,067 m) under normal conditions.

Although the reduction of differential stresses by high pore-fluid pressures probably plays the major role in porosity preservation in the North Sea, other factors also have combined to enhance production. Oil saturation of 70 to 100% reported by Harper and Shaw (1974) for the Maestrichtian chalk at Torfelt implies possible preservation of porosity due to the exclusion of water. If oil entered the section early in its burial history, the presence of the oil itself may have prevented further cementation (Hancock and Scholle, 1975). Extensive fracturing has been reported in chalk reservoirs of the North Sea (Harper and Shaw, 1974; Byrd, 1975) which greatly improves the overall permeability of the chalk. Effective permeabilities (calculated from oil-flow rates) of 12 md have been determined for Ekofisk wells (World Oil, 1971); these values are an order of magnitude greater than the matrix permeabilities measured on core plugs, and presumably largely reflect fracturing. However, elevated pore-fluid pressures also can increase effective permeabilities (Zoback and Byerlee, 1975). Finally, the very low viscosity (36° gravity) of the oil in many of the Ekofisk area fields and the high gas-oil ratio (GOR of 1,000 to 2,500 were reported for the Ekofisk area fields by Harper and Shaw, 1974), and thick pay section (as much as 250 m) all help to improve production.

The time of oil entry into these reservoirs is still unclear. Rickards (1974) reported analyses in the Ekofisk area which indicate a Paleocene source for the oil. Oil may have entered updip but downsection from the flanks of the salt-related structures which localize Ekofisk area production. The preservation of primary porosity in this area indicates either that oil migration was very early or that overpressuring was initiated before extensive burial, and that oil may have bled in from even more strongly overpressured Paleocene shales later in the burial history. Other likely sources for the oil at Ekofisk include the underlying Lower Cretaceous and Jurassic marine sections or even the Upper Cretaceous chalk itself.

Gulf Coast

Oil production from chalks in the Gulf Coast of the United States is far less prolific than in the North Sea. However, several extensive exploration plays are currently underway in the Austin Chalk of Texas (Long, 1976). To date, more than 25 million bbl of oil and significant quantities of gas have been produced from the Austin Chalk in Texas.

Relatively few data are available on the petrophysical characteristics of the Austin and other chalk units in the Gulf Coast area. Cloud (1975) summarized the petrographic and petrophysical features of the Austin Chalk on outcrop. He noted a decrease in porosity from central (the Austin area) to southwestern Texas (Del Rio and Langtry areas). Cloud postulated significantly greater thicknesses of overburden, since removed, in the Del Rio–Langtry region where porosities had been reduced to less than 10% (compared with over 25% near Austin). The same sort of relations can be seen by looking at a series of wells in which the Austin Chalk is buried to progressively greater depths. By 2,000 m of burial the average bulk-matrix porosity of the Austin Chalk is invariably less than 10%. However, greater tectonic stresses and/or higher paleogeothermal gradients in the Mexican border area also may have played a major role in the porosity reduction there.

Most current Austin Chalk exploration plays are in the 1,500 to 2,150-m depth range. As predicted by the data in Figure 5, the porosities encountered range from 6 to 8% with permeabilities of 0.0 to 0.4 md (Doyle, 1955). Almost all production is from fracture porosity; structural position apparently makes little difference in production except for the fact that fracturing is stronger on structural crests. Artificial fracturing of the reservoirs improves production because it connects more oil-filled natural fractures to the well bore. However (unlike the North Sea case), there is either little or very slow oil recharge from the rock matrix, and production drops off rapidly after high initial-production rates. Ultimate recovery per well ranges from 18,000 to 120,000 bbl in the Pearsall field (Long, 1976).

At shallower burial depths, production from the Austin Chalk (or other units such as the Saratoga, Annona, or "Monroe" formations) can be somewhat more encouraging. For example, along the Luling-Mexia fault trend in Central Texas porosities of as much as 24% are present along with as much as 40% oil saturation of matrix pore space (Doyle, 1955). In general, however, exploration for chalk reservoirs is difficult in the Gulf Coast because it relies so completely on the presence of intense fracturing. Overpressures have not been important in Gulf Coast chalk fields, at least at the depths drilled. Thus, only in updip areas which have undergone less than 1,000 m of burial are significant matrix porosity values likely. Exploration should also be directed toward areas of possible overpressures within the chalk section.

SUMMARY AND CONCLUSIONS

Chalks are rock units of major importance in deeper water settings. They have been the predominant carbonate rock type during the past 150 m.y. and are taking on increased significance to the oil industry as exploration moves into deeper water settings. Because of their predictable faunal and mineralogic makeup and their broad, uniform facies, the diagenetic patterns of chalks are far more predictable than are the patterns of alteration of shallow-water carbonate sediments. The following points have been made in this paper.

1. Chalks are subject to diagenetic loss of porosity either through early seafloor cementation or through later burial diagenesis. Aging of sediments or freshwater exposure, alone, do not lead to significant cementation of chalks.

2. The rates of cementation as a function of burial depths are predictable but can vary with pore-water chemistry. The presence of marine-pore fluids (or other fluids containing moderate amounts of magnesium in solution) will retard cement generation. Because porosity can be predicted, other properties such as density and sonic velocity also can be determined and may aid in exploration geophysics. However, although gross

porosity of chalk sections is predictable, there may be considerable bed-to-bed variation due to primary fluctuations of grain size, clay content, or faunal composition.

3. Both mechanical and chemical (solution transfer) cementation are important mechanisms for the reduction of porosity in chalks. In most cases, mechanical compaction predominates during the early (dewatering) phase, but pressure solution and reprecipitation are more important during later diagenesis.

4. Cement is derived internally within chalk sections through dissolution of specific fossil groups or by larger scale dissolution along solution seams and stylolites. Thus, section thickness is reduced as porosity of the remaining section is decreased. This autocannibalization reduces the need for input of large volumes of cement-bearing fluids.

5. Isotopic analysis can be used to confirm the patterns of chalk diagenesis and, in some cases, even can be used as a measure of the maximum burial depth to which chalks have been subjected. Indeed, bulk oxygen-isotopic analyses in chalks commonly are a better indicator of diagenetic history than of primary depositional environments and temperatures. Carbon isotopes, on the other hand, usually are not subject to strong diagenetic alteration and may reveal information about primary depositional conditions.

6. Overpressuring reduces differential stresses within chalks and thereby retards both mechanical and chemical compaction and preserves primary porosity. Early introduction of oil into the rock also may preserve initial porosity through the exclusion of water.

7. Chalks can form significant hydrocarbon reserves if one or more of the following are present: (a) lack of burial (chalks which have not been deeply buried may have 50% or more porosity and permeabilities of 1 to 15 md although it may be difficult to generate hydrocarbons at such shallow depths); (b) burial with overpressuring or early oil emplacement (the prolific production of oil from North Sea chalks falls into this class); (c) burial, cementation, and fracturing (in these cases, as with the Austin Chalk in Texas, production is largely from fractures with little or slow contribution of oil from the rock matrix).

8. Chalks can have a significant role in oil trapping by acting as impermeable cap rocks if they have been buried to depths of more than 2,000 m.

9. The principles and patterns of diagenesis in chalks are applicable, to a considerable degree, to other limestones as well. Shallow-water carbonate sediments, with their unstable mineralogies, are subject to early freshwater alteration. However, they commonly retain considerable porosity after early diagenetic alteration and, therefore, are subject to the same types of subsurface processes as are chalks. Work on initially stable lithologies (such as chalks) may help to clarify the patterns of late diagenesis in other types of limestones as well.

REFERENCES CITED

Adelseck, C. G., Jr., G. W. Geehan, and P. H. Roth, 1973, Experimental evidence for the selective dissolution and overgrowth of calcareous nannofossils during diagenesis: Geol. Soc. America Bull., v. 84, p. 2755-2762.

Anderson, T. F., 1969, Self-diffusion of carbon and oxygen in calcite by isotope exchange with carbon dioxide: Jour. Geophys. Research, v. 74, p. 3918-3932.

——— and N. Schneidermann, 1973, Stable isotope relationships in pelagic limestones from the central Caribbean: leg 15, Deep Sea Drilling Project: Initial Rept. Deep Sea Drilling Project, v. 15, p. 795-803.

Arthur, M. A., and W. M. Roggenthen, 1976, Primary bedding rhythms in pelagic calcareous sediments (abs.): AAPG Bull., v. 60, p. 646.

Atkinson, T. C., and D. I. Smith, 1974, Rapid groundwater flow in fissures in the chalk: an example from South Hampshire: Engineering Geology—Internat. Jour., v. 7, p. 197-205.

Bathurst, R. G. C., 1970, Problems of lithification in carbonate muds: Geologists' Assoc. Proc., v. 81, p. 429-440.

——— 1971, Carbonate sediments and their diagenesis: New York, Elsevier, 620 p.

Beall, A. O., and A. G. Fischer, 1969, Sedimentology: Initial Rept. Deep Sea Drilling Project, v. 1, p. 521-593.

Berger, W. H., 1973, Deep-sea carbonates: evidence for a coccolith lysocline: Deep-Sea Research, v. 20, p. 917-921.

Bignot, G., and S. Defretin, 1967, Remarques sur les spongiaires siliceux de la craie de la région de Dieppe (S.-M): Soc. Géol. Normandie Bull., v. 57, p. 9-14.

Black, M., 1953, The constitution of the chalk: Geol. Soc. London Proc., no. 1499, p. 81-86.

——— 1971, The systematics of coccoliths in relation to the paleontological record, *in* B. M. Funnell and W. R. Riedel, eds., The micropalaeontology of oceans: Cambridge, England, Cambridge Univ. Press , p. 611-620.

——— and B. Barnes, 1959, The structure of coccoliths from the English Chalk: Geol. Mag., v. 96, p. 321-328.

Bond, G., 1976, Evidence for continental subsidence in North America during the Late Cretaceous global submergence: Geology, v. 4, p. 557-560.

Borst, R. L., 1972, Authigenic kaolinite crystals within microfossils of the Danian limestone, North Sea: Internat. Clay Conf., Madrid, Kaolin Symp. Proc., p. 41-48.

Bosellini, A., and E. L. Winterer, 1975, Pelagic limestone and radiolarite of the Tethyan Mesozoic: a genetic model: Geology, v. 3, p. 279-282.

Bowen, R., 1966, Paleotemperature analysis: New York, Elsevier, 265 p.
Bromley, R. G., 1965, Studies in the lithology and conditions of sedimentation of the chalk rock and comparable horizons: PhD thesis, Univ. London, 355 p.
——— 1967, Some observations on burrows of thalassinidean crustacea in chalk hardgrounds: Geol. Soc. London Quart. Jour., v. 123, p. 157-182.
——— 1968, Burrows and boring in hardgrounds: Geol. Soc. Denmark Bull., v. 18, p. 247-250.
——— 1975, Trace fossils at omission surfaces, *in* R. W. Frey, ed., The study of trace fossils: New York, Springer-Verlag, p. 399-428.
——— M. G. Schulz, and N. B. Peake, 1975, Paramoudras: giant flints, long burrows and the early diagenesis of chalks: Kgl. Danske Vidensk. Selsk. Biol. Skr., v. 20, 31 p.
Brown, G., et al, 1969, Partial silicification of chalk fossils from the Chilterns: Geol. Mag., v. 106, p. 583-586.
Bukry, D., 1969, Upper Cretaceous coccoliths from Texas and Europe: Kansas Univ. Paleont. Contrib., Art. 51 (Protista 2), 79 p.
——— 1974, Coccoliths as paleosalinity indicators—evidence from Black Sea, *in* The Black Sea—geology, chemistry, and biology: AAPG Mem. 20, p. 353-363.
Burnaby, T. P., 1961, The paleoecology of the Foraminifera of the Chalk Marl: Palaeontology, v. 4, p. 599-608.
Byrd, W. D., 1975, Geology of the Ekofisk field, offshore Norway, *in* A. W. Woodland, ed., Petroleum and the continental shelf of north-west Europe, v. 1: New York, John Wiley & Sons, p. 439-445.
Carter, P. G., and D. J. Mallard, 1974, A study of the strength, compressibility, and density trends within the chalk of south east England: Engineering Geology Quart. Jour., v. 7, p. 43-55.
Cayeux, L., 1935, Les roches sédimentaires de France; roches carbonatées (calcaires et dolomies): Paris, Masson, 463 p. (Sedimentary rocks; limestones and dolomites; translated and updated by A. V. Carozzi, 1970, Hafner Pub. Co., Darien, Conn., 506 p.)
Childs, F. B., and P. E. C. Reed, 1975, Geology of the Dan field and the Danish North Sea, *in* A. W. Woodland, ed., Petroleum and continental shelf of north-west Europe, v. 1: New York, John Wiley & Sons, p. 429-438.
Cloud, K. W., 1975, The diagenesis of the Austin Chalk: Master's thesis, Univ. Texas—Dallas, 71 p.
Cook, P. J., 1974, Geochemistry and diagenesis of interstitial fluids and associated oozes, Deep Sea Drilling Project, leg 27, site 262, Timor Trough: Initial Rept. Deep Sea Drilling Project, v. 27, p. 463-480.
Cooper, B. S., et al, 1975, Palaeotemperatures in the northern North Sea basin, *in* A. W. Woodland, ed., Petroleum and the continental shelf of north-west Europe, v. 1: New York, John Wiley & Sons, p. 487-492.
Coplen, T. B., and S. O. Schlanger, 1973, Oxygen and carbon isotope studies of carbonate sediments from site 167, Magellan Rise, leg 17: Initial Rept. Deep Sea Drilling Project, v. 17, chapt. 15, p. 505-509.
Craig, H., 1957, Isotopic standards for carbon and oxygen and correction factors for mass-spectrometric analysis of carbon dioxide: Geochim. et Cosmochim. Acta, v. 12, p. 133-149.
Davies, T. A., and P. R. Supko, 1973, Oceanic sediments and their diagenesis: some examples from deep-sea drilling: Jour. Sed. Petrology, v. 43, p. 381-390.
Dessenne, J. L., and P. Duffaut, 1970, Les propriétés rhéologiques de la craie et leur influence sur le percement de galeries: Houille Blanche, v. 25, p. 477-488.
Douglas, R. G., and S. M. Savin, 1975, Oxygen and carbon isotope analyses of Tertiary and Cretaceous microfossils from Shatsky Rise and other sites in the North Pacific Ocean: Initial Rept. Deep Sea Drilling Project, v. 32, p. 509-520.
Doyle, W. M., Jr., 1955, Production and reservoir characteristics of the Austin Chalk in South Texas: Gulf Coast Assoc. Geol. Socs. Trans., v. 5, p. 3-10.
Dunnington, H. V., 1967, Aspects of diagenesis and shape change in stylolitic limestone reservoirs: 7th World Petroleum Cong., Mexico City, Proc., v. 2, p. 339-352.
Durney, D. W., 1972, Solution-transfer, an important geological deformation mechanism: Nature, v. 235, p. 315-317.
Edmunds, W. M., P. E. R. Lovelock, and D. A. Gray, 1973, Interstitial water chemistry and aquifer properties in the Upper and Middle Chalk of Berkshire, England: Jour. Hydrology, v. 19, p. 21-31.
Eicher, D. L., 1969, Paleobathymetry of Cretaceous Greenhorn sea in eastern Colorado: AAPG Bull., v. 53, p. 1075-1090.
Evans, T. R., and N. C. Coleman, 1974, North Sea geothermal gradients: Nature, v. 247, p. 28-30.
Fischer, A. G., S. Honjo, and R. E. Garrison, 1967, Electron micrographs of limestones and their nannofossils: Princeton, New Jersey, Princeton Univ. Press, 141 p.
Foster, S. S. D., 1975, The chalk groundwater tritium anomaly: Jour. Hydrology, v. 19, p. 21-31.
Frey, R. W., 1972, Paleoecology and depositional environment of Fort Hays Limestone member, Niobrara Chalk (Upper Cretaceous), west-central Kansas: Kansas Univ. Paleont. Contrib., Art. 58, 72 p.
Gallois, R. W., 1965, British regional geology: the Wealden district, 4th ed.: London, Her Majesty's Stationery Office, 101 p.
Garlick, G. D., 1969, The stable isotopes of oxygen, *in* K. H. Wedepohl, ed., Handbook of geochemistry, v. 2: New York, Springer-Verlag, p. 8B1-27.
Garrison, R. E., and A. G. Fischer, 1969, Deep-water limestones and radiolarites of the Alpine Jurassic, *in* Depositional environments in carbonate rocks: SEPM Spec. Pub. 14, p. 20-55.
——— and W. J. Kennedy, in press, Origin of solution seams and flaser structure in Upper Cretaceous chalks of southern England: Sed. Geology.
Gross, M. G., 1964, Variations in the O^{18}/O^{16} ratios of diagenetically altered limestones in the Bermuda Islands: Jour. Geology, v. 72, p. 170-194.
Håkansson, E., R. Bromley, and K. Perch-Nielsen, 1974, Maastrichtian chalk of north-west Europe—a pelagic shelf sediment, *in* Pelagic sediments: on land and under the sea: Internat. Assoc. Sedimentology Spec. Pub. 1, p. 211-234.

Hamilton, E. L., 1959, Thickness and consolidation of deep-sea sediments: Geol. Soc. America Bull., v. 70, p. 1399-1424.
Hancock, J. M., 1963, The hardness of the Irish Chalk: Irish Naturalists' Jour., v. 14, p. 157-164.
——— 1975a, The sequence of facies in the Upper Cretaceous of northern Europe compared with that in the Western Interior, *in* The Cretaceous System in the Western Interior of North America: Geol. Assoc. Canada Spec. Paper 13, p. 83-118.
——— 1975b, The petrology of the Chalk: Geologists' Assoc. Proc., v. 86, p. 499-535.
——— and W. J. Kennedy, 1967, Photographs of hard and soft chalks taken with a scanning electron microscope: Geol. Soc. London Proc., no. 1643, p. 249-252.
——— and P. A. Scholle, 1975, Chalk of the North Sea, *in* A. W. Woodland, ed., Petroleum and the continental shelf of north-west Europe, v. 1: New York, John Wiley & Sons, p. 413-425.
Harms, J. C., and P. W. Choquette, 1965, Geologic evaluation of a gamma-ray porosity device, *in* 6th Ann. SPWLA Logging Symp., Trans., v. 2: Soc. Prof. Well Log Analysts, p. C1-C37.
Harper, M. L., and B. B. Shaw, 1974, Cretaceous-Tertiary carbonate reservoirs in the North Sea: Offshore North Sea Technology Conf., Stavanger, Norway, Paper G IV/4, 20 p.
Hattin, D. E., 1975, Petrology and origin of fecal pellets in Upper Cretaceous strata of Kansas and Saskatchewan: Jour. Sed. Petrology, v. 45, p. 686-696.
——— and D. A. Darko, 1971, Technique for determining coccolith abundance in shaly chalk of Greenhorn Limestone (Upper Cretaceous) of Kansas: Kansas Geol. Survey Bull. 202, pt. 2, 11 p.
Havard, D., and N. Steward, 1969, The nature of the Cretaceous/Tertiary unconformity as seen at two exposures in S.E. England: Kingston Geol. Rev., v. 2, p. 23-25.
Hay, W. W., J. R. Southam, and M. R. Noel, 1976, Carbonate mass balance—cycling and deposition on shelves and in deep sea (abs.): AAPG Bull., v. 60, p. 678.
Hays, J. D., and W. C. Pitman, III, 1973, Lithospheric plate motions, sea level changes and climatic and ecological consequences: Nature, v. 246, p. 18-22.
Honjo, S., 1975, Dissolution of suspended coccoliths in the deep-sea water column and sedimentation of coccolith ooze, *in* Dissolution of deep-sea carbonates: Cushman Found. Foram. Research Spec. Pub., v. 13, p. 114-128.
Hudson, J. D., 1967, Speculations on the depth relations of calcium carbonate solution in recent and ancient seas: Marine Geology, v. 5, p. 473-480.
——— 1975, Carbon isotopes and limestone cement: Geology, v. 3, p. 19-22.
Illing, L. V., G. V. Wood, and J. G. C. M. Fuller, 1967, Reservoir rocks and stratigraphic traps in non-reef carbonates: 7th World Petroleum Cong., Mexico City, Proc., v. 2, p. 487-499.
Jeans, C. V., 1968, The origin of the montmorillonite of the European chalk, with special reference to the Lower Chalk of England: Clay Minerals, v. 7, p. 311-329.
Kauffman, E. G., 1967, Coloradoan macroinvertebrate assemblages, central Western Interior, United States, *in* Paleoenvironments of the Cretaceous seaway: Colorado School of Mines, p. 67-143.
Kennedy, W. J., 1967, Burrows and surface traces from the Lower Chalk of southern England: British Mus. (Nat. Hist.) Bull., Geology, v. 15, p. 125-167.
——— 1970, Trace fossils in the Chalk environment, *in* T. P. Crimes and J. C. Harper, eds., Trace fossils: Liverpool, Seel House Press, p. 263-282.
——— and R. E. Garrison, 1975, Morphology and genesis of nodular chalks and hardgrounds in the Upper Cretaceous of southern England: Sedimentology, v. 22, p. 311-386.
——— and P. Juignet, 1974, Carbonate banks and slump beds in the Upper Cretaceous (upper Turonian–Santonian) of Haute Normandie, France: Sedimentology, v. 21, p. 1-42.
Kent, P. E., 1975, Review of North Sea basin development: Geol. Soc. London Jour., v. 131, p. 435-568.
Lawrence, J. R., 1973, Interstitial water studies, leg 15—stable oxygen and carbon isotope variations in water, carbonates, and silicates from the Venezuela basin (site 149) and the Aves Rise (site 148): Initial Rept. Deep Sea Drilling Project, v. 20, p. 891-899.
Leclaire, L., G. Alcayde, and F. Froelich, 1973, La silicification des craies; rôle des spherules de cristobalite-tridymite observées dans les craies des bassins océaniques et dans celles du bassin de Paris: Acad. Sci. Comptes Rendus, ser. D, v. 277, p. 2121-2124.
Long, M., 1976, Austin Chalk drilling surges in south Texas: Oil and Gas Jour., v. 74, no. 36, p. 62-63.
Manus, R. W., and A. H. Coogan, 1974, Bulk volume reduction and pressure-solution derived cement: Jour. Sed. Petrology, v. 44, p. 466-471.
Mapstone, N. B., 1975, Diagenetic history of a North Sea chalk: Sedimentology, v. 22, p. 601-613.
Matter, A., 1974, Burial diagenesis of pelitic and carbonate deep-sea sediments from the Arabian Sea: Initial Rept. Deep Sea Drilling Project, v. 23, p. 421-469.
——— R. G. Douglas, and K. Perch-Nielsen, 1975, Fossil preservation, geochemistry, and diagenesis of pelagic carbonates from Shatsky Rise, northwest Pacific: Initial Rept. Deep Sea Drilling Project, v. 32, p. 891-921.
McIntyre, A., and A. W. H. Be, 1967, Modern Coccolithophoridae of the Atlantic Ocean, 1, Placoliths and cyrtoliths: Deep-Sea Research, v. 14, p. 561-597.
——— and R. McIntyre, 1971, Coccolith concentrations and differential solution in oceanic sediments, *in* B. M. Funnell and W. R. Riedel, eds., The micropalaeontology of oceans: Cambridge, England, Cambridge Univ. Press, p. 253-261.
Milliman, J. D., 1974, Marine carbonates: New York, Springer-Verlag, 375 p.
——— 1975, Dissolution of aragonite, Mg-calcite, and calcite in the North Atlantic Ocean: Geology, v. 3, p. 461-462.
Mimran, Y., 1975, Fabric deformation induced in Cretaceous chalks by tectonic stresses: Tectonophysics, v. 26, p. 309-316.
——— in press, Chalk deformation and large-scale mi-

gration of calcium carbonate: Sedimentology.
Minato, M., S. Honjo, and M. Ishii, 1967, Electron microscopic study of oil bearing calcilutite, Khafji oil field, Neutral Zone: 7th World Petroleum Cong., Mexico City, Proc., v. 2, p. 447-456.
Mossop, G. D., 1972, Origin of the peripheral rim, Redwater reef, Alberta: Bull. Canadian Petroleum Geology, v. 20, p. 238-280.
Müller, A. H., 1951, Diagenetische Untersuchungen in der obersenonen Schreibkreide von Rügen: Germany, Geol. Abh. Dienst. Berlin, H. 228, 29 p.
Neugebauer, J., 1973, The diagenetic problem of chalk—the role of pressure solution and pore fluid: Neues Jahrb. Geologie u. Paläontology Abh., v. 143, p. 223-245.
——— 1974, Some aspects of cementation in chalk, *in* Pelagic sediments: on land and under the sea: Internat. Assoc. Sedimentology Spec. Pub. 1, p. 149-176.
——— 1975, Fossil-Diagenese in der Schreibkreide: Coccolithen: Neues Jahrb. Geologie u. Paläontologie Monatsch., p. 489-502.
Owen, J. D., 1972, A log analysis method for Ekofisk field, Norway: Soc. Prof. Well Log Analysts 13th Ann. Logging Symposium, Proc., Paper 10, 22 p.
Peach, B. N., W. Gunn, and E. T. Newton, 1901, On a remarkable volcanic vent of Tertiary age in the Island of Arran, enclosing Mesozoic fossiliferous rocks: Geol. Soc. London Quart. Jour., v. 57, p. 226-243.
Peake, N. B., and J. M. Hancock, 1961, The Upper Cretaceous of Norfolk, *in* The geology of Norfolk: Norfolk and Norwich Naturalists Soc. Trans., v. 19, p. 293-339.
Reeside, J. B., Jr., 1944, Map showing thickness and general character of the Cretaceous deposits in the Western Interior of the United States: U.S. Geol. Survey Oil and Gas Invs. Prelim. Map 10, scale approx. 1 in. to 218 mi.
Reid, R. E. H., 1968, Bathymetric distribution of Calcarea and Hexactinellida in the present and the past: Geol. Mag., v. 105, p. 546-559.
Rickards, L. M., 1974, The Ekofisk area, discovery to development: Offshore North Sea Technology Conf., Stavanger, Norway, Paper G-IV/3, 16 p.
Rose, P. R., 1970, Stratigraphic interpretation of submarine versus subaerial discontinuity surfaces: an example from the Cretaceous of Texas: Geol. Soc. America Bull., v. 81, p. 2787-2798.
Rothe, P., J. Hoefs, and V. Sonne, 1974, The isotopic composition of Tertiary carbonates from the Mainz basin: an example of isotopic fractionations in "closed basins": Sedimentology, v. 21, p. 373-395.
Saito, T., and J. Van Donk, 1974, Oxygen and carbon isotope measurements of Late Cretaceous and early Tertiary Foraminifera: Micropaleontology, v. 20, p. 152-177.
Savin, S. M., and R. G. Douglas, 1973, Stable isotope and magnesium geochemistry of recent planktonic Foraminifera from the South Pacific: Geol. Soc. America Bull., v. 84, p. 2327-2342.
Sayles, F. L., and F. T. Manheim, 1975, Interstitial solutions and diagenesis in buried marine sediments: results from the Deep Sea Drilling Project: Geochim. et Cosmochim. Acta, v. 39, p. 103-127.
Schlanger, S. O., and R. G. Douglas, 1974, Pelagic ooze-chalk-limestone transition and its implications for marine stratigraphy, *in* Pelagic sediments: on land and under the sea: Internat. Assoc. Sedimentology Spec. Pub. 1, p. 117-148.
——— et al, 1973, Fossil preservation and diagenesis of pelagic carbonates from the Magellan Rise, central North Pacific Ocean: Initial Rept. Deep Sea Drilling Project, v. 17, p. 407-427.
Scholle, P. A., 1971, Diagenesis of deep-water carbonate turbidites, Upper Cretaceous Monte Antola Flysch, Northern Apennines, Italy: Jour. Sed. Petrology, v. 41, p. 233-250.
——— 1974, Diagenesis of Upper Cretaceous chalks from England, Northern Ireland and the North Sea, *in* Pelagic sediments: on land and under the sea: Internat. Assoc. Sedimentology Spec. Pub. 1, p. 177-210.
——— 1975, Application of chalk-diagenetic studies to petroleum-exploration problems (abs.): AAPG Bull., v. 59, p. 2197-2198.
——— 1976, Chalk diagenesis as a burial depth indicator (abs.): Geol. Soc. America Abs. with Programs, v. 8, p. 261-262.
——— and W. J. Kennedy, 1974, Isotopic and petrophysical data on hardgrounds from Upper Cretaceous chalks from western Europe (abs.): Geol. Soc. America Abs. with Programs, v. 6, p. 943.
——— and S. A. Kling, 1972, Southern British Honduras: lagoonal coccolith ooze: Jour. Sed. Petrology, v. 42, p. 195-204.
Schwarcz, H. P., 1969, Carbon isotopes in nature, *in* K. H. Wedepohl, ed., Handbook of geochemistry, v. 2: New York, Springer-Verlag, p. 6B1-6B16.
Shepherd, W., 1975, Flint, its origin, properties and uses: London, Faber and Faber, 255 p.
Shreveport Geological Society, 1968, Stratigraphy and selected gas-field studies of north Louisiana, *in* Natural gases of North America, v. 1: AAPG Mem. 9, p. 1099-1175.
Smith, C. C., 1973, Calcareous nannoplankton of the upper Eagle Ford and lower Austin Formations: PhD thesis, Univ. Texas—Dallas, 336 p.
Smith, D. B., et al, 1976, The age of groundwater in the Chalk of the London basin: Water Resources Res., v. 13, p. 392-404.
Stenestad, E., 1972, Traek af det danske bassins udvikling i Øvre Kridt: Dansk Geol. Foren., Arsskrift f. 1971, p. 63-69.
Thompson, G., 1972, A geochemical study of some lithified carbonate sediments from the deep-sea: Geochim. et Cosmochim. Acta, v. 36, p. 1237-1253.
Thorez, J., et al, 1971, The petrography and origin of deposits filling solution pipes in the Chalk near South Mimms, Hertfordshire: Geol. Mag., v. 108, p. 413-423.
Tiratsoo, E. N., 1976, Oilfields of the world (2d ed.): Beaconsfield, England, Scientific Press, Ltd., 384 p.
Tracey, J. I., Jr., et al, 1971, Initial Reports of the Deep Sea Drilling Project: v. 8, 1037 p.
Valeton, I., 1960, Vulkanische Tuffiteinlagerungen in der nordwestdeutschen Oberkreide: Hamburg, Geol. Staatsinst. Mitt., v. 29, p. 26-41.

Wachs, D., and J. R. Hein, 1974, Petrography and diagenesis of Franciscan limestones: Jour. Sed. Petrology, v. 44, p. 1217-1231.
Weyl, P. K., 1959, Pressure solution and the force of crystallization—a phenomenological theory: Jour. Geophys. Research, v. 64, p. 2001-2025.
Wilson, H. E., 1972, Regional geology of Northern Ireland: Belfast, N.I., Her Majesty's Stationery Office, 113 p.
Wise, D. U., 1972, Freeboard of continents through time, *in* Studies in earth and space sciences (Hess Volume): Geol. Soc. America Mem. 132, p. 87-100.
Wise, S. W., Jr., 1973, Calcareous nannofossils from cores recovered during leg 18, Deep Sea Drilling Project: biostratigraphy and observations of diagenesis: Initial Rept. Deep Sea Drilling Project, v. 18, p. 569-615.
Wolfe, M. J., 1968, Lithification of a carbonate mud: Senonian chalk in Northern Ireland: Sed. Geology, v. 2, p. 263-290.
——— 1970, Dolomitization and dedolomitization in the Senonian chalk of Northern Ireland: Geol. Mag., v. 107, p. 39-49.
Woollard, G. P., 1962, The relation of gravity anomalies to surface elevation, crystal structure and geology: Univ. Wisconsin Research Rept. Ser. 62-9, 329 p.
World Oil, 1971, Massive Danian limestone key to Ekofisk success: World Oil, v. 172, no. 6, p. 51-52.
Yaalon, D. H., and S. Singer, 1974, Vertical variation in strength and porosity of calcrete (nari) on chalk, Shefela, Israel, and interpretation of its origin: Jour. Sed. Petrology, v. 44, p. 1016-1023.
Ziegler, P. A., 1975, Geologic evolution of North Sea and its tectonic framework: AAPG Bull., v. 59, p. 1073-1097.
Zoback, M. D., and J. D. Byerlee, 1975, Permeability and effective stress: AAPG Bull., v. 59, p. 154-158.

The American Association of Petroleum Geologists Bulletin
V. 55, No. 10 (October 1971), p. 1873-1878, 3 figs.

Microporosity in Carbonate Rocks[1]

EDWARD D. PITTMAN[2]
Tulsa, Oklahoma 74102

Abstract Large pores in carbonate rocks hold and transmit fluids, whereas associated micropores may hold irreducible water; *i.e.*, water not available for flow. Analysis of borehole logs of microporous carbonate rocks can result in misleadingly high calculated water saturations and possibly bypassing of a potential oil or gas reservoir.

Microporosity occurs in carbonate rocks that range from friable to well indurated. Intercrystalline micropores may be present in micrite and within ooids, pisolites, micrite intraclasts, pellets, and cryptocrystalline grains. Micropores in ancient rocks appear to result from (1) formation of micrite envelopes, (2) void space present after transition of aragonite to calcite, or (3) incipient weathering and dissolution. Aggrading neomorphism of micrite to microspar tends to destroy microporosity.

Introduction

Choquette and Pray (1970, p. 248) defined micropores as "equant to equant-elongate pores whose average diameter is less than 1/16 mm and for tubular or platy pores whose average cross-sectional diameter or least diameter, respectively, is less than 1/16 mm." The term "micropore" as described and used in this paper is more restrictive and refers to pores and pore throats (interconnections between large voids) well below these size restrictions. In most cases the micropores are tabular with a least diameter of 1 μ or less.

Some carbonate reservoir rocks have a bimodal distribution of pore sizes: (1) large pores and pore throats commonly related to intergranular or moldic porosity; and (2) micropores and associated pore throats commonly related to intercrystalline or intraparticle porosity. Large pores hold and transmit fluids, whereas micropores may hold irreducible water. Microporosity with irreducible water may be so abundant that an electric log analysis indicates high water saturation although much of the water is bound in the rock; that is, the molecular attraction of the solid for the liquid is so great that the liquid tends to remain in the void. This means a potential reservoir may be bypassed, hence the importance of this phenomenon.

Size of the pore throat is critical to the flow of fluids in a rock. If a large pore capable of holding oil has micropore-size throats with irreducible water, no fluid flow is possible.

Procedure and Techniques

Most of the irreducible water is removed from micropores in a core plug during preparation for core analysis. After cores have been cleaned in refluxing toluene (boiling point 110.8°C; 231.4°F), they are dried in an oven at 105°C (221°F). This enables plastic to enter micropores during pressure impregnation of the rock. The samples then may be used to make pore casts and pore-cast thin sections, which are plastic replicas of the void space in a rock and are useful in porosity studies (Pittman and Duschatko, 1970).

Initial investigations of microporosity were limited to routine examination of thin sections, pore casts, and pore-cast thin sections. These techniques provide the best means of visually locating concentrations of micropores. However, micropores are too small to be studied adequately with conventional light microscopes.

Pertinent information on the micropore-crystal relation was gained with the transmission electron microscope (TEM) through the efforts of Fred Rowe and R. E. Garrison. A better understanding of the nature and occurrence of micropores was obtained later with a scanning electron microscope (SEM).

The technique of Honjo and Fischer (1965) was used for replica preparation of samples examined with the TEM. All SEM samples were coated with gold or gold-palladium (40:60 ratio) and examined at 20 KV.

[1] Manuscript received, September 30, 1970; revised and accepted, February 2, 1971.

[2] Amoco Research Center.

Appreciation is expressed to Carol Erwin for typing the manuscript and to Fred Rowe and R. E. Garrison for sample preparation and interpretation of micrographs taken with the transmission electron microscope. H. D. Winland read various drafts of the manuscript and made many helpful comments.

Description of Micropores

Micropores occur in ancient rocks and modern sediments. In rocks, intercrystalline micropores commonly are within coatings on grains (oolites and pisolites), carbonate mud aggregate grains (pellets and intraclasts), and cryptocrystalline grains, as well as in micrite. Intercrystalline micropores also are abundant in comparable modern grains and sediments. Borings into recent grains by blue-green algae result in 5–20 μ diameter tubes commonly filled by micrite. This may lead to the formation of a micrite envelope that contains micropores. In all samples examined with the SEM, micropores are intercrystalline in microcrystalline carbonate.

Micropores are illustrated for the following representative occurrences (terminology from Dunham, 1962): (1) lime grainstone (Fig. 1A-B); (2) lime packstone (Fig. 1C-D); (3) lime wackestone (Fig. 2A-B-C; (4) micrite envelope on Holocene grain (Fig. 2D).

Origin of Microporosity

Modern carbonate grains and sediments contain micropores analogous to those in ancient rocks. The crystal-to-micropore relation is similar and intercrystalline in both situations. Dissimilarities result from the transition of aragonite to calcite. In a carbonate mud this transition causes a change from a mass of aragonite needles that average 3 μ in length to equant 3 μ polyhedrons of calcite. Shoji and Folk (1964) pointed out that it is not a size increase if size is defined by length of crystals. However, if size is defined in terms of volume it would take many aragonite needles to make one calcite polyhedron. Any primary intercrystalline microporosity associated with aragonite needles must be modified substantially if not destroyed during the transition of aragonite to calcite; that is, the pore geometry is changed. Therefore, it is unlikely that any microporosity in ancient rocks truly can be termed primary. Aggrading neomorphism of $< 4\ \mu$ calcite crystals (micrite) to microspar (Folk, 1965) shows a definite tendency to destroy microporosity.

Micropores occur in micrite envelopes, which form on all types of carbonate grains (Hessland, 1949; Swinchatt, 1965; Klement and Toomey, 1967). Micritization of the surficial parts of the grains is accomplished through the boring and perforating action of blue-green algae (Bathurst, 1966; Winland, 1968). These fine tubes subsequently are filled by micrite that contains intercrystalline micropores (Fig. 2D).

Limestones commonly contain micropores in carbonate mud matrix and grains of various types. These micropores occur among calcite polyhedrons and occupy spatial relations among crystals similar to micropores in aggregates of aragonite crystals.

Microporosity appears to be associated with joints and fractures in some rocks, and is probably the result of incipient weathering and dissolution. In capillary pores, material transport is accomplished by diffusion (Heling, 1968, p. 104). Through this process micropores may be enlarged slightly in the vicinity of water flow, as along fractures.

It is possible that some micropores may be formed by expansion of a core resulting from release of overburden pressure as the core is brought to the surface. If this is an important process one would expect all cores from deep wells to be microporous, but they are not.

Example of Reservoir with Microporosity

The Wolfcamp reservoir in the Pan American No. 1 Powell, discovery well for Blalock field, Glasscock County, Texas, is an example of an oil reservoir with high irreducible water content because of microporosity. This Permian reservoir is a light-colored, chalky appearing, fine- to medium-grained, pellet-intraclast-skeletal lime packstone (Fig. 1C). The carbonate mud matrix, pellets, and carbonate mud intraclasts have undergone partial aggrading neomorphism to form microspar. Macropores and micropores are present in this reservoir. Macropores are interparticle and intraparticle, and result from partial dissolution of microspar. Micropores are prominent among calcite crystals within carbonate mud aggregate grains (Fig. 1D).

An encouraging core of a Wolfcampian limestone led to a subsequent drillstem test covering the upper part of the limestone (Fig. 3). During this test the well flowed 60 bbl/hour of oil with no water. The borehole was then drilled to projected total depth and logged. Water saturations of 40–50 percent and porosity values up to 18 percent were computed from logs for the porous and permeable part of the interval tested. The water-free flow of oil despite high water saturations was anomalous for Permian carbonate reservoirs in western Texas. Below the drillstem-tested interval, water saturations up to 80 percent were computed for rock with porosity values comparable to the upper part of the reservoir. The No. 1 Powell was completed from perforations at the base of the porous limestone from an interval with a calculated average porosity and water saturation of

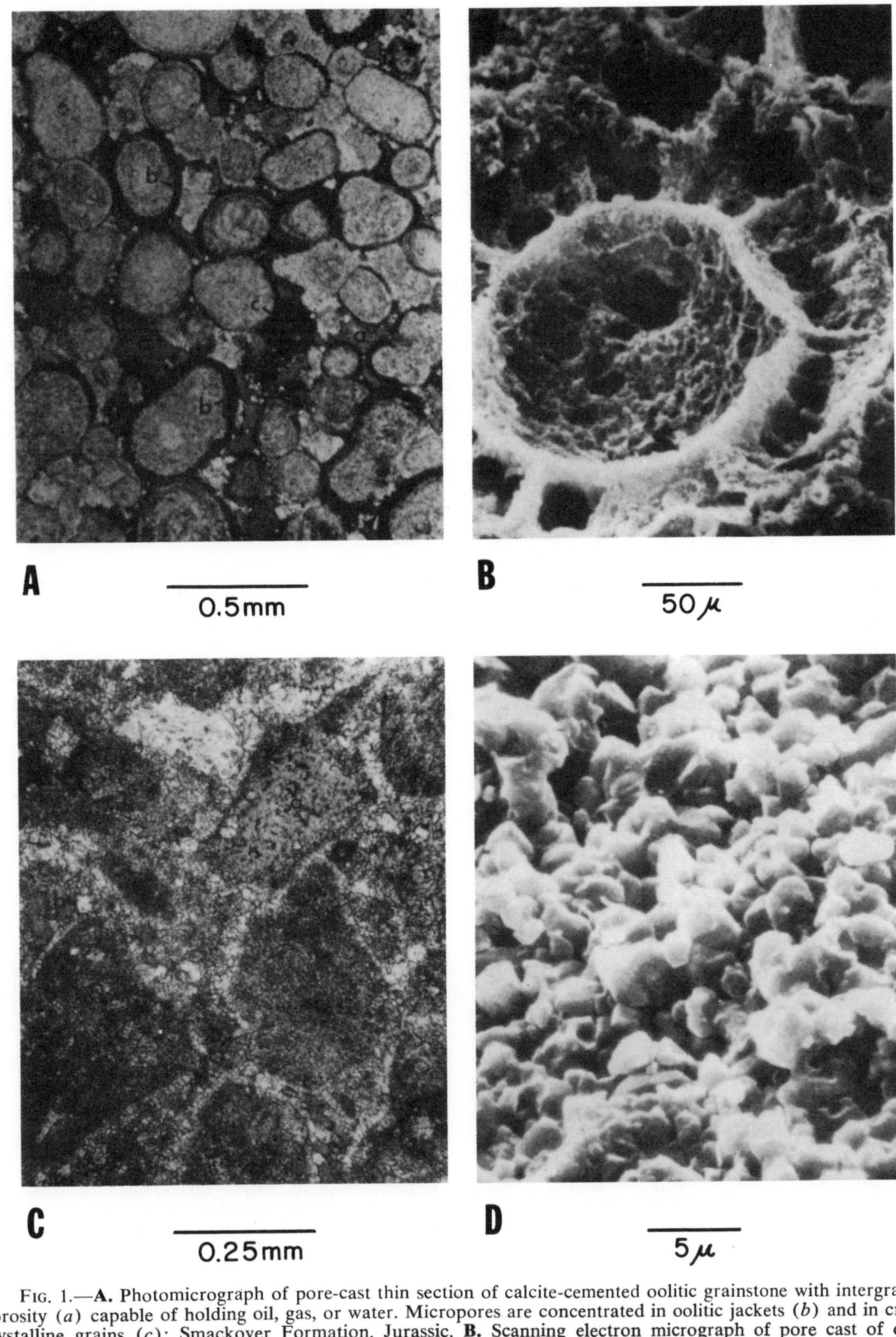

FIG. 1.—**A.** Photomicrograph of pore-cast thin section of calcite-cemented oolitic grainstone with intergranular porosity (*a*) capable of holding oil, gas, or water. Micropores are concentrated in oolitic jackets (*b*) and in cryptocrystalline grains (*c*); Smackover Formation, Jurassic. **B.** Scanning electron micrograph of pore cast of oolitic grainstone showing concentration of micropores forming complex network of interconnected, thin, tabular-shaped pores among crystals in ooid; Smackover Formation, Jurassic. **C.** Photomicrograph of intraclast lime packstone with intragranular (*a*) and intergranular (*b*) porosity that resulted from dissolution. Other samples contain pellets, crinoids, and forams. Micrite matrix has undergone partial aggrading neomorphism to form microspar; unnamed limestone, Permian. **D.** Scanning electron micrograph showing micropores among < 4-μ crystals of calcite that comprise micrite intraclast; unnamed limestone, Permian.

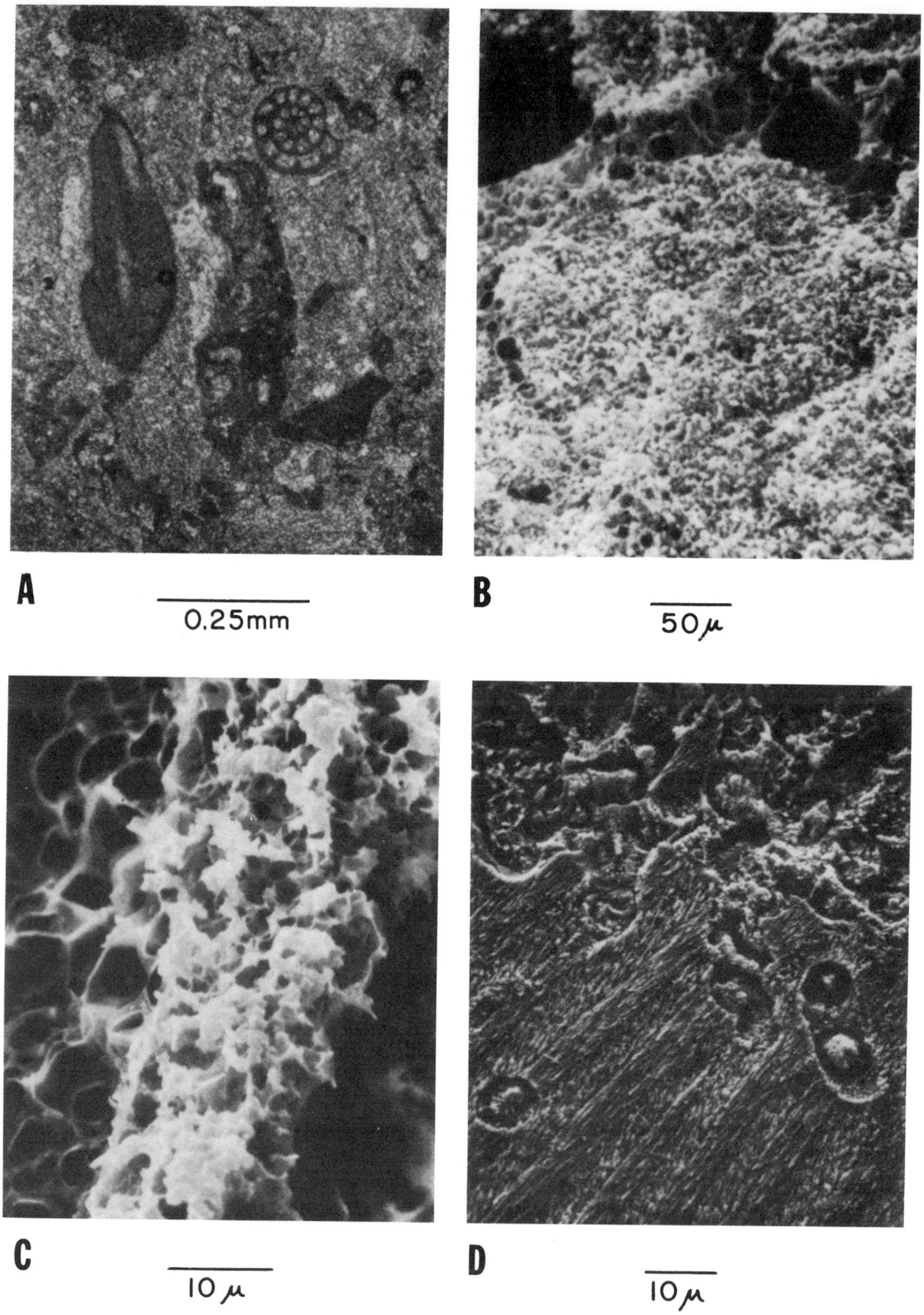

FIG. 2.—**A.** Photomicrograph of skeletal-intraclast lime wackestone. Micropores are within matrix, carbonate mud intraclasts, and walls of fusulines; unnamed limestone, Pennsylvanian. **B.** Scanning electron micrograph of pore cast of micrite intraclast with microporosity. Large holes represent location of microspar and pseudospar in rock; unnamed limestone, Pennsylvanian. **C.** Scanning electron micrograph of pore cast showing micropores within wall of fusuline; unnamed limestone, Pennsylvanian. **D.** Scanning electron micrograph of polished surface showing borings into mollusk shell by blue-green algae. Micrite envelope, formed by extensive borings filled with micrite, is in upper part of micrograph. Isolated micrite-filled borings are visible below micrite envelope. Intercrystalline micropores occur within micrite in borings; recent mollusk from British Honduras (photo courtesy of H. D. Winland).

17.5 percent and 56.1 percent, respectively. On initial potential the well flowed 254 bbl of oil and 63 bbl of water per day. The reason for the high water saturations without a significant initial flow of water is the presence of abundant micropores with irreducible water. If the formation had not been cored and drillstem tested, the reservoir probably would not have been discovered. High water saturations calculated from logs would have discouraged testing.

Conclusions

1. The size of pores can affect the fluid retention capabilities of a reservoir. If large pores are associated with micropores, the rock can yield oil or gas while micropores hold irreducible water. Thus, high water saturations may be calculated with logs, although the rock may not yield a significant flow of water. This may lead to bypassing of a potential reservoir.

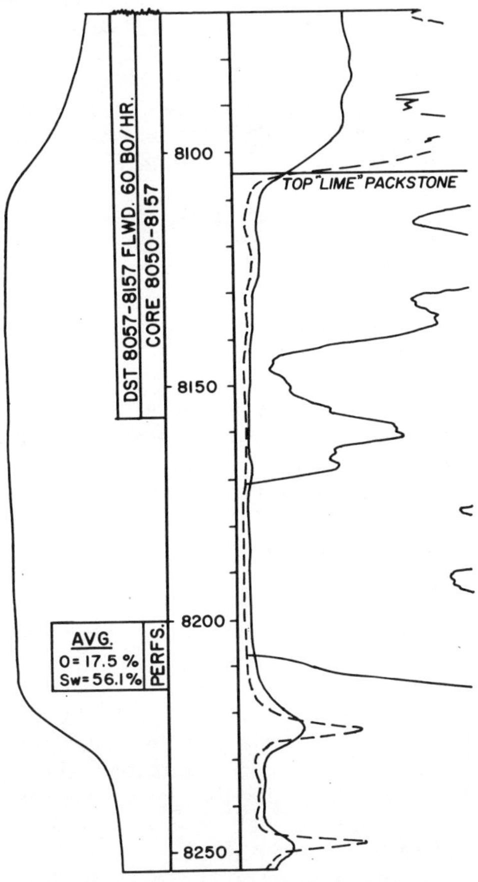

FIG. 3.—Induction electrical log for Wolfcamp limestone reservoir in Pan American No. 1 Powell, Glasscock County, Texas. Note low resistivity caused by irreducible water in micropores. This well was completed from perforations at base of limestone and flowed 254 bbl of oil and 63 bbl of water per day.

2. Micropores are present in carbonate mud matrix, ooids, pisolites, carbonate mud intraclasts, pellets, and cryptocrystalline grains of carbonate rocks that range from friable to well indurated.

3. Micropores in ancient rocks result from (a) algal borings that are subsequently infilled by micrite with microporosity, (b) void space present after transition of aragonite to calcite, but prior to further aggrading neomorphism, or (c) incipient weathering and dissolution.

4. In all samples investigated to date, aggrading neomorphism of micrite to microspar tended to destroy microporosity.

References Cited

Bathurst, R. G. C., 1966, Boring algae, micrite envelopes, and lithification of molluscan biosparites: Geol. Jour., v. 5, pt. 1, p. 15–32.

Choquette, P. W., and L. C. Pray, 1970, Geologic nomenclature and classification of porosity in sedimentary carbonates: Am. Assoc. Petroleum Geologists Bull., v. 54, no. 2, p. 207–250.

Dunham, R. J., 1962, Classification of carbonate rocks according to depositional texture, *in* Classification of carbonate rocks, a symposium: Am. Assoc. Petroleum Geologists Mem. 1, p. 108–121.

Folk, R. L., 1965, Some aspects of recrystallization in ancient limestones, *in* Dolomitization and limestone diagenesis: Soc. Econ. Paleontologists and Mineralogists Spec. Pub. No. 13, p. 124–168.

Heling, D., 1968, Microporosity of carbonate rocks, *in* Recent developments in carbonate sedimentology in central Europe: New York, Springer-Verlag, p. 98–105.

Hessland, I., 1949, Investigation of the Lower Ordovician of the Siljan District, Sweden; II, Lower Ordovician penetrative and enveloping algae from the Siljan District: Uppsala Univ., Geol. Inst., Bull. v. 33, p. 409–426.

Honjo, S., and A. G. Fischer, 1965, Paleontological investigation of limestones by electron microscope, *in* Handbook of paleontological techniques: San Francisco, W. H. Freeman and Co., p. 326–334.

Klement, K. W., and D. F. Toomey, 1967, Role of the blue-green alga *Girvanella* in skeletal grain destruction and lime-mud formation in the Lower Ordovician of West Texas: Jour. Sed. Petrology, v. 37, p. 1045–1051.

Pittman, E. D., and R. W. Duschatko, 1970, Use of pore casts and scanning electron microscope to study pore geometry: Jour. Sed. Petrology, v. 40, p. 1153–1157.

Shoji, R., and R. L. Folk, 1964, Surface morphology of some limestone types as revealed by electron microscope: Jour. Sed. Petrology, v. 34, p. 144–155.

Swinchatt, J. P., 1965, Significance of constituent composition, texture, and skeletal breakdown in some Recent carbonate sediments: Jour. Sed. Petrology, v. 35, p. 71–90.

Winland, H. D., 1968, The role of high Mg-calcite in the preservation of micrite envelopes and textural features of aragonite sediments: Jour. Sed. Petrology, v. 38, p. 1320–1325.

The American Association of Petroleum Geologists Bulletin
V. 67, No. 2 (February, 1983), P.191-200, 7 Figs., 2 Tables

High-Porosity Cenozoic Carbonate Rocks of South Florida: Progressive Loss of Porosity with Depth[1]

ROBERT B. HALLEY and JAMES W. SCHMOKER[2]

ABSTRACT

Porosity measurements by borehole gravity meter in subsurface Cenozoic carbonates of south Florida reveal an extremely porous mass of limestone and dolomite which is transitional in total pore volume between typical porosity values for modern carbonate sediments and ancient carbonate rocks. A persistent decrease of porosity with depth, similar to that of chalks of the Gulf Coast, occurs in these rocks. We make no attempt to differentiate depositional or diagenetic facies which produce scatter in the porosity-depth relationship; the dominant data trends thus are functions of carbonate rocks in general rather than of particular carbonate facies. Carbonate strata with less than 20% porosity are absent from the rocks studied here.

Aquifers and aquicludes cannot be distinguished on the basis of porosity. Although aquifers are characterized by great permeability and well-developed vuggy and even cavernous porosity in some intervals, they are not exceptionally porous when compared to other Tertiary carbonate rocks in south Florida. Permeability in these strata is governed more by the spacial distribution of pore space and matrix than by the total volume of porosity present.

Dolomite is as porous as, or slightly less porous than, limestones in these rocks. This observation places limits on any model proposed for dolomitization and suggests that dolomitization does not take place by a simple ion-for-ion replacement of magnesium for calcium. Dolomitization may be selective for less porous limestone, or it may involve the incorporation of significant amounts of carbonate as well as magnesium into the rock.

The great volume of pore space in these rocks serves to highlight the inefficiency of early diagenesis in reducing carbonate porosity and to emphasize the importance of later porosity reduction which occurs during the burial or late near-surface history of limestones and dolomites.

INTRODUCTION

This paper presents and discusses porosity data obtained from borehole gravity measurements in Cenozoic carbonate rocks of peninsular Florida. Tested depths range from about 6 to 2,700 ft (2 to 830 m), and the study includes Pleistocene to Paleocene rocks that partly span the transition between modern high-porosity carbonate sediments and ancient low-porosity carbonate rocks. This paper is a companion to one that surveys carbonate-rock porosity throughout the south Florida peninsula to depths of 18,000 ft (5,500 m) (Schmoker and Halley, 1982). The present paper emphasizes the significance of very highly porous (approaching 40% porosity on average) carbonate rocks in the upper part of the south Florida carbonate platform.

The diagenetic processes associated with the porosity decrease from modern to ancient carbonates have been the subject of considerable research. Until recently, a lack of evidence in ancient limestones and dolomites for significant physical compaction has led to a focus on cementation as the primary mechanism for porosity loss. The generalities of material transport and carbonate cementation have been discussed by Weller (1959), Dunham (1969), Bathurst (1975, 1976), and Friedman (1975), but difficulties in quantitative evaluation of the roles of mechanical compaction and of pressure solution have been continuing problems (Pratt, 1982; Wanless, 1982). Both processes may be more important than previously thought (Shinn et al, 1977; Wanless, 1979; Buxton and Sibley, 1981).

With increasing detail, carbonate petrographers have documented the processes and products of the transition from carbonate sediments to carbonate rocks and have described the petrographic, mineralogic, and geochemical transformations which occur early in the history of carbonate diagenesis. Ancient carbonate rocks likewise have been extensively studied, and the typically complex series of diagenetic events leading to a given end product has been documented for many formations. However, petrographic studies often do not quantitatively examine porosity, and despite progress toward a general understanding of carbonate diagenesis, attendant volumetric changes in porosity are rather poorly documented.

Borehole-gravity measurements of porosity are derived from in-situ sample sizes on the order of hundreds of cubic meters. Our data suggest that on this scale of measurement, diagenetic processes have reduced the porosity of typical south Florida limestone and dolomite units from roughly 45% near the surface to 25 or 30% at a depth of 2,600 ft (800 m). Carbonate rocks in south Florida retain porosity during mineralogic transformation at shallow depths. The character of the pore space is greatly modified during mineralogical stabilization, resulting mainly in cementation and development of secondary porosity as

[1]Manuscript received, December 21, 1981; accepted, August 26, 1982.

[2]U.S. Geological Survey, Denver Federal Center, Box 25046, MS-921, Denver, Colorado 80225.

The authors thank F. W. Meyer, U.S. Geological Survey, Miami, and Stephen L. Robbins, Donald L. Smith, and Robert J. Martinez, U.S. Geological Survey, Denver, for their assistance during field work for this study. Anthony F. Randazzo, University of Florida, and Louis G. Zachos, Law Engineering, Miami, kindly loaned thin sections which supplemented our sample collections. William L. Irwin of Bass Enterprises Production Co. graciously provided a sonic log from Bass' Pumpkin Bay well to complete our log suite at that locality. Reviews of earlier manuscripts by Robert F. Loucks and Eugene A. Shinn provided significant constructive criticism for which we are grateful.

documented by many workers including Schlanger (1963), Friedman (1964), Land et al (1967), Robinson (1967), Matthews (1968), Gavish and Friedman (1969), Steinen (1974), Harrison (1975), and many others. Considerable porosity is carried into the subsurface after early diagenesis, and depth-burial-dependent cementation and compaction probably account for most of the porosity loss in the limestones and dolomites studied here.

SCALE PROBLEM IN MEASUREMENT OF CARBONATE POROSITY

Pore size in carbonate rocks ranges from micrometers to tens of meters. Linear pore dimensions can thus vary by 6 or 7 orders of magnitude and pore volumes by 18 to 21 orders of magnitude. In addition, the distribution of pores can be heterogeneous at almost any scale. Clearly, adequate estimation of carbonate porosity presents a significant sampling problem.

Table 1 lists many of the methods used to evaluate porosity in carbonate rocks and gives estimates of the sample volume investigated by each. Implicit in the use of a given method are the assumptions that pores are evenly distributed and are small relative to the volume being investigated. In reality, these assumptions often are not satisfied, and a strategy of analyzing a large number of small samples is frequently adopted in the hope that their average porosity will approach that of the whole rock.

The scale problem can be particularly acute when laboratory methods of porosity measurement are applied to highly porous (more than 35% porosity) carbonate rocks. Core recovery in such rocks may be poor, causing a sampling bias toward less porous zones. Pores are likely to be large and irregularly distributed in highly porous carbonate rocks. Laboratory samples are too small to adequately represent pore-space distribution. In addition to these factors, thin-section porosity estimates are subject to systematic measurement errors (Halley, 1978). We believe that measurement inaccuracies associated with the scale problem and thin-section analyses have resulted in an apparent porosity decrease in carbonate sediments during early, freshwater diagenesis as reported by Robinson (1967), Land et al (1967), Pittman (1974), and Friedman (1964). Our porosity measurements, discussed below, suggest little or no porosity loss during freshwater diagenesis. Such porosity preservation during diagenesis has significant implications for understanding the mass transfer of calcite in the subsurface. For example, it eliminates the difficulties of insufficient geological time encountered by Enos and Sawatsky (1981) who modeled cementation of Pleistocene limestones. It avoids problems associated with mass-balance calculations and the source of "excess" cement encountered by Steinen (1974) and Harrison (1975). Finally, early porosity preservation underscores the importance of later diagenetic events in reducing porosity.

In this study of high-porosity carbonate rocks, borehole gravimetry was selected as the best technique to determine the average porosity of samples having vertical dimensions ranging from 10 to 100 ft (3 to 30 m^3) and effective volumes greater than 3,500 ft^3 (100 m^3). This large measurement scale is uncommon in carbonate studies and, compared to other methods of porosity measurement, more closely approaches the size of hydrocarbon reservoirs and aquifers. Our objective was to smooth and average the effect of porosity variations on the scale of centimeters and even meters in order to examine in-situ bulk-rock porosity as a function of depth.

METHODS

The borehole gravity meter has an effective radius of investigation that is measured in meters and is comparable to large carbonate pore spaces (caverns). It is a density logging tool that is not significantly influenced by casing, borehole rugosity, or formation damage caused by drilling. Rocks away from the borehole can be sensed because gravity is not attenuated by intervening material.

The U.S. Geological Survey–LaCoste and Romberg[3] borehole gravity meter (described by McCulloh et al, 1967a, b) was used in early 1979 to conduct borehole gravity surveys in five cased wells penetrating the Cenozoic sediments and rocks of peninsular Florida (Fig. 1). Full details of the borehole gravity surveys are given by Schmoker et al (1979).

A comprehensive listing of references to fundamentals of borehole-gravity logging and interpretation can be found in Robbins (1980). In the absence of complicating structural factors, the relation between formation density and measurements of gravity in a borehole is given by (McCulloh, 1966; Robbins, 1981):

$$\varrho = 39.131(F - \Delta g/\Delta z), \qquad (1)$$

where ϱ is the average formation density (g/cm^3) between two vertically separated points in the borehole, F is the free-air vertical gradient of gravity (mgal/ft), Δg is the measured difference in gravity between the vertically separated points (mgal), and Δz is the vertical separation (ft). In evaluating equation 1, a value for the free-air gradient of 0.09409 mgal/ft, appropriate for the latitudes of south Florida, was used.

Porosity, ϕ, was computed from the borehole-gravity density, ϱ, using the equation:

$$\phi = 100(\varrho_G - \varrho)/(\varrho_G - \varrho_F), \qquad (2)$$

where ϱ_G is the grain density and ϱ_F is the pore-fluid density (g/cm^3). Pores were assumed to be 100% water saturated, with water density ranging between 1.00 and 1.02 g/cm^3 depending upon salinity. Grain density was estimated from lithologic logs prepared from well cuttings, with the assumption that such logs are representative of rocks within the radius of investigation of the borehole gravity meter.

GEOLOGIC SETTING

The strata investigated in this study are primarily shallow-water carbonate rocks that accumulated on the stable Florida platform in a wide range of depositional environments. Secondary lithologies include sulfates,

[3]Use of brand names in this paper is for descriptive purposes only and does not imply endorsement by the U.S. Geological Survey.

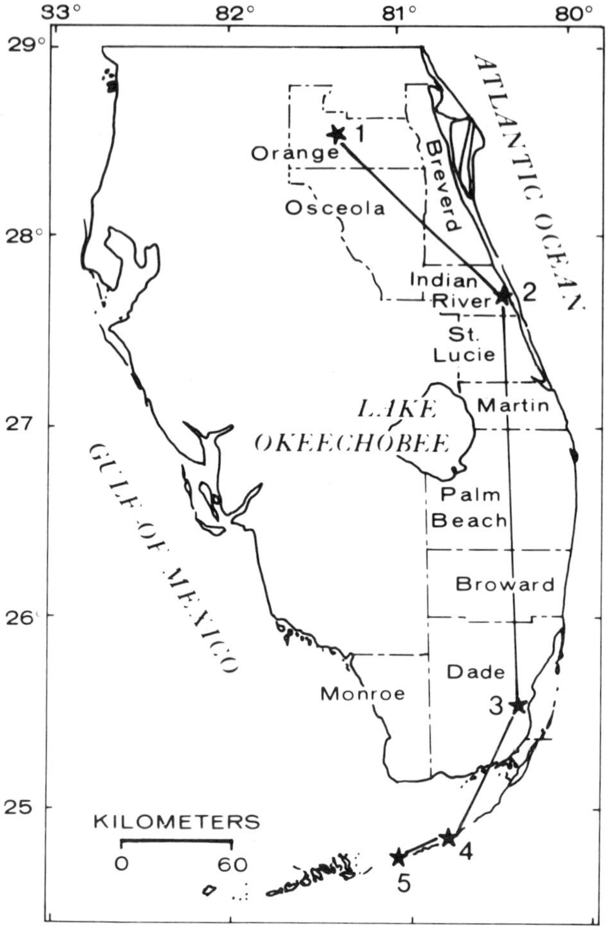

FIG. 1—Index map showing location of wells in south Florida logged for this study.

uncommon in Eocene rocks but widespread in Paleocene strata, and lower Pliocene and Miocene terrigenous clastic deposits. A summary of the formations and depth intervals logged is presented in Table 2. Many of the formation boundaries, particularly within the Eocene rocks, are quite subjective. Puri and Winston (1974, p. 26) described the problem: "Lack of any consistent lithologic horizon in the almost universally porous limestone, as well as the erratic nature of dolostone occurrence, precluded using lithologic parameters for subdividing the section."

Reefs and carbonate-sand bodies are represented in the Pleistocene Key Largo Limestone and Miami Oolite, respectively (Stanley, 1966; Hoffmeister et al, 1967; Halley et al, 1977; Perkins, 1977). Platform-interior facies, well represented in Oligocene and Eocene formations (Table 2), include open-marine grainstones, wackestones and mudstones, lagoonal wackestones and mudstones, and nearshore and tidal-flat limestones and dolomites that were deposited in a cyclic succession interpreted as reflecting transgressions and regressions across central Florida (Randazzo, 1972; Randazzo and Saroop, 1976; Zachos, 1978).

Complex diagenetic facies, spanning a range of typical early-diagenetic alteration styles, also occur in these carbonate rocks. Diagenetic alteration in many of the formations has been described by Fischer (1953), Stanley (1966), Robinson (1967), Hanshaw et al (1971), Randazzo (1972), Randazzo et al (1977), Perkins (1977), Halley and Rose (1977), and Randazzo and Hickey (1978). Alteration has occurred along surfaces of subaerial exposure, in fresh-water vadose zones and fresh-water phreatic zones, and in zones of mixing between fresh and sea water. Dolomite in the aquifers of south Florida may be the result of mixing-zone dolomitization (Hanshaw et al, 1971), and this and other origins for Eocene dolomites are discussed by Puri and Winston (1974). About 40% of the rocks examined in this study are very permeable and occur in the high-permeability zones of Puri and Winston (1974). About half of these rocks, or 20% of the total, are vuggy and cavernous, but porosity measurement by borehole gravimetry was considered a reliable measurement technique.

The climate of south Florida today is humid (about 39 to 59 in. [100 to 150 cm] rainfall per year) but was much more arid in the past as evidenced by evaporites in Paleocene formations of the peninsula. The effect of climate changes on carbonate porosity, if any, is unclear and must await further investigation. A significant study of petrographic, mineralogic, and trace-element changes during progressive diagenesis of carbonate rocks from an arid environment by Gavish and Friedman (1969) did not include quantitative porosity measurements.

Table 1. Techniques for Porosity Measurement and Associated Typical Sample Volumes (cm^3)

Scanning electron microscopy*	$10^{-12}/10^{-4}$
Thin-section analysis	10^{-2}
"Perm" plug analysis (porosimeter)	10
Gamma ray attenuation porosity evaluation (GRAPE)	10
Whole core analysis	10^2
Neutron log analysis	10^5
Borehole gravimetry	10^9

*Typically this method is used only qualitatively.

RESULTS

Porosity data for the five logged wells are shown collectively in Figure 2 and for individual wells in Figure 3. Zones with clear evidence of porosity plugging by sulfate minerals have been eliminated from Figure 2, as have intervals containing more than 50% shale or quartz sand. No attempt, other than separating dolomites and limestones in Figure 4, has been made to differentiate facies that may be petrographically or lithologically distinct, although the depositional and diagenetic variations described in the previous section probably account for much of the porosity scatter at a given depth (Fig. 2). The general data trends thus are functions of carbonate rocks in general, rather than of particular carbonate facies.

In the following paragraphs, Figures 2, 3, and 4 are discussed with respect to six main points:

1. Average porosity of the carbonate sequence is high.
2. Near-surface porosity approaches that of modern

Table 2. Approximate Formation Tops and Logged Intervals (in Meters Below Surface) for Wells Located in Figure 1*

Well Number from Figure 1 Florida Bureau of Geology Number	1 W13287	2 W14167	3 W13768	4 W13752	5 W12799
Unconsolidated quartz sand and clay	0	0	--	--	--
Miami Oolite (Pleistocene)	--	--	0	--	--
Key Largo Limestone (Pleistocene)	--	--	--	0	0
Fort Thompson Formation (Pleistocene)	--	--	10	--	--
Tamiami Formation (Pliocene)	--	30	30	34	53
Hawthorn Formation (Miocene)	33	137	76	84	107
Tampa Limestone (Miocene)	--	--	116	190	174
Suwannee Limestone (Oligocene)	--	152	277	274	
Ocala Limestone (Eocene)	--	236	--		
Avon Park Limestone (Eocene)	53	298	343		
Lake City Limestone (Eocene)	168	451	526		
Oldsmar Limestone (Eocene)	426	610			
Cedar Keys Formation (Paleocene)	700				
Top logged interval	2	2	6	2	2
Base logged interval	733	716	842	296	178
Number of porosity intervals	31	30	65	33	28

*Dashes indicate units are absent.

carbonate sediments.

3. Porosity decreases with depth in a gradual, irregular, but persistent manner.
4. Carbonate intervals with less than 20% porosity are rare.
5. Aquifers and aquicludes cannot be distinguished on the basis of porosity.
6. Dolomite porosity is less than or equal to limestone porosity.

The average porosity of the carbonate intervals shown in Figure 2 is 37%, reflecting an extremely porous mass of limestone and dolomite. For comparison, six south Florida oil reservoirs in the Cretaceous Sunniland Limestone average 18% porosity at a depth of about 11,500 ft (3,500 m). Nine oil reservoirs between 4,600 and 5,900 ft (1,400 and 1,800 m) in the carbonate Jurassic Smackover Formation of Arkansas have an average porosity of 22%. The rocks logged in this study seem unusually porous compared to "average" limestones and dolomites, but are probably not atypical compared to other Cenozoic carbonate accumulations buried to similar depths and deposited in similar tectonic settings. Highly porous sequences such as that of south Florida should occur in other thick Cenozoic carbonate deposits, such as those of the Bahama Islands, Yucatan Peninsula, and some Pacific atolls.

The average porosity of 15 carbonate intervals near the surface (with tops above 82 ft [25 m]) is 44%, a value approaching that of modern carbonate sands. Enos and Sawatsky (1981) found the average porosity of modern grainstone and packstone equivalents to be 44 and 55%, respectively. The porosity of modern ooid sands and Holocene oolites also falls in this range (Halley and Harris, 1979). The high porosities measured in this study support previous studies by Halley and Beach (1979) which indicate that early mineralogical stabilization of shallow-water carbonates does not greatly affect average porosity, although the character of the pore space is greatly altered.

Porosity in each well shows an irregular but persistent decrease with increasing depth (Fig. 3), as does the composite porosity plot of all carbonate intervals (Fig. 2). The composite plot is approximated by the exponential curve:

$$\phi = 43.2e^{-0.000575z}, \tag{3}$$

where ϕ is the porosity in percent and Z is the depth in meters. Porosity as represented by equation 3 decreases about 16% between the surface and 2,600 ft (800 m). Scatter about the primary exponential trend is due to experimental measurement errors and the secondary dependence of porosity upon factors such as age and depositional and diagenetic history. The porosity-loss curve derived here for shallow-water carbonate rocks is similar to that for chalks from the United States Gulf Coast described by Scholle (1977). The rate of porosity loss with depth for these rocks is less than that for Celtic Sea and Berkshire chalks both greater than that of Scotian Shelf and North Sea chalks. The mineralogic and textural homogeneity of chalks aids in establishing pressure solution and cementation (solution transfer of Bathurst, 1975) as the primary process causing porosity loss with depth in pelagic carbonates (Schlanger and Douglas, 1974; Scholle, 1977).

Carbonate intervals with less than 20% porosity are rare in the Cenozoic rocks examined here (Fig. 2). On the sample scale of borehole gravimetry, complete or near-complete porosity destruction does not occur. At a much smaller scale, samples with little or no porosity can be found (Fig. 5a), but such samples are not representative of the whole rock.

The wells near Vero Beach and Miami (Fig. 1, numbers 2 and 3) were drilled into the Floridan aquifer as waste-disposal wells. In well 2, the top of the Floridan aquifer is at about 330 ft (100 m) and confining beds occur from 1,398 to 1,952 ft (426 to 595 m) and from 2,119 ft (646 m) through the deepest measurement interval. In well 3, the top of the aquifer is at about 755 ft (230 m), and confining beds occur from 1,903 (580 m) through the deepest measurement interval. The porosity of these confining beds is not significantly different from that of the more transmissive parts of the Floridan aquifer (Fig. 3). Likewise, five

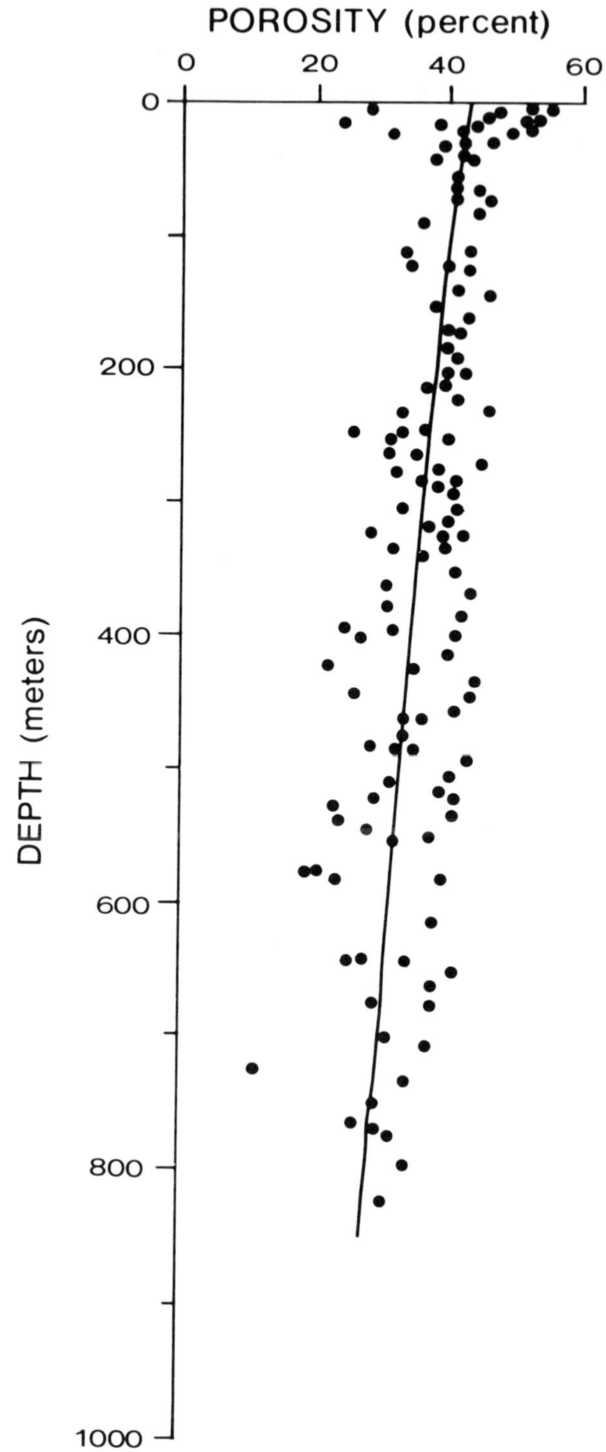

FIG. 2—Composite plot of porosity versus depth for five wells illustrated in Figure 3. Sulfate and terrigenous clastic intervals have been omitted. Line indicates exponential fit to data points.

measurements in the Biscayne aquifer (uppermost 115 ft [35 m] of well 3) yield an average porosity of 38%. The 259 ft (80 m) of confining beds in the Floridan aquifer between 1,916 and 2,175 ft (584 and 663 m) in this well also average 38% porosity. Aquifers and aquicludes cannot be differentiated on the basis of porosity, illustrating that permeability and total porosity do not correlate strongly in the rocks considered here.

The data of Figure 2 are replotted in Figure 4 to show limestone (50 to 100% limestone) and dolomite (50 to 100% dolomite) intervals. About one-third of the intervals is dolomite. (The dolomite interval at 2,380 ft [725 m] with a porosity of 11% is probably partly cemented with anhydrite.) For the rock units and depth range considered here, dolomites are not more porous than limestones. Below about 1,150 ft (350 m), dolomites appear to be less porous than limestones. Because of the limited number of wells and the possible effects on porosity of lateral formation variations, different ages, and different depositional and diagenetic histories, we would be very cautious about generalizing these observations. Nevertheless, within the data set described in this report, the porosity of limestones is greater than or equal to that of dolomites.

DISCUSSION

The porosity of carbonate rocks investigated in this study ranges from about 45% near the surface to 25 or 30% at a depth of 2,600 ft (800 m). The trend of decreasing porosity with depth is present in the data of individual wells (Fig. 3) and in the composite plot of all carbonate intervals (Fig. 2). Carbonate rocks of south Florida retain significant porosity as they are buried in the subsurface during basin subsidence. The grouping of data representing rocks of different ages and significantly different depositional and diagenetic histories does not mask the basic trend of porosity decrease with depth. With the exception of the thin cavernous intervals mentioned above, no particularly unusual physical influences appear to be acting upon the shallow carbonate rocks of peninsular Florida. We speculate that the porosity data presented here may be representative of the "typical" transition from near-surface to deeply buried limestones and dolomites.

Loss of Porosity with Depth

The borehole gravity measurements do not indicate directly the causative processes affecting porosity in the subsurface. Thin sections show evidence of mechanical compaction and associated porosity loss (Fig. 5b). However, laboratory compaction of carbonate sediments over a pressure range equivalent to depths of 0 to 2,950 ft (900 m) (Terzaghi, 1940; Fruth et al, 1966; Hathaway and Robertson, 1961; Robertson, 1967; Morelock and Bryant, 1971; E. A. Shinn, personal commun., 1980) reduces porosity to a minimum of only 35% (Fig. 6). In nature, early cementation strengthens the sediment fabric and inhibits grain readjustment, fracturing, and deformation relative to laboratory results. Thus mechanical compaction is probably not the primary cause of the porosity-depth relations shown in Figures 2 and 3.

A steady deposition of cement as a function of time or depth is another possible reason for the decrease of porosity observed in our data. Thin-section examples of cement deposition are common (Figs. 5c, d), but to directly evalu-

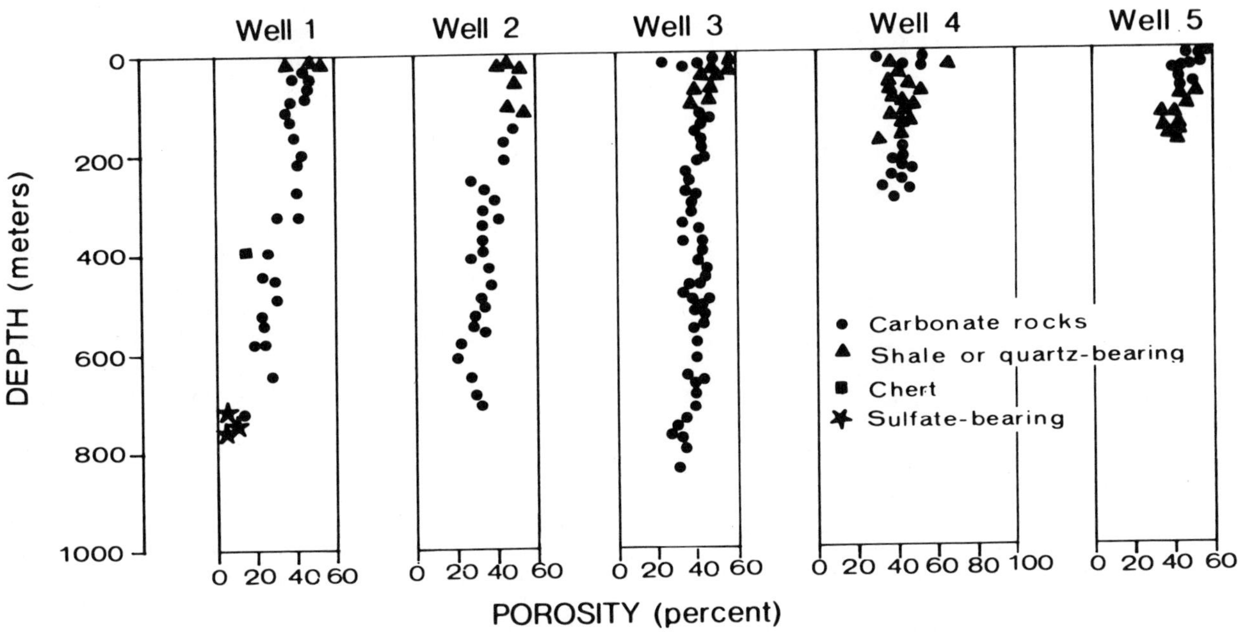

FIG. 3—Porosity versus depth for each of five wells examined in this study.

ate the nature and significance of cementation on a reservoir scale is difficult.

Porosity loss due to the precipitation of externally derived cement carried by the circulation of pore water should be a permeability-dependent process. However, our data show little porosity difference between permeable (aquifer) and impermeable (aquiclude) lithologies. Carbonate aquicludes are not the more cemented, less porous equivalents of carbonate aquifers.

The porosity decrease associated with the accumulation of externally derived cement might be a time-dependent rather than a depth-dependent process, and we can look for evidence of time dependence in our data. No consistent porosity changes are associated with the unconformities penetrated in wells 1, 2, and 3 (Fig. 3, Table 2). The average porosity of the Eocene Avon Park, Lake City, and upper Oldsmar Limestones is plotted as a function of average depth for wells 1, 2, and 3 in Figure 7. Connecting lines paralleling the exponential curve imply depth-dependent porosity; connecting lines trending more to the vertical imply age-dependent porosity. Both examples are present, and the evidence presented by Figure 7 is mixed. Taken as a whole, our indirect evidence bearing on the question of externally derived cement is also mixed, due in large part to lateral formation variations, few available wells, and a limited age and depth range. The importance of this mechanism in the shallow carbonate rocks of south Florida cannot be evaluated here.

Another cement-producing process that is a likely contributor to the porosity trend of Figures 2 and 3 is that of pressure solution. Petrographic evidence of pressure solution occurs as shallow as 390 ft (120 m) in the rocks studied here (Figs. 5e, f). In the high-porosity ranges typical of our data, dissolution at grain contacts is by itself sufficient to reduce porosity significantly; the redeposition of cement derived from pressure dissolution is a relatively unimportant porosity-reducing mechanism. For example, for packed spherical grains it can be shown that more than two-thirds of the porosity loss is due to grains moving closer together. In contrast, less than one-third of the porosity loss is due to the addition of cement derived from pressure solution, until total porosity drops to around 20% (Manus and Coogan, 1974; Mitra and Beard, 1980).

Pressure solution is generally regarded as a significant depth-dependent process in deeply buried carbonate rocks, but published evidence for pressure solution at shallow depths is meager. Schlanger (1964) described microstylolites formed under less than 330 ft (100 m) of overburden. Dunnington (1967) observed pressure-solution features when overburden reached 1,970 to 2,950 ft (600 to 900 m), and Buxton and Sibley (1981) documented extensive pressure-solution features in a Devonian limestone buried to a maximum depth of 4,900 ft (1,500 m).

Theoretical calculations by Neugebauer (1973, 1974) indicate that pressure solution should be a significant process at depths less than 3,300 ft (1,000 m) if pore fluids have low magnesium concentrations. The Cenozoic rocks of south Florida may have been exposed to waters with relatively low magnesium content at various times in their burial history, allowing the process of pressure solution to operate at relatively shallow depths. Pressure-solution contacts between grains in the Avon Park Limestone at 390 ft (120 m) depth (Fig. 5e) may owe their existence to fresh water now present within this formation beneath much of peninsular Florida.

Bearing on Dolomitization

The porosity of dolomites relative to associated limestones has received much attention, especially with respect to hydrocarbon reservoirs at depths greater than those

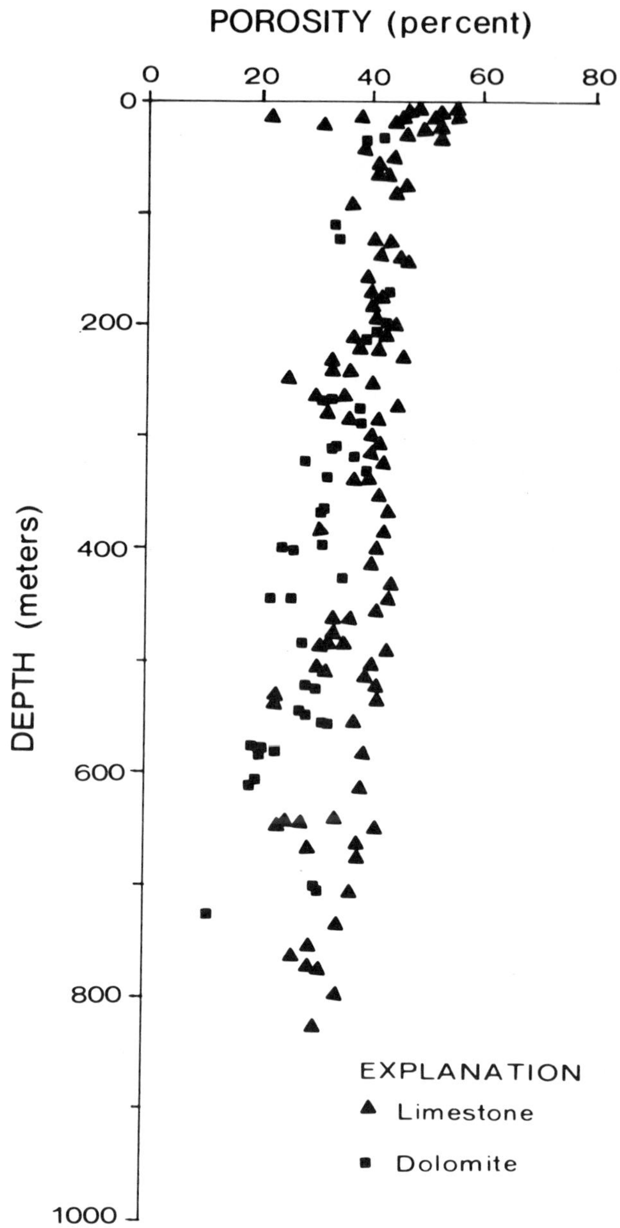

FIG. 4—Composite porosity plot as in Figure 2, both with dolomite and limestone differentiated.

considered here. The general belief, particularly when Paleozoic rocks are considered, is that dolomites are more porous than limestones (Pray and Murray, 1965), although permeability may sometimes be equated with porosity in these comparisons.

The observation that dolomite is not more porous than limestone at shallow depths in southern Florida is perhaps best understood by recalling that dolomite is the replacement product of highly porous limestones or metastable carbonate sediments. During dolomitization, porosity is inherited from preexisting carbonate rock or sediment.

Weyl (1960) pointed out that a 13% volume decrease is associated with the replacement of calcite by dolomite if the replacement is accomplished by an ion-for-ion replacement of calcium for magnesium. This reaction is described by the equation:

$$Mg^{++} + 2CaCO_3 \rightarrow MgCa(CO_3)_2 + Ca^{++}. \quad (4)$$

Because of the high initial porosity of these limestones, this mechanism of dolomitization would result in a porosity increase of only 8 to 9%. Such a porosity increase is not observed and may be overshadowed by compaction and cementation processes. Alternatively, dolomitization may proceed by incorporating new carbonate in addition to magnesium. This reaction is represented by the equation:

$$Mg^{++} + CO_3^{=} + CaCO_3 \rightarrow MgCa(CO_3)_2, \quad (5)$$

and results in a volume increase during dolomitization. If both reactions 4 and 5 contribute to dolomitization, reaction 5 would have to occur twice as often as reaction 4 for there to be no volume change associated with dolomitization. A third possibility is that some dolomite has been added as cement, a process represented by the reaction:

$$Mg^{++} + Ca^{++} + 2CO_3^{=} \rightarrow MgCa(CO_3)_2. \quad (6)$$

Complete porosity loss is possible through the addition of dolomite cement to the rock. Porosity data reported here suggest reaction 4 is not the predominant dolomite-forming mechanism in the Cenozoic of south Florida and that both magnesium and carbonate have been added to the rocks during dolomitization. Such a loss of porosity during dolomitization has been observed elsewhere and is termed "overdolomitization" by F. J. Lucia (personal commun., 1981).

Significance to Early Cementation

The plethora of early-cementation studies during the past 20 years has led to a particularly heavy emphasis on early diagenesis in carbonate rocks. Some authors suggest that most cementation may occur at relatively shallow depths (Friedman, 1975; Longman, 1980). Such work has strongly influenced the study of ancient sequences and resulted in interpretations which may be skewed toward early diagenetic processes at the expense of later burial alterations. This influence is particularly true with reference to porosity loss. Early diagenetic processes undoubtedly exert a strong influence on permeability characteristics and later diagenetic modifications of carbonate rocks. However, from the volume of pore space preserved in south Florida carbonates, we see that early diagenesis (with the exception of marine cementation) is ineffective at reducing porosity. Significant porosity reduction does not begin until carbonate strata are buried several hundred meters below the surface. It is the later porosity reduction during burial that must be documented to construct a workable understanding of porosity evolution of carbonate rocks.

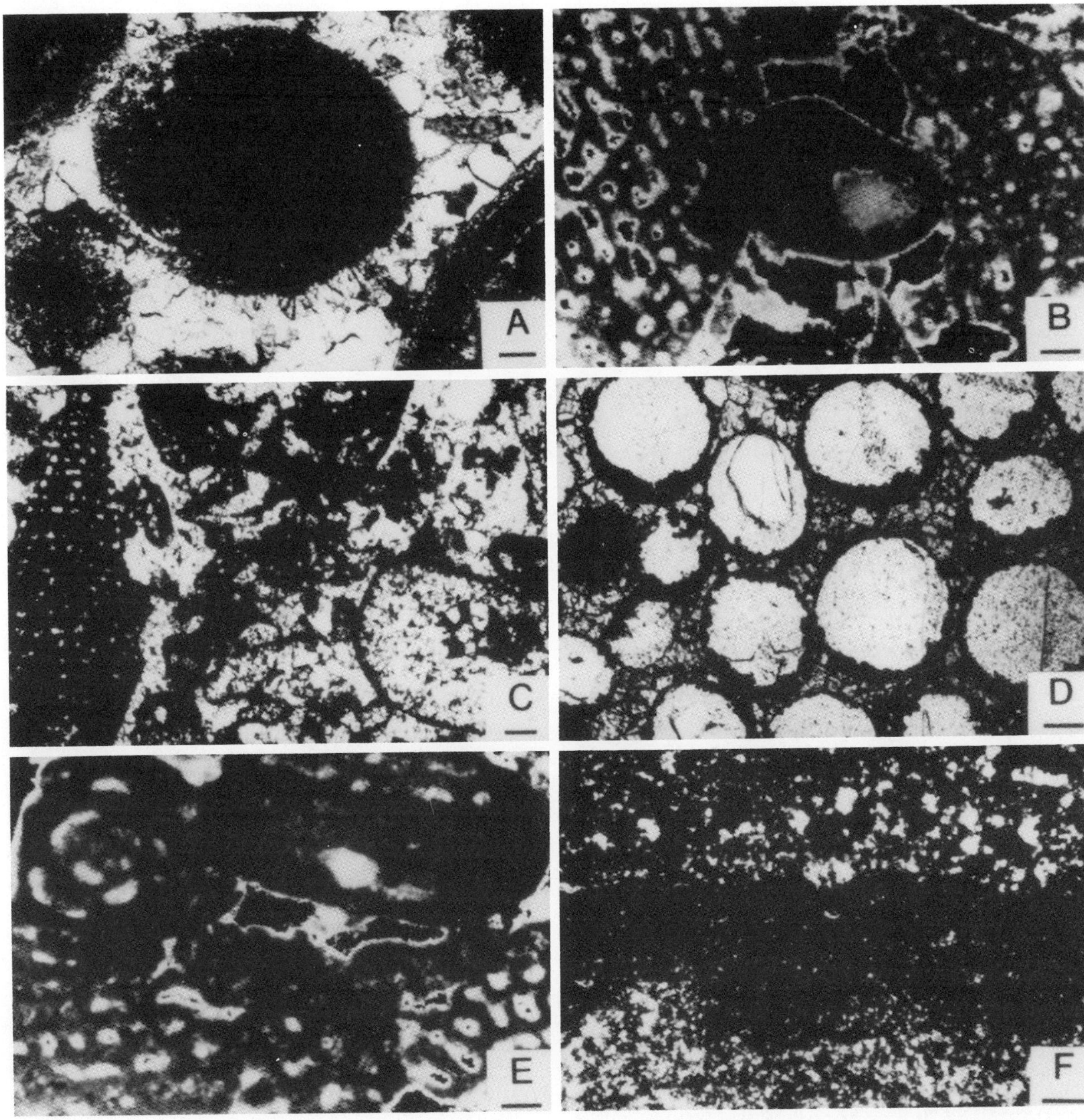

FIG. 5—Photomicrographs illustrating petrographic features associated with porosity modification of Cenozoic rocks of south Florida. The character and distribution of many of these features preclude quantitative evaluation. Nevertheless, their occurrences in these rocks demonstrate several mechanisms discussed in text. Bar scales below figure letters are 100 μm. 5a—Completely cemented part of Miami Oolite, common in thin section but not representative of unit as a whole. 5b—Grain interpenetration, a compaction feature that may or may not involve dissolution of calcite at grain boundaries. Because carbonate grains typically contain intragranular porosity, grain interpenetration may be accommodated by grain deformation rather than dissolution along grain boundaries. 5c—Well-cemented grainstone in Avon Park Limestone. Such early cement may or may not result in net porosity loss depending on volume of secondary porosity developed during early diagenesis. 5d—Cement in Miami Oolite, derived from aragonite dissolved from ooid grains so that total porosity remains almost unchanged. 5e—Stylolitic contacts between grains, from Avon Park Limestone, 390 ft (120 m) below surface, evidence of shallow pressure solution. These contacts are characterized by a dark residue, perhaps organic matter, along stylolitic seams. 5f—Similar seams occur in mudstones but are difficult to evaluate in absence of recognizable grains.

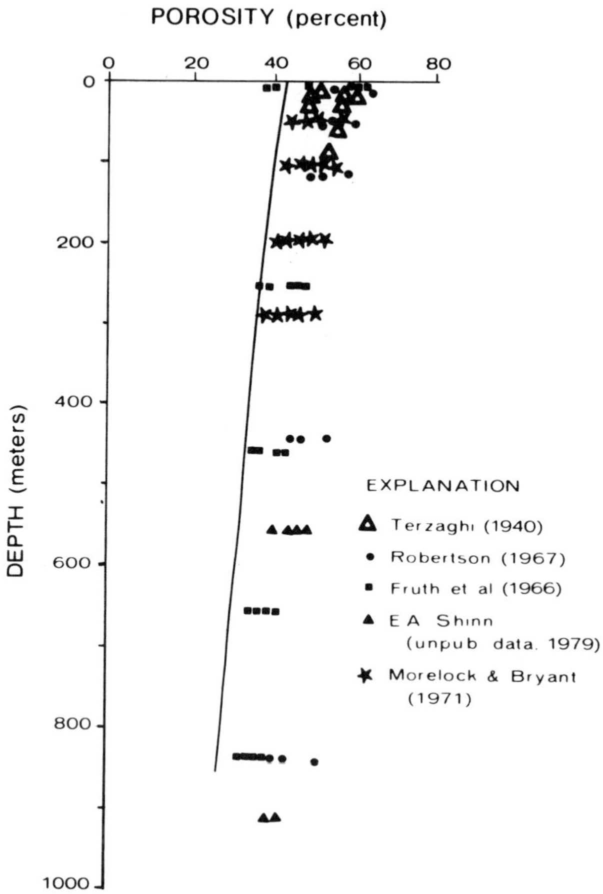

FIG. 6—Comparison of south Florida porosity (line is exponentially fit from Fig. 2) with laboratory compaction data.

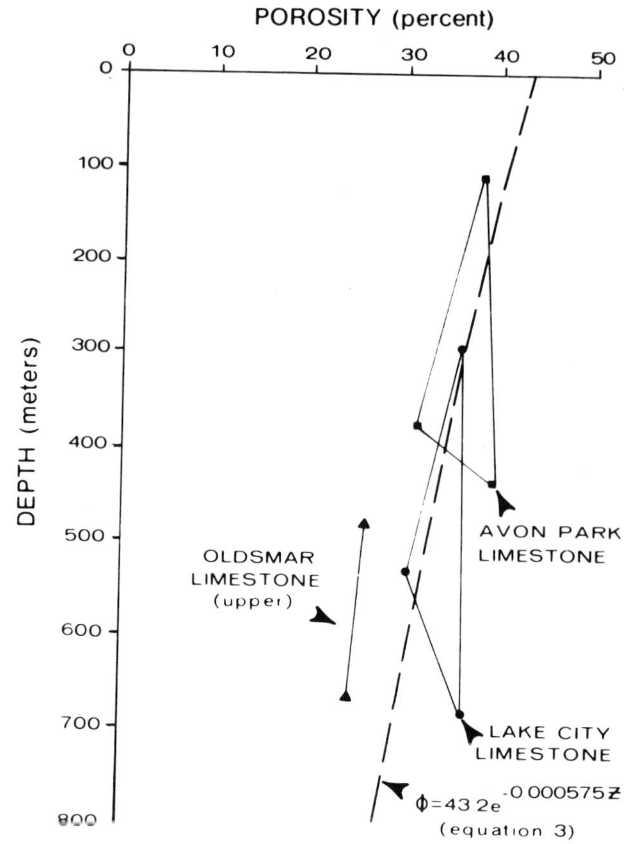

FIG. 7—Age dependence versus depth dependence of porosity from Cenozoic carbonates of south Florida. See text for detailed explanation.

REFERENCES CITED

Bathurst, R. G. C., 1975, Carbonate sediments and their diagenesis (2d ed.): New York, Elsevier Publishing Co., 658 p.

——— 1976, The evolution of porosity in carbonate rocks and the emplacement of cement: Carbonate Seminar, Jakarta, September 12-19, Proc., Spec. volume, Indonesian Petrol. Assoc., p. 112-115.

Buxton, T. M., and D. F. Sibley, 1981, Pressure solution features in a shallow buried limestone: Jour. Sed. Petrology, v. 51, p. 19-26.

Dunham, R. F., 1969, Early vadose silt in Townsend mound (reef), New Mexico, *in* G. M. Friedman, ed., Depositional environments in carbonate rocks: SEPM, Spec. Pub. 14, p. 139-181.

Dunnington, H. V., 1967, Aspects of diagenesis and shape change in stylolitic limestone reservoirs: 7th World Petrol. Congr., Mexico City, Proc., v. 2, p. 339-352.

Enos, Paul, and L. H. Sawatsky, 1979, Pore space in Holocene carbonate sediments (abs.): AAPG Bull., v. 63, p. 445.

——— ——— 1981, Pore networks in Holocene carbonate sediments: Jour. Sed. Petrology, v. 51, p. 961-985.

Fischer, A. G., 1953, Petrology of Eocene limestones in and around the Citrus-Levy County area, Florida: Florida Geol. Survey Rept. Inv. 9, pt. II, p. 41-70.

Friedman, G. M., 1964, Early diagenesis and lithification in carbonate sediments: Jour. Sed. Petrology, v. 34, p. 777-813.

——— 1975, The making and unmaking of limestones or the downs and ups of porosity: Jour. Sed. Petrology, v. 45, p. 379-398.

Fruth, L. S., Jr., G. R. Orme, and F. A. Donath, 1966, Experimental compaction effects in carbonate sediments: Jour. Sed. Petrology, v. 36, p. 747-754.

Gavish, Eliezer, and G. M. Friedman, 1969, Progressive diagenesis in Quaternary to late Tertiary carbonate sediments: Sequence and time scale: Jour. Sed. Petrology, v. 39, p. 980-1006.

Halley, R. B., 1978, Estimating pore and cement volumes in thin section: Jour. Sed. Petrology, v. 48, p. 642-650.

——— and D. K. Beach, 1979, Porosity preservation and early freshwater diagenesis of marine carbonate sands (abs.): AAPG Bull., v. 63, p. 460.

——— and P. R. Harris, 1979, Fresh-water cementation of a 1000-year-old oolite: Jour. Sed. Petrology, v. 49, p. 969-988.

——— and P. R. Rose, 1977, Significance of fresh-water limestones in marine carbonate successions of Pleistocene and Cretaceous age, *in* D. G. Bebout and R. G. Loucks, eds., Cretaceous carbonates of Texas and Mexico: Texas Univ. Bur. Econ. Geology Rept. Inv. 89, p. 206-215.

——— et al, 1977, Pleistocene barrier bar seaward of ooid shoal complex near Miami, Florida: AAPG Bull., v. 61, p. 519-526.

Hanshaw, B. B., William Back, and R. G. Deike, 1971, A geochemical hypothesis for dolomitization by groundwater: Econ. Geology, v. 66, p. 710-724.

Harrison, R. S., 1975, Porosity in Pleistocene grainstones from Barbados: Some preliminary observations: Bull. Canadian Petrol. Geology, v. 23, p. 383-392.

Hathaway, J. C., and E. C. Robertson, 1961, Microtexture of artificially consolidated aragonite mud: U.S. Geol. Survey Prof. Paper 424-C, p. 301-304.

Hoffmeister, J. E., K. W. Stockman, and H. G. Multer, 1967, Miami limestone of Florida and its recent Bahamian counterpart: Geol. Soc. America Bull., v. 78, p. 175-190.

Land, L. S., F. T. MacKenzie, and S. J. Gould, 1967, Pleistocene history of Bermuda: Geol. Soc. America Bull., v. 78, p. 993-1006.

Longman, M. W., 1980, Carbonate diagenetic textures from nearsurface diagenetic environments: AAPG Bull., v. 64, p. 461-487.

Manus, R. W., and A. H. Coogan, 1974, Bulk volume reduction and pressure-solution derived cement: Jour. Sed. Petrology, v. 44, p. 466-471.

Matthews, R. K., 1968, Carbonate diagenesis: Equilibrium of sedimentary mineralogy to the subaerial environment; coral cap of Barbados, West Indies: Jour. Sed. Petrology, v. 38, p. 1110-1119.

Mitra, Shankar, and W. C. Beard, 1980, Theoretical models of porosity reduction by pressure solution for well-sorted sandstone: Jour. Sed. Petrology, v. 50, p. 1347-1360.

McCulloh, T. H., 1966, The promise of precise borehole gravimetry in petroleum exploration and exploitation: U.S. Geol. Survey Circ. 531, 12 p.

——— et al, 1967a, The U.S. Geological Survey-LaCoste and Romberg precise borehole gravimeter system—Instrumentation and support equipment, *in* Geological Survey research 1967: U.S. Geol. Survey Prof. Paper 575-D, p. D92-D100.

——— et al, 1967b, The U.S. Geological Survey-LaCoste and Romberg precise borehole gravimeter system—test results, *in* Geological Survey research 1967: U.S. Geol. Survey Prof. Paper 575-D, p. D101-112.

Morelock, Jack, and W. R. Bryant, 1971, Consolidation of marine sediments, *in* Richard Rezak and H. J. Vernon, eds., Contributions on the geological and geophysical oceanography of the Gulf of Mexico: Texas A&M Univ. Oceanographic Studies, v. 3, p. 181-202.

Neugebauer, Joachim, 1973, The diagenetic problem of chalk—the role of pressure solution and pore fluid: Neues Jahrbuch für Geologie und Paläontology Abhandlungen, v. 143, p. 223-245.

——— 1974, Some aspects of cementation in chalk, *in* K. J. Hsü and H. C. Jenkyns, eds., Pelagic sediments: On land and under the sea: Internat. Assoc. Sedimentologists Spec. Pub. 1, p. 149-176.

Perkins, R. P., 1977, Depositional framework of Pleistocene rocks in south Florida, *in* Paul Enos and R. D. Perkins, eds., Quaternary sedimentation in south Florida: Geol. Soc. America Mem. 147, p. 131-198.

Pittman, E. D., 1974, Porosity and permeability changes during diagenesis of Pleistocene corals, Barbados, West Indies: Geol. Soc. America Bull., v. 85, p. 1811-1820.

Pratt, B. R., 1982, Limestone response to stress: Pressure solution and dolomitization-discussion and examples of compaction in carbonate sediments: Jour. Sed. Petrology, v. 52, p. 323-328.

Pray, L. C., and R. C. Murray, 1965, Dolomitization and limestone diagenesis, a symposium: SEPM Spec. Pub. 13, 180 p.

Puri, H. S., and G. O. Winston, 1974, Geologic framework of the high transmissivity zones in southern Florida: Florida Bur. Geology Spec. Pub. No. 20, 101 p.

Randazzo, A. F., 1972, Petrology of the Suwannee Limestone: Florida Bur. Geology Bull. 54, pt. II, 13 p.

——— and E. W. Hickey, 1978, Dolomitization in the Floridan aquifer: Am. Jour. Science, v. 278, p. 1177-1184.

——— and H. C. Saroop, 1976, Sedimentology and paleoecology of middle and upper Eocene carbonate shoreline sequences, Crystal River, Florida, USA: Sed. Geology, v. 15, p. 259-291.

——— G. C. Stone, and H. C. Saroop, 1977, Diagenesis of middle and upper Eocene carbonate shoreline sequences, central Florida: AAPG Bull., v. 61, p. 492-503.

Robbins, S. L., 1980, Bibliography with abridged abstracts of subsurface gravimetry (especially borehole) and corresponding *in situ* rock density determinations: U.S. Geol. Survey Open-File Rept. 80-710, 47 p.

——— 1981, Reexamination of the values used as constants in calculating rock density from borehole gravity data: Geophysics, v. 46, p. 208-210.

Robinson, R. B., 1967, Diagenesis and porosity development in Recent and Pleistocene oolites from southern Florida and the Bahamas: Jour. Sed. Petrology, v. 37, p. 355-364.

Robertson, E. C., 1967, Laboratory consolidation of carbonate sediment, *in* A. F. Richards, ed., Marine geotechnique: Internat. Research Conf. on Marine Geotechnique, Urbana, Illinois, Univ. Illinois Press, p. 118-127.

Schlanger, S. O., 1963, Subsurface geology of Eniwetok Atoll: U.S. Geol. Survey Prof. Paper 260B, p. 991-1066.

——— 1964, Petrology of the limestones of Guam: U.S. Geol. Survey Prof. Paper 403-D, p. 1-52.

——— and R. G. Douglas, 1974, The pelagic ooze-chalk-limestone transition and its implication for marine stratigraphy, *in* K. J. Hsü and H. C. Jenkyns, eds., Pelagic sediments: on land and under the sea: Internat. Assoc. Sedimentologists Spec. Pub. 1, p. 117-148.

Schmoker, J. W., and R. B. Halley, 1982, Carbonate porosity versus depth: a predictable relation for south Florida: AAPG Bull., v. 66, p. 2561-2570.

——— et al, 1979, Preliminary porosity estimates of south Florida Cenozoic carbonate rocks based on borehole gravity measurements: U.S. Geol. Survey Open-File Rept. 79-1652, 17 p.

Scholle, P. A., 1977, Chalk diagenesis and its relation to petroleum exploration: Oil from chalks, a modern miracle?: AAPG Bull., v. 61, p. 982-1009.

Shinn, E. A., et al, 1977, Limestone compaction: an enigma: Geology, v. 5, p. 21-24.

Stanley, S. M., 1966, Paleoecology and diagenesis of the Key Largo Limestone, Florida: AAPG Bull., v. 50, p. 1927-1947.

Steinen, R. P., 1974, Phreatic and vadose diagenetic modification of Pleistocene limestone: Petrographic observations from the subsurface of Barbados, West Indies: AAPG Bull., v. 58, p. 1008-1024.

Terzaghi, R. D., 1940, Compaction of lime mud as a cause of secondary structure: Jour. Sed. Petrology, v. 10, p. 78-90.

Wanless, H. R., 1979, Limestone response to stress: Pressure solution and dolomitization: Jour. Sed. Petrology, v. 49, p. 437-462.

——— 1982, Limestone response to stress: Pressure solution and dolomitization—Reply: Jour. Sed. Petrology, v. 52, p. 328-333.

Weller, J. M., 1959, Compaction of sediments: AAPG Bull., v. 43, p. 273-310.

Weyl, P. K., 1960, Porosity through dolomitization: Conservation of mass requirements: Jour. Sed. Petrology, v. 30, p. 85-90.

Zachos, L. G., 1978, Stratigraphy and petrology of two shallow wells, Citrus and Levy Counties, Florida: Master's thesis, Univ. Florida, 103 p.

Reprinted by permission of Springer-Verlag New York Inc.
Published in *Carbonate Petroleum Reservoirs,* P.O. Roehl
and P.W. Choquette, editors, pp. 191-206.

11
Glenburn Field

Lee C. Gerhard

RESERVOIR SUMMARY

Location & Geologic Setting	Bottineau & Renville Co.'s, R81 & 82W, North Dakota, northern interior high plains, USA
Tectonics	Gentle structural and paleobathymmetric arch across paleoshelf with underlying salt solution
Regional Paleosetting	SE shelf of cratonic Williston Basin
Nature of Trap	Structural/stratigraphic
Reservoir Rocks	
Age	Mississippian
Stratigraphic Units(s)	Mission Canyon Formation (Frobisher-Alida interval)
Lithology(s)	Limestone
Dep. Environment(s)	Shallow shelf to sabkha, cyclically repeated vertically
Productive Facies	Pisolitic wackestone and grainstone-packstone
Entrapping Facies	Anhydrite caps and lateral seals, structurally assisted
Diagenesis/Porosity	Subaerial and vadose dissolution and pisolite and caliche formation; local dolomitization and cementation by anhydrite
Petrophysics	
Pore Type(s)	Fenestral-vug, interparticle (between pisolites)
Porosity	15–20% (visual estimate in pisolite cores), 17.3% avg
Permeability	23 md avg
Fractures	Uncommon
Source Rocks	
Age	Devonian
Lithology(s)	Bakken Shale
Migration Time	NA
Reservoir Dimensions	
Depth	4464 ft (1360 m)
Thickness	100 ft (30 m) oil column
Areal Dimensions	7 × 4 mi (11.2 × 6.4 km)
Productive Area	11,840 acres (47.6 km^2)
Fluid Data	
Saturations	S_w = 35% avg
API Gravity	27°
Gas-Oil Ratio	100:1
Other	—
Production Data	
Oil and/or Gas in Place	~130 million BO
Ultimate Recovery	26 million BO; URE = 20%
Cumulative Production	13.6 million BO (2.04 million m^3) through November 1979

Remarks: IP 205 BOPD, IP mechanism water drive. Discovered 1958.

11
Porosity Development in the Mississippian Pisolitic Limestones of the Mission Canyon Formation, Glenburn Field, Williston Basin, North Dakota

Lee C. Gerhard

Location and Discovery

The Glenburn Field is located in parts of Renville and Bottineau counties, North Dakota, approximately 100 miles (160 km) north of Bismarck, in the eastern part of the Williston Basin (Fig. 11-1). The field was discovered in February 1958 on a seismically defined structural high, by the Anschutz #1 Einar Christiansen well, which had initial production of 205 barrels of oil per day with 27° API gravity oil and a gas-oil ratio of about 100:1. Glenburn produces oil and associated gas from limestones of the Frobisher-Alida interval of the Mission Canyon (Mississippian). The nature of these limestone reservoirs has been a subject of differing interpretations and is the principal topic of this case study.

General Features of Williston Basin Mississippian Fields

The Frobisher-Alida is one of a series of informally named zones in the Mission Canyon that produce in the North Dakota, Montana, and Canadian portions of the Williston Basin. Certain of these zones are shown in Figure 11-2. Glenburn is one of a long cluster of Mississippian oil fields distributed along an arc-shaped trend extending from southern Saskatchewan and Manitoba southeastward into north-central North Dakota.

Oil production in the North Dakota part of the basin was first discovered in 1951, in Devonian carbonates along the Nesson Anticline, a major north-south structure in the northwestern part of the state. Mississippian production was established soon after the original Devonian discoveries.

Unlike the Nesson Anticline fields, the fields arranged along the northwest–southeast trend previously mentioned have little structural closure and seem to involve very little fracturing. They are located in a series of arcs in which the lowest zones of the Mississippian are productive in the eastward arcs and progressively younger zones in the arcs to the west. This arrangement of productive zones, as well as the distribution and shapes of the fields, suggests a strong influence on reservoir distribution of facies belts arranged generally parallel to depositional strike of the sedimentary basin. Drilling in flush areas soon demonstrated that fields are largely limited to small structural noses with updip evaporite barriers. Evaporites lateral to the reservoirs assist trapping where structures are not sufficient to form effective traps. The Mission Canyon reservoirs include a variety of limestone and dolomitic limestone facies.

Glenburn is the southeasternmost of the Frobisher-Alida fields along the arc-shaped trend in North Dakota. Another field producing from the same zone, Wiley, lies 15 miles (24 km) to the northwest and has similar reservoir lithologies

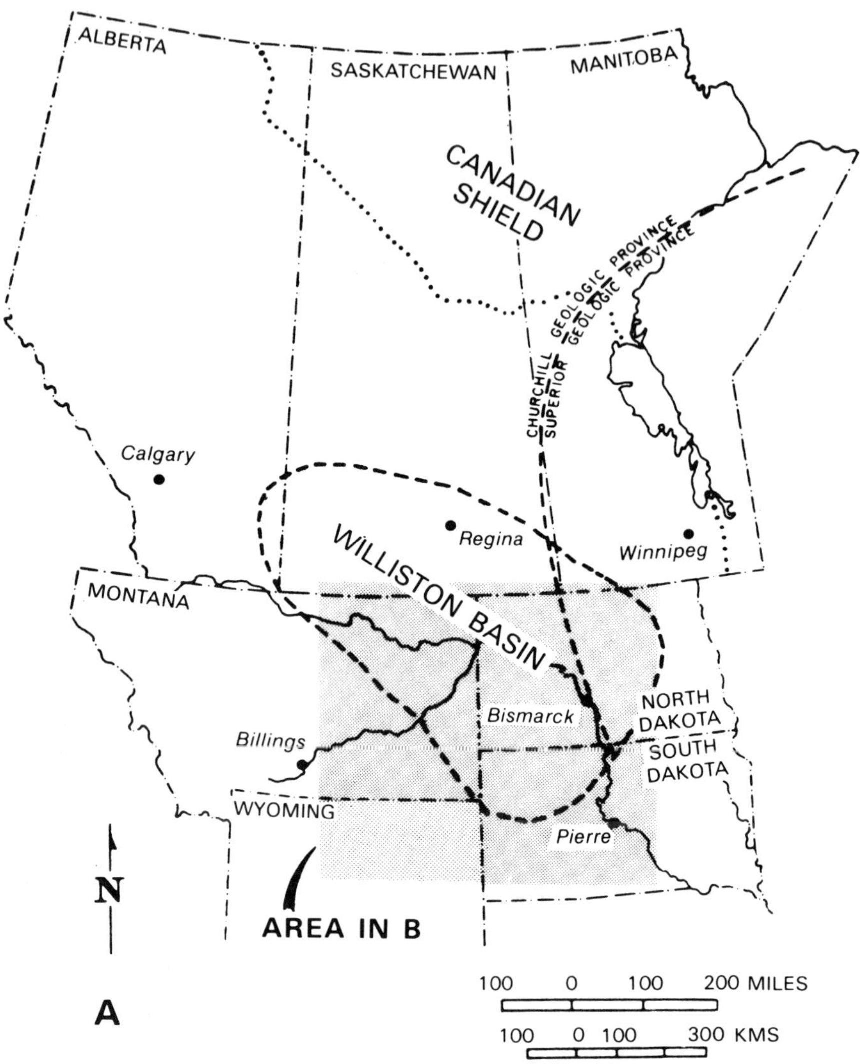

Fig. 11-1. Index map showing location of Glenburn Field. *A*. Regional setting of Williston Basin.

and trapping mechanisms. Glenburn was selected for investigation in this study because of its geographic position just north of extensive untested areas in Ward and McHenry counties, North Dakota, and because relatively good core control was available for study.

Glenburn Reserves and Production

Since its discovery in 1958, Glenburn has been on continuous production from the Frobisher-Alida interval of the Mission Canyon. In 1979 it produced 31,000 barrels of oil from 54 pumping wells and 160,000 barrels of water per month. Cumulative production to December 1, 1979, was 13,597,191 barrels of oil and 33,350,092 barrels of water. Ultimate recoverable reserves were originally estimated at 20 million barrels of oil and have since been revised to 26 million barrels.

Since that time, recognition of the stacked porosity zones deriving from the porosity development processes suggested here and in Gerhard *et al.* (1978) has resulted in a major redrilling and recompletion of the field. Originally spaced on one well per 80 acres, the field now sustains two wells per spacing unit, effectively decreasing the spacing to 40 acres. Production in June 1982 was about 62,000 barrels of oil from 93 wells. Nearly original reservoir pressures have been encountered in porosity zones unconnected to the original completed zone. Stacking of porosity zones implies that oil-producing zones may underlie water-saturated zones because of lack of continuity of the weathering-derived porous intervals. A similar recompletion of the Wiley Field to the north has further substantiated the concepts of porosity development outlined here.

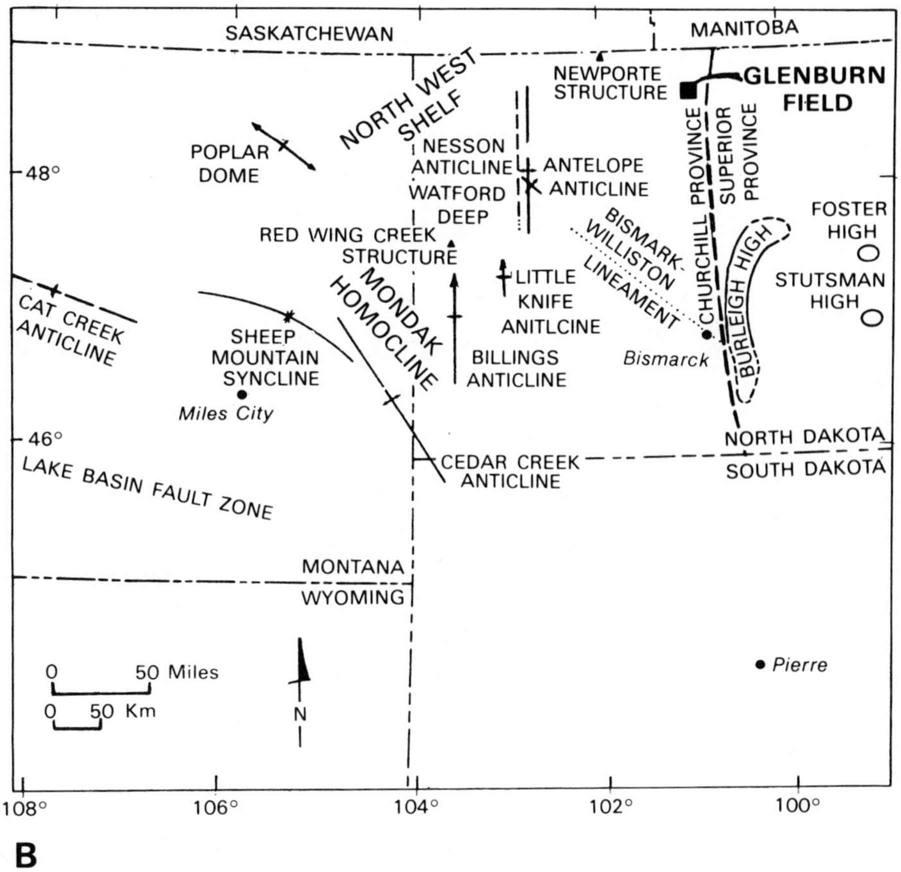

Fig. 11-1B. Structural elements in USA portion of Williston Basin.

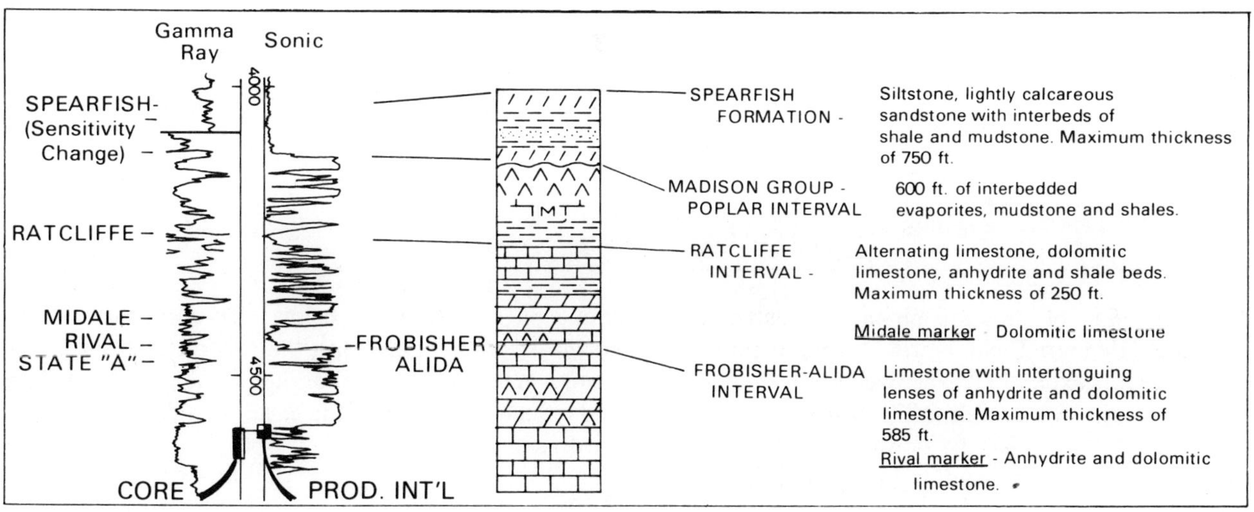

Fig. 11-2. Gamma-ray/laterolog response with a generalized lithologic column of the Frobisher-Alida zone in the Texota Oil Weber #1 well (NDGS #3630), Section 14-T158N-R82W, Renville County.

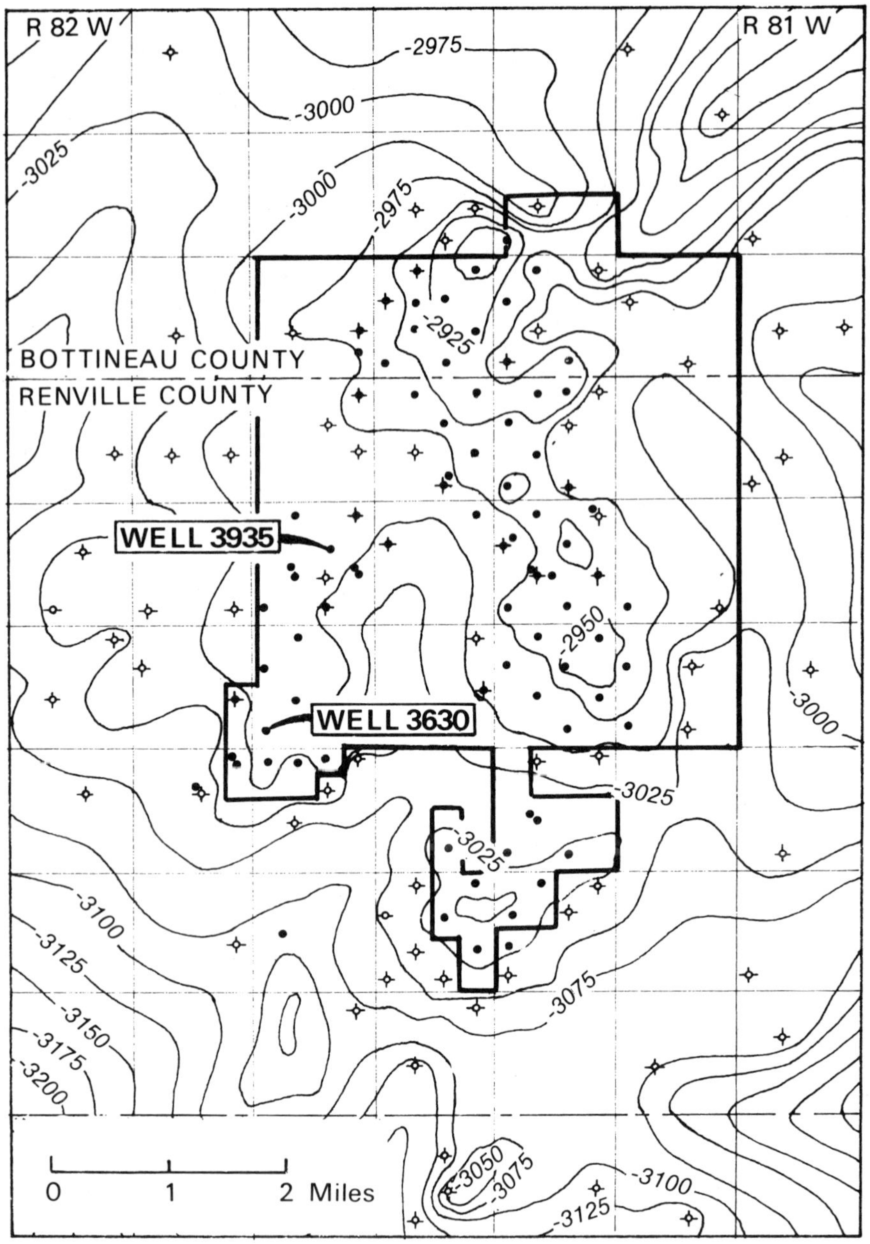

Fig. 11-3. Map of structure on the Frobisher-Alida carbonate, Glenburn Field, North Dakota. Contour interval 25 feet. Texota #1 Weber well (NDGS #3630) and Vaughn #1 Weber well (NDGS #3935), Section 11-T158N-R82W. Note the northeast and northwest trends. The northeast trend is interpreted to be the result of salt solution below the carbonates (e.g., northeast gradient of map); the less distinct northwest trend is interpreted to be parallel to the paleostrandline and reflects the shape of the eastern flank of the Williston Basin.

Reservoir and Trap

The top of the Frobisher-Alida reservoir lies at an average depth of 4464 feet (1361 m) below surface, with a maximum oil column approximately 100 feet (30 m) thick. The field is irregular in shape (Fig. 11-3) covering about 11,840 acres ($47.4\ km^2$). Structure as defined by the top of the Frobisher-Alida has relatively low relief of about 75 feet (23 m). Entrapment is accomplished by updip facies change from porous limestones to anhydrites with some structural assistance. The vertical seal is provided by thin terrigenous shales and massive anhydrites.

Frobisher-Alida reservoir rocks, with a net pay thickness averaging 10 feet (3 m) as originally drilled, are principally pisolitic limestones (grainstones and packstones) with vuggy fenestral and vuggy interparticle porosity. These facies and their fabrics and porosity are chiefly the result of subaerial exposure and a complex of vadose-diagenetic processes. The porous intervals tend to be relatively extensive and continuous areally. Their fabrics vary considerably in detail.

Porosity is in the form of vugs, in both matrix (muddy rocks) and between particles, and range up to 4 millimeters in longest dimension. No detailed quantitative determination of porosity is possible from the well logs available. Porosities by visual estimates range up to 15 to 20 percent, but more typically are lower. For example, in the Texota #1 Weber well, which was cored extensively, the most porous zone is a 5-foot (1.5-m) interval with 8 to 10 percent porosity.

Initial production was assumed to be a strong water drive, but some solution-gas drive is now evident; water saturations in the reservoir are high. Typical completions were open-hole, and Frobisher-Alida rocks were penetrated only a few feet so as to avoid encountering water. Typically, little acid and no fracturing treatment were used. Only in a few wells was the entire Frobisher-Alida section penetrated.

Previous Studies

Early studies of the Mississippian oil fields in North Dakota identified algal pisolites as a major lithology of the Mission Canyon, with updip permeability barriers of anhydrite as major seals for stratigraphic traps (Johnson, 1956). Other studies published around the same time described and illustrated "algal" and "oolitic" structures and lithologies in the MC-3 and MC-5 pay zones of the Mission Canyon, and interpreted these reservoir rocks as marine strandline facies (Fuller, 1956), generally near-shore marine deposits (George, 1962; Kinard, 1964; Moore, 1964), or "bank" facies (Harris *et al.*, 1966). In an earlier study, the producing sequence at Glenburn itself was described as "algal and oolitic vuggy limestone" (George, 1962).

The present study of cores, logs, and samples from Glenburn Field indicates that subaerial processes rather than marine algal growth or carbonate "bank" environments were mainly responsible for creating the limestone fabrics and porosity in the Frobisher-Alida reservoirs of Glenburn. It is probable that Mission Canyon limestones in other oil fields of the Williston Basin had similar origins. Hydrocarbon entrapment in the Glenburn Field is largely accomplished by a structurally assisted updip permeability pinchout into evaporites and evaporite-cemented limestones.

Geologic Characteristics

Upper Mission Canyon rocks of the Glenburn Field include limestone, dolomite, and anhydrite, in varying ratios. Each of these rock types has several fabrics. Cores and core chips from many wells were examined, but core from one well, the Texota #1 Weber (NDGS #3630) was especially important to this study because of its length and the central position of the well in the field (Figs. 11-3 and 11-4). In addition, much lithologic information was obtained from the Vaughn #1 Weber well (NDGS #3935), which is located on the west flank of the field.

Dolomite and Anhydrite

Dolomite and anhydrite fabrics appear to be relatively simple. Dolomite is present in two distinct forms: (1) primary or "supratidal" dolomite, characterized by very small crystals (25 μ) in thin beds and associated with desiccation structures (Gerhard *et al.*, 1978); and (2) dolomite that occurs as distinct rhombs in limestone (and primary dolomite), replacing an earlier carbonate. Dolomitization does not appear to be pervasive; most of the secondary dolomite-rhombs appear in dolomitic limestone.

"Chickenwire" anhydrite, massive bedded anhydrite, and clear, vug-filling anhydrite are present in varying amounts. At Glenburn, the "chickenwire" form is considered to be of sabkha origin, and is commonly associated with primary dolomites and desiccation features. Massive anhydrite is characterized by a subhedral crystalline fabric and is distinctly bedded and interfingered by thin carbonate beds, many of which are fine-grained dolomite beds. The clear anhydrite that fills fenestrae in carbonate beds occurs as large subhedral to anhedral crystals. There are sharp

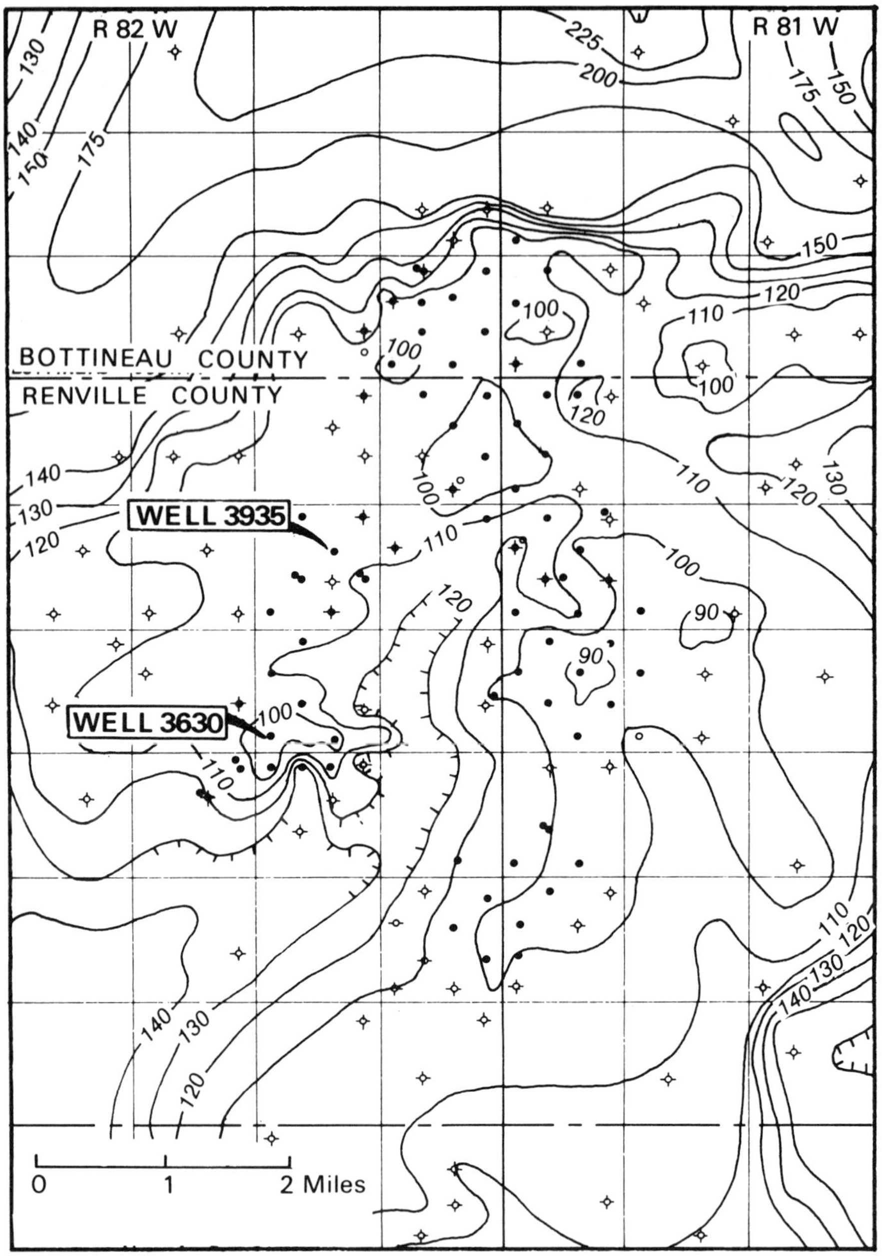

Fig. 11-4. Isopach map of the Frobisher-Alida evaporite, Glenburn Field. Contour interval 10 feet (3.3 m).

contacts between the anhydrite crystals and the walls of the fenestrae.

Limestones

Limestones are more complex texturally than either of the foregoing lithologies. The most obvious features of the limestones are rounded, commonly spherical, concentric bodies averaging a centimeter in diameter. In places these occur randomly distributed in matrix, in other places packed in grain-to-grain contact with high interparticle porosity. These are the bodies that were described as oolitic, pisolitic, and oncolitic by earlier workers, and were attributed to construction by blue-green algae. Porous fine-grained limestone beds, thin nonporous limestone beds, and shale partings are interbedded with rocks bearing these round grains (Fig. 11-5).

A variety of primary sedimentary structures and fabrics can be seen in thin sections of the limestones, and are used as the basis for recognizing three major facies: (1) skeletal pisolitic wackestone, (2) mudstone, and (3) pisolitic packstone or grainstone. It is probable that the pisolitic fabrics are diagenetic because primary fabrics are difficult to recognize. These structures and fabrics suggest a strandline origin for the

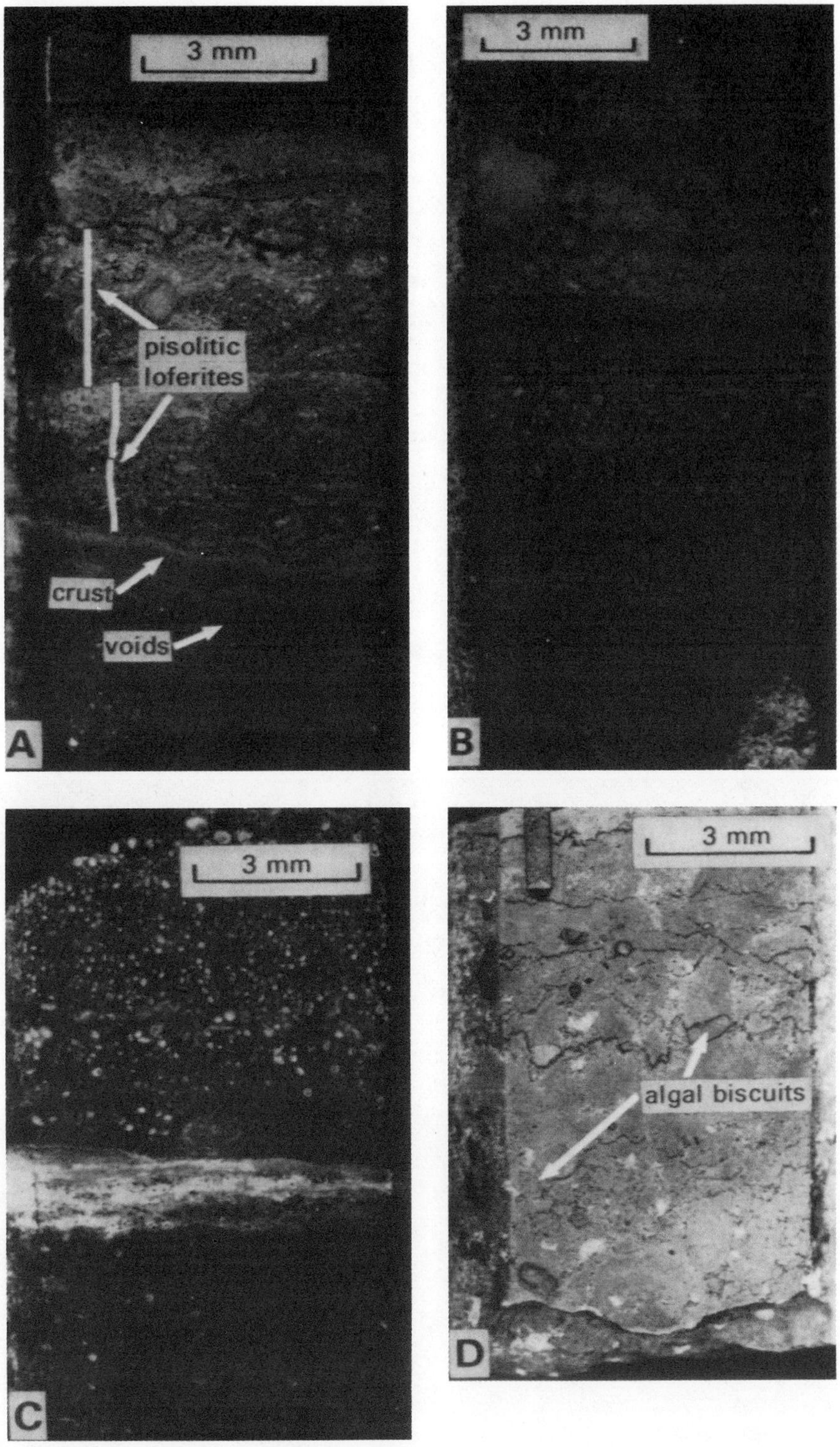

Fig. 11-5. Photograph of core slabs of Frobisher-Alida limestone. *A*. Löferite-type shrinkage or desiccation voids overlain by subaerial laminated crust, in turn overlain by pisolitic löferites. Texota #1 Weber well, 4600 feet (1402 m). From Gerhard *et al.*, 1982. Published by permission, American Association of Petroleum Geologists. *B*. Weathering pisolite in löferite. Texota #1 Weber well, 4596.5 feet (1401 m). *C*. Weathering pisolite overlain by laminated crust, in turn overlain by mixture of coated grains, löferite and weathering pisolite. Texota #1 Weber well. 4614 feet (1406 m). *D*. Algal biscuits with stylolites. Algal biscuits are faint caps on micritic intraclasts. Allochems are hypersaline oolites. Note the abundant stylolites. Texota #1 Weber well, 4623 feet (1409 m).

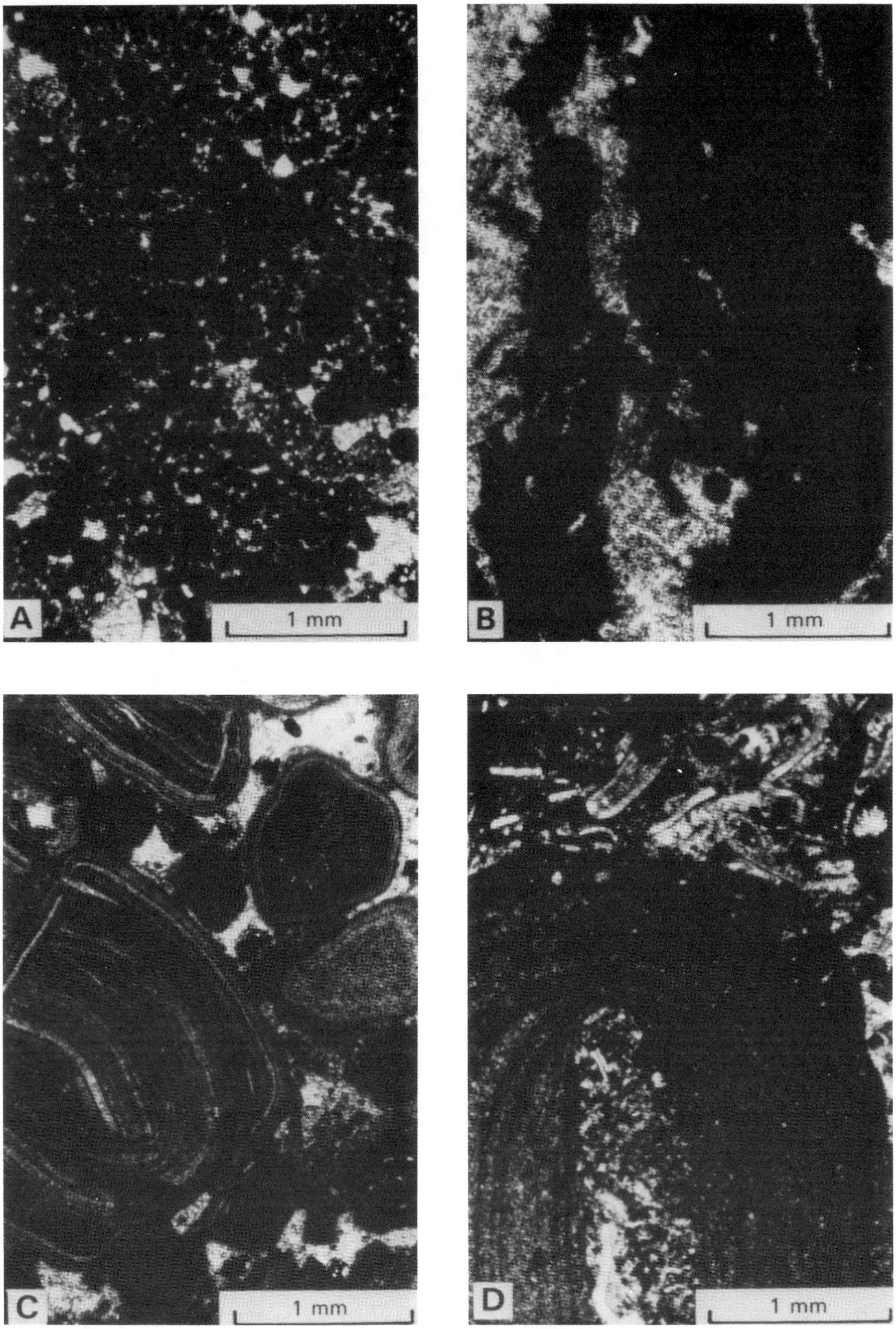

Fig. 11-6. Photomicrographs, cross-polarized light. *A*. Hypersaline oolite. Note the radial fibrous habit of the larger grains, distinct from normal-marine oolites (Bahamian) that are concentrically laminated without the radial structure. Texota #1 Weber well, 4623 feet (1409 m). *B*. Blue-green algal stromatolite interbedded with anhydrite. View is oriented 90° from horizontal. Vaughn #1 Weber well, 4557.3 feet (1389 m). *C*. Vadose pisolite in the Mississippian Frobisher-Alida zone. Note the distinct radial fibrous habit and interrupted laminates. Texota #1 Weber well, 4624 feet (1409 m). *D*. Oncolite in skeletal packstone from the Meagher Limestone (Cambrian), Montana, demonstrating discontinuous and uneven lamination.

Table 11.1. Classification of Spheroidal Sedimentary Structures (Based on average whole rock characteristics, or on n = 10)

I. Positive evidence for algal origin
   Oncolite
II. No evidence for algal origin
   A. Larger than 2 mm diameter
      1. Coated
         a. *vadose pisolite*—radial fibrous w/concentric coats
         b. *incipient weathering pisolite*—grain nucleus w/amorphous or micritic coatings
      2. Uncoated, micritic
         a. *weathering pisolite* (other evidence for weathering present)
         b. *peloid*
   B. Smaller than or equal to 2 mm diameter
      1. Concentric coating
         a. *micropisolite* (occurs w/vadose pisolite)
         b. oolite
            (1) Fully concentric coating w/pseudouniaxial "figure," or radial fibrous habit within concentric coatings; *common oolite*
            (2) Major aspect is radial fibrous, concentricity subtle; *hypersaline oolite*
      2. No coating
         a. Micritic: *peloid*
         b. Allochem: *abraded grain* or *primary grain*

Table 11.2. Characteristics of Oncolites and Subaerial Pisolites.[1]

Subaerial
Vadose Pisolite
Fully concentric banding
Radial fibrous fabric
Multiple generation (nuclei of earlier pisolite)
Gravity cements
Grain-to-grain contacts
Reverse grading common

Weathering Pisolite
No banding or faint banding
No observable nuclei
Mud-supported fabric
Often associated with crusts

Oncolites (Algal)
Anastomosing laminae
Skeletal or intraclastic nuclei
Grain-to-grain contact common
Algal filaments sometimes present
Associated "community" or other organisms
Muddy laminae, *not* radial fibrous

[1] Several of these characteristics are illustrated in Figures 11.5 through 11.7.

rocks, with hypersalinity occurring episodically during deposition. Oolites and stromatolites are present in the Vaughn #1 Weber core. The oolites have radial fibrous structure, however, and are similar to Holocene oolites of the Great Salt Lake, Utah (Eardley, 1938) rather than to the concentric oolites with pseudo-uniaxial optics in marine oolite sands of the Bahamas (Fig. 11-6). It is suggested that the radial fibrous habit is characteristic of hypersalinity (Table 11-1). Stromatolite structures, attributed here to blue-green algae, are interbedded with anhydrite (Fig. 11-6B).

Mudstones and wackestones are virtually devoid of skeletal intraclasts, except for occasional replaced fragments, suggesting that faunas were restricted. Desiccation structures and fenestral porosity also suggest subaerial exposure or supratidal deposition. Pisolites are of several types (Table 11-1).

Vadose Fabrics and Oncolites

Dunham (1969) first demonstrated the similarity in structure and fabric between the pisolite in the Yates Formation (Permian) of the Capitan reef complex, West Texas and New Mexico, and modern caliche. The Yates pisolite had been considered to be oncolitic—that is, the concentrically banded spherical structures formed by blue-green algae accreting mud particles in an agitated aqueous environment. Since Dunham's first paper, more examples have been identified in the geologic record. These include: (1) Holocene and Mississippian exposure surfaces, described by Harrison and Steinen (1978); (2) vadose pisolites (Dunham, 1969); and (3) crusts (Kahle, 1977) and micritized fabrics. These origins are topics of current investigation.

Different environmental interpretations and exploration strategies could result if oncolites were confused with subaerial fabrics. For instance, in the Glenburn Field, part of the trapping mechanism is the filling of interpisolite porosity by anhydrite cement in laterally adjacent rock. Algal pisolites (oncolites) would require an agitated environment that is not compatible with evaporite pore fillings. If a skeletal or other grain nucleus can be clearly demonstrated for these pisolitic bodies, or primary organic structures (filaments) are preserved in lineations, then an algal or at least a subaqueous origin may be inferred (Table 11-2 and Fig. 11-5). However, in most of the Frobisher-Alida zone, these criteria are not met.

Comparison With Other Carbonates

A direct comparison of Frobisher-Alida pisolites with pisolites described by Dunham (1969) clearly demonstrates the subaerial origin of coated multi-generation pisolites if Dunham is correct (Fig. 11-6). However, examples from the literature do not present enough data for comparison to establish an origin for micritic pisolites.

A series of samples of subaerially exposed Holocene (or late Pleistocene) skeletal sands from St. Croix, U.S. Virgin Islands, illustrates one possible sequence of generation of weathering pisolite (Fig. 11-7). Micritic and aphanitic rims are initially formed around skeletal nuclei. Further weathering progressively micritizes grains until only featureless pseudo-pisolites or pseudo-oolites are left. Apparently, this process can also take place in mud-rich fabrics, but the sequence of pisolite formation cannot be demonstrated as clearly.

Porosity

Two major porosity fabrics are present in the Glenburn reservoir. Fenestral porosity is commonly associated with micrites, some of which have isolated micritic pisolites "floating" in micrite. This vuggy fenestral porosity was described as löferites by Fischer (1964) and suggests analogy with desiccation porosity in modern muds, since vugs are elongate parallel to the assumed depositional surface and may have secondary vertical elongation (Fig. 11-5). Vuggy fenestral porosity such as this occurs in sediments on partly desiccated tidal flats in Florida Bay mud islands.

The second major porosity type is interparticle vug porosity, associated with weathering (micritic) pisolites (Figs. 11-5, 11-7). If the interpretation regarding pisolite origin is correct, the porosity is diagenetic and formed by subaerial, near-surface vadose processes. It is this porosity type that is most common in the reservoir.

Table 11.3. Outline of Frobisher-Alida Paragenesis, Glenburn Field, North Dakota

1. Salt solution in buried Prairie Evaporite created surface structures
2. Transgressive episode: Deposition of shallow-marine mud-based carbonate sediments
3. Stable episode: Oscillatory sea stand within small limits. Created marginal evaporite facies, distal sabkha, minor weathering pisolite. Some evaporite filling of pores
4. Regressive episode: Created major weathering pisolite, vadose pisolite, massive evaporite (off-structure), primary dolomite, clastic mud layers, major crusts. Pisolitic porosity developed. Proximal sabkha. (Episodes II, III, and IV are cyclical)
5. Offlap of sabkha environment
6. Trap sealed by lateral evaporites
7. Development of stylolites
8. Oil migration into trap

Origin and Paragenesis

Other observations pertinent to further interpretation:

1. Anhydrite occurs as vug fillings in carbonates, and as massive beds in apparent stratigraphic equivalence to carbonates.
2. Production has been established along topographic (structural) highs without regard for structural closure (Fig. 11-3).
3. Structural noses on carbonates, plunging basinward, are separated by evaporites at the same structural position.

A low depositional slope, arid climate, a slightly unstable sea level, and presence of slight structural deformation either penecontemporaneous with or pre-Frobisher-Alida deposition are assumed. Selective salt solution in older evaporites could easily have accomplished the deformation (Anderson and Hunt, 1964).

In the area of the Glenburn Field during Frobisher-Alida time, carbonate muds and a few muddy skeletal sands were deposited during a

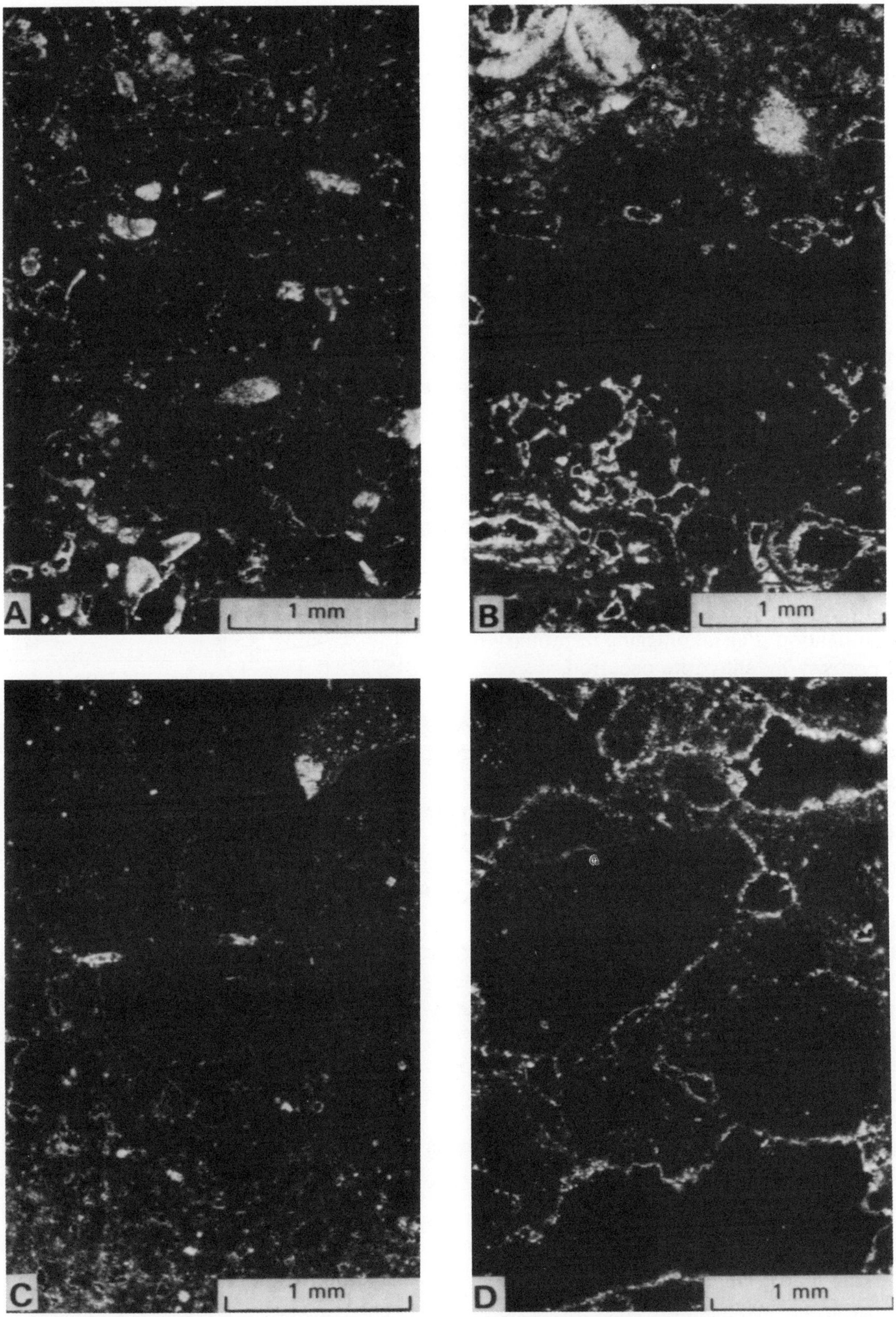

Fig. 11-7. Photomicrographs (in cross-polarized light) comparing some Holocene porosity and pisolites with Mississippian samples. Porosity is black. *A*. Incipient pisolites with coated and partially micritized grains and well-developed porosity, St. Croix, U.S. Virgin Islands. *B*. Weathering pisolites with crust and intrapisolite porosity, Holocene, St. Croix, U.S. Virgin Islands. *C*. Löferite showing micritic pisolites and vertically oriented laminated cracks. Texota #1 Weber well, 4617 feet (1407 m). *D*. Weathering pisolites with interparticle porosity. Texota #1 Weber well, 4595.5 feet (1391.6 m).

transgressive episode. Water depths were probably never more than a few feet. It is possible that a low depositional slope coupled with tidal phenomena could have caused environments to vary rapidly from intertidal to very shallow lagoonal in depth with only minor fluctuations in sea level. Supratidal desiccation may have occurred, creating primary fenestral porosity (löferite).

Regression, with minor fluctuations, exposed the carbonates. Lithification probably occurred by both early hypersaline and later vadose cementation. Flushing of hypersaline lenses by fresh water appears to be a common modern cementation process, as exemplified by case hardening of beachrock (Moore, 1973; Hanor, 1978). Exposed sediment, inferred to have been partly or poorly lithified, was present on structural highs with intervening hypersaline pans (Fig. 11-8). As weathering of the carbonates progressed, evaporite pumping may have occurred along the periphery of the limestone, leading to dolomitization. Partial plugging of primary fenestral porosity by anhydrite occurred.

Porosity enhancement, resulting from the formation of vadose and weathering pisolites, was particularly pronounced in the uppermost parts of the limestone high. Topographic relief was probably less than six or seven feet (2 m) between the top of the limestone mounds and the bottom of the evaporite pans.

Updip to the east, the limestones interfingered with sabkha evaporites, which created impermeable barriers. On the flanks of the limestone mounds, anhydrite that formed by penecontemporaneous evaporitic deposition occludes porosity, creating barriers to lateral oil migration. As mentioned previously, the areas of evaporite pans may be sites of earlier salt solution, either prior to or during Mission Canyon time.

The final seal at the top of the Frobisher-Alida porous zone originated in either of two ways. One was by deposition of a thin shale cap of terrigenous muds, which washed over the evaporite-filled pans and covered the porous limestones. These thin shales are only a fraction of an inch thick, but appear to be effective seals. In conjunction with or independent of this area are calcareous caliche crusts of low permeability, which may form seals as well (Fig. 11-5). The other major type of seal is massive anhydrite, perhaps representing salt-pan deposition during the next

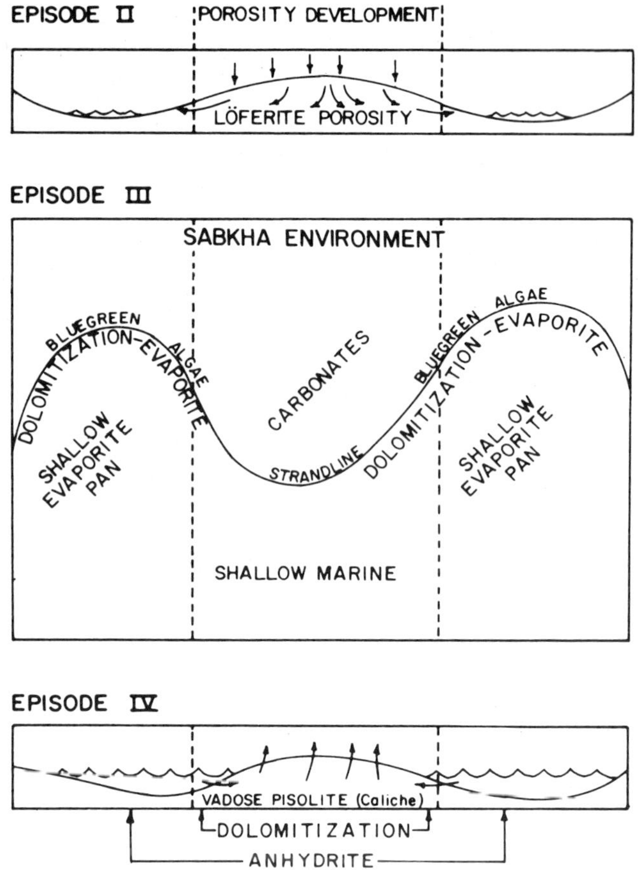

Fig. 11-8. Diagrammatic representation of episodes II to IV of the sequence of porosity development in the Glenburn Field Frobisher-Alida rocks. Refer to Table 11-3 for episode descriptions. From Gerhard *et al.*, 1982. Published by permission, American Association of Petroleum Geologists.

minor transgressive interval. Pre-existing permeability was unaffected by the massive anhydrite development, due to the presence of hypersaline fluid in the pores or earlier anhydrite. This fluid inhibited the mixing with external, sulfate-rich fluids needed for anhydrite formation; early anhydrite has been remobilized.

Post-trap intrastratal solution is indicated by widely distributed stylolites that appear to predate oil migration. The stylolites are inferred to have created lenticular zones of permeability.

Summary and Conclusions

Limestones of the Frobisher-Alida zone of the upper Mission Canyon produce hydrocarbons in the Glenburn Field from porosity largely developed through subaerial diagenetic processes.

Vadose pisolites, carbonate crusts, and solution features are major fabrics in cores of the pay zone. The development of porosity was in part coeval with anhydrite deposition in pans or supratidal sabkhas on the flanks of a very low topographic arch. Flank salt solution or ancestral structural development are plausible mechanisms for the construction of the limestone prominences. Dolomitization along the flanks, coupled with minute interbeds of anhydrite, support contemporaneous porosity development and evaporite deposition.

Early cementation was followed by the formation of vadose pisolite and other solution/desiccation features. The vadose pisolite may have been in part contemporaneous with early cementtation. The last major diagenetic phase was intrastratal solution. Porosity is lenticular in part. The oil-water contact is probably an irregular surface because of varying pore dimensions.

Acknowledgments This study was supported by the North Dakota Geological Survey, the University of North Dakota Carbonate Studies Program funded by industry, and a U.S. Geological Survey grant (#14-08-0001-G-597) for studies of subaerial diagenesis. S. B. Anderson and James Berg were instrumental in producing the earliest report of the project (Gerhard *et al.*, 1978). Published by permission of the North Dakota Geological Survey.

References

ANDERSON, S.B., and J.B. HUNT, 1964, Devonian salt solution in north central North Dakota: Third Internat. Williston Basin Symp., p. 93–104.

DUNHAM, R.J., 1969, Vadose pisolite in the Capitan reef (Permian), New Mexico and Texas: *in* Friedman, G.M., ed., Depositional Environments in Carbonate Rocks—a symposium: Soc. Econ. Paleontolotists and Mineralogists, Spec. Publ. 14, p. 182–191.

EARDLEY, A.J., 1938, Sediments of Great Salt Lake, Utah: Amer. Assoc. Petroleum Geologists Bull., v. 50, p. 1305–1412.

FISCHER, A.G., 1964, The löfer cyclothems of the alpine Triassic: Kansas Geol. Survey Bull. 169, p. 107–149.

FULLER, J.G.C.M., 1956, Mississippian rocks in the Saskatchewan portion of the Williston Basin: First Internat. Williston Basin Symp., p. 29–35.

GEORGE, R.S., 1962, Glenburn Field: *in* Oil and Gas Fields of North Dakota: North Dakota Geol. Soc., p. 109–110.

GERHARD, L.C., S.B. ANDERSON, and J. BERG, 1978, Mission Canyon porosity development, Glenburn Field, North Dakota Williston Basin: *in* Economic Geology of the Williston Basin: Williston Basin Symp., Montana Geol. Soc., p. 177–188.

HANOR, J.S., 1978, Precipitation of beachrock cements: mixing of marine and meteoric waters vs. CO_2-degassing: Jour. Sedimentary Petrology, v. 48, p. 489–501.

HARRIS, S.B., C.B. LAND, JR., and J.H. MCKEEVER, 1966, Mission Canyon stratigraphy, North Dakota: Amer. Assoc. Petroleum Geologists Bull., v. 50, p. 2269–2276.

HARRISON, R.S., and R.P. STEINEN, 1978, Subaerial crusts, caliche profiles, and breccia horizons: Comparison of some Holocene and Mississippian exposure surfaces, Barbados and Kentucky: Geol. Soc. America Bull., v. 89, p. 385–396.

JOHNSON, W., 1956, Mississippian oil fields of northeastern Williston Basin: First Internat. Williston Basin Symp., p. 52–53.

KAHLE, C.F., 1977, Origin of subaerial Holocene calcareous crusts: Role of algae, fungi and sparmicritisation: Sedimentology, v. 24, p. 413–435.

KINARD, J.C., 1964, Recent Mission Canyon discoveries in the Roth-N.E. Landa area, Bottineau County, North Dakota: Third Internat. Williston Basin Symp., p. 241–245.

MOORE, C.H., JR., 1973, Intertidal carbonate cementation Grand Cayman, West Indies: Jour. Sedimentary Petrology, v. 43, p. 591–602.

MOORE, W.L., 1964, Mechanical log and phase relationships in the carbonates of the Mississippian of the Williston Basin: Third Internat. Williston Basin Symp., p. 263–265.

NORTH DAKOTA GEOLOGICAL SOCIETY, 1967, Glenburn Field: *in* Oil and gas fields of North Dakota, 1967 Supplement: North Dakota Geol. Soc., p. 20.

ROEHL, P.O., 1967, Stony Mountain (Ordovician) and Interlake (Silurian) facies analogs of Recent low-energy marine and subaerial carbonates, Bahamas: Amer. Assoc. Petroleum Geologists Bull., v. 51, no. 10, p. 1979–2032.